Automorphic Representations and L-functions

Proceedings of the International Colloquium, Mumbai 2012

TATA INSTITUTE OF FUNDAMENTAL RESEARCH
STUDIES IN MATHEMATICS

1.	M. Hervé	:	*Several Complex Variables*
2.	M.F. Atiyah and others	:	*Differential Analysis*
3.	B. Malgrange	:	*Ideals of Differentiable Functions*
4.	S.S. Abhyankar and others	:	*Algebraic Geometry*
5.	D. Mumford	:	*Abelian Varieties*
6.	L. Schwartz	:	*Radon Measures on Arbitrary Topological Spaces and Cylindrical Measures*
7.	W.L. Baily Jr. and others	:	*Discrete Subgroups of Lie Groups an Applications to Moduli*
8.			*C.P. Ramanujam – A Tribute*
9.	C.L. Siegel	:	*Advanced Analytic Number Theory*
10.	S. Gelbart and others	:	*Automorphic Forms, Representation Theory and Arithmetic*
11.	M.F. Atiyah and others	:	*Vector Bundles on Algebraic Varietie*
12.	R. Askey and others	:	*Number Theory and Related Topics*
13.	S.S. Abhyankar and others	:	*Geometry and Analysis*
14.	S.G. Dani (Editor)	:	*Lie Groups and Ergodic Theory*
15.	T.N. Venkataramana (Editor)	:	*Cohomology of Arithmetic Groups, L-functions and Automorphic Forms*
16.	R. Parimala (Editor)	:	*Algebra, Arithmetic and Geometry*
17.	S.G. Dani and Gopal Prasad (Editors)	:	*Algebraic Groups and Arithmetic*
18.	S.G. Dani and P. Graczyk (Editors)	:	*Probability Measures on Groups: Recent Directions and Trends*
19.	V.B. Mehta (Editor)	:	*Algebraic Groups and Homogeneous Spaces*
20.	N. Saradha (Editor)	:	*Diophantine Equations*
21.	V. Srinivas (Editor)	:	*Cycles, Motives & Shimura Varieties*
22.	D. Prasad and others (Editors)	:	*Automorphic Representations and L-Functions*

Automorphic Representations and L-functions

Proceedings of the International Colloquium, Mumbai 2012

Editors

D. Prasad
C.S. Rajan
A. Sankaranarayanan
J. Sengupta

Published for the
tifr Tata Institute of Fundamental Research
by
HINDUSTAN BOOK AGENCY

International Distribution by
American Mathematical Society, USA

Editors

D. Prasad
C.S. Rajan
A. Sankaranarayanan
J. Sengupta

HINDUSTAN BOOK AGENCY (INDIA)
P 19 Green Park Extension, New Delhi 110 016

email: info@hindbook.com
http://www.hindbook.com

ISBN 978-93-80250-49-6

Contents

International Colloquium
on
Automorphic Representations and L-functions

Mumbai, 3–11 January, 2012

Report

An international colloquium on Automorphic Representations and L-functions was held at the Tata Institute of Fundamental Research, Mumbai, from January 3, 2012 to January 11, 2012. This was the latest in the series of Colloquia held at the School of Mathematics, once every four years, since 1956. We are grateful to the Tata Institute of Fundamental Research for the financial support for hosting this event.

The organising committee of the Colloquium consisted of Professors D. Prasad, C.S. Rajan, A. Sankaranarayanan and J. Sengupta from the Tata Institute of Fundamental Research.

There were thirty nine invited speakers and a total of seventy participants for this event. We were honoured to have a galaxy of foremost experts in the area deliver a one hour lecture each. Besides mathematicians from the School of Mathematics of the Tata Institute of Fundamental Research, several mathematicians from educational and research institutions from India and abroad attended the Colloquium. Our request for contributions to this volume was warmly received and the outcome is the present memorable volume.

The social programme of the Colloquium consisted of an inaugural dinner party on 3rd January, a dinner at Joss restaurant on 6th January and a dinner in the Institute West Lawns on 9th January. There was a classical dance recital on 5th January and a vocal music concert on 10th January. A hike to the Rajamachi hills and Kondana Buddhist caves was organised on 8th January.

List of Talks

Speaker	Title
J. Adams	Automorphisms of spaces of representations
N. Anantharaman	Quantum limits on locally symmetric spaces.
J. Arthur	The Endoscopic classification of representations.
M. Bhargava	The average rank of elliptic curves.
V. Blomer	Applications of the Kuznetsov formula on $GL(3)$.
W.T. Gan	A conjecture of Sakellaridis-Venkatesh on the unitary spectrum of spherical varieties.
E. Ghate	Local semisimplicity over totally real fields.
D. Goldfeld	The distribution of low lying zeros for the family of adjoint L-functions on $GL(3)$.
B. Gross	The average order of the 2-Selmer group for hyperelliptic curves.
M. Hanzer	Explicit construction of automorphic representations of symplectic group with given quadratic unipotent Arthur parameter.
J. Hoffstein	Multiple Dirichlet series and shifted convolutions.
R. Holowinsky	First moments of Rankin-Selberg Convolutions.
A. Ichino	Periods of quaternionic Shimura varieties.
H. Kim	Logarithmic derivatives of Artin L-functions at $s = 1$.
W. Kohnen	Generalized modular functions and their Fourier coefficients.
E. Lapid	Whittaker-Fourier coefficients of cuspidal representations on Mp_{2n}.
X. Li	Voronoi formulas and applications.
W. Luo	Asymptotic variance for Linnik distribution.
P. Michel	On representations of binary forms by quarternary forms.
S. Morel	Mixed l-adic complexes on varieties over Q.
R. Munshi	Bounds for L-functions.
M. Ram Murty	The Fibonacci zeta function.
B-C. Ngo	Kloosterman sheaves for reductive groups.
O. Offen	On representations of $Sp(2n)$ distinguished by $Sp(n) \times Sp(n)$.
A. Raghuram	Cohomology of arithmetic groups and the special values of automorphic L-functions.

D. Ramakrishnan	Comparison of cusp forms on $GL(n)$ agreeing at degree one primes.
M. Rapoport	On the arithmetic fundamental lemma of Wei Zhang.
A. Reznikov	Torus periods of automorphic functions and meromorphic continuation of certain Dirichlet Series.
D. Rohrlich	Self-dual Artin representations.
A. Saha	Determination of modular forms by Fourier coefficients.
Y. Sakellaridis	Beyond endoscopy for the relative trace formula.
G. Savin	Siegel modular forms of half-integral weight: Hecke operators at $p = 2$.
K. Srinivas	On the zeros of Selberg class functions.
M. Tadic	On interaction between harmonic analysis and theory of automorphic forms.
E. Urban	On the rank of Selmer groups for elliptic curves over $\mathbb{Q}$.
J-K. Yu	Epipelagic representations and invariant theory.
W. Zhang	Global Gan-Gross-Prasad conjecture for unitary groups.
C-B. Zhu	Degenerate principal series of real classical groups.

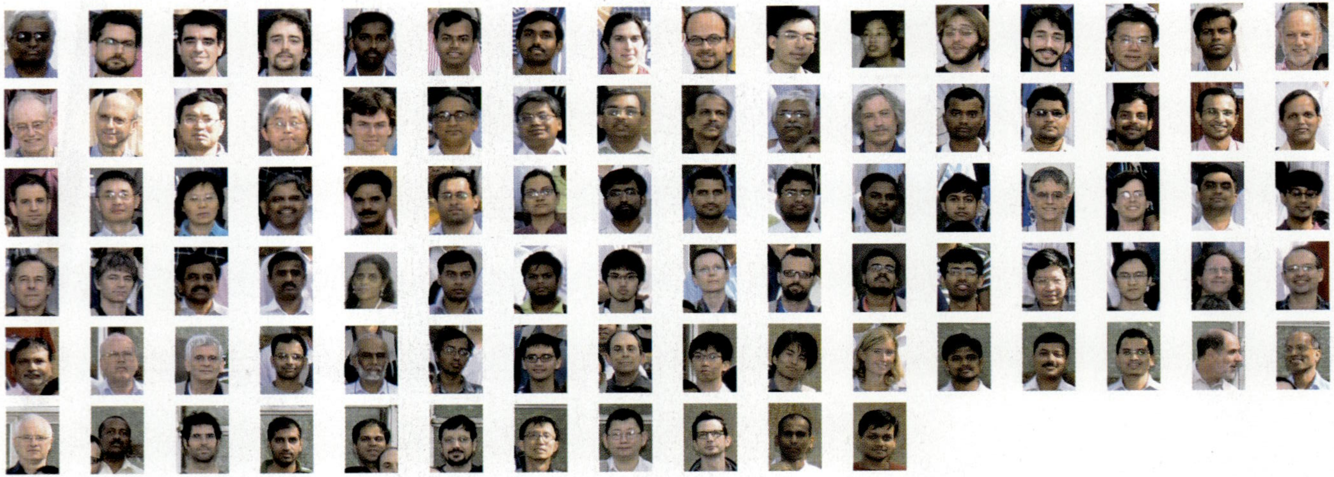

1. B. Ramakrishnan, C. Cunningham, M. Adrian, —, S. Lonka, A. Mondal, P. Kewat, N. Anantharaman, Y. Sakellaridis, C. Kuan, L-M. Lim, T. Hulse, M. Kiral, B-C. Ngo, M. Verma, M. Rapoport
2. J. Arthur, S. Gelbart, H. Kim, M. Furusawa, V. Kala, M. Ram Murty, J. Sengupta, S.M. Garge, T. Aravindakshan, A. Sankaranarayanan, J. Hoffstein, S.K. Singh, A.P. Sawant, S.P. Patel, A. Raghuram, C.S. Rajan
3. R. Holowinsky, W. Luo, X.Q. Li, D.B. Sawant, K. Srinivas, V. Thomas, D. Sharma, A. Gupta, K-D. Shankhadhar, B. Sahu, J. Meher, —, R. Schulze-Pillot, S. Spallone, E. Ghate, M. Bhargava
4. R. Miatello, E. Urban, V. Vengurlekar, D. Salvi, A. Prajapati, S. Das, V. Vaish, K. Morimoto, M. Hanzer, J. Hundley, —, A. Pitale, C-B. Zhu, W. Zhang, E. Lapid, P. Michel
5. P. Bundele, W. Kohnen, D. Goldfeld, R. Raghunathan, R. Tandon, A. Karnataki, R. Gomez, D. Rohrlich A. Ichino, H. Ma Jiajun, C. Schoemann, V. Sailor, M. Mehwala, M. Pilankar, B. Gross, D. Prasad
6. M. Tadić, S.E. Rao, O. Offen, P. Vishe, U. K. Anandavardhanan, A. Reznikov, Wee Teck Gan, J.K. Yu, G. Gotsbacher, M. Krishnamurthy, A. Saha

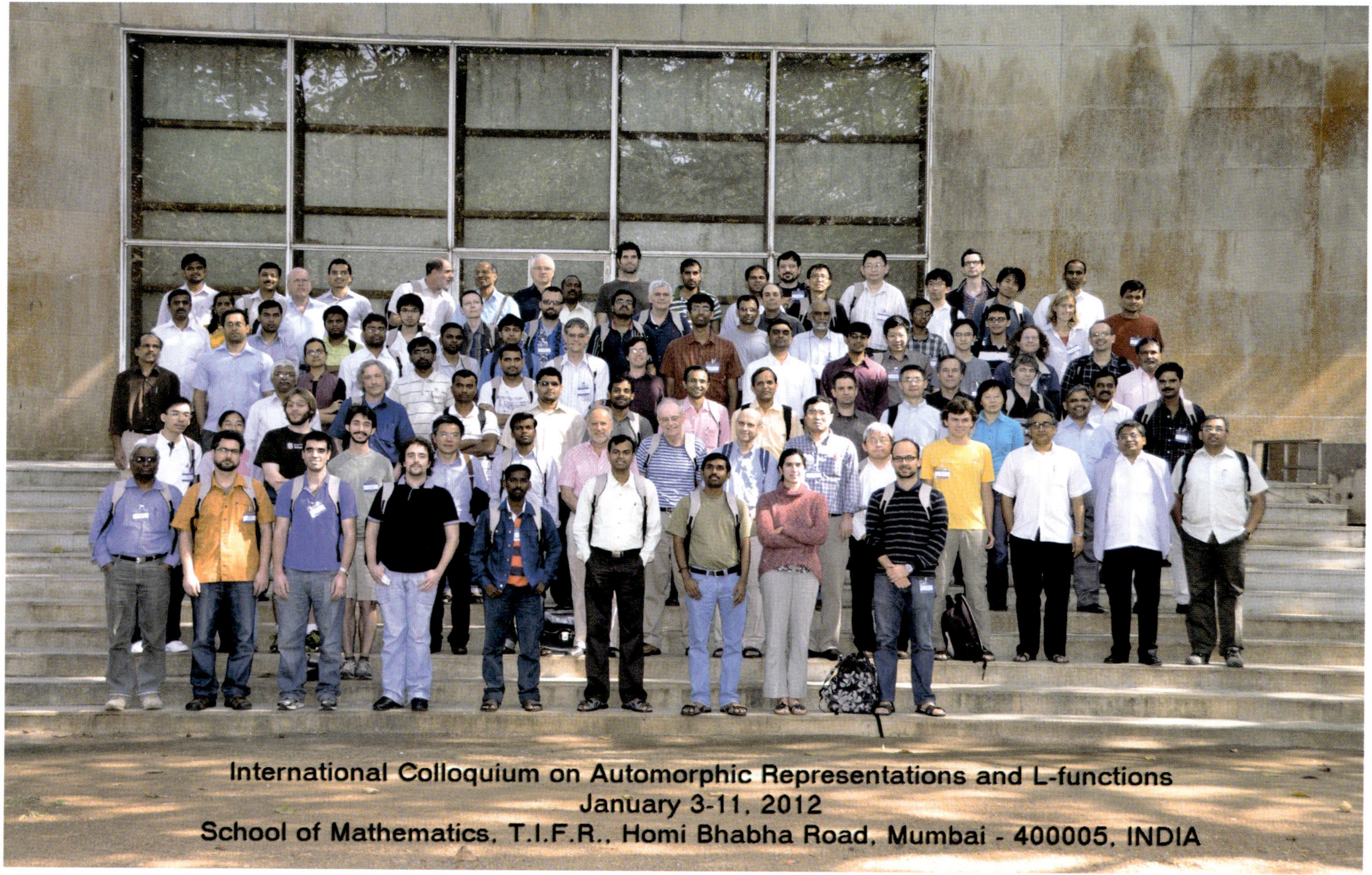

International Colloquium on Automorphic Representations and L-functions
January 3-11, 2012
School of Mathematics, T.I.F.R., Homi Bhabha Road, Mumbai - 400005, INDIA

Automorphic Representations and L-Functions
Editors: D. Prasad, C.S. Rajan, A. Sankaranarayanan, J. Sengupta

Publisher: Hindustan Book Agency, New Delhi, India

The Endoscopic Classification of Representations

James Arthur

We shall outline a classification [A] of the automorphic representations of special orthogonal and symplectic groups in terms of those of general linear groups. This necessarily includes a classification of local L-packets of representations. It also requires a classification of the extended packets that are the local constituents of nontempered automorphic representations. Our description will be brief. In particular, we will restrict it to *quasisplit*[1] orthogonal and symplectic groups G, even though at least some of the results can be extended (not without effort) to inner twists of G.

The methods rest ultimately on two comparisons of trace formulas. One is the spectral identity that is the end product of the stabilization of the trace formula for G. This was established some years ago [A1], under the assumption of the fundamental lemma. It now holds without condition, thanks to the work of Waldspurger [W1][W2], the recent breakthrough by Ngo [N], and the extensions of Chaudouard and Laumon [CL1] [CL2]. The other is the spectral identity given by the stabilization of the twisted trace formula for $GL(N)$. This formula is still conditional. The relevant twisted fundamental lemmas are now known [W3], [W4], at least up to the twisted variants of [CL1] and [CL2]. The problem is to develop twisted generalizations of the techniques of [A1] and related papers. Until this is done, the results we describe here have also to be regarded as conditional.

We take F to be a local or global field of characteristic 0, and G to be a quasisplit, special orthogonal or symplectic group over F. Then G has a complex dual group $\widehat{G}$, and a corresponding L-group

$$^{L}G = \widehat{G} \rtimes \Gamma_{E/F}.$$

We are taking $\Gamma_{E/F} = \mathrm{Gal}(E/F)$ to be the Galois group of a suitable finite extension E/F. If G is split, for example, the absolute Galois group $\Gamma = \Gamma_F = \Gamma_{\overline{F}/F}$ acts trivially on $\widehat{G}$, and we can take $E = F$.

[1]It is understood that G is "classical", in the sense that it is not an outer twist of the split group $SO(8)$ by a triality automorphishm of order 3.

There are three general possibilities for G, which correspond to the three infinite families of simple groups $\mathbf{B}_n$, $\mathbf{C}_n$ and $\mathbf{D}_n$. They are as follows

Type $\mathbf{B}_n$: $G = SO(2n+1)$ is split, and $\widehat{G} = Sp(2n, \mathbb{C}) = {}^L G$.

Type $\mathbf{C}_n$: $G = Sp(2n)$ is split, and $\widehat{G} = SO(2n+1, \mathbb{C}) = {}^L G$.

Type $\mathbf{D}_n$: $G = SO(2n)$ is quasisplit, and $\widehat{G} = SO(2n, \mathbb{C})$. In this case, ${}^L G = SO(2n, \mathbb{C}) \rtimes \Gamma_{E/F}$, where E/F is an arbitrary extension of degree 1 or 2 whose Galois group acts by outer automorphisms on $SO(2n, \mathbb{C})$ (which is to say, by automorphisms that preserve a fixed splitting of $SO(2n, \mathbb{C})$). The nontrivial outer automorphism of $SO(2n, \mathbb{C})$ is induced by conjugation by some element in its complement in $O(2n, \mathbb{C})$.

The other infinite family of simple groups is of course $\mathbf{A}_n$. We will regard the split (reductive) group $GL(N)$, with $N = n+1$, as our representative from this family. Its role is different. For we are treating the representations of $GL(N)$ as known objects, in terms of which we want to classify the representations of G.

1 Statement of the Local Results

Suppose first that F is local. The local Langlands group is the locally compact extension of the local Weil group W_F defined by

$$L_F = \begin{cases} W_F, & \text{if } F \text{ is archimedean,} \\ W_F \times SU(2), & \text{if } F \text{ is } p\text{-adic.} \end{cases}$$

The group G comes with the family

$$\Phi(G) = \{\phi : \ L_F \longrightarrow {}^L G\}$$

of Langlands parameters, and the family

$$\Pi(G) = \{\pi : \ G(F) \longrightarrow GL(V)\}$$

of irreducible (admissible) representations of $G(F)$. We recall that ϕ is an L-homomorphism (which means among other things that it commutes with the projections of L_F and ${}^L G$ onto $\Gamma_{E/F}$), and that it is taken up to conjugacy by $\widehat{G}$ on ${}^L G$. The representation π is of course also taken up to its usual form of equivalence.

We actually work with quotients

$$\widetilde{\Phi}(G) = \Phi(G)/\sim$$

and

$$\widetilde{\Pi}(G) = \Pi(G)/\sim$$

of $\Phi(G)$ and $\Pi(G)$. The equivalence relation $\sim$ is trivial for types $\mathbf{B}_n$ and $\mathbf{C}_n$, but is defined by conjugation by $O(2n)$ (in place of $SO(2n)$) for type $\mathbf{D}_n$. More precisely, if $G = SO(2n)$ is of type D_n, $\widetilde{\Phi}(G)$ is the set of $O(2n,\mathbb{C})/SO(2n,\mathbb{C})$-orbits in $\Phi(G)$ under the action of $O(2n,\mathbb{C})$ by conjugation on ${}^LG$, and $\widetilde{\Pi}(G)$ is the set of $O(2n,F)/SO(2n,F)$ orbits in $\Pi(G)$ under the action of $O(2n,F)$ by conjugation on $G(F)$. It is only these coarser sets that are related to the analogous sets for $GL(N)$. Among them, a special role is played by the smaller sets

$$\widetilde{\Phi}_{\mathrm{bdd}}(G) = \{\phi \in \widetilde{\Phi}(G) : \ \mathrm{im}(\phi) \text{ is bounded}\}$$

of parameters of bounded image, and

$$\widetilde{\Pi}_{\mathrm{temp}}(G) = \{\pi \in \widetilde{\Pi}(G) : \ \pi \text{ is tempered}\}$$

of irreducible representations whose characters (and matrix coefficients) are tempered with respect to Harish-Chandra's Schwartz space $\mathcal{C}(G)$.

Each of the three pairs of sets above has a natural role in the local classification of representations. We have one more pair of local sets to introduce, but its role is primarily global. It consists of the set

$$\widetilde{\Psi}(G) = \{\psi : \ L_F \times SU(2) \longrightarrow {}^LG, \ \mathrm{im}(\psi) \text{ is bounded}\}$$

of (orbits of) L-homomorphisms from the product $L_F \times SU(2)$ to ${}^LG$, and the set

$$\widetilde{\Pi}_{\mathrm{unit}}(G) = \{\pi \in \widetilde{\Pi}(G) : \ \pi \text{ is unitary}\}$$

of (orbits of) irreducible representations of $G(F)$ that are unitary. This pair of sets is especially important for us. For we shall see that it governs the local constituents of automorphic representations.

We can regard $\widetilde{\Psi}(G)$ as an intermediate set between two families of Langlands parameters. The earlier set $\widetilde{\Phi}_{\mathrm{bdd}}(G)$ can obviously be identified with the subset of parameters in $\widetilde{\Psi}(G)$ that are trivial on the second factor $SU(2)$. On the other hand, any $\psi \in \widetilde{\Psi}(G)$ maps to a parameter

$$\phi_\psi(w) = \psi\left(w, \begin{pmatrix} |w|^{\frac{1}{2}} & 0 \\ 0 & |w|^{-\frac{1}{2}} \end{pmatrix}\right), \qquad w \in L_F,$$

in $\widetilde{\Phi}(G)$, where $|w|$ is the pullback to L_F of the absolute value on W_F, and ψ is identified with its analytic extension from $L_F \times SU(2)$ to the group

$L_F \times SL(2,\mathbb{C})$. It follows from the boundedness condition on ψ, and the fact that a homomorphism from $SL(2,\mathbb{C})$ to a complex group is determined by its diagonal weight, that the mapping $\psi \to \phi_\psi$ from $\widetilde{\Psi}(G)$ to $\widetilde{\Phi}(G)$ is injective. There is consequently a chain

$$\widetilde{\Phi}_{\mathrm{bdd}}(G) \subset \widetilde{\Psi}(G) \subset \widetilde{\Phi}(G)$$

of parameter sets. We also have a parallel claim

$$\widetilde{\Pi}_{\mathrm{temp}}(G) \subset \widetilde{\Pi}_{\mathrm{unit}}(G) \subset \widetilde{\Pi}(G)$$

of sets of irreducible representations, since any tempered representation is unitary.

For any $\psi \in \widetilde{\Psi}(G)$, we have the centralizer

$$S_\psi = \mathrm{Cent}\big(\mathrm{im}(\psi), \widehat{G}\big)$$

in $\widehat{G}$ of the image of (a representative of) ψ. This is of course a complex reductive subgroup of $\widehat{G}$. We write

$$\overline{S}_\psi = S_\psi / Z(\widehat{G})^\Gamma$$

for its quotient by the group of Galois invariants in the center of $\widehat{G}$. The group

$$\mathcal{S}_\psi = \pi_0(\overline{S}_\psi) = \overline{S}_\psi / \overline{S}_\psi^0$$

of connected components of $\overline{S}_\psi$ is then a finite abelian 2-group.

Theorem 1.1 [A, Theorem 1.5.1] (F local). (a) *For any $\psi \in \widetilde{\Psi}(G)$, there is a finite set $\widetilde{\Pi}_\psi$ over $\widetilde{\Pi}_{\mathrm{unit}}(G)$, together with a mapping*

$$\pi \in \widetilde{\Pi}_\psi \longrightarrow \langle \cdot, \pi \rangle \in \widehat{\mathcal{S}}_\psi,$$

from $\widetilde{\Pi}_\psi$ to the group of (linear) characters on $\mathcal{S}_\psi$, both of which are canonically determined by endoscopic character relations.

(b) *Suppose that $\phi = \psi$ lies in the subset $\widetilde{\Phi}_{\mathrm{bdd}}(G)$ of $\widetilde{\Psi}(G)$. Then the elements in $\widetilde{\Pi}_\phi$ are tempered and multiplicity free. Moreover, the mapping from $\widetilde{\Pi}_\phi$ to $\widehat{\mathcal{S}}_\phi$ is injective in general, and bijective in case F is p-adic. Finally*

$$\widetilde{\Pi}_{\mathrm{temp}}(G) = \coprod_{\phi \in \widetilde{\Phi}_{\mathrm{bdd}}(G)} \widetilde{\Pi}_\phi.$$

Remarks 1. By a set $\widetilde{\Pi}_\psi$ over $\widetilde{\Pi}_{\text{unit}}(G)$, we understand simply a mapping

$$\widetilde{\Pi}_\psi \longrightarrow \widetilde{\Pi}_{\text{unit}}(G).$$

of sets. The multiplicity free assertion in (b) means that the fibres are trivial if $\phi = \psi$ lies in $\widetilde{\Phi}_{\text{bdd}}(G)$, and that we can therefore regard $\widetilde{\Pi}_\phi$ as a subset of $\widetilde{\Pi}_{\text{temp}}(G)$.

2. The last assertion of (b) is that any element in $\widetilde{\Pi}_{\text{temp}}(G)$ lies in a unique packet $\widetilde{\Pi}_\phi$. This is the local Langlands correspondence for G, or rather a slightly weaker version in the case $G = SO(2n)$ that classifies orbits in $\Pi_{\text{temp}}(G)$ under a group of order 2, rather than individual representations in $\Pi_{\text{temp}}(G)$.

We will need to be precise about the endoscopic character relations that determine the packets and the resulting classification. Since this is slightly more technical, we shall postpone it until after we have stated the global theorems, and listed some of the preliminary applications.

In preparation for the global theorems, we have to introduce a larger set $\widetilde{\Psi}^+(G)$ of local parameters. It will be used to represent the local constituents of automorphic representations in a way that provides for the possible failure of the generalized Ramanujan conjecture for $GL(N)$. We define $\widetilde{\Psi}^+(G)$ in the same way as $\widetilde{\Psi}(G)$, but without the condition that the image of the parameter ψ be bounded. The set so obtained can be regarded as a natural complexification of $\widetilde{\Psi}(G)$. Theorem 1.1 then extends by analytic continuation to parameters ψ in $\widetilde{\Psi}^+(G)$, as will become evident once we have described the endoscopic character formulas on which it is based. The price to pay for this generalization is that the representations in a packet $\widetilde{\Pi}_\psi$ need no longer be irreducible or unitary. However, for the subset of parameters that are local constituents of automorphic representations (in the discrete spectrum, say), the representations are unitary, and most probably also irreducible.

2 Statement of the Global Results

Suppose now that F is global. In this report, we shall state the global theorems in terms of the hypothetical global Langlands group L_F. This group is expected to be a locally compact extension

$$1 \longrightarrow K_F \longrightarrow L_F \longrightarrow W_F \longrightarrow 1$$

of the global Weil group W_F by a compact connected group K_F. It should come with a conjugacy class of local embeddings

$$\begin{array}{ccc} L_{F_v} & \longrightarrow & W_{F_v} \\ \downarrow & & \downarrow \\ L_F & \longrightarrow & W_F \end{array}$$

over the corresponding Weil groups, for any valuation v of F. The essential property of L_F is that its irreducible N-dimensional representations should be in canonical bijection with the cuspidal automorphic representations of the locally compact adelic group $GL(N,\mathbb{A})$.

The structure of L_F is far from known, in contrast to the simple formula we have for the local Langlands group. Its existence is deeper than the theory of endoscopy, to which the results we are describing pertain, and deeper even than the principle of functoriality. However, we shall assume here that L_F does exist in order to simplify the discussion. We cannot of course do this in the volume [A]. The only choice there is to formulate global results for G in terms of self-dual, cuspidal automorphic representations of $GL(N,\mathbb{A})$ rather than irreducible self-dual N-dimensional representations of L_F. This leads to some significant technical complications.

Assuming the existence of L_F, we define the global parameter sets $\widetilde{\Phi}(G)$, $\widetilde{\Phi}_{\mathrm{bdd}}(G)$ and $\widetilde{\Psi}(G)$ as in the local case above. For example,

$$\widetilde{\Psi}(G) = \Psi(G)/\sim,$$

where $\sim$ is the equivalence relation above, and $\Psi(G)$ is the set of $\widehat{G}$-conjugacy classes of L-homomorphisms ψ from the product $L_F \times SU(2)$ to ${}^LG$ such that the image of ψ is bounded. For any $\psi \in \widetilde{\Psi}(G)$, and any valuation v, we obtain a localization ψ_v by pulling ψ back to the local Langlands group L_{F_v}. We may as well treat this object as a local parameter in the larger set $\widetilde{\Psi}^+(G_v)$, following what must be done in [A]. In the context of L_F, this amounts to our taking on the weaker assumption necessitated by the possible failure of the generalized Ramanujan conjecture, that L_{F_v} embeds only in a natural complexification $L_{F,\mathbb{C}}$ of L_F.

For any $\psi \in \widetilde{\Psi}(G)$, we have the centralizer S_ψ in $\widehat{G}$ of the image of ψ, and the 2-group defined by its quotient

$$\mathcal{S}_\psi = S_\psi / S_\psi^0\, Z(\widehat{G})^\Gamma = \overline{S}_\psi / \overline{S}_\psi^0.$$

For any v, we then have a localization mapping

$$x \longrightarrow x_v, \qquad x \in \mathcal{S}_\psi,$$

from $\mathcal{S}_\psi$ to $\mathcal{S}_{\psi_v}$. Given Theorem 1.1, we define the global packet

$$\widetilde{\Pi}_\psi = \Big\{\pi = \bigotimes_v \pi_v : \ \pi_v \in \widetilde{\Pi}_{\psi_v}, \ \langle \cdot, \pi_v \rangle = 1 \text{ for almost all } v\Big\}.$$

Any representation $\pi = \bigotimes \pi_v$ in $\widetilde{\Pi}_\psi$ then has a character

$$\langle x, \pi \rangle = \prod_v \langle x_v, \pi_v \rangle, \qquad x \in \mathcal{S}_\psi,$$

on $\mathcal{S}_\psi$.

Our main global theorem gives a decomposition of the automorphic discrete spectrum

$$L^2_{\text{disc}}\big(G(F)\backslash G(\mathbb{A})\big) \subset L^2\big(G(F)\backslash G(\mathbb{A})\big).$$

The problem is to describe this space explicitly as a module over the global Hecke algebra

$$\mathcal{H}(G) = \widetilde{\bigotimes_v} \mathcal{H}(G_v),$$

which is of course defined as a restricted tensor product of local Hecke algebras on the groups $G_v = G(F_v)$. But because we are dealing with orbits of local parameters in the case $G = SO(2n)$, we have to be content with a slightly weaker result. For any v, we can write $\widetilde{\mathcal{H}}(G_v)$ for the full local Hecke algebra $\mathcal{H}(G_v)$ if G is of type $\mathbf{B}_n$ or $\mathbf{C}_n$. If G is of the type $\mathbf{D}_n$ however, we must take $\widetilde{\mathcal{H}}(G_v)$ to be the proper subalgebra of symmetric functions in $\mathcal{H}(G_v)$ under a suitable fixed automorphism of $G(F_v)$ of order 2 attached to the nontrivial outer automorphism of G. We can then form the locally symmetric subalgebra

$$\widetilde{\mathcal{H}}(G) = \widetilde{\bigotimes_v} \widetilde{\mathcal{H}}(G_v)$$

of the global Hecke algebra $\mathcal{H}(G)$.

Theorem 2.1 [A, Theorem 1.5.2] (F global). *There is an $\widetilde{\mathcal{H}}(G)$-module isomorphism*

$$L^2_{\text{disc}}\big(G(F)\backslash G(\mathbb{A})\big) \cong \bigoplus_{\psi \in \widetilde{\Psi}_2(G)} \ \bigoplus_{\pi \in \widetilde{\Pi}_\psi(\varepsilon_\psi)} m_\psi \pi,$$

where

$$\widetilde{\Psi}_2(G) = \big\{\psi \in \widetilde{\Psi}(G) : \ |S_\psi| < \infty\big\},$$

and m_ψ *equals* 1 *or* 2, *while*

$$\varepsilon_\psi : \mathcal{S}_\psi \longrightarrow \{\pm 1\}$$

is a linear character defined explicitly in terms of symplectic ε*-factors, and*

$$\widetilde{\Pi}_\psi(\varepsilon_\psi) = \big\{\pi \in \widetilde{\Pi}_\psi : \ \langle \cdot, \pi\rangle = \varepsilon_\psi\big\}$$

is the subset of the global packet $\widetilde{\Pi}_\psi$ *attached to* ε_ψ.

The integer m_ψ is the order of the set $\Psi(G,\psi)$ of $\widehat{G}$-orbits of L-homomorphisms from $L_F \times SU(2)$ to ${}^LG$ that map to the (possibly larger) orbit ψ. It equals 2 if G is of type $\mathbf{D}_n$ and the degrees of the irreducible constituents of ψ (as an N-dimensional representation) are all even. In all other cases, it equals 1.

The sign character ε_ψ is more interesting. We define a natural representation

$$\tau_\psi : \ S_\psi \times L_F \times SU(2) \longrightarrow GL(\widehat{\mathfrak{g}})$$

of the product of S_ψ with $L_F \times SU(2)$ on the Lie algebra $\widehat{\mathfrak{g}}$ of $\widehat{G}$ by setting

$$\tau_\psi(s,g,h) = \mathrm{Ad}\big(s \cdot \psi(g,h)\big), \qquad s \in S_\psi,\ (g,h) \in L_F \times SU(2),$$

where Ad is the adjoint representation of ${}^LG$ on $\widehat{\mathfrak{g}}$. Let

$$\tau_\psi = \bigoplus_\alpha \tau_\alpha = \bigoplus_\alpha (\lambda_\alpha \otimes \mu_\alpha \otimes \nu_\alpha)$$

be its decomposition into irreducible representations λ_α, μ_α and ν_α of the groups S_ψ, L_F and $SU(2)$. We then define

$$\varepsilon_\psi(x) = \prod_\alpha{}' \det\big(\lambda_\alpha(s)\big), \qquad s \in S_\psi,$$

where x is the image of s in $\mathcal{S}_\psi$, and $\prod'$ denotes the product over those indices α with μ_α symplectic and

$$\varepsilon\left(\frac{1}{2}, \mu_\alpha\right) = -1.$$

Theorem 2.1 is the main global theorem. However, to prove it, one must also establish the following result.

Theorem 2.2 [A, Theorem 1.5.3] (F global). (a) *Suppose that* ϕ *belongs to the subset*

$$\widetilde{\Phi}_{\mathrm{sim}}(G) = \{\phi \in \widetilde{\Phi}(G) : \ \overline{S}_\psi = S_\psi/Z(\widehat{G})^\Gamma = 1\}$$

of simple generic parameters in $\widetilde{\Psi}(G)$. Then the dual group $\widehat{G}$ is orthogonal if and only if the symmetric square L-function $L(s, \phi, S^2)$ has a pole at $s = 1$, while $\widehat{G}$ is symplectic if and only if it is the skew-symmetric square L-function $L(s, \phi, \Lambda^2)$ that has a pole at $s = 1$.

(b) *Suppose $\phi_1 \in \widetilde{\Phi}_{\mathrm{sim}}(G_1)$ and $\phi_2 \in \widetilde{\Phi}_{\mathrm{sim}}(G_2)$, for two of our groups G_1 and G_2. Then the corresponding Rankin-Selberg ε-factor satisfies*

$$\varepsilon\left(\tfrac{1}{2}, \phi_1 \times \phi_2\right) = 1,$$

if $\widehat{G}_1$ and $\widehat{G}_2$ are either both orthogonal or both symplectic.

The conclusions of Theorem 2.2 are familiar from the work of Cogdell, Kim, Piatetskii-Shapiro, and Shahidi [CKPS1], [CKPS2], Ginzburg, Rallis and Soudry [GRS], and Lapid [Lap] on representations with Whittaker models. We refer the reader also to the recent paper [Ji] of Jiang on the important problem of comparing the global classification here with the explicit constructions provided by the theta correspondence.

3 Initial Applications

Let us describe a few of the initial applications of the three theorems. Some of these are included in the assertions of the theorems, others are proved in [A], while a couple of others will require further thought. We shall list them, with little comment, according to whether F is local or global.

F local

(i) The representations $\pi \in \widetilde{\Pi}_\psi$ in the packets attached to local parameters $\psi \in \widetilde{\Psi}(G)$ are *unitary*. This is included in the assertion of Theorem 1.1(a).

(ii) The *local Langlands correspondence* is valid if the quasisplit group G equals $SO(2n+1)$ or $Sp(2n)$, and in the slightly weaker form given by $O(2n)$-orbits in the remaining case that G equals $SO(2n)$. This is Theorem 1.1(b), together with the local refinement Theorem 4.1 that we will state in the next section. (In the third case $G = SO(2n)$, there is a further refinement in [A, §8.4] that falls just short of the full Langlands correspondence.)

F local or global

(iii) If F is local (respectively global), the packet $\widetilde{\Pi}_\phi$ attached to any parameter ϕ in the subset $\widetilde{\Phi}_{\mathrm{bdd}}(G)$ of $\widetilde{\Psi}(G)$ contains a representation π

that is locally (resp. globally) *generic*, and such that the character $\langle\cdot,\pi\rangle$ on $\mathcal{S}_\phi$ equals 1. This is established in [A, Proposition 8.3.2], using the results of Ginzburg, Rallis and Soudry. (The Whittaker datum with respect to which π is generic is determined by the underlying Langlands-Shelstad transfer factors, a topic we will not review in this report.)

F global

(iv) *Rankin-Selberg L-functions* for orthogonal and symplectic groups (that is, pairs of quasisplit groups (G_1, G_2) of the kind we are considering) have analytic continuation and functional equation. In fact, their analytic behaviour coincides with that of classical Rankin-Selberg L-functions for general linear groups. This is just the *principle of functoriality* for the standard representations of the L-groups ${}^LG_1$ and ${}^LG_2$, which follows from Theorem 2.1 and the refinement of Theorem 1.1 we will state presently. In other words, Rankin-Selberg L-functions for pairs (G_1, G_2) are among the original L-functions studied in [JPS].

(v) Theorem 2.1 will also give us some control of the *symmetric square* and *skew symmetric square L-functions.* On the one hand, these L-functions govern normalizing factors of global intertwining operators for maximal (Siegel) parabolic subgroups of our groups G. On the other, Langlands' theory of Eisenstein series attaches representations in the automorphic discrete spectrum of G to any poles in the right half plane of the intertwining operators. But Theorem 2.1 should rule out such representations, and therefore also such poles. This question needs to be examined more carefully, and is under consideration by Shahidi.

(vi) Our group G has no *embedded eigenvalues*, in the sense of families of Hecke eigenvalues

$$c^S = \{c_v : v \notin S\}$$

(rather than its analogue from mathematical physics for eigenvalues of Laplace operators at archimedean places). As usual, c represents a family of semisimple conjugacy classes in ${}^LG$, taken up to the equivalence relation $(c')^{S'} \sim c^S$ if $c'_v = c_v$ for almost all v. (If $\widehat{G} = SO(N,\mathbb{C})$ for N even, conjugacy is assumed to be taken with respect to the full orthogonal group $O(N,\mathbb{C})$.) The assertion here is that if ψ belongs to the complement of $\widetilde{\Psi}_2(G)$ in $\widetilde{\Psi}(G)$, there is no representation π in the automorphic discrete spectrum of G such that the associated two equivalence classes of families $c(\pi)$ and $c(\psi)$ of Hecke eigenvalues are equal. This follows again from Theorem 2.1, together with the corresponding property [JS] for $GL(N)$.

(vii) Automorphic representations π of $G(\mathbb{A})$ attached to a generic parameter ϕ in the set

$$\widetilde{\Phi}_2(G) = \widetilde{\Phi}_{\mathrm{bdd}}(G) \cap \widetilde{\Psi}_2(G)$$

occur in the discrete spectrum with *multiplicity* 1 or 2. Here π is intended to be an actual representation of $G(\mathbb{A})$, which represents some orbit in the global packet $\widetilde{\Pi}_\phi$. More precisely, π occurs with multiplicity 1 unless

(a) G is of type $\mathbf{D}_n$,

(b) the global integer m_ϕ defined in §2 equals 2, and

(c) each of the corresponding local integers m_{ϕ_v} equals 1,

in which case it occurs with multiplicity 2. (See the remarks surrounding (8.3.8) in [A], which are based on Corollaries 6.6.6 and 6.7.3 of [A].)

(viii) *Symplectic ε-factors* govern which (orbits of) representations in the global packet $\widetilde{\Pi}_\psi$ attached to a general parameter $\psi \in \widetilde{\Psi}_2(G)$ actually do occur in the discrete spectrum. This "application" falls into the "requires further thought" category. What are the implications of the sign character ε_ψ (which is trivial if ψ lies in the subset $\widetilde{\Phi}_{\mathrm{bdd}}(G)$ of $\widetilde{\Psi}(G)$), as it occurs in the multiplicity formula of Theorem 2.1?

4 On Endoscopic Character Relations

We return to the case that F is local, and more specifically, the endoscopic character relations referred to in Theorem 1.1(a). We shall state a more precise supplement to Theorem 1.1(a) that characterizes the packets $\widetilde{\Pi}_\psi$ and the associated families of linear characters $\langle \cdot, \pi \rangle$ on $\mathcal{S}_\psi$.

We write N for the rank of the general linear group attached to the dual $\widehat{G}$ of G. In other words, N equals $2n$, $2n+1$ or $2n$, according to whether the given group G lies in the family $\mathbf{B}_n$, $\mathbf{C}_n$ or $\mathbf{D}_n$. There is then a natural embedding of the L-group ${}^LG$ of G into $GL(N, \mathbb{C})$. It is also convenient to introduce the connected component

$$\widetilde{G}(N) = GL(N) \rtimes \widetilde{\theta}(N)$$

in the semidirect product $\widetilde{G}(N)^+$ of $GL(N)$, for the standard automorphism

$$\widetilde{\theta}(N):\ g \longrightarrow \widetilde{J}\,{}^t g^{-1}\widetilde{J}^{-1},$$

$$g \in GL(N),\ \widetilde{J} = \widetilde{J}(N) = \begin{pmatrix} & & 0 & & 1 \\ & & & -1 & \\ & & \cdot^{\cdot^{\cdot}} & & \\ (-1)^{N+1} & & & & 0 \end{pmatrix},$$

of order 2. This twisted object comes with analogues of the parameter sets $\widetilde{\Psi}(G)$ and $\widetilde{\Phi}(G)$. They consist of the set $\widetilde{\Psi}(N) = \Psi\big(\widetilde{G}(N)\big)$ of (equivalence classes of) self-dual, N-dimensional, unitary representations of the product $L_F \times SU(2)$, and the set $\widetilde{\Phi}(N) = \Phi\big(\widetilde{G}(N)\big)$ of (equivalence classes of) self-dual, N-dimensional, not necessarily unitary representations of L_F.

The embedding of ${}^L G$ into $GL(N,\mathbb{C})$ gives a mapping from $\widetilde{\Psi}(G)$ to $\widetilde{\Psi}(N)$, which can be seen to be injective. There is also an injective mapping from $\widetilde{\Psi}(N)$ to $\widetilde{\Phi}(N)$, obtained by pulling back a representation ψ of $L_F \times SU(2)$ to the representation

$$\phi_\psi(w) = \psi\left(w, \begin{pmatrix} |w|^{\frac{1}{2}} & 0 \\ 0 & |w|^{-\frac{1}{2}} \end{pmatrix}\right), \qquad w \in L_F,$$

of L_F. Finally, we can restrict the local Langlands correspondence for $GL(N)$ (which we recall was established by Harris-Taylor and Henniart, and more recently by Scholze) to the set of self-dual representations of L_F. This gives a bijection from $\widetilde{\Phi}(N)$ to the set $\widetilde{\Pi}(N) = \Pi\big(\widetilde{G}(N)\big)$ of (equivalence classes of) irreducible, self-dual representations of $GL(N,F)$. The composition

$$\widetilde{\Psi}(G) \hookrightarrow \widetilde{\Psi}(N) \hookrightarrow \widetilde{\Phi}(N) \xrightarrow{\sim} \widetilde{\Pi}(N)$$

of the three mappings then provides an injection

$$\psi \longrightarrow \pi_\psi, \qquad \psi \in \widetilde{\Psi}(G),$$

from $\widetilde{\Psi}(G)$ into $\widetilde{\Pi}(N)$. The representation π_ψ of $GL(N,F)$ is called a *Speh representation.* It is known to be unitary.

For any $\psi \in \widetilde{\Psi}(G)$, the self-dual representation π_ψ is θ-stable. It can therefore be extended to the semidirect product

$$\widetilde{G}(N,F)^+ = GL(N,F) \rtimes \langle\widetilde{\theta}(N)\rangle.$$

There are actually two extensions, which differ by a sign on the coset $\widetilde{G}(N,F)$ of $\widetilde{\theta}(N)$. However, the theory of Whittaker models leads to a *canonical* extension $\widetilde{\pi}_\psi$ of π_ψ to the group $\widetilde{G}(N,F)^+$. (One takes $\widetilde{\pi}_\psi\big(\widetilde{\theta}(N)\big)$ to be a quotient of the intertwining operator of order 2 for the standard representation ρ_ψ attached to ψ that stabilizes a Whittaker vector.) This in turn gives us a $GL(N,F)$-invariant linear form on the Hecke bi-module $\widetilde{\mathcal{H}}(N) = \mathcal{H}\big(\widetilde{G}(N)\big)$ of functions on the coset $\widetilde{G}(N,F)$. It is the twisted invariant character

$$\widetilde{f}_N(\psi) = \mathrm{tr}\big(\widetilde{\pi}_\psi(\widetilde{f})\big), \qquad \widetilde{f} \in \widetilde{\mathcal{H}}(N),$$

where $\widetilde{\pi}_\psi(\widetilde{f})$ is the operator

$$\int_{\widetilde{G}(N,F)} \widetilde{f}(x)\widetilde{\pi}_\psi(x)dx.$$

The given group G has its own Hecke algebra, and of more relevance to us here, the subalgebra $\widetilde{\mathcal{H}}(G)$ of symmetric functions in $\mathcal{H}(G)$ defined prior to the statement of Theorem 2.1. It also comes with an essential set of algebraic objects. We write $\mathcal{E}(G)$ for the set of (isomorphism classes of) endoscopic data G' for G [LS, (1.2)], [KS, 2.1].

Here is our supplement to Theorem 1.1.

Theorem 4.1 [A, Theorem 2.2.1] (F local). (a) *For any $\psi \in \widetilde{\Psi}(G)$, there is a unique stable linear form*

$$f \longrightarrow f^G(\psi), \qquad f \in \widetilde{\mathcal{H}}(G),$$

on $\widetilde{\mathcal{H}}(G)$ such that

$$\widetilde{f}^G(\psi) = \widetilde{f}_N(\psi), \qquad \widetilde{f} \in \widetilde{\mathcal{H}}(N),$$

where $\widetilde{f}^G$ is the Kottwitz-Shelstad (twisted) transfer of $\widetilde{f}$ from $\widetilde{G}(N,F)$ to $G(F)$.

(b) *For any $\psi \in \widetilde{\Psi}(G)$, the packet $\widetilde{\Pi}_\psi$ and the mapping $\pi \to \langle \cdot, \pi\rangle$ from $\widetilde{\Pi}_\psi$ to $\widehat{\mathcal{S}}_\psi$ are defined canonically by the relation*

$$f'(\psi') = \sum_{\pi\in\widetilde{\Pi}_\psi} \langle s_\psi x, \pi\rangle\, \mathrm{tr}\big(\pi(f)\big), \qquad s \in S_{\psi,\mathrm{ss}},\ f \in \widetilde{\mathcal{H}}(G),$$

where x is the image in $\mathcal{S}_\psi$ of the semisimple point $s \in S_{\psi,\mathrm{ss}}$, and s_ψ is the ψ-image in S_ψ or $\mathcal{S}_\psi$ of the point $1 \times \begin{pmatrix} -1 & 0 \\ 0 & -1 \end{pmatrix}$ in $L_F \times SU(2)$, while

$$(G', \psi'), \qquad G' \in \mathcal{E}(G),\ \psi' \in \widetilde{\Psi}(G'),$$

is the endoscopic pair that corresponds bijectively with the given pair (ϕ, s), and $f' = g^{G'}$ is the Langlands-Shelstad transfer of f from $G(F)$ to $G'(F)$.

This theorem requires further comment. We have written f^G at the beginning of Part (a) to mean the *equivalence class* of f, relative to the equivalence relation $f_1 \sim f_2$ if $S(f_1) = S(f_2)$ for every stable linear form on $\widetilde{\mathcal{H}}(G)$. (We recall that a stable linear form is one that lies in the linear span of the set of strongly regular, stable orbital integrals, which is to say, orbital integrals over intersections of $G(F)$ with strongly regular geometric conjugacy classes.) The Kottwitz-Shelstad transfer $\widetilde{f}^G$ of $\widetilde{f}$ is defined only as an equivalence class in $\widetilde{\mathcal{H}}(G)$, but its value $\widetilde{f}^G(\psi)$ at the putative stable linear form $f \to f^G(\psi)$ still makes sense. Part (a) asserts that the stable linear form is uniquely determined by the requirement that this value be equal to the linear form $\widetilde{f}_N(\psi)$ on $\widetilde{\mathcal{H}}(N)$ defined above.

Part (b) is predicated on the bijective correspondence

$$(G', \psi') \leftrightarrow (\psi, s), \qquad s \in S_{\psi,\mathrm{ss}},$$

that has always (at least in the case of parameters in the subset $\widetilde{\Phi}_{\mathrm{bdd}}(G)$ of $\widetilde{\Psi}(G)$) been the motivational heart of Langlands' conjectural theory [L] of endoscopy. (See [A, §1.4], for example.) One verifies that the construction of the stable linear form described for (G, ψ) in (a) specializes to an endoscopic pair (G', ψ'). If G' lies in the important subset $\widetilde{\mathcal{E}}_{\mathrm{ell}}(G)$ of elliptic endoscopic data, for example, there is a decomposition

$$(G', \psi') = (G'_1 \times G'_2, \psi'_1 \times \psi'_2)$$

into groups G'_1 and G'_2 of the kind we are considering, and if $f' = f'_1 \times f'_2$ also is decomposable, we obtain a product

$$f'(\psi') = f'_1(\psi'_1)\, f'_2(\psi'_2)$$

of stable linear forms defined as in (a). For every pair (ψ, s), we thus arrive at a linear form

$$f'(\psi'), \qquad f \in \widetilde{\mathcal{H}}(G),$$

on $\widetilde{\mathcal{H}}(G)$. Part (b) is then an explicit definition of the packets $\widetilde{\Pi}_\psi$ and pairings $\langle x, \pi \rangle$ in terms of these linear forms.

Theorem 4.1 thus gives a canonical construction of the objects in Theorem 1.1. They in turn characterize the global objects of Theorems 2.1 and 2.2. Taken together, the theorems thus reduce the representation theory of our groups G to that of general linear groups $GL(N)$.

5 On the Comparison of Trace Formulas

The proof of the theorems we have stated rests ultimately on a comparison of trace formulas. The comparison has many ramifications, which occupy much of the volume [A]. We shall say just a few words, if only to be able to state a supplementary global theorem. For an elementary discussion of the comparison that is a little more comprehensive, the reader can look at the introductory paper [A2] on the embedded eigenvalue problem.

We are assuming now that F is global. The starting point is actually a definition, rather than a formula. It is the linear form

$$I^G_{\mathrm{disc}}(f) = \sum_{\{M\}} |W(M)|^{-1} \sum_{w \in W(M)_{\mathrm{reg}}} |\det(w-1)|^{-1} \operatorname{tr}\big(M_P(w)\,\mathcal{I}_P(f)\big), \qquad f \in \widetilde{\mathcal{H}}(G),$$

on the locally symmetric global Hecke algebra $\widetilde{\mathcal{H}}(G)$, whose terms will be familiar from the trace formula. The outer sum is over conjugacy classes of Levi subgroups M of (parabolic subgroups $P = NM$ of) G. The inner sum is over the set

$$W(M)_{\mathrm{reg}} = \{w \in W(M) :\ \det(w-1) \neq 0\}$$

of regular elements in the Weyl group

$$W(M) = \mathrm{Norm}(A_M, G)/M$$

of (G, A_M), where the determinant refers to the action of w on the Lie algebra of the F-split component A_M of the centre of M. We have written $\mathcal{I}_P$ for the representation of $G(\mathbb{A})$ induced parabolically from the discrete spectrum

$$L^2_{\mathrm{disc}}\big(M(F)\,A^+_{M,\infty}\backslash M(\mathbb{A})\big), \qquad A^+_{M,\infty} = (R_{F/\mathbb{Q}}A_M)(\mathbb{R})^0,$$

of $M(\mathbb{A})$. Finally, $M_P(w)$ is the self-intertwining operator of $\mathcal{I}_P$ attached to w that plays a central role in Langlands' theory of Eisenstein series.

The linear form $I^G_{\mathrm{disc}}(f)$ is known as the *discrete part* of the trace formula. It is composed of those terms on the spectral side that are discrete in the relevant spectral variables. The term with $M = G$ corresponds to the actual discrete spectrum to which Theorem 2.1 applies. However, its contribution to the trace formula is not easily separated from that of the terms with $M \neq G$. The most fundamental spectral variables are the Hecke families $c = \{c_v\}$ mentioned in the application (vi) in §3. They have further properties, which we recall as follows.

The locally symmetric Hecke algebra is a direct limit

$$\widetilde{\mathcal{H}}(G) = \varinjlim_{S} \widetilde{\mathcal{H}}(G, K^S),$$

where S ranges over finite sets of valuations outside of which G is unramified, and $\widetilde{\mathcal{H}}(G, K^S)$ is the subspace of functions in $\widetilde{\mathcal{H}}(G)$ that are bi-invariant under a "hyperspecial" maximal compact subgroup

$$K^S = \prod_{v \notin S} K_v$$

of $G(\mathbb{A}^S)$. For any such S, the *unramified* (locally symmetric) Hecke algebra

$$\widetilde{\mathcal{H}}^S_{\mathrm{un}} = \widetilde{\mathcal{H}}\big(K^S \backslash G(\mathbb{A}^S)/K^S\big)$$

acts by multipliers

$$(f, h) \longrightarrow f_h, \qquad\qquad f \in \widetilde{\mathcal{H}}(G, K^S),\ h \in \widetilde{\mathcal{H}}^S_{\mathrm{un}},$$

on $\widetilde{\mathcal{H}}(G, K^S)$. The function $f_h \in \widetilde{\mathcal{H}}(G, K^S)$ is characterized by property that

$$\pi(f_h) = \widehat{h}\big(c^S(\pi)\big)\pi(f),$$

for any π in the subset $\widetilde{\Pi}(G, K^S)$ of representations in $\widetilde{\Pi}(G)$ with K^S-fixed vectors. The function $\widehat{h}$ is the Satake transform of h, which is defined on the set of S-families

$$c^S = \{c_v : v \notin S\}$$

of Hecke eigenvalues, while $c^S(\pi)$ is the family attached to π.

Since it is discrete in all of the spectral variables, the linear form $I^G_{\mathrm{disc}}(f)$ can be written as a sum of Hecke eigenforms. More precisely, if f belongs to $\widetilde{\mathcal{H}}(G, K^S)$, there is a canonical decomposition

$$I^G_{\mathrm{disc}}(f) = \sum_{c^S} I^G_{\mathrm{disc},c^S}(f)$$

where

$$I^G_{\mathrm{disc},c^S}(f_h) = \widehat{h}(c^S)\, I^G_{\mathrm{disc},c^S}(f),$$

for any $h \in \mathcal{H}^S_{\mathrm{un}}$. Taking a direct limit over S, we obtain a decomposition

$$I^G_{\mathrm{disc}}(f) = \sum_{c} I^G_{\mathrm{disc},c}(f), \qquad\qquad f \in \widetilde{\mathcal{H}}(G),$$

in which c ranges over equivalence classes of families c^S. This sum, incidentally, is infinite, but convergence is assured by the work [Mu] of Muller. (In [A], we indexed $I^G_{\mathrm{disc}}(f)$ by a discrete set of positive numbers t that parametrize the norms of the imaginary parts of the internal archimedean spectral parameters. For any t and f, there are then only finitely many c with $I^G_{\mathrm{disc},t,c}(f) \neq 0$.)

Suppose that ψ lies in the global set $\widetilde{\Psi}(N)$ of self-dual, unitary, N-dimensional representations of the group $L_F \times SU(2)$. Since the global parameter ψ is unramified at almost all places, it comes with a natural equivalence class $c(\psi)$ of families of Hecke eigenvalues. The following proposition from [A] is an important starting point for the comparison.

Proposition [A, Proposition 3.4.1]. *Suppose that c is an equivalence class of families of Hecke eigenvalues. Then*

$$I^G_{\mathrm{disc},c}(f) = 0, \qquad f \in \widetilde{\mathcal{H}}(G),$$

unless $c = c(\psi)$ for some $\psi \in \widetilde{\Psi}(N)$.

We write

$$I^G_{\mathrm{disc},\psi}(f) = I^G_{\mathrm{disc},c(\psi)}(f),$$

for any $\psi \in \widetilde{\Psi}(N)$. The proposition then implies that

$$I^G_{\mathrm{disc}}(f) = \sum_{\psi \in \widetilde{\Psi}(N)} I^G_{\mathrm{disc},\psi}(f), \qquad f \in \widetilde{\mathcal{H}}(G).$$

To study I^G_{disc}, and in particular the term with $M = G$ in I^G_{disc} that corresponds to the automorphic discrete spectrum of G, one has only to study the linear form $I^G_{\mathrm{disc},\psi}$ attached to any ψ. Notice however that the proposition does not assert that ψ belongs to $\widetilde{\Psi}(G)$. It is in fact an important step along the way to show $I^G_{\mathrm{disc},\psi}(f)$ vanishes unless ψ does lie in the subset $\widetilde{\Psi}(G)$ of $\widetilde{\Psi}(N)$.

At this point, our formula for the linear form $I^G_{\mathrm{disc},\psi}$ is only a definition. The stabilization of the trace formula for G provides a separate description of $I^G_{\mathrm{disc},\psi}$. It is a decomposition

$$I^G_{\mathrm{disc},\psi}(f) = \sum_{G' \in \mathcal{E}_{\mathrm{ell}}(G)} \iota(G,G')\, \widehat{S}^{G'}_{\mathrm{disc},\psi}(f'), \qquad f \in \widetilde{\mathcal{H}}(G),$$

of $I^G_{\mathrm{disc},\psi}(f)$ into a linear combination of linear forms parametrized by the set $\mathcal{E}_{\mathrm{ell}}(G)$ of isomorphism classes of (global) elliptic endoscopic data G' for G, with explicit coefficients $\iota(G,G')$. For any G', the linear form

$$f \longrightarrow \widehat{S}^{G'}_{\mathrm{disc},\psi}(f'), \qquad f \in \widetilde{\mathcal{H}}(G'),$$

is the composition of a stable linear form

$$S^{G'}_{\mathrm{disc},\psi} = \sum_{\{\psi' : \psi' \to \psi\}} S^{G'}_{\mathrm{disc},\psi'}$$

on $\widetilde{\mathcal{H}}(G')$ (or rather the associated form $\widehat{S}^{G'}_{\mathrm{disc},\psi}$ on the space of equivalence classes in $\widetilde{\mathcal{H}}(G')$ defined locally in the last section) with the global Langlands-Shelstad transfer

$$f \longrightarrow f' = f^{G'}$$

of f from $\widetilde{\mathcal{H}}(G)$ to $\widetilde{\mathcal{H}}(G')$. The stable linear form $\widehat{S}^{G'}_{\mathrm{disc},\psi'}(f')$ depends only on the elliptic endoscopic group

$$G' = G'_1 \times G'_2$$

and the corresponding global parameter

$$\psi' = \psi'_1 \times \psi'_2,$$

and equals

$$\widehat{S}^{G'}_{\mathrm{disc},\psi'}(f') = \widehat{S}^{G'_1}_{\mathrm{disc},\psi'_1}(f'_1)\,\widehat{S}^{G'_2}_{\mathrm{disc},\psi'_2}(f'_2),$$

if $f' = f'_1 \times f'_2$ is decomposable. It is therefore defined inductively for G as the difference

$$S^G_{\mathrm{disc},\psi}(f) = I^G_{\mathrm{disc},\psi}(f) - \sum_{G' \neq G} \iota(G,G')\,\widehat{S}^{G'}_{\mathrm{disc},\psi}(f'), \qquad f \in \widetilde{\mathcal{H}}(G).$$

For any ψ, the definition of $I^G_{\mathrm{disc},\psi}(f)$ specializes to a formula

$$I^G_{\mathrm{disc},\psi}(f) = \sum_{\{M\}} |W(M)|^{-1} \sum_{w \in W(M)_{\mathrm{reg}}} |\det(w-1)|^{-1}\,\mathrm{tr}\big(M_{P,\psi}(w)\,\mathcal{I}_{P,\psi}(f)\big), \qquad f \in \widetilde{\mathcal{H}}(G),$$

for its ψ-component. Here $\mathcal{I}_{P,\psi}$ is the direct sum of those subrepresentations π of $\mathcal{I}_P$ such that $c(\pi) = c(\psi)$, while $M_{P,\psi}(w)$ is the corresponding restriction of the intertwining operator $M_P(w)$. We must also have a concrete formula for the terms in the stabilization of $I^G_{\mathrm{disc},\psi}(f)$ if we hope to compare the two decompositions. This is provided by the following supplement to Theorem 2.1, in which the identity (a) is called the stable multiplicity formula in [A].

Theorem 5.1 *[A, Theorem 4.1.2 and Corollary 4.1.3] (F global).*
(a) *Given any $\psi \in \widetilde{\Psi}(N)$, we have a formula*

$$S^G_{\mathrm{disc},\psi}(f) = |\mathcal{S}_\psi|^{-1} m_\psi\,\varepsilon_\psi(s_\psi)\,\sigma(\overline{S}^0_\psi)\,f^G(\psi), \qquad f \in \widetilde{\mathcal{H}}(G),$$

with a product

$$f^G(\psi) = \prod_v f^G_v(\psi_v), \qquad f = \prod_v f_v,$$

of local stable linear forms defined in Theorem 4.1(a), and coefficients $|\mathcal{S}_\psi|^{-1}$, m_ψ, $\varepsilon_\psi(s_\psi)$ *and* $\sigma(\overline{S}^0_\psi)$ *that vanish unless* ψ *lies in the subset* $\widetilde{\Psi}(G)$ *of* $\widetilde{\Psi}(N)$.

(b) *The linear form attached to* G' *in the stabilization of* $I^G_{\mathrm{disc},\psi}(f)$ *satisfies*

$$\widehat{S}^{G'}_{\mathrm{disc},\psi}(f') = \sum_{\{\psi' \in \Psi(G') :\, \psi' \to \psi\}} |\mathcal{S}_\psi|^{-1} \varepsilon_{\psi'}(s_{\psi'})\, \sigma(\overline{S}^0_{\psi'})\, f'(\psi').$$

We again need to append some comments. The first three coefficients $|\mathcal{S}_\psi|^{-1}$, m_ψ and $\varepsilon_\psi(s_\psi)$ in Part (a) have already been discussed. We note that prescriptions of m_ψ and ε_ψ following the statement of Theorem 2.1 carry over to general parameters $\psi \in \widetilde{\Psi}(G)$, so the value

$$\varepsilon_\psi(s_\psi) = \varepsilon_\psi\left(\psi\left(1, \begin{pmatrix} -1 & 0 \\ 0 & -1 \end{pmatrix}\right)\right)$$

is defined. The fourth coefficient $\sigma(\overline{S}^0_\psi)$ is something else . It is the value at $\overline{S}^0_\psi$ of a certain $\mathbb{Q}$-valued function σ, whose domain is the set of all isomorphism classes of complex, connected reductive algebraic groups S_1. The function σ is defined inductively by a combinatorial identity, which mimics the identity between the two decompositions of $I^G_{\mathrm{disc},\psi}(f)$. With this analogy, invariant linear forms correspond to the values of a $\mathbb{Q}$-valued function i on the larger domain of connected components S of (isomorphism classes of) general complex reductive groups, while as we have said, stable linear forms correspond to the values of σ on the subset of connected complex groups. We have no call to review the combinatorial construction here, referring the reader instead to [A, Proposition 4.1.1] and [A2, Theorem 4]. Part (b) of the theorem is considerably simpler. It follows directly from the specialization to G' of the formula for G in (a).

Theorem 5.1(a) is one of the main results of [A]. We cannot expect to prove it from the two decompositions of $I^G_{\mathrm{disc},\psi}(f)$ alone. For we have seen that the spectral decomposition serves only as a definition for $I^G_{\mathrm{disc},\psi}(f)$, while the endoscopic decomposition amounts to an inductive definition of the stable linear form $S^G_{\mathrm{disc},\psi}(f)$. To have a chance of proving any of the theorems, one needs to combine these with something else. The extra ingredients are two similar decompositions from the twisted trace formula for $GL(N)$, which is to say, the trace formula for the connected component

$$\widetilde{G}(N) = GL(N) \rtimes \widetilde{\theta}(N)$$

of the group $\widetilde{G}(N)^+$ over F.

For any $\psi \in \widetilde{\Psi}(N)$, the ψ-discrete part of the (twisted) trace formula for $\widetilde{G}(N)$ is the (twisted) invariant linear form

$$\widetilde{I}^N_{\mathrm{disc},\psi}(\widetilde{f}) = \sum_{\{M\}} |W(M)|^{-1} \sum_{w \in W^N(M)_{\mathrm{reg}}} |\det(w-1)|^{-1}\,\mathrm{tr}\big(M_{P,\psi}(\widetilde{f})\,\mathcal{I}_{P,\psi}(\widetilde{f})\big), \qquad \widetilde{f} \in \widetilde{\mathcal{H}}(N),$$

on the Hecke bi-module $\widetilde{\mathcal{H}}(N)$ of functions on $\widetilde{G}(N,\mathbb{A})$. The terms on the right hand side are analogues for $\widetilde{G}(N)$ of terms in the corresponding expression for $I_{\mathrm{disc},\psi}(f)$ ([A, §4.1], [A2, §3]). They serve again simply to define the linear form on the left. However, they are now to be regarded as known, since they pertain to the representations of $GL(N)$ that are to characterize representations of our groups G. The same goes for the stabilization of $\widetilde{I}^N_{\mathrm{disc},\psi}$. It is an endoscopic decomposition

$$\widetilde{I}^N_{\mathrm{disc},\psi}(\widetilde{f}) = \sum_{G \in \widetilde{\mathcal{E}}_{\mathrm{ell}}(N)} \widetilde{\iota}(N,G)\,\widehat{S}^G_{\mathrm{disc},\psi}(\widetilde{f}^G), \qquad \widetilde{f} \in \widetilde{\mathcal{H}}(N),$$

in terms of the stable linear forms $S^G_{\mathrm{disc},\psi}$ that have already been defined. The indexing set $\widetilde{\mathcal{E}}_{\mathrm{ell}}(N)$ consists of the isomorphism classes of (global) elliptic endoscopic data for $\widetilde{G}(N)$. It includes our groups G, which comprise the subset $\widetilde{\mathcal{E}}_{\mathrm{sim}}(N)$ of simple endoscopic data, as well as composite groups

$$G = G_S \times G_O,$$

in which $\widehat{G}_S$ is symplectic and $\widehat{G}_O$ is orthogonal, and for which $\widehat{S}^G_{\mathrm{disc},\psi}(\widetilde{f}^G)$ is expressed in terms of the relevant products.

The twisted trace formula has now been established in complete generality [LW]. Its stabilization is what remains conditional. However, there is reason to be optimistic that the stabilization of the general twisted trace formula will be established in the not too distant future.

The proof of the theorems is complex, even with the two trace formulas and their stabilizations. I hope I have given some idea of its foundations. Let me just reiterate that the technical supplement Theorem 5.1 of Theorem 2.1 is a fundamental part of the argument. It provides a concrete formula for the stable linear forms $S^G_{\mathrm{disc},\psi}$ in the endoscopic expansion of $\widetilde{I}^N_{\mathrm{disc},\psi}(\widetilde{f})$, in addition to the stable forms $S^{G'}_{\mathrm{disc},\psi'}$ in the endoscopic expansion of of $I^G_{\mathrm{disc},\psi}(f)$.

References

[A] J. Arthur, *The Endoscopic Classification of Representations: Orthogonal and Symplectic Groups*, preprint on http://www.claymath.org/cw/arthur/

[A1] ———, *A stable trace formula* III. *Proof of the main theorems*, Annals of Math. **158** (2003), 769–873.

[A2] ———, *The embedded eigenvalue problem for classical groups*, in *On Certain L-functions*, Clay Mathematics Proceedings, vol. **13**, 2011, 19–33.

[CL1] P.-H. Chaudouard and G. Laumon, *Le lemme fondamental pondéré* I: *constructions géométriques*, Compositio Math. **146** (2010), 1416–1506.

[CL2] ———, *Le lemme fondamental pondéré* II: *Énoncés cohomologiques*, Annals of Math. **176** (2012), 1647–1781.

[CKPS1] J. Cogdell, H. Kim, I. Piatetski-Shapiro, and F. Shahidi, *On lifting from classical groups to* GL_N, Publ. Math. Inst. Hautes Études Sci. **93** (2001), 5–30.

[CKPS2] ———, *Functoriality for the classical groups*, Publ. Math. Inst. Hautes Études Sci. **99** (2004, 163–233.

[GRS] D. Ginzburg, S. Rallis, and D. Soudry, *Generic automorphic forms on* $SO(2n+1)$*: functorial lift to* $GL(2n)$*, endoscopy and base change*, Internat. Math. Res. Notices **14** (2001), 729–764.

[JPS] H. Jacquet, I. Piatetski-Shapiro, and J. Shalika, *Rankin-Selberg convolutions*, Amer. J. Math. **105** (1983), 367–464.

[JS] H. Jacquet and J. Shalika, *On Euler products and the classification of automorphic representations* II, Amer. J. Math. **103** (1981), 777–815.

[Ji] D. Jiang, *Automorphic integral transforms for classical groups* I: *endoscopy correspondences*, to appear in Proc. on the *Conference on Automorphic Forms and Related Geometry: Assessing the Legacy of I.I. Piatetskii-Shapiro*, arXiv:1212.6525.

[KS] R. Kottwitz and D. Shelstad, *Foundations of Twisted Endoscopy*, Astérisque, **255**, 1999.

[LW] J.-P. Labesse and J.-L. Waldspurger, *La formule des traces tordues d'apres le Friday Morning Seminar*, preprint, arXiv:1204.2888.

[L] R. Langlands, *Les débuts d'une formule des traces stables*, Publ. Math. Univ. Paris VII **13**, 1983.

[LS] R. Langlands and D. Shelstad, *On the definition of transfer factors*, Math. Ann. **278** (1987), 219–271.

[Lap] E. Lapid, *On the root number of representations of orthogonal type*, Compositio Math. **140** (2004), 274–286.

[Mu] W. Müller, *The trace class conjecture in the theory of automorphic forms*, Ann. of Math. **130** (1989), 473–529.

[N] B.C. Ngo, *Le lemme fondamental pour les algèbres de Lie*, Publ. Math. I.H.E.S. **111** (2010), 1–269.

[W1] J.-L. Waldspurger, *Le lemme fondamental implique le transfer*, Compositio Math. **105** (1997), 153–236.

[W2] ———, *Endoscopie et changement de caractéristique*, J. Inst. Math. Jussieu **5** (2006), 423–525.

[W3] ———, *L'endoscopie tordue n'est pas si tordue*, Memoirs of AMS **908** (2008).

[W4] ———, *A propos du lemme fondamental pondéré tordu*, Math. Ann. **343**, 103–174.

Department of Mathematics, University of Toronto, Toronto, Canada M5S 2E4.

E-mail: arthur@math.toronto.edu

Automorphic Representations and L-Functions
Editors: D. Prasad, C.S. Rajan, A. Sankaranarayanan, J. Sengupta

Publisher: Hindustan Book Agency, New Delhi, India

The Average Size of the 2-Selmer Group of Jacobians of Hyperelliptic Curves Having a Rational Weierstrass Point

Manjul Bhargava and Benedict H. Gross

Abstract

We prove that when all hyperelliptic curves of genus $n \geq 1$ having a rational Weierstrass point are ordered by height, the average size of the 2-Selmer group of their Jacobians is equal to 3. It follows that (the limsup of) the average rank of the Mordell-Weil group of their Jacobians is at most 3/2.

The method of Chabauty can then be used to obtain an effective bound on the number of rational points on most of these hyperelliptic curves; for example, we show that a majority of hyperelliptic curves of genus $n \geq 3$ with a rational Weierstrass point have fewer than 20 rational points.

1 Introduction

A hyperelliptic curve C of genus $n \geq 1$ over $\mathbb{Q}$ with a marked rational Weierstrass point O has an affine equation of the form

$$y^2 = x^{2n+1} + c_2 x^{2n-1} + c_3 x^{2n-2} + \ldots + c_{2n+1} = f(x) \tag{1.1}$$

with rational coefficients c_m. The point O lies above $x = \infty$, the polynomial $f(x)$ is separable, and the ring of functions which are regular outside of O is the Dedekind domain $\mathbb{Q}[x, y] = \mathbb{Q}[x, \sqrt{f(x)}]$. The change of variable $(x' = u^2 x, y' = u^{2n+1} y)$ results in a change in the coefficients $c'_m = u^{2m} c_m$. Hence we may assume the coefficients are all integers. These integers are unique if we assume further that, for every prime p, the integral coefficients c_m are not all divisible by p^{2m}. In this case we say the coefficients are *indivisible*.

To make the next two definitions, we assume that C is given by its unique equation with indivisible integral coefficients. The discriminant of

$f(x)$ is a polynomial $D(c_2, c_3, \ldots, c_{2n+1})$ of weighted homogeneous degree $2n(2n+1)$ in the coefficients c_m, where c_m has degree m. Since $f(x)$ is separable, this discriminant is nonzero. We define the *discriminant* Δ of the curve C by the formula $\Delta(C) := 4^{2n} D(c_2, c_3, \ldots, c_{2n+1})$, and the (naive) *height* H of the curve C by

$$H(C) := \max\{|c_k|^{2n(2n+1)/k}\}_{k=2}^{2n+1}.$$

We include the expression $2n(2n+1)$ in the definition so that the weighted homogeneous degree of the height function H is the same as that of the discriminant Δ. If $\Delta(C)$ is prime to p, the hyperelliptic curve C has good reduction modulo p. The height $H(C)$ gives a concrete way to enumerate all of the hyperelliptic curves of a fixed genus with a rational Weierstrass point: for any real number $X > 0$ there are clearly only finitely many curves with $H(C) < X$.

The discriminant and height extend the classical notions in the case of elliptic curves (which is the case $n = 1$). Any elliptic curve E over $\mathbb{Q}$ is given by a unique equation of the form $y^2 = x^3 + c_2 x + c_3$, where $c_2, c_3 \in \mathbb{Z}$ and for all primes p: $p^6 \nmid c_3$ whenever $p^4 \mid c_2$. The discriminant is then defined by the formula $\Delta(E) := 2^4(-4c_2^3 - 27c_3^2)$ and the naive height by $H(E) := \max\{|c_2|^3, |c_3|^2\}$.

Recall that the 2-Selmer group $S_2(J)$ of the Jacobian $J = \mathrm{Jac}(C)$ of C is a finite subgroup of the Galois cohomology group $H^1(\mathbb{Q}, J[2])$, which is defined by local conditions and fits into an exact sequence

$$0 \to J(\mathbb{Q})/2J(\mathbb{Q}) \to S_2(J) \to Ш_J[2] \to 0,$$

where $Ш_J$ denotes the Tate-Shafarevich group of J over $\mathbb{Q}$.

The purpose of this paper is to prove the following theorem.

Theorem 1.1 *When all hyperelliptic curves of fixed genus $n \geq 1$ over $\mathbb{Q}$ having a rational Weierstrass point are ordered by height, the average size of the 2-Selmer groups of their Jacobians is equal to* 3.

More precisely, we show that

$$\lim_{X \to \infty} \frac{\sum_{H(C)<X} \#S_2(\mathrm{Jac}(C))}{\sum_{H(C)<X} 1} = 3.$$

In fact, we prove that the same result remains true even when we average over any subset of hyperelliptic curves C defined by a finite set of congruence conditions on the coefficients $c_2, c_3, \ldots, c_{2n+1}$.

Since the 2-rank $r_2(S_2(J))$ of the 2-Selmer group of the Jacobian J bounds the rank of the Mordell-Weil group $J(\mathbb{Q})$, and since $2r_2(S_2(J)) \leq 2^{r_2(S_2(J))} = \#S_2(J)$, by taking averages we immediately obtain:

Corollary 1.2 *When all rational hyperelliptic curves of fixed genus $n \geq 1$ with a rational Weierstrass point are ordered by height, the average 2-rank of the 2-Selmer groups of their Jacobians is at most $3/2$. Thus the average rank of the Mordell-Weil groups of their Jacobians is at most $3/2$.*

Another corollary of Theorem 1.1 is that the average 2-rank of the 2-torsion subgroup of the Tate-Shafarevich group $Ш_J[2]$ is also at most $3/2$. The analogous statements for elliptic curves were proven in [5].

We suspect that the average rank of the Mordell–Weil group is equal to $1/2$. However, the fact that the average rank is at most $3/2$ (independent of the genus) already gives some interesting arithmetic applications. Recall that the method of Chabauty [16], as refined by Coleman [17], yields a finite and effective bound on the number of rational points on a curve over $\mathbb{Q}$ whenever its genus is greater than the rank of its Jacobian. Theorem 1.1 implies

Corollary 1.3 *Let δ_n denote the lower density of hyperelliptic curves of genus n with a rational Weierstrass point satisfying Chabauty's condition. Then $\delta_n \to 1$ as $n \to \infty$.*

Indeed, we have $\delta_n \geq 1-2/(2^n-1)$, so $\delta_n \to 1$ quite rapidly as n gets large. Thus for an asymptotic density of 1 of hyperelliptic curves with a rational Weierstrass point, one can effectively bound the number of rational points.

As an explicit consequence, we may use Theorem 1.1, together with the arguments in [49], to prove the following

Corollary 1.4

(a) *For any $n \geq 2$, a positive proportion of hyperelliptic curves of genus n with a rational Weierstrass point have at most 3 rational points.*

(b) *For any $n \geq 3$, a majority (i.e., a proportion of $> 50\%$) of all hyperelliptic curves of genus n with a rational Weierstrass point have fewer than 20 rational points.*

In fact, we will give a lower bound on the proportion in (a) which is independent of the genus n. In (b), when $n = 2$, we may still deduce that more than a quarter of all such curves have fewer than 20 rational points.

The numbers in Corollary 1.4 can certainly be improved with a more careful analysis. Bjorn Poonen and Michael Stoll have recently shown using Theorem 1.1 and Chabauty's method that (provided $n \geq 3$) a positive proportion of such curves have the point at infinity as their only rational point; moreover, the density of such curves having only this one rational point approaches 1 as $n \to \infty$. See [39] for details and further such results.

Note that Corollary 1.4(a) is also true when $n = 1$: it follows from [6] that a positive proportion of elliptic curves have only one rational point (the origin of the group law). Meanwhile, Corollary 1.4(b) is not expected to be true when $n = 1$, since half of all elliptic curves are expected to have rank one and thus infinitely many rational points. We suspect that 100% of all hyperelliptic curves of genus $n \geq 2$ over $\mathbb{Q}$ with a rational Weierstrass point have only one rational point.

2 The Method of Proof

The proof of Theorem 1.1 follows the argument in [5] for elliptic curves. There the main idea was to use the classical parametrization [7, Lemma 2] of 2-Selmer elements of elliptic curves by certain binary quartic forms to transform the problem into one of counting the integral orbits of the group PGL_2 on the space Sym_4 of binary quartics. Over $\mathbb{Q}$, the group PGL_2 is isomorphic to the special orthogonal group of the three-dimensional quadratic space W of 2×2 matrices of trace zero, with quadratic form given by the determinant, and the representation Sym_4 of PGL_2 on the space of binary quartic forms is given by the action of $\mathrm{SO}(W)$ by conjugation on the space V of self-adjoint operators $T : W \to W$ with trace zero [4].

Let W now be a bilinear space of rank $2n+1$ over $\mathbb{Q}$, where the Gram matrix A of the bilinear form $\langle w, u\rangle$ on $\mathbb{Q}^{2n+1}$ consists of 1's on the anti-diagonal and 0's elsewhere:

$$A = \begin{bmatrix} & & & & 1 \\ & & & 1 & \\ & & \cdot^{\cdot^{\cdot}} & & \\ & 1 & & & \\ 1 & & & & \end{bmatrix}. \tag{2.1}$$

Let $\mathrm{SO}(W)$ be the special orthogonal group of W over $\mathbb{Q}$, i.e., the subgroup of $\mathrm{GL}(W)$ defined by the algebraic equations $\langle gw, gu\rangle = \langle w, u\rangle$ and $\det(g) = +1$. (One can define this group scheme over $\mathbb{Z}$, using the unimodular lattice defined by A, but it is only smooth and reductive over $\mathbb{Z}[1/2]$). Let V be the representation of $\mathrm{SO}(W)$ given by conjugation on the self-adjoint operators $T : W \to W$ of trace zero. With respect to the above basis, an operator T is self-adjoint if and only if its matrix M is symmetric when reflected about the anti-diagonal. The matrix $B = AM$ is then symmetric in the usual sense. The symmetric matrix B is the Gram matrix of the associated bilinear form $\langle w, u\rangle_T = \langle w, Tu\rangle$, and has anti-trace zero.

The coefficients of the monic characteristic polynomial of T, defined by

$$f(x) = \det(xI - T) = (-1)^n \det(xA - B) = x^{2n+1} + c_2 x^{2n-1} + \cdots + c_{2n+1}, \tag{2.2}$$

give $2n$ independent invariant polynomials $c_2, c_3, \ldots c_{2n+1}$ on V, having degrees $2, 3, \ldots, 2n+1$ respectively, which generate the (free) ring of $\mathrm{SO}(W)$-invariants [4]. When the discriminant $\mathrm{disc}(f)$ of f is nonzero, we show how classes in the 2-Selmer group of the Jacobian of the hyperelliptic curve C with equation $y^2 = f(x)$ over $\mathbb{Q}$ correspond to certain orbits of $\mathrm{SO}(W)(\mathbb{Q})$ on $V(\mathbb{Q})$ with these polynomial invariants.

More precisely, to each self-adjoint operator T with invariants $c_2, c_3, \ldots, c_{2n+1}$ satisfying $\Delta(T) := 4^{2n} \cdot \mathrm{disc}(f) \neq 0$, we associate a non-degenerate pencil of quadrics in projective space $\mathbb{P}(W \oplus \mathbb{Q}) = \mathbb{P}^{2n+1}$. Two quadrics generating this pencil are $Q(w, z) = \langle w, w\rangle$ and $Q'(w, z) = \langle w, Tw\rangle + z^2$. The discriminant locus $\mathrm{disc}(xQ - x'Q')$ of this pencil is a homogeneous polynomial $g(x, x')$ of degree $2n+2$ satisfying $g(1, 0) = 0$ and $g(x, 1) = f(x)$. The Fano variety F_T of maximal linear isotropic subspaces of the base locus is smooth of dimension n over $\mathbb{Q}$ and forms a principal homogeneous space for the Jacobian J of the curve C (see [21]).

Both the pencil and the Fano variety have an involution τ induced by the involution $\tau(w, z) = (w, -z)$ of $W \oplus \mathbb{Q}$. The involution τ of F_T has 2^{2n} fixed points over an algebraic closure of $\mathbb{Q}$, which form a principal homogeneous space P_T for the 2-torsion subgroup $J[2]$ of the Jacobian J. Indeed, one can define the structure of a commutative algebraic group on the disconnected variety $G = J \cup F_T$ such that the involution -1_G of G induces the involution τ on the non-trivial component F_T (see [51]). We prove that the isomorphism class of the principal homogeneous space P_T over $\mathbb{Q}$ determines the orbit of T. This gives an injective map from the set of rational orbits of $\mathrm{SO}(W)$ on V with characteristic polynomial $f(x)$ to the set of elements in the Galois cohomology group $H^1(\mathbb{Q}, J[2])$, which classifies principal homogeneous spaces for this finite group scheme.

Theorem 2.1 *Let C be the hyperelliptic curve of genus n which is defined by the affine equation $y^2 = f(x)$, where $f(x)$ is a monic, separable polynomial of degree $2n+1$ over $\mathbb{Q}$. Then the classes in the 2-Selmer group of the Jacobian J of C over $\mathbb{Q}$ correspond bijectively to the orbits of $\mathrm{SO}(W)(\mathbb{Q})$ on self-adjoint operators $T : W \to W$ with characteristic polynomial $f(x)$ such that the associated Fano variety F_T has points over $\mathbb{Q}_v$ for all places v.*

We call such special orbits of $\mathrm{SO}(W)(\mathbb{Q})$ on $V(\mathbb{Q})$ *locally soluble.* Since they are in bijection with elements in the 2-Selmer group of $\mathrm{Jac}(C)$, they are finite in number (whereas the number of rational orbits with characteristic polynomial $f(x)$ turns out to be infinite). Local solubility is a subtle concept. For example, for any d in $\mathbb{Q}^*$ there is an obvious bijection ($T \to dT$) between the rational orbits with characteristic polynomial $f(x)$ and the rational orbits whose characteristic polynomial $f^*(x)$ has coefficients $c_k^* = d^k c_k$. But this bijection does not necessarily preserve the locally soluble orbits, as the hyperelliptic curves C and C^* are isomorphic over $\mathbb{Q}$ only when d is a square.

When $n = 1$, the Fano variety $F = F_T$ is the intersection of two quadrics in $\mathbb{P}^3$. This intersection defines a curve of genus 1 whose Jacobian is the elliptic curve $J = C$. The quotient of F by the involution τ is a curve X of genus 0, and the covering $F \to X$ is ramified at the 4 fixed points comprising P_T. Since the curve X injects into the variety of isotropic lines in the split quadratic space W of dimension $2n+1 = 3$, we conclude that X is isomorphic to $\mathbb{P}^1$ over $\mathbb{Q}$. (For $n > 1$, the quotient X of F by τ embeds as singular subvariety of the Lagrangian Grassmannian of W, with 2^{2n} double points, which in turn embeds in projective space of dimension $2^n - 1$ via the spin representation. The composition is the Kummer embedding of X.) In the case where $n = 1$, once we fix an isomorphism of X with $\mathbb{P}^1$, the Fano variety F has an equation of the form $z^2 = T(x, y)$, where $T(x, y)$ is a binary quartic form. The orbit is locally soluble if and only if the quartic form $T(x, y)$ represents a square in $\mathbb{Q}_v$ for all places v, and the homogeneous space P_T is trivial if and only if $T(x, y)$ has a linear factor over $\mathbb{Q}$. This is the point of view taken in [5], where the latter orbits are called reducible.

In order to obtain Theorem 1.1 from Theorem 2.1, we are reduced to counting the $\mathbb{Q}$-orbits in the representation V that are both locally soluble and have bounded integral invariants. To count these locally soluble orbits, we prove that every such orbit of V with integral polynomial invariants has an integral representative having those same invariants (at least away from the prime 2). By a suitable adaptation of the counting techniques of [2] and [5], we first carry out a count of the total number of integral orbits in these representations having bounded height satisfying certain *irreducibility* conditions over $\mathbb{Q}$ (to be defined). The primary obstacle in this counting, as in representations encountered previously, is that the fundamental region in which one has to count points is not compact but instead has a rather complex system of cusps going off to infinity. A priori, it could be difficult to obtain exact counts of points of bounded height in the cusps of these fundamental regions. We show however that, for all n, most of the integer points in the cusps correspond to points that are reducible; meanwhile, most of the points in the main bodies of these fundamental regions are irreducible. The orbits which contain a reducible point turn out to correspond to the identity classes in the corresponding 2-Selmer groups.

Since not all integral orbits correspond to Selmer elements, the proof of Theorem 1.1 requires a sieve to those points that correspond to locally soluble $\mathbb{Q}$-orbits on V, for which the squarefree sieve of [3] is applied. By carrying out this sieve, we prove that the average occurring in Theorem 1.1 arises naturally as the sum of two contributions. One comes from the main body of the fundamental region, which corresponds to the average number of non-identity elements in the 2-Selmer group and which we show is given by the Tamagawa number ($= 2$) of the adjoint group $\mathrm{SO}(W)$ over $\mathbb{Q}$. The other comes from the cusp of the fundamental region, which counts the average number ($= 1$) of identity elements in the 2-Selmer group. The sum $2 + 1 = 3$ then gives us the average size of the 2-Selmer group, as stated in Theorem 1.1.

To obtain Corollary 1.3, we note that by Theorem 1.1 the density δ_n must satisfy

$$\delta_n \cdot 1 + (1 - \delta_n) \cdot 2^n \leq 3 \tag{2.3}$$

yielding $\delta_n \geq 1 - 2/(2^n - 1)$.

To obtain explicit bounds on the number of rational points on these curves, we follow the arguments in [49]. Assume that $n \geq 2$. Using congruence conditions at the prime $p = 3$, we may assume that all of the hyperelliptic curves C of genus n in our sample have good reduction modulo 3, and that the only point on C over $\mathbb{Z}/3\mathbb{Z}$ is the Weierstrass point at ∞. This simply means that the discriminant Δ of the integral polynomial $f(x)$ is prime to 3 and that the three values $f(0)$, $f(1)$, and $f(-1)$ are all $\equiv -1$ modulo 3. Hence any rational point P on C lies in the 3-adic disc reducing to the Weierstrass point.

Since the average rank of the Jacobians in our sample is at most $3/2$, a positive proportion (indeed, at least a third) of these Jacobians have rank zero or one. Assume that this is the case for such a Jacobian J, let T be the (finite) torsion subgroup of $J(\mathbb{Q}_3)$, and let A be the $\mathbb{Z}_3$-submodule of $J(\mathbb{Q}_3)/T$ which is generated by $J(\mathbb{Q})$. Then the $\mathbb{Z}_3$-rank of A is either zero or one. Let Ω be the $\mathbb{Q}_3$-vector space of invariant differentials on J, or equivalently the regular 1-forms on the curve C over $\mathbb{Q}_3$. The logarithm on the formal group gives a $\mathbb{Z}_3$-bilinear pairing $\Omega \times J(\mathbb{Q}_3)/T \to \mathbb{Q}_3$ with trivial left and right kernels. Let Ω_0 be the subspace of Ω which annihilates the submodule A. The codimension of Ω_0 is either zero or one, and any differential ω in the subspace Ω_0 pairs trivially with the class of the divisor $(P) - (O)$, for any rational point P. Since P and O both lie in the disc reducing to the Weierstrass point, this pairing is computed by 3-adic integration. Namely, the function $z = x^n/y$ is a uniformizing parameter on this disc which vanishes at O and is taken to its negative by the hyperelliptic involution. If $z(P)$ is the parameter of the rational point P and we expand $\omega = (a_0 + a_2 z^2 + a_4 z^4 + \cdots)dz$, then the integral of ω from O to P is given by the value of the formal integral $F(z) = a_0 z + (a_2/3)z^3 + (a_4/5)z^5 + \cdots$ at the point $z(P)$ in the maximal ideal of $\mathbb{Z}_3$. We can choose ω integral, and since we are in a subspace Ω_0 of codimension at most one, the reduction of ω has at most a double zero at the Weierstrass point. Hence the coefficients a_{2i} are all integral, and either a_0 or a_2 is a unit. It follows that the formal integral $F(z)$ has at most three zeroes in the disc $3\mathbb{Z}_3$, and hence that there are at most three rational points on the curve C, proving Corollary 1.4(a).

To obtain Corollary 1.4(b), we note that a density of $1 - 1/7$ of hyperelliptic curves with a rational Weierstrass point have good reduction at 7. The congruence version of Theorem 1.1 implies that among these curves, the density of curves having rank ≤ 2 is at least $1 - 2/(2^3 - 1)$. Since $\#C(\mathbb{F}_7) \leq 15$, Stoll's refinement [49, Cor. 6.7] of Coleman's effective bound for Chabauty's method then results in $\#C(\mathbb{Q}) \leq 15 + 2 \cdot 2 = 19$ for such curves. Hence the majority (indeed greater than $(1 - \frac{1}{7})(1 - \frac{2}{2^{2+1}-1}) = 30/49 \approx 61\%$) of our curves have fewer than 20 rational points, as desired.

3 A Representation of the Orthogonal Group

Let W be a lattice of rank $2n+1$ having a non-degenerate integral bilinear form with signature $(n+1, n)$. By this we mean that W is a free abelian group of rank $2n+1$ with a non-degenerate symmetric bilinear pairing $\langle\ ,\ \rangle : W \times W \to \mathbb{Z}$ having signature $(n+1, n)$ over $\mathbb{R}$. It follows that the determinant of W is equal to $(-1)^n$.

Such a lattice is unique up to isomorphism [43, Ch. V]. It therefore has an ordered basis

$$\{e_1, e_2, \ldots, e_n, u, f_n, \ldots, f_2, f_1\}$$

with inner products given by

$$\langle e_i, e_j\rangle = \langle f_i, f_j\rangle = \langle e_i, u\rangle = \langle f_i, u\rangle = 0,$$

$$\langle e_i, f_j\rangle = \delta_{ij},$$

$$\langle u, u\rangle = 1.$$

The Gram matrix A of the bilinear form with respect to this basis (which we will call the *standard basis*) is the anti-diagonal matrix (2.1).

The lattice W gives a split orthogonal space $W \otimes k$ over any field k with $\mathrm{char}(k) \neq 2$. This space has dimension $2n+1$ and determinant $\equiv (-1)^n$ in k^*/k^{*2}. Such an orthogonal space over k is unique up to isomorphism, and has $\mathrm{disc}(W) = (-1)^n \det(W) \equiv 1$ (see [34]).

Let $T : W \to W$ be a $\mathbb{Z}$-linear transformation (i.e., an endomorphism of the abelian group W). We define the *adjoint transformation* T^* uniquely by the formula

$$\langle Tv, w\rangle = \langle v, T^*w\rangle.$$

The matrix M of T^* with respect to our standard basis is obtained from the matrix of T by reflection around the anti-diagonal. In particular, we have the identity $\det(T) = \det(T^*)$. We say that T is *self-adjoint* if $T = T^*$. Any self-adjoint transformation T defines a new symmetric bilinear form on W by the formula

$$\langle v, w \rangle_T = \langle v, Tw \rangle.$$

This bilinear form has symmetric Gram matrix $B = AM$ in our standard basis.

We say that a $\mathbb{Z}$-linear transformation $g : W \to W$ is *orthogonal* if it preserves the bilinear structure on W, i.e., $\langle gv, gw \rangle = \langle v, w \rangle$ for all $w \in W$. In that case g is invertible, with $g^{-1} = g^*$, and $\det(g) = \pm 1$. We define the group scheme $G = \mathrm{SO}(W)$ over $\mathbb{Z}$ by

$$G := \mathrm{SO}(W) = \{g \in \mathrm{GL}(W) : g^* g = 1, \ \det(g) = 1\}.$$

Over $\mathbb{Z}[1/2]$ the group scheme G is smooth and reductive. It defines the split reductive adjoint group with root system of type B_n (see [20]). In particular, for every field k with $\mathrm{char}(k) \neq 2$, the group $G(k)$ gives the points of the split special orthogonal group SO_{2n+1} of the space $W \otimes k$.

We note that the group G is not smooth over $\mathbb{Z}_2$. Indeed, the involution $g \in G(\mathbb{Z}/2\mathbb{Z})$ with matrix

$$\begin{bmatrix} 1 & & & & 1 \\ & 1 & & & \\ & & \ddots & & \\ & & & 1 & \\ & & & & 1 \end{bmatrix} \tag{3.1}$$

does not lift to an element of $G(\mathbb{Z}_2)$, or even to an element of $G(\mathbb{Z}/4\mathbb{Z})$. The theory of Bruhat and Tits provides a smooth model $\mathcal{G}$ over $\mathbb{Z}_2$ with the same points in étale $\mathbb{Z}_2$-algebras (see [12]), namely, $\mathcal{G}$ is the normalizer of a parahoric subgroup P of $G(\mathbb{Q}_2)$. When $n = 1$, the subgroup P is an Iwahori subgroup, and when $n \geq 2$, the subgroup P is a maximal parahoric subgroup. The reductive quotient of P is the split group SO_{2n}. We will not need this smooth model in what follows.

If g is orthogonal and T is self-adjoint, then the transformation $gTg^{-1} = gTg^*$ is also self-adjoint. The self-adjoint operators form a free submodule Y of $\mathrm{End}(W)$, of rank $(2n+1)(n+1)$, and the action of special orthogonal transformations by conjugation $T \to gTg^{-1}$ gives a linear representation $G \to \mathrm{GL}(Y)$. Over $\mathbb{Z}[1/2]$ the representation Y is orthogonal, via the non-degenerate symmetric pairing $\langle T_1, T_2 \rangle = \mathrm{Trace}(T_1 T_2)$. If we identify the module of self-adjoint operators with the module of symmetric matrices, via the map $B = AM$ described above, then Y is isomorphic to the symmetric square of the standard representation W over $\mathbb{Z}[1/2]$.

The trace of an operator gives a nonzero, G-invariant linear form $Y \to \mathbb{Z}$. We define the representation V of G over $\mathbb{Z}$ as its kernel, i.e.,

$$V = \{T \in \mathrm{End}(W) \mid T = T^*,\ \mathrm{Trace}(T) = 0\}.$$

The free $\mathbb{Z}$-module $V(\mathbb{Z})$ has rank $2n^2 + 3n$, and can be identified with the submodule of symmetric matrices having anti-trace zero (where the anti-trace is the sum of the matrix entries on the anti-diagonal). We are going to study the orbits of $G(\mathbb{Z})$ on $V(\mathbb{Z})$.

We note that the representation V of G is algebraically irreducible in all characteristics which do not divide 2 or $2n+1$. If the characteristic divides $2n + 1$, then V contains a copy of the trivial representation (as the scalar multiples of the identity operator I are self-adjoint and have trace zero) and the quotient module is algebraically irreducible. In characteristic 2, there are even more irreducible factors in the Jordan-Hölder series of V.

4 Orbits With a Fixed Characteristic Polynomial

Let k be a field, with $\mathrm{char}(k) \neq 2$. In this section and the next, we consider the orbits of the group $G(k) = \mathrm{SO}(W)(k)$ on the representation $V(k) = V \otimes k$. A discussion of these orbits can also be found in our expository paper [4].

Since the group $G(k)$ acts on the operators T in $V(k)$ by conjugation $T \to gTg^{-1}$, the characteristic polynomial of T is an invariant of its orbit. Since T has trace zero, this monic characteristic polynomial has the form

$$f(x) = \det(xI - T) = x^{2n+1} + c_2 x^{2n-1} + c_3 x^{2n-2} + \cdots + c_{2n+1} \tag{4.1}$$

with coefficients $c_m \in k$. The coefficients c_m are polynomial invariants of the representation, with $\deg(c_m) = m$. These invariants are algebraically independent, and generate the full ring of polynomial invariants on V (cf. [4]). An important polynomial invariant, of degree $2n(2n + 1)$, is the multiple

$$\Delta = \Delta(c_2, c_3, \ldots, c_{2n+1}) = 4^{2n} \operatorname{disc} f(x)$$

of the discriminant of the characteristic polynomial of T, which gives the discriminant Δ of the associated hyperelliptic curve $y^2 = f(x)$ (see [32]). The discriminant $\Delta(T)$ is nonzero precisely when the polynomial $f(x)$ is separable over k, so has $2n + 1$ distinct roots in a separable closure k^s.

Since the characteristic polynomial is an invariant of the $G(k)$-orbit, it makes sense to describe the set of orbits with a fixed characteristic polynomial $f(x)$. We begin by showing that, for any separable $f(x)$, there is a naturally associated orbit of $G(k)$ on $V(k)$, which we call the *distinguished orbit*, that has associated characteristic polynomial $f(x)$.

4.1 The distinguished orbit

We assume throughout that $f(x)$ is *separable* (i.e., $\mathrm{disc}(f) \neq 0$) and is of the form (4.1).

Proposition 4.1 *The group $G(k)$ acts simply transitively on the pairs (T, X), where T is a self-adjoint operator in $V(k)$ with characteristic polynomial $f(x)$ and $X \subset W(k)$ is a maximal isotropic subspace of dimension n with $T(X) \subset X^{\perp}$.*

The action of g on pairs (T, X) is by the formula $(gTg^{-1}, g(X))$. As a corollary, we can identify a *distinguished orbit* of $G(k)$ on the set of self-adjoint operators with characteristic polynomial $f(x)$, namely, the orbit consisting of those T for which a maximal isotropic subspace X exists over k satisfying $T(X) \subset X^{\perp}$. We will frequently use S to denote a self-adjoint operator in this distinguished orbit.

Next, let $L = k[x]/(f(x))$. Since we have assumed that $f(x)$ is separable, L is an étale k-algebra of rank $2n + 1$.

Proposition 4.2 *The stabilizer G_S of a vector S in the distinguished orbit with characteristic polynomial $f(x)$ is (uniquely) isomorphic to the finite commutative étale group scheme D of order 2^{2n} over k, which is the kernel of the norm map $\mathrm{Res}_{L/k}(\mu_2) \to \mu_2$.*

To prove these results, we need a method to construct orbits over k with a fixed separable characteristic polynomial $f(x)$. Let β be the image of x in L, so a power basis of L over k is given by $\{1, \beta, \beta^2, \ldots, \beta^{2n}\}$. We define a symmetric bilinear form $(\ ,\ )$ on L by the formula

$$(\lambda, \nu) := \text{ the coefficient of } \beta^{2n} \text{ in the product } \lambda\nu. \tag{4.2}$$

Since $f(x)$ is separable, it is relatively prime to the derivative $f'(x)$ in $k[x]$ and the value $f'(\beta)$ is a unit in L^*. We then have another formula for this inner product which is due to Euler [45, Ch III, §6]:

$$(\lambda, \nu) = \mathrm{Trace}_{L/k}(\lambda\nu/f'(\beta)).$$

The orthogonal space we have defined on L is non-degenerate, of dimension $2n + 1$ and discriminant $\equiv 1$ over k. Indeed, the entries of its Gram matrix with respect to the power basis are 0 above the anti-diagonal, and 1 on the anti-diagonal. We next observe that the k-subspace M of L spanned by $\{1, \beta, \beta^2, \ldots, \beta^{n-1}\}$ has dimension n and is isotropic for the bilinear form, so the orthogonal space L is split [42]. Hence we have an isometry $\theta : L \to W(k)$ defined over k. This isometry is well-defined up to composition with an orthogonal transformation g of $W(k)$—any other isometry has the form $\theta' = g\theta$ where g is an element of $\mathrm{O}(W) = \mathrm{SO}(W) \times \langle \pm I \rangle$.

The k-linear map $\beta : L \to L$ given by multiplication by β is a self-adjoint transformation with characteristic polynomial equal to $f(x)$. Hence the operator $T = \theta\beta\theta^{-1} : W(k) \to W(k)$ is self-adjoint with the same characteristic polynomial. If we compose the isometry θ with the orthogonal transformation g, we obtain an operator gTg^{-1} in the same $G(k) = \mathrm{SO}(W)(k)$ orbit. This follows from the fact that the dimension of W is odd, so the orthogonal group and its special orthogonal subgroup have the same orbits on the representation V.

This orbit we have just constructed is distinguished, as $\beta(M)$ is contained in $M^{\perp} = \{1, \beta, \ldots, \beta^n\}$. Hence, $X = \theta(M)$ is a maximal isotropic subspace of W over k with $T(X) \subset X^{\perp}$, and (T, X) is a distinguished pair in V, defined over k.

To see that $G(k)$ acts simply-transitively on the pairs (T, X), we first show that this is true when the field $k = k^s$ is *separably closed.* Since $f(x)$ is separable, it is also the minimal polynomial of T on W. By the cyclic decomposition theorem, the centralizer of T in $\mathrm{End}(W)$ is the algebra $k[T]$ generated by T, which is isomorphic to L. When $k = k^s$ is separably closed, this algebra is isomorphic to the product of $2n+1$ copies of the field k. If T' has the same characteristic polynomial, then $T' = hTh^{-1}$, where $h \in \mathrm{GL}(W)(k)$ is well-defined up to right multiplication by $(k[T])^*$. Since both operators T, T' are self-adjoint, the composition h^*h centralizes T, so $\alpha = h^*h$ lies in $(k[T])^*$. In the separably closed case, every unit in the algebra $k[T]$ is a square, so we may modify h by a square root of α to arrange that $h^*h = 1$. Modifying h by $\pm I$ if necessary, we may also arrange that $\det(h) = 1$. It follows that T' is in the same $G(k)$ orbit as T.

The subgroup $G_T(k)$ stabilizing T is the intersection of two subgroups of $\mathrm{GL}(W)(k)$. The first is the centralizer $(k[T])^*$, where all the operators are self-adjoint ($g^* = g$). The second is the subgroup $\mathrm{SO}(W)(k)$, where all the operators are special orthogonal ($g^* = g^{-1}, \ \det(g) = 1$) . Hence $G_T(k)$ is isomorphic to the group

$$D = \{g \in L^* : g^2 = 1, \ N(g) = 1\}.$$

Since L is the product of $2n+1$ copies of k, the group D is an elementary abelian 2-group of order 2^{2n}. For future reference, we record what we have just proved.

Proposition 4.3 *When $k = k^s$ is separably closed, the group $G(k) = \mathrm{SO}(W)(k)$ acts transitively on the set of self-adjoint operators with a fixed separable characteristic polynomial $f(x)$. The stabilizer $G_T(k)$ of an operator T in this orbit is an elementary abelian 2-group of order 2^{2n}.*

We remark that $G_T(k^s)$ embeds as a Jordan subgroup of $\mathrm{SO}(W)(k^s)$, stabilizing a decomposition of W into a direct sum of orthogonal lines. This subgroup is unique up to conjugacy, and is not contained in any maximal torus. It is its own centralizer, and its normalizer is a finite group isomorphic to the Weyl group $W(D_{2n+1}) = 2^{2n}.S_{2n+1}$, where S_{2n+1} is the symmetric group (see [31]).

We continue with the proof of Proposition 4.1 in the case when k is separably closed. The finite stabilizer $G_T(k)$ acts on the set of maximal isotropic subspaces X in V with $T(X)$ contained in $X^{\perp}$, and we must show that it acts simply transitively. There are precisely 2^{2n} such subspaces X defined over an algebraic closure of k: these are the points of the Fano variety of the complete intersection of the two quadrics

$$Q(w) = \langle w, w\rangle, \quad Q_T(w) = \langle w, Tw\rangle$$

in $\mathbb{P}(W)$. For an excellent discussion of the Fano variety of the complete intersection of two quadrics over the complex numbers, see [21]. This material was also treated in the unpublished Cambridge Ph.D. thesis of M. Reid [40]. A full treatment of the arithmetic theory sketched below will appear in the Harvard Ph.D. thesis of X. Wang [51].

The pencil $xQ - x'Q_T$ is non-degenerate. Using the matrices in our standard basis of W we have the formula

$$\mathrm{disc}(xA - B) = (-1)^n \det(xA - AM) = (-1)^n \det(A)\det(xI - M) = f(x)$$

and $f(x)$ is separable. To show that these subspaces X of W are all defined over the separably closed field k, and that $G_T(k)$ acts simply-transitively on them, we consider one such isotropic subspace in the model orthogonal space L, with form (λ, ν) given by (4.2) and the self-adjoint operator given by multiplication by β. The subspace M with basis $\{1, \beta, \ldots, \beta^{n-1}\}$ is maximal isotropic, and $\beta(M)$ is contained in $M^{\perp}$. The subgroup of L^* which fixes M under multiplication is clearly k^*. Hence the subgroup of D which fixes M is trivial and the finite group D must act simply-transitively on the 2^{2n} such subspaces (as it has the same order). This also shows that all of the 2^{2n} points of the Fano variety are defined over the separably closed field k. Translating this to W via the isometry θ completes the proof of Proposition 4.1 when the base field k is separably closed.

For general fields k, we use Galois descent. Let k^s denote a separable closure of k and let $\mathrm{Gal}(k^s/k)$ denote the Galois group. Consider two distinguished pairs (T, X) and (T', X') which are defined over k. By the simple transitivity result we have just proved over k^s, there is a unique element g in the group $\mathrm{SO}(W)(k^s)$ with $g(T, X) = (T', X')$. For any σ in $\mathrm{Gal}(k^s/k)$, the element g^σ has the same property. Hence $g = g^\sigma$ lies in $\mathrm{SO}(W)(k)$, which completes the proof of Proposition 4.1 for general k.

To determine the stabilizer of a vector in the distinguished orbit, as a finite group scheme over k, we again may use the model orthogonal space L with inner product (λ, ν) given by (4.2) and the distinguished self-adjoint operator given by multiplication by β. The centralizer D of β has points

$$D(E) = \{g \in (L \otimes E)^* : g^2 = 1, \; N(g) = 1\}.$$

in any k-algebra E. Hence $D = \mathrm{Res}_{L/k}(\mu_2)_{N=1}$ is the finite group scheme over k described in Proposition 4.2. The choice of isometry $\theta : L \to W$ gives a vector $S = \theta\beta\theta^{-1}$ in the distinguished orbit and an isomorphism of stabilizers $G_S = \theta D \theta^{-1}$. Since θ is well-defined up to left multiplication by G_S, which is a commutative group scheme, the isomorphism with D is uniquely defined. This completes the proof of Proposition 4.2.

Here is a condition on the Gram matrix B of the bilinear form $\langle v, Tw\rangle$ which implies that the orbit of T is distinguished.

Proposition 4.4 *An orbit of* $\mathrm{SO}(W)(k)$ *on the representation* $V(k)$ *having nonzero discriminant is distinguished if and only if it contains a vector* T *whose Gram matrix* $B = AM$ *with respect to the standard basis has the form*

$$\begin{pmatrix} 0 & 0 & \cdots & 0 & 0 & * & * \\ 0 & 0 & \cdots & 0 & * & * & * \\ \vdots & \vdots & \iddots & \iddots & \vdots & \vdots & \vdots \\ 0 & 0 & \iddots & \cdots & \vdots & \vdots & \vdots \\ 0 & * & \cdots & \cdots & * & * & * \\ * & * & \cdots & \cdots & * & * & * \\ * & * & \cdots & \cdots & * & * & * \end{pmatrix}. \tag{4.3}$$

Proof If the Gram matrix B of T has the form (4.3), then the subspace X spanned by the basis elements $e_1, e_2, \ldots, e_n$ is isotropic and $T(X)$ is contained in $X^\perp$, which is spanned by the elements $e_1, e_2, \ldots, e_n, u$. Hence the orbit is distinguished.

Conversely, if the operator T lies in a distinguished orbit, we may identify W with the space $L = k[x]/(f(x))$ having bilinear form (λ, ν) given by (4.2), via an isometry $\theta : L \to W$. The self-adjoint operator $T = \theta\beta\theta^{-1}$ has Gram matrix of the form (4.3) with respect to the basis $1, \beta, \beta^2, \ldots, \beta^{2n}$ of

L. It also has this form with respect to the basis $1, p_1(\beta), p_2(\beta), \ldots, p_{2n}(\beta)$ for any monic polynomials $p_i(x)$ of degree i. But we can choose the monic polynomials $p_i(x)$ so that the inner products $(p_i(\beta), p_j(\beta))$ on L match the inner products of our standard basis elements on W. There is then an isometry $\theta' = g\theta : L \to V$ taking this basis to our standard basis, and the vector $\theta'\beta\theta'^{-1} = gTg^{-1}$ in the orbit of T has Gram matrix B of the desired form.

□

4.2 The remaining orbits via Galois cohomology

To describe the remaining orbits with a fixed separable characteristic polynomial, we use Galois cohomology. Let k^s denote a separable closure of k. If M is a commutative finite étale group scheme over k we let $H^i(k, M) = H^i(\text{Gal}(k^s/k), M(k^s))$ be the corresponding Galois cohomology groups. Similarly, if J is a smooth algebraic group over k, we let $H^1(k, J) = H^1(\text{Gal}(k^s/k), J(k^s))$ be the corresponding pointed set of first cohomology classes [44, Ch I, §5]. If we have a homomorphism $M \to J$ over k, we obtain an induced map of pointed sets $H^1(k, M) \to H^1(k, J)$. The kernel of this map is by definition the subset of classes in $H^1(k, M)$ that map to the trivial class in $H^1(k, J)$.

Let S be a self-adjoint operator with characteristic polynomial $f(x)$ in the distinguished orbit of $G = \text{SO}(W)$ over k, and let G_S be the finite étale subgroup stabilizing S. Let T be any self-adjoint operator with this characteristic polynomial over k. By Proposition 4.3, the vector T is in the same orbit as S over the separable closure k^s, so $T = gSg^{-1}$ for some $g \in \text{SO}(W)(k^s)$. For any $\sigma \in \text{Gal}(k^s/k)$, the element $c_\sigma = g^{-1}g^\sigma$ lies in the subgroup $G_S(k^s)$, and the map $\sigma \to c_\sigma$ defines a 1-cocycle on the Galois group with values in $G_S(k^s)$, whose cohomology class depends only on the $G(k)$-orbit of T. This class clearly lies in the kernel of the map

$$\eta : H^1(k, G_S) \to H^1(k, G). \tag{4.4}$$

Conversely, if we have a cocycle c_σ whose cohomology class lies in the kernel of η, then it has the form $g^{-1}g^\sigma$ and the operator $T = gSg^{-1}$ is defined on W over k. Since T has the same characteristic polynomial as S, we have established the following.

Proposition 4.5 *The map which takes an operator T to the cohomology class of the cocycle c_σ induces a bijection between the set of orbits of $G(k)$ on self-adjoint operators T with characteristic polynomial $f(x)$ and the set of elements in the kernel of the map $\eta : H^1(k, G_S) \to H^1(k, G)$. The trivial class in $H^1(k, G_S)$ corresponds to the distinguished orbit.*

We will make the map η more explicit shortly, via a computation of the pointed cohomology sets of G_S and G. But we can use our discussion of the Fano variety to give an explicit principal homogeneous space P_T for the finite group scheme G_S over k, corresponding to the self-adjoint operator T. Choose $g \in G(k^s)/G_S(k^s)$ with $gSg^{-1} = T$. Since G_S is abelian we obtain a unique isomorphism of the stablizers $G_S \cong G_T$ over k^s, which is given by conjugation by any g in this coset. The unicity implies that this isomorphism is defined over k. It therefore suffices to construct a principal homogeneous space P_T for the finite group scheme G_T. We let P_T be the set of maximal isotropic subspaces $X \subset W$ defined over k^s with the property that $T(X) \subset X^{\perp}$. This set is finite, of cardinality 2^{2n}. The Galois group $\mathrm{Gal}(k^s/k)$ acts, so P_T has the structure of a reduced scheme of dimension zero over k and forms a principal homogeneous space for the group scheme G_T. The isomorphism class of the principal homogeneous space P_T over k depends only on the orbit of T, and corresponds to the cohomology class of the cocycle c_σ. Note that the scheme P_T has a k-rational point if and only if T lies in the distinguished orbit.

We now make the above map η of pointed sets more explicit, using our model orthogonal space L with form (λ, ν) given by (4.2). First, since $2n+1$ is odd, we have a splitting

$$\mathrm{Res}_{L/k}(\mu_2) = \mu_2 \oplus D.$$

This allows us to compute the Galois cohomology groups of D using Kummer theory. We find that

$$\begin{aligned} H^0(k, D) &= L^*[2]_{N=1} \\ H^1(k, D) &= (L^*/L^{*2})_{N\equiv 1} \end{aligned}$$

where we use the subscript $N \equiv 1$ to mean the subgroup of elements in L^*/L^{*2} whose norms are squares in k^*.

Let α be a class in $H^1(k, D)$, viewed as an element of the group L^* whose norm to k^* is a square. The class $\eta(\alpha)$ lies in the pointed set $H^1(k, \mathrm{SO}(L))$, which classifies non-degenerate orthogonal spaces of the same dimension $2n+1$ and discriminant $\equiv 1$ as our model space L over k (see [30]). Unwinding the definitions as in [4], we find that the orthogonal space corresponding to the class $\eta(\alpha)$ is isometric to L with the modified bilinear form $(\lambda, \nu)_\alpha =$ the coefficient of β^{2n} in the product $\alpha\lambda\nu$. Equivalently, we have the formula:

$$(\lambda, \nu)_\alpha = \mathrm{Trace}_{L/k}(\alpha\lambda\nu/f'(\beta)).$$

We note that the isomorphism class of this space depends only on the element α in the quotient group (L^*/L^{*2}), and that the condition that the norm of α is a square in k^* translates into the fact that the discriminant of the form $(\lambda, \nu)_\alpha$ is $\equiv 1$.

If we translate this over to the orthogonal space W via the isometry θ, we see that a class α in the group $H^1(k, G_S) = H^1(k, D) = (L^*/L^{*2})_{N\equiv 1}$ lies in the kernel of the map $\eta : H^1(k, G_S) \to H^1(k, \mathrm{SO}(W))$ precisely when the form $(\lambda, \nu)_\alpha$ on L gives a split orthogonal space over k. In that case α gives an orbit of self-adjoint operators T on W, and the principal homogeneous space P_T splits over the extension of L where α is a square. We will produce examples of non-trivial α where the bilinear form $(\lambda, \nu)_\alpha$ is split in the next section, using the Jacobians of hyperelliptic curves.

5 Hyperelliptic Curves

We now use some results in algebraic geometry, on hyperelliptic curves and the Fano variety of the complete intersection of two quadrics in $\mathbb{P}(W \oplus k)$, to produce classes in the kernel of the map in cohomology $\eta : H^1(k, G_S) \to H^1(k, \mathrm{SO}(W))$, and hence orbits with a fixed separable characteristic polynomial $f(x)$.

Let C be the smooth projective hyperelliptic curve of genus n over k with affine equation $y^2 = f(x)$ and a k-rational Weierstrass point O above $x = \infty$. The functions on C which are regular outside of O form an integral domain:

$$H^0(C - O, \mathscr{O}_{C-O}) = k[x, y]/(y^2 = f(x)) = k[x, \sqrt{f(x)}].$$

The complete curve C is covered by this affine open subset U_1, together with the affine open subset U_2 associated to the equation $w^2 = v^{2n+2}f(1/v) = c_{2n+1}v^{2n+2} + \cdots + c_2v^3 + v$ and containing the point $O = (0, 0)$. The gluing of U_1 and U_2 is by $(v, w) = (1/x, y/x^{n+1})$ and $(x, y) = (1/v, w/v^{n+1})$ wherever these maps are defined.

5.1 The 2-torsion subgroup

Let $J = \mathrm{Pic}^0(C)$ denote the Jacobian of the hyperelliptic curve C over k, which is an abelian variety of dimension n over k, and let $J[2]$ denote the 2-torsion subgroup of J. This is a finite étale group scheme of order 2^{2n}. Recall that we have defined a finite group scheme $D = (\mathrm{Res}_{L/k}\,\mu_2)_{N=1}$ of this order, associated to the étale algebra $L = k[x]/(f(x))$. We begin by constructing an isomorphism $D \cong J[2]$ of finite group schemes over k.

The other Weierstrass points $P_\gamma \neq O$ of $C(k^s)$ correspond bijectively to k-algebra homomorphisms $\gamma : L \to k^s$: we have $x(P_\gamma) = \gamma(\beta)$ and $y(P_\gamma) = 0$. Associated to such a point we have the divisor $d_\gamma = (P_\gamma) - (O)$

of degree zero. The divisor class of d_γ lies in the 2-torsion subgroup $J[2](k^s)$ of the Jacobian, as $2(d_\gamma)$ is the divisor of the function $(x-\gamma(\beta))$. The Riemann-Roch theorem shows that the classes d_γ generate the finite group $J[2](k^s)$, and satisfy the single relation $\sum(d_\gamma)\equiv 0$. Indeed, this sum is the divisor of the function y. This gives an isomorphism $\mathrm{Res}_{L/k}(\mu_2)/\mu_2 \cong J[2]$. We recall that D is defined as the kernel of the map $\mathrm{Res}_{L/k}(\mu_2)\to\mu_2$.

Since the degree of L over k is odd, the composition $D\to\mathrm{Res}_{L/k}(\mu_2)\to\mathrm{Res}_{L/k}(\mu_2)/\mu_2$ is an isomorphism. We summarize what we have proved:

Proposition 5.1 *There is an isomorphism of finite group schemes $G_S\to J[2]$ over k, where G_S is the stabilizer of a point S in the distinguished orbit with characteristic polynomial $f(x)$ and $J[2]$ is the 2-torsion subgroup of the Jacobian of the hyperelliptic curve with equation $y^2=f(x)$. Both group schemes are isomorphic to $D=(\mathrm{Res}_{L/k}(\mu_2))_{N=1}$; their points over any extension field E of k correspond bijectively to the monic factorizations $f(x)=g(x)h(x)$ of $f(x)$ over E.*

The exact sequence $0\to J[2]\to J\to J\to 0$ of commutative group schemes over k^s gives an exact descent sequence in Galois cohomology:

$$0\to J(k)/2J(k)\to H^1(k,J[2])\to H^1(k,J)[2]\to 0.$$

The middle term $H^1(k,J[2])$ can be identified with the group $H^1(k,D)=(L^*/L^{*2})_{N\equiv 1}$ via the above isomorphism. We need an explicit description of the coboundary map

$$\delta: J(k)/2J(k)\to (L^*/L^{*2})_{N\equiv 1}.$$

If $P=(a,b)$ is a point of C over k with $b\neq 0$, and $D=(P)-(O)$ is the corresponding divisor class in $J(k)$, then (cf. [41], [48, p. 7]) we have

$$\delta(D)\equiv(a-\beta).$$

5.2 A subgroup in the kernel of η

For any self-adjoint operator S in the distinguished orbit, we have constructed an isomorphism from D to the stabilizer G_S in $\mathrm{SO}(W)$. Hence we may identify the cohomology groups $H^1(k,J[2])=H^1(k,G_S)$. With this identification, we have the following important result.

Proposition 5.2 *The subgroup $J(k)/2J(k)$ of $H^1(k,J[2])$ lies in the kernel of the map $\eta: H^1(k,D)\to H^1(k,\mathrm{SO}(W))$. The classes α which lie in $J(k)/2J(k)$ correspond bijectively to the orbits of $\mathrm{SO}(W)(k)$ on $V(k)$ with characteristic polynomial $f(x)$, where the homogeneous space F_α for the Jacobian J (to be defined below) has a k-rational point.*

Proof This is proved in [4, Prop. 6], and we briefly recall the argument here. We first need to make the homomorphism from $H^1(k, D) = H^1(k, J[2])$ to $H^1(k, J)[2]$ which arises in the descent sequence more explicit. That is, we need to associate to each class α in the group $H^1(k, D) = (L^*/L^{*2})_{N\equiv 1}$ a principal homogeneous space F_α of order 2 for the Jacobian J over k. The classes in the subgroup $J(k)/2J(k)$ will correspond to the homogeneous spaces with a k-rational point. We will use the existence of k-rational points on F_α to show that the bilinear form $(\ ,\ )_\alpha$ on L is split. Thus every α in $J(k)/2J(k)$ corresponds to a class in the kernel of η, and hence to an orbit of $\mathrm{SO}(W)(k)$ on $V(k)$.

More generally, let α be any class in $H^1(k, D) = (L^*/L^{*2})_{N\equiv 1}$. The vector space $L \oplus k$ has dimension $2n+2$ over k. Consider the two quadrics on $\mathbb{P}(L \oplus k)$:

$$Q(\lambda, a) = (\lambda, \lambda)_\alpha = \mathrm{Trace}(\alpha\lambda^2/f'(\beta)),$$

$$Q'(\lambda, a) = (\lambda, \beta\lambda)_\alpha + a^2 = \mathrm{Trace}(\alpha\beta\lambda^2/f'(\beta)) + a^2.$$

The pencil $uQ - vQ'$ is non-degenerate and contains exactly $2n+2$ singular elements: the quadric Q at $v = 0$ and the $2n+1$ quadrics $\gamma Q - Q'$ at the points $(\gamma, 1)$ where $f(\gamma) = 0$. More precisely, the discriminant of the pencil is given by the binary form

$$\mathrm{disc}(uQ - vQ') = v^{2n+2} f(u/v)$$

of degree $2n+2$ over k. Hence the base locus $Q \cap Q'$ of the pencil is non-singular. The Fano variety $F = F_\alpha$ of this complete intersection, consisting of the n-dimensional subspaces Z of the $2n{+}2$ dimensional space $L\oplus k$ which are isotropic for all of the quadrics in the pencil, is smooth of dimension n (see [21]).

Since the discriminant of the the pencil defines the hyperelliptic curve C over k, a point $c = (x, y)$ on the curve determines both a quadric $Q_x = xQ - Q'$ in the pencil together with a ruling of Q_x. (A *ruling* of Q_x is, by definition, a component of the variety of $n+1$ dimensional subspaces in $L \oplus k$ which are isotropic for Q_x.) Each point of C gives an involution of the corresponding Fano variety $\tau(c) : F \to F$, which has 2^{2n} fixed points over a separable closure k^s, and is defined as follows. A point of F over k^s consists of a common isotropic subspace Z of dimension n for all quadrics in the pencil. The point c then gives a fixed maximal isotropic subspace Y of dimension $n+1$ for the quadric Q_x which contains Z. If we restrict any non-singular quadric other than Q_x in the pencil to the subspace Y, we get a reducible quadric which is the sum of two hyperplanes: Z and another common isotropic subspace $\tau(c)(Z)$. This defines the involution, and hence a morphism $\phi : C \times F \to F$ with $\phi(c, Z) = \tau(c)(Z)$.

The morphism $C \times F \to F$ can be used to define a fixed-point free action of the Jacobian $J = \text{Pic}^0(C)$ on F over k, which gives F the structure of a principal homogeneous space for J. For a general non-degenerate pencil of quadrics, this principal homogeneous space has order 4 in the group $H^1(k, J)$, and its square is the principal homogeneous space $\text{Pic}^1(C)$. In this case, we have a k-rational Weierstrass point O on C, which corresponds to the degenerate quadric Q in the pencil. Hence $\text{Pic}^1(C)$ has a rational point and the homogenous space F_α has order 2 in $H^1(k, J)$.

We can prove that F_α represents the image of α in the descent sequence as follows. The involution $\tau = \tau(O)$ is induced by the linear involution $(\lambda, a) \to (\lambda, -a)$. Hence the fixed points P_α of τ are just the n-dimensional subspaces X of L over k^s which are isotropic for both quadrics Q and Q'. This is the Fano variety $P_T = P_\alpha$ of dimension zero, which is the principal homogeneous space for $G_T = J[2]$ associated to the cohomology class α. It follows that F_α is the image of α in $H^1(k, J)[2]$. For more details on this construction, see the Harvard Ph.D. thesis of X. Wang [51]. Note that the variety F_α has a k-rational point whenever α lies in the subgroup $J(k)/2J(k)$, but F_α only has a k-rational point fixed by the involution τ when α is the trivial class in $H^1(k, J[2])$.

We may now prove Proposition 5.2. Assume that the class α is in the subgroup $J(k)/2J(k)$. Then its image in $H^1(k, J)[2]$ is trivial, and the homogenous space F_α has a k-rational point. Hence there is a k-subspace Z of $L \oplus k$ which is isotropic for both Q and Q'. Since it is isotropic for Q', the subspace Z does not contain the line $0 \oplus k$. Its projection to the subspace L has dimension n and is isotropic for Q. This implies that the orthogonal space L with bilinear form $(\lambda, \nu)_\alpha = \text{Trace}(\alpha\lambda\nu/f'(\beta))$ is split, so the class α is in the kernel of the map η.

□

We call such special $G(k)$-orbits on $V(k)$ (and the elements in them) *soluble*. Thus Proposition 5.2 states that the $\text{SO}(W)(k)$-orbits on the set of soluble elements in $V(k)$ with characteristic polynomial $f(x)$ are in bijection with the elements of $J(k)/2J(k)$.

Remark 5.3 We note that the finite group scheme $D = J[2]$ does not determine the hyperelliptic curve C. Indeed, for any class $d \in k^*/k^{*2}$, one has the hyperelliptic curve C_d with equation $dy^2 = f(x)$ and Jacobian J_d. The determination of $J[2]$ shows that these Jacobians share the same 2-torsion subgroup: $J_d[2] \cong J[2]$. This Jacobian J_d acts simply transitively on the Fano variety $F_{\alpha,d}$ of the complete intersection of the two quadrics

$$Q(\lambda, a) = \text{Trace}(\alpha\lambda^2/f'(\beta)),$$
$$Q'(\lambda, a) = \text{Trace}(\alpha\beta\lambda^2/f'(\beta)) + da^2.$$

Indeed, the discriminant of the quadric $uQ - vQ'$ in the pencil is equal to $dv^{2n+2}f(u/v)$. A similar argument to the one given in the proof of Proposition 5.2 shows that the subgroup $J_d(k)/2J_d(k)$ of $H^1(k, D)$ is also contained in the kernel of the map $\eta : H^1(k, D) \to H^1(k, \mathrm{SO}(W))$.

6 Orbits Over Arithmetic Fields

In this section, we describe the orbits with a fixed characteristic polynomial, when k is a finite, local, or global field.

6.1 Orbits over a finite field

We first consider the case when k is finite, of odd order q. In this case, every non-degenerate quadratic space of odd dimension is split, so $H^1(k, \mathrm{SO}(W)) = 1$, and the orbits with a fixed separable characteristic polynomial $f(x)$ are in bijection with elements of the group $H^1(k, D) = (L^*/L^{*2})_{N\equiv 1}$. This elementary abelian 2-group has order 2^m, where $m + 1$ is the number of irreducible factors of $f(x)$ in $k[x]$. This is also equal to the order of the stabilizer $H^0(k, D) = D(k) = L^*[2]_{N=1}$ of any point in the orbit over k. Hence the number of vectors T in V with any fixed separable polynomial is equal to the order of the finite group $\mathrm{SO}(W)(q)$, and is given by the formula

$$\#\,\mathrm{SO}(W)(q) = q^{n^2}(q^{2n} - 1)(q^{2n-2} - 1)\cdots(q^2 - 1).$$

Since k is perfect, a similar argument applies to all *regular* orbits—those orbits for which the characteristic polynomial of T is equal to its minimal polynomial. Any regular vector T with characteristic and minimal polynomial $f(x)$ has stabilizer D, the finite étale group scheme which is the kernel of the map $\mathrm{Res}_{L/k}\,\mu_2 \to \mu_2$ with $L = k[x]/(f(x))$. This finite group scheme has order 2^r over k, where $r + 1$ is the number of distinct roots of $f(x)$ in the separable closure k^s. (Here $0 \le r \le 2n$, as we do not assume that the minimal polynomial $f(x)$ is separable.) Since the number of regular orbits with this characteristic polynomial is the order of the finite group $H^1(k, D)$, and this order is equal to the order of the stabilizer $H^0(k, D)$ of any point in the orbit, the number of regular vectors T with any fixed characteristic polynomial $f(x)$ is equal to the order of the finite group $\mathrm{SO}(W)(q)$. Since there are q^{2n} choices for the coefficients of the characteristic polynomial $f(x)$, we can count the total number of regular vectors in $V(q)$. This number is

$$q^{2n}\cdot\#\,\mathrm{SO}(W)(q) = q^d - q^{d-2} + \cdots$$

where $d = \dim V$. Since equality of the characteristic and minimal polynomials is an algebraic condition which is independent of field extension, this count shows that irregularity is a codimension 2 condition.

We return to the case where $f(x)$ is separable. By Lang's theorem, we have $H^1(k, J) = 0$, where J is the (connected) Jacobian of the smooth hyperelliptic curve C with equation $y^2 = f(x)$ over k. Hence the homomorphism $J(k)/2J(k) \to H^1(k, D)$ is an isomorphism, and every orbit with characteristic polynomial $f(x)$ comes from a k-rational point on the Jacobian.

6.2 Orbits over a non-archimedean local field

Next, we consider the case when k is a non-archimedean local field, with ring of integers A and finite residue field $A/\pi A$. In this case, the spin cover gives a connecting homomorphism in Galois cohomology, which is an isomorphism by Kneser's theorem [29], cf. [36, Thms. 6.4, 6.10, pp. 356–397]:

$$H^1(k, \mathrm{SO}(W)) \cong H^2(k, \mu_2) \cong \mathbb{Z}/2\mathbb{Z}.$$

One can show that the resulting map $\eta : H^1(k, D) \to H^1(k, \mathrm{SO}(W)) \cong \mathbb{Z}/2\mathbb{Z}$ is an even quadratic form whose associated bilinear form is the cup product on $H^1(k, J[2])$ induced from the Weil pairing $J[2] \times J[2] \to \mu_2$ (see [51], [38]). The subgroup $J(k)/2J(k)$ is a maximal isotropic subspace on which $\eta = 0$. This allows us to count the number of orbits with a fixed separable characteristic polynomial $f(x)$.

We first do this count when the characteristic of the residue field is odd. Let $m + 1$ be the number of irreducible factors of $f(x)$ in $k[x]$, and let A_L be the integral closure of the ring A in L. Then the stabilizer $H^0(k, D) = D(k)$ has order 2^m. This is the order of $H^2(k, D)$ by Tate's local duality theorem, and his Euler characteristic formula then shows that the group $H^1(k, D)$ has order 2^{2m} (see [44, §II.5]). The number of orbits with characteristic polynomial $f(x)$ is equal to the number of zeroes of the even quadratic form η on this vector space, which is $2^{m-1}(2^m + 1) = 2^{2m-1} + 2^{m-1}$. The subgroup $J(k)/2J(k)$ has order 2^m; it is a maximal isotropic subspace for this quadratic form. We note that 2^m is also the order of the subgroup $(A_L^*/A_L^{*2})_{N\equiv 1}$ of units in $(L^*/L^{*2})_{N\equiv 1}$. We will see that these two subgroups of $(L^*/L^{*2})_{N\equiv 1}$ coincide in many cases, such as when the polynomial $f(x)$ has coefficients in A and the order $A[x]/(f(x))$ is equal to the integral closure A_L of A in L.

Next assume that the characteristic of the residue field is even, and for simplicity that k is an unramified extension of $\mathbb{Q}_2$. Let $m + 1$ be the number of irreducible factors of $f(x)$ in $k[x]$, and let A_L be the integral closure of the ring A in L. Then the stabilizer $H^0(k, D) = D(k)$ has order 2^m. This is the order of $H^2(k, D)$ by Tate's local duality theorem, and his Euler characteristic formula now shows that the group $H^1(k, D)$ has order 2^{2m+2n}. The number of orbits with characteristic polynomial $f(x)$ is again equal to the number of zeroes of the even quadratic form η on this vector space. The subgroup $J(k)/2J(k)$ has order 2^{m+n}; it is a maximal isotropic subspace for the quadratic form. Here the order of the subgroup $(A_L^*/A_L^{*2})_{N\equiv 1}$ of units in $(L^*/L^{*2})_{N\equiv 1}$ is equal to 2^{m+2n}, which is strictly larger than the order of $J(k)/2J(k)$.

6.3 Orbits over $\mathbb{C}$ and $\mathbb{R}$

The local field $k = \mathbb{C}$ is algebraically closed, so there is a unique (distinguished) orbit for each separable characteristic polynomial by Proposition 4.3.

The classification of orbits is more interesting for the local field $k = \mathbb{R}$. We now work this out in some detail, as we will need an explicit description of the real orbits in order to use arguments from the geometry of numbers later in this paper. Over the real numbers the pointed set $H^1(\mathbb{R}, \mathrm{SO}(W))$ has $n + 1$ elements, corresponding to the quadratic spaces of signature (p, q) with $p + q = 2n + 1$ and $q \equiv n \pmod 2$. The split space W of dimension $2n + 1$ and discriminant $\equiv 1$ has signature $(n + 1, n)$ over $\mathbb{R}$, so $\mathrm{SO}(W) = \mathrm{SO}(n + 1, n)$.

Let $2m + 1$ be the number of real roots of $f(x)$, with $0 \leq m \leq n$. Then the polynomial $f(x)$ has exactly $(2m+1)+(n-m) = m+n+1$ irreducible factors over $\mathbb{R}$, and $L = \mathbb{R}^{2m+1} \times \mathbb{C}^{n-m}$. The stabilizer of any self-adjoint operator T with this characteristic polynomial is an elementary abelian 2-group of order 2^{n+m}, isomorphic to $H^0(\mathbb{R}, D) = L^*[2]_{N=1}$. The group $H^1(\mathbb{R}, D) = ((\mathbb{R}^*/\mathbb{R}^{*2})^{2m+1} \times (\mathbb{C}^*/\mathbb{C}^{*2})^{n-m})_{N\equiv 1}$ is an elementary abelian 2-group of order 2^{2m}. The real locus of the hyperelliptic curve $y^2 = f(x)$ has $m+1$ components, and the group $J(\mathbb{R})/2J(\mathbb{R}) = J(\mathbb{R})/J(\mathbb{R})^0$ has order 2^m. This is a maximal isotropic subspace of $H^1(\mathbb{R}, J[2])$ for the cup product pairing and each class in $J(\mathbb{R})/2J(\mathbb{R})$ gives a soluble real orbit for the split group.

There are $\binom{2m+1}{m}$ orbits of $\mathrm{SO}(n + 1, n)$ on the self-adjoint operators T with characteristic polynomial $f(x)$. One can identify these orbits as follows. Such an operator T has $2m+1$ distinct real eigenspaces of dimension 1 and $(n-m)$ stable subspaces of dimension 2, which correspond to the pairs of complex conjugate eigenvalues. The orthogonal space $W(\mathbb{R})$ decomposes as an orthogonal direct sum of these eigenspaces, and the 2-dimensional eigenspaces each have signature $(1, 1)$. It follows that the orthogonal sum of the 1-dimensional eigenspaces has signature $(m + 1, m)$, and the subset of m eigenspaces which are negative definite determines the real orbit of T.

Let

$$\lambda_0 < \lambda_1 < \cdots < \lambda_{2m}$$

be the real roots of $f(x)$. The operators S in the distinguished orbit have eigenvectors $S(w_i) = \lambda_i.w_i$ with $\langle w_i, w_i\rangle = (-1)^i$. In other words, the signs of the inner products of the eigenvectors of an operator in the distinguished orbit are

$$+ \quad - \quad + \quad - \quad \cdots \quad + \,. \tag{6.1}$$

Using the explicit description of the map $\delta : J(\mathbb{R})/2J(\mathbb{R}) \to H^1(\mathbb{R}, J[2])$, we see that for a point $P = (a, b)$ on the curve, with $\lambda_{2k} < a < \lambda_{2k+1}$, the class of the divisor $D = (P) - (O)$ gives an orbit with eigenvectors w_i satisfying $\langle w_i, w_i\rangle = (-1)^i$ for $i \leq 2k$ and $\langle w_i, w_i\rangle = (-1)^{i+1}$ for $i \geq 2k+1$. It follows that the signs of the inner products of eigenvectors of an operator in a soluble real orbit have the form

$$+ \quad \epsilon_1 \quad \epsilon_2 \quad \cdots \quad \epsilon_m \tag{6.2}$$

where each ϵ_i is equal to either $(-\,+)$ or $(+\,-)$.

The space $\mathbb{R}^{2n} - \{\Delta = 0\}$ of characteristic polynomials with nonzero discriminant has $n+1$ components $I(m)$ indexed by integers $0 \leq m \leq n$, where $2m+1$ is the number of real roots. We get a similar decomposition of the space $V(\mathbb{R}) - \{\Delta = 0\}$ of self-adjoint operators T with characteristic polynomials in $I(m)$. We label the various components $V^{(m,\tau)}$ in $V(\mathbb{R})$ which map to the component $I(m)$ in $\mathbb{R}^{2n} - \{\Delta = 0\}$ by integers $1 \leq \tau \leq \binom{2m+1}{m}$, and let $V^{(m,1)}$ be the set of operators which lie in the distinguished orbits.

The integer τ indexes a choice of signs $\langle w_i, w_i\rangle = \pm 1$ for the inner products of the $2m+1$ real eigenvectors w_i of T, with m signs equal to -1. We use $\tau = 1$ to designate the distinguished orbit, where we have seen that the signs for the inner products, ordered for increasing real eigenvalues, are given by (6.1). We use the first 2^m indices $\tau = 1, \ldots, 2^m$ to denote the soluble orbits, whose signs are described in (6.2). Note that the stabilizer of every vector T in the component $V^{(m,\tau)}$ has order 2^{m+n}, independent of τ.

6.4 Orbits over a global field

Finally, we consider the case when k is a global field. In this case the group $H^1(k, D) = H^1(k, J[2])$ is infinite. We will see why there are also infinitely many classes in the kernel of η, so infinitely many orbits of self-adjoint operators with characteristic polynomial $f(x)$. First, we show that every class in the 2-Selmer group lies in the kernel of η.

Proposition 6.1 *Every class α in the 2-Selmer group $S_2(J) \subset H^1(k, J[2])$ lies in the kernel of η, so corresponds to an orbit of self-adjoint operators over k with characteristic polynomial $f(x)$.*

Proof Let α be a class in the 2-Selmer group. By definition, these are the classes in $H^1(k, J[2])$ whose restriction to $H^1(k_v, J[2])$ is in the image of $J(k_v)/2J(k_v)$ for every completion k_v. By Proposition 5.2, the orthogonal space U_v associated to the restriction of the class α in $H^1(k_v, \mathrm{SO}(W))$ is split over k_v, for every valuation v of k. By the theorem of Hasse and Minkowski, a non-degenerate orthogonal space U of dimension $2n+1$ is split over k if and only if $U \otimes k_v$ is split over every completion k_v. Hence the orthogonal space U associated to α is split over k, and α lies in the kernel of η.

□

We call the orbits corresponding to the 2-Selmer group *locally soluble*, as the Fano variety F_T associated to T has points over k_v for all completions v. For each $f(x)$, these locally soluble orbits are finite in number, as the 2-Selmer group is finite.

We have already remarked that, for any class $d \in k^*/k^{*2}$, the finite group scheme $J[2]$ is isomorphic to the 2-torsion subgroup $J_d[2]$ of the Jacobian of the hyperelliptic curve C_d with equation $dy^2 = f(x)$. Using similar methods, we can also obtain k-rational orbits with characteristic polynomial $f(x)$ from classes in the 2-Selmer group of J_d over k. Since the 2-Selmer groups of these quadratic twists of C become arbitrarily large finite elementary abelian 2-groups (see [8] for the case $n = 1$), the number of k-rational orbits with characteristic polynomial $f(x)$ is infinite.

We have proven Theorem 2.1, and the remarks following it.

7 Nilpotent Orbits and Vinberg's Theory

The results in this section will not be needed in the proof of the main theorems of this paper; we have included them here to place the orbits in this representation, as well as the hyperelliptic curves which appear in their study, into a larger context.

We say that an orbit T of $\mathrm{SO}(W)$ on V is *nilpotent* if the characteristic polynomial of T is equal to $f(x) = x^{2n+1}$. We say that it is *regular nilpotent* if the minimal polynomial of T is also equal to x^{2n+1}. One can show that the group $\mathrm{SO}(W)(k)$ acts simply-transitively on regular nilpotent elements in $V(k)$, i.e., there is a single orbit with trivial stabilizer. A representative operator E on the standard basis is given by:

$$E : f_1 \to f_2 \to \cdots \to f_n \to u \to e_n \to \cdots \to e_2 \to e_1 \to 0.$$

We say that a nilpotent orbit T is sub-regular if the minimal polynomial of T is equal to x^{2n}. An example of a sub-regular nilpotent operator E' is given on the standard basis by:

$$E' : f_1 \to f_2 \to \cdots \to f_n \to e_n \to \cdots \to e_2 \to e_1 \to 0, \ \ u \to 0.$$

The subregular nilpotent orbits of $\mathrm{SO}(W)(k)$ are indexed by the distinct classes $d \in k^*/k^{*2}$ and represented by the operators $d \cdot E'$.

When the characteristic of k is either 0 or is greater than $2n+1$, the representation V of $\mathrm{SO}(W)$ appears in Vinberg's theory of torsion automorphisms θ, where nilpotent orbits play a special role [35]. We review this connection briefly here, as a model for future work on those automophisms θ which lift regular elliptic classes in the Weyl group.

To do this, we change our notation for the rest of this section to conform with Vinberg, and let $G = \mathrm{SL}(W)$ with W of dimension $2n+1$ over k. This reductive algebraic group has Lie algebra $\mathfrak{g} = \{T \in \mathrm{End}(W) : \mathrm{Trace}(T) = 0\}$. The outer (pinned) involution $\theta(g) = (g^*)^{-1}$ has fixed subgroup $G^\theta = G(0) = \mathrm{SO}(W)$ with Lie algebra $\mathfrak{g}(0) = \{T \in \mathrm{End}(V) : T + T^* = 0\}$. The fixed subgroup $\mathrm{SO}(W)$ acts on the non-trivial eigenspace for θ on the Lie algebra:

$$\mathfrak{g}(1) = \{T \in \mathrm{End}(V) : T^* = T, \ \ \mathrm{Trace}(T) = 0\}$$

and this is precisely the representation V that we have been studying. The invariant polynomials for $G(0)$ on $\mathfrak{g}(1)$ in this case are the restriction of the invariant polynomials $c_2, c_3, \ldots, c_{2n+1}$ for G on the adjoint representation $\mathfrak{g}$. This gives a polynomial map to the geometric quotient:

$$p : \mathfrak{g}(1) \to \mathfrak{g}(1)/\!\!/G(0) = \mathrm{Spec}\, k[c_2, \ldots, c_{2n+1}].$$

The regular nilpotent vector E in $\mathfrak{g}(1)$ can be completed to a principal $\mathfrak{sl}_2$-triple $[E, F, H]$, with F regular nilpotent in $\mathfrak{g}(1)$ and H regular semi-simple in $\mathfrak{g}(0)$. Similarly, the subregular nilpotent vector E' in $\mathfrak{g}(1)$ can be completed to a subregular $\mathfrak{sl}_2$-triple $[E', F', H']$, with F' subregular nilpotent in $\mathfrak{g}(1)$ and H' semi-simple in $\mathfrak{g}(0)$. The centralizer $z(F)$ of F in the Lie algebra $\mathfrak{g}$ is abelian of dimension $2n$ and is contained in $\mathfrak{g}(1)$. If we restrict the invariant polynomials to the affine subspace $E + z(F)$ of $\mathfrak{g}(1)$, the resulting map

$$p : E + z(F) \to \mathrm{Spec}\, k[c_2, \ldots, c_{2n+1}]$$

is a bijection, whose inverse is called the Kostant section [35]. The isotropic subspace X of W spanned by $\{f_1, f_2, \ldots, f_n\}$ satisfies the condition: $T(X)$ is contained in $X^\perp$ for all operators T in the Kostant section. It also contains no T-invariant subspace Y except for $Y = 0$. Hence the Kostant section gives a representative for each distinguished orbit, once a representative E of the regular nilpotent orbit has been chosen.

In his Harvard PhD thesis, J. Thorne [50] studies the centralizer $z(F')$ of a subregular nilpotent vector F' in $\mathfrak{g}$. This has dimension $2n+2$, and the eigenspace $z(F')(1)$ has dimension $2n+1$ in $\mathfrak{g}(1)$. The restriction of the invariant polynomials to the affine subspace $E' + z(F')(1)$ of $\mathfrak{g}(1)$ defines a family

$$p : E' + z(F')(1) \to \operatorname{Spec} k[c_2, \ldots, c_{2n+1}]$$

of affine curves over the geometric quotient, which can be identified with the affine hyperelliptic curves $y^2 = f(x)$. They are smooth over the open subset where $\Delta \neq 0$. If one starts instead with the subregular vector $d \cdot E'$ and forms the corresponding $\mathfrak{sl}_2$-triple, one gets the family of affine curves $dy^2 = f(x)$, with the same 2-torsion in their Picard groups.

8 Integral Orbits

Fix a separable polynomial $f(x) = x^{2n+1} + c_2x^{2n-1} + c_3x^{2n-2} + \cdots + c_{2n+1}$ over $\mathbb{Q}$, let C be the hyperelliptic curve with equation $y^2 = f(x)$ and let J be its Jacobian. We have seen that classes in the 2-Selmer group of J give locally soluble orbits for the split orthogonal group $\mathrm{SO}(W)(\mathbb{Q})$ on $V(\mathbb{Q})$. However, since there are an infinite number of rational orbits and the Selmer group is finite, not all of the orbits over $\mathbb{Q}$ are locally soluble. In this section, we consider those orbits with an integral representative, in a sense to be defined below. We will show that this is a finite set that contains the locally soluble orbits.

8.1 Orbits over $\mathbb{Z}_p$ for $p \neq 2$

We begin by discussing the integral orbits over $\mathbb{Z}_p$, with $p \neq 2$. In this case the group $G = \mathrm{SO}(W)$ is smooth over $\mathbb{Z}_p$, and the lattice $W \otimes \mathbb{Z}_p$ has rank $2n+1$ and discriminant $\operatorname{disc}(W) = (-1)^n \det(W) \equiv 1$ in $\mathbb{Z}_p/\mathbb{Z}_p^{*2}$.

Lemma 8.1 *Assume that $p \neq 2$ and let I be a $\mathbb{Z}_p$-module of rank $2n+1$ equipped with a symmetric bilinear form $I \times I \to \mathbb{Z}_p$ whose discriminant is the square of a unit. Then I is isometric to W over $\mathbb{Z}_p$.*

This follows from the fact that there are, up to isomorphism, only two orthogonal spaces over $\mathbb{Q}_p$ of dimension $2n+1$ and discriminant $\equiv 1$. The non-split orthogonal space does not contain a non-degenerate lattice I over $\mathbb{Z}_p$, and the corresponding lattices in the split orthogonal space are all conjugate. The stabilizer of such a lattice is a hyperspecial maximal compact subgroup of $\mathrm{SO}(W)(\mathbb{Q}_p)$.

Using this lemma, we can give a simple classification of the orbits of $\mathrm{SO}(W)(\mathbb{Z}_p)$ on the self-adjoint operators $T : W \otimes \mathbb{Z}_p \to W \otimes \mathbb{Z}_p$ with a fixed characteristic polynomial $f(x)$ over $\mathbb{Z}_p$, which is assumed to be separable over $\mathbb{Q}_p$. Let $R = \mathbb{Z}_p[x]/(f(x))$, which is a $\mathbb{Z}_p$-order in the étale $\mathbb{Q}_p$-algebra $L = \mathbb{Q}_p[x]/(f(x))$. We say that a $\mathbb{Z}_p$-lattice I in L that spans L over $\mathbb{Q}_p$ is a *fractional ideal for* R if it is stable under multiplication by R. Such an ideal I has a norm $N(I)$, which is an element of $\mathbb{Q}_p^*/\mathbb{Z}_p^*$: if $p^a I$ is contained in R and the order of the finite $\mathbb{Z}_p$-module $R/p^a I$ is p^b, then $N(I) \equiv p^{b-(2n+1)a}$. If α is an element of L^*, then we have the principal fractional ideal (α) consisting of all R-multiples of α. This has norm $\equiv N(\alpha)$.

We will define an equivalence relation on pairs (I, α), where I is a fractional ideal for R, α is an element of L^* whose norm is a square in $\mathbb{Q}_p^*$, the square I^2 of the ideal I is contained in the principal ideal (α), and the square of its norm satisfies $N(I)^2 \equiv N(\alpha) \pmod{\mathbb{Z}_p^{*2}}$. We say that the pair (I', α') is *equivalent* to (I, α) if there is an element c in L^* with $I' = cI$ and $\alpha' = c^2\alpha$.

If R is maximal in L, every fractional ideal I is principal and every equivalence class is represented by a pair (R, α), where α is a unit having square norm to $\mathbb{Z}_p^*$. In this case, the set of equivalence classes form a finite group $(R^*/R^{*2})_{N\equiv 1}$. In general, the equivalence classes form a finite set which contains the finite group $(R^*/R^{*2})_{N\equiv 1}$ corresponding to those pairs where the ideal I is principal. Since the ring R is generated by a single element β over $\mathbb{Z}_p$, it is Gorenstein of dimension one. Hence the ideal I is principal if and only if it is *proper*, i.e., $\mathrm{End}_R(I) = R$ (see [1]).

The set of equivalence classes always maps to the group $(L^*/L^{*2})_{N\equiv 1}$, by taking the class of (I, α) to the class of α. This map need not be injective—for example, when the ring R is not maximal, a unit α that is not a square in R^* may become a square in L^*.

Proposition 8.2 *Assume that* $p \neq 2$ *and that* $f(x) = x^{2n+1} + c_2x^{2n-1} + c_3x^{2n-2} + \cdots + c_{2n+1}$ *is a polynomial with coefficients in* $\mathbb{Z}_p$ *and nonzero discriminant. Then the integral orbits of* $\mathrm{SO}(W)(\mathbb{Z}_p)$ *on self-adjoint operators* T *in* $V(\mathbb{Z}_p)$ *with characteristic polynomial* $f(x)$ *correspond to the equivalence classes of pairs* (I, α) *for the order* $R = \mathbb{Z}_p[x]/(f(x))$*. The stabilizer in* $\mathrm{SO}(W)(\mathbb{Z}_p)$ *of* T *is isomorphic to* $S^*[2]_{N=1}$*, where* $S = \mathrm{End}_R(I)$*. The integral orbit corresponding to the pair* (I, α) *maps to the rational orbit of* $\mathrm{SO}(W)(\mathbb{Q}_p)$ *on* $V(\mathbb{Q}_p)$ *corresponding to the class of* α *in* $(L^*/L^{*2})_{N\equiv 1}$.

Proof Starting with a pair (I, α) we construct an integral orbit as follows. Recall that $L = \mathbb{Q}_p + \mathbb{Q}_p\beta + \cdots + \mathbb{Q}_p\beta^{2n}$. Define the symmetric bilinear pairing $I \times I \to \mathbb{Z}_p$ by $(\lambda, \nu)_{\alpha^{-1}} :=$ the coefficient of β^{2n} in the product $\alpha^{-1}\lambda\nu$.

Since the square I^2 of the ideal I is contained in the principal ideal (α), this product lies in the subring $R = \mathbb{Z}_p + \mathbb{Z}_p\beta + \cdots + \mathbb{Z}_p\beta^{2n}$ and the coefficient of β^{2n} is integral. Since $N(I)^2 \equiv N(\alpha)$, this pairing has discriminant $\equiv 1$. By the above lemma, the bilinear module I is isometric to W over $\mathbb{Z}_p$. Since I is an ideal of R, multiplication by β gives a self-adjoint operator on I with characteristic polynomial $f(x)$. Translating this operator to W by the isometry gives the desired integral orbit. Since a self-adjoint $T : W \to W$ gives W the structure of a torsion-free $\mathbb{Z}_p[T] = R$-module of rank one, every integral orbit arises in this manner. The centralizer of T in $\mathrm{End}(W)$ is equal to $S = \mathrm{End}_R(I)$, and the elements in this centralizer that also lie in $\mathrm{SO}(W)$ are the elements that have order 2 (since they are both self-adjoint and orthogonal) and norm 1 (since this is the condition that the determinant $= 1$). Finally, since the bilinear form defined above is equivalent to the form $(\lambda, \nu)_\alpha$ over $\mathbb{Q}_p$, it maps to the rational orbit of α. This completes the proof of Proposition 8.2.

□

As a corollary, we obtain the following.

Corollary 8.3 *Assume that $f(x)$ has coefficients in $\mathbb{Z}_p$ and that the ring $R = \mathbb{Z}_p[x]/(f(x))$ is maximal in L. Then the integral orbits correspond to classes α in the unit subgroup $(R^*/R^{*2})_{N\equiv 1}$ of $(L^*/L^{*2})_{N\equiv 1}$. Every rational orbit whose class lies in the unit subgroup has a unique integral representative.*

An important special case of the corollary is when the discriminant $\mathrm{disc}(f)$ of the characteristic polynomial is a unit in $\mathbb{Z}_p^*$. Then the hyperelliptic curve C with affine equation $y^2 = f(x)$ over $\mathbb{Z}_p$ has good reduction modulo p. The Jacobian J of C also has good reduction, and the finite group scheme $J[2]$ is étale over $\mathbb{Z}_p$. It is obtained by restriction of scalars from the finite étale extension R of $\mathbb{Z}_p$, i.e., $J[2]$ is isomorphic to the kernel of the map $\mathrm{Res}_{R/\mathbb{Z}_p}(\mu_2) \to \mu_2$. Taking the étale cohomology of the exact sequence of group schemes $0 \to J[2] \to J \to J \to 0$ over $\mathbb{Z}_p$, we see that the image of $J(\mathbb{Q}_p)/2J(\mathbb{Q}_p)$ in $H^1(\mathbb{Q}_p, J[2]) = (L^*/L^{*2})_{N\equiv 1}$ is precisely the unit subgroup $H^1(\mathbb{Z}_p, J[2]) = (R^*/R^{*2})_{N\equiv 1}$.

More generally, we have the following result, which certainly holds (cf. [48, Prop. 4.6]) whenever the discriminant of $f(x)$ is a unit in $\mathbb{Z}_p^*$ or is exactly divisible by p.

Corollary 8.4 *Assume that $f(x)$ has coefficients in $\mathbb{Z}_p$ and that the order $\mathbb{Z}_p[x]/(f(x)) = R$ is maximal in the étale algebra $\mathbb{Q}_p[x]/(f(x)) = L$. Then the integral orbits with characteristic polynomial $f(x)$ are in bijection with the soluble orbits over $\mathbb{Q}_p$, i.e., those whose classes lie in the image of* *the subgroup* $J(\mathbb{Q}_p)/2J(\mathbb{Q}_p)$*. Moreover, the stabilizer in* $\mathrm{SO}(W)(\mathbb{Z}_p)$ *of an element in such an orbit is isomorphic to* $J[2](\mathbb{Q}_p)$*.*

Proof We know from Corollary 8.3 that the integral orbits correspond to classes in the unit subgroup $(R^*/R^{*2})_{N\equiv 1}$ of $(L^*/L^{*2})_{N\equiv 1}$. Hence it suffices to show that the image of $J(\mathbb{Q}_p)/2J(\mathbb{Q}_p)$ lies in this subgroup (as the two subgroups have the same order). Write $L = \prod L_i$ as a product of fields, and let β_i denote the component of β in L_i. Then $R = \prod R_i$, where $R_i = \mathbb{Z}_p[\beta_i]$ is the maximal order in L_i. We will show that for any point $P = (x(P), y(P))$ on the curve C over $\mathbb{Q}_p$, the valuation of $x(P) - \beta_i$ in L_i is even for all i, which will imply the first assertion of the corollary.

If the valuation of $x(P)$ is negative, it must be even. Indeed, $y(P)^2 = x(P)^{2n+1} + c_2 x(P)^{2n-1} + \cdots + c_{2n+1} = f(x(P))$, and $f(x)$ has integral coefficients c_k. Since each β_i is integral, the valuation of $x(P) - \beta_i$ is also negative and even, equal to the valuation of $x(P)$. Next assume that the valuation of $x(P)$ is ≥ 0. If L_i is not totally ramified, β_i must be a unit whose image in the residue field of R_i is not in the prime field $\mathbb{F}_p$. Hence the valuation of $x(P) - \beta_i$ is equal to zero. If L_i is totally ramified and the valuation of $x(P) - \beta_i$ is strictly positive, then the valuation of $x(P) - \beta_j$ in L_j must be zero for all $j \neq i$. Indeed, the images of β_i and β_j in their respective residue fields, when embedded in a common algebraic closure of $\mathbb{F}_p$, cannot be equal or even conjugate over $\mathbb{F}_p$. Otherwise, the discriminant of the polynomial $f(x)$ would be larger than the product of the discriminants of the maximal orders R_i, which would contradict the maximality of the ring R. As L_i is totally ramified, the valuation of $x(P) - \beta_i$ in L_i is equal to the valuation of its norm in $\mathbb{Q}_p$. Since the valuation of $x(P) - \beta_j$ in L_j is zero for all $j \neq i$, the valuation of $x(P) - \beta_i$ in L_i is equal to the valuation of $N(x(P) - \beta) = y(P)^2$ in $\mathbb{Q}_p$, which is clearly even. Finally, to obtain the second assertion of the corollary we simply observe that, because R is maximal in L, we have $S^*[2]_{N=1} \cong R^*[2]_{N=1} \cong L^*[2]_{N=1} \cong J[2](\mathbb{Q}_p)$. This completes the proof.

□

In general, given a p-adic rational orbit with integral characteristic polynomial $f(x)$, we can ask if it is represented by an integral self-adjoint operator $T : W \otimes \mathbb{Z}_p \to W \otimes \mathbb{Z}_p$. That is, given an element α in L^* with square norm to k^*, which represents a class in the quotient group $(L^*/L^{*2})_{N\equiv 1}$ lying in the kernel of the map η, we can ask if there is an ideal I of R with $I^2 \subset \alpha R$ and $(NI)^2 \equiv N(\alpha)$. One situation in which we can prove that an integral orbit exists is when the corresponding p-adic rational orbit is soluble, i.e., for those elements α whose classes are contained in the subgroup $J(\mathbb{Q}_p)/2J(\mathbb{Q}_p)$ of $H^1(\mathbb{Q}_p, J[2]) = (L^*/L^{*2})_{N\equiv 1}$.

We thank Michael Stoll for his help with the proof of the following key proposition:

Proposition 8.5 *Assume that $f(x)$ has coefficients in $\mathbb{Z}_p$ and nonzero discriminant. Let $R = \mathbb{Z}_p[x]/(f(x))$ and $L = \mathbb{Q}_p[x]/(f(x))$. If $\alpha \in L^*$ represents a class in the subgroup $J(\mathbb{Q}_p)/2J(\mathbb{Q}_p)$ of $(L^*/L^{*2})_{N\equiv 1}$, then there is an ideal I of R with $I^2 \subset \alpha R$ and $(NI)^2 \equiv N(\alpha)$. Hence there is an integral orbit representing the rational orbit of α.*

Proof We prove this by an explicit construction of the ideal I. We can represent any class in $J(\mathbb{Q}_p)/2J(\mathbb{Q}_p)$ by a Galois-invariant divisor of degree zero

$$D = (P_1) + (P_2) + \cdots + (P_m) - m(O)$$

with $m \leq n$, where the $P_i = (a_i, b_i)$ are defined over some finite Galois extension K of $\mathbb{Q}_p$. We may assume that the coordinates of all the P_i lie in the integers of K with $b_i \neq 0$, and that the a_i are all distinct. Define the monic polynomial

$$P(x) = (x - a_1)(x - a_2) \cdots (x - a_m)$$

of degree m over $\mathbb{Q}_p$. By our explicit description of the coboundary map from $J(\mathbb{Q}_p)/2J(\mathbb{Q}_p)$ to $H^1(\mathbb{Q}_p, J[2])$, we have $(-1)^m P(\beta) \equiv \alpha$ in $(L^*/L^{*2})_{N\equiv 1}$. Next define $R(x)$ as the unique polynomial of degree $\leq m-1$ over $\mathbb{Q}_p$ with the property that

$$R(a_i) = b_i, \ 1 \leq i \leq m.$$

Then $R(x)^2 - f(x) = h(x)P(x)$ in $\mathbb{Q}_p[x]$. This gives the Mumford representation of the divisor D, which is cut out on the curve by the two equations $P(x) = 0$ and $y = R(x)$.

The polynomial $P(x)$ clearly has coefficients in the ring $\mathbb{Z}_p$. When $R(x)$ also has coefficients in $\mathbb{Z}_p$, we define the ideal I_D of $R = \mathbb{Z}_p[\beta]$ as the R-submodule of L generated by the two elements $P(\beta)$ and $R(\beta)$. For example, when $m = 1$, the divisor D has the form $(P) - (O)$ with $P = (a, b)$, and the ideal I_D is generated by the two elements $(\beta - a)$ and b.

We claim that I_D^2 is contained in the ideal generated by $\alpha = (-1)^m P(\beta)$ and that $N(I_D) \equiv b_1 b_2 \cdots b_m$. Hence $N(I_D)^2$ generates the same ideal as $N(P(\beta)) = (-1)^m b_1^2 b_2^2 \cdots b_m^2$ and we have constructed an integral orbit mapping to the rational orbit of α.

The fact that the ideal $I_D^2 = (P(\beta)^2, P(\beta)R(\beta), R(\beta)^2)$ is contained in the ideal generated by $P(\beta)$ follows from $f(\beta) = 0$, so $R(\beta)^2 = h(\beta)P(\beta)$. On the other hand,

$$R/I_D = \mathbb{Z}_p[\beta]/I_D = \mathbb{Z}_p[x]/(f(x), P(x), R(x)) = \mathbb{Z}_p[x]/(P(x), R(x))$$

and so $N(I_D) \equiv \#\mathbb{Z}_p[x]/(P(x), R(x))$. The latter cardinality may be computed as the determinant (viewed as a power of p) of the $\mathbb{Z}_p$-linear transformation $\times R(x) : \mathbb{Z}_p[x]/(P(x)) \to \mathbb{Z}_p[x]/(P(x))$. This is equivalent to computing the determinant of the K-linear transformation $\times R(x)$: $K[x]/(P(x)) \to K[x]/(P(x))$. But $P(x)$ splits over K, and we have

$$K[x]/(P(x)) \cong K[x]/(x - a_1) \times \cdots \times K[x]/(x - a_m) \cong K^m \tag{8.1}$$

since the a_i are distinct. The image of $R(x)$ in K^m above is given by $(R(a_1), \ldots, R(a_m)) = (b_1, \ldots, b_m)$, and thus the determinant of $\times R(x)$ as a transformation of $K[x]/(P(x)) \cong K^m$ is simply $b_1 \cdots b_m$, as claimed. This completes the construction of an integral orbit when the polynomial $R(x)$ has coefficients in $\mathbb{Z}_p$.

If $R(x)$ is not integral, we consider the $2n + 1$ roots of the monic polynomial $F(x) - R(x)^2$. These contain the roots $a_1, a_2, \ldots, a_m$ of $P(x)$, and a consideration of the Newton polygon shows that at most $m - 2$ of the remaining roots $r_1, r_2, \ldots, r_{m-2}$ have non-negative valuation. Let $s_i = R(r_i)$. Then the function $y - R(x)$ on the curve has principal divisor

$$\{(a_1, b_1) + \cdots + (a_m, b_m) - m(O)\} + \\ \{(r_1, s_1) + \cdots + (r_{m-2}, s_{m-2}) - (m - 2)(O)\} + E$$

where the x-coordinates of the points in the divisor E are the remaining roots. Since these all have negative valuation, the class of E lies in $2J(\mathbb{Q}_p)$. Hence the class of the divisor D is equivalent to the class of the divisor

$$D^* = (r_1, s_1) + \cdots + (r_{m-2}, s_{m-2}) - (m - 2)(O)$$

in the quotient group $J(\mathbb{Q}_p)/2J(\mathbb{Q}_p)$. The proof now concludes by a descent on the number m of points in D. Once $m \leq 1$ the polynomials $P(x)$ and $R(x)$ are both integral.

□

We end with some brief remarks on integral orbits (I, α) in the case when $\mathrm{End}_R(I) = R'$ is strictly larger than R, so I is not a free R-module. When the order R' is Gorenstein, I is a free R'-module, so can be identified with R'. Moreover, the dualizing module $\mathrm{Hom}(R', \mathbb{Z}_p)$ is free, and can be identified with $(1/\gamma)R'$ under the trace pairing, with γ in L^* chosen so that $N(\gamma) \equiv N(f'(\beta))$ in $\mathbb{Q}_p^*/\mathbb{Q}_p^{*2}$. The element γ is well-defined up to multiplication by R'^{*2}, and defines an integral orbit via taking the equivalence class of the pair $(I, \alpha) = (R', \gamma/f'(\beta))$. So the integral orbits with characteristic polynomial $f(x)$ and endomorphisms by a Gorenstein order R' containing R in L form a principal homogeneous space for the group

$(R'^*/R'^{*2})_{N\equiv 1}$. They map to the rational orbits with classes $\gamma/f'(\beta)$ in $(L^*/L^{*2})_{N\equiv 1}$, where $(1/\gamma)$ generates the dualizing module of R' in L, and these rational classes need not lie in the unit subgroup of $(L^*/L^{*2})_{N\equiv 1}$.

8.2 General integral orbits

The lattice $W \otimes \mathbb{Z}_2$ has rank $2n+1$ and discriminant $\mathrm{disc}(W) = (-1)^n \det(W) \equiv 1$ in $\mathbb{Z}_2/\mathbb{Z}_2^{*2}$. In this case we need a bit more to identify those lattices isometric to W over $\mathbb{Z}_2$, as both orthogonal spaces over $\mathbb{Q}_2$ of dimension $2n+1$ and discriminant $\equiv 1$ contain non-degenerate lattices with these invariants.

Lemma 8.6 *Let I be a $\mathbb{Z}_2$-module of rank $2n+1$ equipped with a symmetric bilinear form $I \times I \to \mathbb{Z}_2$ whose discriminant is the square of a unit. Then I is isometric to W over $\mathbb{Z}_2$ if and only if it contains an isotropic sublattice of rank n.*

Indeed, the existence of such an isotropic sublattice is equivalent to the assumption that $I \otimes \mathbb{Q}_2$ is a split orthogonal space, and in this space all non-degenerate odd lattices are conjugate. The stabilizer of such a lattice is the normalizer of a parahoric subgroup of $\mathrm{SO}(W)(\mathbb{Q}_2)$.

If the bilinear lattice I contains a maximal isotropic sublattice over $\mathbb{Z}_2$, then we say that it is *split* [34]. The same argument as the proof of Proposition 8.2 yields the following.

Proposition 8.7 *Assume that $f(x) = x^{2n+1} + c_2x^{2n-1} + c_3x^{2n-2} + \cdots + c_{2n+1}$ is a polynomial with coefficients in $\mathbb{Z}_2$ and nonzero discriminant in $\mathbb{Q}_2$. Then the integral orbits of $G(\mathbb{Z}_2)$ on self-adjoint operators T with characteristic polynomial $f(x)$ correspond to the equivalence classes of pairs (I, α) for the order $R = \mathbb{Z}_2[x]/(f(x))$, with the property that the bilinear form $(\ ,\ )_{\alpha^{-1}}$ on I is split. The integral orbit corresponding to the pair (I, α) maps to the rational orbit of $\mathrm{SO}(W)(\mathbb{Q}_2)$ corresponding to the class of α in $(L^*/L^{*2})_{N\equiv 1}$.*

The condition that the pair (I, α) gives a split orthogonal space is not empty. Consider the case when $f(x)$ has integral coefficients and the discriminant of $f(x)$ is a unit in $\mathbb{Z}_2^*$. Then the ring $R = A_L$ is the maximal order in L, and every pair (I, α) is equivalent to (R, u) with u in the unit subgroup $(A_L^*/A_L^{*2})_{N\equiv 1}$ of $(L^*/L^{*2})_{N\equiv 1}$. This subgroup of has order 2^{m+2n}, where $f(x)$ has $m+1$ irreducible factors over $\mathbb{Q}_2$. Not all of these orthogonal spaces can be split over $\mathbb{Q}_2$, as a maximal isotropic subgroup for the quadratic form η on $(L^*/L^{*2})_{N\equiv 1}$ has order 2^{m+n}. On the other hand, the image of classes in $J(\mathbb{Q}_2)/2J(\mathbb{Q}_2)$ give a subgroup of the units of order 2^{m+n}, which do correspond to orbits.

We suspect that this remains true in general. More precisely, we conjecture that whenever $f(x)$ has coefficients in $\mathbb{Z}_2$ and α is a class in $J(\mathbb{Q}_2)/2J(\mathbb{Q}_2)$, the corresponding orbit over $\mathbb{Q}_2$ is represented by (at least one) integral orbit over $\mathbb{Z}_2$. One can try to imitate the construction of an ideal that we gave in the case where p is odd. The Mumford representative works just as well, but the reduction to a divisor of smaller degree used the fact that when $P = (a, b)$ on C has $\mathrm{ord}(a) < 0$, the class $(P) - (O)$ is divisible by 2 in the Jacobian. For $p = 2$ this is true if we assume that the integral coefficients satisfy: 2^{4k} divides c_k, for $k = 2, 3, \ldots, 2n+1$. So for integral polynomials of this form, each class in $J(\mathbb{Q}_2)/2J(\mathbb{Q}_2)$ is represented by at least one integral orbit.

Similarly, we have the following result on integral orbits over $\mathbb{Z}$.

Proposition 8.8 *Assume that $f(x) = x^{2n+1} + c_2x^{2n-1} + c_3x^{2n-2} + \cdots + c_{2n+1}$ is a polynomial with coefficients in $\mathbb{Z}$ and nonzero discriminant in $\mathbb{Q}$. Then the integral orbits of $G(\mathbb{Z})$ on self-adjoint operators T with characteristic polynomial $f(x)$ correspond to the equivalence classes of pairs (I, α) for the order $R = \mathbb{Z}[x]/(f(x))$, with the property that the bilinear form $(\ ,\ )_{\alpha^{-1}}$ on I is split. The integral orbit corresponding to the pair (I, α) maps to the rational orbit of $\mathrm{SO}(W)(\mathbb{Q})$ corresponding to the class of α in $(L^*/L^{*2})_{N\equiv 1}$.*

Here one can check if the bilinear form is split simply by a calculation of the signature of the orthogonal space over $\mathbb{R}$ (see [43, Chapter V]).

In general, given a rational orbit with integral characteristic polynomial $f(x)$, we can ask if it is represented by an integral self-adjoint operator $T : W \to W$. That is, given an element α in L^* that represents a class in the group $(L^*/L^{*2})_{N\equiv 1}$ lying in the kernel of η, we can ask if there is an ideal I of $R = \mathbb{Z}[x]/(f(x))$ with $I^2 \subset \alpha R$ and $(NI)^2 = N(\alpha)$. In this case, there is no need to check that the bilinear form defined by α on I is split, as it is split over $\mathbb{Q}$. The number of integral orbits mapping to a fixed rational orbit is finite, as there are only finitely many ideals with a given norm.

Fix a representative of the class α in the group L^*, and let $m(\alpha)$ be the number of ideals I of R with the property that $I^2 \subset \alpha R$ and $(NI)^2 = N(\alpha)$. Let $m_p(\alpha)$ be the number of ideals I_p of $R_p := R \otimes \mathbb{Z}_p$ with this property. For all primes p that do not divide Δ and where the element α is a unit, we must have $I_p = R_p$. Hence $m_p(\alpha_p) = 1$ for almost all p, where α_p denotes the image of α in R_p. Since an ideal of R is determined by its localizations, and since the intersection of L with the product $\prod_p I_p$ is an ideal I of R

with the desired properties, we have the product formula

$$m(\alpha) = \prod_p m_p(\alpha).$$

The pair (I, α) gives the same orbit as the pair (cI, α) whenever c is an element of L^* with $c^2 = 1$ and $N(c) = 1$. On the other hand, the ideals I and cI are equal only when c is a unit in the endomorphism ring $R(I)$ of the ideal I. So the number of distinct ideals giving the same integral orbit is the order of the finite quotient group $L^*[2]_{N=1}/R(I)^*[2]_{N=1}$. Note that the stabilizer of an operator in this integral orbit is the finite group $R(I)^*[2]_{N=1}$ and the stabilizer of an operator in the rational orbit α is the finite group $L^*[2]_{N=1}$. Hence we also have the formula

$$m(\alpha) = \sum_{\sigma \in O(\alpha)} \#\,\mathrm{Stab}_{G(\mathbb{Q})}(\alpha)/\#\,\mathrm{Stab}_{G(\mathbb{Z})}(\sigma)$$

where the sum is taken over the set $O(\alpha)$ of integral orbits σ mapping to the rational orbit corresponding to α.

The same argument comparing ideals and integral orbits works locally, so combining the above formulas gives the following

Proposition 8.9 *Assume that the number of integral orbits representing a fixed rational orbit α is nonzero. Then*

$$\#\,\mathrm{Stab}_{G(\mathbb{Q})}(\alpha) \sum_{\sigma \in O(\alpha)} 1/\#\,\mathrm{Stab}_{G(\mathbb{Z})}(\sigma) =$$
$$\prod_p \#\,\mathrm{Stab}_{G(\mathbb{Q}_p)}(\alpha_p) \sum_{\sigma_p \in O_p(\alpha_p)} 1/\#\,\mathrm{Stab}_{G(\mathbb{Z}_p)}(\sigma_p)$$

where almost all factors in the product are equal to 1.

Finally, we have the following result on locally soluble orbits.

Proposition 8.10 *Let $f(x)$ have coefficients in $\mathbb{Z}$, and assume that 2^{4k} divides c_k for all k. Then every class in the 2-Selmer group of J is represented by at least one integral orbit.*

This follows from the construction of an ideal I_p at each prime. The divisibility hypothesis was needed to construct I_2 and an integral orbit over $\mathbb{Z}_2$. It is not necessary when the order R is maximal at 2, and we suspect that it is not necessary in general.

9 Construction of Fundamental Domains

In the previous sections, we have studied the rational and integral orbits of the representation of SO(W) on V, and shown how the locally soluble orbits are naturally related to the 2-Selmer groups of the Jacobians of hyperelliptic curves. In order to prove Theorem 1.1, we now turn to the question of counting these orbits.

In this section, we construct convenient fundamental domains for the action of $G(\mathbb{Z})$ on $V(\mathbb{R})$. In Section 10, we will then count the number of "irreducible" points in $V(\mathbb{Z})$ in these fundamental domains having bounded height, which will then allow us to prove Theorem 1.1 in Sections 11 and 12 via the appropriate sieve methods.

We say that an element in $V(\mathbb{Z})$ is *reducible* if it either lies in a distinguished orbit over $\mathbb{Q}$ or has discriminant zero; we say that it is *irreducible* otherwise. The *height* of an element $B \in V(\mathbb{R})$ and of its associated characteristic polynomial $f(x)$ is defined as follows: if $B \in V(\mathbb{R})$ has associated invariants $c_2, \ldots, c_{2n+1}$ given by

$$f(x) = \det(Ax - B) = x^{2n+1} + c_2 x^{2n-1} + \cdots + c_{2n+1},$$

then the height $H(B)$ of B and the height $H(f)$ of f is defined by

$$H(B) := H(f) := \max\{|c_k|^{2n(2n+1)/k}\}_{k=2}^{2n+1}.$$

To describe our fundamental domains explicitly, we put natural coordinates on V. Recall that V may be identified with the $n(2n+3)$-dimensional space of symmetric $(2n+1) \times (2n+1)$ matrices having anti-trace zero. We use b_{ij} to denote the (i,j) entry of $B \in V$; thus the b_{ij} ($i \leq j$; $(i,j) \neq (n+1, n+1)$) form a natural set of coordinates on V.

9.1 Fundamental sets for the action of $G'(\mathbb{R})$ on $V(\mathbb{R})$

Recall that we have naturally partitioned the set of elements in $V(\mathbb{R})$ with $\Delta \neq 0$ into $\sum_{m=0}^{n} \binom{2m+1}{m}$ components, which we denote by $V^{(m,\tau)}$ for $m = 0, \ldots, n$ and $\tau = 1, \ldots, \binom{2m+1}{m}$. For a given value of m, the component $V^{(m,\tau)}$ in $V(\mathbb{R})$ maps to the component $I(m)$ of characteristic polynomials in $\mathbb{R}^{2n}$ having nonzero discriminant and $2m+1$ real roots. Let us define $G'(\mathbb{R}) := \Lambda \times G(\mathbb{R})$, where $\Lambda = \{\lambda > 0\}$. Then we may view $V(\mathbb{R})$ also as a representation of $G'(\mathbb{R})$, where Λ acts by scaling. Furthermore, $G'(\mathbb{R})$ acts also on each $V^{(m,\tau)}$. Note that the action of $\lambda \in \Lambda$ on $V(\mathbb{R})$ takes c_k to $\lambda^k c_k$, so that $H(\lambda B) = \lambda^{2n(2n+1)} H(B)$.

We first construct fundamental sets $L^{(m,\tau)}$ for the action of $G'(\mathbb{R})$ on $V^{(m,\tau)}$, which will also then end up being fundamental sets for the action of $G(\mathbb{R})$ on $\{B \in V^{(m,\tau)} : H(B) = 1\}$. We begin with the case when $m = n$. Fix a self-adjoint operator $T : W \to W$ in the distinguished orbit having a given characteristic polynomial $f(x)$ with $2n + 1$ real roots. The polynomial $f(x)$ of T is separable, of trace 0 and split over $\mathbb{R}$. Let $r_1 < r_2 < \cdots < r_{2n+1}$ be the roots of $f(x)$ and let $L = \mathbb{R}[T]$ be the algebra centralizing T in $\mathrm{End}(W)$. All of the operators in L are self-adjoint and have the same eigenvectors v_i in W. We thus obtain an isomorphism $L \to \mathbb{R}^{2n+1}$ which maps the polynomial $p(T)$ to the eigenvalues $(x_1, \ldots, x_{2n+1}) = (p(r_1), \ldots, p(r_{2n+1}))$. With these coordinates, consider the cone C defined by the inequalities $x_1 < x_2 < \cdots < x_{2n+1}$ and $\sum x_i = 0$ in the subspace of elements of trace 0 in L. The operators in this cone represent the distinguished orbits with split characteristic polynomial, as the eigenvectors for the eigenvalues $x_1 < x_2 < \cdots < x_{2n+1}$ are the same as the eigenvectors for T. If we scale the elements in C to have height 1 we are in a bounded region. This is the set $L^{(n,1)}$ of $G(\mathbb{R})$-representatives of height 1 that we take for $V^{(n,1)}$. Now take another $V^{(n,\tau)}$ where the characteristic polynomial splits: τ is given by a collection of sign changes α in $L^* = \mathbb{R}^{*2n+1}$. If T' in the region has characteristic polynomial $f(x)$, then we can find an element T in the distinguished orbit and an element $g \in \mathrm{GL}(V)(\mathbb{R})$ with $gTg^{-1} = T'$ and $g^*g = \alpha \in L^*$. The elements $g(v_i)$ are now eigenvectors for T' with the appropriately changed signs, and we obtain a fundamental set $L^{(n,\tau)} = gL^{(n,1)}g^{-1}$. Since we are conjugating by a fixed g (which is a continuous map), this set of $G'(\mathbb{R})$-representatives for $V^{(n,\tau)}$ also lies in a compact region.

Next assume that $m < n$, and choose T in the distinguished orbit with characteristic polynomial $f(x)$. In this case, the algebra $L = \mathbb{R}[T]$ is the product of $2m+1$ copies of $\mathbb{R}$ and $n-m$ copies of $\mathbb{C}$. We take the elements of trace zero in L which lie in the product of the cone defined by the inequalities $x_1 < x_2 < \cdots < x_{2m+1}$ in the real part with a fundamental domain for the action of the symmetric group S_{n-m} on the product of $n-m$ upper half planes minus all the diagonals in the complex part. Again, we scale to obtain a set $L^{(m,1)}$ of orbit representatives of height 1 for the action of $G'(\mathbb{R})$ on $V^{(m,1)}$. As before, there exists $g \in \mathrm{GL}(V)(\mathbb{R})$ such that $L^{(m,\tau)} := gL^{(m,1)}g^{-1}$ is a fundamental set for the action of $G'(\mathbb{R})$ on $V^{(m,\tau)}$, and $L^{(m,\tau)}$ also lies in a compact set.

We have thus chosen our fundamental sets $L^{(m,\tau)}$ so that all elements in these sets have height 1 and all entries of elements in these sets are bounded, i.e., the $L^{(m,\tau)}$ all lie in a compact subset of $V(\mathbb{R})$. Note that for any fixed $h \in G'(\mathbb{R})$ (or for any h lying in a fixed compact subset $G_0 \subset G'(\mathbb{R})$), the set $hL^{(m,\tau)}$ is also a fundamental set for the action of $G'(\mathbb{R})$ on $V^{(m,\tau)}$ that is bounded (independent of $h \in G_0$).

9.2 Fundamental sets for the action of $G(\mathbb{Z})$ on $G'(\mathbb{R})$

In [9], Borel (building on earlier work of Borel–Harish-Chandra [10]) constructs a natural fundamental domain in $G(\mathbb{R})$ for the left action of $G(\mathbb{Z})$ on $G(\mathbb{R})$, which immediately also gives us a fundamental domain $\mathcal{F}$ in $G'(\mathbb{R})$ for the left action of $G(\mathbb{Z})$ on $G'(\mathbb{R})$. This set $\mathcal{F}$ may be expressed in the form

$$\mathcal{F} = \{ut\theta\lambda : u \in N'(t),\ t \in T',\ \theta \in K,\ \lambda \in \Lambda\},$$

where $N'(t)$ is an absolutely bounded measurable set, which depends on $t \in T'$, of unipotent lower triangular real $(2n+1)\times(2n+1)$ matrices; the set T' is the subset of the torus of diagonal matrices with positive entries given by

$$T' = \left\{ \begin{pmatrix} t_1^{-1} & & & & & & \\ & \ddots & & & & & \\ & & t_n^{-1} & & & & \\ & & & 1 & & & \\ & & & & t_n & & \\ & & & & & \ddots & \\ & & & & & & t_1 \end{pmatrix} : t_1/t_2 > c, \ldots, t_{n-1}/t_n > c, t_n > c \right\}; \qquad (9.1)$$

and K is a maximal compact real subgroup of $G(\mathbb{R})$. In the above, c denotes an absolute positive constant. (Note that t_i/t_{i+1} for $i = 1, \ldots, n-1$ and t_n form a set of simple roots for our choice of maximal torus T in G.)

It will be convenient in the sequel to parametrize T' in a slightly different way. Namely, for $k = 1, \ldots, n$, we make the substitution $t_k = s_k s_{k+1} \cdots s_n$. Then we may speak of elements $s = (s_1, \ldots, s_n) \in T'$ as well, and the conditions on s for $s = (s_1, \ldots, s_n)$ to be in T' is simply that $s_k > c$ for all $k = 1, \ldots, n$.

9.3 Fundamental sets for the action of $G(\mathbb{Z})$ on $V(\mathbb{R})$

For any $h \in G'(\mathbb{R})$, by the previous subsection, we see that $\mathcal{F}hL^{(m,\tau)}$ is the union of 2^{m+n} fundamental domains for the action of $G(\mathbb{Z})$ on $V^{(m,\tau)}$; here, we regard $\mathcal{F}hL^{(m,\tau)}$ as a multiset, where the multiplicity of a point x in $\mathcal{F}hL^{(m,\tau)}$ is given by the cardinality of the set $\{g \in \mathcal{F} : x \in ghL^{(m,\tau)}\}$. Thus, as explained in [5, §2.1], a $G(\mathbb{Z})$-equivalence class x in $V^{(m,\tau)}$ is represented in this multiset $\sigma(x)$ times, where $\sigma(x) = \#\operatorname{Stab}_{G(\mathbb{R})}(x)/\#\operatorname{Stab}_{G(\mathbb{Z})}(x)$. In particular, $\sigma(x)$ is always a number between 1 and 2^{m+n}.

For any $G(\mathbb{Z})$-invariant set $S \subset V^{(m,\tau)} \cap V(\mathbb{Z})$, let $N(S; X)$ denote the number of $G(\mathbb{Z})$-equivalence classes of irreducible elements $B \in S$ satisfying $H(B) < X$. Then we conclude that, for any $h \in G'(\mathbb{R})$, the product $2^{m+n} \cdot N(S;X)$ is exactly equal to the number of irreducible integer points in $\mathcal{F}hL^{(m,\tau)}$ having height less than X, with the slight caveat that the (relatively rare—see Proposition 10.8) points with $G(\mathbb{Z})$-stabilizers of cardinality r ($r > 1$) are counted with weight $1/r$.

As mentioned earlier, the main obstacle to counting integer points of bounded height in a single domain $\mathcal{F}hL^{(m,\tau)}$ is that the relevant region is not bounded, but rather has cusps going off to infinity. We simplify the counting in this cuspidal region by "thickening" the cusp; more precisely, we compute the number of integer points of bounded height in the region $\mathcal{F}hL^{(m,\tau)}$ by averaging over lots of such fundamental regions, i.e., by averaging over the domains $\mathcal{F}hL^{(m,\tau)}$ where h ranges over a certain compact subset $G_0 \in G'(\mathbb{R})$. The method, which is an extension of the method of [2], is described next.

10 Counting Irreducible Integral Points of Bounded Height

In this section, we derive asymptotics for the number of $G(\mathbb{Z})$-equivalence classes of irreducible elements of $V(\mathbb{Z})$ having bounded invariants. We also describe how these asymptotics change when we restrict to counting only those elements in $V(\mathbb{Z})$ that satisfy a specified finite set of congruence conditions.

Let $L^{(m,\tau)}$ ($m = 1,\ldots,n$, $\tau = 1,\ldots,\binom{2m+1}{m}$) denote fundamental sets for the action of $G'(\mathbb{R})$ on $V^{(m,\tau)}$, as constructed in §9.1, and let

$$c_{m,\tau} = \frac{\mathrm{Vol}(\mathcal{F}L^{(m,\tau)} \cap \{v \in V(\mathbb{R}) : H(v) < 1\})}{2^{m+n}}.$$

Then in this section we prove the following theorem:

Theorem 10.1 *Fix m,τ. For any $G(\mathbb{Z})$-invariant set $S \subset V(\mathbb{Z})^{(m,\tau)} := V^{(m,\tau)} \cap V(\mathbb{Z})$, let $N(S;X)$ denote the number of $G(\mathbb{Z})$-equivalence classes of irreducible elements $B \in S$ satisfying $H(B) < X$. Then*

$$N(V(\mathbb{Z})^{(m,\tau)};X) = c_{m,\tau}X^{(2n+3)/(4n+2)} + o(X^{(2n+3)/(4n+2)}).$$

10.1 Averaging over fundamental domains

Let G_0 be a compact left K-invariant set in $G'(\mathbb{R})$ that is the closure of a nonempty open set and in which every element has determinant greater than or equal to 1. Then for any m, τ, we may write

$$N(V(\mathbb{Z})^{(m,\tau)}; X) = \frac{\int_{h \in G_0} \#\{x \in \mathcal{F}hL \cap V(\mathbb{Z})^{\text{irr}} : H(x) < X\} dh}{2^{m+n} \int_{h \in G_0} dh} \tag{10.1}$$

for any Haar measure dh on $G'(\mathbb{R})$, where $V(\mathbb{Z})^{\text{irr}}$ denotes the set of irreducible elements in $V(\mathbb{Z})$ and L is equal to $L^{(m,\tau)}$. The denominator of the right hand side of (10.1) is an absolute constant $C_{G_0}^{(m,\tau)}$ greater than zero.

More generally, for any $G(\mathbb{Z})$-invariant subset $S \subset V(\mathbb{Z})^{(m,\tau)}$, let $N(S; X)$ denote again the number of irreducible $G(\mathbb{Z})$-orbits in S having height less than X. Let S^{irr} denote the subset of irreducible points of S. Then $N(S; X)$ can be similarly expressed as

$$N(S; X) = \frac{\int_{h \in G_0} \#\{x \in \mathcal{F}hL \cap S^{\text{irr}} : H(x) < X\} dh}{C_{G_0}^{(m,\tau)}}. \tag{10.2}$$

We use (10.2) to define $N(S; X)$ even for sets $S \subset V(\mathbb{Z})$ that are not necessarily $G(\mathbb{Z})$-invariant.

Now, given $x \in V_{\mathbb{R}}^{(m,\tau)}$, let x_L denote the unique point in L that is $G'(\mathbb{R})$-equivalent to x. We have

$$N(S; X) = \frac{1}{C_{G_0}^{(m,\tau)}} \sum_{\substack{x \in S^{\text{irr}} \\ H(x) < X}} \int_{h \in G_0} \#\{g \in \mathcal{F} : x = ghx_L\} dh. \tag{10.3}$$

For a given $x \in S^{\text{irr}}$, there exist a finite number of elements $g_1, \ldots, g_k \in G'(\mathbb{R})$ satisfying $g_j x_L = x$. We then have

$$\int_{h \in G_0} \#\{g \in \mathcal{F} : x = ghx_L\} dh =$$
$$\sum_j \int_{h \in G_0} \#\{g \in \mathcal{F} : gh = g_j\} dh = \sum_j \int_{h \in G_0 \cap \mathcal{F}^{-1} g_j} dh.$$

As dh is an invariant measure on $G'(\mathbb{R})$, we have

$$\begin{aligned}
\sum_j \int_{h \in G_0 \cap \mathcal{F}^{-1} g_j} dh &= \sum_j \int_{h \in G_0 g_j^{-1} \cap \mathcal{F}^{-1}} dh \\
&= \sum_j \int_{g \in \mathcal{F}} \#\{h \in G_0 : gh = g_j\} dg \\
&= \int_{g \in \mathcal{F}} \#\{h \in G_0 : x = ghx_L\} dg.
\end{aligned}$$

Therefore,

$$N(S;X) = \frac{1}{C_{G_0}^{(m,\tau)}} \sum_{\substack{x\in S^{\rm irr}\\ H(x)<X}} \int_{g\in\mathcal{F}} \#\{h \in G_0 : x = ghx_L\}dg \tag{10.4}$$

$$= \frac{1}{C_{G_0}^{(m,\tau)}} \int_{g\in\mathcal{F}} \#\{x \in S^{\rm irr} \cap gG_0L : H(x) < X\}\, dg \tag{10.5}$$

$$= \frac{1}{C_{G_0}^{(m,\tau)}} \int_{g\in N'(s)T'\Lambda K} \#\{x\in S^{\rm irr} \cap us\lambda\theta G_0L : H(x)<X\}dg. \tag{10.6}$$

Let us now fix our Haar measure dg on $G'(\mathbb{R})$ by setting

$$dg = \prod_{k=1}^{n} t_k^{2k-2n-1} \cdot du\, d^\times t\, d^\times\lambda\, d\theta = \prod_{m=1}^{n} s_k^{k^2-2kn} \cdot du\, d^\times s\, d^\times \lambda\, d\theta\,, \tag{10.7}$$

where du is an invariant measure on the group N of unipotent lower triangular matrices in $G(\mathbb{R})$, and where we normalize the invariant measure $d\theta$ on K so that $\int_K d\theta = 1$.

Let us write $E(u,s,\lambda,X) = us\lambda G_0L \cap \{x \in V^{(m,\tau)} : H(x) < X\}$. As $KG_0 = G_0$ and $\int_K d\theta = 1$, we have

$$N(S;X) = \frac{1}{C_{G_0}^{(m,\tau)}} \int_{g\in N'(s)T'\Lambda} \#\{x \in S^{\rm irr} \cap E(u,s,\lambda,X)\} \prod_{k=1}^{n} s_k^{k^2-2kn} \cdot du\, d^\times s\, d^\times\lambda\,. \tag{10.8}$$

We note that the same counting method may be used even if we are interested in counting both reducible and irreducible orbits in $V(\mathbb{Z})$. For any set $S \subset V^{(m,\tau)}$, let $N^*(S;X)$ be defined by (10.8), but where the superscript "irr" is removed:

$$N^*(S;X) = \frac{1}{C_{G_0}^{(m,\tau)}} \int_{g\in N'(s)T'\Lambda} \#\{x \in S \cap E(u,s,\lambda,X)\} \prod_{k=1}^{n} s_k^{k^2-2kn} \cdot du\, d^\times s\, d^\times\lambda\,. \tag{10.9}$$

Thus for a $G(\mathbb{Z})$-invariant set $S \subset V^{(m,\tau)}$, $N^*(S;X)$ counts the total (weighted) number of $G(\mathbb{Z})$-orbits in S having height less than X (not just the irreducible ones).

The expression (10.8) for $N(S;X)$, and its analogue (10.9) for $N^*(S,X)$, will be useful to us in the sections that follow.

10.2 An estimate from the geometry of numbers

To estimate the number of lattice points in $E(u, s, \lambda, X)$, we have the following proposition due to Davenport [19].

Proposition 10.2 *Let $\mathcal{R}$ be a bounded, semi-algebraic multiset in $\mathbb{R}^n$ having maximum multiplicity m, and that is defined by at most k polynomial inequalities each having degree at most ℓ. Let $\mathcal{R}'$ denote the image of $\mathcal{R}$ under any (upper or lower) triangular, unipotent transformation of $\mathbb{R}^n$. Then the number of integer lattice points (counted with multiplicity) contained in the region $\mathcal{R}'$ is*

$$\mathrm{Vol}(\mathcal{R}) + O(\max\{\mathrm{Vol}(\bar{\mathcal{R}}), 1\}),$$

where $\mathrm{Vol}(\bar{\mathcal{R}})$ denotes the greatest d-dimensional volume of any projection of $\mathcal{R}$ onto a coordinate subspace obtained by equating $n-d$ coordinates to zero, where d takes all values from 1 to $n-1$. The implied constant in the second summand depends only on n, m, k, and ℓ.

Although Davenport states the above lemma only for compact semi-algebraic sets $\mathcal{R} \subset \mathbb{R}^n$, his proof adapts without significant change to the more general case of a bounded semi-algebraic multiset $\mathcal{R} \subset \mathbb{R}^n$, with the same estimate applying also to any image $\mathcal{R}'$ of $\mathcal{R}$ under a unipotent triangular transformation.

10.3 Estimates on reducibility

In this subsection, we describe the relative frequencies with which reducible and irreducible elements sit inside various parts of the fundamental domain $\mathcal{F}hL$, as h varies over the compact region G_0.

We begin by describing some sufficient conditions that guarantee that a point in $V(\mathbb{Z})$ is reducible. Recall that a point $B \in V(\mathbb{Z})$ is called *reducible* if it lies in a distinguished rational orbit or it has discriminant zero.

Lemma 10.3 *Let $B \in V(\mathbb{Z})$ be an element such that, for some $k \in \{1, \ldots, n\}$, the top left $k \times (2n+1-k)$ block of entries of B are all equal to 0. Then $\mathrm{Disc}(\mathrm{Det}(Ax - B)) = 0$, and thus B is reducible.*

Proof If the top left $k \times (2n+1-k)$ block of B is zero, then B (and indeed $Ax_0 - B$ for any $x_0 \in \mathbb{C}$) may be viewed as a block anti-triangular matrix, with square blocks of length k, $2n+1-2k$, and k respectively on the anti-diagonal. Furthermore, the two blocks of length k will be transposes of each other. There then must be a scalar multiple Ax_0 of A such that $Ax_0 + B$ has the property that one and thus both of these square blocks of length k are singular. It follows that x_0 is a double root of $\mathrm{Det}(Ax - B)$, yielding $\mathrm{Disc}(\mathrm{Det}(Ax - B)) = 0$, as desired.

□

We also have the following consequence of Proposition 4.4:

Lemma 10.4 *Let $B = (b_{ij}) \in V(\mathbb{Z})$ be an element such that, for all pairs (i, j) satisfying $i + j < 2n + 1$, we have $b_{ij} = 0$. Then B is reducible.*

We are now ready to give an estimate on the number of irreducible elements $B = (b_{ij}) \in \mathcal{F}hL \cap V(\mathbb{Z})$, on average, satisfying $b_{11} = 0$:

Proposition 10.5 *Let h take a random value in G_0 uniformly with respect to the Haar measure dh. Then the expected number of irreducible elements $B \in \mathcal{F}hL^{(m,\tau)} \cap V(\mathbb{Z})$ such that $H(B) < X$ and $b_{11} = 0$ is $O_\varepsilon\left(X^{\frac{2n+3}{4n+2} - \frac{1}{2n(2n+1)} + \varepsilon}\right)$.*

Proof We follow the method of proof of [2, Lemma 11]. Let U denote the set of all $n(2n + 3)$ variables b_{ij} corresponding to the coordinates on $V(\mathbb{Z})$. Each variable $b \in U$ has a *weight*, defined as follows. The action of $s = (s_1, \ldots, s_n) \cdot \lambda$ on $B \in V(\mathbb{R})$ causes each variable b to multiply by a certain weight which we denote by $w(b)$. These weights $w(b)$ are evidently rational functions in $\lambda, s_1, \ldots, s_n$. If we use $w_1, \ldots, w_{2n}$ to denote the quantities $s_1, s_2, \ldots, s_n, s_n, \ldots, s_2, s_1$, respectively, then we have the explicit formula

$$w(b_{ij}) = \lambda \cdot s_1^{-2} \cdots s_n^{-2} \cdot (w_1 \cdots w_{i-1}) \cdot (w_1 \cdots w_{j-1}). \tag{10.10}$$

In particular, if $i+j = 2n+2$, then we see that $w(b_{ij}) = \lambda$. If $i+j = 2n+1$, with $i \leq n$, then $w(b_{ij}) = \lambda s_i^{-1}$. Finally, note that $\prod_{b \in U} w(b) = \lambda^{n(2n+3)}$.

Let $U_0 \subset U$ be any subset of variables in U containing b_{11}. We now give an upper estimate on the total number of irreducible $B \in \mathcal{F}hL^{(m,\tau)} \cap V(\mathbb{Z})$ such that all variables in U_0 vanish, but all variables in $U \setminus U_0$ do not. To this end, let $V(U_0)$ denote the set of all such $B \in V(\mathbb{Z})$. Furthermore, let U_1 denote the the set of variables having the minimal weights $w(b)$ among the variables $b \in U \setminus U_0$ (by "minimal weight" in $U \setminus U_0$, we mean there is no other variable $u \in U \setminus U_0$ with weight having equal or smaller exponents for all parameters $\lambda, s_1, \ldots, s_n$). As explained in [2], given U_1, it suffices to assume that U_0 in fact consists of all variables in U having weight smaller than that of some variable in U_1.

In that case, if U_1 contains any variable b_{ij} with $i+j > 2n+1$, then every element $B \in V(U_0)$ will be reducible by Lemma 10.3. If U_1 consists exactly of those variables b_{ij} for which $i+j=2n+1$ (i.e., U_0 contains all variables b_{ij} with $i+j<2n+1$), then every element $B\in V(U_0)$ will be reducible by Lemma 10.4. Thus, we may assume that all variables $b_{ij}\in U_1$ satisfy $i+j\le 2n+1$, and at least one variable $b_{ij}\in U_1$ satisfies $i+j<2n+1$.

In particular, there exist variables $\beta_1,\dots,\beta_n\in U_1$ such that, for all k, the weight $w(\beta_k)$ has equal or smaller exponents for all parameters $\lambda,s_1,\dots,s_n$ when compared to the weight $w(b_{k,2n+1-k})=\lambda s_k^{-1}$.

For each subset $U_0\subset U$ as above, we wish to show that $N(V(U_0);X)$, as defined by (10.8), is $O_\varepsilon\Big(X^{\frac{2n+3}{4n+2}-\frac{1}{2n(2n+1)}+\varepsilon}\Big)$. Since the set $N'(s)$ is absolutely bounded, the equality (10.9) implies that

$$N^*(V(U_0);X)\ll\int_{\lambda=c'}^{X^{1/(2n(2n+1))}}\int_{s_1,\dots,s_n=c}^{\infty}\sigma(V(U_0))\prod_{k=1}^{n}s_k^{k^2-2kn}\cdot d^\times s\,d^\times\lambda, \tag{10.11}$$

where $\sigma(V(U_0))$ denotes the maximum possible number of integer points in the region $E(u,s,\lambda,X)$ that also satisfy the conditions

$$b=0 \text{ for } b\in U_0 \text{ and } |b|\ge 1 \text{ for } b\in U_1. \tag{10.12}$$

By our construction of L, all entries of elements $B\in G_0L$ are uniformly bounded. Let C be a constant that bounds the absolute value of all variables $b\in U_1$ over all elements $B\in G_0L$. Then, for an element $B\in E(u,s,\lambda,X)$, we have

$$|b|\le Cw(b) \tag{10.13}$$

for all variables $b\in U_1$; therefore, the number of integer points in $E(u,s,\lambda,X)$ satisfying $b=0$ for all $b\in U_0$ and $|b|\ge 1$ for all $b\in U_1$ will be nonzero only if we have

$$Cw(b)\ge 1 \tag{10.14}$$

for all weights $w(b)$ such that $b\in U_1$. Thus if the condition (10.14) holds for all weights $w(b)$ corresponding to $b\in U_1$, then—by the definition of U_1—we will also have $Cw(b)\gg 1$ for all weights $w(b)$ such that $b\in U\setminus U_0$. In particular, note that we have

$$Cw(\beta_k)\gg 1 \tag{10.15}$$

for all $k=1,\dots,n$.

Therefore, if the region $\mathcal{E}=\{B\in E(u,s,\lambda,X): b=0\ \forall b\in U_0;\ |b|\ge 1\ \forall b\in U_1\}$ contains an integer point, then (10.14) and Proposition 10.2 together imply that the number of integer points in $\mathcal{E}$ is $O(\mathrm{Vol}(\mathcal{E}))$ (where we are computing the volume in the $\mathbb{R}$-subspace of $V(\mathbb{R})$ spanned by the coordinates $b \in U \setminus U_0$); this is because the volumes of all the projections of $u^{-1}\mathcal{E}$ will in that case also be $O(\mathrm{Vol}(\mathcal{E}))$. Now clearly

$$\mathrm{Vol}(\mathcal{E}) = O\Big(\prod_{b\in U\setminus U_0} w(b)\Big),$$

so we obtain

$$N(V(U_0);X) \ll \int_{\lambda=c'}^{X^{1/(2n(2n+1))}} \int_{s_1,\ldots,s_n=c}^{\infty} \prod_{b\in U\setminus U_0} w(b) \prod_{k=1}^{n} s_k^{k^2-2kn} \cdot d^\times s\, d^\times\lambda. \tag{10.16}$$

We wish to show that the latter integral is bounded by

$$O_\varepsilon\Big(X^{\frac{2n+3}{4n+2}-\frac{1}{2n(2n+1)}+\varepsilon}\Big)$$

for every choice of U_0. If, for example, the exponent of s_k in (10.16) is nonpositive for all k in $\{1,\ldots,n\}$, then it is clear that the resulting integral will be at most $O_\varepsilon(X^{(n(2n+3)-|U_0|)/(2n(2n+1))+\varepsilon})$ in value, since each s_k is bounded above by a power of X by (10.15).

For cases where this nonpositive exponent condition does not hold, we observe that, due to (10.15), the integrand in (10.16) may be multiplied by any product π of the variables β_k ($k = 1,\ldots,n$) without harm, and the estimate (10.16) will remain true. Extend the notation w multiplicatively, i.e., $w(ab) = w(a)w(b)$, and for any subset $U' \subset U$, write $w(\prod_{b\in U'} b) = \lambda^{|U'|} \cdot \prod s_k^{-e_k(U')}$. Then we set

$$\pi = \pi(U_0) := \prod_{k=1}^{n} \beta_k^{\max\{0\,,\,e_k(U_0)+k^2-2kn\}}.$$

We may then apply the inequalities (10.13) to each of the variables in π, yielding

$$N(V(U_0);X) \ll \int_{\lambda=c'}^{X^{1/(2n(2n+1))}} \int_{s_1,\ldots,s_n=c}^{\infty} \prod_{b\in U\setminus U_0} w(b)\, w(\pi) \prod_{k=1}^{n} s_k^{k^2-2kn} \cdot d^\times s\, d^\times\lambda. \tag{10.17}$$

By the definition of β_k, we have $w(\beta_k) \ll \lambda s_k^{-1}$, implying

$$N(V(U_0);X) \ll \int_{\lambda=c'}^{X^{1/(2n(2n+1))}} \int_{s_1,\ldots,s_n=c}^{\infty} \prod_{b\in U\setminus U_0} w(b)\, \lambda^{|\pi|} \times \prod_{k=1}^{n} s_k^{k^2-2kn-\max\{0\,,\,e_k(U_0)+k^2-2kn\}} \cdot d^\times s\, d^\times\lambda. \quad (10.18)$$

In the above, we have chosen the factor π so that the exponent of each s_m in the integrand of (10.18) is nonpositive. Thus we obtain from (10.18) that $N(V(U_0);X) = O_\varepsilon(X^{(n(2n+3)-\#U_0+\#\pi)/(2n(2n+1))+\varepsilon})$, where $\#\pi$ denotes the total number of variables of U appearing in π (counted with multiplicity). It thus suffices now to check that we always have $\#U_0 - \#\pi(U_0) > 0$.

Let U^- denote the subset of variables b_{ij} in U such that $i+j < 2n+1$. It then suffices to check the condition $\#U_0 - \#\pi(U_0) > 0$ for all proper nonempty subsets U_0 of U^-, and this is the content of the following combinatorial lemma.

Lemma 10.6 *Let $U_0 \subseteq U^-$. Then*

$$\sum_{k=1}^{n} \max\{0\,,\, e_k(U_0) + k^2 - 2kn\} \le |U_0| \quad (10.19)$$

with equality if and only if $U_0 = \varnothing$ or $U_0 = U^-$.

Proof We prove the lemma by induction on n. The lemma is directly verified for $n = 1$. If $n > 1$, then we note that the weights for s_k $(k > 1)$ agree with the weights for s_{k-1} when n is replaced by $n-1$. Let $U_0(1)$ (resp. $U_0(\ge 2)$) denote the set of all variables $b_{ij} \in U_0$ such that $i = 1$ (resp. $i \ge 2$), and define $U^-(1)$ and $U^-(\ge 2)$ analogously. Then, by the induction hypothesis, we have

$$\sum_{k=2}^{n} \max\{0\,,\, e_k(U_0(\ge 2)) + (k-1)^2 - 2(k-1)(n-1)\} \le |U_0(\ge 2)| \quad (10.20)$$

with equality if and only if $U_0(\ge 2) = \varnothing$ or $U_0(\ge 2) = U^-(\ge 2)$.

Let $|U_0(1)| = r$, so that $r \in \{0,\ldots,2n-1\}$. Then to prove (10.19), we may assume that $U_0(1) = \{b_{11}, b_{12}, \ldots, b_{1r}\}$, for any other choice of $U_0(1) \subset U^-(1)$ of size r could only decrease the sum on the left hand side of (10.19).

We first consider the case where $r < n$; in that case, we have

$$\begin{aligned}
&\max\{0\,,\,e_1(U_0(1)) + 1 - 2n\} + \sum_{k=2}^{n} \max\{0\,,\,e_k(U_0) + k^2 - 2kn\} \\
&\quad = 0 + \sum_{k=2}^{n} \max\{0\,,\,e_k(U_0(\geq 2)) + (k-1)^2 - 2(k-1)(n-1) + \\
&\qquad\qquad\qquad\qquad\qquad\qquad e_k(U_0(1)) + (1-2n)\} \\
&\quad \leq |U_0(\geq 2)| + \sum_{k=2}^{n} \max\{0\,,\,e_k(U_0(1)) + (1-2n)\} \\
&\quad = |U_0(\geq 2)| \leq |U_0|,
\end{aligned}$$

with equality if and only if r is zero and all terms on the left hand sides of (10.19) and (10.20) are zero, i.e., $U_0 = \varnothing$.

Now consider the case where $r \geq n$. In that case, we have similarly

$$\begin{aligned}
&\sum_{k=1}^{n} \max\{0\,,\,e_k(U_0) + k^2 - 2kn\} \\
&\qquad \leq |U_0(\geq 2)| + \sum_{k=1}^{n} \max\{0\,,\,e_k(U_0(1)) + (1-2n)\} \\
&\qquad = |U_0(\geq 2)| + 2(r-n-1) - 1 \leq |U_0|,
\end{aligned}$$

with equality if and only if r is $2n-1$ and we have equality in (10.20) with all terms nonzero, i.e., $U_0 = U^-$.

□

This completes the proof of Proposition 10.5.

□

We now give an estimate on the number of reducible elements $B = (b_{ij}) \in \mathcal{F}hL \cap V(\mathbb{Z})$, on average, satisfying $b_{11} \neq 0$:

Proposition 10.7 *Let h take a random value in G_0 uniformly with respect to the measure dh. Then the expected number of reducible elements $B \in \mathcal{F}hL^{(m,\tau)} \cap V(\mathbb{Z})$ such that $H(B) < X$ and $b_{11} \neq 0$ is $o(X^{(2n+3)/(4n+2)})$.*

We defer the proof of this lemma to the end of the section.

We also have the following proposition which bounds the number of $G(\mathbb{Z})$-equivalence classes of irreducible elements in $V^{(m,\tau)}$ with height less than X that have large stabilizers inside $G(\mathbb{Q})$; we again defer the proof to the end of the section.

Proposition 10.8 *The number of $G(\mathbb{Z})$-equivalence classes of irreducible elements in $V(\mathbb{Z})$ having height less than X whose stabilizer in $G(\mathbb{Q})$ has size greater than 1 is $o(X^{(2n+3)/(4n+2)})$.*

10.4 The main term

Fix again m, τ and let $L = L^{(m,\tau)}$. The work of the previous subsection shows that, in order to obtain Theorem 10.1, it suffices to count those integral elements $B \in \mathcal{F}hL$ of bounded height for which $b_{11} \neq 0$, as h ranges over G_0.

Let $\mathcal{R}_X(hL)$ denote the region $\mathcal{F}hL \cap \{B \in V(\mathbb{R}) : H(B) < X\}$. Then we have the following result counting the number of integral points in $\mathcal{R}_X(hL)$, on average, satisfying $b_{11} \neq 0$:

Proposition 10.9 *Let h take a random value in G_0 uniformly with respect to the Haar measure dh. Then the expected number of elements $B \in \mathcal{F}hL \cap V(\mathbb{Z})$ such that $|H(B)| < X$ and $b_{11} \neq 0$ is* $\mathrm{Vol}(\mathcal{R}_X(L)) + O\Big(X^{\frac{2n+3}{4n+2} - \frac{1}{2n(2n+1)}}\Big)$.

Proof Following the notation in the proof of Proposition 10.5, let $V^{(m,\tau)}(\varnothing)$ denote the subset of $V(\mathbb{Z})^{(m,\tau)}$ such that $b_{11} \neq 0$. We wish to show that

$$N^*(V^{(m,\tau)}(\varnothing); X) = \mathrm{Vol}(\mathcal{R}_X(L)) + O\Big(X^{\frac{2n+3}{4n+2} - \frac{1}{2n(2n+1)}}\Big). \tag{10.21}$$

We have

$$N^*(V^{(m,\tau)}(\varnothing); X) = \frac{1}{C_{G_0}^{(m,\tau)}} \int_{\lambda=c'}^{X^{1/(2n(2n+1))}} \int_{s_1,\ldots,s_n=c}^{\infty} \int_{u\in N'(s)} \sigma(V(\varnothing)) \prod_{k=1}^{n} s_k^{k^2-2kn} \cdot du\, d^\times s\, d^\times\lambda, \tag{10.22}$$

where $\sigma(V(\varnothing))$ denotes the number of integer points in the region $E(u, s, \lambda, X)$ satisfying $|b_{11}| \geq 1$. Evidently, the number of integer points in $E(u, s, \lambda, X)$ with $|b_{11}| \geq 1$ can be nonzero only if we have

$$Cw(b_{11}) = C \cdot \frac{\lambda}{s_1^2 s_2^2 \cdots s_n^2} \geq 1. \tag{10.23}$$

Hence, if the region $\mathcal{E} = \{B \in E(u, s, \lambda, X) : |b_{11}| \geq 1\}$ contains an integer point, then (10.23) and Proposition 10.2 imply that the number of integer points in $\mathcal{E}$ is $\mathrm{Vol}(\mathcal{E}) + O(\mathrm{Vol}(\mathcal{E})/w(b_{11}))$, because all smaller-dimensional projections of $u^{-1}\mathcal{E}$ are clearly bounded by a constant times the projection of $\mathcal{E}$ onto the hyperplane $b_{11} = 0$ (since b_{11} has minimal weight).

Therefore, since $\mathcal{E} = E(u, s, \lambda, X) - \big(E(u, s, \lambda, X) - \mathcal{E}\big)$, we may write

$$N^*(V^{(m,\tau)}(\varnothing); X) =$$

$$\frac{1}{C_{G_0}^{(m,\tau)}} \int\limits_{\lambda=c'}^{X^{1/(2n(2n+1))}} \int\limits_{s_1,\dots,s_n=c}^{\infty} \int\limits_{u\in N'(s)} \big(\mathrm{Vol}\big(E(u,s,\lambda,X)\big) - \mathrm{Vol}\big(E(u,s,\lambda,X) - \mathcal{E}\big)$$

$$+ O(\max\{\lambda^{n(2n+3)-1} s_1^2 s_2^2 \cdots s_n^2, 1\})\big) \prod_{k=1}^{n} s_k^{k^2-2kn} \cdot du\, d^\times s\, d^\times \lambda. \quad (10.24)$$

The integral of the first term in (10.24) is $\int_{h\in G_0} \mathrm{Vol}(\mathcal{R}_X(hL))dh$. Since $\mathrm{Vol}(\mathcal{R}_X(hL))$ does not depend on the choice of $h \in G_0$, the latter integral is simply $C_{G_0}^{(m,\tau)} \cdot \mathrm{Vol}(\mathcal{R}_X(L))$.

To estimate the integral of the second term in (10.24), let

$$\mathcal{E}' = E(u, s, t, X) - \mathcal{E},$$

and for each $|b_{11}| \le 1$, let $\mathcal{E}'(b_{11})$ be the subset of all elements $B \in \mathcal{E}'$ with the given value of b_{11}. Then the $(n(2n+3) - 1)$-dimensional volume of $\mathcal{E}'(b_{11})$ is at most $O\Big(\prod_{b\in U\setminus\{b_{11}\}} w(b)\Big)$, and so we have the estimate

$$\mathrm{Vol}(\mathcal{E}') \ll \int_{-1}^{1} \prod_{b\in U\setminus\{b_{11}\}} w(b)\, db_{11} = O\Big(\prod_{b\in U\setminus\{b_{11}\}} w(b)\Big).$$

The second term of the integrand in (10.24) can thus be absorbed into the third term.

Finally, one easily computes the integral of the third term in (10.24) to be $O\Big(X^{\frac{2n+3}{4n+2} - \frac{1}{2n(2n+1)}}\Big)$. We thus obtain

$$N^*(V(\mathbb{Z})^{(m,\tau)}; X) = \mathrm{Vol}(\mathcal{R}_X(L)) + O\Big(X^{\frac{2n+3}{4n+2} - \frac{1}{2n(2n+1)}}\Big), \quad (10.25)$$

as desired.

□

Propositions 10.5, 10.7, 10.8, and 10.9 now yield Theorem 10.1.

10.5 Computation of the volume

Fix again m, τ. In this subsection, we describe how to compute the volume of $\mathcal{R}_X(L^{(m,\tau)})$.

To this end, let $R^{(m,\tau)} := \Lambda L^{(m,\tau)}$. For each $(c_2, \ldots, c_{2n+1}) \in I(m)$, the set $R^{(m,\tau)}$ contains exactly one point $p^{(m,\tau)}(c_2, \ldots, c_{2n+1})$ having invariants $c_2, \ldots, c_{2n+1}$. Let $R^{(m,\tau)}(X)$ denote the set of all those points in $R^{(m,\tau)}$ having height less than X. Then $\mathrm{Vol}(\mathcal{R}_X(L^{(m,\tau)})) = \mathrm{Vol}(\mathcal{F}^{\mathrm{ss}} \cdot R^{(m,\tau)}(X))$, where $\mathcal{F}^{\mathrm{ss}}$ denotes the fundamental domain $N'T'K$ for the action of $G(\mathbb{Z})$ on $G(\mathbb{R})$ (here N', T', and K are as in (9.1)).

The set $R^{(m,\tau)}$ is in canonical one-to-one correspondence with the set $I(m) \subset \mathbb{R}^{2n}$. We thus obtain a natural measure on each of these sets $R^{(m,\tau)}$, given by $dr = dc_2 \cdots dc_{2n+1}$. Let ω be a differential which generates the rank 1 module of top-degree differentials of G over $\mathbb{Z}$. We begin with the following key proposition, which describes how one can change measure from dv on V to $\omega(g)\,dr$ on $G \times R$:

Proposition 10.10 *Let K be $\mathbb{C}$, $\mathbb{R}$, or $\mathbb{Z}_p$ for any prime p. Let dv be the standard additive measure on $V(K)$. Let R be an open subset of K^{2n} and let $s : R \to V(K)$ be a continuous function such that the invariants of $s(c_2, \ldots, c_{2n+1})$ are given by $c_2, \ldots, c_{2n+1}$. Then there exists a rational nonzero constant $\mathcal{J}$ (independent of K, R, and s) such that, for any measurable function ϕ on $V(K)$, we have*

$$\int_{v \in G(K)\cdot s(R)} \phi(v)dv = |\mathcal{J}| \int_R \int_{G(K)} \phi(g \cdot s(c_2, \ldots, c_n))\, \omega(g)\, dr \qquad (10.26)$$

where we regard $G(K) \cdot s(R)$ as a multiset.

The proof of Proposition 10.10 is identical to that of [5, Prop. 3.11] (where we use the important fact that the dimension $n(2n+3)$ of V is equal to the sum of the degrees of the invariants $c_2, \ldots, c_{2n+1}$ for the action of G on V).

Proposition 10.10 may now be used to give a convenient expression for the volume of the multiset $\mathcal{R}_X(L^{(m,\tau)})$:

$$\begin{aligned}
\int_{\mathcal{R}_X(L^{(m,\tau)})} dv &= \int_{\mathcal{F}^{\mathrm{ss}}\cdot R^{(m,\tau)}(X)} dv = |\mathcal{J}| \cdot \int_{R^{(m,\tau)}(X)} \int_{\mathcal{F}^{\mathrm{ss}}} \omega(g)\, dr \\
&= |\mathcal{J}| \cdot \mathrm{Vol}(G(\mathbb{Z})\backslash G(\mathbb{R})) \cdot \int_{R^{(m,\tau)}(X)} dr \\
&= |\mathcal{J}| \cdot \mathrm{Vol}(G(\mathbb{Z})\backslash G(\mathbb{R})) \cdot \int_{\substack{(c_2,\ldots,c_{2n+1})\in I(m) \\ H(c_2,\ldots,c_{2n+1})<X}} dc_2 \cdots dc_{2n+1}.
\end{aligned} \qquad (10.27)$$

10.6 Congruence conditions

In this subsection, we prove the following version of Theorem 10.1 where we count elements of $V(\mathbb{Z})$ satisfying some finite set of congruence conditions:

Theorem 10.11 *Suppose S is a subset of $V(\mathbb{Z})$ defined by congruence conditions modulo finitely many prime powers. Then*

$$N(S \cap V^{(m,\tau)}; X) \;=\; N(V(\mathbb{Z})^{(m,\tau)}; X) \cdot \prod_p \mu_p(S) + o(X^{(2n+3)/(4n+2)}), \tag{10.28}$$

where $\mu_p(S)$ denotes the p-adic density of S in $V(\mathbb{Z})$.

To obtain Theorem 10.11, suppose S is defined by congruence conditions modulo some integer m. Then S may be viewed as the union of (say) k translates $\mathcal{L}_1, \ldots, \mathcal{L}_k$ of the lattice $m \cdot V(\mathbb{Z})$. For each such lattice translate $\mathcal{L}_j$, we may use formula (10.8) and the discussion following that formula to compute $N(S; X)$, but where each d-dimensional volume is scaled by a factor of $1/m^d$ to reflect the fact that our new lattice has been scaled by a factor of m. For a fixed value of m, we thus obtain

$$N(\mathcal{L}_j \cap V^{(m,\tau)}; X) = m^{-n(2n+3)} \operatorname{Vol}(\mathcal{R}_X(L)) + o(X^{(2n+3)/(4n+2)}). \tag{10.29}$$

Summing (10.29) over j, and noting that $km^{-n(2n+3)} = \prod_p \mu_p(S)$, yields Theorem 10.11.

We will also have occasion to use the following weighted version of Theorem 10.11; the proof is identical.

Theorem 10.12 *Let $p_1, \ldots, p_k$ be distinct prime numbers. For $j = 1, \ldots, k$, let $\phi_{p_j} : V(\mathbb{Z}) \to \mathbb{R}$ be a $G(\mathbb{Z})$-invariant function on $V(\mathbb{Z})$ such that $\phi_{p_j}(x)$ depends only on the congruence class of x modulo some power $p_j^{a_j}$ of p_j. Let $N_\phi(V(\mathbb{Z})^{(m,\tau)}; X)$ denote the number of irreducible $G(\mathbb{Z})$-orbits in $V(\mathbb{Z})^{(m,\tau)}$ having height bounded by X, where each orbit $G(\mathbb{Z}) \cdot B$ is counted with weight $\phi(B) := \prod_{j=1}^k \phi_{p_j}(B)$. Then*

$$N_\phi(V(\mathbb{Z})^{(m,\tau)}; X) =$$
$$N(V(\mathbb{Z})^{(m,\tau)}; X) \prod_{j=1}^{k} \int_{B \in V(\mathbb{Z}_{p_j})} \tilde{\phi}_{p_j}(B)\, dB + o(X^{(2n+3)/(4n+2)}), \tag{10.30}$$

where $\tilde{\phi}_{p_j}$ is the natural extension of ϕ_{p_j} to $V(\mathbb{Z}_{p_j})$ by continuity, dB denotes the additive measure on $V(\mathbb{Z}_{p_j})$ normalized so that $\int_{B \in V(\mathbb{Z}_{p_j})} dB = 1$, and where the implied constant in the error term depends only on the local weight functions ϕ_{p_j}.

10.7 Proof of Propositions 10.7 and 10.8

We may use the results of the previous subsection to prove Propositions 10.7 and 10.8. Indeed, to prove Proposition 10.7, we note that if an element $B \in V(\mathbb{Z})$ is reducible over $\mathbb{Q}$ then it also must be reducible modulo p (i.e., its image in $V(\mathbb{F}_p)$ lies in a distinguished orbit or has discriminant zero) for every p.

Let S^{red} denote the set of elements in $V(\mathbb{Z})$ that are reducible over $\mathbb{Q}$, and let S_p^{red} denote the set of all elements in $V(\mathbb{Z})$ that are reducible modulo p. Then $S^{\text{red}} \subset \cap_p S_p^{\text{red}}$. Let $S^{\text{red}}(Y) = \cap_{p<Y} S_p^{\text{red}}$, and let us use as before $V^{(m,\tau)}(\varnothing)$ to denote the set $B \in V(\mathbb{Z})^{(m,\tau)}$ such that $b_{11} \neq 0$. Then the proof of Theorem 10.11 (without assuming Propositions 10.7 and 10.8!) gives that

$$N^*(S^{\text{red}}(Y) \cap V^{(m,\tau)}(\varnothing); X) \leq 2^{2n} \cdot N^*(V^{(m,\tau)}(\varnothing); X) \cdot \prod_{p<Y} \mu_p(S_p^{\text{red}}) + o(X^{(2n+3)/(4n+2)}). \quad (10.31)$$

To estimate $\mu_p(S_p^{\text{red}})$, we recall from Subsection 6.1 that the number of elements $B \in V(\mathbb{F}_p)$ having any given associated separable polynomial $f(x) = \det(Ax - B)$ is $\#\,\mathrm{SO}(W)(\mathbb{F}_p)$. The number of these elements B that lie in the distinguished orbit is $\#\,\mathrm{SO}(W)(\mathbb{F}_p)/2^m$, where $m+1$ is the number of irreducible factors of f in $\mathbb{F}_p[x]$. For $p > 2n+1$, it is an elementary and well-known calculation that the number of separable polynomials

$$f(x) = x^{2n+1} + c_2x^{2n-1} + \cdots + c_{2n+1}$$

over $\mathbb{F}_p$ that are reducible over $\mathbb{F}_p$ is $p^{2n} \cdot 2n/(2n+1) + O(p^{2n-1})$, where the implied O-constant is independent of p. Since such a reducible polynomial $f(x)$ has at least 2 factors in $\mathbb{F}_p[x]$, we see that at least $\#\,\mathrm{SO}(W)(\mathbb{F}_p)/2$ of the elements B having associated polynomial $f(x)$ do not lie in a distinguished orbit. We conclude that

$$\mu_p(S_p^{\text{red}}) \leq 1 - \frac{2n}{2n+1} \cdot \frac{1}{2} + O(1/p).$$

Combining with (10.31), we see that

$$\lim_{X\to\infty} \frac{N^*(S^{\text{red}} \cap V^{(m,\tau)}(\varnothing); X)}{X^{(2n+3)/(4n+2)}} \ll \prod_{p<Y} \mu_p(S_p^{\text{red}}) \ll \prod_{p<Y} \Big(1 - \frac{n}{2n+1} + O(1/p)\Big).$$

When Y tends to infinity, the product on the right tends to 0, proving Proposition 10.7.

We may proceed similarly with Proposition 10.8. If an element $B \in V(\mathbb{Z})$ with nonzero discriminant has a non-trivial stabilizer, then the associated polynomial $f(x) = \det(Ax - B)$ must be reducible in $\mathbb{Q}[x]$, and thus must be reducible in $\mathbb{F}_p[x]$ for all p.

Let $S^{\text{bigstab}} \subset V(\mathbb{Z})$ denote the elements $B \in V(\mathbb{Z})$ such that $f(x) = \det(Ax - B)$ is reducible in $\mathbb{Q}[x]$, and let $S_p^{\text{bigstab}} \subset V(\mathbb{Z})$ denote the elements $B \in V(\mathbb{Z})$ such that $f(x)$ is reducible in $\mathbb{F}_p[x]$. Let $S^{\text{bigstab}}(Y) = \bigcap_{p<Y} S_p^{\text{bigstab}}$. Then we have by the same argument that

$$\begin{aligned}\lim_{X\to\infty} \frac{N^*(S^{\text{bigstab}} \cap V^{(m,\tau)}(\varnothing); X)}{X^{(2n+3)/(4n+2)}} &\ll \prod_{p<Y} \mu_p(S_p^{\text{bigstab}}) \\ &\ll \prod_{p<Y} \Big(1 - \frac{2n}{2n+1} + O(1/p)\Big).\end{aligned}$$

Letting Y tend to infinity, and noting Proposition 10.5, now proves also Proposition 10.8.

11 Sieving to Selmer Elements

For each prime p, let Σ_p be a closed subset of elements $(c_2, \dots, c_{2n+1})$ in $\mathbb{Z}_p^{2n} \setminus \{\Delta = 0\}$ whose boundary has measure zero. To such a collection $(\Sigma_p)_p$ of local specifications, we may associate the set F_Σ of hyperelliptic curves, where $C(c_2, \dots, c_{2n+1}) \in F_\Sigma$ if and only if $(c_2, \dots, c_{2n+1}) \in \Sigma_p$ for all p. We say then that F_Σ is a family of hyperelliptic curves $C(c_2, \dots, c_{2n+1})$ over $\mathbb{Q}$ that is *defined by congruence conditions*.

We may also impose "congruence conditions at infinity" on elements $(c_2, \dots, c_{2n+1})$ in $\mathbb{Z}^{2n}$. Let additionally Σ_∞ denote a union of components $I(m)$ (as defined in §6.3), where m ranges over the elements of some subset of $\{0, \dots, n\}$. Then we may again associate to the collection $(\Sigma_\nu)_\nu$ of local specifications the set F_Σ of hyperelliptic curves over $\mathbb{Q}$, where $C(c_2, \dots, c_{2n+1}) \in F_\Sigma$ if and only if $(c_2, \dots, c_{2n+1}) \in \Sigma_\nu$ for all ν (including $\nu = \infty$). We will still say in that case that F_Σ is defined by congruence conditions.

If F is a set of hyperelliptic curves (1.1) over $\mathbb{Q}$ defined by local congruence conditions, then we use $\text{Inv}(F)$ to denote the set

$$\{(c_2(C), \dots, c_{2n+1}(C)) : C \in F\} \subset \mathbb{Z}^{2n}.$$

We denote the p-adic closure of $\text{Inv}(F)$ in $\mathbb{Z}_p^{2n} \setminus \{\Delta = 0\}$ by $\text{Inv}_p(F)$. We say that such a set F of hyperelliptic curves over $\mathbb{Q}$ is *large at p* if $\text{Inv}_p(F)$

contains all elements $(c_2, \ldots, c_{2n+1}) \in \mathbb{Z}_p^{2n}$ such that $p^2 \nmid \Delta(c_2, \ldots, c_{2n+1})$. Finally, we say that such a set F of hyperelliptic curves is *large* if it is large at all but finitely many primes p.

In this section we prove the following strengthening of Theorem 1.1:

Theorem 11.1 *When all hyperelliptic curves over $\mathbb{Q}$ of genus n with a rational Weierstrass point in any large family are ordered by height, the average size of the 2-Selmer group is 3.*

Theorem 11.1 is proven via an appropriate sieve applied to the counts of $G(\mathbb{Z})$-orbits on $V(\mathbb{Z})$ having bounded height which we obtained in Section 10.

Note that the set of all hyperelliptic curves in (1.1) with indivisible coefficients is large. So too is the set of such hyperelliptic curves $C : y^2 = f(x)$ having semistable reduction. Any family of hyperelliptic curves in (1.1) defined by finitely many congruence conditions on the coefficients is also a large family. Thus Theorem 11.1 applies to quite general families of hyperelliptic curves.

11.1 A weighted set $U(F)$ in $V(\mathbb{Z})$ corresponding to a large family F

It follows from Theorem 5.2 and Proposition 8.10 that non-identity elements of the 2-Selmer group of the Jacobian of the hyperelliptic curve $C = C(c_2, \ldots, c_{2n+1})$ over $\mathbb{Q}$ defined by (1.1) are in bijective correspondence with $G(\mathbb{Q})$-equivalence classes of irreducible locally soluble elements $B \in V(\mathbb{Z})$ having invariants $2^8 c_2$, $2^{12} c_3$, …, $2^{4(2n+1)} c_{2n+1}$; in this bijection, we have $H(B) = 2^{8n(2n+1)} H(C)$. In Section 10, we computed the asymptotic number of $G(\mathbb{Z})$-equivalence classes of irreducible elements $B \in V(\mathbb{Z})$ having bounded height. In order to use this to compute the number of irreducible locally soluble $G(\mathbb{Q})$-equivalence classes of elements $B \in V(\mathbb{Z})$ having invariants in

$$\{(2^8 c_2, \ldots, 2^{4(2n+1)} c_{2n+1}) : (c_2, \ldots, c_{2n+1}) \in \mathrm{Inv}(F)\} \tag{11.1}$$

and bounded height (where F is any large family), we need to count each $G(\mathbb{Z})$-orbit, $G(\mathbb{Z}) \cdot B$, with a weight of $1/n(B)$, where $n(B)$ is equal to the number of $G(\mathbb{Z})$-orbits inside the $G(\mathbb{Q})$-equivalence class of B in $V(\mathbb{Z})$.

To count the number of irreducible locally soluble $G(\mathbb{Z})$-orbits having invariants in the set (11.1) and bounded height, where each orbit $G(\mathbb{Z}) \cdot B$ is weighted by $1/n(B)$, it suffices to count the number of such $G(\mathbb{Z})$-orbits of bounded height such that each orbit $G(\mathbb{Z}) \cdot B$ is weighted instead by

$1/m(B)$, where

$$m(B) := \sum_{B' \in O(B)} \frac{\#\operatorname{Aut}_{\mathbb{Q}}(B')}{\#\operatorname{Aut}_{\mathbb{Z}}(B')} = \sum_{B' \in O(B)} \frac{\#\operatorname{Aut}_{\mathbb{Q}}(B)}{\#\operatorname{Aut}_{\mathbb{Z}}(B')};$$

here $O(B)$ denotes a set of orbit representatives for the action of $G(\mathbb{Z})$ on the $G(\mathbb{Q})$-equivalence class of B in $V(\mathbb{Z})$, and $\operatorname{Aut}_{\mathbb{Q}}(B)$ (resp. $\operatorname{Aut}_{\mathbb{Z}}(B)$) denotes the stabilizer of B in $G(\mathbb{Q})$ (resp. $G(\mathbb{Z})$). The reason it suffices to weight by $1/m(B)$ instead of $1/n(B)$ is that we have shown in the proof of Proposition 10.8 that all but a negligible number $o(X^{(2n+3)/(4n+2)})$ of irreducible $G(\mathbb{Z})$-orbits having height less than X have trivial stabilizer in $G(\mathbb{Q})$ (and thus also in $G(\mathbb{Z})$), while the number of hyperelliptic curves in F of bounded height is $\gg X^{(2n+3)/(4n+2)}$.

We use $U(F)$ to denote the set of all locally soluble elements in $V(\mathbb{Z})$ having invariants in the set (11.1), i.e.,

$$U(F) := \left\{ \begin{array}{l} \text{loc. sol. elts. in } V(\mathbb{Z}) \text{ having inv'ts.} \\ (2^8 c_2, \ldots, 2^{4(2n+1)} c_{2n+1})\} \end{array} \middle| \; C(c_2, \ldots, c_{2n+1}) \in F \right\}. \tag{11.2}$$

We assign to each element $B \in U(F)$ the weight $1/m(B)$. Then we conclude that the weighted number of irreducible $G(\mathbb{Z})$-orbits of height less than $2^{8n(2n+1)}X$ in $U(F)$ is asymptotically equal to the number of non-identity 2-Selmer elements of all hyperelliptic curves of height less than X in F.

The global weights $m(B)$ assigned to elements $B \in U(F)$ are useful for the following reason. For a prime p and any element $B \in V(\mathbb{Z}_p)$, define the local weight $m_p(B)$ by

$$m_p(B) := \sum_{B' \in O_p(B)} \frac{\#\operatorname{Aut}_{\mathbb{Q}_p}(B)}{\#\operatorname{Aut}_{\mathbb{Z}_p}(B')},$$

where $O_p(B)$ denotes a set of orbit representatives for the action of $G(\mathbb{Z}_p)$ on the $G(\mathbb{Q}_p)$-equivalence class of B in $V(\mathbb{Z}_p)$, and $\operatorname{Aut}_{\mathbb{Q}_p}(B)$ (resp. $\operatorname{Aut}_{\mathbb{Z}_p}(B)$) denotes the stabilizer of B in $G(\mathbb{Q}_p)$ (resp. $G(\mathbb{Z}_p)$). Then by Proposition 8.9, we have the following identity:

$$m(B) = \prod_p m_p(B). \tag{11.3}$$

Thus the global weights of elements in $U(F)$ are products of local weights, so we may express the global weighted density of elements $U(F)$ in $V(\mathbb{Z})$ as products of local weighted densities of the closures of the set $U(F)$ in $V(\mathbb{Z}_p)$. We consider these local weighted densities next.

11.2 Local densities of the weighted sets $U(F)$

Suppose that F is a large family of elliptic curves, and for each place ν of $\mathbb{Q}$, let F_ν denote the resulting family of curves defined by congruence conditions over $\mathbb{Z}_\nu$ (where we follow the usual convention that $\mathbb{Z}_\nu = \mathbb{Q}_\nu = \mathbb{R}$ when $\nu = \infty$). Let $U(F)$ denote the set in $V(\mathbb{Z})$ attached to F, as defined by (11.2), and for any place ν of $\mathbb{Q}$ define $U_\nu(F)$ analogously by

$$U_\nu(F) := \left\{ \begin{array}{l} \text{sol. elts. in } V(\mathbb{Z}_\nu) \text{ having inv'ts.} \\ (2^8 c_2, \ldots, 2^{4(2n+1)} c_{2n+1})\} \end{array} \middle| \; C(c_2, \ldots, c_{2n+1}) \in F_\nu \right\}. \tag{11.4}$$

In the case $\nu = p$ is a finite prime, we assign to each element $B \in U_p(F)$ the weight $1/m_p(B)$. In this subsection, we determine the weighted p-adic density of $U_p(F)$ in $V(\mathbb{Z}_p)$ in terms of a *local (p-adic) mass* $M_p(V, F)$ involving the elements of $\mathrm{Jac}(C)/2\mathrm{Jac}(C)$ for curves C in F_p. To do so, we require the following proposition, which is a reformulation of the change-of-measure assertion of Proposition 10.10, with $\mathbb{Z}_p$ in place of $\mathbb{R}$:

Proposition 11.2 *Let $\mathcal{J}$ be the constant of Proposition* 10.10*, let p be a prime, and let ϕ be a continuous function on $V(\mathbb{Z}_p)$. Then*

$$\int_{V_{\mathbb{Z}_p}} \phi(B)dB = |\mathcal{J}|_p \int_{\substack{(c_2,\ldots,c_{2n+1})\in\mathbb{Z}_p^{2n} \\ \Delta(c_2,\ldots,c_{2n+1})\neq 0}} \Bigg(\sum_{B \in \frac{V_{\mathbb{Z}_p}(c_2,\ldots,c_{2n+1})}{G(\mathbb{Z}_p)}} \frac{1}{\#\,\mathrm{Aut}_{\mathbb{Z}_p}(B)} \times$$

$$\int_{g\in G(\mathbb{Z}_p)} \phi(g\cdot B)\,\omega(g)\Bigg) dc_2 \cdots dc_{2n+1}, \tag{11.5}$$

where $\frac{V_{\mathbb{Z}_p}(c_2,\ldots,c_{2n+1})}{G(\mathbb{Z}_p)}$ denotes a set of representatives for the action of $G(\mathbb{Z}_p)$ on elements in $V_{\mathbb{Z}_p}$ having invariants $c_2, \ldots, c_{2n+1}$.

The proof of Proposition 11.2 is identical to that of [5, Prop. 3.12]. We observe that if ϕ is supported only on elements that have nonzero discriminant and are soluble, and if $\phi(B)$ is additionally weighted by $1/m_p(B)$, then Equation (11.5) takes on a particularly nice form:

Corollary 11.3 *Let p be a prime and let ϕ be a continuous $G(\mathbb{Q}_p)$-invariant function on $V_{\mathbb{Z}_p}$ such that every element $B \in V_{\mathbb{Z}_p}$ in the support of ϕ has nonzero discriminant, is soluble, and satisfies $2^{4j} \mid c_j(B)$ for all*

$j \in \{2, \ldots, 2n+1\}$. Then

$$\int_{V_{\mathbb{Z}_p}} \frac{\phi(B)}{m_p(B)} dB = |\mathcal{J}|_p \operatorname{Vol}(G(\mathbb{Z}_p)) \int_{C=C(c_2,\ldots,c_{2n+1})} \frac{1}{\#\operatorname{Jac}(C)[2](\mathbb{Q}_p)} \times \Big(\sum_{\sigma \in \frac{\operatorname{Jac}(C)(\mathbb{Q}_p)}{2\operatorname{Jac}(C)(\mathbb{Q}_p)}} \phi(B_\sigma) \Big) dc_2 \cdots dc_{2n+1}, \quad (11.6)$$

where B_σ is any element in $V_{\mathbb{Z}_p}$ that corresponds to σ under the correspondence of Proposition 5.2. (The existence of such an $f_\sigma \in V_{\mathbb{Z}_p}$ is guaranteed by Proposition 8.10.)

Proof Proposition 11.2 implies that the left side of (11.6) is equal to

$$\begin{aligned} |\mathcal{J}|_p & \int_{\substack{(c_2,\ldots,c_{2n+1}) \in \mathbb{Z}_p^{2n} \\ \Delta(c_2,\ldots,c_{2n+1}) \neq 0}} \Big(\sum_{B \in \frac{V_{\mathbb{Z}_p}(c_2,\ldots,c_{2n+1})}{G(\mathbb{Z}_p)}} \frac{1}{\#\operatorname{Aut}_{\mathbb{Z}_p}(B)} \times \\ & \int_{g \in G(\mathbb{Z}_p)} \frac{\phi(g \cdot B)}{m_p(B)} \omega(g) \Big) dc_2 \cdots dc_{2n+1} = |\mathcal{J}|_p \operatorname{Vol}(G(\mathbb{Z}_p)) \times \\ & \int_{\substack{(c_2,\ldots,c_{2n+1}) \in \mathbb{Z}_p^{2n} \\ \Delta(c_2,\ldots,c_{2n+1}) \neq 0}} \Big(\sum_{B \in \frac{V_{\mathbb{Z}_p}(c_2,\ldots,c_{2n+1})}{G(\mathbb{Z}_p)}} \frac{\phi(B)}{m_p(B)\#\operatorname{Aut}_{\mathbb{Z}_p}(B)} \Big) dc_2 \cdots dc_{2n+1} \end{aligned} \quad (11.7)$$

since both ϕ and m_p are $G(\mathbb{Z}_p)$-invariant. For $B \in V_{\mathbb{Z}_p}$, let $B = B_1, \ldots, B_k$ be the set of all elements in $\frac{V_{\mathbb{Z}_p}(c_2,\ldots,c_{2n+1})}{G(\mathbb{Z}_p)}$ that are $G(\mathbb{Q}_p)$-equivalent to B. Then, since ϕ and m_p are $G(\mathbb{Q}_p)$-invariant, we have

$$\begin{aligned} \sum_{i=1}^k \frac{\phi(B_i)}{m_p(B_i)\#\operatorname{Aut}_{\mathbb{Z}_p}(B_i)} &= \frac{1}{m_p(B)} \sum_{i=1}^k \frac{\phi(B)}{\#\operatorname{Aut}_{\mathbb{Z}_p}(B_i)} \\ &= \left(\sum_{i=1}^k \frac{\#\operatorname{Aut}_{\mathbb{Q}_p}(B)}{\#\operatorname{Aut}_{\mathbb{Z}_p}(B_i)} \right)^{-1} \sum_{i=1}^k \frac{\phi(B)}{\#\operatorname{Aut}_{\mathbb{Z}_p}(B_i)} \\ &= \frac{\phi(B)}{\#\operatorname{Aut}_{\mathbb{Q}_p}(B)}. \end{aligned}$$

Therefore,

$$\int_{V_{\mathbb{Z}_p}} \frac{\phi(B)}{m_p(B)} dB = |\mathcal{J}|_p \operatorname{Vol}(G(\mathbb{Z}_p)) \times$$
$$\int\limits_{\substack{(c_2,\ldots,c_{2n+1})\in\mathbb{Z}_p^{2n} \\ \Delta(c_2,\ldots,c_{2n+1})\neq 0}} \Bigg(\sum_{B\in \frac{V_{\mathbb{Z}_p}(c_2,\ldots,c_{2n+1})}{G(\mathbb{Q}_p)}} \frac{\phi(B)}{\#\operatorname{Aut}_{\mathbb{Q}_p}(B)} \Bigg) dc_2\cdots dc_{2n+1}, \quad (11.8)$$

where $\frac{V_{\mathbb{Z}_p}(c_2,\ldots,c_{2n+1})}{G(\mathbb{Q}_p)}$ analogously denotes a set consisting of one representative from each $G(\mathbb{Q}_p)$-equivalence class in $V_{\mathbb{Z}_p}$ having invariants $c_2, \ldots, c_{2n+1}$. Proposition 5.2 states that soluble elements in $\frac{V_{\mathbb{Z}_p}(c_2,\ldots,c_{2n+1})}{G(\mathbb{Q}_p)}$ are in bijective correspondence with elements in $\operatorname{Jac}(C)(\mathbb{Q}_p)/2\operatorname{Jac}(C)(\mathbb{Q}_p)$, where $C = C(c_2, \ldots, c_{2n+1})$; meanwhile, Proposition 5.1 states that $\operatorname{Aut}_{\mathbb{Q}_p}(B)$ is isomorphic to $\operatorname{Jac}(C)[2](\mathbb{Q}_p)$. Corollary 11.3 thus follows from (11.8). □

We may now prove the following proposition which determines the necessary local p-adic masses.

Proposition 11.4 *Let $\mathcal{J}$ be the constant of Proposition* 10.10, *and let F be any large family of hyperelliptic curves. Then*

$$\int\limits_{U_p(F)} \frac{1}{m_p(v)} dv = |2^{4n(2n+3)}\mathcal{J}|_p \cdot \operatorname{Vol}(G(\mathbb{Z}_p)) \cdot M_p(V,F),$$

where

$$M_p(V,F) := \int\limits_{C=C(c_2,\ldots,c_{2n+1})\in F_p} \sum_{\sigma\in \frac{\operatorname{Jac}(C)(\mathbb{Q}_p)}{2\operatorname{Jac}(C)(\mathbb{Q}_p)}} \frac{1}{\#\operatorname{Jac}(C)[2](\mathbb{Q}_p)} dc_2\cdots dc_{2n+1}.$$

Proof The set $U(F)$ consists of the soluble elements in $V(\mathbb{Z})$ having invariants

$$2^8 c_2, 2^{12} c_3, \ldots, 2^{4(2n+1)} c_{2n+1}$$

where $(c_2, \ldots, c_{2n+1}) \in \operatorname{Inv}_p(F)$. Proposition 11.4 thus follows directly from Corollary 11.3 since $\operatorname{Jac}(C)(c_2, \ldots, c_{2n+1})(\mathbb{Q}_p)$ is isomorphic to

$$\operatorname{Jac}(C)(2^8 c_2, 2^{12} c_3, \ldots, 2^{4(2n+1)} c_{2n+1})(\mathbb{Q}_p)$$

and the volume of

$$\{(2^8c_2, 2^{12}c_3, \ldots, 2^{4(2n+1)}c_{2n+1}) | (c_2, \ldots, c_{2n+1}) \in \mathrm{Inv}_p(F)\}$$

is equal to $|2^{4n(2n+3)}|_p \cdot \mathrm{Vol}(\mathrm{Inv}_p(F))$.

□

We may also define a local mass at the infinite place:

$$M_\infty(V, F; X) := \int_{\substack{C=C(c_2,\ldots,c_{2n+1})\in F_\infty \\ H(C)<X}} \sum_{\sigma\in \frac{\mathrm{Jac}(C)(\mathbb{R})}{2\mathrm{Jac}(C)(\mathbb{R})}} \frac{1}{\#\mathrm{Jac}(C)[2](\mathbb{R})} dc_2 \cdots dc_{2n+1}. \quad (11.9)$$

In the analogous manner, if F is a large family of hyperelliptic curves, then we may define $M_p(F)$ to be the measure of $\mathrm{Inv}_p(F)$ with respect to the measure $dc_2 \cdots dc_{2n+1}$ on $\mathbb{Z}_p^{2n}$, where the measure dc_m on $\mathbb{Z}_p$ is normalized so that the total measure is 1. That is, we have

$$M_p(F) := \int_{C=C(c_2,\ldots,c_{2n+1})\in F_p} dc_2 \cdots dc_{2n+1}. \quad (11.10)$$

Similarly, we define

$$M_\infty(F; X) := \int_{\substack{C=C(c_2,\ldots,c_{2n+1})\in F_\infty \\ H(C)<X}} dc_2 \cdots dc_{2n+1}. \quad (11.11)$$

In Section 12, we will be interested in comparing the masses $M_p(V, F)$ and $M_p(F)$, and $M_\infty(V, F; X)$ and $M_\infty(F; X)$.

11.3 A key uniformity estimate, and a squarefree sieve

For each prime p, let W_p denote the set of elements B in $V(\mathbb{Z})$ such that $p^2 \mid \Delta(B)$. Then the following proposition is proven in [3]:

Proposition 11.5 *There exists $\delta > 0$ such that, for any $M > 0$, we have*

$$\sum_{p>M} N(W_p; X) = O(X^{(2n+3)/(4n+2)}/M^\delta),$$

where the implied constant is independent of X and M.

Proposition 11.5 allows us to prove a more general congruence version of Theorem 10.11, namely, one which allows appropriate infinite sets of congruence conditions to be imposed and which also allows weighted counts of lattice points (where weights are also assigned by congruence conditions). Specifically, let us say that a function $\phi : V(\mathbb{Z}) \to [0,1] \subset \mathbb{R}$ is *defined by congruence conditions* if, for all primes p, there exist functions $\phi_p : V(\mathbb{Z}_p) \to [0,1]$ satisfying the following conditions:

(1) For all $B \in V(\mathbb{Z})$, the product $\prod_p \phi_p(B)$ converges to $\phi(B)$;
(2) For each prime p, the function ϕ_p is locally constant outside some closed set $S_p \subset V(\mathbb{Z}_p)$ of measure zero.

Such a function ϕ is called *acceptable* if, for sufficiently large primes p, we have $\phi_p(B) = 1$ whenever $p^2 \nmid \Delta(B)$. For example, the characteristic function of the set of elements $B \in V(\mathbb{Z})$ having squarefree discriminant is an acceptable function.

We then have the following version of Theorem 10.12, in which we allow weights to be defined by certain infinite sets of congruence conditions:

Theorem 11.6 *Let $\phi : V(\mathbb{Z}) \to [0,1]$ be an acceptable function that is defined by congruence conditions via the local functions $\phi_p : V(\mathbb{Z}_p) \to [0,1]$. Then, with notation as in Theorem* 10.12, *we have:*

$$N_\phi(V(\mathbb{Z})^{(m,\tau)}; X) = N(V(\mathbb{Z})^{(m,\tau)}; X) \prod_p \int_{B \in V(\mathbb{Z}_p)} \phi_p(B)\, dB + o(X^{(2n+3)/(4n+2)}). \quad (11.12)$$

The proof of Theorem 11.6 is identical to that of [5, Thm. 2.21]; the idea is to use Theorem 10.12 to impose more and more congruence conditions, while using Proposition 11.5 to uniformly bound the error term.

11.4 Weighted count of elements in $U(F)$ having bounded height

For a large family F, we may now describe the asymptotic number of $G(\mathbb{Z})$-orbits in $U(F)$ having bounded height, where as before each element $B \in U(F)$ is counted with weight $1/m(B)$:

Theorem 11.7 *Let F be any large family of hyperelliptic curves. Then $N_{1/m}(U(F); 2^{8n(2n+1)}X)$, the weighted number of $G(\mathbb{Z})$-orbits in $U(F)$ hav-*

ing height less than $2^{8n(2n+1)}X$, *is given by*

$$N_{1/m}(U(F);2^{8n(2n+1)}X) = \sum_{m=0}^{n}\sum_{\tau=1}^{2^m} N(U_\infty(F)\cap V(\mathbb{Z})^{(m,\tau)};2^{8n(2n+1)}X)\cdot \prod_p \int_{U_p(F)} \frac{1}{m_p(v)}dv + o(X^{(2n+3)/(4n+2)}). \quad (11.13)$$

Theorem 11.7 follows from Corollary 8.4, Equation (11.3), and Theorem 11.6.

It remains to evaluate expression (11.13) in terms of the total number of hyperelliptic curves in F having height less than X.

12 Proof of Theorem 1.1

12.1 The number of hyperelliptic curves of bounded height

Theorem 12.1 *Let* F *be a large family of hyperelliptic curves. Then the number of curves* C *in* F *with* $H(C) < X$ *is given by*

$$M_\infty(F;X)\cdot\prod_p M_p(F) + o(X^{(2n+3)/(4n+2)}).$$

This theorem is proven in much the same manner as Theorem 11.7 (but is easier), again using the uniformity estimate of Proposition 11.5.

12.2 Evaluation of the average size of the 2-Selmer group

We now have the following theorem, from which Theorem 11.1 (and thus Theorem 1.1) will be seen to follow.

Theorem 12.2 *Let* F *be a large family of hyperelliptic curves. Then we have*

$$\lim_{X\to\infty}\frac{\displaystyle\sum_{\substack{C\in F\\ H(C)<X}}(\#S_2(J)-1)}{\displaystyle\sum_{\substack{C\in F\\ H(C)<X}}1} = \frac{|\mathcal{J}|\ \mathrm{Vol}(G(\mathbb{Z})\backslash G(\mathbb{R}))M_\infty(V,F;X)}{M_\infty(F;X)}\cdot\prod_p\frac{|\mathcal{J}|_p\,\mathrm{Vol}(G(\mathbb{Z}_p))M_p(V,F)}{M_p(F)}. \quad (12.1)$$

Proof This follows by combining Proposition 11.4, Theorem 11.7, Theorem 10.1, expression (10.27) for the volume

$$\begin{aligned}\mathrm{Vol}(\mathcal{R}_X(L^{(m,\tau)})) &= \mathrm{Vol}(\mathcal{F}L^{(m,\tau)} \cap \{v \in V(\mathbb{R}) : H(v) < X\}) \\ &= 2^{m+n} c_{m,\tau} X^{(2n+3)/(4n+2)}\end{aligned}$$

(noting that $2^{m+n} = \#J[2](\mathbb{R})$, where $J = \mathrm{Jac}(C)(c_2, \ldots, c_{2n+1})$ for any $(c_2, \ldots, c_{2n+1}) \in I(m)$), and Theorem 12.1.

□

In order to evaluate the right hand side of the expression in Theorem 12.2, we use the following fact (see [14, Lemma 3.1]):

Lemma 12.3 *Let J be an abelian variety over $\mathbb{Q}_p$ of dimension n. Then*

$$\#(J(\mathbb{Q}_p)/2J(\mathbb{Q}_p)) = \begin{cases} \#J[2](\mathbb{Q}_p) & \text{if } p \neq 2; \\ 2^n \cdot \#J[2](\mathbb{Q}_p) & \text{if } p = 2. \end{cases}$$

Proof It follows from the theory of formal groups that there exists a subgroup $M \subset J(\mathbb{Q}_p)$ of finite index that is isomorphic to $\mathbb{Z}_p^n$ [33]. Let H denote the finite group $J(\mathbb{Q}_p)/M$. Then by applying the snake lemma to the following diagram

$$\begin{array}{ccccccccc} 0 & \longrightarrow & M & \longrightarrow & J(\mathbb{Q}_p) & \longrightarrow & H & \longrightarrow & 0 \\ \downarrow & & \downarrow{\scriptstyle [2]} & & \downarrow{\scriptstyle [2]} & & \downarrow{\scriptstyle [2]} & & \downarrow \\ 0 & \longrightarrow & M & \longrightarrow & J(\mathbb{Q}_p) & \longrightarrow & H & \longrightarrow & 0 \end{array}$$

we obtain the exact sequence

$$0 \to M[2] \to J(\mathbb{Q}_p)[2] \to H[2] \to M/2M \to J(\mathbb{Q}_p)/2J(\mathbb{Q}_p) \to H/2H \to 0.$$

Since H is a finite group and M is isomorphic to $\mathbb{Z}_p^n$, Lemma 12.3 follows.

□

The expression on the right hand side in Theorem 12.2 thus reduces simply to the Tamagawa number $\tau(G) = \mathrm{Vol}(G(\mathbb{Z})\backslash G(\mathbb{R})) \cdot \prod_p \mathrm{Vol}(G(\mathbb{Z}_p))$ of G, which is 2. This completes the proof of Theorem 11.1, and thus of Theorem 1.1.

In fact, the proof of Theorem 1.1 shows more. For example, we may study the distribution of the nonidentity Selmer elements inside $J(\mathbb{Q}_\nu)/2J(\mathbb{Q}_\nu)$ for any (finite or infinite) place ν of $\mathbb{Q}$. In this regard, we have:

Theorem 12.4 *Fix ν. Let F be a large family of hyperelliptic curves C of genus n such that*

(a) *the cardinality of* $\mathrm{Jac}(C)(\mathbb{Q}_\nu)/2\mathrm{Jac}(C)(\mathbb{Q}_\nu)$ *is a constant k for all C in F; and*

(b) *the set $U_\nu(F) \subset V(\mathbb{Z}_\nu)$, defined by (11.4), can be partitioned into k open sets Ω_i such that:*

  (i) *for all i, if two elements in Ω_i have the same invariants $c_2, \ldots, c_{2n+1}$, then they are $G(\mathbb{Q}_\nu)$-equivalent; and*

  (ii) *for all $i \neq j$, we have $(G(\mathbb{Q}_\nu) \cdot \Omega_i) \cap (G(\mathbb{Q}_\nu) \cdot \Omega_j) = \varnothing$.*

(In particular, the groups $\mathrm{Jac}(C)(\mathbb{Q}_\nu)/2\mathrm{Jac}(C)(\mathbb{Q}_\nu)$ *are naturally identified for all C in F.) Then when the hyperelliptic curves C in F are ordered by height, the images of the non-identity 2-Selmer elements under the canonical map*

$$S_2(\mathrm{Jac}(C)) \to \mathrm{Jac}(C)(\mathbb{Q}_\nu)/2\mathrm{Jac}(C)(\mathbb{Q}_\nu) \tag{12.2}$$

are equidistributed.

We first prove that for any given value $c \in \mathbb{Z}_\nu^{2n} \setminus \{\Delta = 0\}$ for the invariants $(c_2, \ldots, c_{2n+1})$, there always exists a sufficiently small ν-adic neighborhood W of c in $\mathbb{Z}_\nu^{2n}$ such that the corresponding (large) family $F = F(W)$, consisting of all hyperelliptic curves having invariants in W, satisfies both (a) and (b). Indeed, when $\nu = \infty$, if $c \in I(m)$ then we simply let $W = I(m)$, let $k = 2^m = \mathrm{Jac}(C(c))(\mathbb{R})/2\mathrm{Jac}(C(c))(\mathbb{R})$, and let $\Omega_1, \ldots, \Omega_k$ be equal to $V^{(m,1)}, \ldots, V^{(m,k)}$, respectively.

If ν is a finite prime p, let k be the cardinality of

$$\mathrm{Jac}(C(c))(\mathbb{Q}_p)/2\mathrm{Jac}(C(c))(\mathbb{Q}_p).$$

Then the set Y of soluble elements in the inverse image of

$$(2^8 c_2, \ldots, 2^{4(2n+1)} c_{2n+1})$$

under the map $\pi : V(\mathbb{Z}_p) \to \mathbb{Z}_p^{2n}$, given by taking invariants, is the disjoint union of k nonempty compact sets $Y_1, \ldots, Y_k$, namely, the k $G(\mathbb{Q}_p)$-equivalence classes in $V(\mathbb{Z}_p)$ comprising Y.

Let $Z_1, \ldots, Z_k \subset V(\mathbb{Z}_p) \setminus \{\Delta = 0\}$ be disjoint neighborhoods of $Y_1, \ldots, Y_k$, respectively, in $V(\mathbb{Z}_p)$ such that each Z_i consists of soluble elements and is the union of $G(\mathbb{Q}_p)$-equivalence classes in $V(\mathbb{Z}_p)$. Such Z_i can be constructed by noting that if ε is sufficiently small, then the ε-neighborhoods $B_\varepsilon(Y_i)$ of the Y_i's are disjoint and consist only of elements that have nonzero discriminant and are soluble. The set $\{g \in G(\mathbb{Q}_p) \,|\, gB_\varepsilon(Y_i) \cap V(\mathbb{Z}_p) \neq \varnothing\}$ is then compact. Indeed, for a single stable element $v \in V(\mathbb{Z}_p)$, the set

$$\{g \in G(\mathbb{Q}_p) \,|\, gv \in V(\mathbb{Z}_p)\} \tag{12.3}$$

is compact, as it is the union of finitely many right cosets

$$G(\mathbb{Z}_p) \cdot g_1, \ldots, G(\mathbb{Z}_p) \cdot g_m$$

of $G(\mathbb{Z}_p)$; this is because there are only finitely many $G(\mathbb{Z}_p)$-orbits in the $G(\mathbb{Q}_p)$-orbit of v, by Propositions 8.2 and 8.7. Next, note that the set given by (12.3) is constant in a neighborhood of v. This is because the number of $G(\mathbb{Z}_p)$-orbits in $G(\mathbb{Q}_p) \cdot v'$ is a constant for elements v' in a sufficiently small neighborhood of v, e.g., we may take a neighborhood of v small enough so that the associated ring in Proposition 8.2 or 8.7 is a constant. The compactness of

$$\{g \in G(\mathbb{Q}_p) \,|\, gS \cap V(\mathbb{Z}_p) = \varnothing\} \tag{12.4}$$

now follows for any compact set S of stable elements (by covering S with neighborhoods where (12.3) is constant, and then taking a finite subcover). Hence $(G(\mathbb{Q}_p) \cdot B_\varepsilon(Y_i)) \cap V(\mathbb{Z}_p)$ is both open and compact, and so is a bounded distance away from Y_j for all $j \neq i$. By shrinking ε if necessary, we can then ensure that $(G(\mathbb{Q}_p) \cdot B_\varepsilon(Y_i)) \cap B_\varepsilon(Y_j) = \varnothing$ for all $i \neq j$, and therefore $(G(\mathbb{Q}_p) \cdot B_\varepsilon(Y_i)) \cap (G(\mathbb{Q}_p) \cdot B_\varepsilon(Y_j)) = \varnothing$ for all $i \neq j$. We then set $Z_i = (G(\mathbb{Q}_p) \cdot B_\varepsilon(Y_i)) \cap V(\mathbb{Z}_p)$.

Let $W' = \{(b_2, \ldots, b_{2n+1}) \in \mathbb{Z}_p^{2n} \,|\, (2^8 b_2, \ldots, 2^{4(2n+1)} b_{2n+1}) \in \cap_i \pi(Z_i)\}$. Since π is an open mapping on $V(\mathbb{Z}_p) \setminus \{\Delta = 0\}$, we see that W' is an open set in $\mathbb{Z}_p^{2n}$ containing c. Let $W \subset W'$ be an open neighborhood of c small enough so that for all hyperelliptic curves C having invariants in W, we have $\#(\mathrm{Jac}(C)(\mathbb{Q}_p)/2\mathrm{Jac}(C)(\mathbb{Q}_p)) = k$. Such a neighborhood W exists because the size of $\mathrm{Jac}(C)(\mathbb{Q}_p)/2\mathrm{Jac}(C)(\mathbb{Q}_p)$, where C is given by $y^2 = f(x)$, depends only on the shape of the factorization of $f(x)$ over $\mathbb{Q}_p$, i.e., the number of irreducible factors and their degrees, so this size is locally constant as a function of the coefficients of f. Then $F(W)$ satisfies both (a) and (b), with

$$\Omega_i = Z_i \cap \pi^{-1}\{(2^8 b_2, \ldots, 2^{4(2n+1)} b_{2n+1}) \,|\, (b_2, \ldots, b_{2n+1}) \in W\}.$$

To prove Theorem 12.4, on the right side of the equation in Theorem 12.2, we replace the sum over all σ in $\mathrm{Jac}(C)(\mathbb{Q}_\nu)/2\mathrm{Jac}(C)(\mathbb{Q}_\nu)$ in the definition of $M_\nu(V, F)$ by the corresponding sum over σ lying in any subset $\Sigma \subset \mathrm{Jac}(C)(\mathbb{Q}_\nu)/2\mathrm{Jac}(C)(\mathbb{Q}_\nu)$. (By property (b), we are still counting elements in a weighted subset in $V(\mathbb{Z})$ defined by congruence conditions, so Theorem 11.6 again applies.) This gives us the average number of non-identity Selmer elements that map to Σ, and we see that the result is proportional to the size of Σ, proving Theorem 12.4. The same argument also allows one to show equidistribution of non-identity 2-Selmer elements in $\prod_{\nu \in S} \mathrm{Jac}(C)(\mathbb{Q}_\nu)/2\mathrm{Jac}(C)(\mathbb{Q}_\nu)$ for any finite set S of places of $\mathbb{Q}$, provided that our large family F of hyperelliptic curves lies in the intersection of sufficiently small ν-adic discs ($\nu \in S$) so that both (a) and (b) are satisfied for all $\nu \in S$.

There is an alternative interpretation of many of these calculations (and cancellations!) in terms of an adelic geometry-of-numbers argument. This interpretation was introduced by Poonen in [37] and we briefly describe it here. Let $\mathbb{A}$ be the ring of adeles of $\mathbb{Q}$ and let dv be the unique Haar measure on $V(\mathbb{A})$ which is the counting measure on the discrete subgroup $V(\mathbb{Q})$ and gives the compact quotient $V(\mathbb{A})/V(\mathbb{Q})$ volume one. Let S be the vector group over $\mathbb{Q}$ of dimension $2n$ where the invariant polynomials take values, and let ds be the unique Haar measure on $S(\mathbb{A})$ which is counting measure on the discrete subgroup $S(\mathbb{Q})$ and gives the compact quotient $S(\mathbb{A})/S(\mathbb{Q})$ volume one.

Define the compact subset $S(\mathbb{A})_{<X}$ as the product of the elements in $S(\mathbb{Z}_p)$ that define a p-indivisible polynomial having nonzero discriminant with the elements in $S(\mathbb{R})$ that define a polynomial having nonzero discriminant and bounded height $H < X$. Define the subset $V(\mathbb{A})_{<X}$ as the subset of elements in $V(\mathbb{A})$ locally soluble everywhere and having characteristic polynomial in $S(\mathbb{A})_{<X}$. The group $G(\mathbb{Q})$ acts on $V(\mathbb{A})_{<X}$ and the orbit space has finite volume. The adelic volume formula is then

$$\mathrm{vol}(V(\mathbb{A})_{<X}/G(\mathbb{Q}), dv) = 2 \cdot \mathrm{vol}(S(\mathbb{A})_{<X}, ds).$$

The factor 2 in this formula is the Tamagawa number of G: the volume of $G(\mathbb{A})/G(\mathbb{Q})$ with respect to Tamagawa measure $\omega(g)$. The proof of this volume formula involves the compatibility of the three measures ds, dv, and $\omega(g)$ (as in Proposition 10.10) as well as a product formula for the Herbrand quotients $\#(J(\mathbb{Q}_v)/2J(\mathbb{Q}_v))/\#J[2](\mathbb{Q}_v)$ (as in Proposition 12.3). Geometry-of-numbers arguments (as in Sections 9 and 10) and a sieve (as in Section 11) can then be used to estimate (as in Section 12) the average number (= 2) of irreducible locally soluble rational orbits, and so the average number of non-trivial classes in the 2-Selmer group.

Acknowledgments

We are very grateful to John Cremona, Joe Harris, Wei Ho, Bjorn Poonen, Arul Shankar, Michael Stoll, Jack Thorne, and Jerry Wang for their help.

References

[1] H. Bass, *On the ubiquity of Gorenstein rings*, Math Zeit. **82** (1963), 8–28.

[2] M. Bhargava, *The density of discriminants of quintic rings and fields*, Ann. of Math. **172** (2010), 1559–1591.

[3] M. Bhargava, *The geometric sieve, and the density of squarefree values of polynomial discriminants and other invariant polynomials*, preprint.

[4] M. Bhargava and B. Gross, *Arithmetic invariant theory*, `http://arxiv.org/abs/1206.4774`.

[5] M. Bhargava and A. Shankar, *Binary quartic forms having bounded invariants, and the boundedness of the average rank of elliptic curves*, `http://arxiv.org/abs/1006.1002`, Ann. of Math., to appear.

[6] M. Bhargava and A. Shankar, *Ternary cubic forms having bounded invariants, and the existence of a positive proportion of elliptic curves having rank 0*, `http://arxiv.org/abs/1007.0052`.

[7] B. J. Birch and H. P. F. Swinnerton-Dyer, *Notes on elliptic curves I*, J. Reine Angew. Math. **212** (1963), 7–25.

[8] R. Bölling, *Die Ordnung der Schafarewitsch-Tate-Gruppe kann beliebig gross werden*, Math. Nachr. **67** (1975), 157–179.

[9] A. Borel, *Ensembles fondamentaux pour les groupes arithmétiques*, Colloque sur la Théorie des Groupes Algébriques, Bruselles **1962**, 23–40.

[10] A. Borel and Harish-Chandra, *Arithmetic subgroups of algebraic groups*, Ann. of Math. **75** (1962), 485–535.

[11] N. Bourbaki, *Groupes et algèbres de Lie*, Hermann (1982).

[12] F. Bruhat and J. Tits, *Schémas en groupes et immeubles des groupes classiques sur un corps locale*, Bull. Soc. Math. France **112** (1987), 259–301.

[13] A. Brumer, *The average rank of elliptic curves I*, Invent. Math. **109** (1992), no. 3, 445–472.

[14] A. Brumer and K. Kramer, *The rank of elliptic curves*, Duke Math J. **44** (1977), no. 4, 715–743.

[15] J. W. S. Cassels, *Arithmetic on curves of genus 1, IV. Proof of the Hauptvermutung*, J. Reine Angew. Math. **211** (1962), 95–112.

[16] C. Chabauty, *Sur les points rationnels des courbes algébriques de genre supérieur à l'unité*, C. R. Acad. Sci. Paris **212** (1941), 882–885.

[17] R.F. Coleman, *Effective Chabauty*, Duke Math. J. **52** (1985), no. 3, 765–770.

[18] J. Cremona, T. Fisher, and M. Stoll, *Minimisation and reduction of 2-, 3- and 4-coverings of elliptic curves*, Algebra & Number Theory **4** (2010), no. 6, 763–820.

[19] H. Davenport, *On a principle of Lipshitz*, J. London Math. Soc. **26** (1951), 179–183. Corrigendum: "On a principle of Lipschitz", J. London Math. Soc. **39** (1964), 580.

[20] M. Demazure, *Groupes réductifs*, SGA (1963/64).

[21] R. Donagi, *Group law on the intersection of two quadrics*, Annali della Scuola Normale Superiore di Pisa **7** (1980), 217–239.

[22] S. Garibaldi, A. Merkurjev, and J-P. Serre, *Cohomological invariants in Galois cohomology*, AMS University Lecture Series **28**, 2003.

[23] C. G. Gibson, *Elementary geometry of algebraic curves*, Cambridge University Press, Cambridge, 1998.

[24] D. Goldfeld, *Conjectures on elliptic curves over quadratic fields*, Number theory, Carbondale 1979 (Proc. Southern Illinois Conf., Southern Illinois Univ., Carbondale, Ill., 1979), pp. 108–118, Lecture Notes in Math. **751**, Springer, Berlin, 1979.

[25] B. Gross, *On Bhargava's representations and Vinberg's invariant theory*, In: *Frontiers of Mathematical Sciences*, International Press (2011), 317–321.

[26] D. R. Heath-Brown, *The average analytic rank of elliptic curves*, Duke Math. J. **122** (2004), no. 3, 591–623.

[27] N. M. Katz and P. Sarnak, *Random matrices, Frobenius eigenvalues, and monodromy*, American Mathematical Society Colloquium Publications **45**, American Mathematical Society, Providence, RI, 1999.

[28] A. W. Knapp, *Lie groups beyond an introduction*, Second edition, Progress in Mathematics **140**, *Birkhäuser Boston, Inc., Boston, MA*, 2002.

[29] M. Kneser, *Galois-Kohomologie halbeinfacher algebraischer Gruppen über p-adischen Körpern I, II*, Math Z. **88, 89** (1965), 40–47, 250–272.

[30] A. Knus, A. Merkurjev, M. Rost, and J.-P. Tignol, *The book of involutions*, American Mathematical Society Colloquium Publications **44**, American Mathematical Society, Providence, RI, 1998.

[31] A. Kostrikin and P. H. Tiep, *Orthogonal decompositions and integral lattices*, deGruyter Expositions in Mathematics **15**, Berlin, 1994.

[32] P. Lockhart, *On the discriminant of a hyperelliptic curve*, Trans. Amer. Math. Soc. **342** (1994), 729–752.

[33] A. Mattuck, *Abelian varieties over p-adic ground fields*, Ann. of Math. **62** (1955), 92–119.

[34] J. Milnor and D. Husemoller, *Symmetric bilinear forms*, Springer Ergebnisse **73**,1970.

[35] D. Panyushev, *On invariant theory of θ-groups*, J. Algebra **283** (2005), 655–670.

[36] V. Platonov and A. Rapinchuk, *Algebraic groups and number theory*, Translated from the 1991 Russian original by Rachel Rowen, Pure and Applied Mathematics **139**, Academic Press, Inc., Boston, MA, 1994.

[37] B. Poonen, *Average rank of elliptic curves (after Manjul Bhargava and Arul Shankar)*, Séminaire Bourbaki **1049** (2011–2012).

[38] B. Poonen and E. Rains, *Random maximal isotropic subspaces and Selmer groups*, J. Amer. Math. Soc. **25** (2012), 245–269.

[39] B. Poonen and M. Stoll, *Chabauty's method proves that most odd degree hyperelliptic curves have only one rational point*, `http://arxiv.org/abs/1302.0061v1`.

[40] M. Reid, *The complete intersection of two or more quadrics*, Ph.D. Thesis, Trinity College, Cambridge (1972).

[41] E. F. Schaefer, 2-*descent on the Jacobians of hyperelliptic curves*, J. Number Theory **51** (1995) , 219–232.

[42] W. Scharlau, *Zur Pfisterschen Theorie der quadratischen Formen*, Invent. Math. **6** (1969), 327–328.

[43] J-P. Serre, *A course in arithmetic*, Springer Graduate Texts in Math. **7**, 1970.

[44] J-P. Serre, *Galois cohomology*, Springer Monographs in Mathematics, 2002.

[45] J-P. Serre, *Local fields*, Springer Graduate Texts in Math. **67**, 1995.

[46] J. H. Silverman, *The arithmetic of elliptic curves*, Springer Graduate Texts in Math. **106**, 1986.

[47] G. Solomon, *Higher plane curves*, Third edition, 1879, reprinted by Chelsea, NY.

[48] M. Stoll, *Implementing 2-descent for Jacobians of hyperelliptic curves*, Acta Arith. **98** (2001), 245–277.

[49] M. Stoll, *Independence of rational points on twists of a given curve*, Compositio Math. **142** (2006), 1201–1214.

[50] J. Thorne, *The arithmetic of simple singularities*, Ph.D. Thesis, Harvard University, 2012.

[51] X. Wang, *Maximal linear spaces contained in the base loci of pencils of quadrics*, Ph.D. Thesis, Harvard University, 2013, `http://arxiv.org/pdf/1302.2385.pdf`.

M. Bhargava, Department of Mathematics, Fine Hall, Washington Road Princeton NJ 08544-1000 USA.
E-mail: `bhargava@math.princeton.edu`

B.H. Gross, Department of Mathematics, Harvard University, One Oxford Street, Cambridge, MA 02138-2901, USA.
E-mail: `gross@math.harvard.edu`

Automorphic Representations and L-Functions
Editors: D. Prasad, C.S. Rajan, A. Sankaranarayanan, J. Sengupta

Publisher: Hindustan Book Agency, New Delhi, India

Weil's Theorem on Rational Points Over Finite Fields and Artin L-functions

Peter J. Cho and Henry H. Kim[1]

1 Introduction

This paper is a continuation of [7], where we studied extreme values of logarithmic derivatives of $L(s,\rho)$ at $s=1$: Let K be a number field of degree n and $\widehat{K}$, its Galois closure over $\mathbb{Q}$ with $\mathrm{Gal}(\widehat{K}/\mathbb{Q})\simeq G$ for some finite group G. Let $\mathrm{Gal}(\widehat{K}/K)\simeq H$ for a subgroup H of G. Then $\mathrm{Ind}_H^G 1_H = 1_G+\rho$ and ρ is an $(n-1)$-dimensional representation of G. Hence we have

$$\zeta_K(s)=\zeta(s)L(s,\rho).$$

Throughout this article, ρ refers to the above $(n-1)$-dimensional representation of G.

For a finite group G, let $\mathfrak{K}(n,G,r_1,r_2)$ be the set of number fields K of degree n with signature (r_1,r_2), whose Galois closures $\widehat{K}$ are G-Galois extensions over $\mathbb{Q}$. In [7], for several families $\mathfrak{K}(n,G,r_1,r_2)$, we were able to show that there are infinitely many number fields $K\in\mathfrak{K}(n,G,r_1,r_2)$ with

$$\log\log|d_K|+O(\log\log\log|d_K|),\quad -(n-1)\log\log|d_K|+O(\log\log\log|d_K|),$$

as the upper and lower bound of $\frac{L'}{L}(1,\rho)$, respectively. In order to obtain the average result, a refinement of Weil's theorem on rational points of algebraic curves over finite fields was crucial: Namely, let $f_t(x)=x^n+a_1(t)x^{n-1}+\cdots+a_n(t)$ be irreducible, where $a_i(t)\in\mathbb{Z}[t]$ such that the splitting field E of $f_t(x)$ over $\mathbb{Q}(t)$ is regular and the Galois group is G. Let p be a prime and for $t\in\mathbb{Z}$, let $N_t(p)$ be the number of solutions $f_t(x)\equiv 0 \pmod p$, and let $\lambda_t(p)=N_t(p)-1$.

For a specialization t in $\mathbb{Z}$, let $K_t=\mathbb{Q}(\theta_t)$ be a number field of degree n by adjoining a root θ_t of $f_t(x)$ to $\mathbb{Q}$. Then

$$L(s,\rho,t)=\frac{\zeta_{K_t}(s)}{\zeta(s)}=\sum_n \lambda(n,t)n^{-s}$$

[1] Partially supported by an NSERC grant.

with $\lambda(p,t) = \lambda_t(p)$ for an unramified prime p. Note that $\lambda(p,t)$ is the trace of ρ at the conjugacy class of Frob_p.

Let A_i be the number of t (mod p) such that $\lambda_t(p) = i$, i.e., $f_t(x) \equiv 0$ (mod p) has $i+1$ roots. Then we used the fact that $\sum_{i=-1}^{n-1} iA_i = O(\sqrt{p})$. We did not need the precise value of A_i.

In this paper we prove that $A_i = \frac{|C_i|}{|G|}p + O(\sqrt{p})$, where C_i is the union of conjugacy classes C in G such that the trace of ρ at C is equal to i. It is instructive to note that if we fix t, the Chebotarev density theorem says that $\#\{p \le x, \mathrm{tr}(\rho(\mathrm{Frob}_p)) = i\} \sim \frac{|C_i|}{|G|}\frac{x}{\log x}$ as $x \to \infty$.

In the special case $f_t(x) = x^2 - g(t)$, where $g(t) \in \mathbb{Z}[t]$ is square-free (mod p), it is the result of Davenport and others on the distribution of quadratic residues (mod p), namely, $\sum_{t \pmod p} (\frac{g(t)}{p}) = O(\sqrt{p})$.

In the case of simplest cubic fields, Duke [12] computed A_i precisely. We follow his idea to compute A_i in the cases of simplest quartic and sextic fields in section 3.

In section 5, we supplement our earlier result to construct infinite families of number fields with extreme logarithmic derivatives: In section 5.1, we construct a family of number fields in $\mathfrak{K}(5, S_5, 1, 2)$ with extreme logarithmic derivative of Artin L-functions. Here the result of [3] on the modularity of 4-dimensional representation of S_5 is crucial. In section 5.2, we use the family of polynomials in [34] for $\mathfrak{K}(6, A_5, 2, 2)$. Here for the modularity of 5-dimensional representation of A_5, we use the recent result of Khare-Wintenberger [16] and the functoriality of symmetric fourth of cuspidal representations of GL_2 due to the second author [18, 19]. In section 5.3, we use the family of polynomials in [32] for $\mathfrak{K}(4, A_4, 0, 2)$. Spearman showed that the polynomials give rise to monogenic quartic extensions, and hence we can distinguish L-functions. In [7], we used the family of polynomials $x^4 - 8tx^3 + 18t^2x^2 + 1$. In order to distinguish L-functions, we need a difficult folklore conjecture on asymptotics of square free values of $1 + 27t^4$ on arithmetic progressions. Finally in section 5.4, we use the family of polynomials in [35] for $\mathfrak{K}(7, C_7, 7, 0)$. Here we need to assume a difficult folklore conjecture on asymptotics of square free values of certain degree 6 polynomials.

In section 6, we study CM fields with large class numbers, motivated by Kumar Murty's question on whether there are infinitely many CM fields K with class numbers larger than $|d_K|^{\frac{1}{2}}$. We construct unconditionally infinitely many CM fields K with the class number larger than $|d_K|^{\frac{1}{2}} \log\log |d_K|$.

Acknowledgements We thank Wenzhi Luo for his help with zero density result on Theorem 4.4 in the case of isobaric automorphic representations, and the referee for several helpful remarks.

2 A Refinement of Weil's Theorem

First, let us recall Weil's celebrated theorem on rational points of algebraic curves over finite fields. (cf. [27], page 75):

Theorem 2.1 *Let $f(x,y) \in \mathbb{F}_p[x,y]$ be absolutely irreducible and of total degree $d > 0$. Let N be the number of zeros of f in $\mathbb{F}_p \times \mathbb{F}_p$. Then*

$$|N - p| \leq (d-1)(d-2)\sqrt{p} + c(d),$$

for a constant $c(d)$.

A finite extension E of the rational function field $\mathbb{Q}(t)$ is called regular if $\overline{\mathbb{Q}} \cap E = \mathbb{Q}$.

Let $f_t(x)$, $N_t(p)$, $L(s,\rho,t)$, and $\lambda_t(p)$ be as in the introduction. We assume that the splitting field E of $f_t(x)$ over $\mathbb{Q}(t)$ is regular.

Note that since $\lambda(p,t) = \lambda_t(p)$, $-1 \leq \lambda(p,t) \leq n-1$ for an unramified prime p.

Let A_i be the number of t (mod p) such that $\lambda_t(p) = i$, i.e., $f_t(x) \equiv 0$ (mod p) has $i+1$ roots. Then clearly,

$$N = \sum_{i=-1}^{n-1} (i+1)A_i + O_f(1).$$

Here if $K_t/\mathbb{Q}$ is Galois, $A_i = 0$ for $i \neq -1, n-1$. So $N = nA_{n-1} + O_f(1)$.

Theorem 2.2 *Let $f_t(x)$ be as above. Fix a prime p. Let C_i be the union of conjugacy classes C in G such that the trace of ρ at C is equal to i. Then*

$$A_i = \frac{|C_i|}{|G|}p + O(\sqrt{p}).$$

Proof This is essentially Chebotarev density theorem for function field, and is proved by Ree [26]. In [26], Theorem 2, it is stated only for when $f_t(x)$ (mod p) is irreducible. (He needs to assume that $f_t(x)$ gives rise to a regular Galois extension E over $\mathbb{Q}(t)$.) But it is straightforward to generalize it. We follow his exposition in [26]. Let $k = \mathbb{F}_p$, and K be the splitting field of $f_t(x)$ over $k(t)$. Since $f_t(x)$ gives rise to a regular Galois extension over $\mathbb{Q}(t)$, $\mathrm{Gal}(K/k(t)) \simeq \mathrm{Gal}(E/\mathbb{Q}(t))$ for sufficiently large p. Let D be the set of elements $a \in k$ such that the place $\mathfrak{p}_a$ of $k(t)$ corresponding to $t-a$ does not ramify in K. Then the conjugacy class C_a of the Frobenius at $\mathfrak{p}_a$ in K is the same as factorization of $f_a(x)$ (mod p). For any conjugacy class C in G, let $N_n(C)$ be the number of elements $a \in D$ such that $C_a = C$. Then the density theorem of Weil says that there exists a constant α_n, depending only on n, such that

$$\left|N_n(C) - \frac{|C|}{|G|}p\right| < \alpha_n\sqrt{p}.$$

Hence our result follows.

□

The above Weil's theorem implies the following theorem due to Serre, regarding the distribution of Frobenius elements in a regular Galois extension ([30], page 45). Let C be any conjugacy class of G.

Theorem 2.3 *There is a constant $c_f > 0$ depending on $f_t(x)$ such that for any prime $p \geq c_f$, there is a $t_{C,p} \in \mathbb{Z}$ so that for any $t \equiv t_{C,p} \pmod{p}$ with* $\mathrm{Gal}(\widehat{K_t}/\mathbb{Q}) = G$*, p is unramified in $\widehat{K_t}/\mathbb{Q}$, and* $\mathrm{Frob}_p \in C$.

Remark 2.4 Theorem 2.2 can be thought of a refinement of Theorem 2.1. Indeed, Theorem 2.2 implies Theorem 2.1: By class equation, $\sum_{i=-1}^{n-1} |C_i| = |G|$, and so Theorem 2.2 implies

$$\begin{aligned}\sum_{i=-1}^{n-1}(i+1)A_i &= \sum_{i=-1}^{n-1}(i+1)\left(\frac{|C_i|}{|G|}p + O(\sqrt{p})\right)\\ &= \frac{p}{|G|}\sum_{i=-1}^{n-1} i|C_i| + \frac{p}{|G|}\sum_{i=-1}^{n-1}|C_i| + O(\sqrt{p}).\end{aligned}$$

Here $\sum_{i=-1}^{n-1} i|C_i| = 0$. We can prove this as follows: Note that χ_ρ is the sum of irreducible characters $\chi_1, \ldots, \chi_k$, and

$$\sum_{i=-1}^{n-1} i|C_i| = \sum_{g\in G}\chi_{\rho_t}(g) = \sum_{j=1}^{k}\sum_{g\in G}\chi_j(g).$$

By orthogonality of characters, $\sum_{g\in G}\chi_j(g) = 0$ for each $j = 1, \ldots, k$. Therefore, $\sum_{i=-1}^{n-1}(i+1)A_i = p + O(\sqrt{p})$. This implies that $\sum_{i=-1}^{n-1} iA_i = O(\sqrt{p})$. This played a crucial role in our paper [7].

In the special case $f_t(x) = x^2 - g(t)$, where $g(t) \in \mathbb{Z}[t]$ is square free $\pmod{p}$, it is the result of Davenport and others (cf. [2], page 107), namely, $\sum_{t \pmod{p}}\left(\frac{g(t)}{p}\right) = O(\sqrt{p})$.

For several cyclic extensions, we determine the exact value of A_i in section 3.

3 Cyclic Extensions

In the case of cyclic extensions, we need to determine only A_{-1} and A_{n-1}. In the case of simplest cubic fields, Duke [12] already obtained the result. The simplest cubic fields are the cubic fields parameterized by the polynomial

$$f_t(x) = x^3 - tx^2 - (t+3)x - 1, \text{ for } t \in \mathbb{Z}^+.$$

For a prime $p \geq 5$, Duke computed the number of residue classes t modulo p for which $f_t(x)$ splits completely, remains inert or ramifies respectively.

	split	inert	ramified
p ≡ 1 mod 3	$\frac{p-4}{3}$	$\frac{2p-2}{3}$	2
p ≡ 2 mod 3	$\frac{p-2}{3}$	$\frac{2p+2}{3}$	0

This can be paraphrased as

$$A_{-1} = \frac{2p-2}{3}, \text{ and } A_2 = \frac{p-4}{3} \text{ if } p \equiv 1 \text{ mod } 3$$

and

$$A_{-1} = \frac{2p+2}{3}, \text{ and } A_2 = \frac{p-2}{3} \text{ if } p \equiv 2 \text{ mod } 3.$$

With Duke's idea, we obtain the analogous results for simplest quartic and sextic fields. The simplest quartic fields are the fields parameterized by

$$f_t(x) = x^4 - tx^3 - 6x^2 + tx + 1$$

with the discriminant $\text{disc}(f_t(x)) = 4(t^2+16)^3$. In [6], we showed that the splitting field E of $f_t(x)$ over $\mathbb{Q}(t)$ is a C_4 regular Galois extension. Note that for odd primes p, $\text{disc}(f_t(x)) \equiv 0 \text{ mod } p$ has a solution if and only if $p \equiv 1 \text{ mod } 4$.

We consider the polynomial $f_t(x)$ over a finite field $\mathbb{F}_p$. Since

$$t = \frac{x^4 - 6x^2 + 1}{x(x^2-1)} \text{ for a zero } x \text{ of } f_t(x),$$

x should belong to $\mathbb{F}_p \backslash \{0, \pm 1\}$.

If $p \equiv 3 \text{ mod } 4$, $\text{disc}(f_t(x))$ has no root mod p, Hence $4 \times A_3 = p - 3$. Hence we obtain that

$$A_{-1} = \frac{3p+3}{4}, \text{ and } A_3 = \frac{p-3}{4}$$

for primes p with $p \equiv 3 \text{ mod } 4$ and $p \geq 3$.

If $p \equiv 1 \bmod 4$, $\mathrm{disc}(f_t(x))$ has two roots mod p. For each t corresponding to a root of $\mathrm{disc}(f_t(x)) \equiv 0 \pmod p$, namely, $t^2+16 \equiv 0 \pmod p$, $f_t(x)$ has exactly one root mod p: We can see easily that $f_t(x) \equiv (x-a)^4 \pmod p$, where $t \equiv 4a \pmod p$ and $a^2+1 \equiv 0 \pmod p$. Hence $4 \times A_3 = p-5$, and

$$A_{-1} = \frac{3p-3}{4}, \text{ and } A_3 = \frac{p-5}{4}$$

for primes p with $p \equiv 1 \bmod 4$ and $p \geq 5$.

Next, we consider the simplest sextic fields. They are parametrized by

$$f_t(x) = x^6 - 2tx^5 - 5(t+3)x^4 - 20x^3 + 5tx^2 + 2(t+3)x + 1$$

with the discriminant $\mathrm{disc}(f_t(x)) = 2^6 3^6 (t^2+3t+9)^5$. In [6], we showed that the splitting field E of $f_t(x)$ over $\mathbb{Q}(t)$ is a C_6 regular Galois extension. Note that for odd prime $p \geq 5$, $\mathrm{disc}(f_t(x)) \equiv 0 \bmod p$ has a solution if and only if $p \equiv 1 \bmod 3$.

We consider the polynomial $f_t(x)$ over a finite field $\mathbb{F}_p$. Since

$$t = \frac{x^6 - 15x^4 - 20x^3 + 6x + 1}{x(x^2-1)(2x+1)(x+2)} \quad \text{for a zero } x \text{ of } f_t(x),$$

x should belong to $\mathbb{F}_p \backslash \{0, \pm 1, -2, -\frac{1}{2}\}$.

If $p \equiv 2 \bmod 3$, $\mathrm{disc}(f_t(x))$ has no root mod p, Hence $6 \times A_5 = p-5$. Hence we obtain that

$$A_{-1} = \frac{5p+5}{6}. \text{ and } A_5 = \frac{p-5}{6}$$

for primes p with $p \equiv 2 \bmod 3$ and $p \geq 5$.

If $p \equiv 1 \bmod 3$, $\mathrm{disc}(f_t(x))$ has two roots mod p. For each t corresponding to a root of $\mathrm{disc}(f_t(x)) \equiv 0 \pmod p$, namely, $t^2+3t+9 \equiv 0 \pmod p$, $f_t(x)$ has exactly one root mod p: We can see easily that $f_t(x) \equiv (x-a)^6 \pmod p$, where $t \equiv 3a \pmod p$ and $a^2+a+1 \equiv 0 \pmod p$. Hence we obtain that

$$A_{-1} = \frac{5p-5}{6}, \text{ and } A_5 = \frac{p-7}{6}$$

for primes p with $p \equiv 1 \bmod 3$ and $p \geq 7$.

4 Approximation of $\log L(1,\sigma)$ and $\frac{L'}{L}(1,\sigma)$

Let σ be an l-dimensional complex representation of a Galois group. We assume $L(s,\sigma)$ is an entire Artin L-function and N is its conductor. Also $L(s,\sigma)$ has a Dirichlet series

$$L(s,\sigma) = \sum_{n=1}^{\infty} a(n) n^{-s}.$$

When an Artin L-function $L(s,\sigma)$ has a "desired" zero-free region, we can approximate $\log L(1,\sigma)$ and $\frac{L'}{L}(1,\sigma)$ as a sum over small primes.

Proposition 4.1 (Daileda, Proposition 2 in [9]) *Let $L(s,\sigma)$ be as above. Let $\frac{6}{7}<\alpha<1$. Suppose that $L(s,\sigma)$ is zero-free in the rectangle $[\alpha,1]\times[-(\log N)^2,(\log N)^2]$. Here the rectangular region is the product of the real part of s and the imaginary part of s. If N is sufficiently large, then for any $0<k<\frac{16}{1-\alpha}$,*

$$\log L(1,\sigma)=\sum_{p\leq(\log N)^k} a(p)p^{-1}+O_{l,k,\alpha}(1).$$

Proposition 4.2 *Let $L(s,\sigma)$ be as in Proposition 4.1. If N is sufficiently large, then for any constant x with $(\log N)^{\frac{16}{1-\alpha}}\leq x\leq N^{\frac{1}{4}}$,*

$$\frac{L'}{L}(1,\sigma)=-\sum_{p\leq x}\frac{a(p)\log p}{p}+O_{l,\alpha}(1).$$

Because we lack the GRH, we cannot use Proposition 4.1 and 4.2 directly. We extend the result of Kowalski-Michel [21] to isobaric automorphic representations of GL_{n-1}. Let $n-1=n_1+\cdots+n_r$, and let $S(q)$ be a set of isobaric representations $\pi=\pi_1\boxplus\pi_2\cdots\boxplus\pi_r$, where π_j is a cuspidal automorphic representation of $GL(n_j)/\mathbb{Q}$ and satisfies the Ramanujan-Petersson conjecture at the finite places. Moreover, $S(q)$ satisfies the following conditions:

1. There exists $e>0$ such that for $\pi=\pi_1\boxplus\pi_2\cdots\boxplus\pi_r\in S(q)$, $Cond(\pi_1)\cdots Cond(\pi_r)\leq q^e$;
2. There exists $d>0$ such that $|S(q)|\leq q^d$.
3. The Γ factors of π_j are of the form $\prod_{k=1}^{n_j}\Gamma(\frac{s}{2}+\alpha_k)$, where $\alpha_k\in\mathbb{R}$.

In addition, we need to assume

$$\text{Given two representations } \pi,\pi'\in S(q)\text{, for each } j,\ \pi_j \text{ is not equivalent to any } \pi'_k \text{ if } n_j=n_k. \quad (4.1)$$

Remark 4.3 If G is a symmetric or alternating group, the representation ρ is irreducible. Hence the above condition (4.1) is empty. If G is a dihedral or cyclic group, ρ is no longer irreducible. In that case, we can verify the above condition (4.1) by computing their Artin conductors. In section 5.4, we explain this in the case of cyclic extensions of degree 7.

Let, for $\alpha \geq \frac{3}{4}$, $T \geq 2$,

$$N(\pi; \alpha, T) = |\{\rho : \, L(\rho, \pi) = 0, \; Re(\rho) \geq \alpha, \; |\operatorname{Im}(\rho)| \leq T\}|.$$

Here zeros are counted with multiplicity. Clearly, $N(\pi; \alpha, T) = N(\pi_1; \alpha, T) + \cdots + N(\pi_r; \alpha, T)$.

Theorem 4.4 *For some $B \geq 0$,*

$$\sum_{\pi \in S(q)} N(\pi; \alpha, T) \ll T^B q^{c_0 \frac{1-\alpha}{2\alpha - 1}}.$$

One can choose any $c_0 > c_0'$, where $c_0' = \frac{5n'e}{2} + d$ and $n' = max\{n_i\}_{1 \leq i \leq r}$. The implied constant depends only on the family $S(q)$ and the choice of c_0.

5 Logarithmic Derivatives of Artin L-functions

In [7], we studied the extreme Euler-Kronecker constants. Let G, ρ be as in the introduction.

We consider the Laurent expansion of $\zeta_K(s)$ at $s = 1$:

$$\zeta_K(s) = c_{-1}(s-1)^{-1} + c_0 + c_1(s-1) + c_2(s-1)^2 + \cdots$$

Ihara [15] defined $\gamma_K = \frac{c_0}{c_{-1}}$ to be the Euler-Kronecker constant of K.

From the decomposition of $\zeta_K(s)$,

$$\gamma_K = \gamma + \frac{L'}{L}(1, \rho). \tag{5.1}$$

Here γ is the Euler constant. Hence, $\frac{L'}{L}(1, \rho)$ determines γ_K.

Ihara [15] found a upper bound and a lower bound for γ_K under GRH. The main term of his upper and lower bound under GRH are

$$2 \log \log \sqrt{|d_K|}, \quad -2(n-1) \log \left(\frac{\log \sqrt{|d_K|}}{n-1} \right).$$

In [7], we obtained a new upper bound and lower bound of $\frac{L'}{L}(1, \rho)$ under assumption of the Artin conjecture, GRH and certain zero Hypothesis, which are

$$\log \log |d_K| + O(\log \log \log |d_K|), \quad -(n-1) \log \log |d_K| + O(\log \log \log |d_K|),$$

respectively.

Moreover, unconditionally we showed that for several families of number fields, there are infinitely many number fields K such that

$$\frac{L'}{L}(1,\rho) \geq \log\log|d_K| + O(\log\log\log|d_K|),$$

and there are infinitely many number fields K such that

$$\frac{L'}{L}(1,\rho) \leq -(n-1)\log\log|d_K| + O(\log\log\log|d_K|).$$

In the next section, we find several new families of number fields with the extreme logarithmic derivatives.

5.1 Quintic fields with extreme values of Logarithmic derivatives

Consider a quintic polynomial for positive square-free integers t with $t \equiv 1 \pmod 5$,

$$f_t(x) = x^5 + tx + t$$

with the discriminant $\operatorname{disc}(f_t(x)) = t^4(256t + 3125)$. We claim that the Galois groups of $f_t(x)$ over $\mathbb{Q}(t)$ and $\overline{\mathbb{Q}}(t)$ are both S_5. It means that $f_t(x)$ gives rise to an S_5 regular extension over $\mathbb{Q}(t)$. Since $f_t(x)$ is an Eisenstein polynomial for an irreducible element t, it is irreducible over $\mathbb{Q}(t)$ and $\overline{\mathbb{Q}}(t)$ and it is clear that $\operatorname{disc}(f_t(x))$ is not an square in $\mathbb{Q}(t)$ and $\overline{\mathbb{Q}}(t)$. If the sextic resolvent has no root in $\mathbb{Q}(t)$ and $\overline{\mathbb{Q}}(t)$, then the Galois group is S_5 over both fields.

The sextic resolvent of $f_t(x)$ is given by

$$\theta_t(y) = (y^3 + b_2y^2 + b_4y + b_6)^2 - 2^{10}\operatorname{disc}(f_t(x))y$$

where $b_2 = -20t$, $b_4 = 240t^2$, and $b_6 = 320t^3$.

We have to show that $\theta_t(y)$ does not have a root in $\overline{\mathbb{Q}}(t)$: If α is a root of $\theta_t(y)$, then

$$(\alpha^3 + b_2\alpha^2 + b_4\alpha + b_6)^2 = 2^{10}\operatorname{disc}(f_t(x))\alpha.$$

Hence α is a divisor of b_6^2. Since the RHS of the above equation cannot be a square, it is a contradiction.

Let K_t be a quintic field obtained by adjoining a root of $f_t(x)$ to $\mathbb{Q}$. Then the signature of K_t is $(1,2)$. Since $f_t(x)$ is an Eisenstein polynomial for a square-free integer t, the field discriminant d_{K_t} of K_t is divided by t^4. Hence $\log d_{K_t} \gg \log t$.

Recently, Calegari obtained the modularity of S_5 Galois representations in a special case.

Theorem 5.1 (Calegari [3]) *Let $K/\mathbb{Q}$ be a quintic extension with the Galois closure $\widehat{K}$ such that* $\mathrm{Gal}(\widehat{K}/\mathbb{Q}) = S_5$. *Furthermore, we assume that*

1. *The complex conjugation in* $\mathrm{Gal}(\widehat{K}/\mathbb{Q}) = S_5$ *has conjugacy class* $(12)(34)$.
2. *The extension $\widehat{K}/\mathbb{Q}$ is unramified at* 5 *and the Frobenius element* Frob_5 *has conjugacy class* $(12)(34)$.

If $\rho : \mathrm{Gal}(\widehat{K}/\mathbb{Q}) \longrightarrow GL_4(\mathbb{C})$ is an irreducible representation of degree 4, *then ρ is modular.*

□

Remark 5.2 Calegari observed that the 4-dimensional representation ρ is equivalent to a twist of $As(\sigma)$ by a character, where σ is the 2-dimensional icosahedral representation of $\tilde{A}_5$ over the quadratic subextension F and $As(\sigma)$ is the Asai lift of σ, and used the modularity of σ proved by Sasaki [28]. Y. Zhang [36] in his thesis also observed the fact that ρ is twist equivalent to $As(\sigma)$.

Since 5 does not divide $\mathrm{disc}(f_t(x)) = t^4(256t + 3125)$, 5 is unramified in $\widehat{K_t}/\mathbb{Q}$. Because $f_t(x) \equiv x^5 + x + 1 \equiv (x+3)(x^2+x+1)(x^2+x+2) \pmod 5$ and the signature of K_t is (1,2), the Galois extension $\widehat{K_t}/\mathbb{Q}$ satisfies the hypotheses of Theorem 5.1. Hence the Artin L-function $L(s, \rho, t) = \frac{\zeta_{K_t}(s)}{\zeta(s)}$ is a cuspidal automorphic L-function of $GL_4/\mathbb{Q}$.

On the other hand, it is easy to check, for a square-free t,

$$p \text{ is totally ramified in } K_t \Longleftrightarrow p \mid t.$$

(It is known that those primes dividing t totally ramify in K_t. Assume that p totally ramifies but does not divide t. This means that $f_t(x) \equiv x^5$ or $(x + a)^5$ mod p for $p \nmid t$. We can induce a contradiction by comparing coefficients mod p.)

Here we need to show that each t gives rise to a distinct L-function. It is related to arithmetic equivalence. Two number fields K_1 and K_2 are called arithmetically equivalent if $\zeta_{K_1}(s) = \zeta_{K_2}(s)$. We have the following ([20, page 94]):

Theorem 5.3 *Let $K/\mathbb{Q}$ be a number field of degree $n \leq 11$. Let $\widehat{K}$ be the Galois closure and assume that there exists a non-conjugate field K' which is arithmetically equivalent to K. Then up to conjugacy, only the following*

4 cases are possible for $G = \mathrm{Gal}(\hat{K}/\mathbb{Q})$*:*

$$\begin{aligned} &n = 7,\ G = GL_3(2);\\ &n = 8,\ G = \mathbb{Z}/8\mathbb{Z} \rtimes (\mathbb{Z}/8\mathbb{Z})^{\times},\ GL_2(3);\\ &n = 11,\ G = PSL_2(11). \end{aligned}$$

Since K_t's are not isomorphic for all square-free t, they are not arithmetically equivalent and hence $L(s, \rho, t)$'s are distinct.

Since $f_t(x)$ gives rise to a regular extension, by Theorem 2.3, there is a constant $c_f > 5$ such that for every prime $p \geq c_f$, there is an $i_p \in \mathbb{Z}$ (resp. $s_p \in \mathbb{Z}$) so that for every integer $t \equiv i_p \bmod p$ (resp. $s_p \bmod p$) with $\mathrm{Gal}(\widehat{K_t}/\mathbb{Q}) = G$, the Frobenius element at p belongs to a conjugacy class with no fixed point (resp. the trivial conjugacy class). For $X > 0$, define $y = \frac{\log X}{\log\log X}$ and $M = 5\prod_{c_f \leq p \leq y} p$. Let i_M (resp. s_M) be an integer with $i_M \equiv 1$ (resp. $s_M \equiv 1$) mod 5 and $i_M \equiv i_p$ (resp. $s_M \equiv s_p$) mod p for all p with $c_f \leq p \leq y$.

Define two sets of square-free integers;

$$\begin{aligned} L(X)_1 &= \left\{ \frac{X}{2} < t < X \mid t : \text{square-free and } t \equiv i_M \pmod{M} \right\},\\ L(X)_2 &= \left\{ \frac{X}{2} < t < X \mid t : \text{square-free and } t \equiv s_M \pmod{M} \right\}. \end{aligned}$$

Note that for every $t \in L(X)_i$, $i = 1, 2$, we can see that $\mathrm{Gal}(\widehat{K_t}/\mathbb{Q}) = S_5$ by showing that its sextic resolvent has no root in $\mathbb{Q}$. Each $t \in L(X)_1$ (resp. $L(X)_2$) corresponds to a distinct automorphic L-functions $L(s, \rho, t) = \sum_{n=1}^{\infty} \lambda(n, t) n^{-s}$ of GL_4 with $\lambda(p, t) = -1$ (resp. $\lambda(p, t) = 4$) for $c_f \leq p \leq y$.

Let $c_0 = 51$ and choose α with $c_0 \frac{1-\alpha}{2\alpha - 1} < \frac{98}{100}$. Since the disc$(f_t(x))$ is a quintic polynomial of t, there is some constant C with disc$(f_t(x)) \leq Ct^5$ for all $t > 0$. Note that $d_{K_t} \leq$ disc$(f_t(x))$. Apply Theorem 4.4 to $L(X)_i$, $i = 1, 2$ with $e = 5$, $d = 1$ and $T = (\log CX^5)^2$. Then $\sum_{\rho \in L(X)_i} N(\rho, \alpha, T) \ll X^{98/100}$. As in [7], we have $|L(X)_i| \sim c_i \frac{X}{M}$ for some constants c_i, and $\log M = \sum_{c_f \leq p \leq y} \log p \ll y$. Hence $M \ll X^{\epsilon}$ for any $\epsilon > 0$. Therefore every automorphic L-functions in $L(X)_i$, $i = 1, 2$ excluding exceptional $O(X^{\frac{98}{100}})$ L-functions has a zero-free region $[\alpha, 1] \times [-(\log d_{K_t})^2, (\log d_{K_t})^2]$. Let $\widehat{L}(X)_i$, $i = 1, 2$ be the set of the automorphic L-functions with the above zero-free region.

First we consider $\widehat{L}(X)_1$ which gives the extreme positive logarithmic derivatives. Applying Proposition 4.2 to $L(s,\rho,t)$ in $\widehat{L}(X)_1$, we have

$$\begin{aligned}\frac{L'}{L}(1,\rho,t) &= -\sum_{p\le x}\frac{\lambda(p,t)\log p}{p}+O_{n,\alpha}(1)\\ &= \sum_{c_f\le p\le y}\frac{\log p}{p}-\sum_{y<p\le x}\frac{\lambda(p,t)\log p}{p}+O_{n,\alpha}(1)\\ &= \log\log X-\sum_{y<p\le x}\frac{\lambda(p,t)\log p}{p}+O(\log\log\log X).\end{aligned}$$

Here we used the fact that $\sum_{p\le y}\frac{\log p}{p}=\log y+O(1)$, and $y=\frac{\log X}{\log\log X}$.

Now we sum the logarithmic derivative $\frac{L'}{L}(1,\rho,t)$ over $\widehat{L}(X)_1$, namely, consider

$$\sum_{L(s,\rho,t)\in\widehat{L}(X)_1}\frac{L'}{L}(1,\rho,t).$$

We need to deal with the sum

$$\sum_{L(s,\rho,t)\in\widehat{L}(X)_1}\sum_{y<p\le x}\frac{\lambda(p,t)\log p}{p}=\sum_{y<p\le x}\frac{\log p}{p}\sum_{L(s,\rho,t)\in\widehat{L}(X)_1}\lambda(p,t).$$

We showed in [7], by using the fact that $\sum_{i=-1}^{n-1} iA_i=O(\sqrt{p})$,

Proposition 5.4 *For all $y<p\le x$ and $i=1,2$,*

$$\sum_{L(s,\rho,t)\in\widehat{L}(X)_i}\lambda(p,t)\ll\frac{|\widehat{L}(X)_i|}{\sqrt{p}}+\frac{|\widehat{L}(X)_i|}{(\log X)^{\frac{1}{2}}}.$$

where the implied constant is independent of p for $y<p\le x$.

Proposition 5.4 implies

$$\begin{aligned}\sum_{L(s,\rho,t)\in\widehat{L}(X)_1}\sum_{y<p\le x}\frac{\lambda(p,t)\log p}{p} &\ll\\ |\widehat{L}(X)_1|\sum_{y<p\le x}\frac{\log p}{p^{3/2}}&+\frac{|\widehat{L}(X)_1|}{(\log X)^{\frac{1}{2}}}\sum_{y<p\le x}\frac{\log p}{p}\\ &\ll\frac{|\widehat{L}(X)_1|}{y^{1/2}}+\frac{|\widehat{L}(X)_1|\log\log X}{(\log X)^{\frac{1}{2}}}.\end{aligned}$$

Hence we have

$$\sum_{L(s,\rho,t)\in\widehat{L}(X)_1} \frac{L'}{L}(1,\rho,t) = |\widehat{L}(X)_1|\log\log X + O(|\widehat{L}(X)_1|\log\log\log X).$$

If there are only finitely many L-functions with $\frac{L'}{L}(1,\rho,t) \geq \log\log|d_{K_t}| + O(\log\log\log|d_{K_t}|)$, they cannot reach the average value $\log\log X$ as X increases. Hence there exist infinitely many L-functions such that $\frac{L'}{L}(1,\rho,t) \geq \log\log|d_{K_t}| + O(\log\log\log|d_{K_t}|)$.

For the case of $\widehat{L}(X)_2$, we can obtain the extreme negative logarithmic derivatives in the similar way. We summarize our results as follows:

Theorem 5.5 *Let $\mathfrak{K}(5, S_5, 1, 2)$ be the set of quintic number fields with signature $(1, 2)$ whose Galois closures over $\mathbb{Q}$ are S_5-extensions. Then*

1. *There are infinitely many number fields K in $\mathfrak{K}(5, S_5, 1, 2)$ with*

$$\frac{L'}{L}(1,\rho) \geq \log\log d_K + O(\log\log\log d_K).$$

2. *There are infinitely many number fields K in $\mathfrak{K}(5, S_5, 1, 2)$ with*

$$\frac{L'}{L}(1,\rho) \leq -4\log\log d_K + O(\log\log\log d_K).$$

5.2 Sextic fields with $PSL(2, 5)$ Galois closures

Consider a family of sextic polynomials in [34],

$$f_t(x) = x^6 - 4x^5 + 2x^4 - 3tx^3 + x^2 + 2x + 1,$$

with $\mathrm{disc}(f_t(x)) = (729t^3 + 522t^2 + 1788t + 2648)^2 = g(t)^2$. Spearman, Watanabe and Williams showed that the Galois group of the splitting field of $f_t(x)$ over $\mathbb{Q}$ is $PSL(2, 5) \simeq A_5$ for $1 \neq t \in \mathbb{Z}$.

Now we claim that $f_t(x)$ gives rise to $PSL(2, 5)$ regular extension, namely, the Galois group of $f_t(x)$ over $\overline{\mathbb{Q}}(t)$ is $PSL(2, 5)$: By Hilbert's irreducibility theorem, the Galois group of $f_t(x)$ over $\mathbb{Q}(t)$ is $PSL(2, 5)$. Then the Galois group of $f_t(x)$ over $\overline{\mathbb{Q}}(t)$ is a subgroup of $PSL(2, 5)$. Now if $f_t(x)$ is not irreducible over $\overline{\mathbb{Q}}(t)$, it is a product of two cubic polynomials, a product of three quadratic polynomials or a product of a quadratic polynomial and a quartic polynomial because $f_t(x)$ does not have a root in $\overline{\mathbb{Q}}(t)$. Then we can verify that the orders of the possible Galois groups of $f_t(x)$ do not divide the order of $PSL(2, 5)$. Hence it should be irreducible over $\overline{\mathbb{Q}}(t)$. So the Galois group of $f_t(x)$ over $\overline{\mathbb{Q}}(t)$ is transitive. But no proper subgroup of $PSL(2,5)$ is transitive. See FIG. 1 in [13]. Hence the Galois group is $PSL(2,5)$.

For a given specialization $t \in \mathbb{Z}$, let $K_t = \mathbb{Q}(\theta_t)$ be a sextic field obtained by adjoining a root θ_t of $f_t(x)$ to $\mathbb{Q}$ and $\widehat{K_t}$ the Galois closure of K_t. Spearman, Watanabe and Williams showed that when $g(t)$ is square-free, K_t is monogenic.

For a positive integer t, the signature of K_t is $(2,2)$: Since $\mathrm{disc}(f_t(x))$ is a square, the possible signatures are $(6,0)$ and $(2,2)$. We recall the following lemma from [25], [29] to find the location of complex roots of $f_t(x)$.

Lemma 5.6 *Let f be a polynomial of degree m and $f(\alpha) \neq 0$, $f'(\alpha) \neq 0$. Then for every circle C passing through α, $\alpha - \frac{mf(\alpha)}{f'(\alpha)}$, at least one root of f is inside of C, and one root outside of C.*

By applying Lemma 5.6 with $\alpha = 0$, we see that there is a root in a circle of radius 1.25 centered at -1.25. However, we can see easily that $f_t(x)$ does not have a real root between -2.5 and 0. Hence the root should be complex and it follows that the signature of K_t is $(2,2)$.

Then we have $\widehat{K_t}^{D_5} = K_t$ and $Ind_{D_5}^{A_5} 1_{D_5} = 1_{A_5} + \rho$, where ρ be the irreducible 5 dimensional representation of A_5. Then by Lemma 5.2 in [17], ρ is equivalent to a twist of $Sym^4(\sigma) \simeq Sym^4(\sigma^\tau)$ by a character, where σ and σ^τ are the two dimensional representations of $SL(2,5)$. Since K_t is not totally real, σ and σ^τ are odd, and hence they are modular by Khare-Wintenberger [16]. By [18, 19], ρ gives rise to a cuspidal automorphic representation of $GL_5/\mathbb{Q}$. Hence $L(s,\rho,t) = \frac{\zeta_{K_t}(s)}{\zeta(s)}$ is a cuspidal automorphic L-function of $GL_5/\mathbb{Q}$. Since K_t is monogenic if $g(t)$ is square-free, and $g(t)$'s are distinct if $t > t_0$ for some constant t_0, K_t's are distinct for $t > t_0$. Hence for such t's, $L(s,\rho,t)$'s are distinct.

As in Section 5.1, we define two sets which give rise to the Artin L-functions $L(s,\rho,t)$:

$$L(X)_1 = \left\{ \frac{X}{2} < t < X \mid g(t) : \text{square-free and } t \equiv i_M \pmod{M} \right\},$$
$$L(X)_2 = \left\{ \frac{X}{2} < t < X \mid g(t) : \text{square-free and } t \equiv s_M \pmod{M} \right\}.$$

As in [5], we can obtain asymptotics of $|L(X)_i|$ since $g(t)$ is a cubic polynomial, and as in the Section 5.1, we can show that $L(X)_1$ (resp. $L(X)_2$) contains infinitely many Artin L-functions whose logarithmic derivatives at $s = 1$ have the extreme positive (resp. negative) values. It is summarized as follows:

Theorem 5.7 *Let $\mathfrak{K}(6, PSL(2,5), 2, 2)$ be the set of sextic number fields with signature $(2,2)$ whose Galois closures over $\mathbb{Q}$ are $PSL(2,5)$ extensions. Then*

1. *There are infinitely many number fields K in $\mathfrak{K}(6, PSL(2,5), 2, 2)$ with*

$$\frac{L'}{L}(1,\rho) \geq \log\log d_K + O(\log\log\log d_K).$$

2. *There are infinitely many number fields K in $\mathfrak{K}(6, PSL(2,5), 2, 2)$ with*

$$\frac{L'}{L}(1,\rho) \leq -5\log\log d_K + O(\log\log\log d_K).$$

5.3 Quartic fields with A_4 Galois closures

Consider a quartic polynomial for positive integers $t \in \mathbb{Z}^+$,

$$f_t(x) = x^4 + 18x^2 - 4tx + t^2 + 81.$$

with $\text{disc}(f_t(x)) = 2^8 t^2 (t^2 + 108)^2$. This polynomial was first studied by Spearman [32]. Let K_t be the quartic field obtained by adjoining a root θ_t of $f_t(x)$ to the rational numbers. It is easy to check that the signature of K_t is $(0,2)$. In our previous paper [7], we used $x^4 - 8tx^3 + 18t^2x^2 + 1$ which generates also A_4 quartic fields with the signature $(0,2)$. But we were not able to show that the quartic fields are distinct. In order to distinguish L-functions, we need a difficult folklore conjecture on asymptotics of square free values of $1 + 27t^4$ on arithmetic progressions. Spearman showed, under some square-free condition, that K_t is monogenic. Hence they are distinct. More precisely

Theorem 5.8 (Spearman) *Suppose that t is a positive integer and that $t(t^2+108)$ is square-free. Then K_t is a monogenic A_4 quartic extensions of $\mathbb{Q}$. Moreover the fields K_t are distinct.*

We claim that the Galois group of $f_t(x)$ over $\mathbb{Q}(t)$ and $\overline{\mathbb{Q}}(t)$ are both A_4. i.e, the splitting field E of $f_t(x)$ over $\mathbb{Q}(t)$ is a regular extension. It is enough to show that the Galois group of $f_t(x)$ over $\overline{\mathbb{Q}}(t)$ is A_4. First we have to show that $f_t(x)$ is irreducible over $\overline{\mathbb{Q}}[t]$. This can be shown by checking that $f_t(x)$ has no root in $\overline{\mathbb{Q}}[t]$ and it cannot be a product of two quadratic polynomials.

The Ferrari resolvent of $f_t(x)$ is

$$y^3 - 18y^2 - (4t^2 + 324)y + (56t^2 + 5832).$$

Since $\text{disc}(f_t(x))$ is a square in $\overline{\mathbb{Q}}[t]$, if the Ferrari resolvent has no root in $\overline{\mathbb{Q}}[t]$, the Galois group of $f_t(x)$ is A_4. The only possibilities for a root are of the forms $a, b(t \pm i\sqrt{\frac{729}{7}})$ and $c(7t^2+729)$ for some $a,b,c \in \overline{\mathbb{Q}}$. We can check that they are not a root of the Ferrari resolvent. Hence we showed that $f_t(x)$ gives rise to an A_4 regular extension. For the detail of Galois group criterion of a polynomial, we refer to [8].

As in Section 5.1, we define two sets which give rise to the Artin L-functions $L(s,\rho,t)$:

$$L(X)_1 = \left\{ \frac{X}{2} < t < X \mid t(t^2+108) : \text{square-free and } t \equiv i_M \pmod{M} \right\},$$

$$L(X)_2 = \left\{ \frac{X}{2} < t < X \mid t(t^2+108) : \text{square-free and } t \equiv s_M \pmod{M} \right\}.$$

To estimate $|L(X)_i|, i = 1,2$, we need Nair's work [24] since $t(t^2+108)$ is reducible. For a polynomial $f(x) \in \mathbb{Z}[x]$, let $\rho(p^k)$ be the number of solutions for $f(x) \equiv 0 \pmod{p^k}$, and define

$$N_k(x,h) = N_k(x,h) = |\{n : x < n \le x+h \mid f(n) : k\text{-free}\}|.$$

Theorem 5.9 (Nair) *If*

$$f(x) = \prod_{i=1}^{m} (f_i(x))^{\alpha_i} \in \mathbb{Z}[x],$$

where each f_i *is irreducible,* $\alpha = \max_i \alpha_i$ *and* $\deg f_i(x) = g_i$*, then*

$$N_k(x,h) = \prod_p \left(1 - \frac{\rho(p^k)}{p^k}\right) h + O\left(\frac{h}{(\log h)^{k-1}}\right),$$

for $h = x^\theta$ *where* $0 < \theta < 1$ *and* $k \ge \max_i\{\lambda g_i \alpha_i\}$ $(\lambda = \sqrt{2} - \frac{1}{2})$*, provided that at least one* $g_i \ge 2$.

Theorem 5.9 implies that

$$|L(X)_i| = \prod_{p \nmid M} \left(1 - \frac{\rho(p^2)}{p^2}\right) \frac{X}{2M} + O\left(\frac{X}{M \log \frac{X}{M}}\right), \quad i = 1,2.$$

As in Section 5.1, we can show that $L(X)_1$ (resp. $L(X)_2$) contains infinitely many Artin L-functions whose logarithmic derivatives at $s=1$ have the extreme positive (resp. negative) values. It is summarized as follows:

Theorem 5.10 *Let $\mathfrak{K}(4, A_4, 0, 2)$ be the set of quartic number fields with the signature $(0, 2)$ whose Galois closures over $\mathbb{Q}$ are A_4 extensions. Then*

1. *There are infinitely many number fields K in $\mathfrak{K}(4, A_4, 0, 2)$ with*

$$\frac{L'}{L}(1, \rho) \geq \log\log d_K + O(\log\log\log d_K).$$

2. *There are infinitely many number fields K in $\mathfrak{K}(4, A_4, 0, 2)$ with*

$$\frac{L'}{L}(1, \rho) \leq -3\log\log d_K + O(\log\log\log d_K).$$

5.4 Cyclic extensions of degree 7

Thaine [35] considered a parametric polynomial, for $d(t) = t^6 + 7t^5 + 21t^4 + 35t^3 + 49t^2 - 49t + 49$,

$$\begin{aligned} f_t(x) = &\\ & x^7 - d(t)x^5 + (-t^2 - 4t + 2)d(t)x^4 + (-4t^3 - 2t^2 + 5t + 8)d(t)x^3 \\ & + (-5t^4 + 11t^3 + 34t - 21)d(t)x^2 \\ & + (-2t^5 + 16t^4 - 6t^3 + 44t^2 - 42t - 7)d(t)x \\ & + (6t^5 - 3t^4 + 19t^3 - 19t^2 - 35t + 31)d(t), \end{aligned}$$

which generate cyclic extensions of degree 7. (We remark that any cyclic extension K of an odd degree $n > 2$ should be totally real: Since K is Galois, it should be either totally real or totally complex. Since n is odd, it should be totally real.)

The discriminant $\mathrm{disc}(f_t(x)) = d(t)^6 g(t)$, where $g(t)$ is a polynomial of degree 56.

Spearman and Williams [33] studied the discriminant of a cyclic field of odd prime degree: Let $f(X) = X^p + a_{p-2}X^{p-2} + \cdots + a_1 X + a_0 \in \mathbb{Z}[X]$ be a polynomial which satisfies the following two conditions: (1) $\mathrm{Gal}(f) \simeq \mathbb{Z}/p\mathbb{Z}$, and (2) there does not exist a prime q such that $q^{p-i} \mid a_i, i = 0, 1, \cdots, p-2$.

Let θ be a root of $f(X)$ and set $K = \mathbb{Q}(\theta)$. Let f_K be the conductor of K. Then $d_K = f_K^{p-1}$. Spearman and Williams proved

Lemma 5.11 *Let q be a prime $\neq p$. Then*

$$q \text{ ramifies in } K \Leftrightarrow q \mid a_i,\ i = 0, 1, \cdots, p-2.$$

and if p ramifies in K, then $p | a_i$ for $i = 1, \ldots, p-2$. Also

$$f_K = p^{\alpha} \prod_{\substack{q \equiv 1 \ (mod\ p) \\ q \text{ ramifies in } K}} q,$$

where q runs through primes and $\alpha = 0$ or 2.

Now assume the coefficient $(-2t^5+16t^4-6t^3+44t^2-42t-7)d(t) = h(t)d(t)$ of x in $f_t(x)$ is 6th-power free. Then $f_t(x)$ satisfies the conditions on $f(X)$. Let $K_t = \mathbb{Q}(\theta_t)$ for a root θ_t of $f(x,t)$, then by Lemma 5.11, $d_{K_t} = f_{K_t}^6 \geq \left(\prod_{q\neq 7, q|d(t)} q\right)^6 \gg d(t)$. Hence we showed that

Lemma 5.12 *For 6th-power free $h(t)d(t)$,*

$$\log d_{K_t} \gg \log t.$$

We claim that $f_t(x)$ generates a C_7 regular extension. First, $f_t(x)$ is irreducible over both $\mathbb{Q}(t)$ and $\overline{\mathbb{Q}}(t)$ since it is an Eisenstein polynomial with respect to $d(t)$. Moreover, $f_t(x)$ generates a C_7 Galois extension over $\mathbb{Q}$ for each specialization $t \in \mathbb{Z}$. Hence, its Galois group over $\mathbb{Q}(t)$ is also C_7. Since C_7 is the smallest Galois group for an irreducible polynomial of degree 7, the Galois group of $f_t(x)$ over $\overline{\mathbb{Q}}(t)$ equals C_7.

For a given $X > 0$, we define y, M, i_M and s_M as in Section 5.1. We define two sets which give rise to the Artin L-functions $L(s,\rho,t)$:

$$L(X)_1 = \left\{\frac{X}{2} < t < X \mid d(t) : \text{square free and } t \equiv i_M \pmod{M}\right\},$$
$$L(X)_2 = \left\{\frac{X}{2} < t < X \mid d(t) : \text{square free and } t \equiv s_M \pmod{M}\right\}.$$

Here we cannot distinguish K_t's by just assuming that $d(t)$ is 6th-power free.

Since ramified primes are among the divisors of $d(t)$ and $d(t)$'s are distinct for $t > t_0$ for some constant t_0, $d(t)$'s have at least one distinct prime. Hence K_t's are non-isomorphic. By Theorem 5.3, each $t \in L(X)_i$ gives rise to a distinct L-function.

Also in this case, $\rho = \chi_{1t} \oplus \cdots \oplus \chi_{6t}$, where χ_{it}'s are characters of C_7 whose conductor is f_{K_t}. Since $d(t)$ is square free, $7 \nmid d(t)$, and hence by Lemma 5.11, $f_{K_t} = d(t)$. Hence f_{K_t}'s are distinct and the condition (4.1) is satisfied.

However, estimating $|L(X)_i|$,$i = 1,2$, is extremely difficult. The following is a folklore conjecture. From the notation in Theorem 5.9,

Conjecture 5.13 $N_2(f,x,h) = ch + O(h/(\log h)^d)$ *for some constant c and $d \geq 1/2$.*

Conjecture 5.13 implies that

$$|L(X)_i| = c\frac{X}{2M} + O\left(\frac{X}{M(\log\frac{X}{M})^d}\right), i = 1,2.$$

As in Section 5.1, we can show that $L(X)_1$ (resp. $L(X)_2$) contains infinitely many Artin L-functions whose logarithmic derivatives at $s = 1$ have the extreme positive (resp. negative) values. It is summarized as follows:

Theorem 5.14 *Let $\mathfrak{K}(7, C_7, 7, 0)$ be the set of cyclic extensions of degree 7. Then under Conjecture 5.13,*

1. *There are infinitely many number fields K in $\mathfrak{K}(7, C_7, 7, 0)$ with*

$$\frac{L'}{L}(1,\rho) \geq \log\log d_K + O(\log\log\log d_K).$$

2. *There are infinitely many number fields K in $\mathfrak{K}(7, C_7, 7, 0)$ with*

$$\frac{L'}{L}(1,\rho) \leq -6\log\log d_K + O(\log\log\log d_K).$$

6 CM Fields With Large Class Numbers

Let $\mathfrak{K}(n, G, r_1, r_2)$ be as in the introduction. Then by the class number formula, the class number h_K for $K \in \mathfrak{K}(n, G, r_1, r_2)$ is given by

$$h_K = \frac{w_K |d_K|^{\frac{1}{2}}}{2^{r_1}(2\pi)^{r_2} R_K} L(1,\rho),$$

where w_K is the number of roots of unity in K, d_K is the discriminant of K and R_K is its regulator and $L(s,\rho) = \dfrac{\zeta_K(s)}{\zeta(s)}$ is the Artin L-function.

Under the Artin conjecture and GRH for $L(s,\rho)$ along with Silverman's lower bound [31] of R_K, we have an upper bound of h_K;

$$h_K \ll |d_K|^{\frac{1}{2}} \frac{(\log\log|d_K|)^{n-1}}{(\log|d_K|)^{r-r_0}}.$$

where $r = r_1 + r_2 - 1$ and r_0 is the maximum of unit ranks of subfields of K.

Now the question is whether the upper bound is sharp. Namely, are there number fields with the largest possible class number of the size

$$|d_K|^{\frac{1}{2}} \frac{(\log\log|d_K|)^{n-1}}{(\log|d_K|)^{r-r_0}}?$$

For real quadratic fields, this is a classical result of Montgomery and Weinberger [23]. Ankeny, Brauer, and Chowla [1] constructed unconditionally, for any n, r_1, r_2, number fields with arbitrarily large discriminants and

$h_K \gg |d_K|^{\frac{1}{2}-\epsilon}$. Under the GRH and Artin conjecture for $L(s,\rho)$, Duke [11] constructed totally real fields of degree n whose Galois closure has the Galois group S_n with the largest possible class numbers. Daileda [9] showed Duke's result unconditionally when $n = 3$.

We showed that the upper bound is sharp up to a constant for many other cases [4], [5], [6]. However, for the case of CM fields, the upper bound is not sharp at all. A CM field is an imaginary quadratic extension of a totally real field. For the case of CM fields, since $r - r_0 = 0$, we have, under the Artin conjecture and GRH,

$$h_K \ll |d_K|^{\frac{1}{2}} (\log\log |d_K|)^{n-1} \tag{6.1}$$

Let $\mathfrak{K}(n,G)$ denote the set of number fields of degree n whose Galois closures are G. Give a conjugacy class C of G, let $\mathfrak{K}_C(n,G)$ be the subset of $\mathfrak{K}(n,G)$ for which the complex conjugation belongs to C. Daileda, Krishnamoorthy and Malyshev [10] showed

Proposition 6.1 (Proposition 3 in [10]) *If $n \geq 4$ and $\mathfrak{K}_C(n,G)$ contains only CM fields then the bound (6.1) is not sharp in the following sense. For any fixed $c > 0$ there are only finitely many $K \in \mathfrak{K}_C(n,G)$ such that $L(s,\rho)$ is entire, satisfies GRH and*

$$h_K \geq c|d_K|^{\frac{1}{2}} (\log\log |d_K|)^{n-1}.$$

They suggested a new upper bound of h_K for a CM field K. Fix a totally real field F. Let K be an imaginary quadratic field over F. Under the Artin Conjecture and GRH for $L(s,\rho)$, they obtained that

$$h_K \ll |d_K|^{\frac{1}{2}} (\log\log |d_K|)^{1+o(1)}. \tag{6.2}$$

Moreover, they constructed CM fields K whose class numbers are of the extreme size $|d_K|^{1/2}(\log\log |d_K|)^{1+o(1)}$ under the Artin conjecture and GRH for $L(s,\rho)$. Here we note that if we use the effective Chebotarev density theorem, we can remove $o(1)$ term in the exponent.

We will show, unconditionally, that there are infinitely many CM fields K with

$$h_K \gg |d_K|^{\frac{1}{2}} \log\log |d_K|.$$

Let $F/\mathbb{Q}$ be a totally real, solvable Galois extension of degree n. Let $K_t = F(\sqrt{-t})$, where $t > 0$, $-t \equiv 1 \pmod 4$ and square free. Suppose that $gcd(d_F, t) = 1$. Then $K_t/\mathbb{Q}$ is a CM field of degree $2n$, and $d_{K_t} = d_F^2(-t)^n$. We have

$$h_{K_t}/h_F = Q_{K_t} w_{K_t}/(2\pi)^n |d_{K_t}/d_F|^{\frac{1}{2}} L(1,\chi_t),$$

where χ_t is the quadratic Hecke character attached to K_t/F by the class field theory, and w_{K_t} is the number of roots of unity in K_t, and $Q_{K_t} = [U_{K_t} : W_{K_t}U_F]$, and U_{K_t}, U_F are unit groups of K_t, F, resp., and W_{K_t} is the group of roots of unity in K_t. Here $Q_{K_t} = 1, 2$, and if N_t is the norm of the conductor of χ_t, then $N_t = t^n$. Hence

$$h_{K_t} \ll_F t^{\frac{n}{2}} |L(1, \chi_t)|.$$

Now

$$L(s, \chi_t) = \prod_{\mathfrak{p}} (1 - \chi_t(\mathfrak{p}) N(\mathfrak{p})^{-s})^{-1},$$

where $\mathfrak{p}$ runs through the prime ideals of F.

Let t be so large as to apply the effective Chebotarev density theorem [22], namely, $(\log N_t)^{\frac{1}{2}} \geq e^{10n(\log |d_F|)^2}$, i.e., $t \geq e^{\frac{1}{n} e^{20n(\log |d_F|)^2}}$. Then

$$\sum_{p \leq (\log N_t)^{\frac{1}{2}},\, \mathrm{Frob}_p = 1} 1 = \frac{(\log N_t)^{\frac{1}{2}}}{n \log(\log N_t)^{\frac{1}{2}}} + O\left(\frac{(\log N_t)^{\frac{1}{2}}}{(\log \log N_t)^2}\right).$$

Then by partial summation,

$$\sum_{p \leq (\log N_t)^{\frac{1}{2}},\, \mathrm{Frob}_p = 1} p^{-1} = \frac{\log \log \log N_t}{n} + O_{F,n}(1).$$

Choose M, t_M, X so that there exists c_F such that $c_F \leq p \leq (\log N_t)^{\frac{1}{2}}$ splits completely in $\mathbb{Q}(\sqrt{-t})$, and let

$$L(X) = \{\frac{X}{2} < t < X : -t \equiv 1 \pmod 4,\ t \equiv t_M \pmod M,\ gcd(d_F, t) = 1\}.$$

Since $F/\mathbb{Q}$ is a solvable Galois extension, it is a series of cyclic extensions. Hence $\pi_t = Ind_F^{\mathbb{Q}} \chi_t$ is an automorphic representation of $GL_n/\mathbb{Q}$, and $L(s, \pi_t) = L(s, \chi_t)$.

Let $c_0 = \frac{5n^2}{2} + 1$. Choose α with $c_0 \frac{1-\alpha}{2\alpha - 1} < \frac{98}{100}$. Apply Theorem 4.4 to $L(X)$ with $e = n$, $d = 1$ and $T = (\log X^n)^2$. Then every automorphic L-functions in $L(X)$, excluding exceptional $O(X^{98/100})$ L-functions has a zero-free region $[\alpha, 1] \times [-(\log N_t)^2, (\log N_t)^2]$. Let $\widehat{L}(X)$ be the set of the automorphic L-functions with the above zero-free region. For $t \in \widehat{L}(X)$, by Proposition 4.1,

$$
\begin{aligned}
\log L(1,\chi_t) &= \sum_{N(\mathfrak{p})\le(\log N_t)^{\frac12},\, deg(\mathfrak{p})=1} \chi_t(\mathfrak{p})N(\mathfrak{p})^{-1} + O_{F,n}(1) \\
&= n \sum_{p\le(\log N_t)^{\frac12},\, \mathrm{Frob}_p=1} p^{-1} + O_{F,n}(1) \\
&= \log\log\log N_t + O_{F,n}(1).
\end{aligned}
$$

So

$$L(1,\chi_t) \gg_F \log\log N_t.$$

Hence we obtained the extreme class numbers for CM fields and it is summarized as follows:

Theorem 6.2 *Let F be a totally real, solvable Galois extension over $\mathbb{Q}$. Then there are infinitely many CM fields K whose maximal real subfields are F with*

$$h_K \gg_F |d_K|^{\frac12} \log\log|d_K|.$$

References

[1] N.C. Ankeny, R. Brauer, and S. Chowla, *A note on the class-numbers of algebraic number fields*, Amer. J. Math. **78** (1956) 51–61.

[2] D.A. Burgess, *The distribution of quadratic residues and non-residues*, Mathematika **4** (1957), 106–112.

[3] F. Calegari, *The Artin Conjecture for some S_5-extensions*, to appear in Math. Ann., available at http://www.math.northwestern.edu/~fcale/research.html.

[4] P. J. Cho, *The Strong Artin Conjecture and number fields with large class numbers*, To appear in Quart. J. Math.

[5] P. J. Cho and H. H. Kim, *Application of the strong Artin conjecture to class number problem*, To appear in Canad. J. Math.

[6] P. J. Cho and H. H. Kim, *Dihedral and cyclic extensions with large class Numbers*, To appear in J. de Theorie des Nom. de Bordeaux.

[7] P. J. Cho and H. H. Kim, *Logarithmic derivatives of Artin L-functions*, To appear in Compos. Math.

[8] D. Cox, *Galois Theory*, Wiley, 2004.

[9] R.C. Daileda, *Non-abelian number fields with very large class numbers,* Acta Arith. **125** (2006) no.3 , 215-255.

[10] R.C. Daileda, R. Krishnamoorthy, and A. Malyshev, *Maximal class numbers of CM number fields*, J. Number Theory 130 (2010), 936–943.

[11] W. Duke, *Extreme values of Artin L-functions and class numbers,* Compos. Math. **136** (2003), 103-115.

[12] W. Duke, *Number fields with large class groups* Number theory, 117-126, CRM Proc. Lecture Notes, 36, Amer. Math. Soc., Providence, RI, 2004.

[13] T.R. Hagedorn, *General formula for solving solvable sextic equations,* J. Algebra **233** (2000), 704–757.

[14] C. Hooley, *Applications of sieve methods to the theory of numbers,* Cambridge; New York: Cambridge University Press, 1976.

[15] Y. Ihara, *On the Euler-Kronecker constants of global fields and primes with small norms*, Algebraic geometry and number theory, 407–451, Progr. Math., **253**, Birkhäuser Boston, Boston, MA, 2006.

[16] C. Khare and J.P. Wintenberger, *Serre's modularity conjecture. I*, Invent. Math. **178** (2009), no. 3, 485–504.

[17] H.H. Kim, *An example of non-normal quintic automorphic induction and modularity of symmetric powers of cusp forms of icosahedral type,* Inv. Math. **156** (2004), 495–502.

[18] H.H. Kim, *Functoriality for the exterior square of GL_4 and symmetric fourth of GL_2. With appendix 1 by Dinakar Ramakrishnan and appendix 2 by H. Kim and P. Sarnak*, Journal of AMS, **16** (2003), 139-183.

[19] H.H. Kim and F. Shahidi, *Cuspidality of symmetric powers with applications.* Duke Math. J. **112** (2002), 177-197.

[20] N. Klingen, *Arithmetical Similarities*, Oxford Mathematical Monographs, Oxford Science Publications, The Clarendon Press, Oxford University Press, New York, 1998.

[21] E. Kowalski and P. Michel, *Zeros of families of automorphic L-functions close to 1,* Pacific J. Math. **207** (2002), no.2, 411–431.

[22] J.C. Lagarias and A.M. Odlyzko, *Effective versions of the Chebotarev density theorem*, Algebraic number fields: L -functions and Galois properties (Proc. Sympos., Univ. Durham, Durham, 1975), pp. 409–464. Academic Press, London, 1977.

[23] H. L. Montgomery and P. J. Weinberger, *Real quadratic fields with large class numbers*, Math. Ann. **225** (2) (1977), 173–176.

[24] M. Nair, *Power free values of polynomials*, Mathematika **23** (1976), 159-183.

[25] N. Obreschkoff, *Verteilung und Berechnung der Nullstellen reeller Polynome*, Deutscher Verlag der Wissenschaften, Berlin 1963.

[26] R. Ree, *Proof of a conjecture of S. Chowla*, J. Num. Th. **3** (1971), 210–212.

[27] W. Schmidt, *Equations over Finite Fields: An Elementary Approach*, Kendrick Press, 2004

[28] S. Sasaki, *On Artin representations and nearly ordinary Hecke algebras over totally real fields. I*, preprint, 2011.

[29] A.M. Schöpp, *Fundamental units in a parametric family of not totally real quintic number fields*, J. de Theorie des Nom. de Bordeaux, **18** (2006), no. 3, 693–706.

[30] J.P. Serre, *Topics in Galois Theory*, Research Notes in Mathematics, A K Peters, Ltd. 2008.

[31] J.H. Silverman, *An inequality relating the regulator and the discriminant of a number field*, J. Number Theory 19 (1984), no. 3, 437-442.

[32] B.K. Spearman, *Monogenic A_4 quartic fields*, Int. Math. Forum. **1** (2006) no.40, 1969–1974.

[33] B.K. Spearman and K.S. Williams, *The discriminant of a cyclic field of odd prime degree*, Rocky Mountain J. **33** (2003), 1101-1122.

[34] B.K. Spearman, A. Watanabe and K.S. Williams, $PSL(2,5)$ *sextic fields with a power basis*, Kodai Math. J. **29** (2006), no.1 , 5-12.

[35] F. Thaine, *On the construction of families of cyclic polynomials whose roots are units*, Exp. Math. **17** (2008), 315–331.

[36] Y. Zhang, *L-functions in Number Theory.* Thesis at the University of Toronto, 2009.

PETER J. CHO, DEPARTMENT OF MATHEMATICS, UNIVERSITY OF TORONTO, TORONTO, ON M5S 2E4, CANADA.
E-mail: jcho@math.toronto.edu

HENRY H. KIM, DEPARTMENT OF MATHEMATICS, UNIVERSITY OF TORONTO, TORONTO, ON M5S 2E4, CANADA AND KOREA INSTITUTE FOR ADVANCED STUDY, SEOUL, KOREA.
E-mail: henrykim@math.toronto.edu

Automorphic Representations and L-Functions
Editors: D. Prasad, C.S. Rajan, A. Sankaranarayanan, J. Sengupta

Publisher: Hindustan Book Agency, New Delhi, India

Linnik's Ergodic Method and the Distribution of Integer Points on Spheres

Jordan S. Ellenberg, Philippe Michel and Akshay Venkatesh[1]

Abstract

We discuss the distribution of integral solutions to

$$x^2 + y^2 + z^2 = d, \text{ as } d \to \infty.$$

In particular, we prove a refinement of Linnik's theorem that the solutions are uniformly distributed modulo q. The paper is intended in large part as an exposition of Linnik's ideas.

1 Introduction

Let $d > 1$ be an integer which, for simplicity of exposition, we assume to be *squarefree*, and let $\mathscr{R}_3(d)$ be the set of integer points on a 2-dimensional sphere of radius $d^{1/2}$:

$$\mathscr{R}_3(d) := \{\mathbf{x} = (x, y, z) \in \mathbb{Z}^3,\ x^2 + y^2 + z^2 = d\}.$$

The study of $\mathscr{R}_3(d)$ is a classical question of number theory. Surprisingly, there are interesting results about $\mathscr{R}_3(d)$ that have been proved in the last two decades, and simply stated problems that remain unresolved. In increasing order of fineness, one may ask:

1. When is $\mathscr{R}_3(d)$ nonempty?
2. If nonempty, how large is $\mathscr{R}_3(d)$, and how can we generate points in $\mathscr{R}_3(d)$?
3. If $\mathscr{R}_3(d)$ gets large, how is it distributed on the sphere of radius $d^{1/2}$?

[1]We thank agencies that have generously supported our research. J.E. was partially supported by NSF-CAREER Grant DMS-0448750 and a Sloan Research Fellowship; A.V. was partially supported by a Packard fellowship, a Sloan Research Fellowship, and an NSF grant; Ph.M. is partially supported by the Advanced research Grant 228304 from the European Research Council and the SNF grant 200021-125291.

1.1 Existence

The first question was studied by Legendre and the answer is:

Theorem 1 (Legendre/Gauss) *$\mathscr{R}_3(d)$ is nonempty if, and only if, d is not of the form $4^a(8b-1)$ $(a, b \in \mathbb{N})$.*

Put in different terms, this amounts to saying that the quadratic equation $x^2 + y^2 + z^2 = d$ satisfies the *Hasse principle.* Legendre's 1798 proof, however, was incomplete[2], and the first complete proof was given by Gauss three years later in his *Disquisitiones arithmeticae* [11]. An integer satisfying Legendre's condition above will be called *admissible.*

1.2 Size

The second question is somewhat subtler and its resolution is the consequence of the work of several people over the course of more than a century. The initial and fundamental insight comes from Gauss, who showed that $\mathscr{R}_3(d)$ is closely connected with the *set of classes of binary quadratic forms of discriminant $-d$.*

This relation amounts, in more modern terms, to the existence of a natural *action* on the quotient $\mathrm{SO}_3(\mathbb{Z})\backslash\mathscr{R}_3(d)$ of the ideal class group, $\mathrm{Pic}(\mathcal{O}_K)$, of the ring of integers $\mathcal{O}_K$ of the quadratic field $K = \mathbb{Q}(\sqrt{-d})$. This action is, in fact, *transitive* (at least if d is squarefree, which we assume here) and is faithful if and only if $d \equiv 3$ modulo 8. In particular, whereas $\mathrm{SO}_3(\mathbb{Z})\backslash\mathscr{R}_3(d)$ does itself not have a natural group structure (what would the identity be?) the notion of "arithmetic progression" makes sense on $\mathrm{SO}_3(\mathbb{Z})\backslash\mathscr{R}_3(d)$. An exposition of these facts is given in §4.

An immediate consequence of the existence of this action is an exact formula relating $|\mathscr{R}_3(d)|$ to the class number $h_K = |\,\mathrm{Pic}(\mathcal{O}_K)|$ of K. About 40 years after Gauss's work (1838), *Dirichlet's class number formula* provided an analytic expression for the class number:

Theorem 2 (Dirichlet) *The class number equals*

$$h_K = \frac{c}{2\pi} d^{1/2}\, \mathrm{res}_{|s=1}\, \zeta_K(s)$$

with $c = 2, 4, 6, 8$ or 12 and $\zeta_K(s)$ is the Dedekind ζ-function of K.

An immediate consequence of this relation is the *non-vanishing* of the residue at 1 of ζ_K (which we recall is a key step in the proof of Dirichlet's

[2]Legendre assumed the existence of primes in arithmetic progressions which was only proven 40 years later by Dirichlet.

prime number theorem). However, in order to get more precise information on the size of this quantity, one had to wait another century for the work of Landau and Siegel (1936) culminating in

Theorem 3 (Siegel)

$$\mathrm{res}_{|s=1}\,\zeta_K(s) = d^{o(1)}$$

from which one concludes that

$$|\mathscr{R}_3(d)| = d^{1/2+o(1)}. \tag{1.1}$$

1.3 Distribution

The third question is the main focus of this paper; whereas it is not at first clear that is a worthy successor to the first two, its investigation has proved very rich. Progress on it has been entwined with the study of modular L-functions, as well as to the study of dynamics on homogeneous spaces. The first significant answer regarding this question are due to Y. V. Linnik who, in the late 50's, proved — amongst other results — the following:

Theorem 1.1 (Linnik) *As $d \to +\infty$ amongst the admissible, squarefree, integers satisfying $d \equiv \pm 1(5)$, the set*

$$\{\frac{\mathbf{x}}{\sqrt{d}}, \mathbf{x} \in \mathscr{R}_3(d)\} \subset S_2$$

becomes equidistributed on the unit sphere S_2 with respect to the Lebesgue probability measure.

In explicit terms, that means that if

$$\mathrm{red}_\infty : \mathscr{R}_3(d) \mapsto S_2$$

denotes the "scaling" map $\mathbf{x} \mapsto d^{-1/2}.\mathbf{x}$, then for any measurable subset $\Omega \subset S_2$ whose boundary has Lebesgue measure zero,

$$\frac{|\mathrm{red}_\infty^{-1}(\Omega)|}{|\mathscr{R}_3(d)|} = \mathrm{area}(\Omega)(1 + o(1)),\ d \to +\infty; \tag{1.2}$$

(where we normalize by requiring that $\mathrm{area}(S_2) = 1$).

Linnik obtained this by an ingenious technique he called the "ergodic method," which exploited the action of $\mathrm{Pic}(\mathcal{O}_K)$ on $\mathrm{SO}_3(\mathbb{Z})\backslash\mathscr{R}_3(d)$. This method was generalized later (notably by Linnik's student, Skubenko) to establish several remarkable results about the distribution of the representations of large integers by (integral) ternary quadratic forms [18]. Until recently, Linnik's ergodic method remained surprisingly little-known, although simplified treatments were given by a number of authors [33, 21]; one possible reason is that the method did not fit into ergodic theory as the term is now usually understood, i.e. dynamics of a measure-preserving transformation.

The aim of the present paper is to revisit and explain in a slightly different language Linnik's original approach and to present further refinements which do not seem present in Linnik's work and do not seem accessible to other approaches. For this, we will discuss in full detail the following discrete variant to Linnik's equidistribution theorem.

1.4 The discrete sphere

Instead of looking at the position of $\mathscr{R}_3(d)$ on the sphere (archimedean distribution), one could consider the congruence properties of points in $\mathscr{R}_3(d)$ (q-adic distribution). For q an integer coprime with d, let $\mathscr{R}_3(d;q)$ denote the "sphere modulo q"

$$\mathscr{R}_3(d;q) := \{\overline{\mathbf{x}} = (\overline{x}, \overline{y}, \overline{z}) \in (\mathbb{Z}/q\mathbb{Z})^3,\ \overline{x}^2 + \overline{y}^2 + \overline{z}^2 \equiv d \bmod q\}.$$

We have the following

Theorem 1.2 (Linnik) *Let q be a fixed integer, coprime with 30. As $d \to +\infty$ amongst the admissible squarefree integers satisfying $d \equiv \pm 1(5)$, $(d,q) = 1$, the multiset*

$$\{\mathbf{x} \pmod q,\ \mathbf{x} \in \mathscr{R}_3(d)\} \subset \mathscr{R}_3(d;q)$$

becomes equidistributed on $\mathscr{R}_3(d;q)$ with respect to the uniform measure.

In explicit terms, that means that if

$$\mathrm{red}_q : \mathscr{R}_3(d) \mapsto \mathscr{R}_3(d;q)$$

denotes the reduction modulo q map, then, for any $\overline{\mathbf{x}} \in \mathscr{R}_3(d;q)$,

$$\frac{|\mathrm{red}_q^{-1}(\overline{\mathbf{x}})|}{|\mathscr{R}_3(d)|} = \frac{1}{|\mathscr{R}_3(d;q)|}(1 + o(1)), \tag{1.3}$$

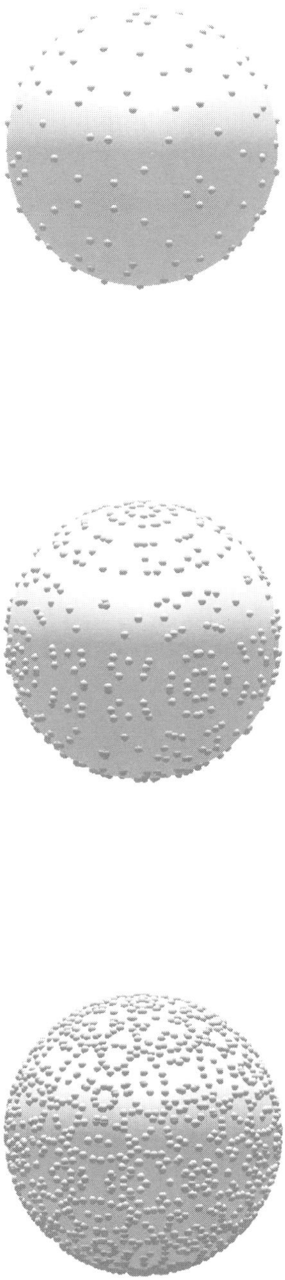

Figure 1 : $d^{-1/2}\mathscr{R}_3(d) \subset S_2$ for $d = 101,\ 8011,\ 104851$

1.5 A refinement of Theorem 1.1

We will focus our attention on proving a sharpened version of Theorem 1.2. For any $\overline{\mathbf{x}} \in \mathscr{R}_3(d;q)$ define the *deviation* at $\overline{\mathbf{x}}$ to be

$$\mathrm{dev}_d(\overline{\mathbf{x}}) = \frac{|\mathrm{red}_q^{-1}(\overline{\mathbf{x}})|}{|\mathscr{R}_3(d)|/|\mathscr{R}_3(d;q)|} - 1.$$

Theorem 1.2 is then equivalent to

$$\mathrm{dev}_d(\overline{\mathbf{x}}) \to 0, \text{ for any } \overline{\mathbf{x}} \text{ as } d \to +\infty$$

(amongst admissible squarefree $d \equiv \pm 1(5)$). We have

Theorem 1.3 *Fix $\nu, \delta > 0$ and suppose that $q^2 \leqslant d^{1/2-\nu}$ and $(q, 30) = 1$. The fraction of $\overline{\mathbf{x}} \in \mathscr{R}_3(d;q)$ for which $|\mathrm{dev}_d(\overline{\mathbf{x}})| > \delta$ tends to zero as $d \to \infty$ with $d \equiv \pm 1(5)$ admissible.*

When q gets large, the volume (i.e. the number of points) of $\mathscr{R}_3(d;q)$ equals $q^{2+o(1)}$,so a qualitative consequence of this theorem (phrased somewhat informally) is:

As long as $\mathscr{R}_3(d)$ is a bit bigger than $\mathscr{R}_3(d;q)$ in size, "almost any solution" to $\bar{x}^2 + \bar{y}^2 + \bar{z}^2 = d$ with $(\bar{x}, \bar{y}, \bar{z}) \in (\mathbb{Z}/q\mathbb{Z})^3$, can be lifted to a solution $x^2 + y^2 + z^2 = d$, $(x, y, z) \in \mathbb{Z}^3$.

It is natural to surmise that this is true for *all* solutions; but this appears to be a very difficult problem. It is also interesting to consider numerics related to this issue.

1.6 An application to mixing

The method of proof of Theorem 1.3 has application regarding the *mixing properties* of the action of $\mathrm{Pic}(\mathcal{O}_K)$ on[3] $\mathscr{R}_3(d)$. Let $[\mathfrak{a}]$ be the ideal class represented by some ideal $\mathfrak{a} \in \mathcal{O}_K$. We would like to understand the distribution of the set of pairs

$$\{(\mathbf{x}, [\mathfrak{a}].\mathbf{x}),\ \mathbf{x} \in \mathscr{R}_3(d)\} \subset \mathscr{R}_3(d) \times \mathscr{R}_3(d)$$

as $d \to +\infty$ (this involves, for each d, the choice of an ideal class $[\mathfrak{a}]$). Here we work in the context of the discrete sphere and consider the behaviour of the multiset

$$\{\mathrm{red}_q(\mathbf{x}, [\mathfrak{a}].\mathbf{x}),\ \mathbf{x} \in \mathscr{R}_3(d)\} \subset \mathscr{R}_3(d;q) \times \mathscr{R}_3(d;q).$$

[3] As we explain below the action on $\mathrm{SO}_3(\mathbb{Z})\backslash\mathscr{R}_3(d)$ can be lifted to an action on $\mathscr{R}_3(d)$

For definiteness, we may and will assume that $\mathfrak{a}$ is *primitive*[4]; let $N = \mathrm{Nr}(\mathfrak{a})$ be its norm. We also assume for simplicity that N is odd and coprime with $15d$.

One has an inclusion

$$\{\mathrm{red}_q(\mathbf{x}, [\mathfrak{a}].\mathbf{x}),\ \mathbf{x} \in \mathscr{R}_3(d)\} \subset \mathscr{R}_3(d; q, N)$$

where $\mathscr{R}_3(d; q, N)$ a specific multiset supported on $\mathscr{R}_3(d;q) \times \mathscr{R}_3(d;q)$: it is the graph of the *Hecke correspondence* T_N whose degree is

$$d_N = \prod_{p^\alpha \| N} (p+1)p^{\alpha-1}.$$

Let $\mu_{d,q,N}$ be the push-forward to $\mathscr{R}_3(d;q) \times \mathscr{R}_3(d;q)$ of the uniform probability measure on $\mathscr{R}_3(d; q, N)$.

When N is fixed, a variant of the proof of Theorem 1.2 gives the following

Let q, N be fixed integers, coprime and coprime with 30. As $d \to +\infty$ amongst the admissible squarefree integers satisfying $d \equiv \pm 1(5)$, $(d, qN) = 1$ and admitting a primitive ideal $\mathfrak{a}$ of norm N exactly[5] *the multiset $\{\mathrm{red}_q(\mathbf{x}, [\mathfrak{a}]\mathbf{x}),\ \mathbf{x} \in \mathscr{R}_3(d)\}$ viewed as a multiset supported on $\mathscr{R}_3(d;q) \times \mathscr{R}_3(d;q)$ becomes equidistributed with respect to the measure $\mu_{d,q,N}$.*

We will now consider the case of $N \to \infty$. Then, $\mathscr{R}_3(d; q, N)$ becomes equidistributed: more precisely the measure $\mu_{d,q,N}$ converge to the uniform probability measure on $\mathscr{R}_3(d;q) \times \mathscr{R}_3(d;q)$. In view of this and of the previous equidistribution result, it is natural to expect that the multiset $\{\mathrm{red}_q(\mathbf{x}, [\mathfrak{a}]\mathbf{x}),\ \mathbf{x} \in \mathscr{R}_3(d)\}$ become equidistributed as well. This is indeed true at least for N in a restricted range:

Theorem 1.4 *Let q be a fixed integer coprime with 30 and $\varepsilon > 0$. For any d squarefree, admissible, satisfying $(d, q) = 1$ and $\equiv \pm 1(5)$ let $\mathfrak{a}$ be a primitive $\mathcal{O}_K$-ideal of norm $N = N_d$. Assume that $N \to +\infty$ as $d \to +\infty$ and that $N \leqslant d^{1/2-\varepsilon}$, then the multiset*

$$\{\mathrm{red}_q(\mathbf{x}, [\mathfrak{a}]\mathbf{x}),\ \mathbf{x} \in \mathscr{R}_3(d)\} \subset \mathscr{R}_3(d;q) \times \mathscr{R}_3(d;q)$$

becomes equidistributed on $\mathscr{R}_3(d;q) \times \mathscr{R}_3(d;q)$ w.r.t. the uniform probability measure.

The idea of the proof is quite simple: a version of Theorem 1.3 (Thm. 3.1) shows that the multiset $\{\mathrm{red}_q(\mathbf{x}, [\mathfrak{a}]\mathbf{x}),\ \mathbf{x} \in \mathscr{R}_3(d)\}$ is almost equidistributed

[4] An integral ideal is primitive, if it is of minimal norm in its $\mathbb{Q}^\times$-homothety class.

[5] This holds if and only if every prime factor of N splits in K

on $\mathscr{R}_3(d;q,N)$ as long as $q^2 d_N \leqslant d^{1/2-\varepsilon}$; then because of the equidistribution of $\mathscr{R}_3(d;q,N)$ on $\mathscr{R}_3(d;q) \times \mathscr{R}_3(d;q)$ (Prop. 8) we can "push" the almost equidistribution on the *varying* space $\mathscr{R}_3(d;q,N)$ to full equidistribution on the *fixed space* $\mathscr{R}_3(d;q) \times \mathscr{R}_3(d;q)$. The fact that $d_N = N^{1+o(1)}$ explains the constraint $N \leqslant d^{1/2-\varepsilon}$.

The proof of Theorem 1.4 can be adapted to the archimedean setting to yield

Theorem 1.5 *Given any $\varepsilon > 0$. For any d squarefree, admissible and $\equiv \pm 1(5)$ let $\mathfrak{a}$ be a primitive $\mathcal{O}_K$-ideal of norm $N = N_d$. Assume that $N \to +\infty$ as $d \to +\infty$ and that $N \leqslant d^{1/2-\varepsilon}$, then the set*

$$\{\mathrm{red}_\infty(\mathbf{x}, [\mathfrak{a}]\mathbf{x}),\ \mathbf{x} \in \mathscr{R}_3(d)\} \subset S^2 \times S^2$$

becomes equidistributed on $S^2 \times S^2$ w.r.t. the product Lebesgue probability measure.

It is then natural to surmise[6] the following

Conjecture (Mixing conjecture) *The above equidistribution results hold without the constraint $N \leqslant d^{1/2-\varepsilon}$ and $d \equiv \pm 1(5)$.*

Notice that by Minkowski's theorem, one has $N \leqslant (4/\pi)d^{1/2}$ and by Siegel's theorem the total number of primitive ideals equals $d^{1/2+o(1)}$ while the number of primitive ideals of norm $\leqslant d^{1/2-\varepsilon}$ is bounded by $\ll d^{1/2-\varepsilon/2}$ so, by comparison with the Mixing conjecture, Theorem 1.4 "misses" a lot of the possible primitive ideals; however, as we discuss below, Theorem 1.4 seem significantly stronger than what could be obtained by different methods, even on very strong hypotheses.

1.6.1 Representation of binary forms by quaternary quadratic forms

The Theorem 1.4 has an interpretation in terms of representations of rank two quadratic forms by a fixed rank four quadratic form[7]: suppose $d \equiv 1, 2 (\mathrm{mod}\, 4)$, then the (classes of) primitive quadratic forms of discriminant $-4d = \mathrm{disc}(\mathcal{O}_K)$ are given by the quadratic lattices $(\mathfrak{a}, \mathrm{Nr}_{K/\mathbb{Q}})$ as $\mathfrak{a}$ ranges

[6]In [24], a similar conjecture was made (in a slightly different context) and some applications of it were described.

[7]a representation of a rank m quadratic form $(\mathbb{Z}^m, q_m)$ by a rank n-quadratic form $(\mathbb{Z}^n, q_n)$, $(n \geqslant m)$ is an isometric linear embedding $\iota : \mathbb{Z}^m \hookrightarrow \mathbb{Z}^n$, ie. $q_n(\iota(\mathbf{x})) = q_m(\mathbf{x})$, $\forall \mathbf{x} \in \mathbb{Z}^m$

over the primitive $\mathcal{O}_K$-ideals; morevoer the various classes of embeddings of such a binary forms by the "four squares" quaternary quadratic form

$$q_4(x, y, z, t) = x^2 + y^2 + z^2 + t^2$$

are precisely described by the set of pairs

$$\{(\mathbf{x}, [\mathfrak{a}]\mathbf{x}),\ \mathbf{x} \in \mathscr{R}_3(d)\}.$$

In particular, Theorem 1.4 and 1.5 translate to q-adic or archimedean equidistribution properties for the set of embeddings of *some* binary quadratic forms of large fundamental discriminant: those associated to a primitive ideal of norm $N(\mathfrak{a}) \leqslant d^{1/2-\varepsilon}$) and the mixing conjecture would establish this equidistribution property for *all* of them.

These techniques and interpretation apply with little changes when q_4 is replaced by an integral anisotropic quaternary form; in particular, if the number of genus classes of this form is greater than 1 (eg. for definite forms), the appropriate analog of Theorem 1.4 or the corresponding mixing conjecture shows (in a way similar to [9]) that the *Hasse principle* holds of the corresponding binary forms when their discriminant gets sufficiently large. Alternatively this technique could be seen as providing non-trivial bounds for *some* Fourier coefficients of Siegel modular forms of genus 2 (Yoshida lifts) and the mixing conjecture would provide such bounds with no restriction. Either of these interpretations should convey the opinion that the mixing conjecture is deep.

1.7 Another approach to Linnik's problem: the works of Duke and Iwaniec

The condition $d \equiv \pm 1(5)$ is stated merely for simplicity. In fact the conclusions of Theorems 1.1, 1.2, 1.3 or 1.4 continue to hold if the condition $d \equiv \pm 1(5)$ is replaced by the more general one: *given $p > 2$ some fixed prime,*

Linnik's condition 1 (at p) *The prime p splits in the quadratic field $\mathbb{Q}(\sqrt{-d})$.*

As we will see, Linnik's condition is genuine to the ergodic method and removing it was considered a major problem: it is only thirty years later that Duke [3] resolved this problem along with other by using very different ideas and techniques (see also the independent work of Fomenko and Golubeva [10]). Since our main aim is to describe Linnik's ergodic method, we will only give a brief account of Duke's approach: for suitable test functions (harmonic homogeneous polynomials) the Weyl sums naturally associated with these equidistribution problems are Fourier coefficients of half-integral weight modular forms –this being a manifestation of the theta correspondence– and the decay of these sums is tantamount to providing non-trivial bounds for these Fourier coefficients. Such bounds were proven by Iwaniec for holomorphic forms in a groundbreaking paper [13] and were generalized to general automorphic forms by Duke, who applied them to solve various equidistribution problems linked to ternary definite or indefinite quadratic forms [3, 4].

1.8 Connections to L-functions

One nice feature of Duke's approach –in addition to removing Linnik's condition entirely– is that is gives explicit (polynomial) control on the rates of equidistribution: there is an absolute constant $\eta > 0$ such that:

- the terms $o(1)$ in (1.2) (at least if Ω is sufficiently regular) and (1.3) take the shape $O(d^{-\eta/2})$:
- In the context of Theorem 1.3: if d is large enough (depending on δ), the number of $\overline{\mathbf{x}} \in \mathscr{R}_3(d;q)$ for which $|\mathrm{dev}_d(\overline{\mathbf{x}})| > \delta$ is *zero*, as long as $q^2 \leqslant d^{\eta/2}$.
- In Theorem 1.4, the multiset $\{\mathrm{red}_q(\mathbf{x}, [\mathfrak{a}]\mathbf{x}),\ \mathbf{x} \in \mathscr{R}_3(d)\}$ becomes equidistributed in $\mathscr{R}_3(d;q) \times \mathscr{R}_3(d;q)$ for $d \to +\infty$ as long as $N \leqslant d^{\eta/2}$.

The exponents η can be explicited and, currently are relatively small.

Alternatively, it follows from the work of Waldspurger [38, 39] that Duke's approach is closely related to analytic properties of some L-functions and in particular to the *subconvexity problem*: the interested reader may consult [14, 22] for a general discussion of the subconvexity problem and [23] for its relation with Linnik's equidistribution problem (this relation is to be discussed in much greater details in [26]); in addition [24] describes –in a slightly different context– an L-function approach to the mixing problem for $\mathrm{Pic}(\mathcal{O}_K)$. It turns out that the validity of the *Generalized Riemann Hypothesis* would solve the subconvexity problem in an optimal way, giving that any $\eta < 1/2$ is admissible.

1.9 Linnik's ergodic method for other quadratic forms

The ergodic method applies to representation of integers by general ternary anisotropic quadratic forms. For definite forms, this shows in particular that the Hasse principle hold for sufficiently large, admissible, integers satisfying Linnik's condition and possibly an additional natural technical assumption (coprimality with the discriminant of the form). The method will be presented in a general setting in [25]. In the isotropic case, the method should work as well but at the price of extra complications linked to non-compactness: in that direction we can mention [8] by M. Einsiedler, E. Lindenstrauss, Ph. M. and A. V., which offers an ergodic theoretic proof of the distribution of the representations of large *positive* integers d by the *discriminant* quadratic form $\mathrm{disc}(a, b, c) = b^2 - 4ac$ (without Linnik's type condition) again largely inspired by the ideas of Linnik and Skubenko.

Going beyond equidistribution, a finer question, is to what extent the set $\mathscr{R}_3(d)$ (when projected to the 2-sphere or the discrete sphere) can be said to exhibit random behavior? This question was investigated recently in the work of Bourgain, Rudnick and Sarnak [2] who tested $d^{-1/2}\mathscr{R}_3(d)$ against various local statistics. Their results point towards a random behavior. Interestingly their methods combine some of Linnik's ideas (surrounding the *basic lemma*) together with Duke's theorem.

1.10 Plan of the paper

In §2 we describe the main steps of the proof of Theorems 1.3 and 1.4. Each of these steps is then explained in detail in the remainder of the paper. The point of working with a discrete sphere $\mathscr{R}_3(d; q)$ rather than on the sphere S_2 is that it makes several technical details slightly simpler while keeping intact the core of the arguments. We definitely hope that the interested reader will be able to use these ideas to shape his own proofs of Theorem 1.1 and of the corresponding analog of Theorem 1.4. Less motivated readers will find the proofs in [25].

- Part 2 concerns those parts of the proof that are most conveniently presented in classical language:
    - In §4, we describe the natural action of the class group on the quotient $\mathrm{SO}_3(\mathbb{Z})\backslash\mathscr{R}_3(d)$ and make it rather explicit. For that purpose, it is particularly useful to express everything in terms of quaternions. Presentations of similar material can be found in [34, 35] and [32]; our approach is slightly different.
    - In §5, we establish the important "basic lemma" of Linnik.
- Part 3 — comprised of §7 to §10 is concerned with transferring some of the key objects — $\mathscr{R}_3(d), \mathscr{R}_3(d; q)$ — in terms of adelic quotients. This will be a key tool in proofs.
- Part 4 is concerned with the part of the proof that is related to expanders.

- In §11 we prove that $\mathscr{R}_3(d;q)$ — endowed with a graph structure that we shall describe — is an expander graph.
- In §12, we recall some basic facts from the theory of random walks on expander graphs. In particular we give a self-contained proof of a *large deviation estimate* (for non-backtracking paths) on such graphs.

1.11 Acknowledgements

The ideas of this paper are based on those of Linnik, and, indeed, this paper should be regarded in considerable part as an exposition of his work. The usefulness of ergodic methods to resolve deep problems in number theory is by now well recognized, at least since the resolution of the Oppenheim conjecture by Margulis in the 80's. Yet, we are still amazed by Linnik's insight which goes back to the 50's.

This paper also draws on the ideas in the work of the latter two authors with M. Einsiedler and E. Lindenstrauss, as well as work of the first- and last- named author [7, 6, 8, 9]. The novel results of the paper were obtained in 2005; we apologize for the delay in bringing them to print.

We also thank P. Sarnak for encouragement of the project, R. Masri for reading an early version, E. Kowalski for reading a later one, J.-P. Serre and the referee for their comments and criticism; finally special thanks are due to G. Harcos who read very carefully the whole manuscript and made numerous corrections and comments on that occasion.

2 An Overview of the Ergodic Method

We now present an overview of the proof of Theorem 1.3. We will try to isolate the main steps of the proof, each one of which contains some key results which may be of more general mathematical interest. In the present section, we treat each of these results as a black box; in the latter part of the paper, we "open the black boxes" one by one and provide complete proofs.

2.1 Assumptions and notations.

We denote "the sum of 3 squares" quadratic form and the associated inner product by

$$q_3(\mathbf{x}) = q_3(x,y,z) = x^2 + y^2 + z^2, \ \mathbf{x}.\mathbf{x}' = xx' + yy' + zz';$$

Throughout the paper we shall make the following assumptions:

1. d will always denote a squarefree integer, not congruent to 7 modulo 8, and congruent to ± 1 modulo 5.
2. q will always denote an integer prime to 30.

Remark As pointed out in the introduction, one could replace 5 by an arbitrary fixed prime p and the condition $d \equiv \pm 1$ modulo 5 by the Linnik's condition at p, ie. *p is split in $\mathbb{Q}(\sqrt{-d})$*. The assumption that q is prime to 30 is for convenience and could be removed entirely.

Write $K = \mathbb{Q}(\sqrt{-d})$; we denote by $\mathscr{O}_K$ the ring of integers of K, by d_K its discriminant (equal to $-d$ or $-4d$) and by $\mathrm{Pic}(\mathscr{O}_K)$ the ideal class group of $\mathscr{O}_K$. We also fix a square root of $-d$ in K, and denote it by $\sqrt{-d}$ without further commentary.

2.2 Some natural quotients

The problems considered in the introduction admit "obvious" *symmetries* owing to the evident action of $\mathrm{SO}_3(\mathbb{Z})$ on $\mathscr{R}_3(d)$, S_2 or $\mathscr{R}_3(d;q)$. We denote the corresponding quotients by

$$\widetilde{\mathscr{R}_3}(d) = \mathrm{SO}_3(\mathbb{Z})\backslash\mathscr{R}_3(d),\ \widetilde{\mathscr{R}_3}(d;q) = \mathrm{SO}_3(\mathbb{Z})\backslash\mathscr{R}_3(d;q),\ \widetilde{S_2} = \mathrm{SO}_3(\mathbb{Z})\backslash S_2.$$

and denote by $[\mathbf{x}]$ the orbit $\mathrm{SO}_3(\mathbb{Z})\mathbf{x}$ of any element in the above sets.

For the purposes of our main theorems, there is no essential difference between working with $\mathscr{R}_3(d)$, S_2 or $\mathscr{R}_3(d;q)$ and working with $\widetilde{\mathscr{R}_3}(d)$, $\widetilde{S_2}$ or $\widetilde{\mathscr{R}_3(d;q)}$ (since $\mathrm{SO}_3(\mathbb{Z})$ is finite). It is conceptually clearer to consider the question of how $\widetilde{\mathscr{R}_3}(d)$ becomes distributed in $\widetilde{\mathscr{R}_3(d;q)}$ or $\widetilde{S_2}$ than the similar question for $\mathscr{R}_3(d)$ and S_2 or $\mathscr{R}_3(d;q)$; this becomes clear when considering these problems for other ternary quadratic forms, especially indefinite ones. However, the latter formulation being slightly more classical, we will make the (slight) extra effort required to state theorems on $\mathscr{R}_3(d)$.

Recall that if G is a group, a *homogeneous space* for G is simply a set X on which G acts transitively (i.e. for any $x, x' \in X$, there is $g \in G$ such that $g.x = x'$). In such a case, the stabilizers of the elements of X under this action are all conjugate.

Proposition 1 ($\widetilde{\mathscr{R}_3}(d)$ is a $\mathrm{Pic}(\mathscr{O}_K)$-homogeneous space) *Let $d > 3$ be a squarefree integer not congruent to 7 mod 8. Then there exists a natural action of $\mathrm{Pic}(\mathscr{O}_K)$ on $\widetilde{\mathscr{R}_3}(d)$, making $\widetilde{\mathscr{R}_3}(d)$ into a homogeneous space for $\mathrm{Pic}(\mathscr{O}_K)$. The stabilizer of any point is trivial if $d \equiv 3$ modulo 4, and is the order 2 subgroup generated by a prime above 2 if $d \equiv 1, 2$ modulo 4. In particular,*

$$|\mathscr{R}_3(d)| = 24|\mathrm{Pic}(\mathscr{O}_K)| \text{ when } d \equiv 3\ (8)$$

and

$$|\mathscr{R}_3(d)| = 12|\operatorname{Pic}(\mathscr{O}_K)| \text{ when } d \equiv 1, 2\ (4).$$

We discuss the precise version of this statement in Proposition 10 in §4.1 and explain its adelic manifestation in §8.1.

Given $\mathfrak{a} \subset \mathscr{O}_K$ an ideal, we denote by $[\mathfrak{a}].[\mathbf{x}]$ the result of the action of the ideal class $[\mathfrak{a}]$ of $\mathfrak{a}$ on an element $[\mathbf{x}] \in \widetilde{\mathscr{R}_3}(d)$.

2.3 The $\operatorname{Pic}(\mathscr{O}_K)$-action

The group $\operatorname{Pic}(\mathscr{O}_K)$ is generated by the classes of prime ideals above primes which are split or ramified in $\mathcal{O}_K$, so to describe the action on $\widetilde{\mathscr{R}_3}(d)$ it is sufficient to describe the action of such classes. In fact we will only need the action of of ideals associated to split primes (ramified primes yield ideal classes of order two). Let $p > 2$ be a split prime, which is to say that the principal ideal $p\mathscr{O}_K$ factors into a product of two prime ideals

$$p\mathscr{O}_K = \mathfrak{p}.\mathfrak{p}'.$$

We note that the prime 5 is split whenever $d \equiv \pm 1(5)$.

We shall now realize explicitly the action of $[\mathfrak{p}]$ and the subgroup of $\operatorname{Pic}(\mathscr{O}_K)$ it generates in terms of the action of SO_3.

Let $\mathcal{A}_p$ be the set of elements δ of $SO_3(\mathbb{Q})$ whose entries have denominator dividing p (that is, $p\delta$ is integral) and such that $\delta \equiv 1$ modulo 3. The significance of the last condition is that it forces $\mathcal{A}_p$ to be a set of representatives for $SO_3(\mathbb{Z})$-cosets on matrices of denominator p. We will see (§4.3) that $|\mathcal{A}_p| = p + 1$. Note that $\mathcal{A}_p$ is *symmetric*, i.e. $\gamma \in \mathcal{A}_p \implies \gamma^{-1} = \gamma^{\mathrm{T}} \in \mathcal{A}_p$.

Proposition 2 (The action of a prime ideal) *Suppose that $p > 3$ splits in $\mathbb{Q}(\sqrt{-d})$. For $\mathbf{x} \in \mathscr{R}_3(d)$, exactly two of*

$$\{\gamma.\mathbf{x},\ \gamma \in \mathcal{A}_p\}$$

belong to $\mathscr{R}_3(d)$. The classes of those two points in $\widetilde{\mathscr{R}_3}(d)$ are $[\mathfrak{p}].[\mathbf{x}]$ and $[\mathfrak{p}'].[\mathbf{x}] = [\mathfrak{p}]^{-1}.[\mathbf{x}]$.

As suggested by Proposition 2, the action of $[\mathfrak{p}]^{\mathbb{Z}}$ can be lifted from $\widetilde{\mathscr{R}_3}(d)$ to $\mathscr{R}_3(d)$. This lifting is rather less canonical than the $\operatorname{Pic}(\mathscr{O}_K)$-action on $\widetilde{\mathscr{R}_3}(d)$.

2.3.1 Example: $p = 5$

Here

$$\mathcal{A}_5 = \{A, A^{-1}, B, B^{-1}, C, C^{-1}\},$$

where A, B, C are, respectively, rotations by angles acos$(-4/5)$ around the x, y, z axes. Explicitly,

$$A = \frac{1}{5}\begin{pmatrix} 5 & 0 & 0 \\ 0 & -4 & 3 \\ 0 & -3 & -4 \end{pmatrix}, B = \frac{1}{5}\begin{pmatrix} -4 & 0 & 3 \\ 0 & 5 & 0 \\ -3 & 0 & -4 \end{pmatrix}, C = \frac{1}{5}\begin{pmatrix} -4 & -3 & 0 \\ 3 & -4 & 0 \\ 0 & 0 & 5 \end{pmatrix}.$$

The first statement of the above proposition can be easily verified directly in this case.

□

2.4 Trajectories

We can use Proposition 2 to determine "distinguished trajectories" in $\mathscr{R}_3(d)$.

To start with, pick any $\mathbf{x} \in \mathscr{R}_3(d)$; now by Lemma 2, there are precisely two matrices in $\mathcal{A}_p$ — say w_1, w_0 — such that $w_1\mathbf{x}$ and $w_0^{-1}\mathbf{x}$ both belong to $\mathscr{R}_3(d)$. Denote $w_1\mathbf{x}$ by $\mathbf{x}_1$, and $w_0^{-1}\mathbf{x}$ by $\mathbf{x}_{-1}$. Similarly, there is a unique choice of matrix $w_2 \in \mathcal{A}_5$ such that $w_2 \neq (w_1)^{-1}$ and $w_2\mathbf{x}_1$ belongs to $\mathscr{R}_3(d)$; we denote $w_2\mathbf{x}_1$ by $\mathbf{x}_2$. In this way, repeated application of Lemma 2 gives rise to a sequence $(\mathbf{x}_i)_{i\in\mathbb{Z}}$ in $\mathscr{R}_3(d)$ such that $\mathbf{x}_0 = \mathbf{x}$ and for any i,

$$\mathbf{x}_i \in \mathscr{R}_3(d),\ \mathbf{x}_{i+1} \in \{\gamma.\mathbf{x}_i,\ \gamma \in \mathcal{A}_p\} - \{\mathbf{x}_{i-1}\}.$$

Alternatively, this string can be represented by the data of $\mathbf{x} = \mathbf{x}_0$ and an infinite (necessarily periodic) word $W_\mathbf{x} = (w_{\mathbf{x},i})_{i\in\mathbb{Z}}$, in the alphabet $\mathcal{A}_p$, satisfying

$$w_{\mathbf{x},i+1} \neq (w_{\mathbf{x},i})^{-1},\ \mathbf{x}_i = w_{\mathbf{x},i}\mathbf{x}_{i-1} \in \mathscr{R}_3(d). \tag{2.1}$$

A word $(w_i)_i$ satisfying the condition $w_{i+1} \neq (w_i)^{-1}$ is called *reduced*; it is easy to see that $W_\mathbf{x}$ is the unique reduced word satisfying (2.1), up to the "switch directions" transformation given by switching w_i and w_{1-i}^{-1}, or, equivalently, switching $\mathbf{x}_i$ and $\mathbf{x}_{-i}$. We refer to the sequence $(\mathbf{x}_i)_{i\in\mathbb{Z}}$ as the *trajectory* of $\mathbf{x}$.

The equivalence of this trajectory to the one defined by the action of $[\mathfrak{p}]^\mathbb{Z}$ is explained in §4.5.

2.5 Example: $p = 5$ continued

Take $d = 101$. In this case, $|\mathscr{R}_3(d)| = 168$ and $|\operatorname{Pic}(\mathscr{O}_K)| = 14$. The points of $|\mathscr{R}_3(d)|$, up to the action of $\mathrm{SO}_3(\mathbb{Z})$, are:

$$(10,1,0), \pm(9,4,2), \pm(8,6,1), \pm(7,6,4)$$

The *trajectory* containing $(10,1,0)$ is:

$$(10,1,0) \overset{B}{\to} (-8,1,-6) \overset{C^{-1}}{\to} (7,4,-6) \overset{B^{-1}}{\to} (-2,4,9) \overset{C^{-1}}{\to}$$
$$(4,-2,9) \overset{A}{\to} (4,7,-6) \overset{C^{-1}}{\to} (1,-8,-6) \overset{A^{-1}}{\to} (1,10,0)\ldots$$
$$W_{(10,1,0)} = \ldots B^{\star}C^{-1}B^{-1}C^{-1}AC^{-1}A^{-1}\ldots$$

with $B^{\star} = w_1$.

We note that after seven steps (and thus, after any multiple of seven steps) the trajectory returns to the $\mathrm{SO}_3(\mathbb{Z})$-orbit of $(10,1,0)$. This periodicity of order 7 reflects the fact that the class of a prime ideal above 5 has order 7 in the class group of $\mathbb{Q}(\sqrt{-101})$, which is cyclic of order 14.

2.6 The lengths of the trajectories

We should emphasize that it is certainly possible, and in some sense probably typical, for $[\mathfrak{p}]^{\mathbb{Z}}$ to be all or most of $\operatorname{Pic}(\mathscr{O}_K)$.

What we know in this direction is rather minimal. On the one hand, it is easy to see that

$$|[\mathfrak{p}]^{\mathbb{Z}}| \gg \log(d)$$

as follows. Suppose that $\mathfrak{p}^k$ is principal for some $k \geqslant 1$; writing $\mathfrak{p}^k = (a + \sqrt{-d}b)\mathcal{O}_K$ and taking norms, one has

$$p^k = a^2 + b^2 d \geqslant b^2 d \geqslant d/4$$

since $b \neq 0$ and $b \in \mathbb{Z}/2$. While this lower bound goes to ∞ with d, it remains quite small compared with the size of $|\operatorname{Pic}(\mathscr{O}_K)| = d^{1/2+o(1)}$. As far as we know, it is unknown whether there exist infinitely many squarefree $d \equiv 1(20)$ such that $|[\mathfrak{p}]^{\mathbb{Z}}|$ is greater than some positive power of d.

These considerations are, of course, related to the question:

for which d does the process described in Proposition 2 traverse all points in $\widetilde{\mathscr{R}_3}(d)$?

This condition on d turns out to be equivalent to the condition that the prime ideal $\mathfrak{p}$ above p and the 2-torsion ideal class above 2 generate the

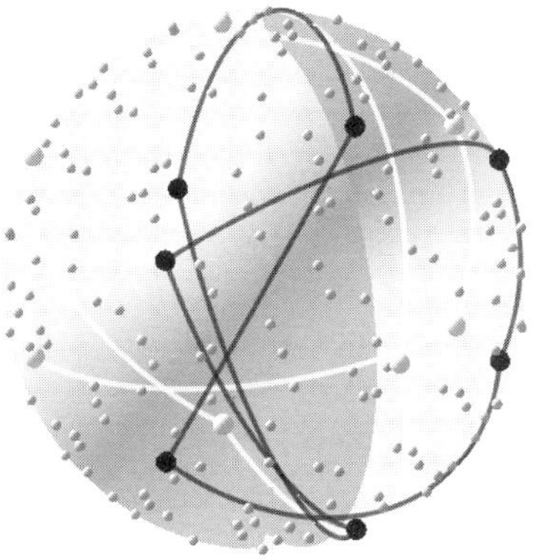

Figure 2: The trajectories of $(10, 1, 0)$ (white) and $(6, 1, 8)$ (black) within $\mathscr{R}_3(d)$

ideal class group $\mathbb{Q}(\sqrt{-d})$. The question of *whether this happens for infinitely many* d is an interesting and seemingly difficult one; heuristically it is reasonable to suppose that this happens a positive fraction of the time. It is analogous to Gauss's famous question:

do there exist infinitely many $d > 0$ *for which* $\mathbb{Q}(\sqrt{d})$ *has class number* 1*?*

The prime above p plays the role in our situation that a real place does in Gauss's problem; the *period* of the process above is analogous to the regulator of a real quadratic field.

2.7 Trajectories on the graph $\mathscr{R}_3(d; q)$

We shall now begin to examine solutions to $x^2 + y^2 + z^2 = d$ modulo a positive integer q, as in Theorem 1.2. We shall suppose, as in that theorem, that q is relatively prime to $6p$.

Under this assumption, the matrices $\mathcal{A}_p = \{\gamma_1, \ldots, \gamma_{p+1}\}$ define elements in $\mathrm{SO}_3(\mathbb{Z}/q\mathbb{Z})$ and act on $\mathscr{R}_3(d; q)$; this endows $\mathscr{R}_3(d; q)$ with a structure of a $p + 1$-regular (undirected) [8] graph by joining each $\overline{x} \in \mathscr{R}_3(d; q)$ to

$$\gamma_1 \overline{x}, \ \ldots \gamma_{p+1} \overline{x}.$$

By an abuse of notation we shall also refer to this graph as $\mathscr{R}_3(d; q)$. More precisely, it is a *multigraph* — we allow multiple edges between pairs of

[8] "undirected" means precisely that in this construction, we consider as the same the edge the link between (say) v to $v' = \gamma.v$ and the link between v' to $v = \gamma^{-1}v'$.

vertices, and edges joining a vertex to itself.

Now to each point $\mathbf{x} \in \mathscr{R}_3(d)$, we may associate a well-defined *marked walk* on the graph $\mathscr{R}_3(d;q)$ in the following way. To a given $\mathbf{x} \in \mathscr{R}_3(d)$, we constructed in §2.4 a word $W_\mathbf{x}$ and a sequence $(\mathbf{x}_i)_{i\in\mathbb{Z}}$ of points in $\mathscr{R}_3(d)$, the *trajectory* of $\mathbf{x}$, which is well-defined up to the substitution $\mathbf{x}_i \mapsto \mathbf{x}_{-i}$. We denote by $\Gamma_\mathbf{x}$ the reduction of this trajectory to $\mathscr{R}_3(d;q)$; the data of $\Gamma_\mathbf{x}$ consists of the sequence of vertices $(\overline{\mathbf{x}}_i)_{i\in\mathbb{Z}}$ of $\mathscr{R}_3(d;q)$ ($\overline{\mathbf{x}}_i = \mathrm{red}_q(\mathbf{x}_i)$), together with a marked basepoint $\overline{\mathbf{x}}_0$ and a choice, for each i, of an edge joining $\overline{\mathbf{x}}_i$ to $\overline{\mathbf{x}}_{i+1}$. For any integer $\ell \geqslant 1$, we denote by $W_\mathbf{x}^{(\ell)}$ and $\Gamma_\mathbf{x}^{(\ell)}$ respectively the truncated word of length 2ℓ and the truncated walk of length 2ℓ centered respectively at 0 and at $\overline{\mathbf{x}}$. In other words,

$$W_\mathbf{x}^{(\ell)} = (w_{\mathbf{x},-\ell+1}, \dots, w_{\mathbf{x},\ell}),\ \Gamma_\mathbf{x}^{(\ell)} = (\overline{\mathbf{x}}, w_{\mathbf{x},-\ell+1}, \dots, w_{\mathbf{x},\ell}).$$

Note that the trajectories arising as $\Gamma_\mathbf{x}$ are not completely arbitrary walks on $\mathscr{R}_3(d;q)$; the condition $w_{i+1} \neq w_i^{-1}$ from (2.1) implies that $\Gamma_\mathbf{x}$ never traverses the same edge twice in succession. (This does not, of course, forbid $\Gamma_\mathbf{x}$ from traveling from $\overline{x}$ to $\overline{y}$ and then back to $\overline{x}$; it just has to use two distinct edges joining $\overline{x}$ and $\overline{y}$.) A *non-backtracking path* in $\mathscr{R}_3(d;q)$ is a (marked) path which never traverses the same edge twice in succession; equivalently, it can be defined by the data of a marked point $\bar{x}_0$, and a word $W_\mathbf{x} = (w_i)_{i\in\mathbb{Z}}$ satisfying $w_{i+1} \neq w_i^{-1}$; the vertices $\bar{x}_i$ can be determined inductively by the rule that $\overline{x}_{i+1}$ is the vertex arrived at by following the edge labeled w_{i+1} from $\overline{x}_i$.

For instance, if $d = 101, q = 7, P = (7,4,-6)$, we have (see §2.5)

$$\begin{aligned}\gamma_P^{(-2,2]} : \cdots\ [(3,1,0)] &\overset{B}{\to} [(-1,1,1)] \overset{C^{-1}}{\to} [(0,4,1)]^\star \\ &\overset{B^{-1}}{\to} [(-2,4,2)] \overset{C^{-1}}{\to} [(4,-2,2)] \cdots\end{aligned}$$

Here $\star$ denotes the marked vertex, and we have used square brackets $[\dots]$ to denote reduction modulo 7.

2.8 Spacing properties of trajectories

Our aim is to show that the integral points $\mathscr{R}_3(d)$ and their associated trajectories $\{\Gamma_\mathbf{x}, \mathbf{x} \in \mathscr{R}_3(d)\}$ are well distributed, in a suitable sense, on the graph $\mathscr{R}_3(d;q)$. (In interpreting this statement, one should imagine that q is fixed, or not increasing too quickly, whereas $d \to \infty$.) The first proposition develops a criterion to measure the "closeness" of two trajectories:

Proposition 3 (Shadowing Lemma) *The truncated words $W_\mathbf{x}^{(\ell)}$ and $W_{\mathbf{x}'}^{(\ell)}$ coincide if and only if $\mathbf{x} \equiv \pm\mathbf{x}' \pmod{p^\ell}$ and the truncated walks*

$\Gamma^{(\ell)}_{\mathbf{x}} = \Gamma^{(\ell)}_{\mathbf{x}'}$ *coincide if and only if, in addition* $\mathbf{x} \equiv \mathbf{x}' (\mathrm{mod}\, q)$. *Suppose this is the case; let* $\mathbf{x}.\mathbf{x}'$ *denote the scalar product associated with the "sum of three squares" quadratic form, then*

$$\mathbf{x}.\mathbf{x}' \equiv d(\mathrm{mod}\, q^2),\ \ \mathbf{x}.\mathbf{x}' \equiv \pm d(p^{2\ell}).$$

This proposition states roughly that if two points have parallel trajectories for a long time then these points have to be p-adically close . This proposition is named after the classical "shadowing lemma" in the theory of the geodesic flow.

Proof The first part of Proposition 3 is proved in §11.4, using geometric properties of the Bruhat-Tits building of $\mathrm{SO}_3(\mathbb{Q}_p) \simeq \mathrm{PGL}_2(\mathbb{Q}_p)$. As for the second part, the equality of the two truncated trajectories is equivalent to $\mathbf{x} \equiv \mathbf{x}'(\mathrm{mod}\, q)$ and $\mathbf{x} \equiv \pm\mathbf{x}'(\mathrm{mod}\, p^{\ell})$. This implies that $(\mathbf{x} - \pm\mathbf{x}').(\mathbf{x} - \pm\mathbf{x}') \equiv 0(\mathrm{mod}\, p^{2\ell})$ and $(\mathbf{x}-\mathbf{x}').(\mathbf{x}-\mathbf{x}') \equiv 0(\mathrm{mod}\, q^2)$ and thus also the two congruences above, since $\mathbf{x}.\mathbf{x} = \mathbf{x}'.\mathbf{x}' = d$ and pq is odd.

□

We shall use Proposition 3 in conjunction with:

Proposition 4 (Linnik's basic Lemma) *Let* $e \in \mathbb{Z}$ *such that* $|e| \neq d$. *The number of pairs* $(\mathbf{x}, \mathbf{x}') \in \mathscr{R}_3(d)^2$ *with dot product* $\mathbf{x} \cdot \mathbf{x}' = e$ *is* $\ll_\varepsilon d^\varepsilon$.

Observe that,by Cauchy-Schwarz, the number of such pairs is zero if $|e| > d$ and equal to $|\mathscr{R}_3(d)|$ if $|e| = d$ (since then $\mathbf{x} = \pm\mathbf{x}'$).

This Lemma is proved in §5. It corresponds to the "basic lemma" in Linnik's ergodic method and is in a sense, a generalization of the well known bounds

$$\tau(d) = \sum_{ab=d} 1 \ll_\varepsilon d^\varepsilon,\ r_2(d) = \sum_{a^2+b^2=d} 1 \ll_\varepsilon d^\varepsilon.$$

Indeed, bounding the divisor function $\tau(d)$, or the number of representations $r_2(d)$ of d as the sum of two squares, amounts to bounding the number of representations of the rank 1 form dX^2 by the binary quadratic forms $Q(x, y) = xy$ and $Q(x, y) = x^2 + y^2$ respectively. By comparison, Proposition 4 concerns the number of ways to represent the rank two form $dX^2 + 2eXY + dY^2$ by the rank three form $x^2 + y^2 + z^2$.

As a consequence of these two results we obtain the following

Corollary 2.1 *For any* $\varepsilon > 0$, *one has*

$$|\{(\mathbf{x}, \mathbf{x}') \in \mathscr{R}_3(d)^2,\ \Gamma^{(\ell)}_{\mathbf{x}} = \Gamma^{(\ell)}_{\mathbf{x}'}\}| \ll_\varepsilon |\mathscr{R}_3(d)| + d^\varepsilon \left(1 + \frac{d}{q^2 p^{2\ell}}\right).$$

Proof We arrange the pairs $(\mathbf{x}, \mathbf{x}')$ according to the value of their inner product $\mathbf{x}.\mathbf{x}' = e$: by the Shadowing Lemma, $|e| \leqslant d$ and $e \equiv \pm d (\operatorname{mod} p^{2\ell})$, $e \equiv d (\operatorname{mod} q^2)$. Hence, by Linnik's basic Lemma, the total number of such pairs is bounded by

$$|\mathscr{R}_3(d)| + \sum_{\substack{|e|<d,\ e\equiv\pm d (\operatorname{mod} p^{2\ell}) \\ e\equiv d (\operatorname{mod} q^2)}} d^{\varepsilon} \ll_{\varepsilon} |\mathscr{R}_3(d)| + d^{\varepsilon}\left(1 + \frac{d}{q^2 p^{2\ell}}\right).$$

□

Remark The above estimate is optimal when the two terms on the right-hand side are roughly equal, that is for ℓ such that

$$q^2 p^{2\ell} \asymp d^{1/2} \asymp |\mathscr{R}_3(d)|.$$

Now $q^2 p^{2\ell} \asymp |\mathscr{R}_3(d;q)|(p+1)p^{2\ell-1}$ is approximately the number of all non-backtracking walks of length 2ℓ on $\mathscr{R}_3(d;q)$, so this Corollary may be interpreted by saying that for this choice of ℓ, the map $\mathbf{x} \mapsto \Gamma_{\mathbf{x}}^{(\ell)}$ from $\mathscr{R}_3(d)$ to the set of such walks is essentially bijective.

2.9 Distribution properties of random walk

The previous section has shown that the deterministic walk $\{\Gamma_{\mathbf{x}},\ \mathbf{x} \in \mathscr{R}_3(d)\}$ behaves roughly like a (non-back tracking) random walk on the graph $\mathscr{R}_3(d;q)$. Here we discuss the distribution of such random walks; these follow from the existence of a *spectral gap*:

Proposition 5 ($\mathscr{R}_3(d;q)$ is an expander) *For any q coprime with $6p$, the graph $\mathscr{R}_3(d;q)$ is connected and non-bipartite. If*

$$\lambda_0 = p+1 \geqslant \lambda_1 \geqslant \dots \lambda_{|\mathscr{R}_3(d;q)|-1} \geqslant -(p+1)$$

denote the eigenvalues of its adjancency matrix $A(\mathscr{R}_3(d;q))$, then

$$|\lambda_j| \leqslant 2\sqrt{p}, \quad j \neq 0.$$

This is explained in Section §11.6 using the adelic description of $\mathscr{R}_3(d;q)$. Indeed, the graphs $\mathscr{R}_3(d;q)$ are closely related to the original Ramanujan graphs of Lubotzky-Phillips-Sarnak [20]. The content of this assertion is equivalent to the (optimal) Ramanujan bound on the p-th Fourier coefficients of weight 2 holomorphic forms of level up to $18.q^2$ (due to Deligne [5]). However, the existence of some spectral gap, which is what we need here, follows from any nontrivial bound for these Fourier coefficients — for instance, those given by Kloosterman or Rankin ($2\sqrt{p}$ replaced by $2p^{5/6}$ and $2p^{3/4}$ respectively).

The bound on the spectral gap in Proposition 5 says that the family of $p+1$-valent graphs $(\mathscr{R}_3(d;q))_{(q,2p)=1}$ form a family of $1-\frac{2\sqrt{p}}{p+1}$ *expander* (in fact Ramanujan) graphs as $q \to \infty$. We refer to [19] for an extensive and motivated discussion of expander graphs, their properties, applications, and their construction via automorphic forms.

We shall use the fact that $\mathscr{R}_3(d;q)$ is an expander through the following proposition.

Proposition 6 (Large deviation estimates) *Fix $\eta, \varepsilon > 0$. For any subset $\mathcal{B} \subset \mathscr{R}_3(d;q)$ with $|\mathcal{B}| \geqslant \eta \mathscr{R}_3(d;q)$, the fraction of non-backtracking walks $\Gamma_{\overline{\mathbf{x}}}$ of length 2ℓ centered at any fixed point $\overline{\mathbf{x}}$ of $\mathscr{R}_3(d;q)$ which satisfy:*

$$\left| \frac{|\Gamma_{\overline{\mathbf{x}}} \cap \mathcal{B}|}{2\ell+1} - \frac{|\mathcal{B}|}{|\mathscr{R}_3(d;q)|} \right| \geqslant \varepsilon$$

is bounded by $c_1 \exp(-c_2 \ell)$, where c_1, c_2 depend only on ε, η.

By $|\Gamma_{\overline{\mathbf{x}}} \cap \mathcal{B}|$ we mean the number of $i \in [-\ell+1, \dots, \ell]$ such that the i-th vertex $\overline{\mathbf{x}}_i$ of Γ is contained in $\mathcal{B}$; in other words, the "amount of time" γ spends in $\mathcal{B}$ if one imagines moving along the trajectory at a constant speed. It is then natural to compare the portion of time spent by a path in $\mathcal{B}$ (i.e. the ratio $|\Gamma_{\overline{\mathbf{x}}} \cap \mathcal{B}|/(2\ell+1)$) with the probability of being in $\mathcal{B}$ (i.e. $|\mathcal{B}|/|\mathscr{R}_3(d;q)|$). Large deviation estimates show that with high probability, long paths spend the right amount of time in any large enough subset $|\mathcal{B}|$.

Such estimates, first proved by Chernoff for complete graphs, are a well-known and useful tool in different contexts; for instance it has been fruitfully applied in computer science (see [12]). We were unable to find a reference in the existing literature for the particular version we need here (a large deviation estimate for non-backtracking walks), and so give a proof from first principles in §12.2 (Proposition 16).

2.10 Conclusion of the proof

We conclude this section by explaining how Propositions 3, 4 and 6 together imply Theorem 1.3.

Let $\mathcal{B}_\delta$ be the set of $\overline{\mathbf{x}} \in \mathscr{R}_3(d;q)$ such that

$$\mathrm{dev}_d(\overline{\mathbf{x}}) = \frac{|\mathrm{red}_q^{-1}(\overline{\mathbf{x}})|}{|\mathscr{R}_3(d)|/|\mathscr{R}_3(d;q)|} - 1 > \delta. \tag{2.2}$$

Suppose that $|\mathcal{B}_\delta| \geqslant \eta|\mathscr{R}_3(d;q)|$. We will derive a contradiction for fixed δ, η and large enough d. A similar bound applies to the set of $\overline{\mathbf{x}}$ for which $\mathrm{dev}_d(\overline{\mathbf{x}}) < -\delta$; taken together these yield Theorem 1.3.

Let $\mathcal{B}$ be a subset of $\mathscr{R}_3(d;q)$; since the action of any $[\mathfrak{p}]^i$ on $\mathscr{R}_3(d)$ permutes $\mathscr{R}_3(d)$, one has for $i \in [-\ell, \ell]$

$$|\mathrm{red}_q^{-1}(\mathcal{B})| = \sum_{\substack{\mathbf{x} \\ \mathrm{red}_q(\mathbf{x}) \in \mathcal{B}}} 1 = \sum_{\substack{\mathbf{x} \\ \mathrm{red}_q([\mathfrak{p}]^i\mathbf{x}) \in \mathcal{B}}} 1$$

$$= \frac{1}{2\ell+1} \sum_{\mathbf{x}} \sum_{\substack{i=-\ell..\ell \\ \mathrm{red}_q([\mathfrak{p}]^i\mathbf{x}) \in \mathcal{B}}} 1 = \frac{1}{2\ell+1} \sum_{\mathbf{x}} |\Gamma_{\mathbf{x}}^{(\ell)} \cap \mathcal{B}| \tag{2.3}$$

We shall take $\mathcal{B} = \mathcal{B}_\delta$ and choose ℓ according to the remark following Corollary 2.1:

$$\frac{1}{p}|\mathscr{R}_3(d)| < q^2 p^{2\ell} \leqslant p|\mathscr{R}_3(d)|. \tag{2.4}$$

This is possible because of the hypothesis that $q^2 \leqslant d^{1/2-\nu}$.

From (2.2) and (2.3), the average value of $\dfrac{|\Gamma_{\mathbf{x}}^{(\ell)} \cap \mathcal{B}_\delta|}{2\ell+1}$ as $\mathbf{x}$ ranges over $\mathscr{R}_3(d)$ exceeds

$$\frac{|\mathcal{B}_\delta|}{|\mathscr{R}_3(d;q)|}(1+\delta) \geqslant \frac{|\mathcal{B}_\delta|}{|\mathscr{R}_3(d;q)|} + \delta\eta.$$

Since $\frac{|\Gamma_{\mathbf{x}}^{(\ell)} \cap \mathcal{B}_\delta|}{2\ell+1} \leqslant 1$ for every $\mathbf{x}$, the number of $\mathbf{x} \in \mathscr{R}_3(d)$ for which

$$\frac{|\Gamma_{\mathbf{x}}^{(\ell)} \cap \mathcal{B}_\delta|}{2\ell+1} > \frac{|\mathcal{B}_\delta|}{|\mathscr{R}_3(d;q)|} + \delta\eta/2 \tag{2.5}$$

is at least

$$\frac{\delta\eta}{2}|\mathscr{R}_3(d)| \gg_\varepsilon \delta\eta d^{1/2-\varepsilon}$$

for any $\varepsilon > 0$ (the last bound following from (1.1)).

Let M be the number of non-backtracking marked paths on $\mathscr{R}_3(d;q)$ satisfying (2.5). Given that ℓ is chosen so that (2.4) is valid, we see from Corollary 2.1 that the number of pairs $(\mathbf{x}, \mathbf{x}')$ yielding the same trajectory $\Gamma_{\mathbf{x}}^{(\ell)}$ on $\mathscr{R}_3(d;q)$ is not much larger than the number of diagonal pairs $|\mathscr{R}_3(d)|$: i.e. is bounded by $\ll_\varepsilon d^{1/2+\varepsilon}$. It follows that

$$M \gg_{\varepsilon,\delta,\eta} d^{1/2-\epsilon}. \tag{2.6}$$

for any $\epsilon > 0$.

On the other hand, we also have an upper bound for M. The total number of non-backtracking marked paths of length 2ℓ on $\mathscr{R}_3(d;q)$ is on order of $p^{2\ell}|\mathscr{R}_3(d;q)| \ll_\varepsilon d^{1/2+\varepsilon}$. Proposition 6 says that, of these, the proportion which satisfy (2.5) is at most $d^{-\tau}$ for some $\tau = \tau(\delta, \eta) > 0$; so

$$M \ll_{\varepsilon,\delta,\eta} d^{1/2-\tau} \tag{2.7}$$

which yields a contradiction for d sufficiently large.

3 Mixing

In this section we prove Theorem 1.4 following the method of proof of Theorem 1.3. Let $d \equiv \pm 1 (\mathrm{mod}\, 5)$ and let $\mathfrak{a}$ be a primitive ideal. To simplify matters slightly we describe the proof when $\mathfrak{a} = \mathfrak{p}$ is a prime ideal of norm $N = p > 5$; in particular, we are requiring that p also splits in K. Then $\mathscr{R}_3(d;q)$ admits two structures of regular (multi)graph: the 6-valent structure defined via $\mathcal{A}_5$ and the $p+1$-valent structure defined via $\mathcal{A}_p$. In the sequel, to differentiate between these two superposed structures, we will use the letter γ to denote an element of the alphabet $\mathcal{A}_p$ and w to denote an element in the alphabet $\mathcal{A}_5$.

Let $\mathscr{R}_3(d;q,p)$, be the set of walks of length 1 on $\mathscr{R}_3(d;q)$ (equivalently the set of *oriented* vertices); that is,

$$\mathscr{R}_3(d;q,p) = \{(\overline{\mathbf{x}}, \gamma),\ \overline{\mathbf{x}} \in \mathscr{R}_3(d;q),\ \gamma \in \mathcal{A}_p\} = \mathscr{R}_3(d;q) \times \mathcal{A}_p.$$

The set $\mathscr{R}_3(d;q,p)$ projects to the $\mathscr{R}_3(d;q) \times \mathscr{R}_3(d;q)$ via the map sending each walk to its extremities:

$$\pi : \begin{array}{ccc} \mathscr{R}_3(d;q,p) & \mapsto & \mathscr{R}_3(d;q) \times \mathscr{R}_3(d;q) \\ (\overline{\mathbf{x}}, \gamma) & \mapsto & (\overline{\mathbf{x}}, \gamma.\overline{\mathbf{x}}) \end{array};$$

we denote by $\mu_{d,q,p}$ the push-forward of the uniform probability measure on $\mathscr{R}_3(d;q,p)$.

$\mathscr{R}_3(d;q,p)$ is evidently a (trivial) degree $p+1$ covering of $\mathscr{R}_3(d;q)$ (via the projection to the first coordinate) and in that way inherits the 6-valent graph structure of $\mathscr{R}_3(d;q)$: for $\gamma \in \mathcal{A}_p$, the neighbors of $(\overline{\mathbf{x}}, \gamma)$ are given by $(w.\overline{\mathbf{x}}, \gamma)$ for w varying over $\mathcal{A}_5$.

For $\mathbf{x} \in \mathscr{R}_3(d)$, let $\gamma_{\mathbf{x}} \in \mathcal{A}_p$ be the matrix such that $[\mathfrak{p}].\mathbf{x} = \gamma_{\mathbf{x}}.\mathbf{x}$ and define

$$\mathrm{red}_{q,p} : \begin{array}{ccc} \mathscr{R}_3(d) & \mapsto & \mathscr{R}_3(d;q,p) \\ \mathbf{x} & \mapsto & (\mathrm{red}_q(\mathbf{x}), \gamma_{\mathbf{x}}) \end{array}$$

We then define for any $(\overline{\mathbf{x}}, \gamma) \in \mathscr{R}_3(d;q,p)$ the *deviation*

$$\mathrm{dev}_d(\overline{\mathbf{x}}, \gamma) = \frac{|\mathrm{red}_{q,p}^{-1}(\overline{\mathbf{x}}, \gamma)|}{|\mathscr{R}_3(d)|/|\mathscr{R}_3(d;q,p)|} - 1.$$

We have the following analog to Theorem 1.3:

Theorem 3.1 *Fix* $\nu, \delta > 0$ *and suppose that* $pq^2 \leqslant d^{1/2-\nu}$ *and* $(q, 30p) = 1$. *The proportion of* $(\overline{\mathbf{x}}, \gamma) \in \mathscr{R}_3(d; q, p)$ *for which* $|\mathrm{dev}_d(\overline{\mathbf{x}}, \gamma)| > \delta$ *tends to zero as* $d \to \infty$ *with* $d \equiv \pm 1(5)$ *admissible.*

Remark If p, q are fixed, this implies that the multiset $\{\mathrm{red}_{q,\mathfrak{p}}(\mathbf{x}),\ \mathbf{x} \in \mathscr{R}_3(d)\}$ becomes equidistributed w.r.t the uniform measure and (by composing with π) that the multiset $\{(\mathbf{x}, [\mathfrak{p}]\mathbf{x}),\ \mathbf{x} \in \mathscr{R}_3(d)\}$ is equidistributed on $\mathscr{R}_3(d; q) \times \mathscr{R}_3(d; q)$ w.r.t $\mu_{d,q,p}$.

Proof The proof is simply an adaptation of the proof of Theorem 1.3 and we will use the Linnik flow at the prime 5: let

$$\Gamma^{(\ell)}_{\mathbf{x},\gamma_{\mathbf{x}}} = ((\overline{\mathbf{x}}, \gamma_{\mathbf{x}}), w_{\mathbf{x},-\ell+1}, \dots, w_{\mathbf{x},\ell}) \in \mathscr{R}_3(d; q, p) \times \mathcal{A}_5^{(\ell)}$$

denote the truncated trajectory of $(\mathrm{red}_q(\mathbf{x}), \gamma_{\mathbf{x}})$ in $\mathscr{R}_3(d; q, p)$ associated with the action of $[\mathfrak{p}_5]^{\mathbb{Z}}$. We have the following version of the shadowing lemma:

Proposition 7 *If the truncated walks* $\Gamma^{(\ell)}_{\mathbf{x},\gamma_{\mathbf{x}}} = \Gamma^{(\ell)}_{\mathbf{x}',\gamma_{\mathbf{x}'}}$ *coincide then*

$$\mathbf{x}.\mathbf{x}' \equiv \pm d (\mathrm{mod}\, pq^2 5^{2\ell}).$$

The congruence $\mathbf{x}.\mathbf{x}' \equiv \pm d (\mathrm{mod}\, q^2 5^{2\ell})$ has already been discussed and the congruence $\mathbf{x}.\mathbf{x}' \equiv \pm d (\mathrm{mod}\, p)$ follows from the equality $\gamma_{\mathbf{x}} = \gamma_{\mathbf{x}'}$ and §11.5. From this we deduce exactly as for Corollary 2.1 that for any $\varepsilon > 0$, one has

$$|\{(\mathbf{x}, \mathbf{x}') \in \mathscr{R}_3(d)^2,\ \Gamma^{(\ell)}_{\mathbf{x},\gamma_{\mathbf{x}}} = \Gamma^{(\ell)}_{\mathbf{x}',\gamma_{\mathbf{x}'}}\}| \ll_\varepsilon |\mathscr{R}_3(d)| + d^\varepsilon \left(1 + \frac{d}{pq^2 5^{2\ell}}\right).$$

Since $pq^2 \leqslant d^{1/2-\nu}$, for d large enough, we may and do choose ℓ so that

$$\frac{1}{5}|\mathscr{R}_3(d)| < pq^2 5^{2\ell} \leqslant 5|\mathscr{R}_3(d)|$$

(compare with (2.4)). Noting that $|\mathscr{R}_3(d; q, p)| = (pq^2)^{1+o(1)}$, we conclude the proof of Theorem 3.1 by repeating the argument of the proof of Theorem 1.3.

□

We now deduce Theorem 1.4; for this, we need the following equidistribution result which is a direct consequence of the fact that the $p+1$-graph $\mathscr{R}_3(d; q)$ has a *uniform* spectral gap when p varies (Proposition 5), and of Lemma 12.1 (with $\ell = 1$ and $\|T_p\| \leqslant \frac{2\sqrt{p}}{p+1}$):

Proposition 8 (Equidistribution of Hecke points) *As $p \to \infty$ the measures $\mu_{\overline{d},q,p}$ converge to the uniform measure on $\mathscr{R}_3(d;q) \times \mathscr{R}_3(d;q)$: for any $(\overline{\mathbf{x}}, \overline{\mathbf{x}}') \in \mathscr{R}_3(d;q) \times \mathscr{R}_3(d;q)$*

$$\frac{|\pi^{-1}(\overline{\mathbf{x}}, \overline{\mathbf{x}}')|}{|\mathscr{R}_3(d;q,p)|} \to \frac{1}{|\mathscr{R}_3(d;q)|^2}.$$

3.1 Proof of Theorem 1.4

Here q is considered as fixed. Given $(\overline{\mathbf{x}}_1, \overline{\mathbf{x}}_2) \in \mathscr{R}_3(d;q) \times \mathscr{R}_3(d;q)$ and any $\varepsilon > 0$, we have

$$\begin{aligned}\frac{|(\mathrm{red}_{q,p} \circ \pi)^{-1}(\overline{\mathbf{x}}_1, \overline{\mathbf{x}}_2)|}{|\mathscr{R}_3(d)|} &= \sum_{\substack{\gamma \in \mathcal{A}_p \\ \gamma.\overline{\mathbf{x}}_1 = \overline{\mathbf{x}}_2}} \frac{|\mathrm{red}_{q,p}^{-1}(\overline{\mathbf{x}}_1, \gamma)|}{|\mathscr{R}_3(d)|} \\ &\geqslant \sum_{\substack{\gamma,\ \gamma.\overline{\mathbf{x}}_1 = \overline{\mathbf{x}}_2 \\ \mathrm{dev}_d(\overline{\mathbf{x}}_1, \gamma) \leqslant \varepsilon}} \frac{|\mathrm{red}_{q,p}^{-1}(\overline{\mathbf{x}}_1, \gamma)|}{|\mathscr{R}_3(d)|} \\ &\geqslant (1-\varepsilon) \frac{|\{\gamma \in \mathcal{A}_p,\ \gamma.\overline{\mathbf{x}}_1 = \overline{\mathbf{x}}_2,\ \mathrm{dev}_d(\overline{\mathbf{x}}_1, \gamma) \leqslant \varepsilon\}|}{|\mathscr{R}_3(d;q,p)|}\end{aligned}$$

Under the assumptions of Theorem 3.1, for d large enough, one has by Theorem 3.1

$$|\{\gamma \in \mathcal{A}_p,\ \gamma.\overline{\mathbf{x}}_1 = \overline{\mathbf{x}}_2,\ \mathrm{dev}_d(\overline{\mathbf{x}}_1, \gamma) > \varepsilon\}| \leqslant \varepsilon |\mathscr{R}_3(d;q,p)|$$

so that

$$\begin{aligned}\frac{|(\mathrm{red}_{q,p} \circ \pi)^{-1}(\overline{\mathbf{x}}_1, \overline{\mathbf{x}}_2)|}{|\mathscr{R}_3(d)|} &\geqslant (1-\varepsilon) \frac{|\{\gamma \in \mathcal{A}_p,\ \gamma.\overline{\mathbf{x}}_1 = \overline{\mathbf{x}}_2\}|}{|\mathscr{R}_3(d;q,p)|} - \varepsilon \\ &= (1-\varepsilon)\mu_{d,q,p}(\overline{\mathbf{x}}_1, \overline{\mathbf{x}}_2) - \varepsilon \geqslant \frac{1}{|\mathscr{R}_3(d;q)|^2} - 3\varepsilon\end{aligned}$$

by Proposition 8, for any $\varepsilon > 0$ and d large enough. Using this lower bound for the points in the complementary set, one deduces that for any $\varepsilon > 0$ and d large enough

$$\frac{|(\mathrm{red}_{q,p} \circ \pi)^{-1}(\overline{\mathbf{x}}_1, \overline{\mathbf{x}}_2)|}{|\mathscr{R}_3(d)|} \leqslant \frac{1}{|\mathscr{R}_3(d;q)|^2} + \varepsilon.$$

□

Part II

Classical Theory

4 The Action of the Class Group on $\widetilde{\mathscr{R}_3}(d)$

We present, in Proposition 9, a precise version of the homogeneous space structure on $\widetilde{\mathscr{R}_3}(d)$ discussed in Proposition 1. As we have remarked, the basic ideas here go back to Venkov [34, 35] and in some sense to Gauss.

It will be convenient to modify, slightly, the definition of $\widetilde{\mathscr{R}_3}(d)$ in the case when $d \equiv 3$ modulo 4. Let $\mathrm{SO}_3(\mathbb{Z})^+$ be the index-2 subgroup of $\mathrm{SO}_3(\mathbb{Z})$ consisting of matrices which act on the coordinate lines via even permutations. Set

$$\widetilde{\mathscr{R}_3}(d)^+ = \begin{cases} \mathrm{SO}_3(\mathbb{Z})^+ \backslash \mathscr{R}_3(d), & \text{if } d \equiv 1, 2 \bmod 4 \\ \mathrm{SO}_3(\mathbb{Z}) \backslash \mathscr{R}_3(d), & \text{if } d \equiv 3 \bmod 4. \end{cases}$$

Thus, $\widetilde{\mathscr{R}_3}(d)^+$ and $\widetilde{\mathscr{R}_3}(d)$ are equal when $d \equiv 3$ modulo 4; otherwise, the former is a double cover of the latter.

Proposition 9 (Torsor structure on $\widetilde{\mathscr{R}_3}(d)^+$) *The set $\widetilde{\mathscr{R}_3}(d)^+$ has the structure of a torsor for* $\mathrm{Pic}(\mathscr{O}_K)$ *(this structure is characterized by the properties given in Proposition 10). This action of* $\mathrm{Pic}(\mathscr{O}_K)$ *descends to $\widetilde{\mathscr{R}_3}(d)$ (obvious if $d \equiv 3 (\mathrm{mod}\, 4)$), and for $d \equiv 1, 2 (\mathrm{mod}\, 4)$ the stabilizer of any point of $\widetilde{\mathscr{R}_3}(d)$ is the order 2 subgroup generated by the prime ideal above 2 (which is ramified).*

Recall that, given a group G, a *G-torsor* or a *principal homogeneous space for G* is a space X endowed with a transitive action of G and for which the stabilizer of some (hence any) point is trivial. Observe that fixing some $x \in X$, the map

$$g \in G \mapsto g.x \in X$$

provides an identification of G with X as G-spaces. However, there is a priori no canonical way of choosing x; thus, one may think of a torsor as a set endowed with many different identifications with G but, in general, with no *canonical* one. For instance, the set of nth roots of 2 is a torsor for the group μ_n of nth roots of unity.

4.1 Quaternions

We first recall the following classical facts.

Let B be the $\mathbb{Q}$-algebra of Hamilton quaternions. For

$$x = u + a.i + b.j + b.k \in \mathrm{B},$$

the canonical involution is noted $\overline{x} = u - a.i - b.j - b.k$, the reduced trace $\mathrm{tr}(x) = x + \overline{x} = 2u$ and the reduced norm $\mathrm{Nr}(x) = x.\overline{x} = u^2 + a^2 + b^2 + c^2$. Let $\mathrm{B}^{(0)}$ denote the space of trace-free quaternions (the kernel of tr) also called the *pure* quaternions. The space $\mathrm{B}^{(0)}$ endowed with the reduced norm is a quadratic space and the map

$$(a, b, c) \mapsto a.i + b.j + b.k \tag{4.1}$$

is an isometry between the quadratic space $(\mathbb{Q}^3, a^2+b^2+c^2)$ and $(\mathrm{B}^{(0)}, \mathrm{Nr})$. In the sequel, we will freely identify $\mathbb{Q}^3$ with $\mathrm{B}^{(0)}$, and, in particular, consider the elements of $\mathscr{R}_3(d)$ as trace-free quaternions.

We denote by $\mathrm{B}^\times$, $\mathrm{B}^{(1)}$ and $\mathrm{PB}^\times = \mathrm{B}^\times/Z(\mathrm{B}^\times)$ respectively , the group of units of B, the subgroup of units of reduced norm one, and the projective group of units; these define $\mathbb{Q}$-algebraic groups, and the action of $\mathrm{B}^\times$ on $\mathrm{B}^{(0)}$ by conjugation: given $x \in \mathrm{B}^\times$

$$\gamma_x : \begin{array}{ccc} \mathrm{B}^{(0)} & \mapsto & \mathrm{B}^{(0)} \\ z & \mapsto & \gamma_x(z) = xzx^{-1} \end{array}$$

induces a covering and an isomorphism of $\mathbb{Q}$-algebraic groups [37, Th. 3.3]:

$$Z(\mathrm{B}^\times) \hookrightarrow \mathrm{B}^\times \twoheadrightarrow \mathrm{PB}^\times \tilde{\to} \mathrm{SO}(a^2 + b^2 + c^2). \tag{4.2}$$

4.2 Integral structures

Let $\mathrm{B}(\mathbb{Z})$ denote the ring of *Hurwitz quaternions*,

$$\mathrm{B}(\mathbb{Z}) = \mathbb{Z}[i, j, k, \frac{1+i+j+k}{2}];$$

It is well known that the ring of Hurwitz quaternions, endowed with the reduced norm Nr, is *euclidean*: for any $y, q \in \mathrm{B}(\mathbb{Z}) - \{0\}$, there is $x, r \in \mathrm{B}(\mathbb{Z})$ such that $\mathrm{Nr}(r) < \mathrm{Nr}(q)$ and $y = qx + r$ ([29, §5.7]). This implies that any left (or right) $\mathrm{B}(\mathbb{Z})$-ideal is a principal ideal: any finitely generated left (resp. right) $\mathrm{B}(\mathbb{Z})$-module $I \subset \mathrm{B}(\mathbb{Q})$ is of the form $\mathrm{B}(\mathbb{Z})q$ (resp. $q\mathrm{B}(\mathbb{Z})$) for some $q \in \mathrm{B}(\mathbb{Q})$; moreover any subring of $\mathrm{B}(\mathbb{Q})$ which is finitely generated (as a $\mathbb{Z}$-module) is conjugate to a subring[9] of $\mathrm{B}(\mathbb{Z})$; in particular $\mathrm{B}(\mathbb{Z})$ is a

[9]Indeed, given R such a subring, $R\mathrm{B}(\mathbb{Z})$ is a right $\mathrm{B}(\mathbb{Z})$ ideal, so of the form $R\mathrm{B}(\mathbb{Z}) = q\mathrm{B}(\mathbb{Z})$ and $q^{-1}Rq \subset q^{-1}RR\mathrm{B}(\mathbb{Z}) = q^{-1}R\mathrm{B}(\mathbb{Z}) = \mathrm{B}(\mathbb{Z})$.

maximal order of $\mathrm{B}(\mathbb{Q})$ and any maximal order in $\mathrm{B}(\mathbb{Q})$ is $\mathrm{B}(\mathbb{Q})^\times$-conjugate to it.

Finally, under (4.1), the lattice $\mathbb{Z}^3 \subset \mathbb{Q}^3$ becomes identified with the trace free integral quaternions,

$$\mathrm{B}^{(0)}(\mathbb{Z}) = \mathrm{B}^{(0)}(\mathbb{Q}) \cap \mathrm{B}(\mathbb{Z}).$$

Let $\mathrm{PB}^\times(\mathbb{Z}) \simeq \mathrm{SO}_3(\mathbb{Z})$ be the set of elements in $\mathrm{PB}^\times(\mathbb{Q})$ leaving $\mathrm{B}^{(0)}(\mathbb{Z})$ invariant. Obviously the rotations associated to integral quaternions of norm 1 are in $\mathrm{PB}^\times(\mathbb{Z})$: in other terms the obvious $\mathrm{B}^\times \to \mathrm{PB}^\times$ induces a map $\mathrm{B}(\mathbb{Z})^\times \to \mathrm{PB}^\times(\mathbb{Z}) \simeq \mathrm{SO}_3(\mathbb{Z})$. That map, however, is not surjective: its image (ie. $\mathrm{B}(\mathbb{Z})^\times/\pm 1$) is the index two subgroup $\mathrm{SO}_3(\mathbb{Z})^+$ and the complementary coset is the coset of the rotations associated with the integral quaternions of norm 2 (for instance $1+i$):

$$\mathrm{SO}_3(\mathbb{Z}) = \mathrm{SO}_3(\mathbb{Z})^+ \sqcup \gamma_{1+i}\mathrm{SO}_3(\mathbb{Z})^+.$$

Similarly, for any prime p, we have a local map $\mathrm{B}(\mathbb{Z}_p)^\times \to \mathrm{SO}_3(\mathbb{Z}_p)$ which is surjective unless $p=2$; in that later case the image has index 2 in $\mathrm{SO}_3(\mathbb{Z}_2)$ and its non-trivial coset is that of γ_{1+i}. Considered in terms of quaternions the quotient $\widetilde{\mathscr{R}}_3(d)^+$ now appears more naturally:

$$\widetilde{\mathscr{R}}_3(d)^+ = \mathrm{B}(\mathbb{Z})^\times \backslash \mathscr{R}_3(d).$$

While our primary interest will be in the Hamilton quaternions and sums of three squares, the reader will observe that much of what we state here and below is valid for more general quaternion algebras endowed with a maximal order.

4.3 The set $\mathcal{A}_p$ in terms of quaternions

Recall that we have defined (for $p \neq 3$)

$$\mathcal{A}_p = \{\delta \in \mathrm{SO}_3(\mathbb{Q}) : p\delta \in 1 + 3M_3(\mathbb{Z})\}. \tag{4.3}$$

Then $\mathcal{A}_p$ is a set of representatives for $\mathrm{SO}_3(\mathbb{Z})$-cosets on $\{\delta \in \mathrm{SO}_3(\mathbb{Q}) : p\delta \in M_3(\mathbb{Z})\}$: this is because the reduction modulo 3 map

$$\mathrm{SO}_3(\mathbb{Z}) \mapsto \mathrm{SO}_3(\mathbb{Z}/3\mathbb{Z})$$

is an isomorphism.

Observe that one may find a set of representatives for (4.3) by taking the image, under $\mathrm{B}(\mathbb{Z}) - \{0\} \to \mathrm{PB}^\times(\mathbb{Z})$, of a set of representatives for Hurwitz-integral quaternions of norm p, under the action of units. In particular, since the number of Hurwitz-integral quaternions of norm p is $24(p+1)$, and there are 24 units, it follows that $|\mathcal{A}_p| = p+1$. (Note, however, that one cannot in general lift elements of $\mathcal{A}_p$ to quaternions of norm p; because of the constraint modulo 3, one can in general lift only to quaternions of norm p or $2p$).

4.3.1 Example: $p = 5$

The set $\mathcal{A}_5$ is the associated with quaternions of norm 10: The integral quaternions of norm 10 can all be expressed in the form ru, where $u \in \mathrm{B}(\mathbb{Z})^\times$ and r is an element of

$$\mathcal{A}_5 = \{1 \pm 3i, 1 \pm 3j, 1 \pm 3k\}. \tag{4.4}$$

Now a direct computation shows that the action of conjugation by the six elements of $\mathcal{A}_5$ yields precisely the action of the six matrices appearing in §2.4. For example, conjugation by $1 - 3i$ acts on $\mathrm{B}^{(0)}$ via the rule

$$\begin{aligned} i &\mapsto i, \\ j &\mapsto \frac{1}{10}(1+3i)j(1-3i) = -\frac{4}{5}j + \frac{3}{5}k, \\ k &\mapsto \frac{1}{10}(1+3i)k(1-3i) = -\frac{3}{5}j - \frac{4}{5}k \end{aligned}$$

which corresponds to the matrix A in §2.3. Observe in particular that this matrix as well as the other elements of $\mathcal{A}_5$ are congruent to the identity modulo 3.

4.4 Construction of representations using ideal classes

As a first example of the usefulness of the quaternions, let us show how to deduce the Gauss-Legendre theorem from the *Hasse-Minkowski local-global principle.* If $d > 0$ is not of the form $4^a(8b-1)$, then, for any prime p, d is representable as a sum of three squares in $\mathbb{Z}_p^3$ and since d is positive, d is also representable over $\mathbb{R}$. By the *Hasse-Minkowski theorem* (cf. [30][Thm. 8, p. 41]), there exists $x = (a, b, c) \in \mathbb{Q}^3$ such that $a^2 + b^2 + c^2 = d$. Let $x = ai + bj + ck$; then $x^2 = -d$ so the ring $\mathbb{Z}[x] = \mathbb{Z} + \mathbb{Z}x$ is finitely generated. By §4.2, there is $q \in \mathrm{B}(\mathbb{Q})$ such that $q\mathbb{Z}[x]q^{-1} \in \mathrm{B}(\mathbb{Z})$, so that $y := qxq^{-1} \in \mathrm{B}^{(0)}(\mathbb{Z})$ is integral and satisfies $y^2 = -d$.

More generally, the above scheme together with the group of $\mathcal{O}_K$-ideal classes makes it possible to generate plenty of *new* integral representations from a given one.

Any element $x \in \mathscr{R}_3(d)$ yields an embedding of $\mathbb{Q}(\sqrt{-d})$ into $\mathrm{B}(\mathbb{Q})$: indeed $x^2 = -d$ and thus $\sqrt{-d} \mapsto x$ defines an embedding

$$\iota_x : K \mapsto \mathbb{Q}[x] \subset \mathrm{B}(\mathbb{Q}).$$

This embedding is *integral* in the following sense: letting

$$\mathscr{O}_x = \mathrm{B}(\mathbb{Z}) \cap \mathbb{Q}[x]$$

we have that $\iota_x^{-1}(\mathscr{O}_x) = \mathscr{O}_K$ is the ring of integers[10] of K.

Now, given such a x and given an $\mathscr{O}_K$-ideal $I \subset K$ (possibly fractional), we can also construct a *new* integral representation $y \in \mathscr{R}_3(d)$ from x and I: the finitely generated $\mathbb{Z}$-module $\mathrm{B}(\mathbb{Z})\iota_x(I)$ is a left $\mathrm{B}(\mathbb{Z})$-ideal, so of the form $\mathrm{B}(\mathbb{Z})q^{-1}$ and then

$$y = q^{-1}xq \subset \mathrm{B}(\mathbb{Z})\iota_x(I)xq = \mathrm{B}(\mathbb{Z})\iota_x(xI)q \subset \mathrm{B}(\mathbb{Z})\iota_x(I)q = \mathrm{B}(\mathbb{Z});$$

moreover if I is replaced by $\lambda.I = I.\lambda\ \lambda \in K^\times$, q may be replaced by $q' = \iota_x(\lambda^{-1})q$ and $q'^{-1}xq' = q^{-1}\iota_x(\lambda)x\iota_x(\lambda^{-1})q = y$. Notice that q is defined only up to multiplication on the right by an element of $\mathrm{B}^\times(\mathbb{Z})$, which is to say that y is well defined up to $\mathrm{B}^\times(\mathbb{Z})$-conjugacy. Since $\mathrm{B}^\times(\mathbb{Z})/\pm 1 \simeq \mathrm{SO}_3(\mathbb{Z})^+$, we obtain for $[x] \in \widetilde{\mathscr{R}_3}(d)^+$ fixed, a well defined map

$$.[x] : \ [I] \in \mathrm{Pic}(\mathscr{O}_K) \mapsto [I].[x] \in \widetilde{\mathscr{R}_3}(d)^+.$$

4.5

In §8, we will give a adelic/group theoretic interpretation of this map and show that it defines a $\mathrm{Pic}(\mathscr{O}_K)$-torsor structure on $\widetilde{\mathscr{R}_3}(d)^+$. In the present section we review this action in more algebraic terms.

For $x, y, q,\ I$ as above, one has

$$\mathbb{Q}[x]q = q\mathbb{Q}[y] = \{\lambda \in \mathrm{B}(\mathbb{Q}),\ x\lambda = \lambda y\}$$

is a 1-dimensional left $\mathbb{Q}[x]$-vector space (resp. right $\mathbb{Q}[y]$-vector space); $\iota_x(I)q = q\iota_y(I)$ is a lattice in $\mathbb{Q}[x]q$ and is clearly a left (resp. right) $\mathscr{O}_x$ (resp. $\mathscr{O}_y$)-module which is locally free of rank 1; moreover its class, $[\iota_x(I)q] = [\iota_x(I)]$ in $\mathrm{Pic}(\mathscr{O}_x) \simeq \mathrm{Pic}(\mathscr{O}_K)$ corresponds to the class $[I]$.

Conversely, for each pair of elements $x, y \in \mathscr{R}_3(d)$ we consider the abelian group

$$\Lambda_{x\to y} = \{\lambda \in \mathrm{B}(\mathbb{Z}) : x\lambda = \lambda y\};$$

[10]It is an order containing $\mathbb{Z}[\sqrt{-d}]$, integrally closed at 2, because the local order $\mathrm{B}(\mathbb{Z}) \otimes_{\mathbb{Z}} \mathbb{Z}_2$ contains all elements of $\mathrm{B} \otimes_{\mathbb{Q}} \mathbb{Q}_2$ with norm in $\mathbb{Z}_2$.

in view of the isomorphism $\mathrm{SO}_3(\mathbb{Q}) \cong \mathrm{PB}^{\times}(\mathbb{Q})$ there exists (by Witt's Theorem) a $q \in \mathrm{B}^{\times}(\mathbb{Q})$ such that $y = q^{-1}xq$. Therefore, $\Lambda_{x\to y}$ is a lattice in $\mathbb{Q}[x]q$ and a locally free of rank 1 left $\mathscr{O}_x$-module (resp. right $\mathscr{O}_y$-module). Indeed

$$\Lambda_{x\to y}q^{-1} \subset \mathbb{Q}[x] \text{ is an } \mathscr{O}_x\text{-ideal, say } \iota_x(I) \subset \iota_x(K). \tag{4.5}$$

In this language a more precise version of Proposition 9 is the following

Proposition 10 (Torsor structure on $\widetilde{\mathscr{R}_3(d)}^+$) *The set $\widetilde{\mathscr{R}_3(d)}^+$ has the structure of a torsor for $\mathrm{Pic}(\mathscr{O}_K)$ in which the unique element of $\mathrm{Pic}(\mathscr{O}_K)$ mapping $[x]$ to $[y]$ is given by the class $[\Lambda_{x\to y}]$.*

More precisely, for any $x, y, z \in \mathscr{R}_3(d)$,

A. *The map*

$$\begin{array}{ccc} \Lambda_{x\to y} \otimes_{\mathscr{O}_y} \Lambda_{y\to z} & \mapsto & \Lambda_{x\to z} \\ \lambda \otimes \mu & \mapsto & \lambda\mu \end{array}$$

is an $(\mathscr{O}_x, \mathscr{O}_z)$-bimodule isomorphism;

B. *The class of $\Lambda_{x\to y}$ is trivial if and only if x and y are identified in $\widetilde{\mathscr{R}_3(d)}^+$;*

C. *For every x and every $[I] \in \mathrm{Pic}(\mathscr{O}_K)$, there exists a y such $[\Lambda_{x\to y}] = [\iota_x(I)]$.*

D. *If $d \equiv 1, 2$ modulo 4, and $x \neq y \in \widetilde{\mathscr{R}_3(d)}^+$ project to the same element of $\widetilde{\mathscr{R}_3(d)}$, then $[\Lambda_{x\to y}]$ is the ideal class of a prime above 2.*

This proposition follows either from the adelic description given in §8 or could be proven directly using the local results of §10.

5 Representations of Binary Quadratic Forms by $x^2 + y^2 + z^2$

We now discuss the proof of Proposition 4, "Linnik's basic lemma." We shall, in fact, discuss two approaches; the first (§5.1) is simply quoting a result of G. Pall, the bound (5.2), which in turn rests on Siegel's mass formula; the second approach, presented in §5.3 after a preliminary discussion, is based on ideas related to §4 and gives a proof of (5.2) at least when the discriminant of the binary form is fundamental. In the related paper [8], we present a self-contained proof of (5.2) for a general integral ternary quadratic form and without any square-freeness condition.

5.1 A result of Venkov and Pall.

Let $(\mathbb{Z}^m, q)$, $(\mathbb{Z}^n, r)$ be two non-degenerate integral quadratic forms with $m \geqslant n$. One says that Q *represents* R if there exists a $\mathbb{Z}$-linear map $\iota : \mathbb{Z}^n \to \mathbb{Z}^m$ such that, for any $\mathbf{x} \in \mathbb{Z}^n$, $Q(\iota(\mathbf{x})) = R(\mathbf{x})$. It follows that ι is an embedding.

In particular, given $\mathbf{x}_1, \mathbf{x}_2 \in \mathscr{R}_3(d)$ with $\mathbf{x}_1.\mathbf{x}_2 = e$, the linear map

$$\iota : \mathbb{Z}^2 \to \mathbb{Z}\mathbf{x}_1 + \mathbb{Z}\mathbf{x}_2$$

defines a representation of the binary quadratic form $R(x, y) = dx^2 + 2exy + dy^2$ by the ternary form $Q(x, y, z) = x^2 + y^2 + z^2$.

Therefore, counting the number of pairs $(\mathbf{x}_1, \mathbf{x}_2) \in \mathscr{R}_3(d)^2$ with $\mathbf{x}_1.\mathbf{x}_2 = e$ is essentially equivalent (up to the action of the finite group $\mathrm{SO}_3(\mathbb{Z})$) to counting the number of representations of R by Q.

The precise computation of the number of embeddings, $r(a, b, c)$ say, of a binary quadratic form $ax^2 + bxy + cy^2$ by the ternary quadratic $x^2 + y^2 + z^2$ was carried out by Venkov [36, p. 168] and (in a slightly more general setting) by Pall [27, Theorem 4, page 359]. His theorem gives a formula

$$r(a, b, c) = 24 \cdot 2^{\nu} \cdot \prod_{p | 2(b^2 - 4ac)} r_p(a, b, c) \tag{5.1}$$

where it follows from Pall's result that:

1. ν is the number of distinct odd primes dividing the discriminant $b^2 - 4ac$;
2. $r_p(a, b, c)$ is bounded by an absolute constant unless $p^2 | (a, b, c)$.

In particular, it follows that

$$|r(a, b, c)| \ll \max(a, b, c)^{\varepsilon}, \tag{5.2}$$

when (a, b, c) has no square factor.

One way to obtain a quantitative result like (5.2) is by use of Siegel's mass formula, combined with a computation of local densities. We will *sketch* in the rest of this section an alternate approach. We must first revisit the material of §4 and describe a "different" (although closely related, as we shall see) connection between $\mathscr{R}_3(d)$ and the class group.

5.2 The orthogonal complement construction

Consider the map

$$x \in \mathscr{R}_3(d) \mapsto (\mathbb{Z}x)^{\perp} = \{\lambda \in \mathbb{Z}^3,\ x.\lambda = 0\}$$

which associates to x the rank 2 lattice of integral vectors orthogonal to x: this is a rank 2-quadratic lattice (when equipped with the restriction of

$q_3(\ ))$ of discriminant $-4d$. Expressed in terms of quaternions (since $x \in \mathrm{B}^0(\mathbb{Z})$)

$$(\mathbb{Z}x)^{\perp} = \{\lambda \in \mathrm{B}^0(\mathbb{Z}),\ \mathrm{tr}(x\overline{\lambda}) = -x\lambda + \lambda x = 0\} = \Lambda_{x\to -x}.$$

Thus we obtain a map

$$\underline{\mathrm{Perp}} : \widetilde{\mathscr{R}_3}(d)^{+} \to \mathrm{Pic}(\mathscr{O}_K). \tag{5.3}$$

Both sides admit an action of $\mathrm{Pic}(\mathscr{O}_K)$; the left-hand side by means of the torsor structure described in section § 4.1, and the right-hand side by multiplication. But $\underline{\mathrm{Perp}}$ is not equivariant for this action; indeed, for x, y in $\mathscr{R}_3(d)$, one has (noting that $[\Lambda_{y\to x}] = [\Lambda_{-x\to -y}]$)

$$\underline{\mathrm{Perp}}(y) = [\Lambda_{y\to -y}] = [\Lambda_{y\to x}][\Lambda_{x\to -x}][\Lambda_{-x\to -y}] = [\Lambda_{x\to y}]^{-2}\underline{\mathrm{Perp}}(x).$$

So $\underline{\mathrm{Perp}}$ intertwines the action of $\mathrm{Pic}(\mathscr{O}_K)$ on the left with the *square* of this action on the right.

It follows that the image of $\underline{\mathrm{Perp}}$ is precisely one coset of $2\,\mathrm{Pic}(\mathscr{O}_K)$ and [11] the cardinality of a fiber of $\underline{\mathrm{Perp}}$ is just the order of the 2-torsion subgroup of $\mathrm{Pic}(\mathscr{O}_K)$.

Remark We have seen that $\widetilde{\mathscr{R}_3}(d)^{+}$ can be placed in bijection with the group $\mathrm{Pic}(\mathscr{O}_K)$; but there is no natural group structure on $\widetilde{\mathscr{R}_3}(d)^{+}$, for there is no natural choice of an identity element of $\widetilde{\mathscr{R}_3}(d)^{+}$. In other words, the torsor structure is natural but admits no natural trivialization. What the orthogonal complement construction supplies is not a trivialization of the torsor $\widetilde{\mathscr{R}_3}(d)^{+}$, but a trivialization of its *square* in the group of $\mathrm{Pic}(\mathscr{O}_K)$-torsors. The square is the torsor T which can be described explicitly as the set of equivalence classes of pairs (x, y) under the relation $(x, y) = (\alpha x, \alpha^{-1} y)$ for all $\alpha \in \mathrm{Pic}(\mathscr{O}_K)$. The action of $\alpha \in \mathrm{Pic}(\mathscr{O}_K)$ sends (x, y) to $(\alpha x, y)$ (or to $(x, \alpha y)$, which is the same.) Then the map sending (x, y) to $[\Lambda_{x\to -y}]$ is a canonical isomorphism between T and $\mathrm{Pic}(\mathscr{O}_K)$. [12]

[11] It is irresistible to ask *which* coset. A quadratic form $Ax^2 + Bxy + Cy^2$ of discriminant d embeds, over $\mathbb{Q}$, into $\mathbb{Z}^3$ with the standard quadratic form only if $(A, d/A)_p = (d, d)_p$ for every p dividing $2d$; here $(a, b)_p$ is the *Hilbert symbol*. This condition in fact defines a coset of squares, and so describes exactly the image of $\underline{\mathrm{Perp}}$.

[12] The situation is similar to one familiar from geometry: If X is a smooth cubic plane curve over a field k with Jacobian E, then the points $X(k)$ form a torsor for the group $E(k)$. This torsor has no canonical trivialization, but the embedding of X into the plane gives a canonical trivialization of the "cube" $X \times_E X \times_E X$.

5.3 Bounds for representations, revisited.

We now sketch how the "orthogonal complement construction" leads us to another proof of Proposition 4.

For simplicity, we consider only the case where $b^2 - 4ac$ is of the form $-4d$, with d squarefree and $d \equiv 1$ modulo 4. Given an embedding

$$\iota : (\mathbb{Z}^2, ax^2 + bxy + cy^2) \hookrightarrow (\mathbb{Z}^3, x^2 + y^2 + z^2).$$

consider the orthocomplement inside $\mathbb{Z}^3$ of $\iota(\mathbb{Z}^2)$. It is generated by a single vector $\mathbf{x} \in \mathbb{Z}^3$, unique up to sign. The quadratic lattice $\mathbf{x}^\perp$ is plainly the same as its sublattice $\iota(\mathbb{Z}^2)$ after tensoring with $\mathbb{Q}$; this means it must have discriminant either $-d$ or $-4d$, and because $d \equiv 1 \bmod 4$, it must be the latter, and the two lattices are identical. This implies that $\mathbf{x}.\mathbf{x} = d$.

The embedding ι is determined up to at most 6 possibilities (6 being the maximal number of automorphisms of a positive definite form in rank two) by $\mathbf{x}$. It suffices, therefore, to count the number of $\mathbf{x} \in \mathscr{R}_3(d)$ so that the quadratic form induced on $\mathbf{x}^\perp$ is isomorphic to $(\mathbb{Z}^2, ax^2 + bxy + cy^2)$. By §5.2, this is at most $24|\operatorname{Pic}(\mathscr{O}_K)[2]|$, where [2] denotes 2-torsion. By genus theory we have $|\operatorname{Pic}(\mathscr{O}_K)[2]| \ll d^\varepsilon$, and the bound (5.2) follows.

Part III

Adelization

In this part, we interpret, in adelic terms, the various classical arithmetic sets discussed so far — i.e. $\mathscr{R}_3(d)$, $\widetilde{\mathscr{R}_3}(d)$, $\operatorname{Pic}(\mathscr{O}_K)$, $\mathscr{R}_3(d;q)$, $\widetilde{\mathscr{R}_3}(d;q)$ — and the various maps between them. This interpretation provide a description of the results described in Part II and more importantly, will used for the proof of Proposition 5.

The principle of adelization goes as follows: firstly we interpret the various sets above in terms of a certain subset of the set of all rational 3-dimensional lattices (the SO_3-genus of $\mathbb{Z}^3$) equipped with additional "structures"; the reduction modulo q map red_q then corresponds to a forgetful map at the level of the structure. These sets of lattices + structures come with a natural action of groups of finite adelic points of SO_3 and suitable subgroups and therefore identified with suitable adelic quotients. We start by recalling how (finite) adèles arise quite naturally in the context of rational lattices.

6 Adelic Actions on Rational Lattices

Given $n \geqslant 1$ an integer, let $\mathcal{L}_n(\mathbb{Q})$ denote the set of lattices in $\mathbb{Q}^n$ (i.e. the finitely generated $\mathbb{Z}$ (resp. $\mathbb{Z}_p$)-modules in $\mathbb{Q}^n$ (resp. $\mathbb{Q}_p^n$) of maximal rank) and, for any prime p, let $\mathcal{L}_n(\mathbb{Q}_p)$ denote the space of lattices in $\mathbb{Q}_p^n$ (i.e. same as above with $\mathbb{Q}$ and $\mathbb{Z}$ replaced by $\mathbb{Q}_p$ and $\mathbb{Z}_p$).

Given $L \in \mathcal{L}_n(\mathbb{Q})$ and p a prime, let L_p be the closure of L in $\mathbb{Q}_p^n$ for the p-adic topology; then $L_p = \mathbb{Z}_p^n$ for a.e. p and (essentially by the Chinese remainder theorem) the map

$$L \mapsto (L_p)_{p \in \mathcal{P}}$$

is a bijection between $\mathcal{L}_n(\mathbb{Q})$ and the (restricted) product

$$\prod_p{}' \mathcal{L}_n(\mathbb{Q}_p) := \{(L_p)_p,\ L_p \in \mathcal{L}_n(\mathbb{Q}_p),\ L_p = \mathbb{Z}_p^n \text{ for a.e. } p\},$$

whose inverse is the map

$$(L_p)_p \mapsto \bigcap_p \mathbb{Q}^n \cap L_p.$$

In particular, for each p, the natural action of the p-adic linear group $\mathrm{GL}_n(\mathbb{Q}_p)$ on $\mathcal{L}_n(\mathbb{Q}_p)$ induces, via this identification, an action of that group on $\mathcal{L}_n(\mathbb{Q})$. All these local actions eventually combine into an action of the restricted product of these groups

$$\begin{aligned}\mathrm{GL}_n(\mathbb{A}_f) &:= \prod_p{}' \mathrm{GL}_n(\mathbb{Q}_p)\\ &= \{(g_p)_p,\ g_p \in \mathrm{GL}_n(\mathbb{Q}_p),\ g_p \in \mathrm{GL}_n(\mathbb{Z}_p) \text{ for a.e. } p\},\end{aligned}$$

which is the group of finite adelic points of GL_n. This action is given explicitely by

$$(g_p)_p.L = \text{ the rational lattice corresponding to the sequence } (g_p.L_p)_p.$$

This action is easily seen to be transitive and the stabilizer of $L_0 := \mathbb{Z}^n$ is the product

$$\mathrm{GL}_n(\widehat{\mathbb{Z}}) := \prod_p \mathrm{GL}_n(\mathbb{Z}_p).$$

In the sequel, we denote by $\mathbb{A}_f$ the ring of finite adèles of $\mathbb{Q}$, i.e. the restricted product

$$\prod_p{}' \mathbb{Q}_p = \{(x_p)_{p \text{ prime}},\ x_p \in \mathbb{Q}_p,\ x_p \in \mathbb{Z}_p \text{ for a.e. } p\}$$

(with respect to the sequence of compact subrings $(\mathbb{Z}_p)_{p\text{ prime}}$) and by $\mathbb{A} = \mathbb{R}\times\mathbb{A}_f$ the full ring of adèles; the field of rationals $\mathbb{Q}$ embed diagonally into $\mathbb{A}_f$ and $\mathbb{A}$, so these are in fact $\mathbb{Q}$-agebras. We also define the subring

$$\widehat{\mathbb{Z}} = \prod_p \mathbb{Z}_p \subset \mathbb{A}_f.$$

$\widehat{\mathbb{Z}}$ is the maximal compact subring of $\mathbb{A}_f$ (and the closure of $\mathbb{Z}$ in $\mathbb{A}_f$), where $\mathbb{A}_f$ has the adelic (i.e. restricted product) topology

Finally, given G, a $\mathbb{Q}$-algebraic group realized as a Zariski-closed subgroup of $\mathrm{GL}_{n,\mathbb{Q}}$ — we we denote by $\mathrm{G}(\mathbb{Q})$, $\mathrm{G}(\mathbb{A}_f)$, $\mathrm{G}(\mathbb{A})$, the groups of points of G in the corresponding rings; we also put $\mathrm{G}(\mathbb{Z}_p) = \mathrm{G}(\mathbb{Q}_p)\cap \mathrm{GL}_n(\mathbb{Z}_p)$ and $\mathbf{G}(\widehat{\mathbb{Z}}) = \prod_p \mathrm{G}(\mathbb{Z}_p)$.

The group of rational points $\mathrm{G}(\mathbb{Q})$ embeds diagonally into $\mathrm{G}(\mathbb{A}_f)$ and $\mathrm{G}(\mathbb{A})$ and in that way is considered a subgroup of these groups.

6.1 Genus

In the sequel, we will consider the restriction of this action on rational lattices to subgroups of $\mathrm{G}(\mathbb{A}_f) \subset \mathrm{GL}_n(\mathbb{A}_f)$, for G varying through certain $\mathbb{Q}$-algebraic subgroups of GL_n. We are particularly interested in the orthogonal group $\mathrm{G} = \mathrm{SO}_3 < \mathrm{GL}_3$ and some subgroups of it. This triggers the following

Definition 6.1 For $\mathrm{G} < \mathrm{GL}_n$ a $\mathbb{Q}$-algebraic subgroup and L a lattice, the G-*genus* of L is the orbit

$$\mathrm{genus}_{\mathrm{G}}(L) := \mathrm{G}(\mathbb{A}_f).L \subset \mathcal{L}_n(\mathbb{Q}).$$

The subgroup of rational points $\mathrm{G}(\mathbb{Q}) < \mathrm{G}(\mathbb{A}_f)$ obviously acts on $\mathrm{genus}_{\mathrm{G}}(L)$: the set of its orbits

$$[\mathrm{genus}_{\mathrm{G}}(L)] := \mathrm{G}(\mathbb{Q})\backslash\mathrm{genus}_{\mathrm{G}}(L)$$

is called the set of G-*genus classes* of L and its cardinality is the *class number of* G *with respect to* L (and for $L = \mathbb{Z}^n$, more simply, the set of genus classes and the the class number of G). In particular we have the identifications

$$\mathrm{genus}_{\mathrm{G}}(\mathbb{Z}^n) \simeq \mathrm{G}(\mathbb{A}_f)/\mathrm{G}(\widehat{\mathbb{Z}}),\ [\mathrm{genus}_{\mathrm{G}}(\mathbb{Z}^n)] := \mathrm{G}(\mathbb{Q})\backslash\mathrm{G}(\mathbb{A}_f)/\mathrm{G}(\widehat{\mathbb{Z}}).$$

We refer to [17] for a more complete introduction to the adelic theory of algebraic groups (in relation with automorphic forms), and to [1, 28] for more extensive treatments.

6.2 Summary

We begin by summarizing §7 to §9:

We set

$$\begin{aligned}\mathcal{P} &= \{(L,\mathbf{x}),\ L\in \mathrm{genus}_{\mathrm{SO}_3}(\mathbb{Z}^3),\ \mathbf{x}\in L,\quad \mathbf{x}.\mathbf{x}=d\}.\\ \mathcal{P}_{(q)} &:= \{(L,\overline{\mathbf{x}}),\ L\in \mathrm{genus}_{\mathrm{SO}_3}(\mathbb{Z}^3),\ \overline{\mathbf{x}}\in L/qL,\ \overline{\mathbf{x}}.\overline{\mathbf{x}}\equiv d(\mathrm{mod}\, q)\}.\end{aligned}$$

For the sequel, we need to fix *base points* $\mathbf{x}_0\in\mathscr{R}_3(d)$ and $\overline{\mathbf{x}}_q\in\mathscr{R}_3(d;q)$; this being done, we define

$$K_f[q] := \ker(\mathrm{SO}_3(\widehat{\mathbb{Z}})\to \mathrm{SO}_3(\mathbb{Z}/q\mathbb{Z})) \tag{6.1}$$

$$K'_f[q] := \{g\in \mathrm{SO}_3(\widehat{\mathbb{Z}}) : g.\overline{\mathbf{x}}_q=\overline{\mathbf{x}}_q\} \tag{6.2}$$

and denote by $\mathrm{SO}_{\mathbf{x}_0}$ the stabilizer of $\mathbf{x}_0$ in SO_3. We will show in the following sections how to describe $\widetilde{\mathscr{R}_3}(d)$ and $\widetilde{\mathscr{R}_3}(d;q)$ adelically, by means of a commutative diagram

$$\begin{array}{ccccc}\widetilde{\mathscr{R}_3}(d) & \xrightarrow{\sim} & \mathrm{SO}_3(\mathbb{Q})\backslash\mathcal{P} & \xleftarrow{\sim} & \mathrm{SO}_{\mathbf{x}_0}(\mathbb{Q})\backslash \mathrm{SO}_{\mathbf{x}_0}(\mathbb{A}_f)/\mathrm{SO}_{\mathbf{x}_0}(\widehat{\mathbb{Z}})\\ \downarrow{\scriptstyle \mathrm{red}_q} & & \downarrow{\scriptstyle \mathrm{red}_q} & & \downarrow\\ \widetilde{\mathscr{R}_3}(d;q) & \xrightarrow{\sim} & \mathrm{SO}_3(\mathbb{Q})\backslash\mathcal{P}_{(q)} & \xleftarrow{\sim} & \mathrm{SO}_3(\mathbb{Q})\backslash \mathrm{SO}_3(\mathbb{A}_f)/K'_f[q].\end{array} \tag{6.3}$$

We will also explain how to describe $\mathscr{R}_3(d)$ and $\mathscr{R}_3(d;q)$ adelically, which is slightly more involved technically (though no different conceptually).

For this we define

$$\begin{aligned}\mathcal{Q} &= \{(L,\mathbf{x},\theta),\ L\in \mathrm{genus}_{\mathrm{SO}_3}(\mathbb{Z}^3),\ \mathbf{x}\in L,\ \mathbf{x}.\mathbf{x}=d, \theta : L/3L\simeq(\mathbb{Z}/3\mathbb{Z})^3\}.\\ \mathcal{P}_{(3,q)} &:= \{(L,\overline{\mathbf{x}},\theta),\ L\in \mathrm{genus}_{\mathrm{SO}_3}(\mathbb{Z}^3),\ \overline{\mathbf{x}}\in L/qL,\ \overline{\mathbf{x}}.\overline{\mathbf{x}}\equiv d(\mathrm{mod}\, q),\\ &\qquad \theta : L/3L\simeq(\mathbb{Z}/3\mathbb{Z})^3\},\end{aligned}$$

and, for any integer a,

$$K_f[a,q] := K_f[a]\cap K'_f[q]\subset K'_f[q]. \tag{6.4}$$

Then we have the following diagram

$$\begin{array}{ccccc}\mathscr{R}_3(d) & \xrightarrow{\sim} & \mathrm{SO}_3(\mathbb{Q})\backslash\mathcal{Q} & \xleftarrow{\sim} & \mathrm{SO}_{\mathbf{x}_0}(\mathbb{Q})\backslash \mathrm{SO}_{\mathbf{x}_0}(\mathbb{A}_f)\times \mathrm{SO}_3(\widehat{\mathbb{Z}}/3\widehat{\mathbb{Z}})/\mathrm{SO}_{\mathbf{x}_0}(\widehat{\mathbb{Z}})\\ \downarrow{\scriptstyle \mathrm{red}_q} & & \downarrow{\scriptstyle \mathrm{red}_q} & & \downarrow\\ \mathscr{R}_3(d;q) & \xrightarrow{\sim} & \mathrm{SO}_3(\mathbb{Q})\backslash\mathcal{P}_{(3,q)} & \xleftarrow{\sim} & \mathrm{SO}_3(\mathbb{Q})\backslash \mathrm{SO}_3(\mathbb{A}_f)/K_f[3,q]\\ \downarrow & & \downarrow & & \downarrow\\ \widetilde{\mathscr{R}_3}(d;q) & \xrightarrow{\sim} & \mathrm{SO}_3(\mathbb{Q})\backslash\mathcal{P}_{(q)} & \xleftarrow{\sim} & \mathrm{SO}_3(\mathbb{Q})\backslash \mathrm{SO}_3(\mathbb{A}_f)/K'_f[q];\end{array} \tag{6.5}$$

where the vertical arrows at the bottom are the evident surjective maps. The horizontal arrows of these diagrams are described in §7 and §8, while the vertical arrows are discussed in §9.

6.3 Convention: orthogonal groups vs. unit group of quaternions

We have already noted in §4.1 that the quadratic spaces $(\mathbb{Q}^3, a^2+b^2+c^2)$, $(\mathrm{B}^{(0)}, \mathrm{Nr})$ are isometric and that the action — by conjugation — of $\mathrm{B}^\times$ on $\mathrm{B}^{(0)}$ induces an isomorphism of $\mathbb{Q}$-algebraic groups

$$\mathrm{PB}^\times \simeq \mathrm{SO}_3$$

with $\mathrm{PB}^\times = Z(\mathrm{B}^\times)\backslash\mathrm{B}^\times$ the projective group of units. In particular the group of adelic points $\mathrm{PB}^\times(\mathbb{A})$ and $\mathrm{SO}_3(\mathbb{A})$ get naturally identified, as are their various respective subgroups. In the sequel, we shall freely use this identification. Moreover we will use the same notations for various subgroups $K_f[q]$, $K'_f[q]$ etc. to denote either some subgroup in $\mathrm{SO}_3(\mathbb{A})$ or its image in $\mathrm{PB}^\times(\mathbb{A})$ under the above isomorphism.

7 Adelic Interpretation of $\mathscr{R}_3(d;q)$

Let $L_0 := \mathbb{Z}^3$. In this section we identify $\mathscr{R}_3(d;q)$ (as well as the sphere S_2), with an adelic quotient of SO_3, verifying the second lines of (6.3) and (6.5).

7.1 The SO_3-genus of $\mathbb{Z}^3$

An adelic consequence of the fact that the ring of Hurwitz quaternions is principal (cf. §4.2) is the following:

Proposition *$\mathbb{Z}^3$ has only one SO_3-genus class:*

$$\mathrm{SO}_3(\mathbb{A}_f) = \mathrm{SO}_3(\mathbb{Q})\mathrm{SO}_3(\widehat{\mathbb{Z}}). \tag{7.1}$$

Proof Let $L \in \mathrm{genus}_{\mathrm{SO}_3}(\mathbb{A}_f).L_0$; in particular the covolume of L equals the covolume of L_0, which is one. We identify L_0 with the traceless integral quaternions $\mathrm{B}^{(0)}(\mathbb{Z})$ and SO_3 with $\mathrm{PB}^\times$; in these terms, we need to show that there is $q \in \mathrm{B}^\times(\mathbb{Q})$ such that $qLq^{-1} = \mathrm{B}^{(0)}(\mathbb{Z})$. By definition, there is $q_f \in \mathrm{B}^\times(\mathbb{A}_f)$ such that $L = q_f\mathrm{B}^{(0)}(\mathbb{Z})q_f^{-1}$ and so $\mathscr{O} := \mathbb{Z} + L = q_f(\mathbb{Z}+\mathrm{B}^{(0)}(\mathbb{Z}))q_f^{-1}$ is a lattice in $\mathrm{B}(\mathbb{Q})$ containing the identity and stable by multiplication, i.e. an order. Hence, by §4.2, there is $q \in \mathrm{B}^\times(\mathbb{Q})$ such that $q\mathscr{O}q^{-1} \subset \mathrm{B}(\mathbb{Z})$ and so $qLq^{-1} \subset \mathrm{B}^{(0)}(\mathbb{Z})$; since qLq^{-1} and $\mathrm{B}^{(0)}(\mathbb{Z})$ have the same covolume they are equal.

$\square$

If we replace $\mathbb{Z}^3$ by another lattice or $x^2+y^2+z^2$ by a different ternary quadratic form Q, the number of classes in the genus will not, in general, be 1.

7.2 Level structure

By construction, the quadratic form $x^2+y^2+z^2$ takes integral values on any lattice $L \in \mathrm{genus}_{\mathrm{SO}_3}(L_0)$. In particular, given $q \geqslant 1$ an integer, the quotient lattice $L/qL \simeq (\mathbb{Z}/q\mathbb{Z})^3$ is naturally a quadratic space for the form $x^2+y^2+z^2$. *A* q*-level structure* on such a lattice is an additional datum related to L/qL. Here, we will consider two type of level q-structures:

- the principal q-structure: this is the datum of an isomorphism of quadratic spaces $\theta : L/qL \simeq (\mathbb{Z}/q\mathbb{Z})^3$. We will use this only for $q = 3$ and mainly for cosmetic purposes.
- a weak q-structure: this is the datum of a point $\overline{\mathbf{x}} \in L/qL$ such that $\overline{\mathbf{x}}.\overline{\mathbf{x}} \equiv d(\mathrm{mod}\, q)$ when $(d, q) = 1$. This will be the main structure considered in the present paper.

Related to these level structures are the open compact subgroups of $\mathrm{SO}_3(\widehat{\mathbb{Z}})$,

$$K_f[q],\ K'_f[q],\ K_f[3,q]$$

defined by (6.1), (6.2), (6.4) relative to the choice of some base point $\overline{\mathbf{x}}_q \in \mathscr{R}_3(d;q)$.

7.3 $\widetilde{\mathscr{R}_3}(d;q)$ as an adelic quotient

Let us consider first the set of pairs

$$\mathcal{P}_{(q)} := \{(L, \overline{\mathbf{x}}),\ L \in \mathrm{genus}_{\mathrm{SO}_3}(L_0),\ \overline{\mathbf{x}} \in L/qL,\ \overline{\mathbf{x}}.\overline{\mathbf{x}} \equiv d(\mathrm{mod}\, q)\}.$$

Since $L/qL \simeq (L \otimes_{\mathbb{Z}} \widehat{\mathbb{Z}})/q(L \otimes_{\mathbb{Z}} \widehat{\mathbb{Z}})$, the group $\mathrm{SO}_3(\mathbb{A}_f)$ (and its subgroup $\mathrm{SO}_3(\mathbb{Q})$) acts diagonally on $\mathcal{P}_{(q)}$.

We start with the action of $\mathrm{SO}_3(\mathbb{Q})$: by (7.1) any $\mathrm{SO}_3(\mathbb{Q})$-orbit in $\mathcal{P}_{(q)}$ contains a pair of the form $(L_0, \overline{\mathbf{x}})$ with $\overline{\mathbf{x}} \in \mathscr{R}_3(d;q)$. Moreover if two pairs $(L_0, \overline{\mathbf{x}})$ and $(L_0, \overline{\mathbf{x}}')$ give rise to the same $\mathrm{SO}_3(\mathbb{Q})$-orbit then $\overline{\mathbf{x}}$ and $\overline{\mathbf{x}}'$ differ by an element of $\mathrm{SO}_3(\mathbb{Q}) \cap \mathrm{SO}_3(\widehat{\mathbb{Z}}) = \mathrm{SO}_3(\mathbb{Z})$. It follows that the map $\overline{\mathbf{x}} \in \mathscr{R}_3(d;q) \mapsto (L_0, \overline{\mathbf{x}}) \in \mathcal{P}_{(q)}$ induces a bijection:

$$\mathrm{SO}_3(\mathbb{Z})\backslash\mathscr{R}_3(d;q) \xrightarrow{\sim} \mathrm{SO}_3(\mathbb{Q})\backslash\mathcal{P}_{(q)}.$$

In the sequel, we will denote by $[L, \overline{\mathbf{x}}]_q$ the $\mathrm{SO}_3(\mathbb{Q})$-orbit of the pair $(L, \overline{\mathbf{x}})$.

Regarding the action of $\mathrm{SO}_3(\mathbb{A}_f)$, one has

Lemma 1 *The group* $\mathrm{SO}_3(\mathbb{A}_f)$ *acts transitively: for any* $\overline{\mathbf{x}}_q \in \mathscr{R}_3(d;q)$, *we have*

$$\mathcal{P}_{(q)} = \mathrm{SO}_3(\mathbb{A}_f).(L_0, \overline{\mathbf{x}}_q) \simeq \mathrm{SO}_3(\mathbb{A}_f)/K'_f[q].$$

Proof Indeed any $(L, \overline{\mathbf{x}}) \in \mathcal{P}_{(q)}$ is in the orbit of a pair of the form $(L_0, \overline{\mathbf{x}}')$ for some $\overline{\mathbf{x}}' \in \mathscr{R}_3(d;q)$. It follows from Lemma 7.1 below that $\mathrm{SO}_3(\widehat{\mathbb{Z}})$ acts transitively on $\mathscr{R}_3(d;q)$ (through its projection to $\mathrm{SO}_3(\mathbb{Z}/q\mathbb{Z})$). Taking $k_{\mathbf{x}'} \in \mathrm{SO}_3(\widehat{\mathbb{Z}})$ such that $k_{\overline{\mathbf{x}}'}\overline{\mathbf{x}}_q = \overline{\mathbf{x}}'$, we have $k_{\overline{\mathbf{x}}'}^{-1}(L_0, \overline{\mathbf{x}}') = (L_0, \overline{\mathbf{x}}_q)$. □

Lemma 7.1 *For any prime* p, *the group* $\mathrm{SO}_3(\mathbb{Z}_p)$ *acts transitively on* $\mathscr{R}_3(d)(\mathbb{Z}_p)$ *(defined in the evident way). Consequently,* $\mathrm{SO}_3(\widehat{\mathbb{Z}})$ *acts transitively on* $\mathscr{R}_3(d)(\widehat{\mathbb{Z}})$.

Proof This — which can be thought of as an analogue of Witt's theorem over $\mathbb{Z}_p$ — is immediate from Proposition 13, using the first case if $p = 2$ and the second case if p is odd. □

We thus have $\mathcal{P}_{(q)} \simeq \mathrm{SO}_3(\mathbb{A}_f)/K'_f[q]$. From the above discussion, we deduce that

$$\widetilde{\mathscr{R}_3}(d;q) \xrightarrow{\sim} \mathrm{SO}_3(\mathbb{Q})\backslash\mathcal{P}_{(q)} \xleftarrow{\sim} \mathrm{SO}_3(\mathbb{Q})\backslash\mathrm{SO}_3(\mathbb{A}_f)/K'_f[q]. \quad (7.2)$$

7.3.1 Lifting to $\mathscr{R}_3(d;q)$

This adelic realization of $\widetilde{\mathscr{R}_3}(d;q)$ may, in fact, be lifted to an adelic realization of $\mathscr{R}_3(d;q)$ itself, at least when q is coprime with 3. For this we add the additional data of the principal level 3-structure discussed in §7.2. This is a little bit artificial, relying on the special fact that the reduction modulo 3 maps yields an isomorphism

$$\mathrm{SO}_3(\mathbb{Z}) \simeq \mathrm{SO}_3(\mathbb{Z}/3\mathbb{Z}).$$

Consider the set of triples

$$\mathcal{P}_{(3,q)} :=$$
$$\{(L, \overline{\mathbf{x}}, \theta),\ L \in \mathrm{genus}_{\mathrm{SO}_3}(L_0),\ \overline{\mathbf{x}} \in L/qL,\ \overline{\mathbf{x}}.\overline{\mathbf{x}} \equiv d(q),\ \theta : L/3L \simeq (\mathbb{Z}/3\mathbb{Z})^3\}.$$

As above, $\mathrm{SO}_3(\mathbb{A}_f)$ (hence its subgroup $\mathrm{SO}_3(\mathbb{Q})$) acts diagonally on $\mathcal{P}_{(3,q)}$: explicitly for $g \in \mathrm{SO}_3(\mathbb{A}_f)$, we have

$$g.(L, \overline{\mathbf{x}}, \theta) = (g.L, g.\overline{\mathbf{x}}, \theta \circ g^{-1}).$$

We consider first the $\mathrm{SO}_3(\mathbb{Q})$-orbits in $\mathcal{P}_{(3,q)}$. From (7.1) any $\mathrm{SO}_3(\mathbb{Q})$-orbit contains a triple of the form $(L_0, \overline{\mathbf{x}}, \theta)$. Moreover, since reduction modulo 3 is an isomorphism $\mathrm{SO}_3(\mathbb{Z}) \simeq \mathrm{SO}_3(\mathbb{Z}/3\mathbb{Z})$, the map $\overline{\mathbf{x}} \mapsto (\overline{\mathbf{x}}, \mathrm{Id})$ yields a bijection between $\mathscr{R}_3(d;q)$ and the set of $\mathrm{SO}_3(\mathbb{Z})$-orbits of pairs $\{(\overline{\mathbf{x}}, \theta),\ \overline{\mathbf{x}} \in \mathscr{R}_3(d;q),\ \theta : (\mathbb{Z}/3\mathbb{Z})^3 \simeq (\mathbb{Z}/3\mathbb{Z})^3\}$. From this, we deduce that the map $\overline{\mathbf{x}} \in \mathscr{R}_3(d;q) \mapsto (L_0, \overline{\mathbf{x}}, \mathrm{Id}) \in \mathcal{P}_{(3,q)}$ induces a bijection:

$$\mathscr{R}_3(d;q) \tilde{\to} \mathrm{SO}_3(\mathbb{Q}) \backslash \mathcal{P}_{(3,q)}.$$

Returning to the action of the whole group $\mathrm{SO}_3(\mathbb{A}_f)$, we have

Lemma 2 *This action is transitive: fixing $\overline{\mathbf{x}}_q \in \mathscr{R}_3(d;q)$, we have*

$$\mathcal{P}_{(3,q)} = \mathrm{SO}_3(\mathbb{A}_f).(L_0, \overline{\mathbf{x}}_q, \mathrm{Id}).$$

Proof The proof is exactly as above: any triple is in the orbit of a triple of the form $(L_0, \overline{\mathbf{x}}, \theta)$, $\overline{\mathbf{x}} \in \mathscr{R}_3(d;q)$, $\theta : (\mathbb{Z}/3\mathbb{Z})^3 \simeq (\mathbb{Z}/3\mathbb{Z})^3$. This follows from the fact that $\prod_{p|q} \mathrm{SO}_3(\mathbb{Z}_p) \times \mathrm{SO}_3(\mathbb{Z}_3)$ acts transitively on the set of pairs $(\overline{\mathbf{x}}, \theta)$; recall (§2.1) that we are assuming that q is prime to 3.

□

We have $\mathcal{P}_{(3,q)} \simeq \mathrm{SO}_3(\mathbb{A}_f)/K_f[3,q]$. (Recall the notation from §7.2). From this and the above discussion, it follows that the map

$$\overline{\mathbf{x}} \in \mathscr{R}_3(d;q) \mapsto (L_0, \overline{\mathbf{x}}, \mathrm{Id}) \in \mathcal{P}_{(3,q)}$$

induces a bijective map:

$$\mathscr{R}_3(d;q) \tilde{\to} \mathrm{SO}_3(\mathbb{Q}) \backslash \mathcal{P}_{(3,q)} \tilde{\leftarrow} \mathrm{SO}_3(\mathbb{Q}) \backslash \mathrm{SO}_3(\mathbb{A}_f)/K_f[3,q]. \tag{7.3}$$

To resume (7.3), the arrow going from the left to the right (7.3) is obtained as follows: for $\mathbf{x}'_q \in \widetilde{\mathscr{R}_3}(d;q)$, let $\kappa' \in \mathrm{SO}_3(\widehat{\mathbb{Z}})$ be such that $\kappa.\mathbf{x}'_q = \mathbf{x}_q$; the arrow is simply

$$\mathbf{x}'_q \to [\kappa]_{q,3}$$

where

$$[.]_{q,3} : g_f \in \mathrm{SO}_3(\mathbb{A}_f) \mapsto \mathrm{SO}_3(\mathbb{Q}) \backslash \mathrm{SO}_3(\mathbb{A}_f)/K'_f[q,3]$$

is the canonical projection.

7.4 The sphere as an adelic quotient

For completeness, we recall the interpretation of sphere S_2 as an adelic quotient; this will not be used here, but it is the way to proceed in order to adapt the proof of Theorems 1.2 and 1.3 to obtain Theorems 1.1 and 1.5 or to obtain hybrid equidistribution theorems. Let $\mathbb{A} = \mathbb{R} \times \mathbb{A}_f$ denote the full ring of adèles and $\mathrm{SO}_3(\mathbb{A}) = \mathrm{SO}_3(\mathbb{R}) \times \mathrm{SO}_3(\mathbb{A}_f)$; by (7.1), we have

$$\mathrm{SO}_3(\mathbb{Z})\backslash\mathrm{SO}_3(\mathbb{R}) \simeq \mathrm{SO}_3(\mathbb{Q})\backslash\mathrm{SO}_3(\mathbb{A})/\mathrm{SO}_3(\widehat{\mathbb{Z}}).$$

Since $\mathrm{SO}_3(\mathbb{R})$ acts transitively on S_2, we obtain — choosing some point $\mathbf{x}_\infty \in S_2$ and letting

$$K_\infty = \mathrm{SO}_{x_\infty}(\mathbb{R}) \simeq \mathrm{SO}_2(\mathbb{R})$$

— the identification

$$\mathrm{SO}_3(\mathbb{Z})\backslash S_2 \simeq \mathrm{SO}_3(\mathbb{Q})\backslash\mathrm{SO}_3(\mathbb{A})/K_\infty.\mathrm{SO}_3(\widehat{\mathbb{Z}}). \tag{7.4}$$

As in the previous section, we may remove the quotient by $\mathrm{SO}_3(\mathbb{Z})$, by adding the principal 3-structure. We obtain

$$\begin{aligned} \mathrm{SO}_3(\mathbb{R}) &\simeq \mathrm{SO}_3(\mathbb{Q})\backslash\mathrm{SO}_3(\mathbb{A})/K_f[3]. \\ S_2 &\simeq \mathrm{SO}_3(\mathbb{Q})\backslash\mathrm{SO}_3(\mathbb{A})/K_\infty.K_f[3]. \end{aligned} \tag{7.5}$$

8 Adelic Interpretation of $\mathscr{R}_3(d)$

In this section, we describe the first line of the diagrams (6.3) and (6.5).

This will be a key tool in the proof of Proposition 3 and Proposition 5. Our presentation follows that in [9], but we emphasize that the material is in essence classical.

8.1

As before, let $\mathcal{P}$ be the set of pairs

$$\mathcal{P} = \{(L, \mathbf{x}),\ \mathbf{x}.\mathbf{x} = d,\ \mathbf{x} \in L,\ L \in \mathrm{genus}_{\mathrm{SO}_3}(L_0)\}.$$

By contrast with $\mathcal{P}_{(q)}$ or $\mathcal{P}_{(3,q)}$, the set $\mathcal{P}$ carries no natural action of $\mathrm{SO}_3(\mathbb{A}_f)$; still, the group $\mathrm{SO}_3(\mathbb{Q})$ acts on $\mathcal{P}$ diagonally. We will denote by $[.]$ the projection $\mathcal{P} \to \mathrm{SO}_3(\mathbb{Q})\backslash\mathcal{P}$. By (7.1), every $\mathrm{SO}_3(\mathbb{Q})$-orbit in $\mathcal{P}$ contains an element of the form $(L_0, \mathbf{x})$ (for some $\mathbf{x} \in \mathscr{R}_3(d)$); two such pairs differ by the action of a unique element of $\mathrm{SO}_3(\mathbb{Z})$. From this is follows that the map $\mathbf{x} \in \mathscr{R}_3(d) \mapsto (L_0, \mathbf{x}) \in \mathcal{P}$ induces a bijective map

$$\widetilde{\mathscr{R}_3}(d) \xrightarrow{\sim} \mathrm{SO}_3(\mathbb{Q})\backslash\mathcal{P}. \tag{8.1}$$

We will now realize $\mathrm{SO}_3(\mathbb{Q})\backslash\mathcal{P}$ as an adelic quotient by realizing it as (the projection of) of an orbit of the adelic points of a *subgroup* of SO_3.

Choose an element $\mathbf{x}_0 \in \mathscr{R}_3(d)$, so $(L_0, \mathbf{x}_0) \in \mathcal{P}$ and let $\mathrm{SO}_{\mathbf{x}_0}$ be its stabilizer in SO_3. Since the quadratic space is 3-dimensional, $\mathrm{SO}_{\mathbf{x}_0}$ is the special orthogonal group of a quadratic plane (namely $\mathbb{Q}\mathbf{x}_0^{\perp}$), and is in fact a 1-dimensional $\mathbb{Q}$-torus.

The group $\mathrm{SO}_{\mathbf{x}_0}(\mathbb{A}_f)$ acts (by multiplication on the first coordinate) on the subset of $\mathcal{P}$ consisting of pairs of the form $(L, \mathbf{x}_0)$; in particular

$$(\mathrm{SO}_{\mathbf{x}_0}(\mathbb{A}_f).L_0, \mathbf{x}_0) \subset \mathcal{P}.$$

This subset is in fact rather big: we have

Proposition 11 *The map*

$$t \in \mathrm{SO}_{\mathbf{x}_0}(\mathbb{A}_f) \to [(t.L_0, \mathbf{x}_0)] \in \mathrm{SO}_3(\mathbb{Q})\backslash\mathcal{P}$$

is surjective and induces an identification

$$\mathrm{SO}_3(\mathbb{Q})\backslash\mathcal{P} = [\mathrm{SO}_{\mathbf{x}_0}(\mathbb{A}_f)(L_0, \mathbf{x}_0)] \simeq \mathrm{SO}_{\mathbf{x}_0}(\mathbb{Q})\backslash\mathrm{SO}_{\mathbf{x}_0}(\mathbb{A}_f)/\mathrm{SO}_{\mathbf{x}_0}(\widehat{\mathbb{Z}}).$$

Proof By Witt's theorem, every $\mathrm{SO}_3(\mathbb{Q})$-orbit in $\mathcal{P}$ is of the form $[L, \mathbf{x}_0]$, $L \in \mathrm{genus}_{\mathrm{SO}_3}(L_0)$ so it suffices to show that $L \in \mathrm{SO}_{\mathbf{x}_0}(\mathbb{A}_f).L_0$ Write $L = g.L_0$ with $g \in \mathrm{SO}_3(\mathbb{A}_f)$. Since $L \otimes \widehat{\mathbb{Z}}$ contains $\mathbf{x}_0$, $L_0 \otimes \widehat{\mathbb{Z}}$ contains $g^{-1}\mathbf{x}_0$; by Lemma 7.1 there is an element k of $\mathrm{SO}_3(\widehat{\mathbb{Z}})$ which sends $\mathbf{x}_0$ to $g^{-1}\mathbf{x}_0$. In particular, $gk = t$ lies in $\mathrm{SO}_{\mathbf{x}_0}(\mathbb{A}_f)$ and $t.L_0 = gk.L_0 = g.L_0 = L$. This show the first equality; The second bijection is induced by the map $t \in \mathrm{SO}_{\mathbf{x}_0}(\mathbb{A}_f) \mapsto (t.L_0, \mathbf{x}_0)$ and follows easily from the equality $\mathrm{SO}_{\mathbf{x}_0}(\mathbb{Q}) = \mathrm{SO}_3(\mathbb{Q}) \cap \mathrm{SO}_{\mathbf{x}_0}(\mathbb{A}_f)$ and the fact that the stabilizer of L_0 in $\mathrm{SO}_{\mathbf{x}_0}(\mathbb{A}_f)$ is $\mathrm{SO}_{\mathbf{x}_0}(\widehat{\mathbb{Z}}) = \mathrm{SO}_{\mathbf{x}_0}(\mathbb{A}_f) \cap \mathrm{SO}_3(\widehat{\mathbb{Z}})$.

□

Hence (8.1) extends to bijections

$$\widetilde{\mathscr{R}_3}(d) \tilde{\to} \mathrm{SO}_3(\mathbb{Q})\backslash\mathcal{P} \tilde{\leftarrow} \mathrm{SO}_{\mathbf{x}_0}(\mathbb{Q})\backslash\mathrm{SO}_{\mathbf{x}_0}(\mathbb{A}_f)/\mathrm{SO}_{\mathbf{x}_0}(\widehat{\mathbb{Z}}). \tag{8.2}$$

8.2 Relationship with the ideal class group

What we have done so far is plainly sufficient to prove the equidistribution of $\widetilde{\mathscr{R}_3}(d)$. However, in our case a bit more is available: since $\mathrm{SO}_{\mathbf{x}_0}(\mathbb{A}_f)$ is commutative (being an orthogonal group in two variables) it acts transitively by multiplication on the righthand side of (8.2), and so also on $\mathrm{SO}_3(\mathbb{Q})\backslash\mathcal{P} \simeq \widetilde{\mathscr{R}_3}(d)$.

On $\mathrm{SO}_3(\mathbb{Q})\backslash\mathcal{P}$, the action is given as follows: for $t \in \mathrm{SO}_{\mathbf{x}_0}(\mathbb{A}_f)$ and $[L, \mathbf{x}_0] \in \mathrm{SO}_3(\mathbb{Q})\backslash\mathcal{P}$,

$$t.[L, \mathbf{x}_0] := [t.L, \mathbf{x}_0].$$

This is well defined since if $[L, \mathbf{x}_0] = [L', \mathbf{x}_0]$ then L and L' differ by an element $\tau \in \mathrm{SO}_{\mathbf{x}_0}(\mathbb{Q})$ and because of the commutativity of $\mathrm{SO}_{\mathbf{x}_0}$,

$$[t\tau L, \mathbf{x}_0] = [\tau t L, \mathbf{x}_0] = [tL, \mathbf{x}_0].$$

Remark This action is very specific to the three dimensional case. For instance, if one studies the representations of an integer d by a quadratic form Q of higher rank, the quotient $\mathrm{SO}_Q(\mathbb{Z})\backslash\mathscr{R}_3(d)$, when non-empty, can always be realized as a finite disjoint union of *projections* of adelic orbits of $\mathrm{SO}_{\mathbf{x}_0}(\mathbb{A}_f)$; so $\mathrm{SO}_Q(\mathbb{Z})\backslash\mathscr{R}_3(d)$ has a description in terms of double cosets of an adelic group but it will not carry a natural *action* of $\mathrm{SO}_{\mathbf{x}_0}(\mathbb{A}_f)$.

We shall now identify $\mathrm{SO}_{\mathbf{x}_0}(\mathbb{Q})\backslash\mathrm{SO}_{\mathbf{x}_0}(\mathbb{A}_f)/\mathrm{SO}_{\mathbf{x}_0}(\widehat{\mathbb{Z}})$ with a quotient of the commutative group $\mathrm{Pic}(\mathscr{O}_K)$. Therefore, the identification (8.2) may be considered as giving an action of $\mathrm{Pic}(\mathscr{O}_K)$ on $\widetilde{\mathscr{R}_3}(d)$.

For this purpose, it is again most convenient to phrase everything in terms of quaternions via the identifications $\mathbb{Q}^3 \simeq \mathrm{B}^{(0)}(\mathbb{Q})$ and $\mathrm{SO}_3 \simeq \mathrm{PB}^\times$. In particular we view $\mathbf{x}_0$ as a trace-free quaternion. Now the stabilizer $\mathrm{SO}_{\mathbf{x}_0} \simeq \mathrm{PB}^\times{}_{\mathbf{x}_0}$ is the multiplicative group generated by (conjugations by) invertible quaternions of the form $a + b\mathbf{x}_0$. In the previous section, we have defined a transitive action of $\mathrm{SO}_{\mathbf{x}_0}(\mathbb{A}_f) \simeq \mathrm{PB}^\times{}_{\mathbf{x}_0}(\mathbb{A}_f)$ on $\widetilde{\mathscr{R}_3}(d)$.

Now, the map $a + b\sqrt{-d} \mapsto a + b\mathbf{x}_0$ defines, via conjugation, an isomorphism of $\mathbb{Q}$-algebraic groups

$$\mathrm{Res}_{K/\mathbb{Q}}\mathbf{G}_m/\mathbf{G}_m \simeq \mathrm{PB}^\times{}_{\mathbf{x}_0}; \tag{8.3}$$

in particular, for any $\mathbb{Q}$-algebra A, $\mathrm{PB}^\times{}_{\mathbf{x}_0}(A) \simeq (A \otimes_\mathbb{Q} K)^\times/A^\times$.

Thus, denoting by $\mathbb{A}^\times_{K,f} = (\mathbb{A}_f \otimes K)^\times$ the group of finite idèles of K — equivalently, the $\mathbb{A}_f$-valued points of $\mathrm{Res}_{K/\mathbb{Q}}$ — (8.3) defines an action of $\mathbb{A}^\times_{K,f}$ on $\widetilde{\mathscr{R}_3}(d)$. The subgroups $\mathbb{A}^\times_f = \mathbf{G}_m(\mathbb{A}_f)$ and $K^\times$ act trivially, as does

$$U = \prod_p U_p \subset \mathbb{A}^\times_{K,f},$$

the preimage of $\mathrm{PB}^\times{}_{\mathbf{x}_0}(\widehat{\mathbb{Z}})$ under (8.3), and we have

$$\widetilde{\mathscr{R}_3}(d) \simeq K^\times\backslash\mathbb{A}^\times_{K,f}/U.$$

Let

$$\widehat{\mathscr{O}}^\times_K = (\mathscr{O}_K \otimes_\mathbb{Z} \widehat{\mathbb{Z}})^\times = \prod_p \mathscr{O}^\times_{K,p} \subset \mathbb{A}^\times_{K,f};$$

clearly $\widehat{\mathscr{O}}_K^\times \subset U$ and more precisely, we have for $p \neq 2$ (since $\mathrm{PB}^\times(\mathbb{Z}_p) \simeq \mathrm{PGL}_2(\mathbb{Z}_p)$)

$$U_p = \mathbb{Q}_p^\times \mathscr{O}_{K,p}^\times, \text{ while for } p = 2,\ U_2 = \begin{cases} \mathbb{Q}_2^\times \mathscr{O}_{K,2}^\times, & \text{if } d \equiv 1, 2 \bmod 4 \\ \mathbb{Q}_2^\times \pi_2^{\mathbb{Z}} \mathscr{O}_{K,2}^\times, & \text{if } d \equiv 3 \bmod 4, \end{cases}$$

where in the latter case, $\pi_2 = a + \sqrt{-d}b \in \mathscr{O}_{K,2}$ is an uniformizer: let us recall that $\mathrm{PB}^\times(\mathbb{Z}_2)$ is the image of $\mathrm{B}(\mathbb{Z}_2)^\times \cup q_2\mathrm{B}(\mathbb{Z}_2)^\times$ in $\mathrm{PB}^\times(\mathbb{Q}_2)$ for $q_2 \in \mathrm{B}(\mathbb{Z}_2)$ any quaternion whose norm has 2-adic valuation 1.

To resume, we have

$$\mathbb{A}_f{}^\times \widehat{\mathscr{O}}_K^\times \subset U \tag{8.4}$$

and in particular

$$\mathrm{Pic}(\mathscr{O}_K) \simeq K^\times \backslash \mathbb{A}_{K,f}^\times / \mathbb{A}_f^\times \widehat{\mathscr{O}}_K^\times$$

acts transitively on $\widetilde{\mathscr{R}_3}(d)$. If $d \equiv 1, 2(4)$, (8.4) is an equality and $\widetilde{\mathscr{R}_3}(d)$ is a $\mathrm{Pic}(\mathscr{O}_K)$-torsor. On the other hand, when $d \equiv 3(4)$, the kernel has order 2 and is generated by (the image of) the element $\pi_2 \in \mathscr{O}_{K,2}$; reproducing what we did for $\widetilde{\mathscr{R}_3}(d)$, we find in that case, that $\widetilde{\mathscr{R}_3}(d)^+$ is naturally identified with $\mathrm{Pic}(\mathscr{O}_K)$ and is in fact a torsor.

A priori, the action we have defined depends on $\mathbf{x}_0$. We verify independence in §8.4.

8.3 Lifting the action to $\mathscr{R}_3(d)$

As before, by introducing an extra level 3-structure, we may replace $\widetilde{\mathscr{R}_3}(d)$ by its covering $\mathscr{R}_3(d)$ and obtain a lift of the action of $\mathrm{SO}_{\mathbf{x}_0}(\mathbb{A}_f)$ and thus of $\mathbb{A}_{K,f}^\times$ on $\widetilde{\mathscr{R}_3}(d)$ to an action on $\mathscr{R}_3(d)$. Notice that this latter action is, in general, not transitive; nor does it factor through the class group, but only through a certain ray class group.

Consider the set of triples

$$\mathscr{Q} = \{(L, \mathbf{x}, \theta),\ L \in \mathrm{genus}_{\mathrm{SO}_3}(L_0),\ \mathbf{x} \in L,\ \mathbf{x}.\mathbf{x} = d, \theta : L/3L \simeq (\mathbb{Z}/3\mathbb{Z})^3\}.$$

Using that the reduction mod 3 map from $\mathrm{SO}_3(\mathbb{Z})$ to $\mathrm{SO}_3(\mathbb{Z}/3\mathbb{Z})$ is an isomorphism, we may verify as above that the map

$$\mathscr{R}_3(d) \to \mathrm{SO}_3(\mathbb{Q}) \backslash \mathscr{Q}, \quad \mathbf{x} \mapsto [L_0, \mathbf{x}, \mathrm{Id}]_3$$

is bijective.

As above, the group $\mathrm{SO}_{\mathbf{x}_0}(\mathbb{A}_f)$ acts on the set of triples in $\mathscr{Q}$ of the form $(L, \mathbf{x}_0, \theta)$ by

$$t.(L, \mathbf{x}_0, \theta) = (t.L, \mathbf{x}_0, \theta \circ t^{-1}). \tag{8.5}$$

Proposition 12 *The set* $\mathrm{SO}_3(\mathbb{Q})\backslash\mathcal{Q}$ *decomposes as a disjoint union of subsets of the form* $[\mathrm{SO}_{\mathbf{x}_0}(\mathbb{A}_f).(L,\mathbf{x}_0,\theta)]_3$. *These subsets are parametrized by the orbits of* $\mathrm{SO}_3(\mathbb{Z}/3\mathbb{Z}) \simeq \mathrm{SO}_3(\widehat{\mathbb{Z}}/3\widehat{\mathbb{Z}})$ *under the action of* $\mathrm{SO}_{\mathbf{x}_0}(\widehat{\mathbb{Z}})$.

Proof By Witt's Theorem every $\mathrm{SO}_3(\mathbb{Q})$-orbit in $\mathcal{Q}$ is of the form $[L,\mathbf{x}_0,\theta]_3$ so $\mathrm{SO}_3(\mathbb{Q})\backslash\mathcal{Q}$ is covered by a union of subsets of the shape

$$[\mathrm{SO}_{\mathbf{x}_0}(\mathbb{A}_f).(L,\mathbf{x}_0,\theta)]_3.$$

This union is disjoint: if $(L',\mathbf{x}_0,\theta') \in [\mathrm{SO}_{\mathbf{x}_0}(\mathbb{A}_f).(L,\mathbf{x}_0,\theta)]_3$ then L' can be written $L' = s_{\mathbb{Q}}s_f L$ with $s_f \in \mathrm{SO}_{\mathbf{x}_0}(\mathbb{A}_f)$ and $s_{\mathbb{Q}} \in \mathrm{SO}_3(\mathbb{Q})$ such that $s_{\mathbb{Q}}\mathbf{x}_0 = \mathbf{x}_0$ so $s_{\mathbb{Q}} \in \mathrm{SO}_{\mathbf{x}_0}(\mathbb{Q})$. Finally the proof of Proposition 11 gives that these sets correspond bijectively with the set of orbits of $\mathrm{SO}_{\mathbf{x}_0}(\widehat{\mathbb{Z}})$-orbits of $\mathrm{SO}_3(\widehat{\mathbb{Z}}/3\widehat{\mathbb{Z}})$.

□

As above, by commutativity, the $\mathrm{SO}_{\mathbf{x}_0}(\mathbb{A}_f)$-action (8.5) descends to an action on $\mathrm{SO}_3(\mathbb{Q})\backslash\mathcal{Q}$ given by

$$t.[L,\mathbf{x}_0,\theta] = [t.L,\mathbf{x}_0,\theta\circ t^{-1}],$$

whose orbits corresponds to the $\mathrm{SO}_{\mathbf{x}_0}(\widehat{\mathbb{Z}})$-orbits of $\mathrm{SO}_3(\mathbb{Z}/3\mathbb{Z})$. In other words, we have a bijection

$$\mathscr{R}_3(d) \xrightarrow{\sim} \mathrm{SO}_3(\mathbb{Q})\backslash\mathcal{Q} \xleftarrow{\sim} \mathrm{SO}_{\mathbf{x}_0}(\mathbb{Q})\backslash\mathrm{SO}_{\mathbf{x}_0}(\mathbb{A}_f)\times \mathrm{SO}_3(\mathbb{Z}/3\mathbb{Z})/\mathrm{SO}_{\mathbf{x}_0}(\widehat{\mathbb{Z}}), \quad (8.6)$$

where $\mathrm{SO}_{\mathbf{x}_0}(\widehat{\mathbb{Z}})$ acts diagonally on the product $\mathrm{SO}_{\mathbf{x}_0}(\mathbb{A}_f) \times \mathrm{SO}_3(\mathbb{Z}/3\mathbb{Z})$. Under this identification, the action of $t \in \mathrm{SO}_{\mathbf{x}_0}(\mathbb{A}_f)$ is the one induced by

$$t.[t',\kappa] = [tt',\kappa],\ (t',\kappa) \in \mathrm{SO}_{\mathbf{x}_0}(\mathbb{A}_f)\times\mathrm{SO}_3(\mathbb{Z}/3\mathbb{Z}).$$

Thus (8.6) gives us the desired way to lift the action of $\mathbb{A}_{K,f}^\times$ to $\mathscr{R}_3(d)$.

8.4 Independence w.r.t. $\mathbf{x}_0$

Let us see that the above defined actions of $\mathbb{A}_{K,f}^\times$ on $\mathscr{R}_3(d)$, $\widetilde{\mathscr{R}_3}(d)$ do not depend on the choice of the base point $\mathbf{x}_0$.

Let $\mathbf{x}_0' \in \mathscr{R}_3(d)$ be another point. By Witt's theorem $\mathbf{x}_0' = \rho.\mathbf{x}_0$, for some $\rho \in \mathrm{PB}^\times(\mathbb{Q})$. Then $\mathrm{PB}^\times{}_{\mathbf{x}_0'} = \rho\mathrm{PB}^\times{}_{\mathbf{x}_0}\rho^{-1}$. Let $u = a + b\sqrt{-d}$ be an element of $\mathbb{A}_K^\times$ $(a,b \in \mathbb{A}_f)$, and let t_u (resp. t_u') denote the corresponding element in $\mathrm{PB}^\times{}_{\mathbf{x}_0}(\mathbb{A}_f)$ (resp. in $\mathrm{PB}^\times{}_{\mathbf{x}_0'}(\mathbb{A}_f)$): that is $a+b\mathbf{x}_0$ (resp. $a+b\mathbf{x}_0'$) modulo scalars. Then $t_u' = \rho t_u\rho^{-1}$. It will suffice to see that

$$t_u[L_0,\mathbf{x}_0,\mathrm{Id}]_3 = t_u'[L_0,\mathbf{x}_0,\mathrm{Id}]_3;$$

the latter expression equals

$$\begin{aligned} t'_u[L_0, \rho^{-1}\mathbf{x}'_0, \mathrm{Id}]_3 = t'_u[\rho L_0, \mathbf{x}'_0, \rho^{-1}]_3 &= [t'_u\rho L_0, \mathbf{x}'_0, \rho^{-1}t'^{-1}_u]_3 \\ &= [\rho^{-1}t'_u\rho L_0, \rho^{-1}\mathbf{x}'_0, \rho^{-1}t'^{-1}_u\rho]_3 \\ &= [t_u L_0, \mathbf{x}_0, t_u^{-1}]_3 = t_u[L_0, \mathbf{x}_0, \mathrm{Id}]_3. \end{aligned}$$

□

9 Adelic Interpretation of $\mathrm{red}_q : \mathscr{R}_3(d) \to \mathscr{R}_3(d;q)$

In this section, we interpret the reduction maps, red_q and red_∞ in terms of the adelic quotients from the previous sections.

Recall that, in (7.4), (7.2), (8.2) (resp. (7.5), (7.3), (8.6)) we have given adelic identifications of $\widetilde{S_2}$, $\widetilde{\mathscr{R}_3}(d;q)$ and $\widetilde{\mathscr{R}_3}(d)$, and of their finite coverings S_2, $\mathscr{R}_3(d;q)$ and $\mathscr{R}_3(d)$.

9.1

First, let us recall the identifications

$$\widetilde{\mathscr{R}_3}(d) \simeq \mathrm{SO}_3(\mathbb{Q})\backslash\mathcal{P},\ \widetilde{\mathscr{R}_3}(d;q) \simeq \mathrm{SO}_3(\mathbb{Q})\backslash\mathcal{P}_{(q)}$$

where

$$\mathrm{SO}_3(\mathbb{Q})\backslash\mathcal{P} = \{[L, \mathbf{x}], L \in \mathrm{SO}_3(\mathbb{A}_f).L_0,\ \mathbf{x} \in L,\ \mathbf{x}.\mathbf{x} = d\}$$

$$\mathrm{SO}_3(\mathbb{Q})\backslash\mathcal{P}_{(q)} = \{[L, \overline{\mathbf{x}}]_q,\ L \in \mathrm{SO}_3(\mathbb{A}_f).L_0,\ \overline{\mathbf{x}} \in L/qL,\ \overline{\mathbf{x}}.\overline{\mathbf{x}} \equiv d(q)\}.$$

The map $\mathrm{red}_q : \widetilde{\mathscr{R}_3}(d) \to \widetilde{\mathscr{R}_3}(d;q)$ is induced by the natural map

$$\mathbf{x} \in L \mapsto \overline{\mathbf{x}} \in L/qL$$

which we also denote red_q.

We now explain how to write red_q in the adelic language. Let t be an element of $\mathrm{SO}_{\mathbf{x}_0}(\mathbb{A}_f)$ and $[t]$ its double coset in $\mathrm{SO}_{\mathbf{x}_0}(\mathbb{Q})\backslash\mathrm{SO}_{\mathbf{x}_0}(\mathbb{A}_f)/\mathrm{SO}_{\mathbf{x}_0}(\widehat{\mathbb{Z}})$. Recall that $\mathbf{x}_0$ and $\overline{\mathbf{x}}_q$ were basepoints in $\widetilde{\mathscr{R}_3}(d)$ and $\widetilde{\mathscr{R}_3}(d;q)$ respectively. We will demonstrate that the reduction map red_q, thought of as a map

$$\mathrm{red}_q : \mathrm{SO}_{\mathbf{x}_0}(\mathbb{Q})\backslash\mathrm{SO}_{\mathbf{x}_0}(\mathbb{A}_f)/\mathrm{SO}_{\mathbf{x}_0}(\widehat{\mathbb{Z}}) \to \mathrm{SO}_3(\mathbb{Q})\backslash\mathrm{SO}_3(\mathbb{A}_f)/K_f[q]$$

is the one sending $[t]$ to $[t.k_{\overline{\mathbf{x}}_0}]_q$, where $k_{\overline{\mathbf{x}}_0} \in \mathrm{SO}_3(\widehat{\mathbb{Z}})$ is a fixed element satisfying
$k_{\overline{\mathbf{x}}_0}.\overline{\mathbf{x}}_q \equiv \mathbf{x}_0 (\mathrm{mod}\, q)$.

Remark Since q is coprime with 3 we may assume that the component of $k_{\overline{\mathbf{x}}_0}$ at 3 is trivial.

To see this, let $t \in \mathrm{SO}_{\mathbf{x}_0}(\mathbb{A}_f)$ be a representative for one of the double cosets in
$\mathrm{SO}_{\mathbf{x}_0}(\mathbb{Q})\backslash\mathrm{SO}_{\mathbf{x}_0}(\mathbb{A}_f)/\mathrm{SO}_{\mathbf{x}_0}(\widehat{\mathbb{Z}})$. We may choose t to have integral coordinates at all primes dividing q. Write $\beta = \gamma k$, with $\gamma \in \mathrm{SO}_3(\mathbb{Q})$ and $k \in \mathrm{SO}_3(\widehat{\mathbb{Z}})$. Then by the definitions of (8.2) and (7.3), one finds that the element of $\mathrm{SO}_3(\mathbb{Z})\backslash\mathscr{R}_3(d)$ corresponding to t is $\gamma^{-1}\mathbf{x}_0$, while the element of $\mathrm{SO}_3(\mathbb{Z})\backslash\mathscr{R}_3(d;q)$ corresponding to $tk_{\overline{\mathbf{x}}_0}$ is $k\bar{\mathbf{x}}_0$. But γk fixes $\mathbf{x}_0$ (whence also $\bar{\mathbf{x}}_0$), so the reduction of $\gamma^{-1}\mathbf{x}_0$ is precisely $k\bar{\mathbf{x}}_0$.

Again, we can lift the situation to $\mathscr{R}_3(d)$: there is a natural map

$$\mathrm{SO}_{\mathbf{x}_0}(\mathbb{Q})\backslash\mathrm{SO}_{\mathbf{x}_0}(\mathbb{A}_f)\times\mathrm{SO}(\mathbb{Z}/3\mathbb{Z})/\mathrm{SO}_{\mathbf{x}_0}(\widehat{\mathbb{Z}}) \mapsto \mathrm{SO}_3(\mathbb{Q})\backslash\mathrm{SO}_3(\mathbb{A}_f)/K_f[3,q]$$

which corresponds to the reduction map $\mathscr{R}_3(d) \to \mathscr{R}_3(d;q)$. Explicitly, choosing for base points the triples $(L_0, \mathbf{x}_0, \mathrm{Id})$, $(L_0, \overline{\mathbf{x}}_q, \mathrm{Id})$ the map is given by

$$[t,\kappa] \mapsto [tk_{\overline{\mathbf{x}}_0}k_3] \tag{9.1}$$

where $k_3 \in \mathrm{SO}_3(\mathbb{Z}_3)$ is a lift of $\kappa \in \mathrm{SO}_3(\mathbb{Z}/3\mathbb{Z})$ and $k_{\overline{\mathbf{x}}_0}$ is as above, but such that its component at the place 3 is trivial.

Remark Much in the same way, the map

$$\mathrm{red}_\infty : \mathscr{R}_3(d) \to S_2$$

corresponds to the map

$$[t,\kappa] \to [tk_{\mathbf{x}_0}k_3] \in \mathrm{SO}_3(\mathbb{Q})\backslash\mathrm{SO}_3(\mathbb{A})/K_\infty K_f[3]$$

with $k_{\mathbf{x}_0} \in \mathrm{SO}_3(\mathbb{R})$ such that $k_{\mathbf{x}_0}\mathbf{x}_\infty = \mathbf{x}_0/|d|^{1/2}$ and $k_3 \in \mathrm{SO}_3(\mathbb{Z}_3)$ is as above.

10 Some Local Analysis

This section is purely local. We obtain here an integral version of Witt's theorem for integral representation of squarefree integers by ternary quadratic forms over $\mathbb{Z}_p$. This is used to prove Lemma 7.1. Here it will prove convenient to work in terms of quaternions.

Let B be a quaternion algebra over $\mathbb{Q}_p$, let $\mathscr{O}$ be a maximal order of B and $\mathscr{O}^{(0)} \subset B^{(0)}$ be the lattice of trace zero elements in $\mathscr{O}$. As above, conjugation by element of $B^\times$ induces a surjective map

$$B^\times \twoheadrightarrow \mathrm{SO}(B^{(0)}, \mathrm{Nr})$$

under which $\mathcal{O}^\times$ maps to $\mathrm{SO}(\mathcal{O}^{(0)}, \mathrm{Nr})$. Let $\mathcal{O}^{(1)} \subset \mathcal{O}^\times$ be the group of norm-1 units. Let $d \in \mathbb{Z}_p$ be squarefree (that is, $\mathrm{ord}_p(d) \leqslant 1$) and let $\mathcal{O}^{(0,d)} \subset \mathcal{O}^{(0)}$ be the set of elements of $\mathcal{O}$ with trace 0 and norm d; $\mathcal{O}^{(0,d)}$ is obviously stable under the action of $\mathcal{O}^\times$ by conjugation.

We fix an element x of $\mathcal{O}^{(0,d)}$ and, as above, write $\Lambda_{x\to y}$ (or, when no confusion is likely, just Λ) for the set of $\lambda \in \mathcal{O}$ such that

$$x\lambda = \lambda y. \tag{10.1}$$

The solutions to (10.1) in B form a vector space of dimension 2, so Λ is a free $\mathbb{Z}_p$-module of rank 2.

Proposition 13 *The action of $\mathcal{O}^{(1)}$ and $\mathcal{O}^\times$ on $\mathcal{O}^{(0,d)}$ can be described as follows.*

- *Suppose B is a division algebra. Then there are two orbits of $\mathcal{O}^{(1)}$ on $\mathcal{O}^{(0,d)}$, which are interchanged by conjugation by any element of $\mathcal{O}^\times$ whose norm is not in $\mathrm{Nr}(\mathbb{Q}_p[x])$. In particular, the special orthogonal group of $\mathcal{O}^{(0)}$ acts transitively on $\mathcal{O}^{(0,d)}$.*
- *Suppose $B = M_2(\mathbb{Q}_p)$. Then:*
  - *If $p \neq 2$ and p does not divide d, the action of $\mathcal{O}^{(1)}$ on $\mathcal{O}^{(0,d)}$ is transitive.*
  - *Otherwise, there are two orbits of $\mathcal{O}^{(1)}$ on $\mathcal{O}^{(0,d)}$; they are interchanged by $\mathcal{O}^\times$, unless $p = 2$ and $d \equiv 3 \bmod 4$.*

Proof Suppose B is a division algebra. Then $\mathcal{O}$ is the unique maximal order, and consists of all elements whose norm lies in $\mathbb{Z}_p$. Let λ be a nonzero element of $\Lambda_{x\to y}$; then $y = \lambda^{-1}x\lambda$ so conjugation by λ is an isometry of $\mathcal{O}^{(0,d)}$ relating x to y. This show that $\mathcal{O}^\times$, hence $\mathrm{SO}(\mathcal{O}^{(0)}; \mathrm{Nr})$ acts transitively on $\mathcal{O}^{(0,d)}$.

Consider λ as above; if there is element $\alpha \in \mathbb{Q}_p[x]$ with $\mathrm{Nr}(\alpha) = \mathrm{Nr}(\lambda)$, then $\alpha^{-1}\lambda$ has norm 1 and to Λ; conversely, any element of Λ of norm 1 is $\alpha^{-1}\lambda$ for some $\alpha \in \mathbb{Q}_p[x]$ whose norm agrees with that of λ. This shows that the orbits of $\mathcal{O}^{(1)}$ on $\mathcal{O}^{(0,d)}$ are naturally identified with $\mathbb{Q}_p^\times/\mathrm{Nr}(\mathbb{Q}_p[x]^\times)$. This quotient is a group of order 2.

Now suppose that $B = M_2(\mathbb{Q}_p)$, so that we can take $\mathcal{O}^\times = \mathrm{GL}_2(\mathbb{Z}_p)$ and $\mathcal{O}^{(1)} = \mathrm{SL}_2(\mathbb{Z}_p)$.

Let $x = \begin{bmatrix} b & a \\ c & -b \end{bmatrix}$ be an element of $\mathcal{O}^{(0,d)}$ (so that $b^2 + ac = -d$) and

$y = \begin{bmatrix} 0 & 1 \\ -d & 0 \end{bmatrix}$; then

$$\Lambda_{x \to y} = \mathbb{Z}_p \begin{bmatrix} b & 1 \\ c & 0 \end{bmatrix} + \mathbb{Z}_p \begin{bmatrix} a & 0 \\ -b & 1 \end{bmatrix}$$

and the elements of $\mathrm{Nr}(\Lambda_{x\to y})$ are those elements of $\mathbb{Z}_p$ represented by the quadratic form $Q = aX^2 + 2bXY - cY^2$, which has discriminant $-4d$. Thus, x and y are in the same orbit of $\mathscr{O}^{(1)}$ if and only if Q represents 1 over $\mathbb{Z}_p$.

For all facts used below about isomorphism classes of binary quadratic forms over $\mathbb{Z}_p$, see [16, §31].

First, suppose p is odd. If p does not divide d, then Q is equivalent to $X^2 + dY^2$, and in particular represents 1. So in this case, $\mathscr{O}^{(1)}$ acts transitively on $\mathscr{O}^{(0,d)}$. If p divides d, then Q is equivalent to either X^2+dY^2 or $\varepsilon X^2 + \varepsilon^{-1}dY^2$, where $\varepsilon \in \mathbb{Z}_p^\times$ is a nonsquare. In the former case, y is in the orbit of x; in the latter case, Q does not represent 1 and

$$y' = \begin{bmatrix} 0 & \varepsilon \\ -\varepsilon^{-1}d & 0 \end{bmatrix}$$

is in the orbit of x. So there are two orbits, as claimed. In both cases, Q represents an element of $\mathbb{Z}_p^\times$, so $\mathscr{O}^\times$ acts transitively on $\mathscr{O}^{(0,d)}$.

Now take $p = 2$. In this case, there are always exactly two equivalence classes of binary forms of discriminant $4d$, one of which represents 1 over $\mathbb{Z}_2$ and the other of which does not. A representative of the non-representing forms is given by:

- $2X^2 + 2XY + (1/2)(d+1)Y^2$ $\qquad (d = 3 \mod 4)$
- $\varepsilon X^2 + \varepsilon^{-1}dY^2$ $\qquad (d = 1, 2 \mod 4)$

where, in the latter case, $\varepsilon \in \mathbb{Z}_2^\times$ is an element which is not a norm from $\mathbb{Q}_2(\sqrt{-d})^\times$. In case $d = 1$ or $2 \mod 4$, we again see that either y or $y' = \begin{bmatrix} 0 & \varepsilon \\ -\varepsilon^{-1}d & 0 \end{bmatrix}$ lies in the orbit of x, so there are two orbits of $\mathscr{O}^{(1)}$ on $\mathscr{O}^{(0,d)}$. And again Q represents an element of $\mathbb{Z}_2^\times$, so some element of $\mathscr{O}^\times$ sends x to y.

In case $d = 3 \mod 4$, take

$$y' = \begin{bmatrix} 1 & -2 \\ (1/2)(d+1) & -1 \end{bmatrix}$$

A direct computation shows that the orbit of x under $\mathscr{O}^{(1)}$ contains either y or y', so again there are two orbits. In this case, the two orbits are not interchanged by $\mathscr{O}^\times$.

□

Part IV

Graphs and Expanders

11 The Graph Structure on $\mathscr{R}_3(d;q)$

Let $p > 3$ be a prime, with $(p,q) = 1$. We shall now replace the role of the finite adèles $\mathbb{A}_f$ in the bijection (7.3) by the much smaller ring $\mathbb{Q}_p$. More precisely, we will show the existence of a bijection:

$$\mathscr{R}_3(d;q) \xleftarrow{\sim} \Gamma_{(3,q)} \backslash \mathrm{SO}_3(\mathbb{Q}_p)/K_p, \tag{11.1}$$

where $\Gamma_{(3,q)} \subset \mathrm{SO}_3(\mathbb{Q}_p)$ is a suitable lattice (i.e. a discrete cofinite subgroup) and K_p is the maximal compact subgroup $\mathrm{SO}_3(\mathbb{Z}_p)$. We have

$$\mathrm{SO}_3(\mathbb{Q}_p) \simeq \mathrm{PB}^\times(\mathbb{Q}_p) \simeq \mathrm{PGL}_2(\mathbb{Q}_p)$$

(since $\mathrm{B}(\mathbb{Q}_p) \simeq M_2(\mathbb{Q}_p)$). Therefore the quotient $\mathrm{SO}_3(\mathbb{Q}_p)/K_p$ is identified with

$$\mathrm{PGL}_2(\mathbb{Q}_p)/\mathrm{PGL}_2(\mathbb{Z}_p) =: \mathcal{T}_p$$

which has the structure of an infinite $p+1$-valent tree (namely, the *Bruhat-Tits* tree of $\mathrm{PGL}_2(\mathbb{Q}_p)$, cf. [31]).

The set $\mathscr{R}_3(d;q)$ has thus a structure of a finite quotient of $\mathcal{T}_p$ and we will see that this graph structure coincides with that described in §2.7. In particular, the latter is connected.

From this viewpoint, it will be possible to prove Proposition 3 and Proposition 5 (i.e. the graph $\mathscr{R}_3(d;q)$ is an expander). The latter relies on the theory of automorphic forms, especially the Jacquet-Langlands correspondence and the work of Eichler-Shimura.

Let us mention that a good part of this section is closely related to the book of Lubotzky [19], especially Chapter 6 and the Appendix, in which the reader will find a motivated discussion of the passage between automorphic forms and expanders.

11.1 $\mathscr{R}_3(d;q)$ as a quotient of a tree

If w is an element of $\mathrm{PB}^\times(\mathbb{Q}_p)$, we denote by $[w]_p$ the element of $\mathrm{PB}^\times(\mathbb{A}_f)$ which projects to w at the p-adic place, and to the identity everywhere else. When no confusion is likely (as in the statement of the following lemma) we will identify $\mathrm{PB}^\times(\mathbb{Q}_p)$ with the subgroup $[\mathrm{PB}^\times(\mathbb{Q}_p)]_p$ of $\mathrm{PB}^\times(\mathbb{A}_f)$.

Lemma 3 *One has*

$$\mathrm{PB}^\times(\mathbb{Q}).\mathrm{PB}^\times(\mathbb{Q}_p).K_f[3,q] = \mathrm{PB}^\times(\mathbb{A}_f).$$

Consequently, the map $g \in \mathrm{PB}^\times(\mathbb{Q}_p) \to [g]_p \in \mathrm{PB}^\times(\mathbb{A}_f)$ *yields a bijective map*

$$\Gamma_{(3,q)}\backslash \mathrm{PB}^\times(\mathbb{Q}_p)/K_p \tilde{\to} \mathrm{PB}^\times(\mathbb{Q})\backslash \mathrm{PB}^\times(\mathbb{A}_f)/K_f[3,q] \tilde{\leftarrow} \mathscr{R}_3(d;q) \quad (11.2)$$

where $\Gamma_{(3,q)}$ *is the lattice* $\mathrm{PB}^\times(\mathbb{Q}) \cap \mathrm{PB}^\times(\mathbb{Q}_p).K_f[3,q]$.

Proof This main ingredient of the proof is the so-called *strong approximation property* (for simply connected semi-simple algebraic groups): we will not discuss this property in any generality and refer to [28, Chap. 7] for a complete treatment. Alternatively, the reader may also refer to [37, Chap. III] for a discussion of the strong approximation property in the context of quaternion algebras. For now, let us merely say that, if $\mathrm{PB}^\times$ satisfied the strong approximation property the assertion would follow immediately; unfortunately, $\mathrm{PB}^\times$ is not simply connected.

One remedies this problem by passing from the $\mathbb{Q}$-algebraic group $\mathrm{PB}^\times$ to its double cover $\mathrm{B}^{(1)}$ (cf. (4.2)). The group $\mathrm{B}^{(1)}$ is simply connected and semisimple. Moreover, $\mathrm{B}^{(1)}(\mathbb{Q}_p)$ is noncompact. Thus, it satisfies a strong approximation property: for any open compact $\Omega \subset \mathrm{B}^{(1)}(\mathbb{A}_f)$, we have

$$\mathrm{B}^{(1)}(\mathbb{Q}).\mathrm{B}^{(1)}(\mathbb{Q}_p).\Omega = \mathrm{B}^{(1)}(\mathbb{A}_f).$$

It follows that $\mathrm{PB}^\times(\mathbb{Q}).\mathrm{PB}^\times(\mathbb{Q}_p).K_f[3,q]$ contains the image

$$\Theta = \mathrm{B}^{(1)}(\mathbb{A}_f)/\{\pm 1\}$$

of $\mathrm{B}^{(1)}(\mathbb{A}_f)$ in $\mathrm{PB}^\times(\mathbb{A}_f)$. It will suffice, then, to verify that

$$\mathrm{PB}^\times(\mathbb{Q}).\Theta.K_f[3,q] = \mathrm{PB}^\times(\mathbb{A}_f); \quad (11.3)$$

this will follow from (7.1) if $(\Theta \cap \mathrm{PB}^\times(\widehat{\mathbb{Z}})).K_f[3,q] = \mathrm{PB}^\times(\widehat{\mathbb{Z}})$; equivalently, if $\Theta \cap \mathrm{PB}^\times(\widehat{\mathbb{Z}})$ acts transitively on $\mathrm{PB}^\times(\widehat{\mathbb{Z}})/K_f[3,q] \simeq \mathscr{R}_3(d;q)$.

In turn, it is equivalent to show that $\Theta_{p'} \cap \mathrm{PB}^\times(\mathbb{Z}_{p'})$ acts transitively on $\mathscr{R}_3(d)(\mathbb{Z}_{p'})$ for each $p'|3q$, where $\Theta_{p'}$ is the image of $\mathrm{B}^{(1)}(\mathbb{Q}_{p'})$ in $\mathrm{PB}^\times(\mathbb{Q}_{p'})$.

Recall that $\mathscr{R}_3(d)(\mathbb{Z}_{p'})$ is identified with $\mathrm{B}^{(0,d)}(\mathbb{Z}_{p'})$, the trace 0 quaternions of norm $d \in \mathbb{Z}_{p'}^\times$; then $\Theta_{p'} \cap \mathrm{PB}^\times(\mathbb{Z}_{p'})$ contains $\mathrm{B}^{(1)}(\mathbb{Z}_{p'})/\{\pm 1\}$ acting by conjugation. The transitivity of this action follows from the second part of Proposition 13, using the fact that $(p', 2d) = 1$.

□

11.2 The graph structure

We can describe the (Bruhat-Tits) graph structure on

$$\mathrm{SO}_3(\mathbb{Q}_p)/\mathrm{SO}_3(\mathbb{Z}_p) \simeq \mathrm{PB}^\times(\mathbb{Q}_p)/\mathrm{PB}^\times(\mathbb{Z}_p)$$

in at least two equivalent ways. We write $K_p = \mathrm{PB}^\times(\mathbb{Z}_p)$ in what follows.

11.2.1

Recall the definition of $\mathcal{A}_p$ from §4.3.

For any fixed $\gamma_p \in \mathcal{A}_p$, let $[\gamma_p]_p \in \mathrm{PB}^\times(\mathbb{Q}_p)$ be the corresponding element considered in the p-adic group (alternatively take any element associated with a quaternion in $\mathrm{B}(\mathbb{Z}_p)$ such that the p-adic valuation of its norm is 1); then

$$K_p[\gamma_p]_p K_p = \bigsqcup_{\gamma \in \mathcal{A}_p} [\gamma]_p K_p. \tag{11.4}$$

For the graph structure, we join any coset gK_p to the $p+1$ cosets $g[\gamma]_p K_p$: $\gamma \in \mathcal{A}_p$. The resulting structure is independent of γ_p.

11.2.2

More intrinsically, we may identify the quotient $\mathrm{SO}_3(\mathbb{Q}_p)/\mathrm{SO}_3(\mathbb{Z}_p)$ with the sublattices of $\mathbb{Q}_p^3$ in the orbit of $\mathrm{SO}_3(\mathbb{Q}_p).\mathbb{Z}_p^3$. Given any such sublattice L, the quadratic form takes values in $\mathbb{Z}_p$ on that lattice and the induced quadratic form on $L/pL \cong (\mathbb{Z}_p/p\mathbb{Z}_p)^3$ takes values in $\mathbb{Z}_p/p\mathbb{Z}_p = \mathbf{F}_p$. Since p is odd, this later quadratic form is isotropic (represent 0 non-trivially over $\mathbf{F}_p$): there are precisely

$$|{\mathbf{F}_p}^\times \backslash \{(x, y, z) \in {\mathbf{F}_p}^3,\ x^2 + y^2 + z^2 = 0\}| = p + 1 (= |\mathbb{P}^1(\mathbf{F}_p)|)$$

isotropic lines. Given such a line, we choose $\overline{\mathbf{v}}$ a vector generating it and $\mathbf{v} \in L$ a lifting of it. For any such isotropic vector $\overline{\mathbf{v}}$ we construct the new lattice:

$$L_{\overline{\mathbf{v}}} = \langle \mathbf{v}/p \rangle + \{\mathbf{z} \in L : \mathbf{v}.\mathbf{z} \equiv 0 \bmod p\}.$$

Then $L_{\overline{\mathbf{v}}}$ depends only on $\overline{\mathbf{v}}$, and belongs to $\mathrm{SO}_3(\mathbb{Q}_p).L = \mathrm{SO}_3(\mathbb{Q}_p).\mathbb{Z}_p^3$ also. In particular, we construct $p+1$ such $L_{\overline{\mathbf{v}}}$, which we declare to be the neighbours of L.

This defines a graph structure on $\mathrm{SO}_3(\mathbb{Q}_p)/\mathrm{SO}_3(\mathbb{Z}_p)$ which is equivalent to the previous one: if $L = g\mathbb{Z}_p^3$

$$\{L_{\overline{\mathbf{v}}},\ \overline{\mathbf{v}} \text{ isotropic}\} = \{g[\gamma]_p\mathbb{Z}_p^3,\ \gamma \in \mathcal{A}_p\},$$

moreover

$$\text{The resulting graph is a tree, i.e., is connected and has no cycles.} \quad (11.5)$$

This later property is established using the following simple facts:

1. $L \cap L_{\overline{\mathbf{v}}} = \{\mathbf{z} \in L : \mathbf{v}.\mathbf{z} \equiv 0 \bmod p\}$ is the preimage under the projection $L \mapsto L/pL$ of the plane $\overline{\mathbf{v}}^{\perp}$. In particular this is an index p sublattice of L and of $L_{\overline{\mathbf{v}}}$.
2. More precisely, for any $\overline{\mathbf{v}}, \overline{\mathbf{v}}'$ generating distinct isotropic lines, let $\overline{\mathbf{w}}$ be a generator the (non-isotropic) line $\langle \overline{\mathbf{v}}, \overline{\mathbf{v}}' \rangle^{\perp}$ and let $\mathbf{w}$ be a lifting of $\overline{\mathbf{w}}$, then
$$L = \mathbb{Z}_p\mathbf{v} + \mathbb{Z}_p\mathbf{v}' + \mathbb{Z}_p\mathbf{w}, \; L_{\overline{\mathbf{v}}} = \mathbb{Z}_p\mathbf{v}/p + \mathbb{Z}_p p\mathbf{v}' + \mathbb{Z}_p\mathbf{w},$$
$$L \cap L_{\overline{\mathbf{v}}} = \mathbb{Z}_p\mathbf{v} + \mathbb{Z}_p p\mathbf{v}' + \mathbb{Z}_p\mathbf{w}.$$
3. In particular for $\overline{\mathbf{v}} \neq \overline{\mathbf{v}}'$
$$L_{\overline{\mathbf{v}}} \cap L_{\overline{\mathbf{v}}'} = \{\mathbf{z} \in L : \mathbf{v}.\mathbf{z} \equiv \mathbf{v}'.\mathbf{z} \equiv 0 \bmod p\} = \mathbb{Z}_p\mathbf{w} + pL$$
is the preimage in L of the line $\langle \overline{\mathbf{v}}, \overline{\mathbf{v}}' \rangle^{\perp}$.
4. Hence given three isotropic vectors $\overline{\mathbf{v}}, \overline{\mathbf{v}}', \overline{\mathbf{v}}''$ generating distinct lines
$$L_{\overline{\mathbf{v}}} \cap L_{\overline{\mathbf{v}}'} \cap L_{\overline{\mathbf{v}}''} = pL.$$

The distance on that tree can be computed via Fact (1) above: the distance between $L, L' \in \mathrm{SO}_3(\mathbb{Q}_p).\mathbb{Z}_p^3$ is given by

$$d(L, L') = p\text{-adic valuation of the index } [L : L \cap L'] = [L' : L \cap L'].$$

Using this one establishes the following generalization of Fact (3):

Proposition 14 *Given $n \geqslant 1$, let L, L', L'' be three lattices such that $d(L, L') = d(L, L'') = n$, $d(L', L'') = 2n$. Then there exists* $\mathbf{v}$ *some isotropic (mod p^n) vector $\mathbf{v} \in L - pL$ (ie. $\mathbf{v}.\mathbf{v} \equiv 0(p^n)$), and $\mathbf{w} \in L$ with $\mathbf{w}.\mathbf{w} \not\equiv 0(p)$ and $\mathbf{v}.\mathbf{w} \equiv 0(p^n)$ so that*

$$L \cap L' = \{\mathbf{x} \in L, \; \mathbf{x}.\mathbf{v} \equiv 0(p^n)\} = \mathbb{Z}_p\mathbf{v} + \mathbb{Z}_p\mathbf{w} + p^n L$$

and

$$L' \cap L'' = \mathbb{Z}_p\mathbf{w} + p^n L$$

is the preimage of a non-isotropic line. under the projection $L \mapsto L/p^n L$.

Using these remarks one can also establish the following

Proposition 15 *Let* $\mathbf{x} \in L$ *be such that* $\mathbf{x}.\mathbf{x}$ *is a square* $\not\equiv 0(p)$. *Then there are exactly two distinct neighbors* $L_{\overline{\mathbf{v}}}$, $L_{\overline{\mathbf{v}}'}$ *of* L *containing* $\mathbf{x}$. *These are the intersection of the orbit* $\mathrm{SO}_{\mathbf{x}}(\mathbb{Q}_p).L$ *with the neighbors of* L *and more generally*

$$\text{The orbit } \mathrm{SO}_{\mathbf{x}}(\mathbb{Q}_p).L \text{ is an infinite geodesic in the tree,} \tag{11.6}$$

(i.e. an isometric embedding of $\mathbb{Z}$ *in the tree w.r.t. the obvious metrics).*

Proof The fact that there are "at most two" neighboring lattices containing $\mathbf{x}$ follows from Fact (4). Now $\mathrm{SO}_{\mathbf{x}}(\mathbb{Q}_p)$ is the set of rotations associated to the invertibles quaternions in the quadratic algebra $\mathbb{Q}_p[\mathbf{x}] = \mathbb{Q}_p + \mathbb{Q}_p\mathbf{x}$. This quadratic algebra is split (ie. $\simeq \mathbb{Q}_p \times \mathbb{Q}_p$) since d is assumed to be square and in particular admits an element, $q_{\mathbf{x}} \in \mathbb{Z}_p[\mathbf{x}]$ whose norm has p-adic valuation 1. Let $t_{\mathbf{x}} \in \mathrm{SO}_{\mathbf{x}}(\mathbb{Q}_p)$ be the associated rotation; then $t_{\mathbf{x}}L$ and $t_{\mathbf{x}}^{-1}L$ are the two neighboring lattices of L containing $\mathbf{x}$. One has $\mathbb{Q}_p[\mathbf{x}]^\times = q_{\mathbf{x}}^{\mathbb{Z}}\mathbb{Z}_p[\mathbf{x}]^\times$ whence

$$\mathrm{SO}_{\mathbf{x}}(\mathbb{Q}_p) = t_{\mathbf{x}}^{\mathbb{Z}}\mathrm{SO}_{\mathbf{x}}(L)$$

($\mathrm{SO}_{\mathbf{x}}(L)$ the stabilizer of L in $\mathrm{SO}_{\mathbf{x}}(\mathbb{Q}_p)$). Therefore,

$$\mathrm{SO}_{\mathbf{x}}(\mathbb{Q}_p).L = t_{\mathbf{x}}^{\mathbb{Z}}.L$$

is an infinite geodesic in $\mathrm{SO}_3(\mathbb{Q}_p).L$.

□

Remark 11.1 If d is not a square in $\mathbb{Z}_p^\times$, then $\mathbb{Q}_p[\mathbf{x}]$ is a quadratic field with uniformizer p, $\mathbb{Q}_p[\mathbf{x}]^\times = p^{\mathbb{Z}}\mathbb{Z}_p[\mathbf{x}]^\times$ and $\mathrm{SO}_{\mathbf{x}}(\mathbb{Q}_p) \subset \mathrm{SO}_3(L)$.

Remark 11.2 There is a further approach to study the tree structure on

$$\mathrm{SO}_3(\mathbb{Q}_p)/\mathrm{SO}_3(\mathbb{Z}_p) \simeq \mathrm{PGL}_2(\mathbb{Q}_p)/\mathrm{PGL}_2(\mathbb{Z}_p):$$

where we interpret this quotient as $[\mathcal{L}_2(\mathbb{Q}_p)]$ the space of homothety classes of (rank 2) lattices in $\mathbb{Q}_p^2$ endowed with the natural transitive action of $\mathrm{PGL}_2(\mathbb{Q}_p)$; this is the viewpoint taken in [31].

11.2.3

Let us now check that the neighbors of $\overline{\mathbf{x}} \in \mathscr{R}_3(d;q)$ for the graph structure defined in §2.7 correspond to the neighbors under T_p of the image of $\overline{\mathbf{x}}$ in (7.3). Thus, the graph structure on $\mathscr{R}_3(d;q)$ coincides with the graph structure on $\Gamma_{(3,q)}\backslash\mathcal{T}_p$.

In the notations of §7.3.1, the neighbors of $\overline{\mathbf{x}}$ i.e. $\{\gamma.\overline{\mathbf{x}},\ \gamma \in \mathcal{A}_p\}$ correspond to the orbits of the triples

$$\{[L_0, \gamma.\overline{\mathbf{x}}, \mathrm{Id}],\ \gamma \in \mathcal{A}_p\} = \{[\gamma^{-1}L_0, \overline{\mathbf{x}}, \mathrm{Id}\circ\gamma],\ \gamma \in \mathcal{A}_p\};$$

now for any $p' \neq p$ and $\gamma \in \mathcal{A}_p$, $[\gamma]_{p'} \in K_{p'}$. Moreover, $[\gamma]_3 \equiv \mathrm{Id}(\mathrm{mod}\, 3)$. Therefore the above set equals $\{[\gamma^{-1}]_p.[L_0, \overline{\mathbf{x}}, \mathrm{Id}],\ \gamma \in \mathcal{A}_p\}$ which manifestly agrees with the graph structure introduced above.

11.3 The action of $[\mathfrak{p}]^{\mathbb{Z}}$

We focus now on the action of a prime ideal $\mathfrak{p}$ above p on $\widetilde{\mathscr{R}_3}(d)$. Fix π a uniformizer of $K_{\mathfrak{p}}$, which we write in the form $a+b\sqrt{-d} \in \mathbb{Z}_p[\sqrt{-d}]$. To any $\mathbf{x} \in \mathscr{R}_3(d)$ we associate the quaternion $q_{\mathbf{x}} = a + b\mathbf{x}$ (the p-adic valuation of $\mathrm{Nr}(q_{\mathbf{x}})$ equals 1) and the corresponding rotation $t_{\mathbf{x}} \in \mathrm{SO}_{\mathbf{x}}(\mathbb{Q}_p)$ induced by conjugation by $q_{\mathbf{x}}$; since $K_{\mathfrak{p}}^{\times} = \pi^{\mathbb{Z}}\mathscr{O}_{K_{\mathfrak{p}}}^{\times}$ we have $\mathrm{SO}_{\mathbf{x}}(\mathbb{Q}_p) = t_{\mathbf{x}}^{\mathbb{Z}}\mathrm{SO}_{\mathbf{x}}(\mathbb{Z}_p)$.

The action of $\mathfrak{p}$ can be interpreted in terms of our adelic viewpoint: We have seen in §8.3, that the group $\mathrm{SO}_{\mathbf{x}}(\mathbb{Q}_p)$ acts on $\mathscr{R}_3(d)$ with $\mathrm{SO}_{\mathbf{x}}(\mathbb{Z}_p)$ acting trivially; by projection this also defines an action on $\widetilde{\mathscr{R}_3}(d)$. Via the map $\pi \mapsto q_{\mathbf{x}} \mapsto t_{\mathbf{x}}$ one obtains an action of the group $\mathbb{Z} \simeq \pi^{\mathbb{Z}}$ which in fact does not depend on the choice of $\mathbf{x}$ (§8.4).

Let us also recall that if $\mathbf{x} \in \mathscr{R}_3(d)$ corresponds to the class $[L_0, \mathbf{x}, \mathrm{Id}] \in \mathrm{SO}_3(\mathbb{Q})\backslash\mathcal{Q}$, the element $\pi.\mathbf{x} \in \mathscr{R}_3(d)$ corresponds to the class $t_{\mathbf{x}}[L_0, \mathbf{x}, \mathrm{Id}] = [t_{\mathbf{x}}L_0, \mathbf{x}, \mathrm{Id}\circ t_{\mathbf{x}}^{-1}] = [t_{\mathbf{x}}L_0, \mathbf{x}, \mathrm{Id}]$, the final equality holding because the 3-component of $t_{\mathbf{x}}$ is trivial. Therefore the trajectory $\pi^{\mathbb{Z}}.\mathbf{x}$ is described by the infinite sequence of lattices

$$\ldots, L_{-2}, L_{-1}, L_0, L_1, L_2, \ldots,\quad L_i = t_{\mathbf{x}}^i L_0.$$

All the L_i contain $\mathbf{x}$. Write $L_{i,p'}$ for $L_{i,p'} = L_i \otimes_{\mathbb{Z}} \mathbb{Z}_{p'}$; then $L_{i,p'} = L_{0,p'}$ for all i and all $p' \neq p$, the p'-th component of $t_{\mathbf{x}}$ being trivial. So the sequence of lattices (L_i) is completely determined by the sequence $(L_{i,p})$ of its p-adic components.

Since the rotation $t_{\mathbf{x}}$ comes from a quaternion whose norm has p-adic valuation equal to 1, $t_{\mathbf{x}}^{\pm}L_{0,p} = L_{\pm1,p}$ are two neighbors of $L_{0,p}$ in the tree. Therefore, they are of the form $\gamma^{\pm}L_{0,p}$ for $\gamma^{\pm} \in \mathcal{A}_p$ two distinct elements; in particular the two elements $\mathbf{x}_{\pm} := \gamma^{\mp}\mathbf{x}$ belong to $L_{0,p}$ and to all other

$L_{0,p'}$ as well, so belong to $\mathscr{R}_3(d)$. In addition by proposition 15, $\gamma^{\pm}$ are the only two elements of $\mathcal{A}_p$ such that $\mathbf{x} \in \gamma L_{0,p}$. So, for $\gamma' \in \mathcal{A}_p - \{\gamma^{\pm}\}$, $\gamma'^{-1}\mathbf{x}$ is not p-integral; this proves Proposition 2.

The sequence $(L_{i,p})_{i \in \mathbb{Z}}$ describes an infinite geodesic passing through $L_{0,p}$ in the tree (cf. 11.2 (1) or more generally (11.6)),

$$\mathrm{SO}_3(\mathbb{Q}_p).L_{0,p} \simeq \mathrm{SO}_3(\mathbb{Q}_p)/\mathrm{SO}_3(\mathbb{Z}_p).$$

Up to orientation, any such geodesic may be encoded by an infinite non-backtracking word in $\mathcal{A}_p$ where the i-th letter connects the $(i-1)$-st element of the geodesic along an edge of the tree to the i-th element. In the present case, the word associated with the sequence $(L_{i,p})_{i \in \mathbb{Z}}$ is the word $(w_i^{-1})_{i \in \mathbb{Z}}$ in the alphabet $\mathcal{A}_p$ satisfying

$$L_{i,p} = g_{i-1} w_i^{-1} L_{0,p}, \text{ for } g_{i-1} \text{ such that } L_{i-1,p} = g_{i-1} L_{0,p};$$

equivalently, $(w_i)_{i \in \mathbb{Z}}$ is the word corresponding to the trajectory of $\mathbf{x}$ defined in §2.4.

11.4 Proof of Proposition 3

Suppose that two points $\mathbf{x}, \mathbf{x}' \in \mathscr{R}_3(d)$ give rise to the same truncated word of length 2ℓ:

$$W^{(\ell)} : [-\ell + 1, \ell] \to \mathcal{A}_p.$$

This means exactly that the geodesics $\mathrm{SO}_{\mathbf{x}}(\mathbb{Q}_p).L_{0,p}$ and $\mathrm{SO}_{\mathbf{x}'}(\mathbb{Q}_p).L_{0,p}$ in the Bruhat-Tits tree coincide "from times $-\ell$ to ℓ " or in other terms

$$L_{i,p} = L'_{i,p}, \; i = -\ell, \ldots, \ell.$$

In particular $\mathbf{x}' \in L_{\ell,p} \cap L_{-\ell,p}$. The last intersection is (Prop. 14) a sublattice of $L_{0,p}$ of index $p^{2\ell}$; more precisely, it is the preimage in $L_{0,p}$ of the line generated by $\mathbf{x} (\mathrm{mod}\, p^{\ell})$ in $L_{0,p}/p^{\ell} L_{0,p}$. Thus $\mathbf{x}$ and $\mathbf{x}'$ are linearly dependent in $L_0/p^{\ell} L_0$, and since both vectors have norm d, we have $\mathbf{x} \equiv \pm\mathbf{x}'$ modulo p^{ℓ}. This concludes the proof of Proposition 3.

11.5 Proof of Proposition 7

We suppose only that the *first* letters coincide

$$w_{\mathbf{x},1} = w_{\mathbf{x}',1}$$

which means precisely that $\mathbf{x}, \mathbf{x}'$ are both contained in (Prop. 14)

$$L_{0,p} \cap L_{1,p} = \{\mathbf{z} \in \mathbb{Z}_p^3, \mathbf{z}.\overline{\mathbf{v}} \equiv 0(p)\},$$

for $\overline{\mathbf{v}}$ a non-zero isotropic vector in $\mathbb{Z}_p^3/p\mathbb{Z}_p^3$. Since $\mathbf{x}.\mathbf{x} = d \neq 0(p)$, $\{\overline{\mathbf{v}}, \overline{\mathbf{x}}\}$ form a basis of $\overline{\mathbf{v}}^{\perp}$ and we can write $\overline{\mathbf{x}}' = \alpha\overline{\mathbf{x}} + \beta\overline{\mathbf{v}}$ with $\alpha \neq 0$, we have

$$\mathbf{x}.\mathbf{x}' \equiv \alpha\mathbf{x}.\mathbf{x} \equiv \alpha d(p).$$

By symmetry

$$\mathbf{x}.\mathbf{x}' \equiv \alpha^{-1}\mathbf{x}'.\mathbf{x}' \equiv \alpha^{-1}d(p) \Rightarrow \alpha = \pm 1,\ \mathbf{x}.\mathbf{x}' \equiv \pm d(p).$$

11.6 Proof of Proposition 5

We assume some familiarity with the theory of automorphic forms; in any case, we refer to Lubotzky's book [19, Chap. 6 & Appendix].

The space $L^2(\mathrm{PB}^\times(\mathbb{Q})\backslash \mathrm{PB}^\times(\mathbb{A}_f)/K_f[3,q])$ admits an orthogonal decomposition into eigenspaces of the commutative algebra generated by Hecke operators. These eigenspaces are the set of $\mathrm{PB}^\times(\mathbb{R})K_f[3,q]$-invariant vectors of automorphic representations on $\mathrm{PB}^\times$. Such representations are of two types:

- one-dimensional representations;
- infinite dimensional representations.

The latter corresponds, via the Jacquet-Langlands correspondence [15] to automorphic representations of GL_2 with trivial central character, which are discrete series of weight 2, unramified outside 2, 3 and outside primes dividing q. (More precisely, their conductor divides $18.q^2$). From the work of Deligne (or rather Eichler/Igusa/Shimura since this is weight 2), the eigenvalue of the standard p-th Hecke operator for such spaces is bounded in absolute value by $2\sqrt{p}$.

As for the former: each such is the representation of $\mathrm{PB}^\times(\mathbb{A})$ on the one-dimensional subspace generated by the function

$$g \in \mathrm{PB}^\times(\mathbb{Q})\backslash \mathrm{PB}^\times(\mathbb{A}) \mapsto \chi(\mathrm{Nr}(g))$$

where $\chi : \mathbb{Q}^\times\backslash\mathbb{A}^\times \mapsto \{\pm 1\}$ is some quadratic character. The action of $\mathrm{PB}^\times(\mathbb{Q})$ on such a representation is trivial, as is the action of $K_f[3,q]$ (by definition); moreover since the elements of Θ come from quaternions of norm 1, the action of Θ is trivial as well; hence from (11.3), such representation has to be the trivial one. It follows from this analysis that $-(p+1)$ does not occur as an eigenvalue of the p-th Hecke operator so the graph, while connected (cf. above) is not bipartite.

Remark The discussion above is also valid for $\widetilde{\mathscr{R}_3}(d;q)$: this follows immediately from the previous discussions by projection. In particular we have

$$\Gamma_{(q)}\backslash \mathrm{PB}^\times(\mathbb{Q}_p)/K_p \xrightarrow{\sim} \mathrm{PB}^\times(\mathbb{Q})\backslash \mathrm{PB}^\times(\mathbb{A}_f)/K'_f[q] \xleftarrow{\sim} \widetilde{\mathscr{R}_3}(d;q)$$

with $\Gamma_{(q)} = \mathrm{PB}^\times(\mathbb{Q}) \cap K'_f[q]\mathrm{PB}^\times(\mathbb{Q}_p)$.

12 Expander Graphs and Random Walks

The contents of this section follow lecture notes of Hoori, Linial and Wigderson [12]. Our goal is to prove Proposition 6.

12.1

Let $\mathscr{G} = (V, E)$ be a (possibly directed) d-regular graph on $|V| = n$ vertices, i.e. the number of incoming edges to each vertex is d, and the number of outgoing edges is also d. We assume $d > 2$. The normalized adjacency matrix T of $\mathscr{G}$ acts on $L^2(V)$ by

$$Tf(x) = \frac{1}{d} \sum_{(x \mapsto y) \in E} f(y)$$

and defines a self-adjoint operator. By an abuse of notation, we will use $L^2(\mathscr{G})$ and $L^2(V)$ interchangeably. More generally, $\mathscr{G}$ may be allowed to have multiple edges and loops, in which case we modify the definition of T in the evident way.

Let $\|T\|$ be the operator norm of T acting on the orthogonal complement of the constants in $L^2(V)$. The graph $\mathscr{G}$ is said to be an α-expander, for some $\alpha < 1$, if $\|T\| \leqslant 1 - \alpha$. In rough terms, the smaller $\|T\|$ is, the more "strongly connected" the graph $\mathscr{G}$.

When we speak of a "random walk on $\mathscr{G}$," we mean that we select a vertex uniformly and randomly from V, and then proceed to walk along directed edges, at each stage choosing one of the adjacent edges one with each choice assigned probability $1/d$.

Lemma 12.1 (Equidistribution of random walks) *Let $x \in V$ be some given point and let Q be a subset of V with density $\mu = |Q|/n$. Then the probability that a random walk on $\mathscr{G}$, originating from x, is in Q at step ℓ equals*

$$\mu + O(\|T\|^\ell \sqrt{\mu |V|}),$$

the implied constant being absolute.

Proof Let 1_x be the Dirac function at x and χ_Q be the characteristic function of Q. Then the probability equals

$$\langle T^\ell 1_x, \chi_Q \rangle = \mu + \langle 1_x, T^\ell(\chi_Q - \mu)\rangle$$

and we conclude using $\|T^\ell(\chi_Q - \mu)\| \leqslant \|T\|^\ell \|\chi_Q - \mu\|$.

□

Lemma 12.2 *Let $Q_1, \ldots, Q_\ell$ be subsets of V, with densities $\mu_i := |Q_i|/n$. The probability that a random walk on $\mathscr{G}$ is in Q_j at step j, for all $1 \leqslant j \leqslant \ell$, is at most*

$$\prod_{i=1}^{\ell-1} \left(\sqrt{\mu_i \mu_{i+1}} + \|T\|\right).$$

Proof Let χ_{Q_i} be the characteristic function of Q_i, and A_i be the endomorphism of $L^2(V)$ defined by $f \mapsto \chi_{Q_i} f$. Let Π denote the projection onto the constants, so that $T\Pi = \Pi$. The endomorphism $A_i.T.A_{i+1}$ may be decomposed:

$$A_i T A_{i+1} = A_i.\Pi.A_{i+1} + A_i T(1 - \Pi)A_{i+1}.$$

The endomorphism $A_i.\Pi.A_{i+1}$ may be written as $f \mapsto \frac{\chi_{Q_i}}{|V|}\langle f, \chi_{Q_{i+1}}\rangle$, and thus has operator norm $\mu_i^{1/2}\mu_{i+1}^{1/2}$. Since the operator norm of $A_i.T(1 - \Pi).A_{i+1}$ is at most $\|T\|$, we conclude that the operator norm of $A_i T A_{i+1}$ is at most $\|T\| + (\mu_i \mu_{i+1})^{1/2}$. The probability that a random walk visits Q_i at step i for every $i \in \{1, \ldots, \ell\}$ is given by

$$|V|^{-1}\langle 1_V, (A_1 T A_2)(A_2 T A_3) \ldots (A_{\ell-1} T A_\ell) 1_V\rangle$$

and the result follows.

□

Lemma 12.3 *Let Q be a subset of V, with $\mu := |Q|/n$, and let γ be a random walk of length ℓ. Then the probability that $|\gamma \cap Q| \geqslant (\mu + \epsilon)\ell$ is at most*

$$c_1 \exp(-c_2 \ell)$$

for positive constants c_1, c_2 depending only on $d, \|T\|, \mu, \epsilon$.

In other words, the number of "bad" walks of length ℓ, with respect to some fixed notion of "bad", decays exponentially with ℓ.

Proof The constants $C_1, C_2, \ldots$ appearing in the proof are all understood to be positive constants depending only on $d, \|T\|, \mu, \epsilon$.

Let S be a subset of $\{1..\ell\}$ of size k. It follows from Lemma 12.2 that the probability that $\gamma_i \in Q$ for $i \in S$ and $\gamma_i \notin Q$ for $i \notin S$ is at most

$$C_1 \mu^k (1-\mu)^{\ell-k} \left(1 + \frac{\|T\|}{\min(\mu, 1-\mu)}\right)^{\ell}.$$

Summing over all choices of S we have that the probability that $|\gamma \cap Q| = k$ is at most

$$C_1 \binom{\ell}{k} \mu^k (1-\mu)^{\ell-k} \left(1 + \frac{\|T\|}{\min(\mu, 1-\mu)}\right)^{\ell}$$

On the other hand, the sum

$$\sum_{k \geqslant (\mu+\epsilon)\ell} \binom{\ell}{k} \mu^k (1-\mu)^{\ell-k}$$

is at most $\exp(-C_2 \ell)$. Thus, we are done if $\|T\|$ is small enough that $(1 + \frac{\|T\|}{\min(\mu, 1-\mu)}) < e^{C_2}$.

If this is not the case, we fix an integer $C_3 \geqslant 1$, and replace the graph $\mathscr{G}$ by the graph $\mathscr{G}^{(C_3)}$ for which a directed edge from x to y corresponds to a directed path of length C_3 in the graph $\mathscr{G}$. (Note that $\mathscr{G}^{(C_3)}$ may have multiple edges and loops.) This improves the spectral gap: if $T^{(C_3)}$ is the normalized adjacency matrix of $\mathscr{G}^{(C_3)}$, we have $T^{(C_3)} = T^{C_3}$. Accordingly, $\|T^{(C_3)}\| = \|T\|^{C_3}$. Choosing C_3 large enough, the argument above shows that the probability that a random walk of length ℓ on the graph $\mathscr{G}^{(C_3)}$ remains within Q for at least $(\mu + \epsilon)\ell$ steps is bounded above by $C_4 \exp(-C_5 \ell)$.

It follows immediately that the probability that a random walk on $\mathscr{G}$ of length $C_3 \ell$ spends time $\geqslant (\mu + \epsilon) C_3 \ell$ inside Q is at most $C_4 \exp(-C_5 \ell)$.

This proves the desired claim.

□

12.2 The arc graph

Lemma 12.3, which tells us that an exponentially small proportion of random walks are poorly distributed in $\mathscr{G}$, will not quite suffice for our purposes; what we need to know is that an exponentially small proportion of *non-backtracking* walks are poorly distributed in $\mathscr{G}$. In this section we explain how to derive such a statement from Lemma 12.3.

We now assume $\mathscr{G}$ to be *symmetric*; that is, E is closed under reversal. For each edge $a \in E$, write a^+ for the target of a and a^- for the source of a. We denote the reversal of a by $\bar{a}$. With these notations, define the arc graph $\mathscr{G}'$ to be a directed graph whose vertices are the directed edges, or arcs, of $\mathscr{G}$. There is an edge from the arc a to the arc b exactly when $a^+ = b^-$ and $a \neq \bar{b}$.

Thus, $\mathscr{G}'$ is regular of degree $d-1$. We denote by T' the normalized adjacency matrix of $\mathscr{G}'$:

$$T'F(a) = \frac{1}{d-1} \sum_{\substack{b^- = a^+ \\ b \neq \bar{a}}} F(b)$$

The key feature of the arc graph $\mathscr{G}'$, for us, is that we have a natural bijection between non-backtracking paths of length ℓ on $\mathscr{G}$, and paths of length $\ell - 1$ on $\mathscr{G}'$.

12.3

Our goal will be to deduce a spectral gap for T' from that for T. This is a simple analogue of Atkin-Lehner theory in the subject of modular forms: when $d = p+1$ for some prime p, we can think of $\mathscr{G}$ as a quotient of the Bruhat-Tits tree attached to $\mathrm{PGL}_2(\mathbb{Q}_p)/\mathrm{PGL}_2(\mathbb{Z}_p)$; then the arc graph of the Bruhat-Tits tree is obtained by replacing $\mathrm{PGL}_2(\mathbb{Z}_p)$ with an Iwahori subgroup, so that the passage from graph to arc graph is much the same as the passage from the congruence subgroup $\Gamma_0(N)$ to the smaller subgroup $\Gamma_0(Np)$.

There are natural maps $B, E : L^2(\mathscr{G}) \to L^2(\mathscr{G}')$ ("beginning" and "end") defined via

$$Bf(a) = f(a^-), \ Ef(a) = f(a^+).$$

Moreover, the orthogonal complement to $\mathrm{im}(B) \oplus \mathrm{im}(E)$ consists of those functions $F \in L^2(\mathscr{G}')$ with the property that

$$\sum_{a^- = v} F(a) = \sum_{a^+ = v} F(a) = 0,$$

for all $v \in V\mathscr{G}$. On this orthogonal complement (the "new space") , the operator T' acts via

$$F \mapsto -\frac{1}{d-1}\bar{F}, \tag{12.1}$$

where $\bar{F}(a) = F(\bar{a})$. Moreover, one checks that

$$\langle Bf_1, Ef_2 \rangle = d\langle Tf_1, f_2 \rangle, \ \ T' \circ B = E.$$

Thus, if $w \in L^2(\mathscr{G})$ is an eigenfunction for T with eigenvalue λ, then the "old space" $\mathbb{C}(Bw) + \mathbb{C}(Ew)$ is stable under T'. From this we see that every eigenvalue of T' on this space is also an eigenvalue of the matrix :

$$\begin{pmatrix} 0 & 1 \\ \frac{-1}{(d-1)} & \frac{d\lambda}{(d-1)} \end{pmatrix}$$

It is easily computed that the eigenvalues are bounded away from 1 if λ is. By (12.1), the eigenvalues of T' on the new space are bounded in absolute value by $1/(d-1) < 1$. We conclude that $\|T'\|$ is bounded away from 1 if $\|T\|$ is.

Proposition 16 *Let $\mathscr{G} = (V, E)$ be an undirected graph with $\|T\| < 1$, and let Q be a subset of V, with $\mu := |Q|/n$. The probability that a random walk* without backtracking *of length ℓ spends more than $(\mu + \epsilon)\ell$ time in Q is at most*

$$c_1 \exp(-c_2 \ell)$$

for constants $c_1, c_2 > 0$ depending only on $d, \|T\|, \mu, \epsilon$.

Proof Let $Q' \subset V\mathscr{G}'$ be the subset of arcs whose initial vertex lies inside Q. Noting that $\frac{|Q'|}{|\mathscr{G}'|} = \frac{|Q|}{|V|}$, we apply Lemma 12.3 to $(\mathscr{G}', Q')$ taking into account that $\|T'\|$ is bounded away from 1 in terms of $\|T\|$.

□

References

[1] A. Borel and G. Mostow, Edts., *Proceed. Symposia Pure Math. Vol. IX: Algebraic groups and discontinuous subgroups*, American Mathematical Society, Providence, R.I., 1966.

[2] J. Bourgain, Z. Rudnick and P. Sarnak, *Local statistics of lattice points on the sphere*, arXiv:1204.0134, 2012.

[3] W. Duke, *Hyperbolic distribution problems and half-integral weight Maass forms*, Invent. Math. **92**, (1988), no. 1 73–90.

[4] W. Duke and R. Schulze-Pillot, *Representation of integers by positive ternary quadratic forms and equidistribution of lattice points on ellipsoids*, Invent. Math. **99** (1990), no. 1 49–57.

[5] P. Deligne, *La conjecture de Weil. I*, Inst. Hautes Études Sci. Publ. Math. **43** (1974), 273–307.

[6] M. Einsiedler, E. Lindenstrauss, P. Michel and A. Venkatesh, *Distribution of periodic torus orbits on homogeneous spaces III: Duke's theorem for cubic fields*, Ann. of Math. **173** no. 2 (2011), 815–885.

[7] M. Einsiedler, E. Lindenstrauss, P. Michel and A. Venkatesh, *The distribution of periodic torus orbits on homogeneous spaces*, Duke Math. J. **148** no. 1, (2009), 119–174.

[8] M. Einsiedler, E. Lindenstrauss, P. Michel and A. Venkatesh, *Distribution of periodic torus orbits on homogeneous spaces II: Duke's theorem for quadratic fields*, to appear in L'Ens. Math., `http://arxiv.org/abs/1109.0413`.

[9] J.S. Ellenberg and A. Venkatesh, *Local-global principles for representations of quadratic forms*, Invent. Math. **171** no. 2 (2008), 257–279.

[10] O.M. Fomenko and E.P. Golubeva, *Asymptotic distribution of lattice points on the three-dimensional sphere*, (Russian) Zap. Nauchn. Sem. Leningrad. Otdel. Mat. Inst. Steklov. (LOMI), **160**, (1987), Anal. Teor. Chisel i Teor. Funktsii. 8, 54–71, 297, translation in J. Soviet Math. **52** (1990), no. 3, 3036–3048.

[11] C.F. Gauss, *Disquisitiones arithmeticae*, Translated and with a preface by Arthur A. Clarke; Revised by William C. Waterhouse, Cornelius Greither and A. W. Grootendorst and with a preface by Waterhouse, Springer-Verlag, New York, 1801.

[12] S. Hoory, N. Linial and A. Wigderson, *Expander graphs and their applications*, Bull. Amer. Math. Soc. (N.S.) **43**, (2006), no. 4, 439–561 (electronic).

[13] H. Iwaniec, *Fourier coefficients of modular forms of half-integral weight*, Invent. Math. **87** (1987), no. 2, 385–401.

[14] H. Iwaniec and P. Sarnak, *Perspectives on the analytic theory of L-functions*, Geom. Funct. Anal. (2000), 705–741.

[15] H. Jacquet and R.P. Langlands, *Automorphic forms on* GL(2), Lecture Notes in Mathematics **114** Springer-Verlag, Berlin, 1970.

[16] B.W. Jones, *The Arithmetic Theory of Quadratic Forms*, Carus Monograph Series, **10**, The Mathematical Association of America, Buffalo, N. Y., 1950,

[17] A.W. Knapp, *Introduction to the Langlands program*, Representation theory and automorphic forms (Edinburgh, 1996), Proc. Sympos. Pure Math. **61**, Amer. Math. Soc., Providence, RI, 1997, 245–302.

[18] Yu. V. Linnik, *Ergodic properties of algebraic fields*, Translated from the Russian by M. S. Keane. Ergebnisse der Mathematik und ihrer Grenzgebiete, Band 45, Springer-Verlag New York Inc., New York, 1968.

[19] A. Lubotzky, *Discrete groups, expanding graphs and invariant measures*, Progress in Mathematics, **125**, With an appendix by Jonathan D. Rogawski, Birkhäuser Verlag, Basel, 1994.

[20] A. Lubotzky, R. Phillips and P. Sarnak, *Ramanujan graphs*, Combinatorica, **8** (1988), no. 3, 261–277.

[21] A.V. Malyshev, *Discrete ergodic method and its applications to the arithmetic of ternary quadratic forms*, Colloq. Math. Soc. János Bolyai, **34**, pp. 1023–1049, North-Holland, Amsterdam, 1984.

[22] P. Michel, *Analytic number theory and families of automorphic L-functions*, IAS/Park City Math. Ser. **12**, Amer. Math. Soc. Providence, RI, (2007), 179–296.

[23] P. Michel and A. Venkatesh, *Equidistribution, L-functions and ergodic theory: on some problems of Yu. Linnik*, International Congress of Mathematicians. Vol. II, Eur. Math. Soc., Zürich, 421–457, 2006.

[24] P. Michel and A. Venkatesh, *Heegner points and non-vanishing of Rankin/Selberg L-functions*, in Analytic number theory, Clay Math. Proc. **7**, Amer. Math. Soc., Providence, RI, (2007), 169–183.

[25] P. Michel and A. Venkatesh, *Lectures on Linnik's ergodic method*, Expanded notes of lectures given in 2010 at the Tata Institute summer school and the CIMPA summer school in Weihai, in preparation.

[26] P. Michel and A. Venkatesh, *Subconvexity, periods and equidistribution*, Book in preparation; expanded version of notes from a 2009 Nachdiplom vorlesung given at ETH Zuerich.

[27] G. Pall, *Representation by quadratic forms*, Canadian J. Math. **1** (1949), 344–364.

[28] V. Platonov and A. Rapinchuk, *Algebraic groups and number theory*, Pure and Applied Mathematics, **139**, Translated from the 1991 Russian original by Rachel Rowen, Academic Press Inc., Boston, MA, 1994.

[29] P. Samuel, *Algebraic theory of numbers*, Translated from the French by Allan J. Silberger, Houghton Mifflin Co., Boston, Mass., 1970.

[30] J.-P. Serre, *A course in arithmetic*, Translated from the French, Graduate Texts in Mathematics, **7**, Springer-Verlag, New York, 1973.

[31] J.-P. Serre, *Trees*, Springer Monographs in Mathematics, Translated from the French original by John Stillwell, Corrected 2nd printing of the 1980 English translation, Springer-Verlag, Berlin, 2003.

[32] T.R. Shemanske, *Representations of ternary quadratic forms and the class number of imaginary quadratic fields*, Pacific J. Math. **122** (1986), no. 1, 223–250

[33] Yu. G. Teterin, *Representation of algebraic integers by ternary quadratic forms*, (Russian), Studies in number theory, **8**, Zap. Nauchn. Sem. Leningrad. Otdel. Mat. Inst. Steklov. (LOMI) **121**, (1983), 157–168.

[34] B.A. Venkov, *On the arithmetic of quaternion algebras*, Izv. Akad. Nauk, **1** (1922), 205–241.

[35] B.A. Venkov, *On the arithmetic of quaternion algebras*, Izv. Akad. Nauk, (1929), 489–509, 532–562, 607–622.

[36] B.A. Venkov, *Elementary number theory*, Translated from the Russian and edited by Helen Alderson, Wolters-Noordhoff Publishing, Groningen, 1970.

[37] M.-F. Vignéras, *Arithmétique des algèbres de quaternions*, (French), Lecture Notes in Mathematics, **800**, Springer, Berlin, 1980.

[38] J.-L. Waldspurger, *Sur les valeurs de certaines fonctions L automorphes en leur centre de symétrie*, (French), Compositio Math. **54** (1985), no. 2, 173–242.

[39] J.-L. Waldspurger, *Correspondances de Shimura et quaternions*, (French), Forum Math. **3** (1991), no. 3, 219–307. doi=10.1515/form.1991.3.219.

J.S. Ellenberg, Department of Mathematics, 480 Lincoln Drive, Madison, WI 53706-1325, USA.
E-mail: ellenber@math.wisc.edu

P. Michel, EPFL, Section de mathématiques EFPL, FSB-SMA Station 8 - Bâtiment MA, CH-1015 Lausanne, Switzerland.
E-mail: philippe.michel@epfl.ch

A. Venkatesh, Department of Mathematics, Stanford University, Stanford, CA, 94305, USA.
E-mail: akshay.venkatesh@gmail.com

Automorphic Representations and L-Functions
Editors: D. Prasad, C.S. Rajan, A. Sankaranarayanan, J. Sengupta

Publisher: Hindustan Book Agency, New Delhi, India

Arithmeticity for Periods of Automorphic Forms

Wee Teck Gan[1] and A. Raghuram[2]

1 Introduction

Let G be a connected reductive algebraic group over a number field F, and let (π, V_π) be a cuspidal automorphic representation of $G(\mathbb{A}_F)$. Let H be an algebraic F-subgroup of G, and let χ be an automorphic character of $H(\mathbb{A}_F)$. We say that π has a non-vanishing (H, χ)-period if the functional

$$\phi \mapsto \ell_\chi(\phi) := \int_{[H]} \chi(h)^{-1}\phi(h)\, dh, \quad \phi \in V_\pi, \tag{1.1}$$

is nonzero, where $[H] := H(F)\backslash H(\mathbb{A}_F)$ or sometimes $[H] := Z_G(\mathbb{A}_F)H(F)\backslash H(\mathbb{A}_F)$. Let us now suppose that we are in an arithmetic situation, in as much as that we can talk of the automorphic representation ${}^\sigma\pi$ for any $\sigma \in \mathrm{Aut}(\mathbb{C})$. For example, if π is a cohomological cuspidal automorphic representation of $\mathrm{GL}_n(\mathbb{A}_F)$ then, by a result of Clozel (see Theorem 3.1 below), we know that so is ${}^\sigma\pi$. In this paper we study, mostly by the way of presenting a lot of examples, the dictum: *π has a non-vanishing (H, χ)-period if and only if ${}^\sigma\pi$ has a non-vanishing $(H, {}^\sigma\chi)$-period.* It is this dictum that we call 'arithmeticity for periods of automorphic forms.'

Let us remind the reader that automorphisms of $\mathbb{C}$, with the exceptions of the identity automorphism and complex-conjugation, are discontinuous; in particular, it is almost never the case that $\sigma(\ell_\chi(\phi)) = \ell_{\sigma\circ\chi}(\sigma(\phi))$. So one cannot naively take the σ inside the integral sign. In every example that we study, the dictum holds, and the argument is always indirect via some characterization of existence of such periods.

Let us also observe at the outset that the problem is a distinctly global problem. The corresponding local problem, at any finite place v, is trivial.

[1]Partially supported by a startup grant at the National University of Singapore.
[2]Partially supported by the National Science Foundation (NSF), award number DMS-0856113, and an Alexander von Humboldt Research Fellowship.

If π_v is (H_v, χ_v)-distinguished, i.e., there exists a nonzero functional $\ell : \pi_v \to \mathbb{C}$ such that $\ell(\pi_v(h)v) = \chi_v(h)\ell(v)$ for all $h \in H(F_v)$. For any $\sigma \in \mathrm{Aut}(\mathbb{C})$, it is easy to see that $\sigma \circ \ell$ gives a $(H_v, {}^\sigma\chi_v)$-distinguishing functional for the conjugated representation ${}^\sigma\pi_v$.

The above local observation says that the problem of arithmeticity of automorphic periods is a consequence of a positive solution of the classical local-to-global problem: 'If τ is an automorphic representation, and suppose at every place v, τ_v is (H_v, χ_v)-distinguished, then does τ have a nonzero global (H, χ)-period?' Here, take τ to be ${}^\sigma\pi$.

Given a cuspidal automorphic representation π of $G(\mathbb{A})$, and a $\sigma \in \mathrm{Aut}(\mathbb{C})$ we need to discuss when the representation ${}^\sigma\pi$ makes sense. This will be possible when the representation π contributes to the cohomology of a locally symmetric space of G with coefficients in a sheaf attached to a finite-dimensional coefficient system. In Section 2 we briefly discuss the appropriate cohomological preliminaries needed to talk about the Galois-conjugated representation ${}^\sigma\pi$ for a general G, and in Section 3 we explicate the case of $\mathrm{GL}(n)$ and recall the basic Theorem 3.1 due to Clozel which says that ${}^\sigma\pi$ is also a cuspidal cohomological representation. In Theorem 9.5, we give the natural generalization of Clozel's theorem to certain classical groups, exploiting the recent results of Arthur [2].

In Section 4, we begin by looking at two of the easiest nontrivial examples when the ambient group $G = \mathrm{GL}_2/F$. In particular, in the GL_2 context, we look at the question of arithmeticity for Whittaker periods which boils down to every cuspidal automorphic representation being globally generic and that the space of cuspidal cohomology having a rational structure; indeed, the same ingredients give arithmeticity for Whittaker periods when $G = \mathrm{GL}_n/F$. The other GL_2 example we analyze is (GL_1, χ)-periods for Hecke characters χ; here, arithmeticity is a consequence of Manin's and Shimura's classical algebraicity results on the critical values of L-functions for GL_2.

In the rest of the paper we analyze the following situations for arithmeticity problems, which are various generalizations of the GL_2 cases considered in Section 4:

1. Shalika period integrals for representations of GL_{2n}. See Theorem 5.3. The nonvanishing of Shalika period integrals is characterized in terms of functorial transfers from $\mathrm{GSpin}(2n+1)$.
2. $\mathrm{GL}(n)/F$-periods for representations of $\mathrm{GL}(n)/E$, for a quadratic extension E/F. See Theorem 6.3. The nonvanishing of such period integrals is characterized in terms of functorial transfers from the unitary groups $\mathrm{U}(n)$.

3. $\mathrm{GL}(n-1)$-periods for representations of $\mathrm{GL}(n) \times \mathrm{GL}(n-1)$. See Theorem 7.1. These are the Gross-Prasad periods for the general linear groups.
4. $\mathrm{GL}(n) \times \mathrm{GL}(n)$-periods for representations of $\mathrm{GL}(2n)$ over a totally real field. See Theorem 8.1.
5. Whittaker and Gross-Prasad periods for classical groups. See Theorem 10.1.

It is clear that these examples are pointing toward some general motivic interpretation of period integrals. Automorphic representations with a nonzero (H, χ)-period are usually characterized in terms of functorial transfers and/or in terms of some L-function attached to π having a pole or (not having) a zero at a certain point. In terms of L-values, the situation is very similar to a conjecture of Gross on motivic L-functions; see [10, Conjecture 2.7 (ii)]. This says that for a critical motive M, the order of vanishing of the critical L-value $L(\sigma, M, 0)$ is independent of the conjugating automorphism σ. In our situation, suppose π corresponds to a motive, and suppose having a non-vanishing (H, χ)-period corresponds to the (non-)vanishing of an L-value attached to π which happens to be a critical L-value, then Gross's conjecture would predict the validity of the dictum. For example, the situation in (3), respectively (4), above exactly ties up with critical L-values of the underlying Rankin-Selberg L-function, respectively the standard L-function.

Acknowledgements We thank Jeff Adams for several helpful discussions concerning the proof of the results in Section 9, and Harald Grobner for many comments on a first draft of this paper which corrected some errors and greatly improved the exposition. It is also a pleasure to thank Dipendra Prasad, C.S. Rajan, A. Sankaranarayanan and Jyoti Sengupta for organizing a very memorable international colloquium at TIFR during which this article took shape. This article was completed during the first author's visit at the IHES at Bures-sur-Yvette in July 2012; the first author thanks the IHES for its support and for providing a peaceful yet stimulating working environment.

2 Cohomological Preliminaries

We will often talk about a 'cohomological cuspidal automorphic representation'. In this section, we will briefly review its definition and discuss some of its very basic properties that we need later.

Let $G/\mathbb{Q}$ be a connected reductive algebraic group over $\mathbb{Q}$ and let S be the maximal $\mathbb{Q}$-split torus in the center Z of G. Let C_∞ be a maximal compact subgroup of $G(\mathbb{R})$ and let $K_\infty = C_\infty S(\mathbb{R})$. The connected component of the identity of K_∞ is denoted K_∞°. For any open-compact subgroup $K_f \subset G(\mathbb{A}_f)$, define the locally symmetric space

$$S^G_{K_f} := G(\mathbb{Q})\backslash G(\mathbb{A})/K_\infty^\circ K_f.$$

Allowing K_f to vary, one has an inverse system $S^G = \text{proj}\lim S^G_{K_f}$ of locally symmetric spaces.

Fix a maximal torus $T/\mathbb{Q} \subset G/\mathbb{Q}$, with associated absolute Weyl group $W(G,T) = (N_G(T)/T)(\mathbb{C})$. Set $X(T) = \text{Hom}_{\mathbb{C}}(T, \mathbb{G}_m)$, so that $W(G,T)$ acts naturally on $X(T)$. By the theory of highest weight, each $W(G,T)$-orbit μ in $X(T)$ corresponds to an irreducible algebraic representation E_μ of $G(\mathbb{C})$ with highest weight μ. Let $\mathcal{E}_\mu$ be the associated (inverse system of) sheaf on S^G. We are interested in the sheaf cohomology groups

$$H^\bullet(S^G, \mathcal{E}_\mu) := \text{inj}\lim_{K_f} H^\bullet(S^G_{K_f}, \mathcal{E}_\mu)$$

on which the finite adelic group $G(\mathbb{A}_f)$ acts naturally.

We can compute the above sheaf cohomology via the de Rham complex, and then reinterpreting the de Rham complex in terms of the complex computing relative Lie algebra cohomology, we get the isomorphism:

$$H^\bullet(S^G, \mathcal{E}_\mu) \simeq H^\bullet(\mathfrak{g}_\infty, K_\infty^\circ;\ C^\infty(G(\mathbb{Q})\backslash G(\mathbb{A})) \otimes E_\mu).$$

The inclusion $C^\infty_{\text{cusp}}(G(\mathbb{Q})\backslash G(\mathbb{A})) \hookrightarrow C^\infty(G(\mathbb{Q})\backslash G(\mathbb{A}))$ of the space of smooth cusp forms in the space of all smooth functions induces, via well-known results of Borel [5], an injection in cohomology; this defines cuspidal cohomology:

$$\begin{array}{ccc} H^\bullet(S^G, \mathcal{E}_\mu) & \longrightarrow & H^\bullet(\mathfrak{g}_\infty, K_\infty^\circ; C^\infty(G(\mathbb{Q})\backslash G(\mathbb{A})) \otimes E_\mu) \\ \uparrow & & \uparrow \\ H^\bullet_{\text{cusp}}(S^G, \mathcal{E}_\mu) & \longrightarrow & H^\bullet(\mathfrak{g}_\infty, K_\infty^\circ; C^\infty_{\text{cusp}}(G(\mathbb{Q})\backslash G(\mathbb{A})) \otimes E_\mu) \end{array}$$

Using the usual decomposition of the space of cusp forms into a direct sum of cuspidal automorphic representations, we get the following fundamental decomposition

$$H^\bullet_{\text{cusp}}(S^G, \mathcal{E}_\mu) = \bigoplus_{\Pi} H^\bullet(\mathfrak{g}_\infty, K_\infty^\circ; \Pi_\infty \otimes E_\mu) \otimes \Pi_f \qquad (2.1)$$

of $\pi_0(G_\infty) \times G(\mathbb{A}_f)$-modules.

We say that Π *is a cohomological cuspidal automorphic representation* if Π has a nonzero contribution to the above decomposition for some μ, or equivalently, if Π is a cuspidal automorphic representation whose representation at infinity Π_∞, after twisting by E_μ, has nontrivial relative Lie algebra cohomology. In this situation, we write $\Pi \in \mathrm{Coh}(G, \mu)$.

One may also consider cohomology with compact supports $H_c^\bullet(S^G, \mathcal{E}_\mu)$. *Inner cohomology* is defined as the image of compactly supported cohomology in global cohomology:

$$H_!^\bullet(S^G, \mathcal{E}_\mu) := \mathrm{Image}\left(H_c^\bullet(S^G, \mathcal{E}_\mu) \longrightarrow H^\bullet(S^G, \mathcal{E}_\mu)\right).$$

In the literature, inner cohomology is also called *interior* or *parabolic* cohomology. It is a fundamental fact (which comes from analyzing the long-exact sequence arising from the Borel-Serre compactification; see, for example, Li–Schwermer [32]) that

$$H_{\mathrm{cusp}}^\bullet(S^G, \mathcal{E}_\mu) \subset H_!^\bullet(S^G, \mathcal{E}_\mu). \tag{2.2}$$

On the other hand, since any compactly supported function is square-integrable, we also have that inner cohomology sits inside the cohomology group whose elements are represented by square-integrable automorphic forms, i.e.,

$$H_!^\bullet(S^G, \mathcal{E}_\mu) \subset H_{(2)}^\bullet(S^G, \mathcal{E}_\mu). \tag{2.3}$$

Now let us briefly recall the action of $\mathrm{Aut}(\mathbb{C})$ on the various objects introduced above. First, observe that $\mathrm{Aut}(\mathbb{C})$ acts naturally on $X(T)$ by

$$({}^\sigma\chi)(t) = \sigma(\chi(\sigma^{-1}(t))), \quad \text{for } \sigma \in \mathrm{Aut}(\mathbb{C}),\ t \in T(\mathbb{C}) \text{ and } \chi \in X(T).$$

Similarly, $\mathrm{Aut}(\mathbb{C})$ acts naturally on the set of equivalence classes of irreducible algebraic representations (E, ρ) of $G(\mathbb{C})$ (where $\rho : G \to \mathrm{GL}(E)$) by

$${}^\sigma\rho(g) = \sigma(\rho(\sigma^{-1}(g))) \quad \text{for } g \in G(\mathbb{C}),$$

where, the $\mathrm{Aut}(\mathbb{C})$-action on $\mathrm{GL}(E) \cong \mathrm{GL}_n(\mathbb{C})$ is with respect to the canonical Chevalley structure over $\mathbb{Q}$. In particular, it follows that if $E = E_\mu$ has highest weight μ, then ${}^\sigma E_\mu$ has highest weight ${}^\sigma\mu$.

Another description of the σ-conjugated algebraic representation $({}^\sigma E, {}^\sigma\rho)$ is as follows. Set $E_\sigma = E \otimes_{\mathbb{C},\sigma} \mathbb{C}$. Then $({}^\sigma E, {}^\sigma\rho)$ is isomorphic to the representation of $G(\mathbb{C})$ on E_σ with $g \in G(\mathbb{C})$ acting by

$$v \mapsto \rho(\sigma^{-1}(g))(v).$$

In particular, when restricted to $G(\mathbb{Q})$, ${}^\sigma E$ is realized on $E_\sigma = E \otimes_{\mathbb{C},\sigma} \mathbb{C}$ with $G(\mathbb{Q})$ acting via its action on the first component E of the tensor product. Thus, there is a natural σ-linear, $G(\mathbb{Q})$-equivariant map

$$E \longrightarrow {}^\sigma E.$$

In an analogous way, $\mathrm{Aut}(\mathbb{C})$ acts naturally on the smooth representations (W, Π_f) of $G(\mathbb{A}_f)$. Namely, we define ${}^\sigma\Pi_f$ to be the action of $G(\mathbb{A}_f)$ on $W \otimes_{\mathbb{C},\sigma} \mathbb{C}$ with $G(\mathbb{A}_f)$ acting on the first component W of the tensor product.

Now, passing to sheaves on the locally symmetric space S^G, the above considerations lead to a commutative diagram, where the horizontal arrows are σ-linear $G(\mathbb{A}_f)$-equivariant isomorphisms

$$\begin{array}{ccc} H^\bullet(S^G, \mathcal{E}_\mu) & \longrightarrow & H^\bullet(S^G, {}^\sigma\mathcal{E}_\mu) \\ \uparrow & & \uparrow \\ H_!^\bullet(S^G, \mathcal{E}_\mu) & \longrightarrow & H_!^\bullet(S^G, {}^\sigma\mathcal{E}_\mu) \end{array}$$

Thus we have the following

Proposition 2.1 *Suppose that π is a cohomological cuspidal automorphic representation of $G(\mathbb{A})$. Then for any $\sigma \in \mathrm{Aut}(\mathbb{C})$, there exists an automorphic representation τ_σ appearing in the automorphic discrete spectrum of $G(\mathbb{A})$ such that:*

- $\tau_{\sigma,f} = {}^\sigma\pi_f$;
- $\tau_{\sigma,\infty}$ *has nonzero Lie algebra cohomology with respect to* ${}^\sigma\mu$.

Proof By assumption, π_f occurrs as a $G(\mathbb{A}_f)$-summand in $H^\bullet_{\mathrm{cusp}}(S^G, \mathcal{E}_\mu)$ for some $\mathcal{E}_\mu$. We deduce by (2.2), (2.3) and the above commutative diagram that ${}^\sigma\pi_f$ occurs in $H_!^\bullet(S^G, {}^\sigma\mathcal{E}_\mu)$ and hence in $H^\bullet_{(2)}(S^G, {}^\sigma\mathcal{E}_\mu)$. This proves the proposition.

□

Let us now suppose that G is a connected reductive group defined over a number field F with ring of adeles $\mathbb{A}_F$ and let $T \subset G$ be a maximal torus defined over F. We may apply the above discussion to the reductive group $G_0 := \mathrm{Res}_{F/\mathbb{Q}} G$ over $\mathbb{Q}$ containing the torus $T_0 = \mathrm{Res}_{F/\mathbb{Q}} T$. In this case, one has

$$G_0 \times_{\mathbb{Q}} \mathbb{C} \cong \prod_{\tau \in \mathrm{Hom}(F,\mathbb{C})} G_\tau \qquad \text{with} \qquad G_\tau := G \times_{F,\tau} \mathbb{C}$$

and

$$T_0 \times_{\mathbb{Q}} \mathbb{C} \ \cong \prod_{\tau \in \mathrm{Hom}(F,\mathbb{C})} T_\tau \qquad \text{with} \qquad T_\tau := T \times_{F,\tau} \mathbb{C},$$

so that

$$X(T_0) \ \cong \bigoplus_{\tau \in \mathrm{Hom}(F,\mathbb{C})} X(T_\tau), \quad \text{with} \qquad X(T_\tau) = \mathrm{Hom}_{\mathbb{C}}(T_\tau, \mathbb{G}_m).$$

Note that $X(T_\tau)$ comes equipped with a natural action of $\mathrm{Aut}(\mathbb{C}/\tau(F))$. Thus, an irreducible algebraic representation E of $G_0(\mathbb{C})$ is of the form $E = \bigotimes_\tau E_{\mu_\tau}$ where E_{μ_τ} is an irreducible algebraic representation of G_τ.

Let us explicate the action of $\mathrm{Aut}(\mathbb{C})$ on $X(T_0)$. For $\sigma \in \mathrm{Aut}(\mathbb{C})$ and $\tau \in \mathrm{Hom}(F, \mathbb{C})$, with $\tau' = \sigma \circ \tau$, the automorphism σ induces:

(a) a natural isomorphism

$$\mathrm{Aut}(\mathbb{C}/\tau(F)) \longrightarrow \mathrm{Aut}(\mathbb{C}/\tau'(F))$$

sending ϕ to $\sigma \circ \phi \circ \sigma^{-1}$;

(b) a natural equivariant isomorphism

$$\sigma_* : X(T_\tau) \longrightarrow X(T_{\tau'})$$

given by:

$$\sigma_* : \chi \mapsto \sigma \circ \chi \circ p_\sigma,$$

where

$$p_\sigma : T_{\tau'} = T_\tau \times_{\mathbb{C},\sigma} \mathbb{C} \longrightarrow T_\tau$$

is the natural projection. Here, the equivariance of σ_* is with respect to the action of $\mathrm{Aut}(\mathbb{C}/\tau(F))$ on the source and the action of $\mathrm{Aut}(\mathbb{C}/\tau'(F))$ on the target, via the isomorphism in (a).

Then the action of $\sigma \in \mathrm{Aut}(\mathbb{C})$ on $X(T_0)$ is via the bijections σ_* described above. In particular, for any fixed $\tau \in \mathrm{Hom}(F, \mathbb{C})$,

$$X(T_0) \cong \mathrm{Ind}_{\mathrm{Aut}(\mathbb{C}/\tau(F))}^{\mathrm{Aut}(\mathbb{C})} X(T_\tau)$$

as $\mathrm{Aut}(\mathbb{C})$-modules.

As examples, consider the following cases:

- when G is F-split, the action of $\mathrm{Aut}(\mathbb{C}/\tau(F))$ on $X(T_\tau)$ is trivial, so that $\sigma_* : X(T_\tau) \longrightarrow X(T_{\tau'})$ is independent of σ (subject to $\sigma \circ \tau = \tau'$). In other words, one has a canonical identification $X(T_\tau) \leftrightarrow X(T_{\tau'})$. Thus, in this case, the action of $\mathrm{Aut}(\mathbb{C})$ on $X(T_0)$ is via the permutation of the components $X(T_\tau)$, so that if $E = \bigotimes_\tau E_{\mu_\tau}$, then

$$^{\sigma}E = \bigotimes_\tau E_{\mu_{\sigma^{-1}\circ\tau}}.$$

- when G is an inner form of a split group over F, the action of $\mathrm{Aut}(\mathbb{C}/\tau(F))$ on $X(T_\tau)$ factors through the action of the Weyl group $W_\tau := W(G_\tau, T_\tau)$. Thus, one still has a canonical bijection $X(T_\tau)/W_\tau \leftrightarrow X(T_{\tau'})/W_{\tau'}$, i.e., between the sets of highest weights. As in the split case, one still has

$$^{\sigma}E = \bigotimes_{\tau} E_{\mu_{\sigma^{-1}\circ\tau}}.$$

Finally, note that the set S_∞ of infinite places of F is the set of orbits of complex conjugation on $\mathrm{Hom}(F,\mathbb{C})$. If we write $G_\infty = \mathrm{Res}_{F/\mathbb{Q}}(G)(\mathbb{R}) = \prod_{v\in S_\infty} G_v$, then $\mu \in X(T_0)$ defines a finite dimensional representation $E = \otimes_v E_{\mu_v}$ via:

- if $v = \tau \in \mathrm{Hom}(F,\mathbb{R})$ is real, then $E_{\mu_v} = E_{\mu_\tau}$;
- if $v = \{\tau, \overline{\tau}\} \subset \mathrm{Hom}(F,\mathbb{C})$ is complex, then $E_{\mu_v} = E_{\mu_\tau} \otimes E_{\mu_{\overline{\tau}}}$ as a representation of $G_v = \{(g,\overline{g}) : g \in G_\tau\} \subset G_\tau \times G_{\overline{\tau}}$.

3 Cohomological Representations of GL_n

In this section, we discuss in greater depth the case when $G = \mathrm{GL}_n/F$ and recall a fundamental result due to Clozel [7, Théorème 3.13], which refines Proposition 2.1.

Theorem 3.1 (Clozel) *Let Π be a cuspidal automorphic representation of $\mathrm{GL}_n(\mathbb{A}_F)$ such that $\Pi \in \mathrm{Coh}(\mathrm{GL}_n/F,\mu)$. For any $\sigma \in \mathrm{Aut}(\mathbb{C})$, there is a cuspidal automorphic representation ${}^{\sigma}\Pi \in \mathrm{Coh}(\mathrm{GL}_n/F, {}^{\sigma}\mu)$ whose finite part is ${}^{\sigma}\Pi_f$.*

Thus, the extra information contained here is that the representation τ_σ in Proposition 2.1 is cuspidal. For the precursor to this result of Clozel, see Shimura [44, Section 2] for the classical situation of Hilbert modular forms; for representations of GL_2, see Harder [21] and Waldspurger [47]. We also remark that in [7], Clozel works with the notion of the "infinity type $p(\Pi)$" of a cohomological cuspidal representation Π, which consists of the exponents appearing in the Langlands parameter of Π_v for $v \in S_\infty$. The infinity type $p(\Pi) = \{p_\tau : \tau \in \mathrm{Hom}(F,\mathbb{C})\}$ is basically the infinitesimal character of the finite-dimensional representation $E_\mu = \otimes_{\tau\in\mathrm{Hom}(F,\mathbb{C})} E_{\mu_\tau}$ with highest weight μ. More precisely, if we assume that μ is dominant, then for each $\tau \in \mathrm{Hom}(F,\mathbb{C})$,

$$\mu_\tau + \rho_n = p_\tau + \Big(\frac{n-1}{2}, \ldots, \frac{n-1}{2}\Big),$$

where $\rho_n = (\frac{n-1}{2}, \frac{n-3}{2}, \cdots, -\frac{n-1}{2})$ is the usual half sum of positive roots for GL_n. We also refer the reader to [7, Defn. 3.6 on p. 107] where the action of $\sigma \in \mathrm{Aut}(\mathbb{C})$ on the infinity type $p(\Pi)$ is defined; it agrees with the action of σ on μ explicated in the previous section.

Note that in Theorem 3.1, the archimedean component of ${}^\sigma\Pi$ is not precisely specified: one only knows the finite dimensional representation $E_{{}^\sigma\mu_v}$ with respect to which ${}^\sigma\Pi_v$ has nonzero relative Lie algebra cohomology. Thus, one is naturally led to ask: to what extent does the highest weight μ_v determine the cohomological representation Π_v? Let us examine this issue more closely for GL_n.

Assume first that $v = \{\tau, \overline{\tau}\}$ is a complex place. Then suppose that

$$\mu_\tau + \rho_n = (a_1, a_2, \ldots, a_n) \quad \text{and} \quad \mu_{\overline{\tau}} + \rho_n = (b_1, b_2, \ldots, b_n),$$

with a_i and b_i decreasing. By the purity lemma of Clozel [7, Lemma 4.9], there is a $\mathsf{w} \in \mathbb{Z}$ (independent of v) such that $a_i + b_{n+1-i} = \mathsf{w}$ for all i. Then

$$\Pi_v = \mathrm{Ind}_B^{\mathrm{GL}_n(\mathbb{C})} z^{a_1}\overline{z}^{b_n} \times \cdots \times z^{a_n}\overline{z}^{b_1}.$$

In other words, Π_v is completely determined by $\mu_v = \{\mu_\tau, \mu_{\overline{\tau}}\}$ if v is complex.

Assume now that v is a real place. Then $\mu_v = (\mu_{v,1}, \ldots, \mu_{v,n})$ where $\mu_{v,j} \in \mathbb{Z}$ and $\mu_{v,1} \geq \mu_{v,2} \geq \cdots \geq \mu_{v,n}$. Further, one knows that the highest weight μ is pure (because only pure weights support nonzero cuspidal cohomology); namely there exists a $\mathsf{w} \in \mathbb{Z}$ (independent of v) such that

$$\mu_{v,j} + \mu_{v,n-j+1} = \mathsf{w}.$$

Note that if n is odd, then $\mathsf{w} = 2\mu_{v,(n-1)/2}$ is even. Now put

$$\ell_v = 2\mu_v - \mathsf{w} + 2\rho_n.$$

Then

$$\ell_{v,j} = 2\mu_{v,j} - \mathsf{w} + n - 2j + 1 = \mu_{v,j} - \mu_{v,n-j+1} + n - 2j + 1.$$

Observe that:

$$\begin{cases} \ell_{v,1} > \cdots > \ell_{v,[n/2]} > 0; \\ \ell_{v,n-j+1} = -\ell_{v,j}; \\ \ell_{v,j} \equiv \mathsf{w} + n - 1 \pmod 2). \end{cases}$$

In particular, $\ell_{v,j} \equiv 0 \pmod 2$ if n is odd.

We define an irreducible representation J_{μ_v} as the representation induced from the $(2, \ldots, 2)$-parabolic if $n = 2m$ is even:

$$J_{\mu_v} = D_{\ell_{v,1}} |\ |^{-\mathsf{w}/2} \times \cdots \times D_{\ell_{v,m}} |\ |^{-\mathsf{w}/2}, \tag{3.1}$$

and if $n = 2m + 1$ is odd then it is induced from the $(2, \ldots, 2, 1)$-parabolic subgroup:

$$J_{\mu_v} = D_{\ell_{v,1}} |\ |^{-\mathsf{w}/2} \times \cdots \times D_{\ell_{v,m}} |\ |^{-\mathsf{w}/2} \times |\ |^{-\mathsf{w}/2}, \tag{3.2}$$

where D_l is the 'discrete series' representation of $\mathrm{GL}_2(\mathbb{R})$ of lowest weight $l+1$ and central character sgn^{l+1}. Given $\Pi \in \mathrm{Coh}(\mathrm{GL}_n/F, \mu)$, we know, when n is even, that

$$\Pi_v \ \simeq \ J_{\mu_v}, \tag{3.3}$$

and when n is odd, we know that

$$\Pi_v \ \simeq \ J_{\mu_v} \otimes \mathrm{sgn}^{\epsilon(\Pi_v)}, \tag{3.4}$$

where $\epsilon(\Pi_v) \in \{0, 1\}$ is defined by

$$(-1)^{\epsilon(\Pi_v)} = (-1)^{(n-1)/2} \omega_{\Pi_v}(-1).$$

(See, for example, [39, Section 5.1].) Thus, when v is real, μ_v completely determines Π_v when n is even; however, when n is odd, we need not only μ_v but also the parity of the central character at the real place v to pin down Π_v.

Now we bring in the $\mathrm{Aut}(\mathbb{C})$-action. As we have noted in the previous section, for $\sigma \in \mathrm{Aut}(\mathbb{C})$, we have ${}^\sigma\mu_\tau = \mu_{\sigma^{-1}\tau}$ for $\tau \in \mathrm{Hom}(F, \mathbb{C})$. The above discussion implies that when n is even or when v is complex, the local component ${}^\sigma\Pi_v$ is completely determined by ${}^\sigma\mu_v$. We explicate the situation in two cases:

Proposition 3.2 (i) *Assume first that F is a totally complex quadratic extension of a totally real F^+. Then for any $\sigma \in \mathrm{Aut}(\mathbb{C})$,*

$${}^\sigma\Pi_v = \Pi_{\sigma^{-1}v}.$$

(ii) *Assume that F is totally real. When n is even, we have:*

$${}^\sigma\Pi_\infty \ = \ \bigotimes_{v \in S_\infty} \Pi_{\sigma^{-1}v} \ = \ \bigotimes_{v \in S_\infty} J_{\mu_{\sigma^{-1}v}}. \tag{3.5}$$

When n is odd, we have:

$${}^\sigma\Pi_\infty \ = \ \bigotimes_{v \in S_\infty} \left(J_{\mu_{\sigma^{-1}v}} \otimes \mathrm{sgn}^{\epsilon(\Pi_v)} \right), \tag{3.6}$$

where $\epsilon(\Pi_v)$ is as defined in (3.4). In particular, if the sign $\omega_{\Pi_v}(-1)$ is independent of the infinite place v, then

$${}^\sigma\Pi_\infty \ = \ \bigotimes_{v \in S_\infty} \Pi_{\sigma^{-1}v} \ = \ \bigotimes_{v \in S_\infty} J_{\mu_{\sigma^{-1}v}}$$

as in the case when n is even.

Proof (i) For each place v^+ of F^+, let $v = \{\tau, \overline{\tau}\}$ be the place of F over v^+, so that τ and $\overline{\tau}$ are the two elements of $\mathrm{Hom}(F, \mathbb{C})$ whose restriction to F^+ is v^+. Then for any $\sigma \in \mathrm{Aut}(\mathbb{C})$, $\sigma^{-1} \circ v := \{\sigma^{-1} \circ \tau, \sigma^{-1} \circ \overline{\tau}\}$ are the two elements which restrict to $\sigma^{-1} \circ v^+$. Thus, ${}^{\sigma}\mu_v = \mu_{\sigma^{-1} \cdot v}$ and so we have:

$$ {}^{\sigma}\Pi_v = \Pi_{\sigma^{-1} \circ v}. $$

(ii) Now assume that F is totally real, so that $S_\infty = \mathrm{Hom}(F, \mathbb{C})$. The case when n is even follows from our discussion above. When n is odd, the situation is a little more tricky and we need to consider central characters. Let ω_Π be the global central character of Π. It is of the form:

$$ \omega_\Pi = \omega_\Pi^\circ \otimes |\ |^{-n\mathsf{w}/2} $$

with ω_Π° a Hecke character of finite order. Let us simplify notations and write this as: $\omega = \omega^\circ |\ |^{\mathsf{m}}$ where ω° is a finite-order Hecke character and $\mathsf{m} \in \mathbb{Z}$. Then, for any $\sigma \in \mathrm{Aut}(\mathbb{C})$, we have:

$$ ({}^{\sigma}\omega)_v = \omega_v, \quad \forall v \in S_\infty. \tag{3.7} $$

This follows from the following two observations:

- $({}^{\sigma}\omega^\circ) = \sigma \circ \omega^\circ$; this is because the latter is still a continuous character of $F^\times \backslash \mathbb{A}_F^\times$ as ω° takes value in a finite group. In particular, for real v, $\sigma \circ \omega_v^\circ = \omega_v^\circ$ since ω_v° takes value in $\{\pm 1\}$.
- $\sigma|\ | = |\ |$; this is because at all finite places w, $|\ |_w$ takes value in $\mathbb{Q}$ and so $\sigma|\ |_w = |\ |_w$ (and a Hecke character is determined by almost all its local components, by weak approximation).

Next, the global central character of ${}^{\sigma}\Pi$ satisfies:

$$ \omega_{{}^{\sigma}\Pi} = {}^{\sigma}\omega_\Pi, $$

which can be seen by checking equality of local characters at all finite unramified places. Hence the parity that is needed in pinning down the representations at infinity as in (3.4) is given by $\epsilon({}^{\sigma}\Pi_v) = \epsilon(\Pi_v)$, since

$$ (-1)^{\epsilon({}^{\sigma}\Pi_v)} = (-1)^{(n-1)/2}\omega_{{}^{\sigma}\Pi_v}(-1) \text{ and } \omega_{{}^{\sigma}\Pi_v}(-1) = {}^{\sigma}\omega_{\Pi_v}(-1) = \omega_{\Pi_v}(-1) $$

by (3.7). This proves the proposition.

□

Remark 3.3 When F is totally real and n is odd, the hypothesis in the above theorem that the sign $\omega_{\Pi_v}(-1)$ is independent of v is arithmetically interesting because it is a necessary condition for the standard L-function of Π to have a critical point. This will also be the case if the rank n Grothendieck motive $M = M_\Pi$ over F that is conjecturally attached to Π is *special*, i.e., has the property that complex conjugation acts via the 'same' scalar on the middle Hodge type for every real embedding v of F; see, for example, Blasius [6, M3].

4 The GL_2-examples

After the preliminary discussions of the previous two sections, we can now begin the consideration of periods. In this section, we illustrate the question we will study for the case of GL_2. Let us, for the sake of simplicity, take $G = \mathrm{GL}_2/\mathbb{Q}$, although everything discussed in this section works for GL_2 over any number field.

4.1 Whittaker periods

For the subgroup H we take

$$H = U_2 = U = \left\{ \begin{pmatrix} 1 & x \\ 0 & 1 \end{pmatrix} \; : \; x \in \mathbb{G}_a \right\},$$

i.e., U is the unipotent radical of the standard Borel subgroup of upper triangular matrices in G. Fix a nontrivial additive character $\psi : \mathbb{Q}\backslash\mathbb{A} \to \mathbb{C}^\times$. Then, as usual, ψ gives a character $\psi : U(\mathbb{Q})\backslash U(\mathbb{A}) \to \mathbb{C}^\times$ by $\psi\left(\begin{smallmatrix} 1 & x \\ 0 & 1 \end{smallmatrix}\right) = \psi(x)$. Using the same symbol ψ for both the characters will cause no confusion. In this situation, the linear functional ℓ_ψ defined in (1.1) is called a global Whittaker functional.

Given a cuspidal automorphic representation (π, V_π) of $\mathrm{GL}_2(\mathbb{A})$ we can define for each $\phi \in V_\pi$ the associated Whittaker vector

$$W_\phi(g) \;:=\; \int_{U(\mathbb{Q})\backslash U(\mathbb{A})} \phi(ug)\psi(u)^{-1}\, du. \tag{4.1}$$

Observe that $W_\phi(1) = \ell_\psi(\phi)$. Using the action of $\mathrm{GL}_2(\mathbb{A})$ we see that $\ell_\psi(\phi) \neq 0$ for some ϕ if and only if $W_\phi \neq 0$ for some ϕ. A fundamental fact at the heart of the GL_2-theory of automorphic forms is that W_ϕ determines ϕ. (See, for example, [8, Lecture 4, Section 1].) Indeed, we have a Fourier expansion of the form

$$\phi(g) \;=\; \sum_{\gamma \in \mathbb{Q}^\times} W_\phi\left(\begin{pmatrix} \gamma & 0 \\ 0 & 1 \end{pmatrix} g \right), \tag{4.2}$$

In particular, every cuspidal automorphic representation has a nonvanishing Whittaker period.

Now let us suppose that π is a cohomological cuspidal automorphic representation of $\mathrm{GL}_2(\mathbb{A})$, and in particular, π has nonvanishing Whittaker periods. For any $\sigma \in \mathrm{Aut}(\mathbb{C})$ Theorem 3.1 says that ${}^\sigma\pi$ is also a (cohomological) cuspidal automorphic representation of $\mathrm{GL}_2(\mathbb{A})$. Hence, by the above discussion once again, ${}^\sigma\pi$ also has nonvanishing Whittaker periods, i.e, we have arithmeticity for Whittaker periods for GL_2.

The main ingredients in arithmeticity for Whittaker periods are (4.2) and Theorem 3.1. Both these ingredients, which are nontrivial assertions, are valid for GL_n/F over any number field F after suitable modification; for example, the Fourier expansion takes the form:

$$\phi(g) \;=\; \sum_{\gamma \in \mathrm{GL}_{n-1}(F)/U_{n-1}(F)} W_\phi\left(\begin{pmatrix} \gamma & 0 \\ 0 & 1 \end{pmatrix} g\right), \tag{4.3}$$

(See, for example, [8, loc. cit.].) Hence we get arithmeticity for Whittaker periods for GL_n/F. In Section 10 we consider the case of classical groups where the analysis is far more complicated.

Reverting to $\mathrm{GL}_2/\mathbb{Q}$, let us go through the analysis for arithmeticity for Whittaker periods in the classical context of modular forms. Fix a positive integer N and consider the space $S_k(N)$ consisting of all holomoprhic cusp forms of weight k on the upper half plane for the discrete subgroup $\Gamma_1(N)$ of $\mathrm{SL}_2(\mathbb{R})$. A cusp form $\varphi \in S_k(N)$ has a Fourier expansion

$$\varphi(z) = \sum_{n=1}^{\infty} a_n(\varphi) e^{2\pi i n z}.$$

Now define $S_k(N, \mathbb{Q})$ to be the $\mathbb{Q}$-subspace of the $\mathbb{C}$-vector space $S_k(N)$ consisting of all φ such that $a_n(\varphi) \in \mathbb{Q}$ for all $n \geq 1$. One has the following *nontrivial* fact:

$$S_k(N) \;=\; S_k(N, \mathbb{Q}) \otimes_{\mathbb{Q}} \mathbb{C}. \tag{4.4}$$

(See, for example, Shimura [45, Theorem 3.52].) This may be stated as the fact that the space of cusp forms of weight k and level N has a basis of cusp forms all of whose Fourier coefficients are in $\mathbb{Q}$. Indeed, there is a deeper integrality statement which says that the above is true with $\mathbb{Z}$ instead of $\mathbb{Q}$; however, for our purposes, a $\mathbb{Q}$-basis is sufficient. Let us note that (4.4) is the classical analogue of the statement (see Clozel [7, Théorème 3.19]) that cuspidal cohomology for GL_n/F admits a suitable rational structure. Now, given $\varphi \in S_k(N)$ and $\sigma \in \mathrm{Aut}(\mathbb{C})$ we can define a function ${}^\sigma\varphi$ via a

q-expansion.

$$^{\sigma}\varphi(z) := \sum_{n=1}^{\infty} \sigma(a_n(\varphi))e^{2\pi inz}.$$

It follows from (4.4) that $^{\sigma}\varphi \in S_k(N)$. This is the classical analogue of Theorem 3.1. Arithmeticity for Whittaker models takes the form:

$$a_n(\varphi) \neq 0 \implies a_n(^{\sigma}\varphi) \neq 0,$$

which is built into the definition of the Galois conjugate $^{\sigma}\varphi$. The depth of the phenomenon lies in the rationality statement in (4.4).

4.2 GL_1-periods

We continue with $G = \mathrm{GL}_2/\mathbb{Q}$ and now we take

$$H = \left\{ \begin{pmatrix} x & 0 \\ 0 & 1 \end{pmatrix} \; : \; x \in \mathbb{G}_m \right\} \simeq \mathrm{GL}_1.$$

Take a Hecke character $\chi : \mathbb{Q}^\times\backslash\mathbb{A}^\times \to \mathbb{C}^\times$, which gives a character $\chi : H(\mathbb{Q})\backslash H(\mathbb{A}) \to \mathbb{C}^\times$ by $\chi\left(\begin{smallmatrix} x & 0 \\ 0 & 1 \end{smallmatrix}\right) = \chi(x)$. Using the same symbol χ for both the characters will cause no confusion. Consider a cuspidal automorphic representation π of $\mathrm{GL}_2(\mathbb{A})$. Suppose there is a $\phi \in V_\pi$ such that

$$\ell_\chi(\phi) = \int_{x\in\mathbb{Q}^\times\backslash\mathbb{A}^\times} \phi\left(\begin{pmatrix} x & 0 \\ 0 & 1 \end{pmatrix}\right) \chi(x)\, dx \neq 0.$$

To analyze these integrals, and to relate them to L-values, following Jacquet-Langlands [24], fix a nontrivial additive character ψ as in the previous subsection and consider the Whittaker model $\mathcal{W}(\pi) = \mathcal{W}(\pi, \psi)$ of π. Let the cusp form ϕ correspond to $W_\phi \in \mathcal{W}(\pi)$ where W_ϕ is defined in (4.1). Then for a complex variable s such that $\Re(s) \gg 0$, the classical unfolding argument gives:

$$\begin{aligned} \ell(s, \phi, \chi) &:= \int_{x\in\mathbb{Q}^\times\backslash\mathbb{A}^\times} \phi\left(\begin{pmatrix} x & 0 \\ 0 & 1 \end{pmatrix}\right) \chi(x)|x|^{s-\frac{1}{2}}\, dx \\ &= \int_{x\in\mathbb{A}^\times} W_\phi\left(\begin{pmatrix} x & 0 \\ 0 & 1 \end{pmatrix}\right) \chi(x)|x|^{s-\frac{1}{2}}\, dx. \end{aligned}$$

Denote the *zeta integral* on the right hand side by $Z(s, W_\phi, \chi)$, and note that all the ingredients in that integral are factorizable. Changing notation if necessary, there is a cusp form ϕ so that $\ell_\chi(\phi) \neq 0$ and the associated Whittaker vector W_ϕ is a pure-tensor $W_\phi = \otimes_p W_p$. Outside a finite set of primes S containing the infinite prime and all the primes where π or χ is ramified, one knows that W_p is the spherical vector normalized so that $W_p(1) = 1$, and the local zeta-integral computes the local L-function:

$$Z(s, W_p, \chi_p) = \int_{x_p \in \mathbb{Q}_p^\times} W_p\left(\begin{pmatrix} x_p & 0 \\ 0 & 1 \end{pmatrix}\right) \chi_p(x_p)|x_p|_p^{s-\frac{1}{2}}\, dx_p \;=\; L(s, \pi_p \otimes \chi_p).$$

(See, for example, Gelbart [15, Prop. 6.17].) Let

$$L^S(s, \pi \otimes \chi) := \prod_{p \notin S} L_p(s, \pi_p \otimes \chi_p)$$

denote the partial L-function. So far we have:

$$\ell(s, \phi, \chi) = Z(s, W_\phi, \chi) = \left(\prod_{p \in S} Z(s, W_p, \chi_p)\right) \cdot L^S(s, \pi \otimes \chi).$$

Now multiply and divide the right hand side by the local L-factors at $p \in S$ to get:

$$\ell(s, \phi, \chi) = \left(\prod_{p \in S} \frac{Z(s, W_p, \chi_p)}{L_p(s, \pi_p \otimes \chi_p)}\right) \cdot L(s, \pi \otimes \chi). \tag{4.5}$$

The integral $\ell(s, \phi, \chi)$ converges for all s since ϕ is rapidly decreasing. On the right hand side, one knows from Jacquet-Langlands that each of the ratios $Z(s, W_p, \chi_p)/L_p(s, \pi_p \otimes \chi_p)$, *a priori* defined only for $\Re(s) \gg 0$, in fact have an analytic continuation to all of s (see, for example, [15, Theorem 6.12 (ii)]), and that the completed L-function $L(s, \pi \otimes \chi)$ is an entire function of s (see, for example, [15, Theorem 6.18]). We can now prove the following characterization of the existence of (GL_1, χ)-periods and nonvanishing of a certain L-value:

Proposition 4.1 *Let π be a cuspidal automorphic representation of $\mathrm{GL}_2(\mathbb{A})$, and χ a Hecke character of $\mathbb{Q}$. Then, the following are equivalent:*

1. *There exists a cusp form $\phi \in V_\pi$ such that $\ell_\chi(\phi) \neq 0$.*
2. $L(\frac{1}{2}, \pi \otimes \chi) \neq 0$.

Proof For (1) $\Rightarrow$ (2), put $s = 1/2$ in (4.5) to get:

$$0 \neq \ell_\chi(\phi) = r \cdot L(\tfrac{1}{2}, \pi \otimes \chi)$$

where r is an ad-hoc notation for the product

$$\prod_{p\in S} Z(\tfrac{1}{2}, W_p, \chi_p)/L_p(\tfrac{1}{2}, \pi_p \otimes \chi_p).$$

Hence the right hand side is not zero.

For (2) $\Rightarrow$ (1), given $L(\frac{1}{2}, \pi \otimes \chi) \neq 0$, to construct a cusp form ϕ with non-vanishing period, we construct the associated Whittaker vector W_ϕ as a pure-tensor. Outside a finite set S as above, take W_p to be the normalized spherical vector. For places in S, given any s_0 (for us $s_0 = 1/2$), we are always guaranteed the existence of a Whittaker vector W_p such that the ratio $Z_p(s_0, W_p, \chi_p)/L(s_0, \pi_p \otimes \chi_p) \neq 0$. See [15, (6.29)]. (**Note:** Indeed, for GL_2 there is a W_p for each place p so that the local zeta integral computes the local L-factor, and so the ratio is in fact 1. We deliberately stated it in a weaker form of just nonvanishing of that ratio as that is the way it will generalize to $\mathrm{GL}_n \times \mathrm{GL}_{n-1}$.) Now put all the local Whittaker vectors to get a global Whittaker vector, and take ϕ to be the associated cusp form. The proof follows again from (4.5) at $s = 1/2$.

□

Remark 4.2 Observe that it is possible for $L(\frac{1}{2}, \pi \otimes \chi) \neq 0$ and yet $L_f(\frac{1}{2}, \pi \otimes \chi) = 0$. (We use $L(s, \dots)$ for the completed L-function, and $L_f(s, \dots)$ for the finite part.) Such a phenomenon will happen when the infinite part $L_\infty(s, \pi \otimes \chi)$ has a pole at $s = 1/2$. Here is an easy example: Let $\Delta \in S_{12}(\mathrm{SL}_2(\mathbb{Z}))$ be the Ramanujan Δ-function which is a weight 12 cusp form of full level. Let $\pi := \pi(\Delta) \otimes |\ |^{-6}$ and take χ to be the trivial character. (For us, cuspidal automorphic representations need not be unitary, and indeed, π is not unitary.) Anyway, let $L(s, \pi)$ be the Jacquet-Langlands L-function, and $L(s, \Delta)$ be the classical Hecke L-function; then $L(s, \pi) = L(s - 6, \pi(\Delta)) = L(s - 1/2, \Delta)$. Using the classical functional equation $L(s, \Delta) = L(12 - s, \Delta)$ we get

$$L(\tfrac{1}{2}, \pi) \;=\; L(0, \Delta) \;=\; L(12, \Delta) \;\neq\; 0,$$

The L-factor at infinity is given by:

$$L_\infty(s, \pi_\infty) \;=\; L_\infty(s - 6, \pi(\Delta)_\infty) \;=\; 2\,(2\boldsymbol{\pi})^{-s+\frac{1}{2}}\,\Gamma\left(s - \tfrac{1}{2}\right)$$

(For the last equation, see, for example, [40, 4.4]; the presence of an additional factor of 2 makes no difference to the discussion at hand.) Hence, $L_\infty(s, \pi_\infty)$ has a pole at $s = 1/2$, in other words, *nonvanishing of the global L-function at a (seemingly interesting) point does not guarantee that the point is a critical point.*

Now, given a cuspidal representation π as above with a nonvanishing (GL_1, χ)-period, and consequently with $L(\frac{1}{2}, \pi \otimes \chi) \neq 0$, we want to analyze the dictum of arithmeticity, whence we take π to be of cohomological type. But, even if π is of cohomological type with $L(\frac{1}{2}, \pi \otimes \chi) \neq 0$, it is not guaranteed that $s = 1/2$ is a critical point. The same counterexample as in the above remark will work for this. Henceforth, we assume in addition that $s = 1/2$ is a critical point; i.e., by definition, $L_\infty(s, \pi \otimes \chi)$ and $L_\infty(s, \pi^\vee \otimes \chi^{-1})$ are regular at $s = 1/2$. Since local L-values are always nonzero, under the additional assumption of criticality of $s = 1/2$, we get

$$L(\tfrac{1}{2}, \pi \otimes \chi) \neq 0 \iff L_f(\tfrac{1}{2}, \pi \otimes \chi) \neq 0. \tag{4.6}$$

Now one can prove arithmeticity, for which we need the following algebraicity theorem due to Manin [34] in certain special cases, and more generally due to Shimura [43]; for the version stated below, see [40].

Proposition 4.3 *Let π be a cohomological cuspidal automorphic representation of $\mathrm{GL}_2(\mathbb{A})$. There exists two nonzero complex number $p^\pm(\pi)$ such that if $s = \frac{1}{2}$ is critical then for any algebraic Hecke character χ, and any $\sigma \in \mathrm{Aut}(\mathbb{C})$ we have*

$$\sigma\left(\frac{L_f(\frac{1}{2}, \pi \otimes \chi)}{p^{\epsilon_\chi}(\pi)\mathcal{G}(\chi)(2\pi i)^{d_\infty}}\right) = \frac{L_f(\frac{1}{2}, {}^\sigma\pi \otimes {}^\sigma\chi)}{p^{\epsilon_\chi}({}^\sigma\pi)\mathcal{G}({}^\sigma\chi)(2\pi i)^{d_\infty}},$$

where $\mathcal{G}(\chi)$ is the Gauß sum attached to χ, ϵ_χ is a sign keeping track of the parity of χ, and d_∞ is an integer determined entirely by π_∞. (For more details see [40].)

A trivial corollary to the above deep proposition is that

$$L_f(\tfrac{1}{2}, \pi \otimes \chi) \neq 0 \iff L_f(\tfrac{1}{2}, {}^\sigma\pi \otimes {}^\sigma\chi) \neq 0. \tag{4.7}$$

The reader should compare this with Gross's conjecture mentioned in the introduction.

Theorem 4.4 (Arithmeticity for (GL_1, χ)-periods for GL_2) *Let π be a cohomological cuspidal automorphic representation of $\mathrm{GL}_2(\mathbb{A}_\mathbb{Q})$. Suppose that π has a nonvanishing (GL_1, χ)-period for an algebraic Hecke character of $\mathbb{Q}$. Suppose, further, that $s = 1/2$ is a critical point for the L-function $L(s, \pi \otimes \chi)$. Then ${}^\sigma\pi$ has a nonvanishing $(\mathrm{GL}_1, {}^\sigma\chi)$-period.*

Proof Follows from Proposition 4.1, (4.6) and (4.7). □

Before closing this section, let us note that the above discussion is equivalent to taking:

$$G = \mathrm{GL}_2 \times \mathrm{GL}_1, \quad \text{and} \quad H = \Delta\mathrm{GL}_1 := \left\{ \left(x, \begin{pmatrix} x & 0 \\ 0 & 1 \end{pmatrix} \right) \; : \; x \in \mathrm{GL}_1 \right\}.$$

It is from this perspective that it generalizes readily to the context of GL_n and GL_{n-1} which we discuss in Section 7.

5 Arithmeticity of Shalika Models for GL_{2n}

In the remainder of the paper, we shall consider various generalizations of the results in the previous section. One generalization of the Whittaker model for GL_2 to GL_n is the so-called Shalika model. We will first define the notion of a *Shalika model* of a cuspidal automorphic representation Π of $\mathrm{GL}_{2n}(\mathbb{A})$ where $\mathbb{A} = \mathbb{A}_F$ is the adele ring of a number field F; this particular situation was our original motivation to consider arithmeticity questions for periods. Let

$$\mathcal{S} := \left\{ s = \begin{pmatrix} h & 0 \\ 0 & h \end{pmatrix} \begin{pmatrix} 1 & X \\ 0 & 1 \end{pmatrix} \,\middle|\, \begin{matrix} h \in \mathrm{GL}_n \\ X \in \mathrm{M}_n \end{matrix} \right\} \subset \mathrm{GL}_{2n} =: G.$$

It is traditional to call $\mathcal{S}$ the Shalika subgroup of G. A character $\eta : F^\times\backslash\mathbb{A}^\times \to \mathbb{C}^\times$ and a character $\psi : F\backslash\mathbb{A} \to \mathbb{C}^\times$ can be extended to a character of $\mathcal{S}(\mathbb{A})$:

$$s = \begin{pmatrix} h & 0 \\ 0 & h \end{pmatrix} \begin{pmatrix} 1 & X \\ 0 & 1 \end{pmatrix} \mapsto (\eta \otimes \psi)(s) := \eta(\det(h))\psi(Tr(X)).$$

We will also denote $\eta(s) = \eta(\det(h))$ and $\psi(s) = \psi(Tr(X))$. For a cusp form $\varphi \in \Pi$, and a character η with $\eta^n = \omega_\Pi$, consider the integral

$$\mathcal{S}_\psi^\eta(\varphi)(g) := \int_{Z_G(\mathbb{A})\mathcal{S}(F)\backslash\mathcal{S}(\mathbb{A})} (\Pi(g)\cdot\varphi)(s)\eta^{-1}(s)\psi^{-1}(s)ds, \quad g \in \mathrm{GL}_{2n}(\mathbb{A}).$$

When $n = 1$, observe that S_ψ^η is simply the ψ-Whittaker period of GL(2), since η is forced to be the central character of Π.

The following theorem, due to the works of many people (Jacquet–Shalika [26], Asgari–Shahidi [3, 4], Hundley–Sayag [23]) gives a necessary and sufficient condition for $\mathcal{S}_\psi^\eta$ to be non–zero.

Theorem 5.1 *Let* Π *be a cuspidal automorphic representation of* $\mathrm{GL}_{2n}(\mathbb{A}_F)$. *For a pair of characters* (η, ψ), *the following are equivalent:*

(i) *There is a* $\varphi \in \Pi$ *and* $g \in G(\mathbb{A})$ *such that* $\mathcal{S}_\psi^\eta(\varphi)(g) \neq 0$.

(ii) $\mathcal{S}_\psi^\eta$ *defines an injection of* $G(\mathbb{A})$*-modules*

$$\Pi \hookrightarrow \mathrm{Ind}_{\mathcal{S}(\mathbb{A})}^{G(\mathbb{A})}[\eta \otimes \psi].$$

(iii) *Let* S *be any finite set of places containing* $S_{\Pi,\eta}$. *The twisted partial exterior square* L*-function*

$$L^S(s, \Pi, \wedge^2 \otimes \eta^{-1}) := \prod_{v \notin S} L(s, \Pi_v, \wedge^2 \otimes \eta_v^{-1})$$

has a pole at $s = 1$.

(iv) Π *is the transfer of a globally generic cuspidal automorphic representation* π *of* $\mathrm{GSpin}_{2n+1}(\mathbb{A})$ *whose central character* $\omega_\pi = \eta$.

Moreover, when these conditions hold, the transfer in (iv) is strong at all archimedean places, in the sense that it respects L-parameters.

If Π satisfies any one, and hence all, of the equivalent conditions of Theorem 5.1, then we say that Π *has an* (η, ψ)*-Shalika model*, and we call the isomorphic image $\mathcal{S}_\psi^\eta(\Pi)$ of Π under $\mathcal{S}_\psi^\eta$ a *global* (η, ψ)*-Shalika model* of Π.

There is a companion theorem to Theorem 5.1 which considers the Sym^2 L-function (see [3, 4] and [23]):

Theorem 5.2 *Let* Π *be a cuspidal automorphic representation of* $\mathrm{GL}_{2n}(\mathbb{A}_F)$. *Then the following are equivalent:*

(i) *Let* S *be any finite set of places containing* $S_{\Pi,\eta}$. *The twisted partial symmetric square* L*-function*

$$L^S(s, \Pi, \mathrm{Sym}^2 \otimes \eta^{-1}) := \prod_{v \notin S} L(s, \Pi_v, \mathrm{Sym}^2 \otimes \eta_v^{-1})$$

has a pole at $s = 1$.

(ii) Π *is the transfer of a globally generic cuspidal automorphic representation* π *of* $\mathrm{GSpin}_{2n}(\mathbb{A})$ *with connected central character* $\omega_\pi^0 = \eta$.

Moreover, when these conditions hold, the transfer in (ii) is strong at all archimedean places, in the sense that it respects L-parameters.

Finally, here is the main result of this section:

Theorem 5.3 *(Arithmeticity of Shalika periods.) Suppose that F is totally real. Let Π be a cohomological cuspidal automorphic representation of $\mathrm{GL}_{2n}(\mathbb{A}_F)$ which has an (η, ψ)-Shalika model. Then, for any $\sigma \in \mathrm{Aut}(\mathbb{C})$, ${}^\sigma\Pi$ is a cohomological cuspidal automorphic representation with a $({}^\sigma\eta, \psi)$-Shalika model.*

Proof The proof is the content of the appendix of [19]; but for the sake of completeness we give a brief sketch here, elaborating upon certain important points. The hypothesis that Π is a cohomological cuspidal representation having an (η, ψ)-Shalika model imposes certain restrictions on η. In particular, we claim that η_v *is independent of* $v \in S_\infty$.

Recall from Sections 2 and 3 that the representation Π being cohomological means that there is a highest weight $\mu = (\mu_v)_{v \in S_\infty}$, with $\mu_v = (\mu_{v,1}, \ldots, \mu_{v,2n})$ where $\mu_{v,j} \in \mathbb{Z}$ and $\mu_{v,1} \geq \mu_{v,2} \geq \cdots \geq \mu_{v,2n}$, etc., and, from (3.1) and (3.3) we have

$$\Pi_v = J_{\mu_v} = D_{\ell_{v,1}} |\ |^{-\mathsf{w}/2} \times \cdots \times D_{\ell_{v,n}} |\ |^{-\mathsf{w}/2}. \tag{5.1}$$

Then the central character of Π_v is given by

$$\omega_{\Pi_v} = (\mathrm{sgn}^{\ell_{v,1}+1} |\ |^{-\mathsf{w}}) \cdots (\mathrm{sgn}^{\ell_{v,n}+1} |\ |^{-\mathsf{w}}) = \mathrm{sgn}^{n\mathsf{w}} |\ |^{-n\mathsf{w}}.$$

Now let us invoke the hypothesis that Π has an (η, ψ)-Shalika model. Then $\eta^n = \omega_\Pi$. Hence for all $v \in S_\infty$, there is an $e_v \in \{0, 1\}$ such that $\eta_v = \mathrm{sgn}^{\mathsf{w}} |\ |^{-\mathsf{w}} \mathrm{sgn}^{e_v}$, with $ne_v \equiv 0 \pmod 2$. In fact, a stronger statement is true: for all $v \in S_\infty$ one has

$$\eta_v \ = \ \mathrm{sgn}^{\mathsf{w}} |\ |^{-\mathsf{w}}. \tag{5.2}$$

To prove (5.2), note that by Theorem 5.1, Π is a Langlands functorial transfer of a cuspidal representation of $\mathrm{GSpin}_{2n+1}(\mathbb{A})$ with central character η. Moreover, the lift is strong at the archimedean places, i.e., for each archimedean place, the L-parameter ϕ_v of Π_v factors through the dual group $\mathrm{GSp}_{2n}(\mathbb{C})$ of GSpin_{2n+1} with similitude character η_v. For $v \in S_\infty$, let (ϕ_v, U_v) be the L-parameter of Π_v, i.e., ϕ_v is a representation of $W_{F_v} = W_\mathbb{R}$ on a $2n$-dimensional $\mathbb{C}$-vector space U_v. From (5.1) we know that

$$(\phi_v, U_v) \ = \ \bigoplus_{i=1}^{n} (\phi_{v,i}, U_{v,i}),$$

where each $\phi_{v,i}$ is an irreducible 2-dimensional representation of W_{F_v}. To say that the L-parameter ϕ_v factors through $\mathrm{GSp}(2n, \mathbb{C})$, means that there is a skew-symmetric non-degenerate bilinear form B_v on U_v such that

$$B_v \in \mathrm{Hom}_{W_{F_v}}(U_v \otimes U_v, \eta_v).$$

Further, from (5.1) one knows that

$$\phi_{v,j} = \mathrm{Ind}_{\mathbb{C}^\times}^{W_\mathbb{R}}\left(re^{i\theta} \mapsto e^{i\ell_{v,j}\theta}\right) \otimes |\ |^{-\mathsf{w}/2} =: I(\ell_{v,j}) \otimes |\ |^{-\mathsf{w}/2}.$$

In particular, the dual $\phi_{v,j}^\vee$ is $I(\ell_{v,j}) \otimes |\ |^{\mathsf{w}/2}$. Hence if $i \neq j$ then $\phi_{v,i}$ is not twist-equivalent to the dual of $\phi_{v,j}$. This implies that $B_v = \bigoplus_i B_{v,i}$ and each $B_{v,i} := B|_{U_{v,i}}$ is a non-degenerate skew-symmetric bilinear form on $U_{v,i}$:

$$B_{v,i} \in \mathrm{Hom}_{W_{F_v}}(U_{v,i} \otimes U_{v,i}, \eta_v) = \mathrm{Hom}_{W_{F_v}}(\phi_{v,i} \otimes \phi_{v,i}, \eta_v).$$

But $B_{v,i}$ is skew-symmetric, hence

$$0 \neq B_{v,i} \in \mathrm{Hom}_{W_{F_v}}(\wedge^2 U_{v,i}, \eta_v) = \mathrm{Hom}_{W_{F_v}}(\det(\phi_{v,i}), \eta_v) = \mathrm{Hom}_{W_{F_v}}(\mathrm{sgn}^{\mathsf{w}}|\ |^{-\mathsf{w}}, \eta_v),$$

since $\det(\phi_{v,i}) = \mathrm{sgn}^{\ell_{v,i}+1}|\ |^{-\mathsf{w}} = \mathrm{sgn}^{\mathsf{w}}|\ |^{-\mathsf{w}}$. This proves (5.2). Together with (3.7), we conclude that

$$({}^\sigma\eta)_v = \eta_v = \eta_{\sigma^{-1}v}$$

for all $\sigma \in \mathrm{Aut}(\mathbb{C})$.

Next, it follows by Theorem 5.1 that $L^S(s, \Pi, \wedge^2 \otimes \eta^{-1})$ has a pole at $s = 1$, and thus $\Pi^\vee \cong \Pi \otimes \eta^{-1}$. For $\sigma \in \mathrm{Aut}(\mathbb{C})$, we see, by checking locally almost everywhere, that

$${}^\sigma\Pi^\vee \cong {}^\sigma\Pi \otimes {}^\sigma\eta^{-1},$$

and thus

$$L^S(s, {}^\sigma\Pi \otimes {}^\sigma\Pi \otimes {}^\sigma\eta^{-1}) = L^S(s, {}^\sigma\Pi, \mathrm{Sym}^2 \otimes {}^\sigma\eta^{-1}) \cdot L^S(s, {}^\sigma\Pi, \wedge^2 \otimes {}^\sigma\eta^{-1})$$

has a pole at $s = 1$. To prove the theorem, we need to show that the Sym^2 L-function does not have a pole at $s = 1$.

Let us suppose, for the sake of contradiction, that $L^S(s, {}^\sigma\Pi, \mathrm{Sym}^2 \otimes {}^\sigma\eta^{-1})$ has a pole at $s = 1$. Then by Theorem 5.2 , one knows that ${}^\sigma\Pi$ is a Langlands functorial transfer from a cuspidal representation of $\mathrm{GSpin}_{2n}(\mathbb{A})$ with connected central character ${}^\sigma\eta$, and this lift is strong at the archimedean places. Fix any $v \in S_\infty$ and put $w = \sigma^{-1}v$. The L-parameter of ${}^\sigma\Pi_v = \Pi_{\sigma^{-1}v} = \Pi_w$ (see (3.5)) preserves a symmetric non-degenerate bilinear form C_w with similitude character $({}^\sigma\eta)_v = \eta_v = \mathrm{sgn}^{\mathsf{w}}|\ |^{-\mathsf{w}}$. In this case, we get a non-degenerate symmetric bilinear form $C_{w,i}$ on $U_{w,i}$ and W_{F_w} preserves this form up to similitude character $\mathrm{sgn}^{\mathsf{w}}|\ |^{-\mathsf{w}}$, i.e.,

$$C_{w,i} \in \mathrm{Hom}_{W_{F_w}}(U_{v,i} \otimes U_{v,i},\ \mathrm{sgn}^{\mathsf{w}}|\ |^{-\mathsf{w}}).$$

But each $\phi_{w,i}$ being irreducible, by Schur's Lemma, we know that the space

$$\mathrm{Hom}_{W_{F_w}}(U_{v,i} \otimes U_{v,i},\ \mathrm{sgn}^{\mathsf{w}}|\ |^{-\mathsf{w}}) = \mathrm{Hom}_{W_{F_w}}(\phi_{v,i},\ \phi_{v,i}^{\vee} \otimes \mathrm{sgn}^{\mathsf{w}}|\ |^{-\mathsf{w}}),$$

is one-dimensional. Hence, the non-degenerate symmetric form $C_{w,i}$ is a multiple of the non-degenerate skew-symmetric form $B_{w,i}$; a contradiction!

□

Remark 5.4 The hypothesis that F is totally real is rather artificial. One expects the arithmeticity result to hold even without this hypothesis, however, the above proof would not go through. Suppose, for example, F is an imaginary quadratic extension, then we need to consider cohomological representations of $\mathrm{GL}_{2n}(\mathbb{C})$. For the infinite place v, the parameter of the representation Π_v, which is a $2n$-dimensional representation of $W_{\mathbb{C}} = \mathbb{C}^\times$, is of the form:

$$\phi_v = \bigoplus_{j=1}^{2n} (z \mapsto z^{a_j} \bar{z}^{b_j}).$$

(Here the a_j and b_j are half-integers; see Clozel [7, p.112].) If Π has a Shalika model, then the image of the Langlands parameter ϕ_v is inside a split torus in $\mathrm{Sp}(2n, \mathbb{C})$. But this split torus may also be viewed as sitting inside $\mathrm{SO}(2n, \mathbb{C})$. Hence, from information of $\Pi_\infty = \Pi_v$ it is not possible to deduce that the parameter of ${}^\sigma\Pi_v$ is not of orthogonal type.

Remark 5.5 The proof of Theorem 5.3 amounts to showing that

$$L^S(s, \Pi, \wedge^2 \otimes \eta^{-1}) \text{ has a pole at } s = 1 \Longrightarrow$$
$$L^S(s, {}^\sigma\Pi, \wedge^2 \otimes {}^\sigma\eta^{-1}) \text{ has a pole at } s = 1$$

under the conditions in the theorem. The same proof shows that when F is totally real,

$$L^S(s, \Pi, \mathrm{Sym}^2 \otimes \eta^{-1}) \text{ has a pole at } s = 1 \Longrightarrow$$
$$L^S(s, {}^\sigma\Pi, \mathrm{Sym}^2 \otimes {}^\sigma\eta^{-1}) \text{ has a pole at } s = 1$$

when Π is cuspidal cohomological.

6 Arithmeticity of $\mathrm{GL}(n)/F$-periods for Representations of $\mathrm{GL}(n)/E$

The argument of the previous section can be applied to prove the arithmeticity of $\mathrm{GL}_n(F)$-periods for representations of $\mathrm{GL}_n(E)$, where E is a quadratic extension of F.

More precisely, let c be the nontrivial element in $\mathrm{Gal}(E/F)$ and let $\omega_{E/F}$ be the quadratic Hecke character associated to E/F by global class field theory. Let χ be a Hecke character of $\mathbb{A}_E^\times$ whose restriction to $\mathbb{A}_F^\times$ is equal to $\omega_{E/F}$. Then for $\epsilon = \pm$, we set

$$\omega_{E/F}^\epsilon = \begin{cases} 1, & \text{if } \epsilon = +; \\ \omega_{E/F} & \text{if } \epsilon = -. \end{cases} \qquad \chi^\epsilon = \begin{cases} 1 & \text{if } \epsilon = +; \\ \chi & \text{if } \epsilon = -. \end{cases}$$

For a cuspidal representation Π of $\mathrm{GL}_n(\mathbb{A}_E)$ and $\epsilon = \pm$, we shall consider the period integral

$$\mathcal{P}^\epsilon(\varphi) = \int_{Z_F(\mathbb{A}_F)\mathrm{GL}_n(F)\backslash \mathrm{GL}_n(\mathbb{A}_F)} \varphi(h) \cdot \omega_{E/F}^\epsilon(\det(h))\, dh,$$

where $\varphi \in \Pi$. For the period integral $\mathcal{P}^\epsilon$ to have a chance to be nonvanishing, it is necessary that the central character ω_Π of Π is equal to $(\omega_{E/F}^\epsilon)^n$ when restricted to the center $Z_F(\mathbb{A}_F) = \mathbb{A}_F^\times$ of $\mathrm{GL}_n(\mathbb{A}_F)$.

Associated to Π is a pair of partial L-functions $L^S(s, \Pi, \mathrm{Asai}^\pm)$, known as the $\mathrm{Asai}^\pm$ (or twisted tensor) L-function (see [14, Section 7]). One has

$$L^S(s, \Pi, \mathrm{Asai}^-) = L^S(s, \Pi \otimes \chi, \mathrm{Asai}^+)$$

and

$$L^S(s, \Pi \times \Pi^c) = L^S(s, \Pi, \mathrm{Asai}^+) \cdot L^S(s, \Pi, \mathrm{Asai}^-)$$

where c acts on the representations of $\mathrm{GL}_n(\mathbb{A}_E)$ by

$$\Pi^c(g) = \Pi(c(g)).$$

The following theorem is a consequence of the works of many people (Kim-Krishnamurthy [30, 31], Flicker [11, 12], Ginzburg-Rallis-Soudry [16]).

Theorem 6.1 *For a cuspidal automorphic representation Π of $\mathrm{GL}_n(\mathbb{A}_E)$, the following are equivalent:*

(i) *There is a $\varphi \in \Pi$ such that $\mathcal{P}^\epsilon(\varphi) \neq 0$.*

(ii) *For a sufficiently large finite set S of places of F, the partial Asai^ϵ L-function $L^S(s, \Pi, \mathrm{Asai}^\epsilon)$ has a pole at $s = 1$.*

(iii) $\Pi \otimes \chi^{\epsilon \cdot (-1)^{n-1}}$ *is the transfer (standard base change) of a globally generic cuspidal automorphic representation π of the quasi-split $\mathrm{U}_n(\mathbb{A})$.*

Moreover, when these conditions hold, the transfer in (iii) is strong at all archimedean places of F, in the sense that it respects L-parameters.

One has a local analog of the above global theorem, which is due to the works of many people (A. Kable [28], Anandavardhanan-Kable-Tandon [1], N. Matringe [35, 36]):

Theorem 6.2 *Let v be a non-archimedean place of F which is inert in E and let Π_v be a generic representation of $\mathrm{GL}_n(E_v)$. Then the following are equivalent:*

(i) *Π_v is $(\mathrm{GL}_n(F_v), \omega^{\epsilon}_{E_v/F_v})$-distinguished.*

(ii) *The local Asai L-function $L(s, \Pi_v, \mathrm{Asai}^{\epsilon})$ has an "exceptional" pole at $s = 0$.*

(iii) *The L-parameter ϕ_v of Π_v is conjugate-self-dual with sign ϵ (in the sense of [14, Section 3]).*

In analogy with Theorem 5.3, one has the following theorem.

Theorem 6.3 *Let Π be a cohomological cuspidal automorphic representation of $\mathrm{GL}_n(\mathbb{A}_E)$ which is globally distinguished by $(\mathrm{GL}_n(\mathbb{A}_F), \omega^{\epsilon}_{E/F})$. Assume that one of the following conditions hold:*

(1) *n is odd; or*

(2) *E is a totally complex and F is totally real; or*

(3) *there is a finite place v of F which is inert in E where Π_v is discrete series.*

Then, for any $\sigma \in \mathrm{Aut}(\mathbb{C})$, ${}^{\sigma}\Pi$ is a cohomological cuspidal representation which is globally distinguished by $(\mathrm{GL}_n(\mathbb{A}_F), \omega^{\epsilon}_{E/F})$.

Proof This is similar to the proof of Theorem 5.3, exploiting Theorems 6.1 and 6.2:

$$\begin{aligned}
&\Pi \text{ is globally distinguished by } (\mathrm{GL}_n(\mathbb{A}_F), \omega^{\epsilon}_{E/F}) \\
\Longrightarrow & L^S(s, \Pi, \mathrm{Asai}^{\epsilon}) \text{ has a pole at } s = 1 \\
\Longrightarrow & \Pi^c \cong \Pi^{\vee} \\
\Longrightarrow & {}^{\sigma}\Pi^c \cong {}^{\sigma}\Pi^{\vee} \\
\Longrightarrow & L^S_E(s, {}^{\sigma}\Pi \times {}^{\sigma}\Pi^c) \text{ has a pole at } s = 1 \\
\Longrightarrow & L^S(s, {}^{\sigma}\Pi, \mathrm{Asai}^{+}) \text{ or } L^S(s, {}^{\sigma}\Pi, \mathrm{Asai}^{-}) \text{ has a pole at } s = 1.
\end{aligned}$$

Suppose that $L^S(s, {}^{\sigma}\Pi, \mathrm{Asai}^{-\epsilon})$ has a pole at $s = 1$, rather than $L^S(s, {}^{\sigma}\Pi, \mathrm{Asai}^{\epsilon})$, so that ${}^{\sigma}\Pi$ is distinguished by $(\mathrm{GL}_n(\mathbb{A}_F), \omega^{-\epsilon}_{E/F})$. We shall obtain a contradiction under one of the hypotheses (1), (2) or (3).

Under hypothesis (1), so that n is odd, we note that the central character of Π is equal to $\omega_{E/F}^{\epsilon}$ when restricted to the center of $\mathrm{GL}_n(\mathbb{A}_F)$, whereas that of ${}^{\sigma}\Pi$ is equal to $\omega_{E/F}^{-\epsilon}$. In particular, one restriction is the trivial character of $\mathbb{A}_F^{\times}$ whereas the other is the quadratic character $\omega_{E/F}$. However, at all finite places, it is clear that the central characters of Π_v and ${}^{\sigma}\Pi_v$ have the same restriction to the center of $\mathrm{GL}_n(F_v)$ since this restriction is at most a quadratic character. This gives the desired contradiction under hypothesis (1).

Now assume hypothesis (2), so that E is a totally complex extension of the totally real field F. By Theorem 6.1, $\Pi \otimes \chi^{\epsilon\cdot(-1)^n}$ is lifted from $U_n(\mathbb{A})$ and the lift is strong at archimedean places. Thus for each infinite place v of E, the L-parameter ϕ_v of Π_v is of the form

$$\phi_v = \bigoplus_i z^{a_i} \cdot \overline{z}^{-a_i}$$

with $a_1 > a_2 > \cdots > a_n$ half-integers and with each character $z \mapsto z^{a_i} \cdot \overline{z}^{-a_i}$ conjugate-self-dual with sign ϵ, i.e.,

$$2a_i = \begin{cases} \text{even, if } \epsilon = +1; \\ \text{odd, if } \epsilon = -1. \end{cases}$$

On the other hand, consider the L-parameter ϕ'_v of ${}^{\sigma}\Pi_v$. By Proposition 3.2, ${}^{\sigma}\Pi_v = \Pi_{\sigma^{-1}\cdot v}$, so that $\phi'_v = \phi_{\sigma^{-1}v}$. Thus ϕ'_v is the direct sum of characters which are conjugate-self-dual of sign ϵ. But if ${}^{\sigma}\Pi_v$ is distinguished by $(\mathrm{GL}_n(\mathbb{A}_F), \omega_{E/F}^{-\epsilon})$, Theorem 6.1 implies that ϕ'_v is the direct sum of characters which are conjugate-self-dual with sign $-\epsilon$. This gives the desired contradiction.

Finally, assume hypothesis (3). For all finite places v of F, ${}^{\sigma}\Pi_v$ is locally distinguished by $(\mathrm{GL}_n(F_v), \omega_{E_v/F_v}^{-\epsilon})$ and so its L-parameter ϕ'_v is conjugate-self-dual with sign $-\epsilon$. On the other hand, the L-parameter ϕ_v of Π_v is conjugate self-dual of sign ϵ. When Π_v is discrete series, $\phi'_v = {}^{\sigma}\phi_v$ up to the quadratic character $x \mapsto \sigma(|x|_{E_v}^{1/2})/|x|_{E_v}^{1/2}$ of $W_{E_v}^{ab} \cong E_v^{\times}$. Observe that this character is trivial on $F_v^{\times}$, so it is conjugate orthogonal in the sense of [14, Section 3]. In particular, ϕ_v and ϕ'_v are conjugate-self-dual of the same sign; this gives the desired contradiction when Π_v is discrete series.

□

7 Arithmeticity of GL_{n-1} Periods on $\mathrm{GL}_n \times \mathrm{GL}_{n-1}$

In this section, we consider the GL_{n-1}-period for cuspidal representations of $\mathrm{GL}_n \times \mathrm{GL}_{n-1}$ over $\mathbb{Q}$. This context is a very nice generalization of the example in subsection 4.2 where we studied (GL_1, χ)-periods for representations of GL_2. The nonvanishing of periods is equivalent to a certain central L-value being nonzero. If we further impose the condition that the central value is a critical value then an appropriate algebraicity theorem for this critical value gives arithmeticity. The situation is analogous to Gross's conjecture concerning order of vanishing of critical motivic L-values as discussed in the introduction.

Theorem 7.1 *Let Π be a cohomological cuspidal automorphic representation of $\mathrm{GL}_n(\mathbb{A})$, say $\Pi \in \mathrm{Coh}(\mathrm{GL}_n, \mu)$. Here $\mathbb{A}$ is the adele ring of $\mathbb{Q}$. Similarly, let $\Sigma \in \mathrm{Coh}(\mathrm{GL}_{n-1}, \lambda)$. Suppose that $\Pi \otimes \Sigma$, as a representation of $(\mathrm{GL}_n \times \mathrm{GL}_{n-1})(\mathbb{A})$, has a non-vanishing period with respect to the diagonally embedded subgroup $\mathrm{GL}_{n-1}(\mathbb{A})$. Suppose further that the coefficient systems E_μ and E_λ satisfy:*

$$\mathrm{Hom}_{\mathrm{GL}_{n-1}}(E_\mu \otimes E_\lambda,\ 1\!\!1) \neq 0.$$

Then for any $\sigma \in \mathrm{Aut}(\mathbb{C})$, the representation ${}^\sigma\Pi \times {}^\sigma\Sigma$ also has a non-vanishing period with respect to $\mathrm{GL}_{n-1}(\mathbb{A})$ under the assumption that [37, Hypothesis 3.10] holds.

Proof Every step of the proof is a suitable generalization of the proof of arithmeticity of (GL_1, χ)-periods for representations of GL_2 as in subsection 4.2.

To begin, the generalization of Proposition 4.1 goes like this: $\Pi \otimes \Sigma$ as a representation of $(\mathrm{GL}_n \times \mathrm{GL}_{n-1})(\mathbb{A})$ has a non-vanishing $\mathrm{GL}_{n-1}(\mathbb{A})$ period if and only if $L(\frac{1}{2}, \Pi \times \Sigma) \neq 0$. This follows from using the integrals studied by Jacquet, Piatetskii-Shapiro and Shalika [25] as follows. For cusp forms $\phi \in V_\Pi$ and $\phi' \in V_\Sigma$, define

$$\ell(s, \phi, \phi') = \int_{\mathrm{GL}_{n-1}(\mathbb{Q}) Z_{n-1}(\mathbb{A}) \backslash \mathrm{GL}_{n-1}(\mathbb{A})} \phi \begin{pmatrix} g & 0 \\ 0 & 1 \end{pmatrix} \phi'(g) |\det(g)|^{s-\frac{1}{2}}\, dg,$$

where Z_{n-1} is the center of GL_{n-1}. Our assumption on $\Pi \otimes \Sigma$ is that $\ell(\frac{1}{2}, \phi, \phi') \neq 0$ for some ϕ and ϕ'. Let W_ϕ and $W_{\phi'}$ be the corresponding Whittaker vectors; we may and will take ϕ and ϕ' so that W_ϕ and $W_{\phi'}$ are pure-tensors: $W_\phi = \otimes W_p$ and $W_{\phi'} = \otimes W'_p$. The unfolding argument gives

$\ell(s, \phi, \phi') = Z(s, W_\phi, W_{\phi'})$ for $\Re(s) \gg 0$, where

$$Z(s, W_\phi, W_{\phi'}) = \int_{N_{n-1}(\mathbb{A})\backslash \mathrm{GL}_{n-1}(\mathbb{A})} W_\phi \begin{pmatrix} g & 0 \\ 0 & 1 \end{pmatrix} W_{\phi'}(g)|\det(g)|^{s-\frac{1}{2}}\, dg;$$

here N_{n-1} is the subgroup of all upper triangular unipotent elements in GL_{n-1}. The analogue of (4.5) takes the form:

$$\ell(s, \phi, \phi') = \left(\prod_{p \in S} \frac{Z(s, W_p, W_p')}{L_p(s, \Pi_p \otimes \Sigma_p)} \right) \cdot L(s, \Pi \otimes \Sigma). \tag{7.1}$$

Using [25, Theorem 2.7] we get that both sides and especially both the factors on the right hand side are entire functions. Evaluating at $s = 1/2$ gives

$$\Pi \otimes \Sigma \text{ has a non-vanishing } \mathrm{GL}_{n-1}(\mathbb{A}) \text{ period} \iff L(\tfrac{1}{2}, \Pi \times \Sigma) \neq 0. \tag{7.2}$$

Next, the hypothesis that $\Pi \in \mathrm{Coh}(\mathrm{GL}_n, \mu)$ and $\Sigma \in \mathrm{Coh}(\mathrm{GL}_{n-1}, \lambda)$ puts us in an arithmetic context; however, this doesn't guarantee that $s = \frac{1}{2}$ is critical. With the foresight of wanting to appeal to algebraicity results, we impose the condition:

$$\mathrm{Hom}_{\mathrm{GL}_{n-1}}(E_\mu \otimes E_\lambda,\ 1\!\!1) \neq 0.$$

It was observed by Kasten and Schmidt [29, Theorem 2.3] that this condition implies $s = 1/2$ is critical for the Rankin-Selberg L-function $L(s, \Pi \times \Sigma)$. The same condition is also needed for an algebraicity result for critical values due the second author; see [37]. We get the analogue of (4.6) which looks like

$$L(\tfrac{1}{2}, \Pi \otimes \Sigma) \neq 0 \iff L_f(\tfrac{1}{2}, \Pi \otimes \Sigma) \neq 0. \tag{7.3}$$

Under an additional nonvanishing hypothesis involving only representations at infinity as in [37, Hypothesis 3.10], the main result of that paper, [37, Theorem 1.1], says that

$$\sigma \left(\frac{L_f(\frac{1}{2}, \Pi \times \Sigma)}{p^\epsilon(\Pi) p^\eta(\Sigma) \mathcal{G}(\omega_{\Sigma_f})\, p_\infty(\mu, \lambda)} \right) = \frac{L_f(\frac{1}{2}, \Pi^\sigma \times \Sigma^\sigma)}{p^\epsilon(\Pi^\sigma)\, p^\eta(\Sigma^\sigma)\, \mathcal{G}(\omega_{\Sigma_f^\sigma})\, p_\infty(\mu, \lambda)},$$

where $p^\epsilon(\Pi)$ and $p^\eta(\Sigma)$ are nonzero complex numbers, $\mathcal{G}(\omega_{\Sigma_f})$ is the Gauss sum of the central character of Σ, and $p_\infty(\mu, \lambda)$ is a nonzero complex number determined by μ and λ. The analogue of (4.7) follows easily:

$$L_f(\tfrac{1}{2}, \Pi \otimes \Sigma) \neq 0 \iff L_f(\tfrac{1}{2}, {}^\sigma\Pi \otimes {}^\sigma\Sigma) \neq 0. \tag{7.4}$$

The reader may easily verify that $s = 1/2$ is critical for $L(s, \Pi \otimes \Sigma)$ if and only if $s = 1/2$ is critical for $L(s, {}^\sigma\Pi \otimes {}^\sigma\Sigma)$.

Arithmeticity follows from first applying (7.2), (7.3) and (7.4) to $\Pi \otimes \Sigma$ and then applying (7.3) and (7.2) to ${}^\sigma\Pi \otimes {}^\sigma\Sigma$.

□

Remark 7.2 The hypothesis on the coefficient systems as in Theorem 7.1, which is itself a nonvanishing period like condition, is crucial for the methods of [37] to apply. Let us note that it is possible to have a pair of cohomological representations Π and Σ for which $s = 1/2$ is critical but for which that condition on the coefficients is not satisfied. For example, take $n = 3$, $\mu = (0,0,0)$ and $\lambda = (1,-1)$; then E_μ is the trivial representation of GL_3. Take $\Pi \in \mathrm{Coh}(\mathrm{GL}_3, \mu)$ and $\Sigma \in \mathrm{Coh}(\mathrm{GL}_2, \lambda)$. Then, we leave it to the reader to check that $s = 1/2$ is critical for $L(s, \Pi \times \Sigma)$, but $\mathrm{Hom}_{\mathrm{GL}_2}(E_\mu \otimes E_\lambda, \ 1\!\!1) = 0$. Now in such a situation, suppose the representation $\Pi \times \Sigma$ of $\mathrm{GL}_3 \times \mathrm{GL}_2$ has a nonvanishing GL_2 period, then the above proof is not applicable; however, we still believe that one should have arithmeticity.

Remark 7.3 In a certain work in progress [38], the second author is studying algebraicity theorems for critical values of L-functions for $\mathrm{GL}_n \times \mathrm{GL}_{n-1}$ over any number field. This would then generalize Theorem 7.1 from $\mathbb{Q}$ to any number field.

Remark 7.4 The assumption [37, Hypothesis 3.10] is a certain limitation of the technique used in that paper. We note that this hypothesis is of a purely local nature and depends only the representations Π_∞ and Σ_∞ at infinity. For $n = 2$ the validity of this hypothesis follows from an explicit calculculation; see [40]; it is this calculation that gives the term $(2\pi i)^{d_\infty}$ in Proposition 4.3. For $n = 3$ the validity of the hypothesis has been proved by Kasten and Schmidt [29].

8 Arithmeticity of $\mathrm{GL}_n \times \mathrm{GL}_n$ Periods for Cusp Forms on GL_{2n}

In this section, we discuss yet another generalization of the example in subsection 4.2 where we studied (GL_1, χ)-periods for representations π of GL_2. Indeed, in that example, we could have carried through the entire discussion by replacing π by $\pi \otimes \chi$ and taking the trivial character of $H = \mathrm{GL}_1 \times \mathrm{GL}_1$ sitting as the diagonal torus in GL_2. (This imposes the condition that the central character of $\pi \otimes \chi$ is trivial.)

Now we take $G = \mathrm{GL}_{2n}$ over a totally real number field F. Take $H = \mathrm{GL}_n \times \mathrm{GL}_n$ sitting as block diagonal matrices in G. Let Π be a cuspidal automorphic representation of $G(\mathbb{A}_F)$. We would like to analyze arithmeticity for the periods:

$$\ell(\phi) := \int_{[H]} \phi \begin{pmatrix} g_1 & 0 \\ 0 & g_2 \end{pmatrix} dg_1 dg_2, \quad \phi \in V_\Pi,$$

where $[H] = Z_G(\mathbb{A}_F)H(F)\backslash H(\mathbb{A}_F)$. Arithmeticity in this context follows from certain zeta integrals studied by Jacquet-Shalika [26] and Friedberg-Jacquet [13], and an algebraicity result due to Grobner and the second author [19].

Theorem 8.1 *Let* Π *be a cohomological cuspidal automorphic representation of* $\mathrm{GL}_{2n}(\mathbb{A}_F)$ *where* F *is a totally real number field. Suppose that*

1. Π *has a nonvanishing* H*-period;*
2. *the point* $s = \frac{1}{2}$ *is critical for* $L(s, \Pi)$.

Then for any $\sigma \in \mathrm{Aut}(\mathbb{C})$, *the representation* ${}^\sigma\Pi$ *also has a non-vanishing* H*-period.*

Proof Again, we follow the proof of arithmeticity of (GL_1, χ) for representations of GL_2 as in subsection 4.2.

To begin, by a theorem of Friedberg-Jacquet [13], Π has nonzero H-period if and only if Π admits a Shalika model and $L(1/2, \Pi) \neq 0$. Now, suppose Π has a Shalika model, then Π having a nonzero H-period being equivalent to nonvanishing of $L(1/2, \Pi)$ maybe seen as follows. For a cusp form $\phi \in V_\Pi$ define

$$\ell(s, \phi) = \int_{H(F)Z_{2n}(\mathbb{A}_F)\backslash H(\mathbb{A}_F)} \phi \begin{pmatrix} g_1 & 0 \\ 0 & g_2 \end{pmatrix} \left| \frac{\det(g_1)}{\det(g_2)} \right|^{s-\frac{1}{2}} dg_1 dg_2,$$

where Z_{2n} is the center of GL_{2n}. Our assumption on Π is that $\ell(\frac{1}{2}, \phi) \neq 0$ for some ϕ. Let $\mathcal{S}_\phi$ be the corresponding vector in the Shalika model of Π; as before, we may and will take ϕ so that $\mathcal{S}_\phi$ is a pure-tensor: $\mathcal{S}_\phi = \otimes \mathcal{S}_p$. (For details concerning Shalika models and related matters we refer the reader to [19], and recommend that any serious reader of this section should have that paper by one's side.)

An unfolding argument ([19, Proposition 3.1.5]) gives $\ell(s, \phi) = Z(s, \mathcal{S}_\phi)$ where

$$Z(s, \mathcal{S}_\phi) = \int_{\mathrm{GL}_n(F)\backslash \mathrm{GL}_n(\mathbb{A}_F)} \mathcal{S}_\phi \begin{pmatrix} g & 0 \\ 0 & 1 \end{pmatrix} |\det(g)|^{s-\frac{1}{2}} \, dg.$$

The analogue of (7.1) takes the form:

$$\ell(s,\phi) = \left(\prod_{p \in S} \frac{Z(s,\mathcal{S}_p)}{L_p(s,\pi_p)} \right) \cdot L(s,\pi). \tag{8.1}$$

Using [19, Proposition 3.3.1] we get that the left hand side and both the factors on the right hand side are entire functions. Evaluating at $s = 1/2$ gives

$$\Pi \text{ has a non-vanishing } H\text{-period} \iff L(\tfrac{1}{2},\Pi) \neq 0. \tag{8.2}$$

Next, the hypothesis that $\Pi \in \mathrm{Coh}(\mathrm{GL}_{2n},\mu)$ puts us in an arithmetic context, however, as before, this doesn't guarantee that $s = \frac{1}{2}$ is critical. So, we now need the assumption that $s = \frac{1}{2}$ is critical for $L(s,\Pi)$. (In the $\mathrm{GL}_n \times \mathrm{GL}_{n-1}$ case, we needed a stronger condition on the coefficient system, but in the current context [19, Proposition 6.3.1] guarantees that.) The analogue of (7.3) is:

$$L(\tfrac{1}{2},\Pi) \neq 0 \iff L_f(\tfrac{1}{2},\Pi) \neq 0. \tag{8.3}$$

Under the assumption that $\Pi \in \mathrm{Coh}(\mathrm{GL}_{2n},\mu)$ has a Shalika model, the algebraicity result in [19, Theorem 7.1.2] says

$$\sigma\left(\frac{L_f(\frac{1}{2},\Pi\otimes\chi)}{\omega^{\epsilon_\chi}(\Pi)\mathcal{G}(\chi)\,\omega_\infty(\mu)}\right) = \frac{L_f(\frac{1}{2},{}^\sigma\Pi\otimes{}^\sigma\chi)}{\omega^{\epsilon_{\sigma\chi}}({}^\sigma\Pi)\,\mathcal{G}({}^\sigma\chi)\,\omega_\infty(\mu)},$$

where χ is an algebraic Hecke character, ϵ_χ its parity, $\mathcal{G}(\chi)$ its Gauß sum, $\omega^{\epsilon_\chi}(\Pi)$ is a nonzero complex number, and $\omega_\infty(\mu)$ is a nonzero complex number determined by μ. The analogue of (7.4) follows easily:

$$L_f(\tfrac{1}{2},\Pi) \neq 0 \iff L_f(\tfrac{1}{2},{}^\sigma\Pi) \neq 0. \tag{8.4}$$

The reader may easily verify that $s = 1/2$ is critical for $L(s,\Pi)$ if and only if $s = 1/2$ is critical for $L(s,{}^\sigma\Pi)$.

Arithmeticity follows from first applying (8.2), (8.3) and (8.4) to Π and then noting that ${}^\sigma\Pi$ has a Shalika model by Theorem 5.3 and then applying (8.3) and (8.2) to ${}^\sigma\Pi$.

□

9 Arithmeticity for Classical Groups

In this section, we consider the possibility of extending Theorem 3.1 for GL(N) to the case of classical groups. By the recent work [2] of Arthur and others, one now has a classification of square-integrable automorphic representations for quasi-split classical groups, in terms of automorphic representations of $\mathrm{GL}(N)$. In view of this, it is natural to ask if arithmeticity results for $\mathrm{GL}(N)$ can be transferred to these classical groups.

More precisely, let G be a quasi-split symplectic, special orthogonal or unitary group over the number field F and let π be a cuspidal automorphic representation of $G(\mathbb{A}_F)$. By Arthur [2], one can attach a discrete A-parameter to π and this is a multiplicity-free formal sum

$$\Psi = \Pi_1 \boxtimes S_{r_1} \boxplus \cdots \boxplus \Pi_k \boxtimes S_{r_k},$$

where Π_i is a cuspidal automorphic representation of $\mathrm{GL}(n_i)$ (over F or a quadratic extension E) satisfying some symmetry conditions and S_{r_i} is the r_i-dimensional irreducible representation of $\mathrm{SL}_2(\mathbb{C})$. Moreover, the set of all π's with a given discrete A-parameter Ψ is a full near equivalence class in the automorphic discrete spectrum of G. If $r_i = 1$ for all i, Ψ is called a tempered A-parameter.

Now suppose further that π is cohomological. Recall from Proposition 2.1 that for any $\sigma \in \mathrm{Aut}(\mathbb{C})$, there is a square-integrable automorphic representation τ_σ of $G(\mathbb{A}_F)$ such that

$$\tau_{\sigma,f} \cong {}^\sigma\pi_f.$$

Note that, for fixed $\sigma \in \mathrm{Aut}(\mathbb{C})$, τ_σ may not be uniquely determined, but any two such candidates are nearly equivalent to each other and thus have the same A-parameter. It is thus natural to ask:

Question: How is the A-parameter of τ_σ related to that of π?

In the remainder of this section, we shall consider this question for the *quasi-split* classical groups G when the A-parameter Ψ of π is *tempered*. In this case, the A-parameter Ψ of π is equal to the L-parameter of π. We may also regard Ψ as the representation

$$\Pi = \boxplus_i \Pi_i := \mathrm{Ind}_P^{\mathrm{GL}(N)}(\otimes_i \Pi_i) \quad \text{of } \mathrm{GL}(N).$$

One knows moreover that this induced representation is irreducible. In the following, we shall use Ψ and Π interchangeably for the A-parameter of π. In particular, for each v, the L-parameter of π_v is precisely the L-parameter of the generic representation $\Psi_v = \Pi_v = \Pi_{1,v} \boxplus \cdots \boxplus \Pi_{k,v}$.

Let us also explicate the symmetry condition satisfied by the summands Π_i in Ψ in the various cases:

- if $G = \mathrm{Sp}(2n)$ or $\mathrm{SO}(2n)$, then $L^S(s, \Pi_i, \mathrm{Sym}^2)$ has a pole at $s = 1$ for each i;
- if $G = \mathrm{SO}(2n+1)$, then $L^S(s, \Pi_i, \wedge^2)$ has a pole at $s = 1$;

- if $G = \mathrm{U}(n)$, then $L^S(s, \Pi_i, \mathrm{Asai}^{(-1)^{n-1}})$ has a pole at $s = 1$.

Remark 9.1 Henceforth, when $G = \mathrm{U}(n)$, we shall assume that the underlying Hermitian space is defined with respect to a totally complex quadratic extension E of the totally real base field F. Moreover, the target group of the functorial lifting is $\mathrm{GL}(n)$ over the CM field E. In the rest of this section, in order to simplify the exposition, we shall only give proofs for the symplectic and orthogonal groups, even though the results are stated for $\mathrm{U}(n)$ as well.

It is natural to first investigate if the functorial transfer of unramified representations from classical groups to $\mathrm{GL}(N)$ is $\mathrm{Aut}(\mathbb{C})$-equivariant.

Lemma 9.2 *Let k be a p-adic field. Let G be an unramified classical group over k and for an unramified representation π of $G(k)$, let $\Sigma(\pi)$ be its functorial transfer to the appropriate* $\mathrm{GL}(N)$.

(i) *If* $G = \mathrm{SO}(2n+1)$, $\mathrm{Sp}(2n)$ *or* $\mathrm{U}(n)$, *then*

$$^\sigma\Sigma(\pi) = \Sigma(^\sigma\pi).$$

(ii) *If* $G = \mathrm{SO}(2n)$, *then*

$$^\sigma(\Sigma(\pi) \otimes |\ |^{-1/2}) \otimes |\ |^{1/2} = \Sigma(^\sigma\pi).$$

Proof Let $T \subset B \subset G$ be a maximal torus contained in a Borel subgroup of G, both defined over k. Similarly, let $T^* \subset B^* \subset \mathrm{GL}(N)$, where $\mathrm{GL}(N)$ is the target of the functorial transfer from G. Now Langlands functoriality gives rise to a transfer $\chi \mapsto \chi^*$ from the set of unramified characters of T to that of T^*. Moreover, this transfer from T to T^* is $\mathrm{Aut}(\mathbb{C})$-equivariant.

Now $\pi \subset I(\chi) := \mathrm{Ind}_B^G\chi$ for some unramified character χ of T (where the induction is normalized here). The functorial transfer of $I(\chi)$ is then equal to $I(\chi^*)$. For $\sigma \in \mathrm{Aut}(\mathbb{C})$, $^\sigma\pi$ is still an unramified representation and

$$^\sigma\pi \subset {}^\sigma I(\chi) = I(^\sigma\chi \cdot \delta_{G,\sigma})$$

where $\delta_{G,\sigma} = {}^\sigma\delta_B^{1/2}/\delta_B^{1/2}$ and $\delta_B^{1/2}$ is the modulus character.

Thus, we are interested in whether

$$I(^\sigma\chi \cdot \delta_{G,\sigma}) \mapsto I(^\sigma\chi^* \cdot \delta_{\mathrm{GL}(N),\sigma})$$

under functorial transfer. Since $^\sigma\chi \mapsto {}^\sigma\chi^*$, it suffices to verify whether $\delta_{G,\sigma} \mapsto \delta_{\mathrm{GL}(N),\sigma}$.

Now we note:

- when $G = \mathrm{Sp}(2n)$, $\mathrm{SO}(2n)$, $\mathrm{U}(2n+1)$ or $\mathrm{GL}(2n+1)$, $\delta_B^{1/2}$ takes value in $\mathbb{Q}^\times$ and so $\delta_{G,\sigma} = 1$.
- when $G = \mathrm{SO}(2n+1)$, $\mathrm{U}(2n)$ or $\mathrm{GL}(2n)$, $\delta_B^{1/2}$ takes value in $q^{\frac{1}{2}\mathbb{Z}}$ where q is the size of the residue field of k, so that $\delta_{G,\sigma}$ is a quadratic character (possibly trivial).

In the context of (i), one sees that the transfer of $\delta_{G,\sigma}$ to T^* is $\delta_{\mathrm{GL}(N),\sigma}$. On the other hand, when $G = \mathrm{SO}(2n)$, $\delta_{G,\sigma}$ is trivial whereas $\delta_{\mathrm{GL}(2n),\sigma}$ is not necessarily trivial. A short computation then gives the result in (ii).

□

Lemma 9.2 already implies that when $G \neq \mathrm{SO}(2n)$, the transfer (in the sense of Arthur) of $\tau_{\sigma,f}$ to $\mathrm{GL}(N)$ is nearly equivalent to the abstract representation ${}^\sigma\Pi_f$. Next, we note the following crucial proposition:

Proposition 9.3 *Let G be a classical group over a totally real F. Assume that π is a cohomological cuspidal representation of $G(\mathbb{A})$ with tempered A-parameter. Then for all infinite places v, π_v is "as close to being a discrete series representation as possible". More precisely,*

(i) *if $G(F_v)$ has discrete series representations, then π_v is discrete series. This is the case precisely when $G(F_v)$ is not of the form $\mathrm{SO}(2a+1, 2b+1)$ with $a, b \in \mathbb{Z}$.*
(ii) *If $G(F_v) = \mathrm{SO}(2a+1, 2b+1)$, then $\pi_v = \mathrm{Ind}_P^G \chi \boxtimes \pi_{0,v}$ where P is a maximal parabolic subgroup with Levi factor $\mathrm{GL}_1(F_v) \times \mathrm{SO}(2a, 2b)$, $\chi = 1$ or sgn, and $\pi_{0,v}$ is a discrete series representation of $\mathrm{SO}(2a, 2b)$.*

In particular, π_v is tempered in all cases. When $G \neq \mathrm{SO}(2n)$, the functorial transfer of π_v to $\mathrm{GL}(N)$ is a cohomological representation.

Proof We shall treat only symplectic and orthogonal groups; the case of unitary groups is similar. We make the following observations:

(a) Since π is cohomological, the infinitesimal character of π_v (for any infinite place v) is (strongly) regular and integral.
(b) Since π has tempered A-parameter, the results of Luo-Rudnick-Sarnak [33] imply that π_v is "close to being tempered".

Now we can express π_v as a quotient of an induced representation $I = \mathrm{Ind}_{P_v}^{G_v} \Sigma$, where

- P_v is a parabolic subgroup with Levi factor

$$\mathrm{GL}_1(F_v)^a \times \mathrm{GL}_2(F_v)^b \times G_{0,v}$$

where $G_{0,v}$ is a classical group of the same type as G_v;

- The representation Σ has the form

$$\Sigma = \left(\bigotimes_{i=1}^{a} \chi_i | - |^{s_i} \right) \bigotimes \left(\bigotimes_{j=1}^{b} \tau_j | - |^{t_i} \right) \bigotimes \pi_{0,v}$$

  with χ_i either the trivial or the sign character, τ_j are discrete series representations of $\mathrm{GL}_2(\mathbb{R})$ whose central character are trivial or sign, $\pi_{0,v}$ is a discrete series representation of $G_{0,v}$ and $s_i, t_j \in \mathbb{C}$.

The integrality of the infinitesimal character of π_v implies that the numbers s_i and t_j are half-integers. The "close to temperedness" of π_v in (b) above implies that the numbers s_i and t_j are close to the imaginary axis. Taken together, they imply that $s_i = t_j = 0$. In particular, we deduce that π_v is tempered.

We have yet to make use of the regularity condition in (a). Let us fix a maximal torus T_v of G_v with associated dual complexified Lie algebra $\mathfrak{t}^*$. The character group $X(T_v)$ endows $\mathfrak{t}^*$ with an integral structure. The infinitesimal character of π_v can be regarded as an element in $\mathfrak{t}^*$ up to conjugacy by the absolute Weyl group $W(G_v, T_v)$. Now we observe:

- there are no GL_2-factors in P_v.

  To see this, note that if there were a GL_2-factor in P_v, then the infinitesimal character of π_v will have the form $(\cdots, k, -k, \cdots)$, with k a half integer. But for the classical groups, there is a nontrivial element of the absolute Weyl group which fixes such an element, namely "exchanging the coordinates k and $-k$, followed by changing the signs of both coordinates". This contradicts regularity.

- there is at most one GL_1-factor in P_v. Indeed, there can be a GL_1-factor if and only if $G_v = \mathrm{SO}(2a+1, 2b+1)$.

  To see this, note that if there were two GL_1-factors in P_v, then the infinitesimal character of π_v will have the form $(\cdots, 0, 0, \cdots)$ which is fixed by a non-trivial element of the absolute Weyl group. Further, if there is a GL_1-factor, then the infinitesimal character has the form $(\cdots, 0, \cdots)$ and this is fixed by a nontrivial element of the absolute Weyl group of type B and C, namely "changing the sign a coordinate". Thus, a GL_1-factor can only occur in type D, so that $G = \mathrm{SO}(a, b)$ with $a + b$ even. If a GL_1-factor does occur, then $\mathrm{SO}(a-1, b-1)$ must have a discrete series representation, so that a and b must both be odd. Conversely, if a and b are both odd, there must be a GL_1-factor, since $\mathrm{SO}(a, b)$ does not have discrete series representations.

Summarizing the above observations, we see that unless $G_v = \mathrm{SO}(2a+1, 2b+1)$, there are no GL_1 or GL_2 factors in P_v, so that $G_{0,v} = G_v$ and π_v is discrete series. When $G_v = \mathrm{SO}(2a+1, 2b+1)$, G_v does not have discrete series and π_v is of the form given in (ii).

To prove the last assertion of the proposition, let us explicate the discrete series L-parameter Ψ_v of π_v when $G \neq \mathrm{SO}(2n)$:

- if $G = \mathrm{Sp}(2n)$, then $\Psi_v = \oplus_{j=1}^n \phi_j \oplus \chi$, where $\chi = 1$ or sign, and the ϕ_j's are pairwise distinct orthogonal representations of the Weil group $W_{\mathbb{R}}$ of $\mathbb{R}$, which correspond to discrete series representations of $\mathrm{GL}_2(\mathbb{R})$ with central character sign. Moreover, $(-1)^n \cdot \chi(-1) = 1$.
- if $G = \mathrm{SO}(2n+1)$, then $\Psi_v = \oplus_{j=1}^n \phi_j$ where the ϕ_j's are pairwise distinct symplectic representations of $W_{\mathbb{R}}$, which correspond to discrete series representations of $\mathrm{GL}_2(\mathbb{R})$ with trivial central character.
- if $G = \mathrm{U}(n)$, then $\Psi_v = \oplus_{j=1}^n \chi_j$, where the χ_j's are conjugate dual character of $W_{\mathbb{C}} = \mathbb{C}^\times$ of the form $\chi_j(z) = (z/\overline{z})^{\frac{a_j}{2}}$ with $a_j \equiv n+1 \mod 2$.

From this and the description of cohomological representations of $\mathrm{GL}(N)$ given in Section 3, we observe that the representation Π_v with L-parameter Ψ_v is cohomological. The proposition is proved.

□

Recall that the A-parameter of π is $\Pi := \Pi_1 \boxplus \cdots \boxplus \Pi_k$. As noted by Clozel, however, it is better to work with a Tate-twisted isobaric sum $\overset{T}{\boxplus}$:

$$\Pi = \boxplus_i \Pi_i = \overset{T}{\boxplus}_i \, \Sigma_i$$

where

$$\Sigma_i = \Pi_i \otimes |\ |^{\frac{n_i - N}{2}}.$$

Then the following corollary follows from Proposition 9.3:

Corollary 9.4 (i) *Suppose that* $G = \mathrm{Sp}(2n)$, $\mathrm{SO}(2n+1)$ *or* $\mathrm{U}(n)$. *Then* $\Pi = \overset{T}{\boxplus}_i \, \Sigma_i$ *is a cohomological representation of* $\mathrm{GL}(N)$ *and for each* i, Σ_i *is a cohomological cuspidal representation of* $\mathrm{GL}(n_i)$.

(ii) *When* $G = \mathrm{SO}(2n)$, *then* $\Pi \otimes |\ |^{-1/2}$ *is an algebraic representation of* $\mathrm{GL}(N)$ *in the sense of [7], and for each* i, $\Sigma_i \otimes |\ |^{-1/2}$ *is an algebraic representation of* $\mathrm{GL}(n_i)$.

Now we come to the main result of this section.

Theorem 9.5 *Let F be totally real and let G be a quasi-split classical group over F with $G = \mathrm{Sp}(2n)$, $\mathrm{SO}(2n+1)$ or $\mathrm{U}(n)$ (c.f. Remark 9.1). Let π be a cohomological cuspidal representation of G with a tempered A-parameter $\Psi = \boxplus_i \Pi_i = \overset{T}{\boxplus}_i \Sigma_i$. Then we have:*

(i) *For $\sigma \in \mathrm{Aut}(\mathbb{C})$, let τ_σ be a square-integrable automorphic representation such that $\tau_{\sigma,f} \cong {}^\sigma\pi_f$. Then the A-parameter of τ_σ is*

$$
{}^\sigma\Psi := {}^\sigma\Sigma_1 \overset{T}{\boxplus} \cdots \overset{T}{\boxplus} {}^\sigma\Sigma_k.
$$

(ii) *For each infinite place v, the L-parameter ${}^\sigma\Psi_v$ of $\tau_{\sigma,v}$ is equal to*

$$
\Psi_{\sigma^{-1}v} = \Pi_{1,\sigma^{-1}v} \boxplus \cdots \boxplus \Pi_{k,\sigma^{-1}v}.
$$

In particular, $\tau_{\sigma,v}$ is a discrete series representation for each infinite place v and thus τ_σ is cuspidal.

Proof Again, we shall treat only symplectic and orthogonal groups in the proof; the case of unitary groups is similar.

(i) Let us first check that ${}^\sigma\Psi$ is actually an A-parameter for G. By Corollary 9.4 and Theorem 3.1, one has the cuspidal automorphic representations ${}^\sigma\Sigma_i$ for each i. By Remark 5.5 and its analog for $\mathrm{U}(n)$ (cf. Theorem 6.3), ${}^\sigma\Sigma_i$ has the same symmetry type as Σ_i (as detected by the poles of the twisted exterior square, symmetric square or the Asai L-function in the respective cases) and hence as Ψ. Thus ${}^\sigma\Psi$ is a bona-fide A-parameter for G and has an associated near equivalence class of square-integrable automorphic representations of G.

To prove (i), we need to show that the representation τ_σ is contained in the near equivalence class associated to ${}^\sigma\Psi$. This follows immediately by Lemma 9.2, since

$$
{}^\sigma\Pi_v \;\cong\; \overset{T}{\boxplus}_i \, {}^\sigma\Sigma_{i,v}.
$$

(ii) In the context of the theorem, one knows by Proposition 3.2 that

$$
{}^\sigma\Psi_v \;=\; \overset{T}{\boxplus}_i \, {}^\sigma\Sigma_{i,v} \;=\; \overset{T}{\boxplus}_i \, \Sigma_{i,\sigma^{-1}v} \;=\; \boxplus_i \Pi_{i,\sigma^{-1}v}.
$$

To be more precise, we have used Proposition 3.2 to deduce that ${}^\sigma\Sigma_{i,v} = \Sigma_{i,\sigma^{-1}v}$. This follows immediately from Proposition 3.2 if n_i is even or if $F_v = \mathbb{C}$. Thus, the only issue is when n_i is odd and $F_v = \mathbb{R}$. Such a situation only occurs when $G = \mathrm{Sp}(2n)$. But it follows by Proposition 9.3

(or rather the last paragraph of its proof) that there is a unique i_0 such that n_{i_0} is odd. However, for this particular i_0, one has $(-1)^n \cdot \epsilon(\Pi_{i_0,v}) = 1$ for each infinite place v, where $\epsilon(\Pi_{i_0,v})$ is the sign in Proposition 3.2. In particular, $\epsilon(\Pi_{i_0,v})$ is independent of v, so that ${}^{\sigma}\Pi_{i_0,v} = \Pi_{i_0,\sigma^{-1}v}$.

Finally, since the archimedean component ${}^{\sigma}\Psi_v = \Psi_{\sigma^{-1}v}$ is a discrete series L-parameter for G_v and is the L-parameter of $\tau_{\sigma,v}$, we deduce that $\tau_{\sigma,v}$ is a discrete series representation. Since τ_σ is a square-integrable automorphic representation, it then follows by a well-known result of Wallach [48] that τ_σ is cuspidal. This proves (ii).

□

Corollary 9.6 *In the context of the theorem, suppose that $\Psi = \Pi$ is a cuspidal representation of* GL(N). *Then ${}^{\sigma}\pi := (\otimes_{v\in S_\infty} \pi_{\sigma^{-1}v}) \otimes {}^{\sigma}\pi_f$ is a cohomological cuspidal automorphic representation of G.*

Proof Under the hypothesis of the corollary, the A-packet associated to Ψ is stable, in the sense that every member of the abstract global A-packet is automorphic. In this case, one may replace $\tau_{\sigma,\infty}$ by ${}^{\sigma}\pi_\infty$ and thus take τ_σ to be ${}^{\sigma}\pi$.

□

10 Arithmeticity of Periods for Classical Groups

After Theorem 9.5, it makes sense to consider the question of arithmeticity of periods for classical groups. We consider two examples here.

10.1 Whittaker periods

For a quasi-split group G over F, if we fix a maximal F-torus T contained in a Borel subgroup $B = T \cdot N$ over F, then for any generic automorphic character ψ of $N(\mathbb{A}_F)$, we may consider the ψ-Whittaker period of an automorphic representation π of G. If this period is nonzero on π, we say that π is ψ-generic.

The group $T(F)$ acts naturally on the set of generic automorphic caracters of $N(\mathbb{A}_F)$ and the notion of being ψ-generic depends only on the $T(F)$-orbit of ψ. In particular, if $T(F)$ acts transitively on the set of generic automorphic characters of $N(\mathbb{A}_F)$, then there is no harm in suppressing ψ. This is the case when $G = \mathrm{SO}(2n+1)$ and $\mathrm{U}(2n+1)$. When $G = \mathrm{Sp}(2n)$, the set of $T(F)$-orbits of generic automorphic characters is a torsor of $F^\times/F^{\times 2}$, whereas if $G = \mathrm{U}(2n)$, it is a torsor of $F^\times/\mathbb{N}_{E/F}E^\times$.

Now we have:

Theorem 10.1 *Let F be a totally real number field, and let $G = \mathrm{SO}(2n+1)$ or $\mathrm{U}(2n+1)$. Let π be a cohomological cuspidal automorphic representation of $G(\mathbb{A}_F)$ which is globally generic. Then for any $\sigma \in \mathrm{Aut}(\mathbb{C})$, the conjugated representation ${}^\sigma\pi$ (as defined in Corollary 9.6) is also a cohomological cuspidal automorphic representation of $G(\mathbb{A}_F)$ which is globally generic.*

Proof Since π is globally generic, it follows by [2] and [9] that the A-parameter Ψ of π is tempered. By Theorem 9.5, we know that ${}^\sigma\pi$ belongs to the global A-packet associated to the tempered parameter ${}^\sigma\Psi$. Moreover, ${}^\sigma\pi_v$ is locally ${}^\sigma\psi_v$-generic for all finite places v. But since all generic characters of $N(F_v)$ are in the same $T(F_v)$-orbit, we deduce that ${}^\sigma\pi_v$ is ψ_v-generic as well. The same holds at the infinite places, since ${}^\sigma\pi_v = \pi_{\sigma^{-1}v}$. Thus, we see that ${}^\sigma\pi$ is abstractly ψ-generic.

Now in a local L-packet of $G(F_v)$, there can be at most one ψ_v-generic representation; this is a consequence of the theory of local descent (see Jiang-Soudry [27] for the case of $G = \mathrm{SO}(2n+1)$). Thus, ${}^\sigma\pi$ is the only member of its A-packet which could be globally ψ-generic. However, the theory of global descent [16] says that a tempered A-packet must contain a globally ψ-generic cuspidal automorphic representation. Thus we conclude that ${}^\sigma\pi$ is a globally generic cohomological cuspidal representation.

□

Remark 10.2 When $G = \mathrm{Sp}(2n)$ or $\mathrm{U}(2n)$, it is still true that the representation ${}^\sigma\pi$ is abstractly generic with respect to the generic character $(\otimes_{v\in S_\infty}\psi_{\sigma^{-1}v}) \otimes (\otimes_{v\notin S_\infty}{}^\sigma\psi_v)$. However, we do not know whether the $T(\mathbb{A}_F)$-orbit of this generic character contains an automorphic character of $N(\mathbb{A}_F)$. Nevertheless, if $\sigma \in \mathrm{Aut}(\mathbb{C}/\mathbb{Q}^{\mathrm{ab}})$, then ${}^\sigma\psi_v = \psi_v$ for all finite v, and so we conclude as in the theorem that ${}^\sigma\pi$ is globally ψ-generic again. One may avoid this complication by working with similitude groups, for example $\mathrm{GSp}(2n)$; however, one needs to await the generalization of Arthur's results [2] to the context of similitude groups.

10.2 Gross-Prasad period

In this speculative final subsection, we consider the Gross-Prasad periods for the classical groups. To be concrete, let us consider the Gross-Prasad period for unitary groups. Let $\pi = \pi_1 \boxtimes \pi_2$ be a tempered cuspidal representation of $G = \mathrm{U}(n) \times \mathrm{U}(n-1)$. Then a recent preprint of Wei Zhang [49] establishes the global Gross-Prasad conjecture under some local hypotheses. In particular, he shows that the period of π over the diagonally embedded $\mathrm{U}(n-1)$ is nonzero if and only if π_v is $\Delta U(n-1)$-distinguished for all v, and

$$L_E(\tfrac{1}{2}, \Pi_1 \times \Pi_2) \neq 0.$$

Here, Π_i denotes the transfer of π to $\mathrm{GL}(n)$ or $\mathrm{GL}(n-1)$ over E.

Assume now that π is stable and cohomological, say $\pi \in \mathrm{Coh}(G, \mu)$. Suppose that π has a nonvanishing period over the diagonally embedded $\mathrm{U}(n-1)$, and that $\mathrm{Hom}_{U(n-1)}(\mu, \mathbb{C}) \neq 0$. Then by Theorem 9.5 and its corollary, one knows that ${}^{\sigma}\pi$ is also a cohomological cuspidal automorphic representation of $\mathrm{U}(n) \times \mathrm{U}(n-1)$. Now one may apply the same argument as in the proof of Theorem 7.1 (with the hypotheses stated there) to deduce that ${}^{\sigma}\pi$ also has nonvanishing period over the diagonally embedded $\mathrm{U}(n-1)$. We will perhaps leave the detailed treatment of this to a future occasion.

References

[1] U. K. Anandavardhanan, A. Kable and R. Tandon, *Distinguished representations and poles of twisted tensor L-functions*, Proc. Amer. Math. Soc. **132** (2004), no. 10, 2875–2883.

[2] J. Arthur, *The endoscopic classification of representations: orthogonal and symplectic groups*, Book in preparation; preliminary version available from `http://www.claymath.org/cw/arthur/`.

[3] M. Asgari and F. Shahidi, *Generic transfer for General Spin Groups*, Duke Math. J., **132**, (2006) 137–190.

[4] M. Asgari and F. Shahidi, *Functoriality for General Spin Groups*, `arXiv:1101.3467v1`, preprint (2011).

[5] A. Borel, *Regularization theorems in Lie algebra cohomology. Applications*, Duke Math. J. **50** (1983), no. 3, 605–623.

[6] D. Blasius, *Period relations and critical values of L-functions*, Olga Taussky-Todd: in memoriam. Pacific J. Math., (1997), Special Issue, 53–83.

[7] L. Clozel, *Motifs et Formes Automorphes: Applications du Principe de Fonctorialité.* in: *Automorphic forms, Shimura varieties, and L-functions*, Vol. I, Perspect. Math., vol. 10, eds. L. Clozel and J. S. Milne, (Ann Arbor, MI, 1988) Academic Press, Boston, MA, 1990, pp. 77–159.

[8] J. Cogdell, *Lectures on L-functions, converse theorems, and functoriality for GL_n*, Lectures on automorphic L-functions, 1–96, Fields Inst. Monogr., 20, Amer. Math. Soc., Providence, RI, 2004.

[9] J. Cogdell, H. Kim, Piatetski-Shapiro and F. Shahidi, *On lifting from classical groups to* GL_n, Publ. Math. Inst. Hautes Études Sci. No. 93 (2001), 5–30.

[10] P. Deligne, *Valeurs de fonctions L et périodes d'intégrales. With an appendix by N. Koblitz and A. Ogus*, In: Proc. Sympos. Pure Math., Vol. XXXIII, part II, AMS, Providence, R.I., (1979) pp. 313–346.

[11] Y. Flicker, *Twisted tensors and Euler products*, Bull. Soc. Math. France **116** (1988) no. 3, 295–313.

[12] Y. Flicker, *On distinguished representations*, J. Reine Angew. Math. **418** (1991), 139–172.

[13] S. Friedberg and H. Jacquet, *Linear periods*, J. Reine Angew. Math. **443** (1993), 91–139.

[14] W.T. Gan, B.H. Gross and D. Prasad, *Symplectic local root numbers, central critical L-values and restriction problems in the representation theory of classical groups*, to appear in Astérisque.

[15] S. Gelbart, *Automorphic forms on adele groups*, Annals of Math. Studies, Number 83, Princeton University Press (1975).

[16] D. Ginzburg, S. Rallis and D. Soudry, *The descent map from automorphic representations of GL(n) to classical groups*, World Scientific Publishing Co. Pte. Ltd., Hackensack, NJ, 2011. x+339 pp.

[17] G. Gotsbacher and H. Grobner, *On the Eisenstein Cohomology of Odd Orthogonal Groups*, to appear in Forum Math.

[18] H. Grobner and A. Raghuram, *On some arithmetic properties of automorphic forms of GL_m over a division algebra*, `arXiv:1102.1872v2` preprint (2011).

[19] H. Grobner and A. Raghuram, *On the arithmetic of Shalika models and the critical values of L-functions for* GL_{2n}, *with an appendix by Wee Teck Gan*, to appear in Amer. J. Math.

[20] B. Gross and M. Reeder, *From Laplace to Langlands via representations of orthogonal groups*, Bulletin AMS, **43** (2006), 163–205.

[21] G. Harder, *General aspects in the theory of modular symbols.* Seminar on number theory, Paris 1981–82 (Paris, 1981/1982), 73–88, Progr. Math., 38, Birkhauser Boston, Boston, MA, 1983.

[22] G. Henniart, *Sur la conjecture de Langlands locale pour* GL_n, 21st Journées Arithmétiques (Rome, 2001). J. Théor. Nombres Bordeaux 13, no. 1, 167–187 (2001).

[23] J. Hundley and E. Sayag, *Descent construction for GSpin groups: Main results and applications*, Electronic Research Announcements in Mathematical Sciences, **16** (2009) 31–36.

[24] H. Jacquet and R. Langlands, *Automorphic forms on GL(2).* Lecture Notes in Mathematics, Vol. 114. Springer-Verlag, Berlin-New York, 1970.

[25] H. Jacquet, I.I. Piatetskii-Shapiro, J.A. Shalika, *Rankin–Selberg Convolutions*, American J. of Math., **105** No. 2, (1983), 367–464.

[26] H. Jacquet, J. Shalika, *Exterior square L-functions*, in: *Automorphic forms, Shimura varieties, and L-functions*, Vol. II, Perspect. Math., vol. 10, eds. L. Clozel and J. S. Milne, (Ann Arbor, MI, 1988) Academic Press, Boston, MA, 1990, 143–226.

[27] D. H. Jiang and D. Soudry, *The local converse theorem for* $\mathrm{SO}(2n+1)$ *and applications*, Ann. of Math. (2) **157** (2003), no. 3, 743–806.

[28] A. Kable, *Asai L-functions and Jacquet's conjecture*, Amer. J. Math. **126** (2004), no. 4, 789–820.

[29] H. Kasten and C.-G. Schmidt, *On critical values of Rankin–Selberg convolutions*, to appear in International J. of Number Theory.

[30] Henry H. Kim and M. Krishnamurthy, *Base change lift for odd unitary groups*, Functional analysis VIII, 116–125, Various Publ. Ser. (Aarhus), 47, Aarhus Univ., Aarhus, 2004.

[31] Henry H. Kim and M. Krishnamurthy, *Stable base change lift from unitary groups to* GL_n, Int. Math. Res. Pap. **2005**, no. 1, 1–52.

[32] Jian-Shu Li and Joachim Schwermer, *On the Eisenstein cohomology of arithmetic groups*, Duke Math. J. **123** (2004), no. 1, 141–169.

[33] W. Z. Luo, Z. Rudnick and P. Sarnak, *On the generalized Ramanujan conjecture for* $\mathrm{GL}(n)$, in *Automorphic forms, automorphic representations, and arithmetic (Fort Worth, TX, 1996)*, 301–310, Proc.

Sympos. Pure Math., 66, Part 2, Amer. Math. Soc., Providence, RI, 1999.

[34] Y. Manin, *Periods of cusp forms, and p-adic Hecke series*, (Russian) Mat. Sb. (N.S.) **92** (134) (1973), 378–401.

[35] N. Matringe, *Distinguished representations and exceptional poles of the Asai L-function*, Manuscripta Math. **131** (2010), no. 3-4, 415–426.

[36] N. Matringe, *Distinguished generic representations of GL(n) over p-adic fields*, Int. Math. Res. Not., 2011, no. 1, 74–95.

[37] A. Raghuram, *On the special values of certain Rankin-Selberg L-functions and applications to odd symmetric power L-functions of modular forms*, Int. Math. Res. Not., Vol. (2010), 334–372, doi:10.1093/imrn/rnp127.

[38] A. Raghuram, *On the critical values of certain Rankin-Selberg L-functions for* $\mathrm{GL}_n \times \mathrm{GL}_{n-1}$, in preparation.

[39] A. Raghuram and F. Shahidi, *Functoriality and special values of L-functions*, Eisenstein series and applications, 271–293, Progr. Math. **258**, Birkhäuser Boston, Boston, MA, 2008.

[40] A. Raghuram and N. Tanabe, *Notes on the arithmetic of Hilbert modular forms*, J. Ramanujan Math. Soc. **26** (2011), no. 3, 261–319.

[41] S. Salamanca-Riba, *On the unitary dual of real reductive groups and the* $A_{\mathfrak{q}}(\lambda)$ *modules: the strongly regular case*, Duke Math. J. **96** No. 3 (1998), 521–546.

[42] F. Shahidi, *Arthur packets and the Ramanujan conjecture*, Kyoto J. Math. **51** (2011), no. 1, 1–23.

[43] G. Shimura, *On the periods of modular forms*, Math. Ann. **229** no. 3, (1977), 211–221.

[44] G. Shimura, *The special values of the zeta functions associated with Hilbert modular forms*, Duke Math. J. **45** (1978), no. 3, 637–679.

[45] G. Shimura, *Introduction to the arithmetic theory of automorphic functions*, Reprint of the 1971 original. Publications of the Mathematical Society of Japan, 11. Kanô Memorial Lectures, 1. Princeton University Press, Princeton, NJ, 1994.

[46] D. Soudry, *On Langlands functoriality from classical groups to* GL_n. *Automorphic forms. I*, Astérisque **298** (2005), 335–390.

[47] J.-L. Waldspurger, *Quelques propriétés arithmétiques de certaines formes automorphes sur* GL(2), Compositio Math. **54** no. 2, (1985), 121–171.

[48] N. Wallach, *On the constant term of a square integrable automorphic form*, Operator algebras and group representations, Vol. II (Neptun, 1980), Monogr. Stud. Math., 18, Pitman, Boston, MA, (1984) 227–237.

[49] W. Zhang, *Fourier transform and the global Gan-Gross-Prasad conjecture for unitary groups*, preprint, available at http://www.math.columbia.edu/~wzhang/math/online/transfer.pdf.

Wee Teck Gan, Department of Mathematics, National University of Singapore, 10 Lower Kent Ridge Road Singapore 119076.
E-mail: matgwt@nus.edu.sg

A. Raghuram, Indian Institute of Science Education and Research (IISER), First floor, Central Tower, Sai Trinity Building Garware Circle, Sutarwadi, Pashan Pune, Maharashtra 411021, India.
E-mail: raghuram@iiserpune.ac.in

Automorphic Representations and L-Functions
Editors: D. Prasad, C.S. Rajan, A. Sankaranarayanan, J. Sengupta

Publisher: Hindustan Book Agency, New Delhi, India

Control Theorems for Ordinary 2-adic Families of Modular Forms

Eknath Ghate and Narasimha Kumar

Abstract

We prove a control theorem for Hida's ordinary Hecke algebra for the prime $p = 2$, thereby establishing a uniqueness result for ordinary 2-adic families of cusp forms. As a consequence we show that the possibly finitely many exceptions that arise, in showing that the local Galois representations attached to the arithmetic members of a non-CM ordinary 2-adic family of cuspidal eigenforms are all non-split, contain no CM forms.

1 Introduction

Hida theory for the prime p is the theory [Hid86a], [Hid86b] that deals with p-ordinary families of elliptic modular forms. While generalizations are now available, for automorphic forms on groups other than GL_2 and base fields other than $\mathbb{Q}$, many authors tend to shy away from the prime $p = 2$.

The goal of this paper is to check that some of the basic results in the literature stated for odd primes p (and originally for $p \geq 5$ in [Hid86a], [Hid86b]) remain valid for the prime $p = 2$. In particular, we shall check that Hida's control theorem (Theorem 8.1) for the ordinary Λ-adic Hecke algebra holds in this setting. Such control theorems are of fundamental importance since

(i) they are connected to the fact that the dimension of the space of ordinary cusp forms $S_k^0(\Gamma_1(Np^r), \mathbb{Z}_p)$ of tame level N coprime to p, and r fixed, is bounded independent of the weight k, even though the dimension of the ambient space of cusp forms grows large with k, and

(ii) they can be used to prove existence and uniqueness results for p-ordinary cusp forms in Hida families.

Both these applications are well-known when p is odd. As a result of our work here, we similarly deduce that a 2-stabilized ordinary cuspidal newform of weight at least 2 lives in a unique primitive 2-ordinary cuspidal family, up to Galois conjugacy (cf. Section 9). As a consequence, we are able to separate primitive 2-adic CM families from primitive non-CM families.

The proof of the control theorem in the case $p = 2$ given here uses a melange of techniques from several of Hida's papers. However, since some key facts needed from the theory of mod 2 modular forms do not still seem to be known, we have had to replace these with other ingredients; see in particular the proof of Theorem 5.3.

The reason we decided to embark on this project was to understand to what extent a recent application of Hida theory for odd primes p, namely to understanding the local splitting behaviour of p-ordinary modular Galois representations, continues to hold for the prime $p = 2$. As explained in [Gha04], a direct geometric approach to this problem using the motive attached to the underlying form only seems possible when the weight is 2. Following instead the Hida theoretic approach in [GV04] for odd primes p, and assuming that a modularity result of Buzzard [Buz03] for Artin-like representations holds in sufficient generality for the prime $p = 2$ (see [All12] for some progress on this front), we show that almost all arithmetic members of a primitive non-CM 2-ordinary family of cusp forms have locally non-split Galois representations. The uniqueness result mentioned above implies that none of these possibly finitely many exceptions are CM forms. Thus, as for odd primes p, if an exception occurs in a 2-ordinary non-CM family, it would give a genuine counterexample to the natural guess of Greenberg that p-ordinary modular Galois representations tend to be locally split only if the underlying form has CM.

2 Preliminaries

We recall some background and notation.

Let Φ be a torsion-free congruence subgroup of $\mathrm{SL}_2(\mathbb{Z})$. Then Φ acts freely on the upper half complex plane $\mathbb{H}$ by linear fractional transformations. Let Y be the complex open manifold associated to Φ, i.e., $Y = \Phi\backslash\mathbb{H}$. Let $C(\Phi)$ denote a finite set of representatives for the Φ-equivalence classes of cusps. Let X be the smooth compactification of Y obtained by adjoining the cusps in $C(\Phi)$.

For any discrete $\mathbb{Z}[\Phi]$-module M, let $F(M) = \Phi\backslash(\mathbb{H} \times M)$, where $\alpha \in \Phi$ acts on $\mathbb{H} \times M$ by $\alpha(z, m) = (\alpha z, \alpha m)$ for $(z, m) \in \mathbb{H} \times M$. We denote by the same symbol $F(M)$ the sheaf of continuous sections of the natural covering map $F(M) \to Y$. The sheaf cohomology group is denoted by $\mathrm{H}^i(Y, F(M))$ and compactly supported cohomology is denoted by

$\mathrm{H}^i_c(Y, F(M))$. The parabolic sheaf cohomology group $\mathrm{H}^1_{\mathrm{p}}(Y, F(M))$ is the image of $\mathrm{H}^1_c(Y, F(M))$ in $\mathrm{H}^1(Y, F(M))$.

We relate the sheaf cohomology group with group cohomology. For $n \geq 0$, let $L_n(\mathbb{Z}) = \mathbb{Z}^{n+1}$ be the n-th symmetric power representation of $\mathrm{SL}_2(\mathbb{Z})$. For an abelian group A, let $L_n(A) := L_n(\mathbb{Z}) \otimes_{\mathbb{Z}} A$. It is naturally an $\mathrm{SL}_2(\mathbb{Z})$-module through its action on the left factor.

It is well-known that

$$\mathrm{H}^i(Y, F(L_n(A))) \simeq \mathrm{H}^i(\Phi, L_n(A)),$$

where $\mathrm{H}^i(\Phi, L_n(A))$ denotes the i-th group cohomology of the $\mathbb{Z}[\Phi]$-module $L_n(A)$. This isomorphism is compatible with the action of the Hecke operators. Moreover, we have

$$\mathrm{H}^1_{\mathrm{p}}(Y, F(L_n(A))) \simeq \mathrm{H}^1_{\mathrm{p}}(\Phi, L_n(A)),$$

where the right hand side is the first parabolic group cohomology group and this isomorphism is again equivariant for the action of the Hecke operators.

When we study the ordinary parts of cohomology groups of $\Gamma_1(Np^r)$ with $(p, N) = 1$, for different r's, we will also need to consider the ordinary parts of cohomology groups of Φ^s_r, for $r \geq s \geq 0$, where

$$\Phi^s_r := \Gamma_1(Np^s) \cap \Gamma_0(p^r) = \\ \left\{ \begin{pmatrix} a & b \\ c & d \end{pmatrix} \in \mathrm{SL}_2(\mathbb{Z}) \mid c \equiv 0 \pmod{Np^r},\ a \equiv 1 \pmod{Np^s} \right\}.$$

From now on $p = 2$, $q = 4$ and $(2, N) = 1$. Let $\Gamma_0 = \Gamma_1 = \mathbb{Z}_p^\times$ and for $r \geq 2$, let Γ_r denote the subgroup $1 + p^r\mathbb{Z}_p$ of Γ, where $\Gamma = \Gamma_2 = 1 + q\mathbb{Z}_p$. There is a short exact sequence of groups

$$0 \to \Gamma_1(N2^r) \to \Phi^s_r \to \Gamma_s/\Gamma_r \to 0,$$

induced by $\Phi^s_r \ni \left(\begin{smallmatrix} a & b \\ c & d \end{smallmatrix}\right) \mapsto \bar{d} \in \Gamma_s/\Gamma_r$. For convenience write Φ_r for Φ^0_r, for $r \geq 0$. In Hida theory, for primes $p \geq 3$, the congruence subgroup Φ_1 plays an important role, but when $p = 2$, the role of this group is played by the congruence subgroup $\Phi_2 = \Gamma_0(4) \cap \Gamma_1(N)$. Note that this last group is torsion-free if $N > 1$.

Let K be a finite extension of $\mathbb{Q}_p$ and $\mathcal{O}_K$ be the integral closure of $\mathbb{Z}_p$ in K. Let Z denote the group $\varprojlim_r (\mathbb{Z}/Np^r\mathbb{Z})^\times = \mathbb{Z}_p^\times \times (\mathbb{Z}/N\mathbb{Z})^\times$. We may consider $\Gamma = \Gamma_2 = 1 + 4\mathbb{Z}_p$ as a subgroup of Z; let u denote a generator of Γ_2, often taken to be $1 + q$. By definition, there is a tautological character $\iota : \Gamma \hookrightarrow \Lambda_K = \mathcal{O}_K[[\Gamma]]$, which takes u to itself in Λ_K. For each character

$\chi : \Gamma \to \mathcal{O}_K^\times$, the element $P_\chi = \iota(u) - \chi(u)$ is a prime element, and $\Lambda_K/P_\chi\Lambda_K \simeq \mathcal{O}_K$, so that $\iota(u)$ corresponds to $\chi(u)$. If $\chi(u) = \epsilon(u)u^k$, where ϵ is a finite order character of Γ, we write $P_{k,\epsilon}$ for P_χ and simply P_k if ϵ is trivial. We may identify Λ_K with $\mathcal{O}_K[[X]]$ sending u to $1+X$ and in this case, $P_{k,\epsilon}$ is nothing but $(1+X) - \epsilon(u)u^k$. When $K = \mathbb{Q}_p$, we denote $\Lambda_{\mathbb{Q}_p}$ by Λ.

Finally, let ω denote the mod 4 cyclotomic character, that is, ω is the mod 4 character defined by $\omega(x) = \pm 1$, for $x \equiv \pm 1 \pmod 4$.

3 Main Theorems

Recall that $p = 2$, and N is odd. We only use the congruence subgroup $\Gamma_1(Np^r)$ with $r \geq 2$, which is a torsion-free group. We denote the corresponding complex Riemann surface by Y_r and its compactification by X_r. Let $\mathrm{H}^i(Y_r, M)$ and $\mathrm{H}^i(X_r, M)$ denote the corresponding sheaf cohomology groups for each constant sheaf M of $\mathbb{Z}$-modules. It is well-known that

$$S_2(\Gamma_1(Np^r)) \simeq \mathrm{H}^1(X_r, \mathbb{R}),$$

where the right hand side of the isomorphism, the sheaf cohomology with $\mathbb{R}$ coefficients, can be identified with de Rham cohomology. The above isomorphism is invariant under the action of the Hecke algebra. The Hecke algebra $\mathfrak{h}_2(\Gamma_1(Np^r), \mathbb{Z})$ acts on $\mathrm{H}^1(X_r, \mathbb{Z})$ and therefore $\mathfrak{h}_2(\Gamma_1(Np^r), \mathbb{Z}_p)$ acts on $\mathrm{H}^1(X_r, \mathbb{Z}_p)$, $\mathrm{H}^1(X_r, \mathbb{Q}_p)$, $\mathrm{H}^1(X_r, \mathbb{T}_p)$, where $\mathrm{H}^1(X_r, M) = \mathrm{H}^1(X_r, \mathbb{Z}) \otimes_{\mathbb{Z}} M$ and $\mathbb{T}_p := \mathbb{Q}_p/\mathbb{Z}_p$.

For every positive integer $r \geq 2$, we simply write

$$\mathcal{V}_r = \mathrm{H}^1(X_r, \mathbb{T}_p), \quad \mathcal{W}_r = \mathrm{H}^1(Y_r, \mathbb{T}_p).$$

Since $\mathrm{H}^1(X_r, \mathbb{T}_p) \simeq \mathrm{H}^1(X_r, \mathbb{Q}_p)/\mathrm{H}^1(X_r, \mathbb{Z}_p)$, we see that $\mathcal{V}_r$ and $\mathcal{W}_r$ are p-divisible modules of finite $\mathbb{Z}_p$-corank. Therefore $\mathrm{End}(\mathcal{V}_r)$ and $\mathrm{End}(\mathcal{W}_r)$ are free of finite rank. Hence one can define Hida's idempotent operator e_r attached to the Hecke operator T_p in $\mathrm{End}(\mathcal{V}_r)$ and $\mathrm{End}(\mathcal{W}_r)$. Define the ordinary parts of $\mathcal{V}_r$ to be $\mathcal{V}_r^0 = e_r\mathcal{V}_r$ and similarly for $\mathcal{W}_r$. $\mathcal{V}_r^0$ is a module for $\mathfrak{h}_2^0(\Gamma_1(Np^r), \mathbb{Z}_p)$, the ordinary part of $\mathfrak{h}_2(\Gamma_1(Np^r), \mathbb{Z}_p)$. By abuse of notation, we will use the same notation e for the various e_r's.

There is also an action of $\Gamma_0(Np^r)/\Gamma_1(Np^r)$ on $\mathcal{V}_r^0$ and $\mathcal{W}_r^0$. Let $\mathcal{V}$ denote the direct limit of $\mathcal{V}_r$ and define similarly $\mathcal{W}$, $\mathcal{V}^0$ and $\mathcal{W}^0$. Since $(\mathbb{Z}/Np^r\mathbb{Z})^\times$ acts on $\mathcal{V}_r^0$ and $\mathcal{W}_r^0$, hence Z acts on $\mathcal{V}^0$ and $\mathcal{W}^0$. In particular $\mathcal{V}^0$ and $\mathcal{W}^0$ become continuous modules over the Iwasawa algebra $\Lambda = \mathbb{Z}_p[[\Gamma]]$ if we equip them with the discrete topology. Let $V^0 = \mathrm{Hom}_{\mathbb{Z}_p}(\mathcal{V}^0, \mathbb{T}_p)$, respectively W^0, be the Pontryagin dual module of $\mathcal{V}^0$, respectively $\mathcal{W}^0$. Then V^0 and W^0 are compact Λ-modules. We can now state one of the main theorems of this article.

Theorem 3.1 *Let $p = 2$. We have:*

1. *For each positive integer $r \geq 2$, the restriction morphism of cohomology groups induces an isomorphism of $\mathcal{V}_r^0$ onto $(\mathcal{V}^0)^{\Gamma_r}$. The same result also holds for $\mathcal{W}^0$.*
2. *Let $N > 1$. The modules V^0 and W^0 are free modules of finite rank over Λ.*

The first part of Theorem 3.1 gives control of the ordinary parts of the cohomology modules associated with the decreasing sequence of congruence subgroups $\Gamma_1(Np^r)$, for $r \geq 2$, and we refer to such a result as a control theorem (for cohomology).

4 Control Theorem for Cohomology

In this section, we prove part (1) of Theorem 3.1.

When studying the action of the Hecke operators on cohomology groups or on parabolic cohomology groups, often one needs to decompose certain double coset spaces into a disjoint union of left cosets with a clever choice of coset representatives. Such decompositions can be found in [Hid86b, Lem. 4.3]. We recall with proof only part (ii) of that lemma, since the hypotheses of the original statement are mildly misstated. We refer the reader to [Hid86b] for the other parts, especially since the lemma is only used implicitly below, in the proof of Proposition 4.6.

Lemma 4.1 *Let r, $m \geq 1$, $r \geq s$. For every integer $u \in \mathbb{Z}$, let $\alpha_u \in \mathrm{M}_2(\mathbb{Z})$ be such that*

$$\alpha_u \equiv \left(\begin{smallmatrix} 1 & u \\ 0 & p^m \end{smallmatrix}\right) \pmod{Np^{\max(m,r)}} \text{ and } \det(\alpha_u) = p^m.$$

Then we have a disjoint decomposition

$$\Phi_r^s \left(\begin{smallmatrix} 1 & 0 \\ 0 & p^m \end{smallmatrix}\right) \Phi_r^s = \bigcup_{u \bmod p^m} \Phi_r^s \alpha_u.$$

Proof Suppose that $m \geq r$. The proof in the other case is similar. The group Γ' in (3.3.2) of [Shi71, p. 67], is just Φ_r^s, for $r \geq s$, if we take $t = 1$, $\mathfrak{h}$ to be the kernel of $(\mathbb{Z}/Np^r)^\times \to (\mathbb{Z}/Np^s)^\times$ and N to be Np^r. Now by [Shi71, Prop. 3.33], we have that $\Phi_r^s \left(\begin{smallmatrix} 1 & 0 \\ 0 & p^m \end{smallmatrix}\right) \Phi_r^s = \{\beta \in \Delta' | \det(\beta) = p^m\}$, where Δ' is as in [Shi71, p. 68]. The elements α_u belongs to the right hand side of the above equality. Observe that $u \equiv u' \pmod{p^m}$ if and only if $\alpha_u \equiv \alpha_{u'} \pmod{p^m}$. By the same proposition, the number of left cosets of Φ_r^s is p^m. Thus, the α_u, for $u \pmod{p^m}$, are candidates for the coset representatives.

$\square$

Before we start the proof of the control theorem, let us state another lemma.

Lemma 4.2 *Let $\{M_r\}_{r\geq 2}$ be an inductive system of compatible modules over $\mathbb{Z}_p[\Gamma/\Gamma_r]$, respectively. Assume that for all $r \geq t \geq 2$, $M_r^{\Gamma_t} = M_t$. Then $(\varinjlim_r M_r)^{\Gamma_t} = M_t$.*

Proof Clearly, $\varinjlim_r M_r$ is a module over $\mathbb{Z}_p[[\Gamma]]$. For every integer $q \geq 0$, one knows $\mathrm{H}^q(\Gamma_t, \varinjlim_r M_r) = \varinjlim_r \mathrm{H}^q(\Gamma_t, M_r)$. In particular, when $q = 0$, we have that $(\varinjlim_r M_r)^{\Gamma_t} = \varinjlim_r (M_r)^{\Gamma_t} = M_t$, where the last equality follows from the assumption.

$\square$

Now, we start the proof of the control theorem for cohomology.

Lemma 4.3 *If $\Phi_r^s/\Gamma_1(Np^r)$ acts on $\mathbb{T}_p$ trivially, then*

$$\mathrm{H}^2(\Phi_r^s/\Gamma_1(Np^r), \mathbb{T}_p) = 0.$$

Proof Since $\Phi_r^s/\Gamma_1(Np^r)$ is a finite cyclic group, $\mathrm{H}^2(\Phi_r^s/\Gamma_1(Np^r), \mathbb{T}_p) = \mathbb{T}_p/\mathcal{N}\mathbb{T}_p$, where $\mathcal{N}$ denote the norm map from $\mathbb{T}_p$ to itself. Since $\Phi_r^s/\Gamma_1(Np^r)$ acts on $\mathbb{T}_p$ trivially, $\mathcal{N}$ is multiplication by the index of $\Gamma_1(Np^r)$ in Φ_r^s, hence is surjective.

$\square$

Lemma 4.4 *For each $r \geq s \geq 2$, $e\mathrm{H}^1(\Gamma_1(Np^s), \mathbb{T}_p) \simeq e\mathrm{H}^1(\Phi_r^s, \mathbb{T}_p)$, where e is the idempotent operator attached to T_p on the respective groups.*

Proof Since $\Phi_r^s \subseteq \Gamma_1(Np^s)$, there is a restriction map $\mathrm{H}^1(\Gamma_1(Np^s), \mathbb{T}_p) \to \mathrm{H}^1(\Phi_r^s, \mathbb{T}_p)$. We have the following commutative diagram

$$\begin{array}{ccc}
\mathrm{H}^1(\Gamma_1(Np^s), \mathbb{T}_p) & \xrightarrow{\text{res}} & \mathrm{H}^1(\Phi_r^s, \mathbb{T}_p) \\
\big\downarrow T_p^{r-s} & \swarrow_{[\Phi_r^s \left(\begin{smallmatrix}1 & 0\\ 0 & p^{r-s}\end{smallmatrix}\right)\Phi_s^s]} & \big\downarrow T_p^{r-s} \\
\mathrm{H}^1(\Gamma_1(Np^s), \mathbb{T}_p) & \xrightarrow{\text{res}} & \mathrm{H}^1(\Phi_r^s, \mathbb{T}_p).
\end{array}$$

By applying the idempotent operator, we get that the vertical morphisms are isomorphisms and hence the diagonal map is an isomorphism.

□

For each $r \geq s \geq 2$, we have the inflation-restriction sequence

$$0 \to \mathrm{H}^1(\Phi_r^s/\Gamma_1(Np^r), \mathbb{T}_p) \xrightarrow{\iota} \mathrm{H}^1(\Phi_r^s, \mathbb{T}_p) \to \mathrm{H}^1(\Gamma_1(Np^r), \mathbb{T}_p)^{\Gamma_s} \to \mathrm{H}^2(\Phi_r^s/\Gamma_1(Np^r), \mathbb{T}_p) = 0$$

where the last term vanishes by Lemma 4.3. The image of the group $\mathrm{H}^1(\Phi_r^s/\Gamma_1(Np^r), \mathbb{T}_p)$ inside $\mathrm{H}^1(\Phi_r^s, \mathbb{T}_p)$ is annihilated by the idempotent e attached to T_p by [Hid86b, Lem. 6.1]. Therefore,

$$e\mathrm{H}^1(\Phi_r^s, \mathbb{T}_p) \simeq e\mathrm{H}^1(\Gamma_1(Np^r), \mathbb{T}_p)^{\Gamma_s} = (\mathcal{W}_r^0)^{\Gamma_s}.$$

By Lemma 4.4, we have that $\mathcal{W}_s^0 = e\mathrm{H}^1(\Gamma_1(Np^s), \mathbb{T}_p) \simeq e\mathrm{H}^1(\Phi_r^s, \mathbb{T}_p)$. By combining these isomorphisms, we get

$$\mathcal{W}_s^0 \simeq (\mathcal{W}_r^0)^{\Gamma_s}.$$

By Lemma 4.2, for any $r \geq 2$, we have that $(\mathcal{W}^0)^{\Gamma_r} \simeq \mathcal{W}_r^0$. This finishes the proof of the control theorem for $\mathcal{W}^0$. Note that so far the proof works for all primes $p \geq 2$.

Now, we shall prove the control theorem for $\mathcal{V}^0$, concentrating on what changes need to be made in Hida's original proof when $p = 2$. For a torsion-free congruence subgroup Φ, let $P(\Phi)$ denote a set of generators of Φ_s, the stabilizer of Φ at s, for $s \in C(\Phi)$. Then, for $r \geq 2$, $\mathcal{W}_r$ is given by

$$\left\{\varphi \in \mathrm{Hom}(\Gamma_1(Np^r), \mathbb{T}_p) \mid \sum_{\pi \in P(\Gamma_1(Np^r))} \varphi(\pi) = 0\right\},$$

and $\mathcal{V}_r$ is the submodule of $\mathcal{W}_r$ given by

$$\mathcal{V}_r = \left\{\varphi \in \mathrm{Hom}(\Gamma_1(Np^r), \mathbb{T}_p) \mid \varphi(\pi) = 0, \text{ for } \pi \in P(\Gamma_1(Np^r))\right\},$$

since the group $\Gamma_1(Np^r)$ acts trivially on $\mathbb{T}_p$. For a detailed proof of the above equalities, see [Hid86b, p. 583].

By Lemma 4.2, it is enough to prove that $(\mathcal{V}_r^0)^{\Gamma_s} = \mathcal{V}_s^0$ for $r \geq s \geq 2$. It is clear that there is a map $\mathcal{V}_s \to \mathcal{V}_r$. By taking the ordinary parts of Γ_s-invariants, we have an inclusion $\mathcal{V}_s^0 \hookrightarrow (\mathcal{V}_r^0)^{\Gamma_s}$. Therefore it is enough to prove the surjectivity of this last map. Since $(\mathcal{W}_r^0)^{\Gamma_s} = \mathcal{W}_s^0$, given a homomorphism $\varphi : \Gamma_1(Np^r) \to \mathbb{T}_p$ invariant under Γ_s and satisfying $\varphi|_e = \varphi$, there exists a homomorphism $\psi : \Gamma_1(Np^s) \to \mathbb{T}_p$ with $\psi|_e = \psi$

such that $\psi = \varphi$ on $\Gamma_1(Np^r)$. Thus, we need to show that $\psi(\pi) = 0$, for all $\pi \in P(\Gamma_1(Np^s))$, i.e., $\psi \in e\mathrm{H}^1_{\mathrm{p}}(\Gamma_1(Np^s), \mathbb{T}_p)$, assuming the same holds for φ with r instead of s.

Let $[\psi]$ denote the equivalence class of ψ in the module

$$\mathcal{G}(\Gamma_1(Np^s), \mathbb{T}_p) := \mathrm{H}^1(\Gamma_1(Np^s), \mathbb{T}_p)/\mathrm{H}^1_{\mathrm{p}}(\Gamma_1(Np^s), \mathbb{T}_p).$$

We need to show that $[\psi] = 0$. We know that $[\psi]|_e = [\psi]$. If $[\psi]|_{1-e} = [\psi]$, then $[\psi] = 0$, since e is an idempotent. Hence, it is enough to show that

$$[\psi]|_{1-e} = [\psi]$$

holds. By following the strategy in [Hid86b], this reduces to proving [Hid86b, Thm. 5.8], which characterizes the elements of $(1 - e)\mathcal{G}(\Gamma_1(Np^s), \mathbb{T}_p)$ as elements of the set

$$V(\mathbb{T}_p) := \{\psi \in \mathrm{Hom}(\Gamma_1(Np^s)^{\infty}_{\mathrm{ab}}, \mathbb{T}_p) \mid \psi(\pi) = 0, \text{ for all } \pi \in P(\Gamma_1(Np^s)) \text{ corresponding to the unramified cusps}\},$$

where the module $\Gamma_1(Np^s)^{\infty}_{\mathrm{ab}}$ is the free submodule of $\Gamma_1(Np^s)_{\mathrm{ab}}$ generated by the elements of $P(\Gamma_1(Np^s))$. Under the above equality, if $\psi \in V(\mathbb{T}_p)$, then $[\psi] = (1 - e)[\psi']$ and hence $[\psi]|_{1-e} = [\psi]$ holds. Thus it suffices to show that $\psi \in V(\mathbb{T}_p)$. Since every unramified cusp of X_s over X_0 is under an unramified cusp of X_r over X_0, the elements of $P(\Gamma_1(Np^s))$ corresponding to unramified cusps in X_s can be taken to be among the elements of $P(\Gamma_1(Np^r))$ corresponding to unramified cusps in X_r. Then $\psi(\pi) = \varphi(\pi) = 0$, for all $\pi \in P(\Gamma_1(Np^s))$, which corresponds to unramified cusps of $\Gamma_1(Np^s)$, as desired.

The proof of [Hid86b, Thm. 5.8] depends, firstly, on various relations between the dimensions of the space of Eisenstein series for $\Gamma_1(Np^r)$ with coefficients in A and the boundary cohomology $\mathrm{G}^1(\Gamma_1(Np^r), L_n(A)) = \oplus_{t \in C(\Gamma_1(Np^r))} \mathrm{H}^1(\Gamma_1(Np^r)_t, L_n(A))$, for any subalgebra A of $\mathbb{C}$ or $\mathbb{C}_p$, and secondly, on the validity of [Hid86b, Prop. 5.7]. The results on the dimensions of the space of Eisenstein series and the space $\mathrm{G}^1(\Gamma_1(Np^r), L_n(A))$ also holds for the prime $p = 2$. But, in the proof of [Hid86b, Prop. 5.7], one crucially uses the fact that $p \neq 2$. Thus to finish the proof of the control theorem for $\mathcal{V}^0$ when $p = 2$, it suffices to check that the proposition holds. Before we do that, let us introduce the notion of regular and irregular cusps.

The stabilizer Φ_s at a cusp $s \in C(\Phi)$ of a torsion-free congruence subgroup Φ is an infinite cyclic group. We fix an element $\alpha = \alpha_s$ in $\mathrm{SL}_2(\mathbb{Z})$ for each $s \in C(\Phi)$ such that $\alpha(\infty) = s$. We can choose a generator $\pi = \pi_s$ of Φ_s so that $\alpha^{-1}\pi\alpha = \pm\left(\begin{smallmatrix} 1 & u \\ 0 & 1 \end{smallmatrix}\right)$ with $u > 0$.

Definition 4.5 When $\alpha^{-1}\pi\alpha = -\left(\begin{smallmatrix} 1 & u \\ 0 & 1 \end{smallmatrix}\right)$, we say that the cusp s is irregular, and otherwise, we say that s is regular.

This definition makes sense since $-1 \notin \Phi$, by assumption. We also remark that some authors define irregularity by the condition $\alpha^{-1}\pi\alpha = \left(\begin{smallmatrix} -1 & u \\ 0 & -1 \end{smallmatrix}\right)$, with $u > 0$. This is easily seen to be equivalent to the above by taking the inverse of the generator π.

Coming back to the proof, the difference between [Hid86b, Prop. 5.7] and the analogous result for $p = 2$ (Proposition 4.6 below) is that in the former case, i.e., when $\Phi = \Gamma_1(Np^r)$ for $p \geq 5$ and $r \geq 1$, the group Φ is torsion-free with regular cusps, whereas in the latter case, i.e., when $\Phi = \Gamma_1(Np^r)$, for $p = 2$ and $r \geq 2$, the group Φ is torsion-free, but its cusps are not necessarily regular. For example, when $N = 1$ and $r = 2$, the group $\Gamma_1(4)$ has both regular and irregular cusps. These irregular cusps create problems in the proof given in [Hid86b, Prop. 5.7] when $p = 2$.

Proposition 4.6 *Let $\Phi = \Gamma_1(Np^r)$ with $r \geq 2$. Let A be either $\mathbb{Z}_p$, $\mathbb{Z}_p/p^i\mathbb{Z}_p$ or any field of characteristic 0. Let s be any* ***unramified*** *cusp of Φ and $\rho_s : \mathrm{G}^1(\Phi, A) := \oplus_{t\in C(\Phi)}\mathrm{H}^1(\Phi_t, A) \to \mathrm{H}^1(\Phi_s, A)$ be the natural projection map. Then for any $c \in \mathrm{G}^1(\Phi, A)$, we have that $\rho_s(c|e) = \rho_s(c)$, where e is the idempotent operator attached to T_p.*

Proof For any positive integer $M \geq 5$, all the cusps of $\Gamma_1(M)$ are regular. Hence, when $p = 2$, all cusps of $\Gamma_1(Np^r)$, for $N \geq 3$ or $r \geq 3$ are regular (so in particular are the unramified ones). So, it is enough to consider the case when $\Phi = \Gamma_1(4)$. By [Hid86b, Lem. 5.1], $\Gamma_1(4)$ has a unique unramified cusp, namely ∞. We see that this cusp is also regular since otherwise we would have

$$\pi_s = \pi_\infty = -\left(\begin{smallmatrix} 1 & u \\ 0 & 1 \end{smallmatrix}\right)$$

for some $u > 0$, which is not an element of $\Gamma_1(4)$. Hence, when $N = 1$ and $r = 2$, the unique unramified cusp is also regular. Thus Hida's original proof of [Hid86b, Prop. 5.7] for regular cusps applies when $p = 2$ as well. □

Remark 4.7 Subsequently we will only need to consider $N \geq 3$, for applications to p-adic families. Also, in the odd prime case, the statement of [Hid86b, Prop. 5.7] also treats the groups $\Phi = \Phi_r^s$ for $r > s \geq 0$. Since we do not need this part of the proposition when $p = 2$, we ignore it. In fact, if Φ has elements of finite order (this happens for small values of N and s when $p = 2$), it is not clear that the proposition holds for such Φ.

This finishes the proof of the control theorem for cohomology.

5 Freeness

In this section, we prove that the modules V^0 and W^0 are free of finite rank over $\mathbb{Z}_p[[X]]$, completing the proof of Theorem 3.1. We restate this formally as:

Theorem 5.1 *Let $p = 2$, $N > 1$ be odd, and $\Lambda = \mathbb{Z}_p[[X]]$. The modules V^0 and W^0 are free modules of finite rank over Λ.*

The freeness of W^0 follows from [Kum, Thm. 5.3], by Poincare duality. The claim for V^0 is more subtle and requires more machinery and results. We start by recalling without proof a lemma [Hid86b, Lem. 6.3] which is useful in proving the freeness of V^0.

Lemma 5.2 *A compact continuous Λ-module M is free of finite rank r over Λ if and only if there is a subset I of positive integers and infinitely many elements $\{P_n\}_{n\in I}$ in Λ such that $\mathcal{M}[P_n] \simeq \mathbb{T}_p^r$, for all $n \in I$, where $\mathcal{M}$ is the Pontryagin dual of M and $\mathcal{M}[P_n] = \{m \in \mathcal{M} | P_n.m = 0\}$.*

We know that the group $\mathbb{Z}_p^\times$ (recall $p = 2$) acts on $\mathcal{V}^0$. In particular, $\mu_2 = (\mathbb{Z}/q\mathbb{Z})^\times$ acts on $\mathcal{V}^0$ (recall $q = 4$). Write

$$\mathcal{V}^0 = \mathcal{V}^0(0) \oplus \mathcal{V}^0(1),$$

where $\mathcal{V}^0(a) = \{v \in \mathcal{V}^0 |\ v|\zeta = \zeta^a v, \text{ for } \zeta \in \mu_2\}$. Since the action of Γ commutes with the action of μ_2, $\mathcal{V}^0(a)$ is also a Λ-module, for $a = 0, 1$. Let $V^0(a)$ denote the Pontryagin dual of $\mathcal{V}^0(a)$. We shall show $V^0(a)$ is a free module of rank $2r(a)$ over Λ, where $r(a)$ is the rank of the Hecke algebra $\mathfrak{h}_2^0(\Phi_2, \omega^a, \mathbb{Z}_p)$.

By part (1) of Theorem 3.1, we have that $V^0(a)/\mathfrak{a}_2 V^0(a)$, where $\mathfrak{a}_2$ is the augmentation ideal of $\mathbb{Z}_p[[\Gamma]]$, is a free module of rank $2r(a)$ over $\mathbb{Z}_p$. By Nakayama's lemma, we see that $V^0(a)$ is a finitely generated Λ-module with minimal number of generators $2r(a)$. Hence there is a surjection from $\Lambda^{2r(a)} \twoheadrightarrow V^0(a)$. Hence, by duality, we have

$$\mathcal{V}^0(a)[P_n] \hookrightarrow \mathbb{T}_p^{2r(a)},$$

where P_n is the prime ideal of Λ defined in section 2. Now define

$$\begin{aligned} &\mathrm{H}^1(\Gamma_1(Np^r), n; \mathbb{Z}_p/p^r\mathbb{Z}_p) := \\ &\qquad\qquad \{v \in \mathrm{H}^1(\Gamma_1(Np^r), \mathbb{Z}_p/p^r\mathbb{Z}_p) \mid v|z = z^n v \text{ for } z \in \mathbb{Z}_p^\times\}, \\ &\mathrm{H}^1_{\mathrm{p}}(\Gamma_1(Np^r), n; \mathbb{Z}_p/p^r\mathbb{Z}_p) := \\ &\qquad\qquad \mathrm{H}^1(\Gamma_1(Np^r), n; \mathbb{Z}_p/p^r\mathbb{Z}_p) \cap \mathrm{H}^1_{\mathrm{p}}(\Gamma_1(Np^r), \mathbb{Z}_p/p^r\mathbb{Z}_p). \end{aligned}$$

Suppose that the following inclusions and isomorphisms are true for $r \geq 2$:

$$\begin{aligned} eH^1_p(\Phi_2, L_n(\mathbb{Z}_p)) \otimes \mathbb{Z}_p/p^r\mathbb{Z}_p &\underset{(1)}{\hookrightarrow} eH^1_p(\Phi_2, L_n(\mathbb{Z}_p/p^r\mathbb{Z}_p)) \underset{(2)}{\simeq} \\ &eH^1_p(\Phi_r, L_n(\mathbb{Z}_p/p^r\mathbb{Z}_p)) \underset{(3)}{\simeq} eH^1_p(\Phi_r, \mathbb{Z}_p/p^r\mathbb{Z}_p(n)) \\ &\underset{(4)}{\hookrightarrow} eH^1_p(\Gamma_1(Np^r), n; \mathbb{Z}_p/p^r\mathbb{Z}_p) \underset{(5)}{\hookrightarrow} \mathcal{V}^0(a)[P_n], \end{aligned} \tag{5.1}$$

where the last inclusion holds only if $n \equiv a \pmod 2$. Then we have

$$eH^1_p(\Phi_2, L_n(\mathbb{Z}_p)) \otimes \mathbb{Z}_p/p^r\mathbb{Z}_p \hookrightarrow \mathcal{V}^0(a)[P_n] \hookrightarrow \mathbb{T}_p^{2r(a)}.$$

Taking direct limits with respect to r, we have that

$$eH^1_p(\Phi_2, L_n(\mathbb{Z}_p)) \otimes \mathbb{T}_p \hookrightarrow \mathcal{V}^0(a)[P_n] \hookrightarrow \mathbb{T}_p^{2r(a)}. \tag{5.2}$$

In the next section, we prove that the module $eH^1_p(\Phi_2, L_n(\mathbb{Z}_p))$ is $\mathbb{Z}_p$-free (see Lemma 6.2). More precisely, we prove in Theorem 6.1 that:

Theorem 5.3 *The $\mathbb{Z}_p$-rank of the module $eH^1_p(\Phi_2, L_n(\mathbb{Z}_p))$ is $2r(a)$, for $n \equiv a \pmod 2$.*

Proof For $p \geq 5$, the theorem is proved in [Hid86a, Thm. 3.1 and Cor. 3.2]. In his proof of [Hid86a, Thm. 3.1], Hida uses results from the theory of Katz modular forms, the theory of mod p modular forms and the fact that $p \geq 5$. For the prime $p = 2$, we need different arguments to prove the theorem and we postpone the proof to the next section.

□

We complete the proof of Theorem 5.1, assuming Theorem 5.3.

Proof By Theorem 5.3, we have that $\mathbb{T}_p^{2r(a)} \simeq eH^1_p(\Phi_2, L_n(\mathbb{Z}_p)) \otimes \mathbb{T}_p$. Hence, $\mathcal{V}^0(a)[P_n] \simeq \mathbb{T}_p^{2r(a)}$, for all $n \equiv a \pmod 2$. The theorem now follows from Lemma 5.2.

□

Now, we shall show that the inclusions and isomorphisms in (5.1) hold. This is the content of the next few lemmas and propositions. The following proposition proves the inclusion (1) in (5.1).

Proposition 5.4 *For all $r \geq 1$ and $n \geq 0$, we have*

$$eH^1_p(\Phi_2, L_n(\mathbb{Z}_p)) \otimes \mathbb{Z}_p/p^r\mathbb{Z}_p \hookrightarrow eH^1_p(\Phi_2, L_n(\mathbb{Z}_p/p^r\mathbb{Z}_p)).$$

Proof For any $\mathbb{Z}_p$-module A, the short exact sequence of modules

$$0 \to e\mathrm{H}^1_{\mathrm{p}}(\Phi_2, L_n(\mathbb{Z}_p)) \to e\mathrm{H}^1(\Phi_2, L_n(\mathbb{Z}_p)) \to e\mathrm{H}^1(\Phi_2, L_n(\mathbb{Z}_p))/e\mathrm{H}^1_{\mathrm{p}}(\Phi_2, L_n(\mathbb{Z}_p)) \to 0$$

induces the long exact sequence

$$\mathrm{Tor}(e\mathrm{H}^1(\Phi_2, L_n(\mathbb{Z}_p))/e\mathrm{H}^1_{\mathrm{p}}(\Phi_2, L_n(\mathbb{Z}_p)), A) \to e\mathrm{H}^1_{\mathrm{p}}(\Phi_2, L_n(\mathbb{Z}_p)) \otimes A \to e\mathrm{H}^1(\Phi_2, L_n(\mathbb{Z}_p)) \otimes A.$$

If $e\mathrm{H}^1(\Phi_2, L_n(\mathbb{Z}_p))/e\mathrm{H}^1_{\mathrm{p}}(\Phi_2, L_n(\mathbb{Z}_p))$ is $\mathbb{Z}_p$-free, then the first term in the above exact sequence is zero, and in particular this is so when $A = \mathbb{Z}_p/p^r\mathbb{Z}_p$. So the second map above is injective. Now, for any congruence subgroup Φ,

$$e\mathrm{H}^1(\Phi, L_n(\mathbb{Z}_p)) \otimes A \xrightarrow{\sim} e\mathrm{H}^1(\Phi, L_n(A)),$$

and under this identification, this second map above preserves parabolic classes, proving the theorem. The $\mathbb{Z}_p$-freeness of the module $e\mathrm{H}^1(\Phi_2, L_n(\mathbb{Z}_p))/e\mathrm{H}^1_{\mathrm{p}}(\Phi_2, L_n(\mathbb{Z}_p))$ follows from [Hid88a, Prop. 2.3]. Although this proposition was proved there for $\Gamma_1(Np^r)$, for $p \geq 5$, the same proof works for Φ_2 for $p = 2$ and $N \geq 3$ with $(2, N) = 1$.

□

The isomorphisms (2) and (3) in (5.1) follows from [Hid86b, Cor. 4.5], noting that the argument given there works for $p = 2$ and for Φ_2, instead of p odd and the Φ_1 there. The following lemma proves the inclusion (4).

Lemma 5.5 *For $r \geq 2$, we have an inclusion*

$$e\mathrm{H}^1_{\mathrm{p}}(\Phi_r, \mathbb{Z}_p/p^r\mathbb{Z}_p(n)) \hookrightarrow e\mathrm{H}^1_{\mathrm{p}}(\Gamma_1(Np^r), n; \mathbb{Z}_p/p^r\mathbb{Z}_p).$$

Proof We have the following inflation-restriction sequence for the groups $\Gamma_1(Np^r) \subseteq \Phi_r$:

$$0 \to \mathrm{H}^1(\Phi_r/\Gamma_1(Np^r), \mathbb{Z}_p/p^r\mathbb{Z}_p(n)) \to \mathrm{H}^1(\Phi_r, \mathbb{Z}_p/p^r\mathbb{Z}_p(n)) \to \mathrm{H}^1(\Gamma_1(Np^r), \mathbb{Z}_p/p^r\mathbb{Z}_p(n))^{\Phi_r/\Gamma_1(Np^r)}.$$

Since

$$\mathrm{H}^1(\Gamma_1(Np^r), \mathbb{Z}_p/p^r\mathbb{Z}_p(n))^{\Phi_r/\Gamma_1(Np^r)} \hookrightarrow \mathrm{H}^1(\Gamma_1(Np^r), n; \mathbb{Z}_p/p^r\mathbb{Z}_p),$$

we have the following exact sequence

$$0 \to \mathrm{H}^1(\Phi_r/\Gamma_1(Np^r), \mathbb{Z}_p/p^r\mathbb{Z}_p(n)) \to \mathrm{H}^1(\Phi_r, \mathbb{Z}_p/p^r\mathbb{Z}_p(n)) \to \mathrm{H}^1(\Gamma_1(Np^r), n; \mathbb{Z}_p/p^r\mathbb{Z}_p).$$

By [Hid86b, Lem. 6.1], we have

$$eH^1(\Phi_r, \mathbb{Z}_p/p^r\mathbb{Z}_p(n)) \hookrightarrow eH^1(\Gamma_1(Np^r), n; \mathbb{Z}_p/p^r\mathbb{Z}_p).$$

Hence, we have the required claim.

□

The following lemma proves inclusion (5) in (5.1).

Lemma 5.6 $eH^1_{\mathrm{p}}(\Gamma_1(Np^r), n; \mathbb{Z}_p/p^r\mathbb{Z}_p) \hookrightarrow \mathcal{V}^0(a)[P_n]$, *if* $n \equiv a \pmod 2$.

Proof Observe that, $eH^1_{\mathrm{p}}(\Gamma_1(Np^r), n; \mathbb{Z}_p/p^r\mathbb{Z}_p)$ is the subspace of $eH^1_{\mathrm{p}}(\Gamma_1(Np^r), \mathbb{Z}_p/p^r\mathbb{Z}_p)$ on which $\mathbb{Z}_p^\times$ act by $v|z = z^n v$, where v is a cohomology class. The group $\mathcal{V}^0(a)[P_n]$ is also the subspace of $\mathcal{V}^0$ such that $\mathbb{Z}_p^\times$ acts by $v|z = z^n v$, where $v \in \mathcal{V}^0$. This is true because μ_2 acts by $\zeta_2^a = \zeta_2^n$ and $\gamma \in \Gamma$ acts by $v|\gamma = \gamma^n v$.

By [Hid86b, p. 584, (5.4)], we have that $eH^1_{\mathrm{p}}(\Gamma_1(Np^r), \mathbb{Z}_p/p^t\mathbb{Z}_p) \simeq eH^1_{\mathrm{p}}(\Gamma_1(Np^r), \mathbb{Z}_p) \otimes \mathbb{Z}_p/p^t\mathbb{Z}_p$. Since tensor product commutes with direct limits, we have that

$$\varinjlim_t eH^1_{\mathrm{p}}(\Gamma_1(Np^r), \mathbb{Z}_p/p^t\mathbb{Z}_p) \simeq eH^1_{\mathrm{p}}(\Gamma_1(Np^r), \mathbb{Z}_p) \otimes_{\mathbb{Z}_p} \mathbb{T}_p \simeq eH^1_{\mathrm{p}}(\Gamma_1(Np^r), \mathbb{T}_p).$$

By part (1) of Theorem 3.1 (i.e., $(\mathcal{V}^0)^{\Gamma_r} \simeq \mathcal{V}^0_r$, for every $r \geq 2$), we have that

$$eH^1_{\mathrm{p}}(\Gamma_1(Np^r), \mathbb{Z}_p/p^r\mathbb{Z}_p) \hookrightarrow \mathcal{V}^0.$$

Now the lemma follows, since this map respects the action of $\mathbb{Z}_p^\times$.

□

6 Constant Rank

In this section, we prove that the ranks of certain cuspidal ordinary 2-adic Hecke algebras of different weights are all equal to the rank of a weight 2 cuspidal ordinary Hecke algebra. As in the previous section, $p = 2$ and $N > 1$ is odd.

For $a = 0$ or 1, recall that $r(a)$ is the rank of the Hecke algebra $\mathfrak{h}_2^0(\Phi_2, \omega^a, \mathbb{Z}_p)$, where ω denotes the mod 4 cyclotomic character. For simplicity, we write $A(\omega^n)$ for the sheaf with twisted action $L_0(\omega^n, A)$, for any $\mathbb{Z}_p$-module A.

Theorem 6.1 *For each positive integer* $n \equiv a \pmod 2$,

$$\mathrm{rank}_{\mathbb{Z}_p}\ \mathfrak{h}^0_{n+2}(\Phi_2, \mathbb{Z}_p) = r(a).$$

Before proving this theorem, we need to gather some results, which we do now.

Lemma 6.2 *For $r > s \geq 0$, the module $e\mathrm{H}^1(\Phi_r^s, L_n(\mathbb{Z}_p))$ is $\mathbb{Z}_p$-free, for $n \geq 0$.*

Proof The short exact sequence

$$0 \to L_n(\mathbb{Z}_p) \to L_n(\mathbb{Q}_p) \to L_n(\mathbb{T}_p) \to 0$$

induces a long exact sequence of cohomology groups for the group Φ_r^s

$$\mathrm{H}^0(\Phi_r^s, L_n(\mathbb{Q}_p)) \xrightarrow{\alpha} \mathrm{H}^0(\Phi_r^s, L_n(\mathbb{T}_p)) \xrightarrow{\beta} \mathrm{H}^1(\Phi_r^s, L_n(\mathbb{Z}_p)) \xrightarrow{\gamma} \mathrm{H}^1(\Phi_r^s, L_n(\mathbb{Q}_p)).$$

If $n = 0$, then the map α is surjective and hence the map β is zero. Therefore the map γ is injective and $\mathrm{H}^1(\Phi_r^s, L_n(\mathbb{Z}_p))$ is $\mathbb{Z}_p$-free. Assume that $n > 0$. If we can show that $e\mathrm{H}^0(\Phi_r^s, L_n(\mathbb{T}_p)) = 0$, then the lemma follows. The operator T_p acts on $L_n(\mathbb{T}_p)$ by $x|T_p = \sum_{i=0}^{p-1} \left(\begin{smallmatrix} 1 & -i \\ 0 & p \end{smallmatrix}\right)^{\iota} x$, where $A^{\iota} = \mathrm{Adj}(A)$. We see that T_p acts on any p-torsion element of $\mathrm{H}^0(\Phi_r^s, L_n(\mathbb{T}_p))$ by the matrix $\left(\begin{smallmatrix} 0 & * \\ 0 & 0 \end{smallmatrix}\right)$ and hence T_p^2 acts trivially on such elements, hence the idempotent e annihilates $\mathrm{H}^0(\Phi_r^s, L_n(\mathbb{T}_p))$.

□

Corollary 6.3 *For any integer $n \geq 0$, the module $e\mathrm{H}^1(\Phi_2, \mathbb{Z}_p(\omega^n))$ is $\mathbb{Z}_p$-free.*

Proof If n is even, then $\omega^n = 1$, hence this follows from the lemma and when n is odd, the proof is similar to the proof of the lemma.

□

Lemma 6.4 $e\mathrm{H}^1_{\mathrm{p}}(\Phi_2, L_n(\mathbb{Z}_p)) \otimes \mathbb{Z}_p/q\mathbb{Z}_p \simeq e\mathrm{H}^1_{\mathrm{p}}(\Phi_2, L_n(\mathbb{Z}_p/q\mathbb{Z}_p))$.

Proof By Proposition 5.4 with $r = 2$, the map

$$0 \to e\mathrm{H}^1_{\mathrm{p}}(\Phi_2, L_n(\mathbb{Z}_p)) \otimes \mathbb{Z}_p/q\mathbb{Z}_p \to e\mathrm{H}^1_{\mathrm{p}}(\Phi_2, L_n(\mathbb{Z}_p/q\mathbb{Z}_p)).$$

is injective. For the surjectivity, we work with sheaf cohomology instead of group cohomology. Let Y be the complex open manifold associated with Φ_2. Observe that we have the following commutative diagram:

$$\begin{array}{ccccc}
e\mathrm{H}^1_{\mathrm{c}}(Y,F(L_n(\mathbb{Z}_p)))/q & \twoheadrightarrow & e\mathrm{H}^1_{\mathrm{p}}(Y,F(L_n(\mathbb{Z}_p)))/q & \hookrightarrow & e\mathrm{H}^1(Y,F(L_n(\mathbb{Z}_p)))/q \\
\downarrow \wr & & \downarrow & & \downarrow \wr \\
e\mathrm{H}^1_{\mathrm{c}}(Y,F(L_n(\mathbb{Z}_p/q\mathbb{Z}_p))) & \twoheadrightarrow & e\mathrm{H}^1_{\mathrm{p}}(Y,F(L_n(\mathbb{Z}_p/q\mathbb{Z}_p))) & \hookrightarrow & e\mathrm{H}^1(Y,F(L_n(\mathbb{Z}_p/q\mathbb{Z}_p))).
\end{array}$$

By [Hid88a, Cor. 2.2], the first vertical map is an isomorphism. As a result, we get that the middle vertical map is surjective and the lemma follows. □

Lemma 6.5 *For any* $n \geq 0$,

$$e\mathrm{H}^1(Y, F(\mathbb{Z}_p(\omega^n)))/q \simeq e\mathrm{H}^1(Y, F(\mathbb{Z}_p/q\mathbb{Z}_p(\omega^n))).$$

Proof The short exact sequence

$$0 \to \mathbb{Z}_p(\omega^n) \xrightarrow{q} \mathbb{Z}_p(\omega^n) \to \mathbb{Z}_p/q\mathbb{Z}_p(\omega^n) \to 0$$

induces another short exact sequence

$$0 \to \mathrm{H}^1(Y, F(\mathbb{Z}_p(\omega^n))) \otimes \mathbb{Z}_p/q\mathbb{Z}_p \to \mathrm{H}^1(Y, F(\mathbb{Z}_p/q\mathbb{Z}_p(\omega^n))) \to \mathrm{H}^2(Y, F(\mathbb{Z}_p(\omega^n)))[q] \to 0,$$

where $\mathrm{H}^2(Y, F(\mathbb{Z}_p(\omega^n)))[q] = \{x \in \mathrm{H}^2(Y, F(\mathbb{Z}_p(\omega^n))) \mid q.x = 0\}$. This last group vanishes, since the cohomological dimension of Φ_r^s is 1. □

Proposition 6.6 *The module* $e(\mathrm{H}^1(\Phi_2, \mathbb{Z}_p(\omega))/\mathrm{H}^1_{\mathrm{p}}(\Phi_2, \mathbb{Z}_p(\omega)))$ *is* $\mathbb{Z}_p$*-free.*

Proof For $i = 0$, 1, define

$$\mathrm{G}^i(\Phi_2, M) = \bigoplus_{s \in C(\Phi_2)} \mathrm{H}^i((\Phi_2)_s, M),$$

for any Φ_2-module M. For each $s \in C(\Phi_2)$ and $x \in \mathrm{G}^i(\Phi_2, M)$, we write x_s for the component of x in $\mathrm{H}^i((\Phi_2)_s, M)$. The module $\mathrm{G}^i(\Phi_2, M)$ has a natural action of the Hecke operators and we have an exact sequence of abelian groups for which the maps are compatible with the action of the Hecke operators:

$$0 \to \mathrm{H}^1_{\mathrm{p}}(\Phi_2, M) \to \mathrm{H}^1(\Phi_2, M) \to \mathrm{G}^1(\Phi_2, M).$$

From the exact sequence above we see that if the module $e\mathrm{G}^1(\Phi_2, \mathbb{Z}_p(\omega))$ is $\mathbb{Z}_p$-free, then the proposition follows. Consider the long exact sequence of cohomology groups

$$\mathrm{G}^0(\Phi_2, \mathbb{Q}_p(\omega)) \xrightarrow{\beta} \mathrm{G}^0(\Phi_2, \mathbb{T}_p(\omega)) \to \mathrm{G}^1(\Phi_2, \mathbb{Z}_p(\omega)) \to \mathrm{G}^1(\Phi_2, \mathbb{Q}_p(\omega)),$$

induced by the short exact sequence $0 \to \mathbb{Z}_p(\omega) \to \mathbb{Q}_p(\omega) \to \mathbb{T}_p(\omega) \to 0$.

Since the image of β is p-divisible, it is sufficient to know that for all $x \in \mathrm{G}^0(\Phi_2, \mathbb{T}_p(\omega))[p]$, $x|T_p$ belongs to $\beta(\mathrm{G}^0(\Phi_2, \mathbb{Q}_p(\omega)))$. Then a small computation shows that

$$e(\mathrm{G}^0(\Phi_2, \mathbb{T}_p(\omega))/\beta(\mathrm{G}^0(\Phi_2, \mathbb{Q}_p(\omega)))) = 0,$$

and hence $e\mathrm{G}^1(\Phi_2, L_n(\mathbb{Z}_p(\omega)))$ is $\mathbb{Z}_p$-free. We now prove, for all $x \in \mathrm{G}^0(\Phi_2, \mathbb{T}_p(\omega))[p]$, the element $x|T_p$ belongs to $\beta(\mathrm{G}^0(\Phi_2, \mathbb{Q}_p(\omega)))$.

Since $(2, N) = 1$ and $N \geq 3$, all the cusps of Φ_2 are regular, because irregularity for Φ_2 implies the irregularity for $\Gamma_1(N)$, but there are no irregular cusps for $\Gamma_1(N)$. Let $s \in C(\Phi_2)$ be a cusp of Φ_2. Let $\alpha_s = \left(\begin{smallmatrix} a & b \\ c & d \end{smallmatrix}\right) \in \mathrm{SL}_2(\mathbb{Z})$ such that $\alpha_s(\infty) = s$. If π_s denotes a generator for $(\Phi_2)_s$, we may write

$$\pi_s = \alpha_s \left(\begin{smallmatrix} 1 & u \\ 0 & 1 \end{smallmatrix}\right) \alpha_s^{-1} = \left(\begin{smallmatrix} 1-cau & a^2u \\ -c^2u & 1+cau \end{smallmatrix}\right) \in \Phi_2, \text{ with } u \neq 0. \tag{6.1}$$

The structure of $\mathrm{G}^0(\Phi_2, M(\omega))$ depends on the action π_s on M. In order to study this, let us divide the cusps into two types. If $p \mid u$, then we refer to this cusp as being of type 1, otherwise of type 2. We assume that Φ_2 acts trivially on M, because we are only interested in the cases when $M = \mathbb{Z}_p$, $\mathbb{Q}_p$ or $\mathbb{T}_p$. Let x be an element of $\mathrm{G}^0(\Phi_2, \mathbb{T}_p(\omega))[p]$.

If s is a cusp of type 1, then we see that $\mathrm{H}^0((\Phi_2)_s, M(\omega)) = M$ by (6.1) and moreover the map β_s is surjective, where $\beta_s : \mathrm{H}^0((\Phi_2)_s, \mathbb{Q}_p(\omega)) \to \mathrm{H}^0((\Phi_2)_s, \mathbb{T}_p(\omega))$. Hence $(x|T_p)_s \in \beta(\mathrm{H}^0((\Phi_2)_s, \mathbb{Q}_p(\omega)) = \mathbb{Q}_p)$.

Suppose s is a cusp of type 2. If π_s acts trivially on M, then

$$\mathrm{H}^0((\Phi_2)_s, M(\omega)) = M$$

and if π_s does not act trivially on M, then $\mathrm{H}^0((\Phi_2)_s, M(\omega)) = M[2]$. In the former case, again $(x|T_p)_s \in \beta(\mathbb{Q}_p)$. In the latter case,

$$(x|T_p)_s = \sum_{i=0}^{p-1} \left(\gamma \left(\begin{smallmatrix} 1 & 0 \\ 0 & p \end{smallmatrix}\right) \pi_s^i\right)^{\iota} \cdot x_t,$$

where $\gamma \in \Phi_2$ such that $t = \gamma \left(\begin{smallmatrix} 1 & 0 \\ 0 & p \end{smallmatrix}\right)(s) \in C(\Phi_2)$. Since x is 2-torsion, we see that

$$(x|T_p)_s = \sum_{i=0}^{p-1} (\pm 1) \cdot x_t = \sum_{i=0}^{p-1} x_t = 2x_t = 0 \in \beta(\mathrm{H}^0((\Phi_2)_s, \mathbb{Q}_p(\omega))) = 0.$$

Hence we have that for any $x \in \mathrm{G}^0(\Phi_2, \mathbb{T}_p(\omega))[p]$, the element $x|T_p$ belongs to $\beta(\mathrm{G}^0(\Phi_2, \mathbb{Q}_p(\omega)))$.

□

Remark 6.7 In the above proof, we have used the fact that $p = 2$.

Corollary 6.8 $e\mathrm{H}^1_{\mathrm{p}}(\Phi_2, \mathbb{Z}_p(\omega^n))/q \hookrightarrow e\mathrm{H}^1_{\mathrm{p}}(\Phi_2, \mathbb{Z}_p/q\mathbb{Z}_p(\omega^n))$.

Proof When n is even, this follows from Proposition 5.4 with $r = 2$. When n is odd, the injectivity of the first vertical map follows from the following diagram

$$\begin{array}{ccc} e\mathrm{H}^1_{\mathrm{p}}(Y, F(\mathbb{Z}_p(\omega)))/q & \xhookrightarrow{\alpha} & e\mathrm{H}^1(Y, F(\mathbb{Z}_p(\omega)))/q \\ \big\downarrow & & \big\downarrow \wr \\ e\mathrm{H}^1_{\mathrm{p}}(Y, F(\mathbb{Z}_p/q\mathbb{Z}_p(\omega))) & \hookrightarrow & e\mathrm{H}^1(Y, F(\mathbb{Z}_p/q\mathbb{Z}_p(\omega))), \end{array}$$

since α is injective by Proposition 6.6, and the second vertical map is an isomorphism by Lemma 6.5.

□

Now we shall give a proof Theorem 6.1.

Proof It is enough to prove that the $\mathbb{Z}_p$-rank of $e\mathrm{H}^1_{\mathrm{p}}(\Phi_2, L_n(\mathbb{Z}_p))$ is the same as the $\mathbb{Z}_p$-rank of $e\mathrm{H}^1_{\mathrm{p}}(\Phi_2, \mathbb{Z}_p(\omega^a))$ (the modules are $\mathbb{Z}_p$-free by Lemma 6.2 and by its corollary). By (5.2), we see that the rank of $e\mathrm{H}^1_{\mathrm{p}}(\Phi_2, L_n(\mathbb{Z}_p))$ is less than or equal to the rank of $e\mathrm{H}^1_{\mathrm{p}}(\Phi_2, \mathbb{Z}_p(\omega^a))$.

Again by Lemma 6.2 and by its corollary, it is enough to show the $\mathbb{Z}_p/q\mathbb{Z}_p$-rank of the module $e\mathrm{H}^1_{\mathrm{p}}(\Phi_2, L_n(\mathbb{Z}_p)) \otimes \mathbb{Z}_p/q\mathbb{Z}_p$ is greater than or equal to that of $e\mathrm{H}^1_{\mathrm{p}}(\Phi_2, \mathbb{Z}_p(\omega^a)) \otimes \mathbb{Z}_p/q\mathbb{Z}_p$. We have the following

$$\begin{aligned} e\mathrm{H}^1_{\mathrm{p}}(\Phi_2, L_n(\mathbb{Z}_p)) \otimes \mathbb{Z}_p/q\mathbb{Z}_p \underset{(1)}{\simeq} e\mathrm{H}^1_{\mathrm{p}}(\Phi_2, L_n(\mathbb{Z}_p/q\mathbb{Z}_p)) \underset{(2)}{\simeq} \\ e\mathrm{H}^1_{\mathrm{p}}(\Phi_2, \mathbb{Z}_p/q\mathbb{Z}_p(\omega^a)) \underset{(3)}{\hookleftarrow} e\mathrm{H}^1_{\mathrm{p}}(\Phi_2, \mathbb{Z}_p(\omega^a)) \otimes \mathbb{Z}_p/q\mathbb{Z}_p, \end{aligned} \tag{6.2}$$

where the isomorphisms (1), (2) and the inclusion (3) follow from Lemma 6.4, the isomorphism (3) in (5.1) with $r = 2$, and Corollary 6.8, respectively. Hence the theorem is proved.

□

7 Λ-adic Hecke Algebras

Recall that our aim is to prove a control theorem for Hida's ordinary Hecke algebra, which we now introduce.

Each element f in $S_k(\Gamma_1(Np^r))$, for $r \geq 0$, has the Fourier expansion $f(z) = \sum_n a_n(f)q^n$, for complex constants $a_n(f)$. By means of this, we may embed $S_k(\Gamma_1(Np^r))$ into the power series $\mathbb{C}[[q]]$. One may then give a rational structure on $S_k(\Gamma_1(Np^r))$ by defining the A-rational subspace $S_k(\Gamma_1(Np^r), A)$ for each subalgebra A of $\mathbb{C}$ by $S_k(\Gamma_1(Np^r), A) = S_k(\Gamma_1(Np^r)) \cap A[[q]]$.

For any $r \geq s \geq 1$, we have a commutative diagram for all n:

$$\begin{array}{ccc} S_k(\Gamma_1(Np^s), A) & \longrightarrow & S_k(\Gamma_1(Np^r), A) \\ \big\downarrow T_n & & \big\downarrow T_n \\ S_k(\Gamma_1(Np^s), A) & \longrightarrow & S_k(\Gamma_1(Np^r), A), \end{array}$$

where the horizontal arrows are the natural inclusion. Then the restriction of each Hecke operator in $\mathfrak{h}_k(\Gamma_1(Np^r), A)$ to the subspace $S_k(\Gamma_1(Np^s), A)$ is again contained in the algebra $\mathfrak{h}_k(\Gamma_1(Np^s), A)$. Thus, we have surjective A-algebra homomorphism, $\mathfrak{h}_k(\Gamma_1(Np^r), A) \to \mathfrak{h}_k(\Gamma_1(Np^s), A)$ and since $T_p \mapsto T_p$, we have that $\mathfrak{h}^0_k(\Gamma_1(Np^r), A) \to \mathfrak{h}^0_k(\Gamma_1(Np^s), A)$ for each $r \geq s \geq 1$, where the ordinary part is defined by using Hida's idempotent attached to T_p.

Now, take limits and set:

$$\begin{aligned} \mathfrak{h}_k(\Gamma_1(Np^\infty), A) &:= \varprojlim_r \mathfrak{h}_k(\Gamma_1(Np^r), A), \\ \mathfrak{h}^0_k(\Gamma_1(Np^\infty), A) &:= \varprojlim_r \mathfrak{h}^0_k(\Gamma_1(Np^r), A), \\ S_k(Np^\infty, A) &:= \cup_{r=1}^\infty S_k(\Gamma_1(Np^r), A). \end{aligned}$$

In Lemma 7.2 below we show there is a surjection $\mathfrak{h}_{k_1}(\Gamma_1(Np^\infty), A) \twoheadrightarrow \mathfrak{h}_{k_2}(\Gamma_1(Np^\infty), A)$, for weights $k_1 \geq k_2 \geq 2$, and hence on the ordinary parts. Before we state it, we need to define a pairing between certain Hecke algebras and certain spaces of modular forms. Recall that K is a finite extension of $\mathbb{Q}_p$ and $\mathcal{O}_K$ is the integral closure of $\mathbb{Z}_p$ in K. Put

$$S_k(Np^r, K/\mathcal{O}_K) = S_k(\Gamma_1(Np^r), K)/S_k(\Gamma_1(Np^r), \mathcal{O}_K).$$

By definition, one can embed this space via q-expansion into the module of formal series $K/\mathcal{O}_K[[q]]$. We take the injective limit:

$$S_k(Np^\infty, K/\mathcal{O}_K) = \varinjlim_r S_k(Np^r, K/\mathcal{O}_K) \to K/\mathcal{O}_K[[q]].$$

Then

$$S_k(Np^\infty, K/\mathcal{O}_K) \simeq S_k(Np^\infty, K)/S_k(Np^\infty, \mathcal{O}_K).$$

The algebra $\mathfrak{h}_k(\Gamma_1(Np^\infty), \mathcal{O}_K)$ acts on $S_k(Np^\infty, K/\mathcal{O}_K)$. Define the pairing

$$(,) : \mathfrak{h}_k(\Gamma_1(Np^\infty), \mathcal{O}_K) \times S_k(Np^\infty, K/\mathcal{O}_K) \to K/\mathcal{O}_K,$$

by $(h, f) = a(1, f|h)$. Then $(h, f|g) = (hg, f)$, for all $h, g \in \mathfrak{h}_k(\Gamma_1(Np^\infty), \mathcal{O}_K)$. Equip the space $S_k(Np^\infty, K/\mathcal{O}_K)$ with the discrete topology. We have (cf. [Hid86b, Lem. 7.1]):

Lemma 7.1 *The pairing above shows that*

$$\mathfrak{h}_k(\Gamma_1(Np^r), \mathbb{Z}_p) \text{ and } S_k(\Gamma_1(Np^r), \mathbb{T}_p)$$

(respectively, $\mathfrak{h}_k^0(\Gamma_1(Np^r), \mathbb{Z}_p)$ and $S_k^0(\Gamma_1(Np^r), \mathbb{T}_p)$), for $r = 1, 2, \ldots, \infty$, are Pontryagin duals.

Lemma 7.2 *For $k_1 \geq k_2 \geq 2$, there exists a surjection*

$$\mathfrak{h}_{k_1}(\Gamma_1(Np^\infty), \mathcal{O}_K) \twoheadrightarrow \mathfrak{h}_{k_2}(\Gamma_1(Np^\infty), \mathcal{O}_K).$$

Proof The proof is similar to the proof of [Hid86b, Lem. 7.2]. For $p = 2$, we need to work with a different Eisenstein series than the one given in that lemma. For $r \geq 2$, define a formal q-expansion for each $t \in (\mathbb{Z}/p^r\mathbb{Z})^\times$ by

$$G(r, t) = -t_0 p^{-r} + \frac{1}{2} + \sum_{n=1}^{\infty} \left(\sum_{\substack{d|n \\ d \equiv t \pmod{p^r}}} \operatorname{sgn}(d) \right) q^n,$$

where t_0 is an integer satisfying $0 \leq t_0 < p^r$ and $t_0 \equiv t \bmod p^r$. Then, as shown by Hecke, $G(r, t)$ gives the q-expansion of an element of $\mathcal{M}_1(\Gamma_1(Np^r), \mathbb{Q})$ and satisfies

$$G(r, t)|_1 = G(r, at) \text{ for } \left(\begin{smallmatrix} a & b \\ c & d \end{smallmatrix}\right) \in \Gamma_0(Np^r).$$

Put $E(r, t) = -p^r G(r, t)$. For odd primes p, the congruence $E(r, t) \equiv t \pmod{p^r}$ holds. For the even prime $p = 2$, the congruence that holds is $E(r, t) \equiv t \pmod{p^{r-1}}$. Multiplication by the Eisenstein series $E(r, 1)$ gives an injective morphism

$$\iota_r : S_{k-1}(Np^\infty, \mathbb{T}_p)[p^{r-1}] \to S_k(Np^\infty, \mathbb{T}_p)[p^{r-1}].$$

Using the injective limit of the maps ι_r and Lemma 7.1, we can finish the proof of the lemma along the lines of the proof of [Hid86b, Lem. 7.2].

□

It is known, by [Hid88b, Thm. 3.2], that the map in the above lemma is an isomorphism. This theorem is stated adelically, but includes the case of $p = 2$. Thus, the Hecke algebra $\mathfrak{h}_k^0(\Gamma_1(Np^\infty), \mathcal{O}_K)$ is independent of the weight, for all $k \geq 2$. Denote this Hecke algebra by $\mathfrak{h}^0(N, \mathcal{O}_K)$.

8 Control Theorem for Ordinary Hecke Algebras

In this section, we prove a control theorem for Hida's ordinary Hecke algebras for the prime $p = 2$. Recall K is a finite extension of $\mathbb{Q}_p$ and $\mathcal{O}_K$ is integral closure of $\mathbb{Z}_p$ in K. Let ϵ be a character of Γ/Γ_r with values in $\mathcal{O}_K$, with $r \geq 2$. In this section, we write Λ for Λ_K and $Q(\Lambda)$ for the field of fractions of Λ_K.

We know that $\mathfrak{h}^0(N, \mathcal{O}_K)$ acts on the finite free Λ-module $\mathcal{V}^0$. Hence the Λ-module $\mathfrak{h}^0(N, \mathcal{O}_K)$ is finitely generated and torsion-free, since the action is faithful on $\mathcal{V}^0$. By abuse of notation, let $P_{k,\epsilon}$ also denote the prime ideal generated by the prime element $P_{k,\epsilon} = \iota(u) - \epsilon(u)u^k$. By the independence of weight of $\mathfrak{h}^0(N, \mathcal{O}_K)$, there is a surjective homomorphisms of $\mathcal{O}_K$-algebras, respectively, of Λ-algebras:

$$\rho : \mathfrak{h}^0(N, \mathcal{O}_K) \twoheadrightarrow \mathfrak{h}^0_k(\Phi^2_r, \epsilon, \mathcal{O}_K) \text{ and } \Lambda_{P_{k,\epsilon}} \twoheadrightarrow \Lambda_{P_{k,\epsilon}}/P_{k,\epsilon}\Lambda_{P_{k,\epsilon}} = K, \quad (8.1)$$

inducing the map

$$\tilde{\rho}_{k,\epsilon} : \mathfrak{h}^0(N, \mathcal{O}_K) \otimes_\Lambda \Lambda_{P_{k,\epsilon}} \twoheadrightarrow \mathfrak{h}^0_k(\Phi^2_r, \epsilon, \mathcal{O}_K) \otimes_{\mathcal{O}_K} K,$$

which in turn factors via $P_{k,\epsilon}$, to give the map:

$$\rho_{k,\epsilon} : \mathfrak{h}^0(N, \mathcal{O}_K) \otimes_\Lambda \Lambda_{P_{k,\epsilon}}/P_{k,\epsilon} \twoheadrightarrow \mathfrak{h}^0_k(\Phi^2_r, \epsilon, \mathcal{O}_K) \otimes_{\mathcal{O}_K} K \simeq \mathfrak{h}^0_k(\Phi^2_r, \epsilon, K),$$

where $\Lambda_{P_{k,\epsilon}}/P_{k,\epsilon}\Lambda_{P_{k,\epsilon}}$ is identified with K with $\iota(u)$ corresponding to $u^k\epsilon(u)$.

Theorem 8.1 *The natural map*

$$\rho_{k,\epsilon} : \mathfrak{h}^0(N, \mathcal{O}_K) \otimes_\Lambda \Lambda_{P_{k,\epsilon}}/P_{k,\epsilon} \twoheadrightarrow \mathfrak{h}^0_k(\Phi^2_r, \epsilon, K)$$

is an isomorphism.

Proof Since the module $\mathfrak{h}^0(N, \mathcal{O}_K)$ is finitely generated and torsion-free over Λ, so is $\mathfrak{h}^0(N, \mathcal{O}_K)_{P_{k,\epsilon}}$ over $\Lambda_{P_{k,\epsilon}}$. Since any finitely generated torsion-free module over a discrete valuation ring is free, the module $\mathfrak{h}^0(N, \mathcal{O}_K)_{P_{k,\epsilon}}$ is free and hence it makes sense to speak of its rank. Let $S(k, \epsilon)$ (respectively, $R(k, \epsilon)$) denote the rank of $\mathfrak{h}^0(N, \mathcal{O}_K)_{P_{k,\epsilon}}$ (respectively, $\mathfrak{h}^0_k(\Phi^2_r, \epsilon, K)$). A priori the number $S(k, \epsilon)$ depends on k and ϵ. Since

$$\mathfrak{h}^0(N, \mathcal{O}_K)_{P_{k,\epsilon}} \otimes_{\Lambda_{P_{k,\epsilon}}} Q(\Lambda) \simeq \mathfrak{h}^0(N, \mathcal{O}_K) \otimes_\Lambda Q(\Lambda),$$

we see that $S(k, \epsilon)$ is independent of k and ϵ and we denote this common value by R.

We first prove the theorem for weights $k > 2$ by assuming that it holds for $k = 2$. The Eisenstein series $E(2,1)$ above has the property that $E(2,1) \equiv 1 \pmod 2$. Multiplication by $E(2,1)^{k-2}$ induces an injection

$$S_2^0(\Gamma_1(Np^r), \mathbb{T}_p)[p] \to S_k^0(\Gamma_1(Np^r), \mathbb{T}_p)[p].$$

By duality, we have a surjection

$$\mathfrak{h}_k^0(\Gamma_1(Np^r), \mathbb{Z}_p) \otimes \mathbb{Z}_p/p\mathbb{Z}_p \twoheadrightarrow \mathfrak{h}_2^0(\Gamma_1(Np^r), \mathbb{Z}_p) \otimes \mathbb{Z}_p/p\mathbb{Z}_p.$$

Then

$$\begin{aligned} R[\Gamma : \Gamma_r] \geq \sum_{\epsilon} R(k,\epsilon) &= \mathrm{rank}_{\mathcal{O}_K}(\mathfrak{h}_k^0(\Gamma_1(Np^r), \mathcal{O}_K)) \\ &\geq \mathrm{rank}_{\mathcal{O}_K}(\mathfrak{h}_2^0(\Gamma_1(Np^r), \mathcal{O}_K)) = R[\Gamma : \Gamma_r], \end{aligned}$$

where the last equality follows by assumption. This can happen only if $R(k,\epsilon) = R$ for all k, ϵ, showing $\rho_{k,\epsilon}$ is an isomorphism.

Now, we shall prove the result for $k = 2$. By Theorem 3.1, we have that the $\mathbb{Z}_p$-rank of $e\mathrm{H}^1_{\mathrm{p}}(\Gamma_1(Np^r), \mathbb{Z}_p)$ is equal to $2[\Gamma : \Gamma_r]\ \mathrm{rank}_{\mathbb{Z}_p}\mathfrak{h}_2^0(\Gamma_1(Nq), \mathbb{Z}_p)$. Hence,

$$\mathrm{rank}_{\mathbb{Z}_p}\mathfrak{h}_2^0(\Gamma_1(Np^r), \mathbb{Z}_p) = [\Gamma : \Gamma_r]\ \mathrm{rank}_{\mathbb{Z}_p}\mathfrak{h}_2^0(\Gamma_1(Nq), \mathbb{Z}_p).$$

Since $\mathfrak{h}_2^0(\Gamma_1(Np^r), K) = \oplus_\epsilon \mathfrak{h}_2^0(\Phi_r^2, \epsilon, K)$, the left hand side of the equality above is also $\sum_\epsilon R(2,\epsilon)$. If $\mathrm{rank}_{\mathbb{Z}_p}\mathfrak{h}_2^0(\Gamma_1(Nq), \mathbb{Z}_p) = R$, then $[\Gamma : \Gamma_r]R = \sum_\epsilon R(2,\epsilon)$. Since $R \geq R(2,\epsilon)$, we get $R = R(2,\epsilon)$, for each ϵ, as desired. Thus, we need to show that $R = \mathrm{rank}_{\mathbb{Z}_p}\mathfrak{h}_2^0(\Gamma_1(Nq), \mathbb{Z}_p)$. This is proved in Theorem 8.3 below.

□

The following lemma is well-known; for the proof refer to [Hid86b, Lem. 6.4].

Lemma 8.2 *For any subfield K of $\mathbb{C}$ or $\mathbb{C}_p$, $\mathrm{H}^1_{\mathrm{p}}(\Gamma_1(M), L_n(K))$ is free of rank 2 over the Hecke algebra $\mathfrak{h}_{n+2}(\Gamma_1(M), K)$ for each positive integer M.*

Set $\epsilon := \left(\begin{smallmatrix} 1 & 0 \\ 0 & -1 \end{smallmatrix}\right)$. The matrix ϵ normalizes $\Gamma_1(Np^r)$ for $r \geq 1$. Let M be a module over the Hecke algebra $\mathfrak{h}_2^0(\Gamma_1(Np^r), \mathbb{Z})$. Let $M^\pm$ denote the subspaces of M defined by $\{m \pm [\epsilon]m \mid m \in M\}$. Since ϵ normalizes $\Gamma_1(Np^r)$, the action of $[\epsilon] = [\Gamma_1(Np^r)\epsilon\Gamma_1(Np^r)]$ commutes with that of the Hecke algebra $\mathfrak{h}_2^0(\Gamma_1(Np^r), \mathbb{Z})$ on M. Therefore, the modules $M^\pm$ are stable under the action of $\mathfrak{h}_2^0(\Gamma_1(Np^r), \mathbb{Z})$. For simplicity, we write $\mathfrak{h}_2^0(N, \mathbb{Z}_p)$ for the weight-2 Λ-adic Hecke algebra $\mathfrak{h}_2^0(\Gamma_1(Np^\infty), \mathbb{Z}_p)$.

Theorem 8.3 *The surjective map*

$$\rho_{2,\mathrm{triv}} : \mathfrak{h}_2^0(N, \mathbb{Z}_p) \otimes_\Lambda \Lambda_{P_2}/P_2 \twoheadrightarrow \mathfrak{h}_2^0(\Gamma_1(Nq), \mathbb{Q}_p)$$

is an isomorphism.

Proof By Theorem 3.1, we have $(\mathcal{V}^0)^{\Gamma_r} = \mathcal{V}_r^0$, for $r \geq 2$, and in particular $(\mathcal{V}^0)^{\Gamma_2} = \mathcal{V}_2^0$, i.e., $\mathcal{V}^0[P_2] = e\mathrm{H}^1(X_2, \mathbb{T}_p) = e\mathrm{H}^1_{\mathrm{p}}(\Gamma_1(Nq), \mathbb{T}_p)$, where the last equality follows from [Hid86b, p. 583 (5.3)]. Again by the same theorem, we have

$$V^0/P_2V^0 \simeq \mathrm{Hom}_{\mathbb{Z}_p}(e\mathrm{H}^1_{\mathrm{p}}(\Gamma_1(Nq), \mathbb{Z}_p), \mathbb{Z}_p). \tag{8.2}$$

Since $\mathcal{V}^0$ is direct limit over $\mathcal{V}_r^0$, we see that $[\epsilon]$ acts on $\mathcal{V}^0$ and the action commutes with that of the Hecke algebra $\mathfrak{h}_2^0(N, \mathbb{Z}_p)$. There is a map $\mathcal{V}^{0+} \oplus \mathcal{V}^{0-} \to \mathcal{V}^0$, which is an isomorphism if p is odd. Since $p = 2$, we tensor this with Λ_{P_2} so that we have an isomorphism $(\mathcal{V}^{0+})_{P_2} \oplus (\mathcal{V}^{0-})_{P_2} \simeq \mathcal{V}^0_{P_2}$.

Let $V^{0\pm}$ denote the Pontryagin dual of $\mathcal{V}^{0\pm}$. Then $(V^{0+})_{P_2} \oplus (V^{0-})_{P_2} \simeq V^0_{P_2}$. We can think of $\mathfrak{h}_2^0(N, \mathbb{Z}_p)_{P_2}$ as a subalgebra of the endomorphism algebra of $(V^{0+})_{P_2}$ and hence we shall restrict ourselves to the module $(V^{0+})_{P_2}$. We now prove that

$$(V^{0+})_{P_2}/P_2(V^{0+})_{P_2} = \mathfrak{h}_2^0(\Gamma_1(Nq), \mathbb{Q}_p).$$

We remark that since $p = 2$ and we work with $(V^{0+})_{P_2}$, the above isomorphism is with $\mathbb{Q}_p$-coefficients, otherwise we would have worked with V^{0+} and the above isomorphism would have been with $\mathbb{Z}_p$-coefficients. Since the functor $\mathrm{Hom}_{\mathbb{Z}_p}(-, \mathbb{T}_p)$ commutes with the $\pm$-action after tensoring with Λ_{P_2}, we see that $V^{0\pm} \otimes_\Lambda \Lambda_{P_2} \simeq (V^0_{P_2})^\pm$ holds. Hence $(V^{0\pm})_{P_2}/P_2(V^{0\pm})_{P_2} \simeq (V^0_{P_2})^\pm/P_2(V^0_{P_2})^\pm \simeq (V^0_{P_2}/P_2V^0_{P_2})^\pm$, where the last isomorphism is an easy check. We have that

$$\begin{aligned}(V^{0+})_{P_2}/P_2(V^{0+})_{P_2} &= (V^0_{P_2}/P_2V^0_{P_2})^+ \\ &\underset{(8.2)}{=} (\mathrm{Hom}_{\mathbb{Z}_p}(e\mathrm{H}^1_{\mathrm{p}}(\Gamma_1(Nq), \mathbb{Z}_p), \mathbb{Z}_p) \otimes_{\mathbb{Z}_p} \mathbb{Q}_p)^+ \\ &= \mathrm{Hom}_{\mathbb{Q}_p}(e\mathrm{H}^1_{\mathrm{p}}(\Gamma_1(Nq), \mathbb{Q}_p)^+, \mathbb{Q}_p) \\ &= \mathfrak{h}_2^0(\Gamma_1(Nq), \mathbb{Q}_p),\end{aligned} \tag{8.3}$$

where the last equality follows from Lemma 8.2 and the third equality follows from the fact that for any $\mathbb{Q}_p$-module M,

$$\mathrm{Hom}_{\mathbb{Q}_p}(M, \mathbb{Q}_p)^\pm \simeq \mathrm{Hom}_{\mathbb{Q}_p}(M^\pm, \mathbb{Q}_p).$$

Let v denote the vector in $(V^{0+})_{P_2}$ corresponding to 1 in $\mathfrak{h}_2^0(\Gamma_1(Nq), \mathbb{Q}_p)$ in (8.3). Therefore, we have a map $\mathfrak{h}_2^0(N, \mathbb{Z}_p)_{P_2} \to (V^{0+})_{P_2}$ defined by mapping $h \to hv$. This map is a surjective map by Nakayama's lemma and by (8.3). The map is injective since the Hecke action is faithful on $(V^{0+})_{P_2}$. Therefore, we have $\mathfrak{h}_2^0(N, \mathbb{Z}_p)_{P_2} \simeq (V^{0+})_{P_2}$. Tensoring this isomorphism with Λ_{P_2}/P_2 and using (8.3), we obtain the theorem.

□

9 Uniqueness

In this section, we prove a uniqueness result for Hida families. Let f be a p-stabilized newform. Let P_f denote the unique height one prime ideal, induced by f, via the isomorphism in Theorem 8.1. Suppose $Q = P_f$ lies over the prime ideal $P_{k,\epsilon}$, where the integer k and the character ϵ depend on f.

First we show that, for the prime $P = P_{k,\epsilon}$ of Λ, the localized Hecke algebra $\mathfrak{h}^0(N, \mathcal{O}_K)_Q$ is étale over $\Lambda_{P_{k,\epsilon}}$. We deduce the uniqueness result as a consequence. For simplicity, let us denote $\mathfrak{h}^0(N, \mathcal{O}_K)$ by $\mathfrak{h}^0(N)$.

Proposition 9.1 *The localized Hecke algebra $\mathfrak{h}^0(N)_Q$ is étale over Λ_P and $Q\mathfrak{h}^0(N)_Q = P\mathfrak{h}^0(N)_Q$, i.e., $\mathfrak{h}^0(N)_Q$ is a regular local ring.*

Proof We apply [Nek06, Lem. 12.7.6], with $A = \Lambda$, $B = \mathfrak{h}^0(N)$ and $J = 0$ (and also by switching the roles of P and Q). The first condition of that lemma, namely the Hecke algebra $\mathfrak{h}^0(N)$ is finitely generated and torsion-free over Λ follows, as mentioned earlier, from Theorem 3.1. By Theorem 8.1, the short exact sequence in the second part of that lemma reduces to

$$0 \to \mathcal{P} \to \mathfrak{h}_k^0(\Phi_r^2, \epsilon, K) \overset{\alpha}{\to} \mathbb{Q}_p(a_n(f))_{n=1}^{\infty} \to 0,$$

where the last map is given by $T_n \to a_n(f)$ and $\mathcal{P}$ denote the kernel of α. By analyzing the proof of that lemma, we see that if $\mathcal{P}_{\mathcal{P}} = 0$, where $\mathcal{P}_{\mathcal{P}}$ denote the localisation, then the proposition follows. From the theory of newforms, one knows that $\mathfrak{h}_k^0(\Phi_r^2, \epsilon, K)_{\mathcal{P}} \overset{\sim}{\to} \mathbb{Q}_p(a_n(f))_{n=1}^{\infty}$, hence $\mathcal{P}_{\mathcal{P}} = 0$.

□

Now we recall the definition of a 2-adic Λ-adic form. Let $p = 2$. Let L denote the integral closure of Λ in a finite extension of $Q(\Lambda)$. Let ζ denote a p^{r-2}-th root of unity in $\bar{\mathbb{Q}}_p$, the algebraic closure of $\mathbb{Q}_p$, with $r \geq 2$, and let $k \geq 1$ be a positive integer. The assignment $X \to \zeta(1+q)^k - 1$ yields a $\mathbb{Z}_p$-algebra homomorphism $\varphi_{k,\zeta} : \Lambda \to \bar{\mathbb{Q}}_p$. We shall say that a height one prime $P \in \mathrm{Spec}(L)(\bar{\mathbb{Q}}_p)$ has weight k if the corresponding Λ-algebra homomorphism $P : L \to \bar{\mathbb{Q}}_p$ extends $\varphi_{k,\zeta}$ on Λ for some $k \geq 1$ and for some ζ. In addition we say that P is arithmetic if P has weight $k \geq 2$.

Recall N is an integer prime to p. We need some notation for certain Dirichlet characters. Let:

- ψ be a Dirichlet character of level Nq,
- ω be the mod 4 cyclotomic character,
- ϵ be the character χ_ζ mod 2^r for each root of unity ζ of order 2^{r-2} with $r \geq 2$ defined by first decomposing

$$(\mathbb{Z}_p/2^r\mathbb{Z}_p)^\times = (\mathbb{Z}_p/q\mathbb{Z}_p)^\times \times \mathbb{Z}/2^{r-2},$$

where the second factor is generated by $1+q$, and then by setting

$$\chi_\zeta = 1 \text{ on } (\mathbb{Z}_p/q\mathbb{Z}_p)^\times \quad \text{and} \quad \chi_\zeta(1+q) = \zeta.$$

Definition 9.2 Let $\mathcal{F} = \sum_{n=1}^{\infty} a(n,\mathcal{F})q^n \in L[[q]]$ be a formal q-expansion with coefficients $a(n,\mathcal{F}) \in I$. We say $\mathcal{F}$ is a Λ-adic form of tame level N and character ψ if for each arithmetic point $P \in \mathrm{Spec}(L)(\bar{\mathbb{Q}}_p)$ lying over $\varphi_{k,\zeta}$, with $k \geq 2$ and ζ of order $2^{r-2}, r \geq 2$, the specialization

$$P(\mathcal{F}) \in \bar{\mathbb{Q}}_p[[q]]$$

of $\mathcal{F}$ at P is the q-expansion of a classical cusp form $f \in S_k(N2^r, \chi)$, where $\chi = \psi\omega^{-k}\chi_\zeta$.

The notion of primitive, ordinary, p-distinguished for Λ-adic forms can be defined similar to the classical case. For definitions, refer to [GV04, §3].

We remark that there is no 2-adic Hida theory when the tame level is 1, showing that our assumption that $N > 1$ in several previous sections loses no generality. Indeed, we have that:

Proposition 9.3 *There are no ordinary Λ-adic eigenforms of tame level* 1.

Proof If such a Λ-adic eigenform were to exist, then for every integer $k \geq 2$ and $r \geq 2$, its' specialization at $P_{k,\zeta}$, where ζ is a 2^{r-2}-th root of unity, would be an element of $S_k(p^r, \omega^{a-k}\chi_\zeta)$, for some $a \in \mathbb{N}$. For parity reasons, $(-1)^{a-k} = (-1)^k$, hence a is even. But, if $r = 2$ and k is even, then are no 2-ordinary, 2-stabilized Hecke eigenforms in $S_k(4, \mathrm{triv})$. For newforms this follows from [Miy89, Thm. 4.6.17] and for oldforms from *loc. cit.* and the fact that $X_0(2)$ has genus 0, and from Hatada [Hat79]. Since there are no ordinary specializations in even weight, there are no 2-ordinary Λ-adic eigenforms of tame level 1.

□

Corollary 9.4 *For odd integers k, the space $eS_k^{2\text{-new}}(4,\omega)$ is zero.*

Proof This follows immediately from the proposition noting that every 2-ordinary eigenform in the above space must live in a 2-ordinary Hida family of tame level 1.

□

Remark 9.5 It can be checked independently that the dimensions of $eS_k^{2\text{-new}}(4,\omega)$ for $k = 3, 5, 7, 9, 11, 13, 15, 17$, are indeed all zero, whereas the dimensions of $S_k^{2\text{-new}}(4,\omega)$ for $k = 3, 5, 7, 9, 11, 13, 15, 17$ are $0, 1, 2, \ldots, 7$, respectively.

We now turn to the uniqueness result for 2-adic families.

Theorem 9.6 *Any p-ordinary elliptic p-stabilized newform is an arithmetic specialization of a* ***unique*** *Hida family, up to Galois conjugacy.*

Proof By (8.1), we know that any p-ordinary eigenform lives in a Hida family. We want to show that such a family is unique, up to Galois conjugacy (this last caveat is necessary since if a form lies in F by specialization under $P : L \to \bar{\mathbb{Q}}_p$, then it also lies in the conjugate family F^σ, by specializing under $\sigma^{-1} \circ P : L^\sigma \to \bar{\mathbb{Q}}_p$. Note that F and F^σ correspond to the same minimal prime ideal of $\mathfrak{h}^0(N)$).

Assume the contrary. Let λ_1 and λ_2 denote the algebra homomorphisms from $\mathfrak{h}^0(N)$ to L and L' respectively, where L, L' are finite integral extensions of Λ. Let P_1 and P_2 denote the minimal prime ideals of $\mathfrak{h}^0(N)$ which are the respective kernels of these homomorphisms. Since λ_1 and λ_2 have one arithmetic specialization in common, there are two algebra homomorphisms $P : L \to \bar{\mathbb{Q}}_p$ and $P' : L' \to \bar{\mathbb{Q}}_p$ such that $P \circ \lambda_1 = P' \circ \lambda_2 = \lambda_{P,P'}$, say. Then the kernel of $\lambda_{P,P'}$ is a height one prime of $\mathfrak{h}^0(N)$, denote by Q, containing both P_1 and P_2 and lying over $P = P_{k,\zeta}$ for some $k \geq 2$, ζ.

By Proposition 9.1, $\mathfrak{h}^0(N)_Q$ is a regular local ring. But, a regular a local ring is a domain, hence the prime ideals P_1 and P_2 have to be equal.

□

As an application of the last result we now show that the notion of CM-ness is pure with respect to families.

Proposition 9.7 *Let $\mathcal{F}$ be a primitive 2-adic Hida family. Then either all arithmetic specializations are CM forms or no arithmetic specialization is a CM form.*

Proof The proof is the same as for odd prime p, once one has the uniqueness result for $p = 2$. Indeed a CM family is defined to be one which is obtained as the theta series of a Λ-adic Hecke character of an imaginary quadratic field. Clearly all its arithmetic specializations are CM forms. Now start with an arbitrary CM form. Assume it lives in a non-CM family (one which is not a theta series). Then explicit interpolation allows us to also construct a CM family passing through this CM form. Clearly the non-CM family and the CM family are not Galois conjugate, which is a contradiction by Theorem 9.6.

□

In view of this result from now on we may and do speak of CM and non-CM 2-adic Hida families.

10 Applications to Galois Representations

In [GV04], the splitting of the local Galois representations associated to ordinary eigenforms was studied for odd primes p. We carry out the same analysis for the case of $p = 2$, assuming that the relevant result of Buzzard continues to hold for $p = 2$ in the residually dihedral setting. That is, under this assumption, we prove that in a non-CM 2-adic Hida family, all arithmetic specializations have non-split local Galois representation, except for a possible finite set of exceptions. By Proposition 9.7, we are able to exclude CM forms from this finite exceptional set, but we do not yet know if this set is empty.

Recall p denotes the prime 2 and $q = 4$. We recall some preliminaries on ordinary eigenforms and their associated Galois representations. Let $f = \sum_{n=1}^{\infty} a_n(f)q^n$ be a primitive elliptic modular Hecke eigenform of weight $k \geq 2$ and nebentypus $\chi : (\mathbb{Z}/Np^r)^\times \to \mathbb{C}^\times$, for some $r \geq 0$. (The two usages of q, that in the q-expansion and the natural number 4, should be clear from the context!) Let K_f denote the number field generated by the Fourier coefficients of the cusp form f. Fix an embedding i_p of $\bar{\mathbb{Q}}$ into $\bar{\mathbb{Q}}_p$. Let $\wp$ be the prime of $\bar{\mathbb{Q}}$ determined by this embedding. Let $\wp$ also denote the induced prime of K_f, and let $K_{f,\wp}$ be the completion of K_f at $\wp$. Let G_p denote the absolute Galois group of $\mathbb{Q}_p$ and also the decomposition group at $\wp$. There is a Galois representation

$$\rho_f = \rho_{f,\wp} : \mathrm{Gal}(\bar{\mathbb{Q}}/\mathbb{Q}) \to \mathrm{GL}_2(K_{f,\wp}),$$

associated to f (and $\wp$) which has the property that for all primes $\ell \nmid Np$,

$$\mathrm{trace}(\rho_f(\mathrm{Frob}_\ell)) = a_\ell(f) \quad \text{and} \quad \det(\rho_f(\mathrm{Frob}_\ell)) = \chi(\ell)\ell^{k-1}.$$

Recall f is ordinary at $\wp$ (or $\wp$-ordinary), if $a_p(f)$ is $\wp$-adic unit. If f is ordinary at $\wp$, then the result of Wiles [Wil88] shows that the restriction of ρ_f to the decomposition group G_p is upper-triangular, i.e.,

$$\rho_f|_{G_p} \sim \begin{pmatrix} \delta & \psi \\ 0 & \epsilon \end{pmatrix},$$

where $\delta, \epsilon : G_p \to K_{f,\wp}^{\times}$ are characters with ϵ unramified and $\psi : G_p \to K_{f,\wp}$ is a continuous function. We say that the ordinary representation $\rho_f|_{G_p}$ splits, if the representation space of ρ_f can be written as direct sum of two G_p-invariant lines.

10.1 Buzzard's result

We shall assume that a slight strengthening of a result of Buzzard holds. Let $\mathcal{O}$ denote the ring of integers in a finite extension L of $\mathbb{Q}_p$. Let $\rho : \mathrm{Gal}(\bar{\mathbb{Q}}/\mathbb{Q}) \to \mathrm{GL}_2(\mathcal{O})$ be a continuous representation. Let λ denote the maximal ideal of $\mathcal{O}$ and let $\bar{\rho}$ denote the mod λ reduction of ρ. The following result is proved in [Buz03], and we refer to that paper for a detailed explanation of all the hypotheses.

Theorem 10.1 (Buzzard) *Assume that*

1. *ρ is ramified at finitely many primes and $\bar{\rho}$ is modular,*
2. *$\bar{\rho}$ is absolutely irreducible when restricted to* $\mathrm{Gal}(\bar{\mathbb{Q}}/\mathbb{Q}(i))$,
3. *$\rho|_{G_p}$ is the direct sum of two 1-dimensional characters α and β : $G_p \to \mathcal{O}^{\times}$, such that $\alpha(I_p)$ and $\beta(I_p)$ are finite, and (α/β) mod λ is non-trivial,*
4. $\bar{\rho}(c) \neq 1$,
5. *$\bar{\rho}(c)$ is both α-modular and β-modular, in the sense that there are eigenforms f_α with T_p-eigenvalue $\bar{\alpha}(\mathrm{Frob}_p)$ and f_β with T_p-eigenvalue $\bar{\beta}(\mathrm{Frob}_p)$ giving rise to $\bar{\rho}$,*
6. ***The projective image of $\bar{\rho}$ is not dihedral.***

Then ρ is modular, in the sense that there exists an embedding $i : L \hookrightarrow \mathbb{C}$ and a classical weight 1 *cuspidal eigenform f such that the composite $i \circ \rho$ is isomorphic to the representation associated to f by Deligne and Serre.*

Let us comment on the assumption (6). There is no restriction (6) on the projective image of $\bar{\rho}$ for odd primes p in [Buz03]. For $p = 2$, this assumption was made due to the unavailability of '$R^{\mathrm{red}} = T$ theorems' in the residually dihedral setting. In his recent thesis, Allen [All12] has proved such a theorem, deducing the modularity of nearly ordinary 2-adic residually dihedral Galois representations. He works under some assumptions, the most crucial for us being that he assumes that the prime 2 does not split in the quadratic extension of $\mathbb{Q}$ corresponding to the (dihedral) residual representation. However, for the application of Buzzard's theorem we have in mind below, the prime 2 does split in this extension. It appears that extending Allen's result to the split case might not be possible without a new idea. From now on we therefore **assume** that Theorem 10.1 holds without condition (6).

Remark 10.2 In [All12], the splitting assumption on the prime 2 is made in order to ensure that the dihedral locus is small in the Hecke algebra. This guarantees the existence of certain 'nice' primes that are needed in order to use a connectivity result of Raynaud in the course of the proof.

10.2 Λ-adic Galois representations

We state a few facts about Λ-adic Galois representations. Let $\mathcal{F} \in I[[q]]$ be a primitive Λ-adic form of level N and with character ψ. Let $K_{\mathcal{F}}$ denote the quotient field of I. Then there exists a Galois representation attached to $\mathcal{F}$, constructed by Hida, and Wiles in the case of $p = 2$,

$$\rho_{\mathcal{F}} : \mathrm{Gal}(\bar{\mathbb{Q}}/\mathbb{Q}) \to \mathrm{GL}_2(K_{\mathcal{F}}),$$

such that for each arithmetic point P of I, $P(\rho_{\mathcal{F}})$, the specialization of $\rho_{\mathcal{F}}$ at P, is isomorphic to the representation ρ_f attached to $f = P(\mathcal{F})$ by Deligne. Note that if ℓ is a prime number such that $\ell \nmid Np$, then

$$\mathrm{trace}(\rho_{\mathcal{F}}(\mathrm{Frob}_\ell)) = a(\ell, \mathcal{F}) \in I, \ \det(\rho_{\mathcal{F}}(\mathrm{Frob}_\ell)) = \psi(\ell)\kappa(\mathrm{Frob}_\ell)\ell^{-1},$$

where $\kappa : \mathrm{Gal}(\bar{\mathbb{Q}}/\mathbb{Q}) \to \Lambda^\times$ is the 'Λ-adic cyclotomic character'.

The restriction of $\rho_{\mathcal{F}}$ to G_p also turns out to be 'upper-triangular'. More precisely, the representation $\rho_{\mathcal{F}}|_{G_p}$ has the following shape

$$\rho_{\mathcal{F}}|_{G_p} \sim \begin{pmatrix} \delta_{\mathcal{F}} & u_{\mathcal{F}} \\ 0 & \epsilon_{\mathcal{F}} \end{pmatrix},$$

where $\delta_{\mathcal{F}}, \epsilon_{\mathcal{F}} : G_p \to K_{\mathcal{F}}^\times$ are characters with $\epsilon_{\mathcal{F}}$ unramified, and $u_{\mathcal{F}} : G_p \to K_{\mathcal{F}}$ is a continuous map. Let

$$c_{\mathcal{F}} = \epsilon_{\mathcal{F}}^{-1}.u_{\mathcal{F}} \in \mathrm{Z}^1(G_p, K_{\mathcal{F}}(\delta_{\mathcal{F}}\epsilon_{\mathcal{F}}^{-1}))$$

be the associated cocycle. Then the representation

$$\rho_{\mathcal{F}}|_{G_p} \text{ splits if and only if } [c_{\mathcal{F}}] = 0 \text{ in } \mathrm{H}^1(G_p, K_{\mathcal{F}}(\delta_{\mathcal{F}}\epsilon_{\mathcal{F}}^{-1})).$$

We shall shortly show that for a primitive 2-adic family $\mathcal{F}$ whose residual representation satisfies some technical conditions (cf. conditions (1), (2), (3) below), the corresponding representation $\rho_{\mathcal{F}}$ splits at p if and if $\mathcal{F}$ is a CM family. As a consequence, standard descent arguments allow us to conclude the following partial result towards Greenberg's question on the local splitting of ordinary 2-adic modular Galois representations.

Theorem 10.3 *Let $\mathcal{F}$ be a primitive non-CM 2-ordinary Hida family of eigenforms with the property that*

1. *$\bar{\rho}_{\mathcal{F}}$ is p-distinguished,*
2. *$\bar{\rho}_{\mathcal{F}}$ is absolutely irreducible, when restricted to* $\mathrm{Gal}(\bar{\mathbb{Q}}/\mathbb{Q}(i))$,
3. *$\bar{\rho}_{\mathcal{F}}(c) \neq 1$ and $\bar{\rho}_{\mathcal{F}}(c)$ is both α-modular and β-modular.*

Then, for all but except possibly finitely many arithmetic members $f \in \mathcal{F}$, the representation $\rho_f|_{G_p}$ is non-split. Moreover the possible exceptions are necessarily non-CM forms.

For the definitions of primitive and p-distinguished, refer to [GV04, §2]. We remark again that the last statement in the theorem is a consequence of uniqueness for 2-adic families, proved in the last section.

10.3 Local splitting for Λ-adic eigenforms

Proposition 10.4 *Let $\mathcal{F}$ be a primitive 2-adic Λ-adic eigenform of fixed tame level N satisfying conditions* (1)-(3) *above. Then $\rho_{\mathcal{F}}|_{G_p}$ splits if and only if $\mathcal{F}$ is of CM type.*

Proof The proof is very similar to that for odd primes given in [GV04, Prop. 14]. One shows that the following statements are equivalent.

1. $\rho_{\mathcal{F}}|_{G_p}$ splits.
2. $\mathcal{F}$ has infinitely many weight one classical specializations.
3. $\mathcal{F}$ has infinitely many weight one classical CM specializations.
4. $\mathcal{F}$ is of CM type.

For the readers convenience, we prove the implications (1) $\Longrightarrow$ (2), to show how the strengthened version of Buzzard's result is used. For the remaining implications, we refer the reader to [GV04, Prop. 14], although a shorter proof of the implication (3) $\Longrightarrow$ (4) can be found in [DG12].

(1) $\Longrightarrow$ (2): Recall that we have the following characters:

$$\psi : \mathrm{Gal}(\bar{\mathbb{Q}}/\mathbb{Q}) \to \bar{\mathbb{Q}}_p^\times \quad \text{the character of } \mathcal{F} \text{ of conductor } Nq,$$
$$\kappa : \mathrm{Gal}(\bar{\mathbb{Q}}/\mathbb{Q}) \to \Lambda^\times, \quad \text{the } \Lambda\text{-adic cyclotomic character},$$
$$\nu : \mathrm{Gal}(\bar{\mathbb{Q}}/\mathbb{Q}) \to \mathbb{Z}_p^\times \quad \text{the 2-adic cyclotomic character.}$$

We know that $\det(\rho_{\mathcal{F}}) = \psi\kappa\nu^{-1}$. The specialization of $\det(\rho_{\mathcal{F}})$ at $\varphi_{k,\zeta}$ is $\chi\nu^{k-1}$, where $\chi = \psi\omega^{-k}\chi_\zeta$. By assumption $\rho_{\mathcal{F}}|_{G_p}$ splits, i.e.,

$$\rho_{\mathcal{F}}|_{I_p} \sim \begin{pmatrix} \psi\kappa\nu^{-1} & 0 \\ 0 & 1 \end{pmatrix}.$$

Let P be a weight one point of L extending $\varphi_{1,\zeta} : \Lambda \to \bar{\mathbb{Q}}_p$. It follows that $P(\rho_{\mathcal{F}}) = \rho_{P(\mathcal{F})}$ has the following shape on I_p:

$$\rho_{P(\mathcal{F})}|_{I_p} \sim \begin{pmatrix} \psi\omega^{-1}\chi_\zeta & 0 \\ 0 & 1 \end{pmatrix},$$

noting that the characters on the diagonal have finite order. Now by Theorem 10.1, we have that

$$\rho_{P(\mathcal{F})} \sim \rho_f,$$

where f is a primitive weight 1 form of level $N2^r$, with character $\psi\omega^{-1}\chi_\zeta$ where ζ is exactly of order 2^{r-2}, $r \geq 2$. As we vary the point P, and therefore $r \geq 2$, we obtain infinitely many classical weight 1 specializations of $\mathcal{F}$ as required.

We remark that elementary arguments (cf. [GV04, (2) $\Longrightarrow$ (3) of Prop. 14]), allow us to conclude that infinitely many of these must be of CM type, and in particular the residual representation must necessarily be of dihedral type. Moreover, by ordinariness, the prime 2 will split in the corresponding imaginary quadratic field. This explains why it is crucial to assume that Buzzard's result holds in this case as well.

□

Acknowledgements. We wish to dedicate this paper to Haruzo Hida, on his 60th birthday. The second author would like to thank Hausdorff Research Institute for Mathematics and Institute of Mathematical Sciences, where parts of this work were carried out during the trimester program "Algebra and Number Theory" and as a Postdoctoral Fellow, respectively.

References

[All12] P. Allen, *Modularity of nearly ordinary 2-adic residually dihedral Galois representations*, UCLA Ph. D. Thesis, 2012.

[Buz03] K. Buzzard, *Analytic continuation of overconvergent eigenforms*, J. Amer. Math. Soc. **16** (1) (2003), 29–55.

[DG12] M. Dimitrov and E. Ghate, *On classical weight one forms in Hida families*, J. Théor. Nombres Bordeaux, **24** (3) (2012), 639–660.

[Gha04] E. Ghate, *On the local behavior of ordinary modular Galois representations*, *Modular curves and abelian varieties*, 105–124, Progr. Math. **224**, Birkhäuser, Basel, 2004.

[GV04] E. Ghate and V. Vatsal, *On the local behaviour of ordinary Λ-adic representations*, Ann. Inst. Fourier (Grenoble) **54** (7) (2004), 2143–2162.

[Kum] N. Kumar, *A 2-adic control theorem for modular curves*, Available at `https://sites.google.com/site/chnarasimhakumar/preprints`.

[Hat79] K. Hatada, *Eigenvalues of Hecke operators on $SL_2(\mathbb{Z})$*, Math. Ann. **239** (1) (1979), 75–96.

[Hid86a] H. Hida, *Iwasawa modules attached to congruences of cusp forms*, Ann. Sci. École Norm. Sup. (4) **19** (2) (1986), 231–273.

[Hid86b] H. Hida, *Galois representations into $GL_2(\mathbf{Z}_p[[X]])$ attached to ordinary cusp forms*, Invent. Math. **85** (3) (1986), 545–613.

[Hid88a] H. Hida, *Modules of congruence of Hecke algebras and L-functions associated with cusp forms*, Amer. J. Math. **110** (2) (1988), 323–382.

[Hid88b] H. Hida, *On p-adic Hecke algebras for* GL_2 *over totally real fields*, Ann. of Math. (2) **128** (2) (1988), 295–384.

[Miy89] T. Miyake, *Modular Forms*. Springer-Verlag, Berlin, 1989.

[Nek06] J. Nekovář, *Selmer complexes*, Astérisque **(310)**, 2006.

[Shi71] G. Shimura, *Introduction to the arithmetic theory of automorphic functions*. Princeton University Press, 1971.

[Wil88] A. Wiles, *On ordinary Λ-adic representations associated to modular forms*, Invent. Math. **94** (3) (1988), 529–573.

E. Ghate, School of Mathematics, Tata Institute of Fundamental Research, Homi Bhabha Road, Mumbai 400005, India.
E-mail: `eghate@math.tifr.res.in`

Narasimha Kumar, Ruprecht-Karls-Universität Heidelberg, Interdisziplinäres Zentrum für Wissenschaftli-ches Rechnen, Im Neuenheimer Feld 368, 69120 Heidelberg, Germany.
E-mail: `kumar@mathi.uni-heidelberg.de`

Automorphic Representations and L-Functions
Editors: D. Prasad, C.S. Rajan, A. Sankaranarayanan, J. Sengupta

Publisher: Hindustan Book Agency, New Delhi, India

On the GL(3) Kuznetsov Formula with Applications to Symmetry Types of Families of L-functions

Dorian Goldfeld[1] and Alex Kontorovich[2]

Abstract

We present an explicit approach to the GL(3) Kuznetsov formula. As an application, for a restricted class of test functions, we obtain the low-lying zero densities for the following three families: cuspidal GL(3) Maass forms ϕ, the symmetric square family $\mathrm{sym}^2 \phi$ on GL(6), and the adjoint family Adϕ on GL(8). Hence we can identify their symmetry types; they are: unitary, unitary, and symplectic, respectively.

1 Introduction

1.1 Symmetry Types

In [KS99], Katz and Sarnak introduced the notion of symmetry type for a family of L-functions. Since then there has been a slew of activity regarding the following problem: given a family of L-functions, determine its symmetry type. A common approach to this determination, as outlined in [Sar08], is to analyze the density of low-lying zeros in the specified family, for test functions whose Fourier transforms have restricted support. Such an analysis has been carried out in many places, including [ILS00, Roy01, Gul05, You06, HM07, AIL$^+$11] for GL(2) and [DM06] for some GL(4) and GL(6) families; see also [DM09]. In all cases in the literature (going beyond $GL(1)$ or abelian methods), the analysis involves a version of the GL(2) Petersson/Kuznetsov formula.

The purpose of this paper is to carry out a similar analysis, for the first time using the GL(3) Kuznetsov formula. Assuming the generalized

[1]Partially supported by NSF grant DMS-1001036.
[2]Partially supported by NSF grants DMS-1209373, DMS-1064214 and DMS-1001252.

Riemann hypothesis (to interpret the low-lying zero densities) and the generalized Ramanujan conjectures (for ease of exposition), we will determine symmetry types for the families of

1. cuspidal $\mathrm{GL}(3,\mathbb{Z})$ automorphic forms ϕ,
2. the GL(6) symmetric square family $\mathrm{sym}^2\,\phi$, and
3. the GL(8) adjoint family $\mathrm{Ad}\phi$.

We will show that the symmetry types are:

1. unitary,
2. unitary, and
3. symplectic,

respectively.

The methods presented here are capable of wide generalization, in particular, it should be possible to determine the symmetry types of families associated to pairs of cuspidal automorphic representations on $GL(n)$ for any $n \geq 2$. We hope to return to this topic in a future publication.

To state our results more precisely, we need some background.

1.2 Hecke-Maass forms

Let $G = \mathrm{GL}(3,\mathbb{R})$ with maximal compact $K = O(3)$ and center $Z = \mathbb{R}^\times$, let $\mathfrak{h}^3 = G/(K\cdot Z)$ be the generalized upper half plane, and take the lattice $\Gamma := \mathrm{GL}(3,\mathbb{Z})$ in G.

The algebra of G-invariant differential operators acts on $\mathcal{H} := L^2(\Gamma\backslash\mathfrak{h}^3)$. The Hecke-Maass forms ϕ_j constitute an orthogonal (Hecke normalized) basis for

$$\mathcal{H}_0 := \bigoplus_{j=1}^{\infty} \mathbb{C}\phi_j \subset \mathcal{H},$$

where $\mathcal{H}_0$ is the cuspidal subspace in the Langlands spectral decomposition [Gol06, Prop. 10.13.1]

$$\mathcal{H} = \mathbb{C}\mathbf{1} \oplus \mathcal{H}_0 \oplus \mathcal{H}_{min} \oplus \mathcal{H}_{max} \oplus \mathcal{H}_{res}.$$

Here $\mathcal{H}_{min}$, $\mathcal{H}_{max}$, and $\mathcal{H}_{res}$ are, respectively, the spans of integrals of the minimal and maximal parabolic Eisenstein series, and the residual spectrum.

Let the Hecke-Maass form ϕ_j have spectral parameters $\nu^{(j)} := (\nu_1^{(j)}, \nu_2^{(j)})$. When discussing a fixed form ϕ, we drop the superscripts (j). Our normalization[3] is such that for a tempered form, ν_1 and ν_2 are purely imaginary.

[3]Note that our normalization differs from that used in [Gol06] by $\nu_j \mapsto 1/3 + \nu_j$.

It is convenient to also introduce the spectral parameters

$$\nu_3 := \nu_1 + \nu_2,$$

and

$$\alpha_1 := \nu_1 + \nu_3, \quad \alpha_2 := -\nu_1 + \nu_2, \quad \alpha_3 := -\nu_2 - \nu_3.$$

Writing λ_ϕ for the Laplace eigenvalue of ϕ, we have

$$\lambda_\phi = 1 - 3(\nu_1^2 + \nu_2^2 + \nu_3^2) = 1 - (\alpha_1^2 + \alpha_2^2 + \alpha_3^2).$$

Weyl's Law in this setting [Mil01] states that

$$\#\{\phi : \lambda_\phi < T^2\} \sim cT^5,$$

as $T \to \infty$, for some constant $c > 0$.

1.3 The GL(3) Kuznetsov Formula

We will state and use the GL(3) Kuznetsov formula with some naturally occurring weights, defined as follows. For $j = 1, 2, 3, \ldots$, let

$$\mathcal{L}_j := \operatorname*{Res}_{s=1} L\big(s, \phi_j \times \tilde{\phi}_j\big)$$

be the residue at the edge of the critical strip of the L-function attached to $\phi_j \times \tilde{\phi}_j$; generically this is the value at $s = 1$ of $L(s, \mathrm{Ad}\phi_j)$.

We introduce an absolute constant

$$R \geq 10,$$

which is needed for certain technical reasons, see the estimates in §4.2. For $T \gg 1$, we define

$$h_{T,R}(\nu) := e^{\left(\alpha_1^2+\alpha_2^2+\alpha_3^2\right)/T^2} \frac{\left(\prod\limits_{1\leq j\leq 3} \Gamma\left(\frac{2+R+3\nu_j}{4}\right)\Gamma\left(\frac{2+R-3\nu_j}{4}\right)\right)^2}{\prod\limits_{1\leq j\leq 3} \Gamma\left(\frac{1+3\nu_j}{2}\right)\Gamma\left(\frac{1-3\nu_j}{2}\right)}. \tag{1.1}$$

In fact, R is needed to enable us later to pull contours in certain integrals with respect to the ν_j's without passing through poles of the numerator in (1.1). Note that $h_{T,R}(\nu) > 0$, and is essentially supported on $\lambda < T^2$, or $|\nu_1|, |\nu_2|, |\nu_3| \ll T$. In this range, one sees from Stirling's formula that if ν is tempered, then

$$h_{T,R}(\nu) \sim c_R\big[(1 + |\nu_1|)(1 + |\nu_2|)(1 + |\nu_3|)\big]^R, \tag{1.2}$$

for some $c_R > 0$. The non-tempered forms constitute a zero density set [Mil01].

Let $A_j(n_1, n_2)$ denote the coefficients of ϕ_j in the Fourier-Whittaker expansion, see §2.3.

Theorem 1.1 *With the above notation and assuming the Ramanujan conjecture at the infinite place, we have the "Weyl Law", that for some $c > 0$,*

$$\sum_j \frac{h_{T,R}(\nu^{(j)})}{\mathcal{L}_j} \sim c\, T^{5+3R}. \tag{1.3}$$

Moreover for fixed $\varepsilon > 0$, $R \geq 10$, $n_1, n_2, m_1, m_2 \in \mathbb{Z}_{\geq 1}$, and $T \gg 1$, we have

$$\sum_j A_j(m_1, m_2)\overline{A_j(n_1, n_2)}\, \frac{h_{T,R}(\nu^{(j)})}{\mathcal{L}_j} =$$

$$\begin{cases} \sum_j \frac{h_{T,R}(\nu^{(j)})}{\mathcal{L}_j} + \mathcal{O}_{R,\varepsilon}\left(T^{3+3R+\varepsilon}\,|m_1 m_2 n_1 n_2|^2\right), & \text{if } \begin{smallmatrix} m_1 = n_1, \\ m_2 = n_2, \end{smallmatrix} \\ \mathcal{O}_{R,\varepsilon}\left(T^{3+3R+\varepsilon}\,|m_1 m_2 n_1 n_2|^2\right), & \text{otherwise.} \end{cases} \tag{1.4}$$

Remark 1.2 In light of the asymptotic formula (1.2), the analytic weight $h_{T,R}$ can be removed with a modicum of effort; we have chosen to leave the weight for ease of exposition. The same is done for GL(2) in [AIL$^+$11].

Remark 1.3 The weight $\mathcal{L}_j$ is more subtle; it is shown in [Blo11, (1.4)] that

$$C_{\nu^{(j)}}^{-1} \ll \mathcal{L}_j \ll_\varepsilon C_{\nu^{(j)}}^{\varepsilon},$$

where

$$C_\nu = (1 + |\nu_1|)(1 + |\nu_2|)(1 + |\nu_3|).$$

Moreover if one assumes the functorial transfer predicting $\phi \times \tilde{\phi}$ is automorphic on GL(9), then using the non-existence of Siegel zeros for the corresponding L-function [HR95], one can improve the lower bound above to $C_\nu^{-\varepsilon}$. With this assumption, the weight can be removed completely, as in [Luo01], giving rise to a clean cut-off.

Remark 1.4 We have not made any attempt to obtain the best possible error terms in (1.4). In particular, we have made no use of stationary phase, nor have we even invoked Deligne's bounds for Kloosterman sums (see e.g. [BFG88, Larsen's appendix]). We tried to present as simple a method as we could, keeping in mind the eventual goal of generalizing these techniques to GL(n) with $n \geq 2$.

Remark 1.5 A similar result is obtained in [Blo11]. Blomer first chooses a test function on the geometric side, and then executes a delicate analysis to obtain implications on the spectral side. In our approach, we choose the test function on the spectral side first, making the asymptotic formula (1.2) immediately visible. In a private communication, Blomer has informed us that from the methods in [Blo11], he can also obtain (1.3) and (1.4) (with a better error term) for a range of test functions.

1.4 Low-Lying Zeros

For a Hecke-Maass form ϕ on GL(3), let $\rho(\phi)$ be one of

$$\rho(\phi) = \begin{cases} \phi \\ \operatorname{sym}^2 \phi \\ \operatorname{Ad}\phi, \end{cases}$$

and let $L(s, \rho(\phi))$ be the corresponding L-function. Let $\alpha_1, \alpha_2, \alpha_3$ be the spectral parameters associated to ϕ. If the Laplace eigenvalue $\lambda_\phi = 1 - (\alpha_1^2 + \alpha_2^2 + \alpha_3^2)$ is sufficiently large, then we define the analytic conductor $c_{\rho(\phi)}$ of $\rho(\phi)$ as follows.

$$c_{\rho(\phi)} = \begin{cases} \pi^{-3} \cdot \prod\limits_{\substack{1 \leq k \leq 3 \\ |\alpha_k| \geq \frac{1}{2}}} \frac{|\alpha_k|}{2}, & \text{if } \rho(\phi) = \phi, \\ \pi^{-5} \cdot \prod\limits_{\substack{1 \leq j \leq k \leq 3 \\ |\alpha_j + \alpha_k| \geq \frac{1}{2}}} \frac{|\alpha_j + \alpha_k|}{2}, & \text{if } \rho(\phi) = \operatorname{sym}^2 \phi, \\ \pi^{-9} \cdot \prod\limits_{\substack{j=1 \\ |\alpha_j - \alpha_k| \geq \frac{1}{2}}}^{3} \prod\limits_{k=1}^{3} \frac{|\alpha_j - \alpha_k|}{2}, & \text{if } \rho(\phi) = \phi \times \bar{\phi}. \end{cases} \tag{1.5}$$

Remark 1.6 Note that this is off by a constant from the more standard Iwaniec-Sarnak definition of "conductor", for which see e.g. [IK04, p. 95]. The constants are crucial in our applications, see specifically (7.6), so we make our definition as above.

We are interested in the weighted average value, denoted C_ρ, of the conductor $c_{\rho(\phi)}$ with respect to the weighting function $h_{T,R}$ defined in (1.1).

Then C_ρ is defined by

$$\sum_j \log c_{\rho(\phi_j)} \frac{h_{T,R}(\nu^{(j)})}{\mathcal{L}_j} \sim \log C_\rho \sum_j \frac{h_{T,R}(\nu^{(j)})}{\mathcal{L}_j}, \qquad (T \to \infty), \tag{1.6}$$

and satisfies

$$C_\rho \asymp \begin{cases} T^3 & \text{if } \rho(\phi) = \phi, \\ T^6 & \text{if } \rho(\phi) = \operatorname{sym}^2 \phi, \\ T^6 & \text{if } \rho(\phi) = \operatorname{Ad}\phi. \end{cases}$$

The weighted average value of the conductor in a family is introduced to normalize the low-lying zeros for comparison between the different families and the different matrix ensembles.

Let ψ be an even test function of Schwartz class on $\mathbb{R}$ and define the low-lying zeros sum

$$D(\rho(\phi); \psi) := \sum_\gamma \psi\left(\gamma \frac{\log C_\rho}{2\pi}\right),$$

where γ runs over the ordinates of nontrivial zeros of $L(s, \rho(\phi))$, counted with multiplicity. To interpret this as capturing the low-lying zeros, we must assume GRH for the corresponding L-functions. As ψ has rapid decay, this sum localizes to those γ which are within $1/\log C_\rho$ of the origin (corresponding to the central point $s = 1/2$ of the L-function).

Theorem 1.7 *Assume the Fourier transform $\widehat{\psi}$ of ψ has support in $(-\delta, \delta)$, where*

$$\delta = \begin{cases} 4/15, & \text{if } \rho(\phi) = \phi, \\ 2/27, & \text{if } \rho(\phi) = \operatorname{sym}^2 \phi, \\ 2/27, & \text{if } \rho(\phi) = \operatorname{Ad}\phi. \end{cases}$$

Assume the Ramanujan conjectures, and GRH for the corresponding L-functions. Then we have the asymptotic formula

$$\frac{1}{\sum_j \frac{h_{T,R}(\nu^{(j)})}{\mathcal{L}_j}} \cdot \sum_j D(\rho(\phi_j); \psi) \frac{h_{T,R}(\nu^{(j)})}{\mathcal{L}_j} =$$

$$\int_{\mathbb{R}} \psi(x) W_{\rho(\phi)}(x)\,dx + \mathcal{O}\left(\frac{\log\log T}{\log T}\right), \tag{1.7}$$

as $T \to \infty$, with the limiting density function W above given by

$$W_{\rho(\phi)}(x) = \begin{cases} 1, & \text{if } \rho(\phi) = \phi, \\ 1, & \text{if } \rho(\phi) = \operatorname{sym}^2 \phi, \\ 1 - \frac{\sin(2\pi x)}{2\pi x}, & \text{if } \rho(\phi) = \operatorname{Ad}\phi. \end{cases} \tag{1.8}$$

That is, the family $\rho(\phi)$ has symmetry type: unitary, unitary, and symplectic, respectively.

Remark 1.8 The exterior square L-function on $GL(3)$ is the same as the contragredient L-function (see [BF90, JS90, Kon10]). So the symmetry type for the exterior square family is unitary.

Remark 1.9 Note that (1.8) is consistent with a recent conjecture by Shin and Templier [ST12].

Remark 1.10 As in Remark 1.4, the range of δ above can also be improved, and is intimately tied to the error terms in (1.4).

Remark 1.11 The Ramanujan conjectures are assumed to make the exposition of Theorem 1.7 as simple as possible. They can easily be removed by decreasing the size of δ in Theorem 1.7.

1.5 Outline

The rest of the paper is organized as follows. In §2, we collect various preliminaries on automorphic forms on $\mathrm{GL}_3(\mathbb{Z})$ (their Fourier development and L-functions), and the Kontorovich-Lebedev-Whittaker transform, as explicated by the authors in [GK11]. In §3, we collect the GL(3) Kuznetsov formula, explicating all the terms which appear.

The careful definition of the choice of test function is given in §4, where we also analyze its growth/decay properties; this is the most important and involved section. We note that, though the argument is a bit complicated (four-dimensional integrals of 12 Gamma factors in the numerator and 7 Gamma factors in the denominator), the analysis uses nothing more than Stirling's asymptotics for the Gamma function. In §5, we input the estimates of §4 into the Kloosterman integrals appearing on the geometric side of the Kuznetsov formula, giving bounds for these, as well as estimating away the contribution from the Eisenstein spectrum. Combining all the above estimates, we prove Theorem 1.1 in §6.

Next we turn our attention to the application to low-lying zeros. In §7, we develop the Explicit Formula for the various L-functions of interest, and analyze the local Langlands-Satake parameters in §8. Having done so, we apply Theorem 1.1 to the low-lying zeros sum in §9 to prove Theorem 1.7.

Acknowledgements

The authors wish to thank Peter Sarnak for suggesting the application of our work on the Kontorovich-Lebedev transform to low-lying zeros. We are grateful to him and Valentin Blomer for many discussions, comments and suggestions regarding this work. Thanks also to Steve J. Miller and Matt Young for comments on an earlier draft. Much of this work was carried out during the 2009-2010 special year in analytic number theory at IAS, and it is a pleasure for the authors to acknowledge the fantastic working conditions. This work has its roots dating back to the AIM workshop "Analytic theory of GL(3) automorphic forms and applications" in November 2008, and we also thank the organizers of this meeting.

2 Preliminaries on Automorphic Forms on $\mathrm{GL}_3(\mathbb{Z})$

2.1 Jacquet's Whittaker Function

Let

$$\mathfrak{h}^3 := \mathrm{GL}_3(\mathbb{R})/(O_3(\mathbb{R}) \times \mathbb{R}^\times)$$

denote the generalized upper half plane. For $z \in \mathfrak{h}^3$ we use Iwasawa coordinates:

$$z = xy = \begin{pmatrix} 1 & x_2 & x_3 \\ 0 & 1 & x_1 \\ 0 & 0 & 1 \end{pmatrix} \begin{pmatrix} y_1y_2 & 0 & 0 \\ 0 & y_1 & 0 \\ 0 & 0 & 1 \end{pmatrix},$$

where $x_1, x_2, x_3 \in \mathbb{R}$ and $y_1, y_2 > 0$. We will frequently abuse notation, not distinguishing between y as above and $y = (y_1, y_2)$. Equip $\mathfrak{h}^3$ with the Haar measure

$$dz = \frac{dx_1dx_2dx_3dy_1dy_2}{(y_1y_2)^3}.$$

With this measure, the group $\Gamma = \mathrm{GL}_3(\mathbb{Z})$ is a lattice, that is, the quotient $\Gamma\backslash\mathfrak{h}^3$ has finite volume. In fact, the volume is

$$\int\limits_{\Gamma\backslash\mathfrak{h}^3} dz = \frac{3\zeta(3)}{2\pi}. \tag{2.1}$$

For the pair $\nu = (\nu_1, \nu_2) \in \mathbb{C}^2$, set

$$\nu_3 := \nu_1 + \nu_2. \tag{2.2}$$

Then we have the I-function, defined by

$$I_\nu(z) = (y_1y_2)^{1+\nu_3} y_1^{\nu_2} y_2^{\nu_1}, \qquad (z = xy \in \mathfrak{h}^3). \tag{2.3}$$

We now define Jacquet's Whittaker function for $GL_3(\mathbb{R})$.[4]

[4]Throughout we use the *completed* Whittaker function, in the terminology of [Gol06].

Definition 2.1 (Whittaker function) For $\nu \in \mathbb{C}^2$ and $z \in \mathfrak{h}^3$, set

$$W_\nu^\pm(z) := \pi^{-3\nu_3} \prod_{j=1}^{3} \Gamma\left(\frac{1+3\nu_j}{2}\right)$$
$$\times \iiint_{\mathbb{R}^3} I_\nu\left(\begin{pmatrix} & & 1 \\ & 1 & \\ 1 & & \end{pmatrix}\begin{pmatrix} 1 & u_2 & u_3 \\ & 1 & u_1 \\ & & 1 \end{pmatrix} z\right) e(-u_1 \mp u_2)\, du_1 du_2 du_3.$$

This function, originally defined for $\mathfrak{Re}\nu_1, \mathfrak{Re}\nu_2 \gg 1$, has analytic continuation to all $\nu \in \mathbb{C}^2$. For $z = y$, the value of $W_\nu^\pm(y)$ is independent of the sign, so we drop the $\pm$.

It is convenient to define the parameters α, given in terms of ν, by the following linear relation:

$$\alpha_1 = 2\nu_1 + \nu_2, \quad \alpha_2 = -\nu_1 + \nu_2, \quad \alpha_3 = -\nu_1 - 2\nu_2. \tag{2.4}$$

Then $\alpha_1 + \alpha_2 + \alpha_3 = 0$ and there is an action of the Weyl group which permutes the parameters $\alpha_1, \alpha_2, \alpha_3$. We say that a function of ν_1, ν_2 is symmetric under the action of the Weyl group if it is invariant under all reorders of the triple $(\alpha_1, \alpha_2, \alpha_3)$.

Consider the representation of the Whittaker function as a double inverse Mellin transform [Sta01]

$$W_\nu(y) = \frac{y_1 y_2 \pi^{3/2}}{(2\pi i)^2} \int_{(C_1)} \int_{(C_2)} \frac{\prod_{j=1}^{3} \Gamma\left(\frac{s_1+\alpha_j}{2}\right)\Gamma\left(\frac{s_2-\alpha_j}{2}\right)}{4\pi^{s_1+s_2}\Gamma\left(\frac{s_1+s_2}{2}\right)} y_1^{-s_1} y_2^{-s_2}\, ds_1 ds_2, \tag{2.5}$$

for any $C_1, C_2 > 0$. Here we use the standard convention that for $C \in \mathbb{R}$, the symbol (C) denotes the line $C + i\mathbb{R}$. Note that W_ν is symmetric under the action of the Weyl group.

For $s \in \mathbb{C}$, Stade's formula [Sta02] gives

$$\iint_{\mathbb{R}_+^2} W_\nu(y)\overline{W_\mu(y)}(\det y)^s \frac{dy_1 dy_2}{(y_1 y_2)^3} = \frac{\pi^{3(1-s)}}{\Gamma\left(\frac{3s}{2}\right)} \prod_{1 \le j,k \le 3} \Gamma\left(\frac{s + \alpha_j + \overline{\beta_k}}{2}\right), \tag{2.6}$$

where $\mu_3 = \mu_1 + \mu_2$ and $\beta_1, \beta_2, \beta_3$ are defined in terms of μ_1, μ_2 as in (2.4). The left side above is originally only defined for $\mathfrak{Re}(s)$ sufficiently large; of course the right side gives its meromorphic continuation.

2.2 Kontorovich-Lebedev transform

Next, we give the analogue of the Kontorovich-Lebedev transform for GL(3), often referred to as the Lebedev-Whittaker transform [GK11, Wal92]. Let $f : \mathbb{R}_+^2 \to \mathbb{C}$ and define $f^\sharp : \mathbb{C}^2 \to \mathbb{C}$ by

$$f^\sharp(\nu) := \iint\limits_{\mathbb{R}_+^2} f(y) W_\nu(y) \, \frac{dy_1 dy_2}{(y_1 y_2)^3}, \tag{2.7}$$

provided the integral converges absolutely. Then $f^\sharp$ is termed the Lebedev-Whittaker transform of f. Note that $f^\sharp$ inherits the property that it is symmetric under the action of the Weyl group.

The inverse transform is given as follows. Assuming g is invariant under the action of the Weyl group and has sufficient decay, we define

$$g^\flat(y) := \frac{1}{(\pi i)^2} \int\limits_{-i\infty}^{i\infty} \int\limits_{-i\infty}^{i\infty} g(\nu) \overline{W_\nu(y)} \frac{d\nu_1 d\nu_2}{\prod\limits_{j=1}^{3} \Gamma\left(\frac{3\nu_j}{2}\right) \Gamma\left(\frac{-3\nu_j}{2}\right)} . \tag{2.8}$$

A sufficient condition on the test functions above (see [GK11]) is that $g(\nu)$ have holomorphic extension to a strip $-\eta < |\mathfrak{Re}(\nu_1)|, |\mathfrak{Re}(\nu_2)| < \eta$ (for some $\eta > 0$) and in this strip satisfy

$$|g(\nu)| < \exp\left(-\frac{3\pi}{4} \sum_{k=1}^{3} |\nu_k|\right) \prod_{k=1}^{3} (1 + |\nu_k|)^{-10}. \tag{2.9}$$

Then under these growth assumptions we have

$$g = f^\sharp \qquad \Longleftrightarrow \qquad f = g^\flat,$$

and the Parseval-type relation:

$$\iint\limits_{\mathbb{R}_+^2} f_1(y) \overline{f_2(y)} \, \frac{dy_1 dy_2}{(y_1 y_2)^3} = \frac{1}{(\pi i)^2} \int\limits_{-i\infty}^{i\infty} \int\limits_{-i\infty}^{i\infty} f_1^\sharp(\nu) \overline{f_2^\sharp(\nu)} \frac{d\nu_1 d\nu_2}{\prod\limits_{j=1}^{3} \Gamma\left(\frac{3\nu_j}{2}\right) \Gamma\left(\frac{-3\nu_j}{2}\right)}. \tag{2.10}$$

2.3 Cusp Forms

Take a Hecke-normalized basis of Maass cusp forms $\{\phi_j\}_{j=1,2,\ldots}$ for $\mathcal{H}_0$, the cuspidal subspace of $L^2(\Gamma \backslash \mathfrak{h}^3)$. The form ϕ_j is of type $(\frac{1}{3} + \nu_1^{(j)}, \frac{1}{3} + \nu_2^{(j)})$

where $\nu^{(j)} = (\nu_1^{(j)}, \nu_2^{(j)}) \in \mathbb{C}^2$ denote the spectral parameters. When speaking of a fixed Maass form ϕ, we drop the superscript (j). For tempered forms, the spectral parameters ν_1 and ν_2 are purely imaginary. Then with $\nu_3 = \nu_1 + \nu_2$, the Laplace eigenvalue λ_ϕ is related to ν by

$$\lambda_\phi = 1 - 6(\nu_1^2 + \nu_2^2 + \nu_3^2).$$

Each such ϕ has the Fourier-Whittaker development given by [Sha73, PŠ75, Gol06]:

$$\phi(z) = \sum_{\gamma \in U_2(\mathbb{Z}) \backslash \mathrm{SL}_2(\mathbb{Z})} \sum_{k_1 \geq 1} \sum_{k_2 \neq 0} \frac{A_\phi(k_1, k_2)}{k_1 |k_2|} W_\nu^{\mathrm{sgn}(k_2)} \times \left(\begin{pmatrix} k_1|k_2| & & \\ & k_1 & \\ & & 1 \end{pmatrix} \begin{pmatrix} \gamma & \\ & 1 \end{pmatrix} z \right), \tag{2.11}$$

with the Hecke normalization $A_\phi(1,1) = 1$.

The L-function attached to ϕ is given by

$$L(s, \phi) := \sum_{n \geq 1} \frac{A(1,n)}{n^s},$$

where $A(1,n) = A_\phi(1,n)$, i.e., we have dropped the ϕ from the notation. This constitutes a degree 3 L-function, which in completed form has Euler product

$$\Lambda(s, \phi) := \prod_p L_p(s, \phi)$$

with local factors for $p < \infty$ of type

$$L_p(s, \phi) := \prod_{k=1}^{3} \left(1 - \frac{\alpha_k(p)}{p^s}\right)^{-1} = \left(1 - \frac{A(p,1)}{p^s} + \frac{A(1,p)}{p^{2s}} - \frac{1}{p^{3s}}\right)^{-1}$$

and for $p = \infty$,

$$L_\infty(s, \phi) := \pi^{-\frac{3s}{2}} \prod_{k=1}^{3} \Gamma\left(\frac{s + \alpha_k}{2}\right)^{-1}.$$

The Rankin-Selberg L-function is

$$L(s, \phi \times \tilde{\phi}) := \zeta(3s) \sum_{k_1, k_2} \frac{|A(k_1, k_2)|^2}{(k_1^2 k_2)^s}.$$

This L-function has a pole at $s = 1$. Standard Rankin-Selberg theory, together with Stade's formula (2.6) shows that the j-th Maass form ϕ_j has L^2 norm given by

$$\|\phi_j\|^2 = 6\ \mathcal{L}_j \cdot \prod_{k=1}^{3} \Gamma\left(\frac{1+3\nu_k^{(j)}}{2}\right)\Gamma\left(\frac{1-3\nu_k^{(j)}}{2}\right), \tag{2.12}$$

where

$$\mathcal{L}_j := \mathrm{Res}_{s=1}\, L(s, \phi_j \times \tilde{\phi}_j). \tag{2.13}$$

3 The GL(3) Kuznetsov Formula

The following equation is the GL(3) Kuznetsov formula, as compiled from [BFG88] and [Blo11]:

$$\boxed{\mathcal{C} + \mathcal{E}_{min} + \mathcal{E}_{max} = \mathcal{M} + \mathcal{K} + \widetilde{\mathcal{K}} + \widetilde{\mathcal{K}}^{\vee},} \tag{3.1}$$

where each component is explicated below. Let $p : \mathbb{R}_+^2 \to \mathbb{C}$ be a test function with suitable decay properties; a sufficient condition is that

$$|p(y_1, y_2)| \ll (y_1 y_2)^{2+\varepsilon}, \tag{3.2}$$

as $y_1, y_2 \to 0$, and that p is otherwise bounded. Fix positive integers n_1, n_2, m_1, m_2.

The left hand side of (3.1), called the spectral side, consists of cuspidal and Eisenstein contributions. The cuspidal contribution is given by

$$\mathcal{C} = \sum_j A_j(m_1, m_2)\overline{A_j(n_1, n_2)} \frac{|p^\sharp(\nu_1^{(j)}, \nu_2^{(j)})|^2}{6\ \mathcal{L}_j\ \prod\limits_{k=1}^{3} \Gamma\left(\frac{1+3\nu_k^{(j)}}{2}\right)\Gamma\left(\frac{1-3\nu_k^{(j)}}{2}\right)}, \tag{3.3}$$

where the sum on j is over cuspidal Hecke-Maass forms ϕ_j on $\mathrm{GL}(3, \mathbb{R})$. The minimal Eisenstein series contributes

$$\mathcal{E}_{min} =$$

$$\frac{1}{(4\pi i)^2} \int\limits_{-i\infty}^{i\infty} \int\limits_{-i\infty}^{i\infty} A_\nu(m_1, m_2)\overline{A_\nu(n_1, n_2)} \frac{|p^\sharp(\nu_1, \nu_2)|^2}{\prod\limits_{k=1}^{3} \left|\zeta(1+3\nu_k)\Gamma\left(\frac{1+3\nu_k}{2}\right)\right|^2}\, d\nu_1 d\nu_2,$$

where the minimal Eisenstein coefficients satisfy

$$|A_\nu(n_1, n_2)| \ll_\varepsilon (n_1 n_2)^\varepsilon. \tag{3.4}$$

Lastly, the maximal Eisenstein contribution is

$$\mathcal{E}_{max} = \frac{c}{2\pi i} \sum_{j=1}^{\infty} \int_{-i\infty}^{i\infty} \frac{B_{\nu,r_j}(m_1, m_2)\overline{B_{\nu,r_j}(n_1, n_2)}}{L(1, \mathrm{Ad}\ u_j)|L(1+3\nu, u_j)|^2} \frac{\left|p^\sharp_{T,R}\left(\nu - \frac{ir_j}{3}, \frac{2ir_j}{3}\right)\right|^2}{\left|\Gamma\left(\frac{1+3\nu-ir_j}{2}\right)\Gamma\left(\frac{1+2ir_j}{2}\right)\Gamma\left(\frac{1+3\nu+ir_j}{2}\right)\right|^2} d\nu,$$

where c is an absolute constant, and $\{u_j\}$ is a basis of Hecke-Maass forms for $\mathrm{GL}(2, \mathbb{Z})$, each of eigenvalue $1/4 + r_j^2$. The trivial bound for these Fourier coefficients is

$$|B_{\nu,r_j}(n_1, n_2)| \ll_\varepsilon (n_1 n_2)^{1/2+\varepsilon}. \tag{3.5}$$

Note that the residual spectrum does not contribute, having only degenerate terms in its Fourier expansion.

For functions $p, q : \mathbb{R}^2_+ \to \mathbb{C}$ let

$$\langle p, q\rangle = \iint_{\mathbb{R}^2_+} p(y_1, y_2)\,\overline{q(y_1, y_2)}\,\frac{dy_1 dy_2}{(y_1 y_2)^3}.$$

Let $\mathbf{1}_C$ denote the indicator function, which is 1 if the condition C holds and 0 otherwise. The right-hand side of (3.1), called the arithmetic side of the Kuznetsov formula, consists of a main term and Kloosterman contributions given by

$$\mathcal{M} = \mathbf{1}_{\left\{ \substack{n_1 = m_1 \\ n_2 = m_2} \right\}} \langle p, p\rangle, \tag{3.6}$$

$$\widetilde{\mathcal{K}} = \sum_{\epsilon = \pm 1} \sum_{\substack{D_1 | D_2 \\ m_2 D_1^2 = n_1 D_2}} \frac{\widetilde{S}(\epsilon m_1, n_1, n_2, D_1, D_2)}{D_1 D_2} \widetilde{\mathcal{J}}_\epsilon\left(\sqrt{\frac{n_1 n_2 m_1}{D_1 D_2}}\right),$$

$$\widetilde{\mathcal{K}}^\vee = \sum_{\epsilon = \pm 1} \sum_{\substack{D_2 | D_1 \\ m_1 D_2^2 = n_2 D_1}} \frac{\widetilde{S}(\epsilon m_2, n_2, n_1, D_2, D_1)}{D_1 D_2} \widetilde{\mathcal{J}}_\epsilon\left(\sqrt{\frac{n_1 n_2 m_2}{D_1 D_2}}\right),$$

$$\mathcal{K} = \sum_{\epsilon_1, \epsilon_2 = \pm 1} \sum_{D_1, D_2} \frac{S(\epsilon_1 m_1, \epsilon_2 m_2, n_1, n_2, D_1, D_2)}{D_1 D_2} \times \mathcal{J}_{\epsilon_1, \epsilon_2}\left(\frac{\sqrt{m_1 n_2 D_1}}{D_2}, \frac{\sqrt{m_2 n_1 D_2}}{D_1}\right).$$

Here $S, \widetilde{S}, \mathcal{J}, \widetilde{\mathcal{J}}$ are certain GL(3) Kloosterman sums and integrals corresponding to various elements of the Weyl group.

Let $e(x) := e^{2\pi i x}$. The Kloosterman sums are given explicitly by:

$$\widetilde{S}(m_1, n_1, n_2, D_1, D_2) := \mathbf{1}_{D_1|D_2} \sum_{\substack{C_1 \,(\mathrm{mod}\, D_1), C_2 \,(\mathrm{mod}\, D_2) \\ (C_1, D_1)=1=(C_2, D_2/D_1)}} \sum e\left(\frac{m_1 C_1 + n_1 \bar{C}_1 C_2}{D_1}\right) e\left(\frac{n_2 \bar{C}_2}{D_2/D_1}\right),$$

and

$$S(m_1, m_2, n_1, n_2, D_1, D_2) := \sum_{\substack{B_1, C_1 \,(\mathrm{mod}\, D_1) \\ B_2, C_2 \,(\mathrm{mod}\, D_2) \\ (B_1, C_1, D_1)=1=(B_2, C_2, D_2) \\ B_1 B_2 + C_2 D_2 + C_2 D_1 \equiv 0 \,(\mathrm{mod}\, D_1 D_2)}} \sum\sum\sum e\left(\frac{m_1 B_1 + n_1 (Y_1 D_2 - Z_1 B_2)}{D_1}\right) \times e\left(\frac{m_2 B_2 + n_2 (Y_2 D_1 - Z_2 B_1)}{D_2}\right),$$

where Y_1, Y_2, Z_1, Z_2 are determined by

$$Y_1 B_1 + Z_1 C_1 \equiv 1 \,(\mathrm{mod}\, D_1) \qquad \text{and} \qquad Y_2 B_2 + Z_2 C_2 \equiv 1 \,(\mathrm{mod}\, D_2).$$

The Kloosterman integrals are given by:

$$\widetilde{\mathcal{J}}_\epsilon(A) = A^{-2} \iint\limits_{\mathbb{R}_+^2} \iint\limits_{\mathbb{R}^2} \overline{p(Ay_1, y_2)}\, e(-\epsilon A x_1 y_1)\, p\left(y_2 \cdot \frac{\sqrt{1+x_1^2+x_2^2}}{1+x_1^2}, \frac{A}{y_1 y_2} \cdot \frac{\sqrt{1+x_1^2}}{1+x_1^2+x_2^2}\right) \times e\left(y_2 \cdot \frac{x_1 x_2}{1+x_1^2} + \frac{A}{y_1 y_2} \cdot \frac{x_2}{1+x_1^2+x_2^2}\right) dx_1 dx_2 \frac{dy_1 dy_2}{y_1 y_2^2},$$

and

$$\mathcal{J}_{\epsilon_1,\epsilon_2}(A_1, A_2) = (A_1 A_2)^{-2} \iint\limits_{\mathbb{R}_+^2} \iiint\limits_{\mathbb{R}^3} \overline{p(A_1 y_1, A_2 y_2)} e(-\epsilon_1 A_1 x_1 y_1 - \epsilon_2 A_2 x_2 y_2) \times p\left(\frac{A_2}{y_2} \cdot \frac{\sqrt{(x_1 x_2 - x_3)^2 + x_1^2 + 1}}{x_3^2 + x_2^2 + 1}, \frac{A_1}{y_1} \cdot \frac{\sqrt{x_3^2 + x_2^2 + 1}}{(x_1 x_2 - x_3)^2 + x_1^2 + 1}\right)$$
$$\times e\left(-\frac{A_2}{y_2} \cdot \frac{x_1 x_3 + x_2}{x_3^2 + x_2^2 + 1} - \frac{A_1}{y_1} \cdot \frac{x_2(x_1 x_2 - x_3) + x_1}{(x_1 x_2 - x_3)^2 + x_1^2 + 1}\right) dx_1 dx_2 dx_3 \frac{dy_1 dy_2}{y_1 y_2}.$$

4 Choice of Test Function and Bounds

We now make a specific choice for the test function $p(y_1, y_2)$. By Lebedev-Whittaker inversion (2.8), we can just as well choose the transform $p^\sharp(\nu_1, \nu_2)$. Let $R \geq 10$ and $T \gg 1$. We define

$$p^\sharp_{T,R}(\nu_1, \nu_2) := \sqrt{6}\, e^{\frac{\alpha_1^2+\alpha_2^2+\alpha_3^2}{2T^2}} \prod_{1\leq j\leq 3} \Gamma\left(\frac{2+R+3\nu_j}{4}\right)\Gamma\left(\frac{2+R-3\nu_j}{4}\right). \tag{4.1}$$

This choice is motivated by the fact that we need $p^\sharp = p^\sharp_{T,R}$ to be invariant under the action of the Weyl group, while also requiring cancellation of the exponential growth of the Γ-factors in the denominator on the right side of (3.3) (cuspidal contribution to the Kuznetsov formula). The variable $R \geq 10$ is introduced to obtain absolute convergence of the sum (3.3), and to pull certain contours without passing through poles, see (4.15). Note first that $p^\sharp_{T,R}$ easily satisfies the requisite bounds (2.9) for Lebedev-Whittaker inversion. It will be shown below that the inverse transform $p_{T,R}$ satisfies (3.2), see (4.17).

Observe then that the cuspidal contribution (3.3) becomes

$$\mathcal{C} = \sum_j A_j(m_1, m_2)\overline{A_j(n_1, n_2)}\frac{h_{T,R}(\nu_1^{(j)}, \nu_2^{(j)})}{\mathcal{L}_j},$$

exactly as desired in (1.1).

4.1 Some Auxiliary Bounds

We collect here some bounds coming from Stirling's asymptotic formula:

$$|\Gamma(\sigma + it)| \sim \sqrt{2\pi}\, |t|^{\sigma-\frac{1}{2}}\, e^{-\frac{\pi|t|}{2}}, \qquad (t \to \pm\infty), \tag{4.2}$$

for fixed $\sigma \in \mathbb{R}$.

There are three types of integrals which we will need to estimate $p_{T,R}$. Throughout we have $y_1, y_2 > 0$, $R \geq 10$, $T \gg 1$.

The First Integral: For any $C_1, C_2 \in \mathbb{R} \setminus \{-2, -4, -6, \dots\}$, let

$$\mathcal{I}^{(1)}_{T,R}(C_1, C_2; y_1, y_2) :=$$

$$\int\limits_{(0)}\int\limits_{(0)}\int\limits_{(C_2)}\int\limits_{(C_1)} \frac{e^{\frac{\alpha_1^2+\alpha_2^2+\alpha_3^2}{2T^2}} \prod\limits_{1\le j\le 3} \Gamma\left(\frac{2+R+3\nu_j}{4}\right)\Gamma\left(\frac{2+R-3\nu_j}{4}\right)}{\prod\limits_{j=1}^{3}\Gamma\left(\frac{3\nu_j}{2}\right)\Gamma\left(\frac{-3\nu_j}{2}\right)}$$

$$\times \frac{\prod\limits_{j=1}^{3}\Gamma\left(\frac{s_1-\alpha_j}{2}\right)\Gamma\left(\frac{s_2+\alpha_j}{2}\right)}{4\pi^{s_1+s_2}\Gamma\left(\frac{s_1+s_2}{2}\right)} y_1^{1-s_1} y_2^{1-s_2}\, ds_1 ds_2 d\nu_1 d\nu_2. \quad (4.3)$$

The Second Integral: Similarly, for any $\kappa_1, \kappa_2, C_1 \in \mathbb{R}$ so that the integrand below doesn't pass through poles of Γ, let

$$\mathcal{I}^{(2)}_{T,R}(\kappa_1, \kappa_2, C_1; y_1, y_2) :=$$

$$\int\limits_{(\kappa_2)}\int\limits_{(\kappa_1)}\int\limits_{(C_1)} \frac{e^{\frac{\alpha_1^2+\alpha_2^2+\alpha_3^2}{2T^2}} \prod\limits_{1\le j\le 3} \Gamma\left(\frac{2+R+3\nu_j}{4}\right)\Gamma\left(\frac{2+R-3\nu_j}{4}\right)}{\prod\limits_{j=1}^{3}\Gamma\left(\frac{3\nu_1}{2}\right)\Gamma\left(\frac{3\nu_2}{2}\right)\Gamma\left(\frac{3\nu_3}{2}\right)\Gamma\left(\frac{-3\nu_2}{2}\right)}$$

$$\times \Gamma\left(\frac{s_1-\alpha_2}{2}\right)\Gamma\left(\frac{s_1-\alpha_3}{2}\right) y_1^{1-s_1} y_2^{1+\alpha_1}\, ds_1 d\nu_1 d\nu_2.$$

The Third Integral: Lastly, for any $\kappa_1, \kappa_2 \in \mathbb{R}$ passing through no poles, let

$$\mathcal{I}^{(3)}_{T,R}(\kappa_1, \kappa_2; y_1, y_2) :=$$

$$\int\limits_{(\kappa_2)}\int\limits_{(\kappa_1)} \frac{e^{\frac{\alpha_1^2+\alpha_2^2+\alpha_3^2}{2T^2}} \prod\limits_{1\le j\le 3} \Gamma\left(\frac{2+R+3\nu_j}{4}\right)\Gamma\left(\frac{2+R-3\nu_j}{4}\right)}{\Gamma\left(\frac{3\nu_1}{2}\right)\Gamma\left(\frac{3\nu_3}{2}\right)\Gamma\left(\frac{-3\nu_2}{2}\right)}$$

$$\times\, y_1^{1-\alpha_2} y_2^{1+\alpha_1}\, d\nu_1 d\nu_2.$$

Define $\kappa_1', \kappa_2', \kappa_3'$ to be related to κ's in the same way that α's are related to ν's, that is,

$$\kappa_1' = 2\kappa_1 + \kappa_2, \quad \kappa_2' = -\kappa_1 + \kappa_2, \quad \kappa_3' = -\kappa_1 - 2\kappa_2. \quad (4.4)$$

Theorem 4.1 *Fix $R \geq 10$ and $\varepsilon > 0$. For any $y_1, y_2 > 0$ and $T \gg 1$, we have the bound*

$$|\mathcal{I}_{T,R}^{(1)}(C_1, C_2; y_1, y_2)| \ll_{\varepsilon, C_1, C_2, R} (y_1 y_2) T^{9/2+3R/2} \left(\frac{y_1}{T}\right)^{-C_1} \left(\frac{y_2}{T}\right)^{-C_2} T^{\varepsilon}. \tag{4.5}$$

Moreover,

$$|\mathcal{I}_{T,R}^{(2)}(\kappa_1, \kappa_2, C_1; y_1, y_2)| \ll_{\varepsilon, C_1, \kappa_1, \kappa_2, R} (y_1 y_2) T^{4+3R/2} \left(\frac{y_1}{T}\right)^{-C_1} \left(\frac{y_2}{T}\right)^{\kappa_1'} T^{\varepsilon}. \tag{4.6}$$

And finally,

$$|\mathcal{I}_{T,R}^{(3)}(\kappa_1, \kappa_2; y_1, y_2)| \ll_{\varepsilon, \kappa_1, \kappa_2, R} (y_1 y_2) T^{7/2+3R/2} \left(\frac{y_1}{T}\right)^{-\kappa_2'} \left(\frac{y_2}{T}\right)^{\kappa_1'} T^{\varepsilon}. \tag{4.7}$$

We give separate treatments of each statement.

Proof (Proof of (4.5)**)** Write $\nu_j = it_j$ and $s_j = C_j + iu_j$. The first exponential in the integrand gives arbitrary decay once $|t_j| > T^{1+\varepsilon}$ for any ε. Bringing the absolute values inside and applying Stirling's asymptotic formula gives

$$|\mathcal{I}_{T,R}^{(1)}(C_1, C_2; y_1, y_2)| \ll_{C_1, C_2, R, \varepsilon} y_1^{1-C_1} y_2^{1-C_2} \iint\limits_{|t_1|, |t_2| \leq T^{1+\varepsilon}} \iint\limits_{\mathbb{R}^2} \mathcal{P} \cdot \exp\left(\frac{\pi}{4} \cdot \mathcal{E}\right) du_1 du_2 dt_1 dt_2,$$

where $\mathcal{E} = \mathcal{E}(t_1, t_2, u_1, u_2)$ is the exponential factor:

$$\mathcal{E} = 3\sum_{k=1}^{3} |t_k| - \sum_{k=1}^{3} |\alpha_k - iu_1| - \sum_{k=1}^{3} |\alpha_k + iu_2| + |u_1 + u_2|,$$

and $\mathcal{P} = \mathcal{P}_{C_1, C_2, R}(t_1, t_2, u_1, u_2)$ is the polynomial factor:

$$\mathcal{P} = \left(\prod_{k=1}^{3}(1 + |t_k|)\right)^{(R+2)/2} \left(\prod_{k=1}^{3}(1 + |\alpha_k - iu_1|)\right)^{(C_1-1)/2} \times \left(\prod_{k=1}^{3}(1 + |\alpha_k + iu_2|)\right)^{(C_2-1)/2} \left(1 + |u_1 + u_2|\right)^{(1-C_1-C_2)/2}.$$

Note that we always have

$$\mathcal{E} \leq 0,$$

with equality only when

$$t_2 - t_1 \le u_1 \le 2t_1 + t_2 \qquad \text{and} \qquad t_1 - t_2 \le u_2 \le t_1 + 2t_2$$

or

$$-t_1 - 2t_2 \le u_1 < t_2 - t_1 \qquad \text{and} \qquad -2t_1 - t_2 \le u_2 \le t_1 - t_2.$$

Hence, there is arbitrary decay outside of this range. Both inequalities have the same contribution, so we only deal with the second.

Make a linear change variables

$$u_1 \mapsto u_1 - t_1 - 2t_2 \qquad \text{and} \qquad u_2 \mapsto u_2 - 2t_1 - t_2,$$

so the new range is

$$0 \le u_1 < 3t_2 \qquad \text{and} \qquad 0 \le u_2 \le 3t_1, \tag{4.8}$$

and the $\mathcal{P}$ factor becomes

$$\begin{aligned} \mathcal{P}_1 \;\; := \;\; & \left((1+|t_1|)\,(1+|t_2|)\,(1+|t_1+t_2|)\right)^{(R+2)/2} \\ & \times\left((1+|u_1|)\,(1+|3t_1+3t_2-u_1|)\,(1+|u_1-3t_2|)\right)^{(C_1-1)/2} \\ & \times\left((1+|u_2|)\,(1+|3t_1+3t_2-u_2|)\,(1+|u_2-3t_1|)\right)^{(C_2-1)/2} \\ & \times\left(1+|-3t_1-3t_2+u_1+u_2|\right)^{(1-C_1-C_2)/2}. \end{aligned}$$

The integral of $\mathcal{P}_1$ over (4.8) in u_1, u_2 is bounded up to constant by

$$\mathcal{P}_2 \;\; := \;\; (1+|t_1|)^{(R+2)/2+C_2}\,(1+|t_2|)^{(R+2)/2+C_1}\,(1+|t_1+t_2|)^{(R+1)/2}.$$

Integrating $\mathcal{P}_2$ over the range $|t_j| < T^{1+\varepsilon}$ gives (4.5), as claimed.

□

Next we give a

Proof (Proof of (4.6)**)** Again by Stirling's formula, we have

$$|\mathcal{I}^{(2)}_{T,R}(\kappa_1,\kappa_2,C_1;y_1,y_2)| \ll_{C_1,\kappa_1,\kappa_2,R,\varepsilon} y_1^{1-C_1}y_2^{1+\kappa_1'} \iint\limits_{|t_1|,|t_2|\le T^{1+\varepsilon}} \int\limits_{\mathbb{R}} \mathcal{P}\cdot\exp\left(\frac{\pi}{4}\cdot\mathcal{E}\right)du_1dt_1dt_2,$$

where $\mathcal{E} = \mathcal{E}(t_1,t_2,u_1)$ is now the exponential factor:

$$\mathcal{E} = -|t_1 - t_2 + u_1| - |t_1 + 2t_2 + u_1| + 3|t_2|,$$

and $\mathcal{P} = \mathcal{P}_{C_1,C_2,\kappa_1,\kappa_2,R}(t_1, t_2, u_1)$ is now the polynomial factor:

$$\mathcal{P} = (1+|t_2|)^{\frac{R}{2}+1} (1+|t_1|)^{\frac{1}{2}(-3\kappa_1+R+1)} (1+|t_1+t_2|)^{\frac{1}{2}(-3\kappa_1-3\kappa_2+R+1)}$$
$$\times (1+|t_1-t_2+u_1|)^{\frac{1}{2}(C_1+\kappa_1-\kappa_2-1)} (1+|t_1+2t_2+u_1|)^{\frac{1}{2}(C_1+\kappa_1+2\kappa_2-1)}.$$

Note that we always have

$$\mathcal{E} \le 0,$$

with equality only when

$$-t_1 - 2t_2 \le u_1 \le t_2 - t_1,$$

so we may restrict the u_1 integral to this range.

Make a linear change variables

$$u_1 \mapsto u_1 - t_1 - 2t_2,$$

so the new range is

$$0 \le u_1 < 3t_2,$$

and the $\mathcal{P}$ factor becomes

$$\mathcal{P}_1 :=$$
$$(1+|t_2|)^{(R+2)/2} (1+|t_1|)^{(-3\kappa_1+R+1)/2} (1+|t_1+t_2|)^{(-3\kappa_1-3\kappa_2+R+1)/2}$$
$$\times (|u_1|+1)^{(C_1+\kappa_1+2\kappa_2-1)/2} (|u_1-3t_2|+1)^{(C_1+\kappa_1-\kappa_2-1)/2}.$$

The integral of $\mathcal{P}_1$ over the u_1 range is bounded up to constant by

$$\begin{aligned}\mathcal{P}_2 &= (1+|t_1|)^{(-3\kappa_1+R+1)/2} (1+|t_2|)^{(R+2C_1+2\kappa_1+\kappa_2+2)/2} \\ &\quad \times (1+|t_1+t_2|)^{(-3\kappa_1-3\kappa_2+R+1)/2}.\end{aligned}$$

Integrating $\mathcal{P}_2$ over $|t_j| < T^{1+\varepsilon}$ gives the claim.

□

Finally, we give a

Proof (Proof of (4.7)**)**

As before, we have

$$|\mathcal{I}^{(3)}_{T,R}(\kappa_1,\kappa_2;y_1,y_2)| \ll_{\kappa_1,\kappa_2,R,\varepsilon} y_1^{1-\kappa_2'} y_2^{1+\kappa_1'} \iint\limits_{|t_1|,|t_2| \le T^{1+\varepsilon}} \mathcal{P}\, dt_1 dt_2,$$

where $\mathcal{P} = \mathcal{P}_{\kappa_1,\kappa_2,R}(t_1,t_2)$ is the polynomial factor

$$\mathcal{P} =$$
$$(1+|t_1|)^{(-3\kappa_1+R+1)/2}(1+|t_2|)^{(3\kappa_2+R+1)/2}(1+|t_1+t_2|)^{(-3\kappa_1-3\kappa_2+R+1)/2}.$$

(Note that the exponential terms exactly cancel.) Integrating $\mathcal{P}$ gives the claim.

□

4.2 Estimating $p_{T,R}$

We use the bounds of the previous section to give an estimate for $p_{T,R}$. Among other things, we must verify that the inverse Lebedev-Whittaker transform $p_{T,R}$ satisfies (3.2). This will follow from the bound (4.17).

By Lebedev inversion (2.8), we define

$$p_{T,R}(y) := \frac{1}{(\pi i)^2} \int\limits_{(0)} \int\limits_{(0)} p^\sharp_{T,R}(\nu)\, \overline{W_\nu(y)} \frac{d\nu}{\prod\limits_{j=1}^{3} \Gamma\left(\frac{3\nu_j}{2}\right)\Gamma\left(\frac{-3\nu_j}{2}\right)}. \tag{4.9}$$

Recall the double inverse Mellin transform formula for the Whittaker function (2.5), and that $\overline{W_\nu(y)} = W_{-\nu}(y)$ for ν tempered.

Then putting (2.5) into (4.9) and comparing with (4.3) gives

$$p_{T,R}(y) = \frac{\sqrt{6\pi^3}}{(\pi i)^2} \cdot \mathcal{I}^{(1)}_{T,R}(C_1, C_2; y_1, y_2), \tag{4.10}$$

and the equality holds for any $C_1, C_2 > 0$. An immediate application of (4.5) proves that for any $y_1, y_2, C_1, C_2 > 0$,

$$p_{T,R}(y_1, y_2) \ll_{C_1,C_2,R,\varepsilon} y_1 y_2\, T^{9/2+3R/2} \left(\frac{y_1}{T}\right)^{-C_1} \left(\frac{y_2}{T}\right)^{-C_2} T^{\varepsilon}. \tag{4.11}$$

4.2.1 Pull past one set of poles

The above bound is insufficient for our purposes, so we return to the definition of $\mathcal{I}^{(1)}_{T,R}$, and pull the s_2 integral from the vertical line (C_2) with $C_2 > 0$ to the vertical line $(-\mathcal{C}_2)$, with $\mathcal{C}_2 = -C_2$, $0 < \mathcal{C}_2 < 2$. In so doing, we pass through simple poles at $s_2 = -\alpha_1, -\alpha_2, -\alpha_3$ (generically the α_j are distinct). Then we can write

$$p_{T,R} = \mathfrak{M} + \mathcal{R}_1 + \mathcal{R}_2 + \mathcal{R}_3, \tag{4.12}$$

where $\mathfrak{M}$ is the remaining 4-dimensional integral (that is, a constant times $\mathcal{I}^{(1)}_{T,R}(C_1, -\mathcal{C}_2; y_1, y_2)$), and the $\mathcal{R}_j$ are the 3-dimensional contributions from the residues at $s_2 = -\alpha_j$. Note that $\mathcal{R}_1$ is exactly equal to a constant times $\mathcal{I}^{(2)}_{T,R}(0, 0, C_1; y_1, y_2)$. In this integral, we pull the ν_1, ν_2 integrals from the vertical lines with $\mathfrak{Re}\,\nu_j = 0$ to the lines $(\kappa_1), (\kappa_2)$ respectively, so that it becomes $\mathcal{I}^{(2)}_{T,R}(\kappa_1, \kappa_2, C_1; y_1, y_2)$. To ensure that we haven't passed any new poles, we require that the κ's satisfy:

$$\begin{aligned} |\kappa_j| &< \frac{R+2}{3} \\ \kappa_2', \kappa_3' &< C_1. \end{aligned} \tag{4.13}$$

Recall that here, as always, the κ' are related to κ by (4.4).

The estimate (4.6) bounds $\mathcal{R}_1$ by

$$y_1 y_2\, T^{4+3R/2} \left(\frac{y_1}{T}\right)^{-C_1} \left(\frac{y_2}{T}\right)^{\kappa_1'} T^{\varepsilon},$$

whereas the term $\mathfrak{M}$ is dominated using (4.5) by

$$y_1 y_2\, T^{9/2+3R/2} \left(\frac{y_1}{T}\right)^{-C_1} \left(\frac{y_2}{T}\right)^{\mathcal{C}_2} T^{\varepsilon}.$$

To make these the same in y_2, we would like to take $\kappa_1' = 2\kappa_1 + \kappa_2$ as large as $\mathcal{C}_2$, subject to (4.13), which requires $-\kappa_1 + \kappa_2 < C_1$ and $-\kappa_1 - 2\kappa_2 < C_2$. This is easily achieved by, say, setting $\kappa_2 = 0, \kappa_1 > 0$; then we can take κ_1 as large as 1, so that κ_1' can be as large as 2. We can take κ_1 as large as 1, as needed, since $R = 10$. So under these conditions, we have dominated $\mathcal{R}_1$ by the bound we already have on $\mathfrak{M}$. The same can be done with $\mathcal{R}_2$ and $\mathcal{R}_3$, by pulling κ's to different ranges.

We have thus given our second intermediate bound: for any y_1, y_2, $C_1 > 0$, and any $0 < \mathcal{C}_2 < 2$, we have

$$p_{T,R}(y_1, y_2) \ll_{C_1, \mathcal{C}_2, R, \varepsilon} y_1 y_2\, T^{9/2+3R/2} \left(\frac{y_1}{T}\right)^{-C_1} \left(\frac{y_2}{T}\right)^{\mathcal{C}_2} T^{\varepsilon}. \tag{4.14}$$

By symmetry, we have the same result with the subscripts "1" and "2" reversed.

4.2.2 Pull past two sets of poles

The above is still insufficient, so we return to (4.12). In the $\mathfrak{M}$ integral, we now also pull the s_1 integral from (C_1) with $C_1 > 0$ to $(-\mathcal{C}_1)$, where $\mathcal{C}_1 = -C_1$, with $0 < \mathcal{C}_1 < 2$, passing through poles at $s_1 = \alpha_j$, giving

$$\mathfrak{M} = \tilde{\mathfrak{M}} + \tilde{\mathcal{R}}_1 + \tilde{\mathcal{R}}_2 + \tilde{\mathcal{R}}_3.$$

The integral $\tilde{\mathfrak{M}}$ is exactly equal to an absolute constant times $\mathcal{I}^{(1)}_{T,R}(-\mathcal{C}_1, -\mathcal{C}_2; y_1, y_2)$, and hence we apply (4.5), giving

$$\tilde{\mathfrak{M}} \ll_{\mathcal{C}_1,\mathcal{C}_2,R,\varepsilon} y_1 y_2\, T^{9/2+3R/2} \left(\frac{y_1}{T}\right)^{\mathcal{C}_1} \left(\frac{y_2}{T}\right)^{\mathcal{C}_2} T^{\varepsilon}.$$

The integrals $\tilde{\mathcal{R}}_j$ are of the same form as $\mathcal{I}^{(2)}_{T,R}(0, 0, \mathcal{C}_2; y_1, y_2)$, and after pulling to appropriate κ's, we can dominate the $\tilde{\mathcal{R}}_j$ integrals by $\tilde{\mathfrak{M}}$, exactly as before.

In the integral $\mathcal{R}_1$, which is exactly equal to a constant times $\mathcal{I}^{(2)}_{T,R}(0, 0, C_1; y_1, y_2)$, we can pull the s_1 integral from (C_1) to $(-\mathcal{C}_1)$, passing through poles at $s_1 = \alpha_2, \alpha_3$. Hence we can write correspondingly

$$\mathcal{R}_1 = \mathcal{R}_1' + \mathcal{P}_{1,2} + \mathcal{P}_{1,3}.$$

Here $\mathcal{R}_1'$ is a triple integral, exactly equal to a constant times $\mathcal{I}^{(2)}_{T,R}(0, 0, -\mathcal{C}_1; y_1, y_2)$, and $\mathcal{P}_{1,2}$ is a double integral, exactly equal to a constant times $\mathcal{I}^{(3)}_{T,R}(0, 0; y_1, y_2)$. The term $\mathcal{P}_{1,3}$ is similar to $\mathcal{P}_{1,2}$.

In the double integral $\mathcal{P}_{1,2}$, we can pull contours in ν_j to any (κ_j) with

$$|\kappa_j| < (2 + R)/3, \tag{4.15}$$

without passing new poles. Apply the estimate (4.7) to bound $\mathcal{P}_{1,2}$ by

$$y_1 y_2\, T^{7/2+3R/2} \left(\frac{y_1}{T}\right)^{-\kappa_2'} \left(\frac{y_2}{T}\right)^{\kappa_1'} T^{\varepsilon}.$$

Elementary linear algebra shows from (4.4) that if we choose

$$\kappa_1 = -\frac{1}{3}(\mathcal{C}_1 + \mathcal{C}_2), \qquad \kappa_2 = \frac{1}{3}(-\mathcal{C}_1 + 2\mathcal{C}_2),$$

then $-\kappa_1' = \mathcal{C}_1$ and $\kappa_2' = \mathcal{C}_2$. Since $0 < \mathcal{C}_1, \mathcal{C}_2 < 2$, the condition $R \geq 10$ is more than sufficient to ensure that (4.15) is satisfied. So the $\mathcal{P}_{1,2}$ contribution is dominated by that from $\tilde{\mathfrak{M}}$. The same (by a different pull in κ's) holds for $\mathcal{P}_{1,3}$.

Lastly, consider the triple integral $\mathcal{R}_1' = \mathcal{I}^{(2)}_{T,R}(0, 0, -\mathcal{C}_1; y_1, y_2)$. Since $-\mathcal{C}_1 = C_1 < 0$, the next poles in ν arise when $\kappa_2', \kappa_3' = 2 + C_1$. Hence we can pull the ν variables to any κ_1, κ_2, satisfying

$$\begin{aligned} |\kappa_j| &< \frac{R+2}{3} \\ \kappa_2', \kappa_3' &< 2 - \mathcal{C}_1, \end{aligned} \tag{4.16}$$

without passing more poles. The estimate (4.6) bounds $\mathcal{R}_1'$ by

$$y_1 y_2 \, T^{4+3R/2} \left(\frac{y_1}{T}\right)^{\mathcal{C}_1} \left(\frac{y_2}{T}\right)^{\kappa_1'} T^{\varepsilon}.$$

Again, taking $\kappa_2 = 0$ and $\kappa_1 > 0$, the second inequalities in (4.16) are satisfied, and we can take $\kappa_1 = \mathcal{C}_2/2 < 1$, so that $\kappa_1' = \mathcal{C}_2$. There are no new constraints on R.

Hence we see that the contribution by $\mathcal{R}_1$ is dominated by that of $\mathfrak{M}$. The same holds for $\mathcal{R}_2$ and $\mathcal{R}_3$ by symmetry, and we have established the following crucial bound.

Proposition 4.2 *Fix $R \geq 10$ and $\varepsilon > 0$. For any $y_1, y_2 > 0$, $T \gg 1$, and any $0 < \mathcal{C}_1, \mathcal{C}_2 < 2$, we have*

$$p_{T,R}(y_1, y_2) \ll_{\mathcal{C}_1,\mathcal{C}_2,R,\varepsilon} y_1 y_2 \, T^{9/2+3R/2} \left(\frac{y_1}{T}\right)^{\mathcal{C}_1} \left(\frac{y_2}{T}\right)^{\mathcal{C}_2} T^{\varepsilon}. \tag{4.17}$$

In particular, $p_{T,R}$ satisfies (3.2)*, as needed in the Kuznetsov formula.*

Remark 4.3 One can pull further and analyze contributions from higher poles. In so doing, the power in T increases, so that the residual contributions $\mathcal{R}$ dominate the contribution from $\mathfrak{M}$. It may still be possible to get further improvements from such an analysis, but the above is sufficient for our purposes, so we stop here.

5 Bounds for the Kloosterman and Eisenstein Contributions

Since we showed in the previous section that our choice $p_{T,R}$ of test function satisfies the requisite bound (3.2), we now invoke Kuznetsov's formula with this choice, and estimate the resulting components.

5.1 Bounds for the Kloosterman integrals $\mathcal{J}$ and $\tilde{\mathcal{J}}$

We shall apply the estimates obtained in the previous section to bound the Kloosterman integrals $\tilde{\mathcal{J}}$ and $\mathcal{J}$ defined in §3. We begin with an analysis of the more difficult case of $\mathcal{J}$. For $\epsilon_1, \epsilon_2 \in \pm 1$, recall $\mathcal{J}_{\epsilon_1,\epsilon_2}$ is given by

$$\mathcal{J}_{\epsilon_1,\epsilon_2}(A_1,A_2) := (A_1A_2)^{-2} \iint_{\mathbb{R}_+^2} \iiint_{\mathbb{R}^3} e(-\epsilon_1 A_1 x_1 y_1 - \epsilon_2 A_2 x_2 y_2)\,\overline{p_{T,R}}(A_1y_1, A_2y_2)$$
$$\times e\left(-\frac{A_2}{y_2}\frac{x_1x_3+x_2}{x_3^2+x_2^2+1} - \frac{A_1}{y_1}\frac{x_2(x_1x_2-x_3)+x_1}{(x_1x_2-x_3)^2+x_1^2+1}\right)$$
$$\times p_{T,R}\left(\frac{A_2}{y_2}\frac{\sqrt{(x_1x_2-x_3)^2+x_1^2+1}}{x_3^2+x_2^2+1}, \frac{A_1}{y_1}\frac{\sqrt{x_3^2+x_2^2+1}}{(x_1x_2-x_3)^2+x_1^2+1}\right) dx\frac{dy}{y}. \tag{5.1}$$

Here $dx = dx_1 dx_2 dx_3$.

We put absolute values inside the integral and note that the resulting bounds are then independent of ϵ_1, ϵ_2, so it is convenient to drop ϵ_1, ϵ_2 from the notation. Since $p = p_{T,R}$, it is convenient to recall the dependence of $\mathcal{J}$ on T and R, so we write henceforth $\mathcal{J}_{T,R}(A_1, A_2)$, etc. We have

$$|\mathcal{J}_{T,R}(A_1,A_2)| \leq \frac{1}{(A_1A_2)^2} \iint\limits_{\mathbb{R}_+^2} \iiint\limits_{\mathbb{R}^3} |p_{T,R}(A_1y_1, A_2y_2)| \left| p_{T,R}\left(\frac{A_2}{y_2}\frac{\xi_1^{1/2}}{\xi_2}, \frac{A_1}{y_1}\frac{\xi_2^{1/2}}{\xi_1}\right)\right| dx\frac{dy_1dy_2}{y_1y_2},$$

where

$$\xi_1 = 1 + x_1^2 + (x_1x_2 - x_3)^2, \qquad \xi_2 = 1 + x_2^2 + x_3^2.$$

For $i = 1, 2$, break the y_i integrals according to $y_i > 1$ or $y_i < 1$; this gives

$$\mathcal{J}_{T,R} \leq \mathcal{J}_0 + \mathcal{J}_1 + \mathcal{J}_1' + \mathcal{J}_2,$$

where the y integral in $\mathcal{J}_0$ has $y_1, y_2 < 1$, the $\mathcal{J}_1, \mathcal{J}_1'$ integrals have one bigger and one smaller, and $\mathcal{J}_2$ has $y_1, y_2 > 1$.

We first estimate $\mathcal{J}_0$. Apply the bound in (4.17) to the second appearance of $p_{T,R}$, choosing the largest possible values $\mathcal{C}_1 = \mathcal{C}_2 = 2 - \varepsilon$:

$$\mathcal{J}_0 \ll_\varepsilon \frac{T^{9/2+3R/2+\varepsilon}}{(A_1A_2)^2}\left(\frac{A_1A_2}{T^2}\right)^{2-\varepsilon} \int_0^1\int_0^1 |p_{T,R}(A_1y_1,A_2y_2)| \frac{A_1A_2}{(y_1y_2)^{3-\varepsilon}}$$
$$\times \iiint \frac{1}{(\xi_1\xi_2)^{\frac{3-\varepsilon}{2}}} dx \frac{dy_1dy_2}{y_1y_2}$$
$$\ll T^{1/2+3R/2+3\varepsilon}(A_1A_2)^{1-\varepsilon} \int_0^1\int_0^1 \frac{|p_{T,R}(A_1y_1,A_2y_2)|}{(y_1y_2)^{3-\varepsilon}}\frac{dy_1dy_2}{y_1y_2},$$

since the x integral converges absolutely. Now estimate the first $p_{T,R}$ again using (4.17), with $\mathcal{C}_1 = \mathcal{C}_2 = 2 - \varepsilon/2$:

$$\begin{aligned}\mathcal{J}_0 &\ll T^{1/2+3R+3\varepsilon}\left(A_1A_2\right)^{1-\varepsilon}\int_0^1\int_0^1 A_1A_2y_1y_2T^{9/2+3R/2+\varepsilon}\\ &\qquad\times\left(\frac{A_1A_2y_1y_2}{T^2}\right)^{2-\varepsilon/2}\frac{1}{(y_1y_2)^{3-\varepsilon}}\frac{dy_1dy_2}{y_1y_2}\\ &\ll T^{1+3R+\varepsilon}\left(A_1A_2\right)^{4-\varepsilon}\int_0^1\int_0^1 (y_1y_2)^{\varepsilon/2}\frac{dy_1dy_2}{y_1y_2} \ll T^{1+3R+\varepsilon}\left(A_1A_2\right)^{4-\varepsilon},\end{aligned}$$

since the y-integral converges absolutely.

To bound $\mathcal{J}_1$, $\mathcal{J}_1'$, and $\mathcal{J}_2$, we simply follow the same procedure as above with minimal changes to ensure convergence, as follows. For $\mathcal{J}_2$, in the second application of (4.17), we choose $\mathcal{C}_1 = \mathcal{C}_2 = 2 - 2\varepsilon$, say, so the final y-integral converges absolutely. Similarly, for $\mathcal{J}_1$ and $\mathcal{J}_1'$, we do the same as before, except in the second application of (4.17), we choose $\mathcal{C}_1 = 2 - \varepsilon/2$, $\mathcal{C}_2 = 2 - 2\varepsilon$ (or vice versa), so that the y-integral converges absolutely. We have thus proved that

$$\mathcal{J}_{T,R}(A_1, A_2) \ll_{R,\varepsilon} T^{1+3R+\varepsilon}(A_1A_2)^{4-\varepsilon}.$$

Next, we want a similar bound for $\widetilde{\mathcal{J}}_{T,R}(A)$. Recall that we have

$$\begin{aligned}\widetilde{\mathcal{J}}_{T,R}(A) \;\le\; A^{-2}\iint\limits_{\mathbb{R}_+^2}\iint\limits_{\mathbb{R}^2} &\left|p_{T,R}(Ay_1, y_2)\right|\\ \times&\left|p_{T,R}\left(y_2\cdot\frac{\sqrt{1+x_1^2+x_2^2}}{1+x_1^2},\ \frac{A}{y_1y_2}\cdot\frac{\sqrt{1+x_1^2}}{1+x_1^2+x_2^2}\right)\right| dx_1dx_2\frac{dy_1dy_2}{y_1y_2^2}.\end{aligned}$$

Note that here the integral involves dy_2/y_2^2, whereas in $\mathcal{J}$ the integral has dy_2/y_2. This will result in a weaker final bound.

As before, for $i = 1, 2$, break the y_i integrals according to $y_i > 1$ or $y_i < 1$; this gives

$$\widetilde{\mathcal{J}}_{T,R} \le \widetilde{\mathcal{J}}_0 + \widetilde{\mathcal{J}}_1 + \widetilde{\mathcal{J}}_1' + \widetilde{\mathcal{J}}_2,$$

where the y integral in $\widetilde{\mathcal{J}}_0$ has $y_1, y_2 < 1$, the $\widetilde{\mathcal{J}}_1, \widetilde{\mathcal{J}}_1'$ integrals have one bigger and one smaller, and $\widetilde{\mathcal{J}}_2$ has $y_1, y_2 > 1$.

We first bound $\widetilde{\mathcal{J}}_0$. Set $\xi_1 = 1 + x_1^2$ and $\xi_2 = 1 + x_1^2 + x_2^2$. Replace the second $p_{T,R}$ by its bound in (4.17), with the choice $\mathcal{C}_1 = \varepsilon$ and $\mathcal{C}_2 = 2 - \varepsilon$:

$$\widetilde{\mathcal{J}}_0 \ll_\varepsilon \frac{T^{9/2+3R/2+\varepsilon}}{A^2} \int_0^1 \int_0^1 \int_{-\infty}^{\infty} \int_{-\infty}^{\infty} |p_{T,R}(Ay_1, y_2)| \frac{A}{y_1\sqrt{\xi_1\xi_2}} \left(\frac{y_2\sqrt{\xi_2}}{T\xi_1}\right)^\varepsilon \times \left(\frac{A\sqrt{\xi_1}}{y_1y_2\xi_2 T}\right)^{2-\varepsilon} dx_1 dx_2 \frac{dy_1 dy_2}{y_1 y_2^2}$$

$$\ll T^{5/2+3R/2+\varepsilon} A^{1-\varepsilon} \int_0^1 \int_0^1 \frac{|p_{T,R}(Ay_1, y_2)|}{(y_1y_2)^{3-\varepsilon}} \frac{dy_1 dy_2}{y_1 y_2},$$

since again the x integral converges absolutely. Here we used that for fixed $x_1 > 1$ and $Z < -1/2$,

$$\int_{\mathbb{R}} (1 + x_1^2 + x_2^2)^Z dx_2 \ll x_1^{1+2Z}.$$

Now apply (4.17) to $p_{T,R}(Ay_1, y_2)$ in the above integral with $\mathcal{C}_1 = \mathcal{C}_2 = 2 - \varepsilon/2$. It follows that

$$\widetilde{\mathcal{J}}_0 \ll$$

$$T^{5/2+3R+\varepsilon} A^{1-\varepsilon} \int_0^1 \int_0^1 Ay_1y_2 T^{9/2+3R/2+\varepsilon} \left(\frac{Ay_1y_2}{T^2}\right)^{2-\varepsilon/2} (y_1y_2)^{-3+\varepsilon} \frac{dy_1 dy_2}{y_1 y_2}.$$

Since $\varepsilon > 0$, the above y-integral converges, and we obtain the bound

$$\widetilde{\mathcal{J}}_0 \ll T^{3+3R+\varepsilon} A^{4-\varepsilon}.$$

Then we bound $\widetilde{\mathcal{J}}_1, \widetilde{\mathcal{J}}_1'$, and $\widetilde{\mathcal{J}}_2$, by the same procedure as above, with suitable modifications, as before. We record the previous computations in the following.

Proposition 5.1 *Fix $R \geq 10$, and any small $\varepsilon > 0$.*
For any $A_1, A_2 > 0$ and $T \gg 1$, we have

$$\mathcal{J}_{T,R}(A_1, A_2) \ll_{R,\varepsilon} T^{1+3R+\varepsilon}(A_1A_2)^{4-\varepsilon}. \tag{5.2}$$

For any $A > 0$ and $T \gg 1$, we have

$$\widetilde{\mathcal{J}}_{T,R}(A) \ll_{R,\varepsilon} T^{3+3R+\varepsilon} A^{4-\varepsilon}. \tag{5.3}$$

5.2 Bounds for the Kloosterman contributions

We shall now apply the bounds in Proposition 5.1 to estimate the Kloosterman contributions $\mathcal{K}, \widetilde{\mathcal{K}}$, and $\widetilde{\mathcal{K}}^{\vee}$ appearing in the geometric side of the GL(3) Kuznetsov formula (3.6). We need estimates for these contributions using our choice of test function (or its transform) given by (4.1).

Let us begin by bounding the long element Kloosterman contribution $\mathcal{K}$. We only use the trivial bound for the Kloosterman sum:

$$|S(m_1, n_1, m_2, n_2, D_1, D_2)| \ll_{\varepsilon} (D_1 D_2)^{1+\varepsilon}. \tag{5.4}$$

It immediately follows from Proposition 5.1 and the trivial bound (5.4) that

$$\begin{aligned}
\mathcal{K} &\ll \sum_{D_1=1}^{\infty} \sum_{D_2=1}^{\infty} \frac{|S(m_1, m_2, n_1, n_2, D_1, D_2)|}{D_1 D_2} \left| \mathcal{J}_{T,R}\left(\frac{\sqrt{m_1 n_2 D_1}}{D_2}, \frac{\sqrt{m_2 n_1 D_2}}{D_1} \right) \right| \\
&\ll_{R,\varepsilon} |m_1 m_2 n_1 n_2|^2 \, T^{1+3R+\varepsilon} \sum_{D_1=1}^{\infty} \sum_{D_2=1}^{\infty} |D_1 D_2|^{\varepsilon-2} \\
&\ll |m_1 m_2 n_1 n_2|^2 \, T^{1+3R+\varepsilon}.
\end{aligned} \tag{5.5}$$

Next, we obtain a similar proposition for the lower rank Kloosterman contributions $\widetilde{\mathcal{K}}$ and $\widetilde{\mathcal{K}}^{\vee}$. In this case, we have the trivial bound

$$\widetilde{S}(m_1, n_1, n_2, D_1, D_2) \ll_{\varepsilon} (D_1 D_2)^{1+\varepsilon}.$$

It immediately follows from Proposition 5.1 that

$$\begin{aligned}
\widetilde{\mathcal{K}} &\ll \sum_{D_2=1}^{\infty} \sum_{\substack{D_1 | D_2 \\ m_2 D_1^2 = n_1 D_2}} \frac{|\widetilde{S}(m_1, n_1, n_2, D_1, D_2)|}{D_1 D_2} \left| \widetilde{\mathcal{J}}_{T,R}\left(\sqrt{\frac{n_1 n_2 m_1}{D_1 D_2}} \right) \right| \\
&\ll_{R,\varepsilon} |m_1 n_1 n_2|^2 \, T^{3+3R+\epsilon} \sum_{D_2=1}^{\infty} \sum_{\substack{D_1 | D_2 \\ m_2 D_1^2 = n_1 D_2}} |D_1 D_2|^{2\epsilon-2} \\
&\ll |m_1 n_1 n_2|^2 \, T^{3+3R+\epsilon}.
\end{aligned} \tag{5.6}$$

The same argument applies to $\widetilde{\mathcal{K}}^{\vee}$. We have proved the following.

Proposition 5.2 *Fix $R \geq 10$, $T \gg 1$, and $\varepsilon > 0$. Then the Kloosterman contributions $\mathcal{K}$, $\widetilde{\mathcal{K}}$ and $\widetilde{\mathcal{K}}^\vee$ in (3.6) satisfy the bounds:*

$$\begin{aligned}
\mathcal{K} &\ll_{R,\varepsilon} |m_1 m_2 n_1 n_2|^2 \, T^{1+3R+\varepsilon}, \\
\widetilde{\mathcal{K}} &\ll_{R,\varepsilon} |m_1 n_1 n_2|^2 \, T^{3+3R+\varepsilon}, \\
\mathcal{K}^\vee &\ll_{R,\varepsilon} |m_2 n_2 n_1|^2 \, T^{3+3R+\varepsilon}.
\end{aligned}$$

5.3 Bounds on the continuous spectrum

Next we obtain bounds for the terms $\mathcal{E}_{min}$, $\mathcal{E}_{max}$ coming from the continuous spectrum in the Kuznetsov formula (3.1). We begin with the term coming from the minimal parabolic Eisenstein series:

$$\mathcal{E}_{min} = \frac{1}{(4\pi i)^2} \int_{-i\infty}^{i\infty} \int_{-i\infty}^{i\infty} A_\nu(n_1, n_2) \overline{A_\nu(m_1, m_2)} \frac{|p^\sharp_{T,R}(\nu_1, \nu_2)|^2}{\prod_{k=1}^{3} \left|\zeta(1+3\nu_k)\Gamma\left(\frac{1+3\nu_k}{2}\right)\right|^2} \, d\nu_1 d\nu_2, \tag{5.7}$$

where the minimal Eisenstein coefficients satisfy (3.4). Inserting the choice of test function (4.1) into (5.7) and using the de la Vallée Poussin bound (Prime Number Theorem)

$$|\zeta(1+it)| \gg \frac{1}{\log(2+|t|)},$$

we get from Stirling's formula (4.2) that

$$\begin{aligned}
\mathcal{E}_{min} \ll \frac{1}{(4\pi i)^2} \int_{-iT^{1+\varepsilon}}^{iT^{1+\varepsilon}} \int_{-iT^{1+\varepsilon}}^{iT^{1+\varepsilon}} & |m_1 m_2 n_1 n_2|^\varepsilon \\
& \times \frac{\prod_{j=1}^{3} \left|\Gamma\left(\frac{2+R+3\nu_j}{4}\right)\Gamma\left(\frac{2+R-3\nu_j}{4}\right)\right|^2}{\prod_{k=1}^{3} \left|\zeta(1+3\nu_k)\Gamma\left(\frac{1+3\nu_k}{2}\right)\right|^2} \, d\nu_1 d\nu_2 \\
\ll T^{2+3R+\varepsilon} & |m_1 m_2 n_1 n_2|^\varepsilon.
\end{aligned} \tag{5.8}$$

Next we consider the term coming from the maximal parabolic Eisenstein series:

$$\mathcal{E}_{max} = \frac{c}{2\pi i} \sum_{j=1}^{\infty} \int_{-i\infty}^{i\infty} \frac{B_{\nu,r_j}(n_1,n_2)\overline{B_{\nu,r_j}(m_1,m_2)}}{L(1,\mathrm{Ad}u_j)|L(1+3\nu,u_j)|^2}$$

$$\times \frac{\left|p^\sharp_{T,R}\left(\nu - \frac{ir_j}{3}, \frac{2ir_j}{3}\right)\right|^2}{\left|\Gamma\left(\frac{1+3\nu-ir_j}{2}\right)\Gamma\left(\frac{1+2ir_j}{2}\right)\Gamma\left(\frac{1+3\nu+ir_j}{2}\right)\right|^2} \, d\nu,$$

where c is an absolute constant, $\{u_j\}$ is a basis of Hecke-Maass forms for $\mathrm{GL}(2,\mathbb{Z})$ (each of eigenvalue $1/4 + r_j^2$), and the Fourier coefficients satisfy (3.5). Here we have the lower bounds

$$L(1,\mathrm{Ad}u_j) \gg_\varepsilon (1+|r_j|)^{-\varepsilon}, \qquad L(1+3\nu,u_j) \gg_\varepsilon (1+|\nu|+|r_j|)^{-\varepsilon}.$$

These lower bounds follow from [HL94, HR95, JS77] and [GLS04]. Combining the above lower bounds with Stirling's formula (4.2), it follows that

$$\mathcal{E}_{max} \ll |m_1m_2n_1n_2|^{1/2+\varepsilon} \sum_{r_j \ll T^{1+\varepsilon}} \int_{-iT^{1+\varepsilon}}^{iT^{1+\varepsilon}} \frac{\left|\Gamma\left(\frac{2+R+3\nu-ir_j}{4}\right)\right|^8}{L(1,\mathrm{Ad}u_j)|L(1+3\nu,u_j)|^2}$$

$$\times \frac{\left|\Gamma\left(\frac{2+R+2ir_j}{4}\right)\right|^4}{\left|\Gamma\left(\frac{1+3\nu-ir_j}{2}\right)\Gamma\left(\frac{1+2ir_j}{2}\right)\Gamma\left(\frac{1+3\nu+ir_j}{2}\right)\right|^2} \, d\nu$$

$$\ll |m_1m_2n_1n_2|^{1/2+\varepsilon} \sum_{r_j \ll T^{1+\varepsilon}} \int_{-iT^{1+\varepsilon}}^{iT^{1+\varepsilon}} (1+|3\nu-ir_j|)^{\frac{R}{4}\cdot 8} \cdot (1+|r_j|)^{\frac{R}{4}\cdot 4} \, |d\nu|$$

$$\ll |m_1m_2n_1n_2|^{1/2+\varepsilon}\, T^{3+3R+\varepsilon}, \tag{5.9}$$

using Weyl's Law for GL(2) and the Ramanujan conjectures at infinity (for GL(2)). In summary, we have proved the following.

Proposition 5.3 *Fix $R \geq 10$ and $\varepsilon > 0$. For any $T \gg 1$, we have*

$$\mathcal{E}_{min} \ll_{R,\varepsilon} |m_1m_2n_1n_2|^{\varepsilon}\, T^{2+3R+\varepsilon}, \quad \mathcal{E}_{max} \ll_{R,\varepsilon} |m_1m_2n_1n_2|^{1/2+\varepsilon} T^{3+3R+\varepsilon}.$$

6 Proof of Theorem 1.1

6.1 The Main Term

We begin by computing an asymptotic formula for the main term $\mathcal{M}$ in the Kuznetsov formula (3.1), (3.6). It follows from (2.10) and Stirling's asymptotic formula (4.2) that the inner product in the main term (3.6) becomes

$$\langle p_{T,R}, p_{T,R} \rangle = \frac{1}{(\pi i)^2} \int_{-i\infty}^{i\infty} \int_{-i\infty}^{i\infty} |p^{\sharp}_{T,R}(\nu_1, \nu_2)|^2 \frac{d\nu_1 d\nu_2}{\prod\limits_{1 \le j \le 3} \Gamma\left(\frac{3\nu_j}{2}\right) \Gamma\left(\frac{-3\nu_j}{2}\right)} \, d\nu_1 \nu_2 \tag{6.1}$$

$$\sim c' \iint\limits_{|\nu_1|, |\nu_2| \ll T} \frac{\left| \prod\limits_{1 \le j \le 3} \Gamma\left(\frac{2+R+3\nu_j}{4}\right) \Gamma\left(\frac{2+R-3\nu_j}{4}\right) \right|^2}{\prod\limits_{1 \le j \le 3} \Gamma\left(\frac{3\nu_j}{2}\right) \Gamma\left(\frac{-3\nu_j}{2}\right)} d\nu_1 d\nu_2$$

$$\sim c T^{5+3R},$$

for some constants $c, c' > 0$. This gives the T^{5+3R} main term as claimed in (1.3).

6.2 Completion of the proof of Theorem 1.2

It follows from the Kuznetsov formula (3.1), together with the choice of test function $p = p_{T,R}$ (with $p^{\#}_{T,R}$ given by (4.1)), that

$$\begin{aligned} \mathcal{C} &= \sum_j A_j(m_1, m_2) \overline{A_j(n_1, n_2)} \frac{h_{T,R}(\nu_1^{(j)}, \nu_2^{(j)})}{\mathcal{L}_j} \\ &= \mathcal{M} + \mathcal{K} + \widetilde{\mathcal{K}} + \widetilde{\mathcal{K}}^{\vee} - \mathcal{E}_{min} - \mathcal{E}_{max}, \end{aligned} \tag{6.2}$$

with $\mathcal{M} = \mathbf{1}_{\left\{ \substack{n_1 = m_1 \\ n_2 = m_2} \right\}} \langle p_{T,R}, p_{T,R} \rangle$.

Then Theorem 1.1 is an immediate consequence of the estimates

$$\begin{aligned} \langle p_{T,R}, p_{T,R} \rangle &\sim c T^{5+3R} \\ |\mathcal{E}_{min}|, |\mathcal{E}_{max}| &\ll_\varepsilon |m_1 m_2 n_1 n_2|^{1/2+\varepsilon} \, T^{3+3R+\varepsilon} \\ |\mathcal{K}| + |\widetilde{\mathcal{K}}| + |\widetilde{\mathcal{K}}^{\vee}| &\ll_{R,\varepsilon} |m_1 m_2 n_1 n_2|^2 \, T^{3+3R+\varepsilon} \end{aligned}$$

given in (6.1), and Propositions 5.3, 5.2, respectively.

7 The Explicit Formula

For a Hecke-Maass form ϕ on GL(3), let $\rho(\phi)$ be one of

$$\rho(\phi) = \begin{cases} \phi \\ \operatorname{sym}^2 \phi \\ \operatorname{Ad}\phi, \end{cases}$$

and let $L(s, \rho(\phi))$ be the corresponding L-function. Note that the Maass form dual to ϕ is just the complex conjugate $\bar{\phi}$. In view of the identity

$$L(s, \operatorname{Ad}\phi) = \frac{L(s,\, \phi \times \bar{\phi})}{\zeta(s)},$$

it is easier to work with the Rankin-Selberg convolution of ϕ and $\bar{\phi}$ instead of the adjoint L-function.

Define

$$\Lambda(s,\, \rho(\phi)) := \begin{cases} \pi^{-\frac{3s}{2}} \prod_{k=1}^{3} \Gamma\left(\frac{s+\alpha_k}{2}\right)^{-1} \prod_p \prod_{k=1}^{3} \left(1 - \frac{\alpha_k(p)}{p^s}\right)^{-1}, \\ \qquad\qquad \text{if } \rho(\phi) = \phi, \\ \pi^{-\frac{5s}{2}} \prod_{1\le j\le k\le 3} \Gamma\left(\frac{s+\alpha_j+\alpha_k}{2}\right)^{-1} \prod_p \prod_{1\le j\le k\le 3} \left(1 - \frac{\alpha_j(p)\alpha_k(p)}{p^s}\right)^{-1}, \\ \qquad\qquad \text{if } \rho(\phi) = \operatorname{sym}^2 \phi, \\ \pi^{-\frac{9s}{2}} \prod_{j=1}^{3} \prod_{k=1}^{3} \Gamma\left(\frac{s+\alpha_j-\alpha_k}{2}\right)^{-1} \prod_p \prod_{j=1}^{3} \prod_{k=1}^{3} \left(1 - \frac{\alpha_j(p)\bar{\alpha}_k(p)}{p^s}\right)^{-1}, \\ \qquad\qquad \text{if } \rho(\phi) = \phi \times \bar{\phi}. \end{cases}$$

Then, in all the above cases, we have the functional equation

$$\Lambda(s,\, \rho(\phi)) = \Lambda(1-s, \widetilde{\rho(\phi)}),$$

where $\tilde{\pi}$ is the contragredient representation of π; its L-function has Dirichlet coefficients which are complex conjugates of the original. This follows from [GJ72, BG92, JPSS83], respectively.

We shall use the functional equation for $\Lambda(s,\, \rho(\phi))$ to determine the so-called "explicit formula" relating zeros and poles of $\Lambda(s,\, \rho(\phi))$ with sums over prime power Fourier coefficients of $L(s, \rho(\phi))$.

Let G be any holomorphic function in the region $-1 \le \mathfrak{Re}(s) \le 2$ satisfying

$$G(s) = G(1-s), \quad |s^2 G(s)| \ll 1.$$

Let $\rho_i = \frac{1}{2} + i\gamma_i$ $(i = \pm 1, \pm 2, \ldots)$ run over the zeros of $\Lambda(s, \rho(\phi))$ with corresponding multiplicity. As we have assumed GRH, the ordinates form a real increasing sequence

$$\cdots \le \gamma_{-2} \le \gamma_{-1} \le 0 \le \gamma_1 \le \gamma_2 \le \cdots$$

By the functional equation and standard shifts of contours, together with the fact (first proved by [BG92]) that $\frac{\Lambda'(s)}{\Lambda(s)}$ has at most simple poles at $s = 0, 1$, with residue

$$r_{\rho(\phi)} = \begin{cases} 0 & \text{if } \rho(\phi) = \phi, \\ 0 & \text{if } \rho(\phi) = \operatorname{sym}^2 \phi \text{ and } \phi \text{ not self dual}, \\ 1 & \text{if } \rho(\phi) = \phi \times \bar{\phi}, \end{cases} \tag{7.1}$$

we have

$$\sum_\rho G(\rho) - r_{\rho(\phi)}\big(G(0) + G(1)\big) =$$

$$\frac{1}{2\pi i} \int\limits_{2-i\infty}^{2+i\infty} G(s) \left[\frac{\Lambda'}{\Lambda}\big(s, \rho(\phi)\big) + \frac{\Lambda'}{\Lambda}\big(s, \widetilde{\rho(\phi)}\big)\right] ds = \sum_{p \le \infty} H_{\rho(\phi)}(p). \tag{7.2}$$

For a finite prime $p < \infty$, the function $H_{\rho(\phi)}(p)$ is defined by

$$H_{\rho(\phi)}(p) = \begin{cases} -\sum\limits_{\ell=1}^{\infty} \left(\sum\limits_{k=1}^{3} \left(\alpha_k(p)^\ell + \overline{\alpha_k(p)^\ell} \right) \right) F\left(p^\ell\right) \log p, \\ \qquad\qquad \text{if } \rho(\phi) = \phi, \\ -\sum\limits_{\ell=1}^{\infty} \left(\sum\limits_{1 \le j \le k \le 3} \left(\alpha_j(p)^\ell \alpha_k(p)^\ell + \overline{\alpha_j(p)^\ell \alpha_k(p)^\ell} \right) \right) F\left(p^\ell\right) \log p, \\ \qquad\qquad \text{if } \rho(\phi) = \operatorname{sym}^2 \phi, \\ -2 \sum\limits_{\ell=1}^{\infty} \left(\sum\limits_{j=1}^{3} \sum\limits_{k=1}^{3} \alpha_j(p)^\ell \, \overline{\alpha_k(p)^\ell} \right) F(p^\ell) \log p, \\ \qquad\qquad \text{if } \rho(\phi) = \phi \times \bar{\phi}. \end{cases}$$

Here $F(y)$ is the inverse Mellin transform of $G(s)$,

$$F(y) = \frac{1}{2\pi i} \int\limits_{(1/2)} G(s) y^{-s}\, ds.$$

For $p = \infty$, we have that $H_{\rho(\phi)}(\infty)$ equals

$$\begin{cases} -3F(1)\log\pi + \frac{1}{4\pi i}\sum_{k=1}^{3}\int_{\frac{1}{2}-i\infty}^{\frac{1}{2}+i\infty}\left(\frac{\Gamma'}{\Gamma}\left(\frac{s+\alpha_k}{2}\right)+\frac{\Gamma'}{\Gamma}\left(\frac{s-\alpha_k}{2}\right)\right)G(s)\,ds, \\ \qquad\qquad \text{if } \rho(\phi)=\phi, \\ 5F(1)\log\pi + \frac{1}{4\pi i}\sum_{1\le j\le k\le 3}\int_{\frac{1}{2}-i\infty}^{\frac{1}{2}+i\infty}\left(\frac{\Gamma'}{\Gamma}\left(\frac{s+\alpha_j+\alpha_k}{2}\right)+\frac{\Gamma'}{\Gamma}\left(\frac{s-\alpha_j-\alpha_k}{2}\right)\right)G(s)\,ds, \\ \qquad\qquad \text{if } \rho(\phi)=\operatorname{sym}^2\phi, \\ -9F(1)\log\pi + \frac{1}{2\pi i}\sum_{j=1}^{3}\sum_{k=1}^{3}\int_{\frac{1}{2}-i\infty}^{\frac{1}{2}+i\infty}\frac{\Gamma'}{\Gamma}\left(\frac{s+\alpha_j-\alpha_k}{2}\right)G(s)\,ds, \\ \qquad\qquad \text{if } \rho(\phi)=\phi\times\bar{\phi}. \end{cases}$$

Fix an even test function ψ of Schwartz class whose Fourier transform has compact support, and apply the above formulae making the choice

$$G(s) = \psi\left((s-1/2)\,\frac{\log C_\rho}{2\pi i}\right),$$

where we recall (see (1.6)) that C_ρ is the weighted average value of the conductor of $L(s,\rho(\phi))$.

Then

$$F(y) = \frac{1}{\sqrt{y}\log C_\rho}\widehat{\psi}\left(\frac{\log y}{\log C_\rho}\right).$$

It follows that $H_{\rho(\phi)}(\infty)$ equals

$$\begin{cases} -\frac{3\widehat{\psi}(0)\log\pi}{\log C_\rho} + \sum_{k=1}^{3}\int_{-\infty}^{\infty}\left(\frac{\frac{\Gamma'}{\Gamma}\left(\frac{\pi i x}{\log C_\rho}+\frac{1+2\alpha_k}{4}\right)+\frac{\Gamma'}{\Gamma}\left(\frac{\pi i x}{\log C_\rho}+\frac{1-2\alpha_k}{4}\right)}{2\log C_\rho}\right)\psi(x)\,dx, \\ \qquad\qquad \rho(\phi)=\phi, \\ -\frac{5\widehat{\psi}(0)\log\pi}{\log C_\rho} + \frac{1}{2\log C_\rho}\sum_{1\le j\le k\le 3}\int_{-\infty}^{\infty}\left(\frac{\Gamma'}{\Gamma}\left(\frac{\pi i x}{\log C_\rho}+\frac{1+2(\alpha_j+\alpha_k)}{4}\right)\right. \\ \qquad\qquad \left.+\frac{\Gamma'}{\Gamma}\left(\frac{\pi i x}{\log C_\rho}+\frac{1-2(\alpha_j+\alpha_k)}{4}\right)\right)\psi(x)\,dx, \\ \qquad\qquad \rho(\phi)=\operatorname{sym}^2\phi, \\ -\frac{9\widehat{\psi}(0)\log\pi}{\log C_\rho} + \frac{1}{\log C_\rho}\sum_{j=1}^{3}\sum_{k=1}^{3}\int_{-\infty}^{\infty}\frac{\Gamma'}{\Gamma}\left(\frac{\pi i x}{\log C_\rho}+\frac{1+2(\alpha_j-\alpha_k)}{4}\right)\psi(x)\,dx, \\ \qquad\qquad \rho(\phi)=\phi\times\bar{\phi}. \end{cases}$$

Recall that

$$D(\rho(\phi);\,\psi) := \sum_{\Lambda\left(\frac{1}{2}+i\gamma_i,\,\rho(\phi)\right)=0} \psi\left(\gamma_i\,\frac{\log C_\rho}{2\pi}\right). \tag{7.3}$$

Consequently, (7.2) becomes

$$D(\rho(\phi);\,\psi) = \begin{cases} B_{\rho(\phi)} - \sum\limits_{p<\infty}\sum\limits_{\ell\geq 1} \widehat{\psi}\left(\frac{\ell\,\log p}{\log C_\rho}\right) \frac{\log p}{\log C_\rho} \sum\limits_{k=1}^{3} \frac{\alpha_k(p)^\ell + \overline{\alpha_k(p)^\ell}}{p^{\ell/2}}, \\ \qquad\qquad \rho(\phi) = \phi, \\ B_{\rho(\phi)} - \sum\limits_{p<\infty}\sum\limits_{\ell\geq 1} \widehat{\psi}\left(\frac{\ell\,\log p}{\log C_\rho}\right) \frac{\log p}{\log C_\rho} \times \\ \qquad \sum\limits_{1\leq j\leq k\leq 3} \frac{\alpha_j(p)^\ell\alpha_k(p)^\ell + \overline{\alpha_j(p)^\ell\alpha_k(p)^\ell}}{p^{\ell/2}}, \\ \qquad\qquad \rho(\phi) = \mathrm{sym}^2\,\phi, \\ B_{\rho(\phi)} - 2\sum\limits_{p<\infty}\sum\limits_{\ell\geq 1} \widehat{\psi}\left(\frac{\ell\,\log p}{\log C_\rho}\right) \frac{\log p}{\log C_\rho} \sum\limits_{j=1}^{3}\sum\limits_{k=1}^{3} \frac{\alpha_j(p)^\ell\,\overline{\alpha_k(p)^\ell}}{p^{\ell/2}}, \\ \qquad\qquad \rho(\phi) = \phi\times\bar{\phi}. \end{cases} \tag{7.4}$$

with

$$B_{\rho(\phi)} = 2r_{\rho(\phi)}\,\psi\left(\frac{\log C_\rho}{4\pi i}\right) + \frac{A_{\rho(\phi)}}{\log C_\rho}, \tag{7.5}$$

and where

$$A_{\rho(\phi)} = \begin{cases} -3\widehat{\psi}(0)\log\pi + \frac{1}{2}\sum\limits_{k=1}^{3}\int\limits_{-\infty}^{\infty} \left(\frac{\Gamma'}{\Gamma}\left(\frac{\pi i x}{\log C_\rho} + \frac{1+2\alpha_k}{4}\right) + \right. \\ \qquad \left. \frac{\Gamma'}{\Gamma}\left(\frac{\pi i x}{\log C_\rho} + \frac{1-2\alpha_k}{4}\right)\right)\psi(x)\,dx, \qquad \rho(\phi) = \phi, \\ -5\widehat{\psi}(0)\log\pi + \frac{1}{2}\sum\limits_{1\leq j\leq k\leq 3}\int\limits_{-\infty}^{\infty} \left(\frac{\Gamma'}{\Gamma}\left(\frac{\pi i x}{\log C_\rho} + \frac{1+2(\alpha_j+\alpha_k)}{4}\right) + \right. \\ \qquad \left. \frac{\Gamma'}{\Gamma}\left(\frac{\pi i x}{\log C_\rho} + \frac{1-2(\alpha_j+\alpha_k)}{4}\right)\right)\psi(x)\,dx, \quad \rho(\phi) = \mathrm{sym}^2\,\phi, \\ -9\widehat{\psi}(0)\log\pi + \sum\limits_{j=1}^{3}\sum\limits_{k=1}^{3}\int\limits_{-\infty}^{\infty} \frac{\Gamma'}{\Gamma}\left(\frac{\pi i x}{\log C_\rho} + \frac{1+2(\alpha_j-\alpha_k)}{4}\right)\psi(x)\,dx, \\ \qquad\qquad \rho(\phi) = \phi\times\bar{\phi}. \end{cases}$$

Now, for $\mathfrak{Re}(\alpha) = 0$, we have

$$\int_{-\infty}^{\infty} \frac{\Gamma'}{\Gamma}\left(\frac{\pi i x}{\log C_\rho} + \alpha + \frac{1}{4}\right)\psi(x)\,dx =$$
$$\widehat{\psi}(0)\frac{\Gamma'}{\Gamma}\left(\alpha + \frac{1}{4}\right) + \mathcal{O}\left(\left(\left|\alpha + 1/4\right| \log C_\rho\right)^{-2}\right).$$

If we combine this with the fact that $\frac{\Gamma'}{\Gamma}\left(\alpha + \frac{1}{4}\right) = \log\alpha + \mathcal{O}(1)$ for $|\alpha| \geq \frac{1}{4}$, it follows that

$$A_{\rho(\phi)} = \begin{cases} \widehat{\psi}(0)\log\left(\pi^{-3}\cdot \prod\limits_{\substack{1\leq k\leq 3\\ |\alpha_k|\geq \frac{1}{2}}} \frac{|\alpha_k|}{2}\right) + \mathcal{O}(1), & \rho(\phi) = \phi, \\ \widehat{\psi}(0)\log\left(\pi^{-5}\cdot \prod\limits_{\substack{1\leq j\leq k\leq 3\\ |\alpha_j+\alpha_k|\geq \frac{1}{2}}} \frac{|\alpha_j+\alpha_k|}{2}\right) + \mathcal{O}(1), & \rho(\phi) = \operatorname{sym}^2\phi, \\ \widehat{\psi}(0)\log\left(\pi^{-9}\cdot \prod\limits_{\substack{j=1\\ |\alpha_j-\alpha_k|\geq \frac{1}{2}}}^{3}\prod\limits_{k=1}^{3} \frac{|\alpha_j-\alpha_k|}{2}\right) + \mathcal{O}(1), & \rho(\phi) = \phi\times\bar{\phi}. \end{cases}$$

Thus, in all cases, we have

$$A_{\rho(\phi)} = \widehat{\psi}(0)\log c_{\rho(\phi)} + \mathcal{O}(1), \tag{7.6}$$

where the analytic conductor $c_{\rho(\phi)}$ is given by (1.5).

We combine the above formula for $A_{\rho(\phi)}$ with (7.4) and (7.5). The contribution to (7.4) from $\ell \geq 3$ is negligible (using the Ramanujan bound $|\alpha_j(p)| \leq 1$), so we have

$$\boxed{D(\rho(\phi);\,\psi) = \widehat{\psi}(0) - \Sigma^1_{\rho(\phi)} - \Sigma^2_{\rho(\phi)} + 2r_{\rho(\phi)}\,\psi\left(\frac{\log c_{\rho(\phi)}}{4\pi i}\right) + \mathcal{O}\left(\frac{1}{\log c_{\rho(\phi)}}\right),} \tag{7.7}$$

where $\Sigma^\ell_{\rho(\phi)}$ is the contribution from $\ell = 1, 2$, namely

$$\Sigma^1_{\rho(\phi)} = \begin{cases} -\sum_{p<\infty} \widehat{\psi}\left(\frac{\log p}{\log C_\rho}\right) \frac{\log p}{\log C_\rho} \sum_{k=1}^{3} \frac{\alpha_k(p) + \overline{\alpha_k(p)}}{p^{1/2}}, & \text{if } \rho(\phi) = \phi, \\ -\sum_{p<\infty} \widehat{\psi}\left(\frac{\log p}{\log C_\rho}\right) \frac{\log p}{\log C_\rho} \sum_{1\le j\le k\le 3} \frac{\alpha_j(p)\alpha_k(p) + \overline{\alpha_j(p)\alpha_k(p)}}{p^{1/2}}, & \text{if } \rho(\phi) = \operatorname{sym}^2 \phi, \\ -2\sum_{p<\infty} \widehat{\psi}\left(\frac{\log p}{\log C_\rho}\right) \frac{\log p}{\log C_\rho} \sum_{j=1}^{3}\sum_{k=1}^{3} \frac{\alpha_j(p)\,\overline{\alpha_k(p)}}{p^{1/2}}, & \text{if } \rho(\phi) = \phi \times \bar{\phi}, \end{cases} \tag{7.8}$$

and

$$\Sigma^2_{\rho(\phi)} = \begin{cases} -\sum_{p<\infty} \widehat{\psi}\left(\frac{2\log p}{\log C_\rho}\right) \frac{\log p}{\log C_\rho} \sum_{k=1}^{3} \frac{\alpha_k(p)^2 + \overline{\alpha_k(p)^2}}{p}, & \text{if } \rho(\phi) = \phi, \\ -\sum_{p<\infty} \widehat{\psi}\left(\frac{2\log p}{\log C_\rho}\right) \frac{\log p}{\log C_\rho} \sum_{1\le j\le k\le 3} \frac{\alpha_j(p)^2\alpha_k(p)^2 + \overline{\alpha_j(p)^2\alpha_k(p)^2}}{p}, & \text{if } \rho(\phi) = \operatorname{sym}^2 \phi, \\ -2\sum_{p<\infty} \widehat{\psi}\left(\frac{2\log p}{\log C_\rho}\right) \frac{\log p}{\log C_\rho} \sum_{j=1}^{3}\sum_{k=1}^{3} \frac{\alpha_j(p)^2\,\overline{\alpha_k(p)^2}}{p}, & \text{if } \rho(\phi) = \phi \times \bar{\phi}. \end{cases} \tag{7.9}$$

8 Local Analysis

Recall the Fourier expansion of ϕ:

$$\phi(z) = \sum_{\gamma\in U_2(\mathbb{Z})\backslash \mathrm{SL}_2(\mathbb{Z})} \sum_{k_1\ge 1} \sum_{k_2\ne 0} \frac{A_\phi(k_1,k_2)}{k_1|k_2|} W_\nu^{\operatorname{sgn}(k_2)} \times \left(\begin{pmatrix} k_1|k_2| & & \\ & k_1 & \\ & & 1 \end{pmatrix} \begin{pmatrix} \gamma & \\ & 1 \end{pmatrix} z \right), \tag{8.1}$$

where ϕ is normalized so that $A_\phi(1,1) = 1$. Then the Fourier coefficients $A_\phi(k_1, k_2)$ satisfy the Hecke relations

$$A_\phi(n,1)A_\phi(k_1,k_2) = \sum_{\substack{d_0 d_1 d_2 = n \\ d_1 | k_1, d_2 | k_2}} A_\phi\left(\frac{k_1 d_0}{d_1}, \frac{k_2 d_1}{d_2}\right), \tag{8.2}$$

as well as the conjugation relation

$$\overline{A(k_1, k_2)} = A(k_2, k_1).$$

Furthermore, the L-function associated to ϕ is given by

$$L(s,\phi) = \prod_p L_p(s,\phi) = \sum_{n=1}^{\infty} \frac{A_\phi(1,n)}{n^s},$$

with local factor

$$L_p(s,\phi) := \prod_{j=1}^{3} \left(1 - \frac{\alpha_j(p)}{p^s}\right)^{-1} = \sum_{m \geq 0} \frac{A_\phi(1,p^m)}{p^{ms}}.$$

It follows that

$$A_\phi(1,p^m) = \sum_{u+v+w=m} \alpha_1(p)^u \alpha_2(p)^v \alpha_3(p)^w.$$

In particular,

$$A_\phi(1,p) = \alpha_1(p) + \alpha_2(p) + \alpha_3(p),$$

and

$$\begin{aligned} A_\phi(1,p^2) &= \alpha_1(p)^2 + \alpha_2(p)^2 + \alpha_3(p)^2 + \alpha_1(p)\alpha_2(p) + \alpha_2(p)\alpha_3(p) \\ &\quad + \alpha_1(p)\alpha_3(p) \\ &= \alpha_1(p)^2 + \alpha_2(p)^2 + \alpha_3(p)^2 + \overline{A_\phi(1,p)}. \end{aligned}$$

Next, we analyze the local contribution at a prime p occurring in the sum $\Sigma^1_{\rho(\phi)}$ given in (7.8). There are 3 cases to consider. In the second and third case we apply the Hecke relations (8.2) to remove the product of two

Fourier coefficients:

$$\sum_{k=1}^{3} \alpha_k(p) = \boxed{A_\phi(1,p),}$$

$$\begin{aligned}\sum_{1\le j\le k\le 3} \alpha_j(p)\alpha_k(p) &= A_\phi(1,p)^2 - \alpha_1(p)\alpha_2(p) - \alpha_1(p)\alpha_3(p) - \alpha_2(p)\alpha_3(p)\\ &= A_\phi(1,p)^2 - \overline{A_\phi(1,p)} = A_\phi(1,p)^2 - A_\phi(p,1)\\ &= \boxed{A_\phi(1,p^2),}\end{aligned}$$

$$\sum_{j=1}^{3}\sum_{k=1}^{3} \alpha_j(p)\overline{\alpha_k(p)} = A_\phi(1,p)A_\phi(p,1) = \boxed{1 + A_\phi(p,p).}$$

Putting the above identities into (7.8) yields

$$\Sigma^1_{\rho(\phi)} = \begin{cases} -\sum\limits_{p<\infty} \widehat{\psi}\left(\frac{\log p}{\log C_\rho}\right) \frac{\log p}{\log C_\rho} \frac{A_\phi(1,p) + A_\phi(p,1)}{p^{1/2}}, & \text{if } \rho(\phi) = \phi,\\[2ex] -\sum\limits_{p<\infty} \widehat{\psi}\left(\frac{\log p}{\log C_\rho}\right) \frac{\log p}{\log C_\rho} \frac{A_\phi(1,p^2) + A_\phi(p^2,1)}{p^{1/2}}, & \text{if } \rho(\phi) = \mathrm{sym}^2\,\phi,\\[2ex] -2\sum\limits_{p<\infty} \widehat{\psi}\left(\frac{\log p}{\log C_\rho}\right) \frac{\log p}{\log C_\rho} \frac{1+A_\phi(p,p)}{p^{1/2}}, & \text{if } \rho(\phi) = \phi\times\bar{\phi},\end{cases} \tag{8.3}$$

Next, we do the same for $\Sigma^2_{\rho(\phi)}$ given in (7.9). As before, there are three cases to consider. We require the following Hecke relations obtained from (8.2):

$$\begin{aligned} A_\phi(p,1)^2 &= A_\phi(p^2,1) + A_\phi(1,p),\\ A_\phi(1,p^2)^2 &= A_\phi(p^2,1) + A_\phi(p,p^2) + A_\phi(1,p^4),\\ A_\phi(p,1)A_\phi(1,p^2) &= A_\phi(p,p^2) + A_\phi(1,p^3),\\ A_\phi(p^2,1)A_\phi(1,p^2) &= A_\phi(1,1) + A_\phi(p,p) + A_\phi(p^2,p^2),\\ -A_\phi(p^2,1)A_\phi(p,1) &= -A_\phi(p,p) - A_\phi(p^3,1),\\ -A_\phi(1,p^2)A_\phi(1,p) &= -A_\phi(p,p) - A_\phi(1,p^3),\\ A_\phi(p,1)A_\phi(1,p) &= 1 + A_\phi(p,p). \end{aligned}$$

It follows from the above Hecke relations that:

$$\sum_{k=1}^{3} \alpha_k(p)^2 = \boxed{A_\phi(1,p^2) - A_\phi(p,1),}$$

$$\sum_{1\le j\le k\le 3} \alpha_j(p)^2\alpha_k(p)^2 = \left(A_\phi(1,p^2) - A_\phi(p,1)\right)^2 - A_\phi(p^2,1) - A_\phi(p,1)$$

$$= A_\phi(1,p^2)^2 - 2A_\phi(p,1)A_\phi(1,p^2) + A_\phi(p,1)^2 - A_\phi(p^2,1) + A_\phi(p,1)$$

$$= \boxed{A_\phi(p^2,1) + A_\phi(1,p^4) - A_\phi(p,p^2) - 2A_\phi(1,p^3) + A_\phi(1,p) + A_\phi(p,1),}$$

$$\sum_{j=1}^{3}\sum_{k=1}^{3} \alpha_j(p)^2\overline{\alpha_k(p)^2} = \left(A_\phi(1,p^2) - A_\phi(p,1)\right)\left(A_\phi(p^2,1) - A_\phi(1,p)\right)$$

$$= A_\phi(p^2,1)A_\phi(1,p^2) - A_\phi(p,1)A_\phi(p^2,1) - A_\phi(1,p^2)A_\phi(1,p) + A_\phi(p,1)A_\phi(1,p)$$

$$= \boxed{2 + A_\phi(p^2,p^2) - A_\phi(1,p^3) - A_\phi(p^3,1).}$$

Putting the above identities into (7.9) yields

$$\Sigma^2_{\rho(\phi)} = \begin{cases} -\sum\limits_{p<\infty} \widehat{\psi}\left(\frac{2\log p}{\log C_\rho}\right)\frac{\log p}{\log C_\rho}\left(\frac{A_\phi(1,p^2)-A_\phi(p,1) + A_\phi(p^2,1)-A_\phi(1,p)}{p}\right), \\ \qquad\qquad \rho(\phi) = \phi, \\ -\sum\limits_{p<\infty} \widehat{\psi}\left(\frac{2\log p}{\log C_\rho}\right)\frac{\log p}{\log C_\rho}\Bigg(\frac{A_\phi(p^2,1)+A_\phi(1,p^4)-A_\phi(p,p^2)-2A_\phi(1,p^3)+A_\phi(1,p)-A_\phi(p,1)}{p} \\ \qquad + \frac{A_\phi(1,p^2)+A_\phi(p^4,1)-A_\phi(p^2,p)-2A_\phi(p^3,1)+A_\phi(p,1)-A_\phi(1,p)}{p}\Bigg), \\ \qquad\qquad \rho(\phi) = \mathrm{sym}^2\,\phi, \\ -2\sum\limits_{p<\infty} \widehat{\psi}\left(\frac{2\log p}{\log C_\rho}\right)\frac{\log p}{\log C_\rho}\left(\frac{2+A_\phi(p^2,p^2)-A_\phi(1,p^3)-A_\phi(p^3,1)}{p}\right), \\ \qquad\qquad \rho(\phi) = \phi\times\bar{\phi}. \end{cases} \tag{8.4}$$

Remark 8.1 It is much simpler to apply the Kuznetsov formula to the sums $\Sigma^1_{\rho(\phi)}$, $\Sigma^2_{\rho(\phi)}$, given in (8.3) and (8.4) since the product of Fourier coefficients has been removed by the use of the Hecke relations.

9 Proof of Theorem 1.7

By Parseval's Theorem,

$$\int_{\mathbb{R}} \psi(x) W_{\rho(\phi)}(x) dx = \int_{\mathbb{R}} \widehat{\psi}(y) \widehat{W}_{\rho(\phi)}(y) dy,$$

where $\widehat{\psi}$ is the Fourier transform of ψ, and $\widehat{W}_{\rho(\phi)}$ can be explicitly computed from (1.8) as a distribution:

$$\widehat{W}_{\rho(\phi)}(y) = \begin{cases} \delta_0(y), & \text{if } \rho(\phi) = \phi \text{ or } \operatorname{sym}^2 \phi, \\ \delta_0(y) - \begin{cases} 1/2, & |y|<1, \\ 1/4, & |y|=\pm 1, \\ 0, & |y|>1,. \end{cases} & \text{if } \rho(\phi) = \mathrm{Ad}\phi. \end{cases}$$

As the support of $\widehat{\psi}$ will be restricted well inside $(-1, 1)$, it follows that

$$\int_{\mathbb{R}} \widehat{\psi}(y) \widehat{W}_{\rho(\phi)}(y) dy = \begin{cases} \widehat{\psi}(0), & \text{if } \rho(\phi) = \phi, \\ \widehat{\psi}(0), & \text{if } \rho(\phi) = \operatorname{sym}^2 \phi \\ \widehat{\psi}(0) - \frac{1}{2}\psi(0), & \text{if } \rho(\phi) = \mathrm{Ad}\phi. \end{cases} \tag{9.1}$$

Recall the asymptotic formula (7.7) for the low-lying zeros sum:

$$\boxed{D(\rho(\phi); \psi) = \widehat{\psi}(0) - \Sigma^1_{\rho(\phi)} - \Sigma^2_{\rho(\phi)} + 2r_{\rho(\phi)} \psi\left(\frac{\log c_{\rho(\phi)}}{4\pi i}\right) + \mathcal{O}\left(\frac{1}{\log c_{\rho(\phi)}}\right),} \tag{9.2}$$

where by (7.1), we have

$$r_{\rho(\phi)} = \begin{cases} 0 & \text{if } \rho(\phi) = \phi, \\ 0 & \text{if } \rho(\phi) = \operatorname{sym}^2 \phi \text{ and } \phi \text{ not self dual}, \\ 1 & \text{if } \rho(\phi) = \phi \times \bar{\phi}, \end{cases}$$

and $\Sigma^1_{\rho(\phi)}$, $\Sigma^2_{\rho(\phi)}$ are given by (8.3), and (8.4), respectively.

We will prove Theorem 1.7 where the limiting density $\int_{\mathbb{R}} \widehat{\psi}(y) \widehat{W}(y) dy$ is in the form (9.1). Note that $\widehat{\psi}(0)$ already appears in (9.2).

The case when $\rho(\phi) = \phi \times \bar{\phi}$:

The main contribution to $\Sigma^1_{\phi\times\bar{\phi}} + \Sigma^2_{\phi\times\bar{\phi}}$ comes from

$$2\sum_p \frac{1}{\sqrt{p}}\, \widehat{\psi}\left(\frac{\log p}{\log C_\rho}\right) \frac{\log p}{\log C_\rho} + 4\sum_p \frac{1}{p}\, \widehat{\psi}\left(\frac{2\log p}{\log C_\rho}\right) \frac{\log p}{\log C_\rho}. \tag{9.3}$$

This may be computed in two steps.

Step 1: We apply the explicit formula (7.2) to the function

$$\Lambda(s) = \pi^{-\frac{s}{2}}\Gamma\left(\frac{s}{2}\right)\zeta(s).$$

Note that for $\log C_\rho$ sufficiently large, we have

$$\sum_{\Lambda(\rho)=0} G(\rho) = \sum_{\Lambda(\frac{1}{2}+i\gamma)=0} \psi\left(\frac{\gamma \log C_\rho}{2\pi}\right) = 0,$$

since there are no low-lying (with $|\gamma| \ll 1/\log C_\rho$) non-trivial zeros of the Riemann zeta function. Further $G(0)+G(1) = 2\psi\left(\frac{\log C_\rho}{4\pi i}\right)$. It follows that the explicit formula for the Riemann zeta function takes the form:

$$2\psi\left(\frac{\log C_\rho}{4\pi i}\right) = 2\sum_{\ell\geq 1} \frac{1}{p^{\ell/2}} \cdot \widehat{\psi}\left(\frac{\ell \log p}{\log C_\rho}\right)\frac{\log p}{\log C_\rho} + \frac{\widehat{\psi}(0)}{\log C_\rho}\frac{\log \pi}{2}$$
$$- \frac{1}{\log C_\rho}\int_{-\infty}^{\infty} \frac{\Gamma'}{\Gamma}\left(\frac{2\pi i x}{\log C_\rho} + \frac{1}{4}\right)\psi(x)\, dx$$

$$= 2\sum_{\ell=1}^{2} \frac{1}{p^{\ell/2}} \cdot \widehat{\psi}\left(\frac{\ell \log p}{\log C_\rho}\right)\frac{\log p}{\log C_\rho} + \mathcal{O}\left(\frac{1}{\log C_\rho}\right).$$

Note that the above accounts for a large part of (9.3) and almost cancels the term $2\psi\left(\frac{\log c_{\rho(\phi)}}{4\pi i}\right)$ in (9.2). In fact, we get

$$D(\phi\times\bar{\phi};\psi) = \widehat{\psi}(0) - \Sigma^2_{\phi\times\bar{\phi}} + 2\psi\left(\frac{\log c_{\phi\times\bar{\phi}}}{4\pi i}\right) - 2\psi\left(\frac{\log C_\rho}{4\pi i}\right)$$
$$- 2\sum_p \frac{1}{p}\widehat{\psi}\left(\frac{2\log p}{\log C_\rho}\right)\frac{\log p}{\log C_\rho} + \mathcal{O}\left(\frac{1}{\log c_{\rho(\phi)}}\right). \tag{9.4}$$

What is really happening here is that every Rankin-Selberg L-function $L(s,\phi\times\bar{\phi})$ in the family is divisible by the Riemann zeta function which has a pole at $s=1$. But the Riemann zeta function does not contribute low lying zeros, so the contribution from the pole at $s=1$ is cancelled.

Step 2: In the second step we make use of the classical Riemann hypothesis which implies that $\Psi(x) = x + \mathcal{O}\left(x^{\frac{1}{2}+\epsilon}\right)$. It follows that

$$2\sum_p \frac{1}{p}\,\widehat{\psi}\left(\frac{2\log p}{\log C_\rho}\right)\frac{\log p}{\log C_\rho} = \frac{2}{\log C_\rho}\int_1^\infty \widehat{\psi}\left(\frac{2\log x}{\log C_\rho}\right)x^{-1}d\Psi(x)$$

$$= \frac{2}{\log C_\rho}\int_1^\infty \widehat{\psi}\left(\frac{2\log x}{\log C_\rho}\right)x^{-2}\left(x+\mathcal{O}\left(x^{\frac{1}{2}+\epsilon}\right)\right)dx$$
$$- \frac{2}{\log C_\rho}\int_1^\infty \frac{2}{x\,\log C_\rho}\,\widehat{\psi}'\left(\frac{2\log x}{\log C_\rho}\right)x^{-1}\left(x+\mathcal{O}\left(x^{\frac{1}{2}+\epsilon}\right)\right)dx$$

$$= \int_0^\infty \widehat{\psi}\,(u)\,du \;+\; \mathcal{O}\left(\frac{1}{\log C_\rho}\right)$$
$$= \frac{1}{2}\psi(0) \;+\; \mathcal{O}\left(\frac{1}{\log C_\rho}\right).$$

If we now combine the computation in Step 2 with equation (9.4), it follows that

$$D(\phi\times\bar{\phi};\,\psi) \;=\; \widehat{\psi}(0) - \frac{1}{2}\psi(0) + 2\psi\left(\frac{\log c_{\phi\times\bar{\phi}}}{4\pi i}\right) - 2\psi\left(\frac{\log C_\rho}{4\pi i}\right)$$
$$+\Sigma^3_{\phi\times\bar{\phi}} + O\left(\frac{1}{\log c_{\phi\times\bar{\phi}}}\right), \tag{9.5}$$

where

$$\Sigma^3_{\phi\times\bar{\phi}} = 2\sum_p\left[\frac{A_\phi(p,p)}{\sqrt{p}}\,\widehat{\psi}\left(\frac{\log p}{\log C_\rho}\right)\right.$$
$$\left.+\left(\frac{A_\phi(p^2,p^2) - A_\phi(1,p^3) - A_\phi(p^3,1)}{p}\right)\widehat{\psi}\left(\frac{2\log p}{\log C_\rho}\right)\right]\frac{\log p}{\log C_\rho}. \tag{9.6}$$

To prove Theorem 1.7 for the family of Rankin Selberg L-functions, we make use of (9.5) and the decay properties of $h_{T,R}$ to obtain

$$\sum_j D(\phi_j\times\phi_j;\,\psi)\,\frac{h_{T,R}(\nu^{(j)})}{\mathcal{L}_j} =$$

$$\sum_j\left[\widehat{\psi}(0) - \frac{1}{2}\psi(0) + \Sigma^3_{\phi_j\times\bar{\phi}_j} + O\left(\frac{\log\log T}{\log T}\right)\right]\frac{h_{T,R}(\nu^{(j)})}{\mathcal{L}_j}.$$

as we average over Maass forms ϕ_j $(j = 1, 2, \ldots)$. The term $\log\log T/\log T$ arises after breaking the sum into two pieces corresponding to ϕ_j with conductor $c_{\phi_j} \ll T^3/\log T$ and $c_{\phi_j} \gg T^3/\log T$. To evaluate the above sum, it remains to estimate

$$\sum_j \Sigma^3_{\phi_j\times\bar{\phi}_j} \frac{h_{T,R}(\nu^{(j)})}{\mathcal{L}_j}.$$

Say the support of $\widehat{\psi}$ is in $(-\delta, \delta)$. It immediately follows from (9.6) that

$$\sum_j \Sigma^3_{\phi_j\times\bar{\phi}_j} \frac{h_{T,R}(\nu^{(j)})}{\mathcal{L}_j} = 2\sum_j \left[\sum_{p\ll T^{6\delta}} \frac{A_\phi(p,p)}{\sqrt{p}}\widehat{\psi}\left(\frac{\log p}{\log C_\rho}\right)\frac{\log p}{\log C_\rho}\right.$$
$$\left. + \sum_{p\ll T^{3\delta}} \left(\frac{A_\phi(p^2,p^2) - A_\phi(1,p^3) - A_\phi(p^3,1)}{p}\right)\widehat{\psi}\left(\frac{2\log p}{\log C_\rho}\right)\frac{\log p}{\log C_\rho}\right].$$
$$\frac{h_{T,R}(\nu^{(j)})}{\mathcal{L}_j}.$$

To finish the estimation, we apply Theorem 1.1 to obtain

$$\sum_j \Sigma^3_{\phi_j\times\bar{\phi}_j} \frac{h_{T,R}(\nu^{(j)})}{\mathcal{L}_j} \ll$$
$$T^{3+3R+\varepsilon}\left[\sum_{p\ll T^{6\delta}} p^{4-\frac{1}{2}}\frac{\log p}{\log C_\rho} + \sum_{p\ll T^{3\delta}} p^7\frac{\log p}{\log C_\rho}\right] \ll T^{3+3R+27\delta+\varepsilon}.$$

So we need

$$\delta < \frac{2}{27}.$$

With this choice of δ we obtain

$$\sum_j D(\phi_j\times\phi_j;\psi)\frac{h_{T,R}(\nu^{(j)})}{\mathcal{L}_j} =$$
$$\left[\widehat{\psi}(0) - \frac{1}{2}\psi(0) + O\left(\frac{\log\log T}{\log T}\right)\right]\cdot\sum_j \frac{h_{T,R}(\nu^{(j)})}{\mathcal{L}_j},$$

as claimed.

Remark 9.1 The family of Rankin-Selberg L-functions for $GL(3)$ has the same symmetry type as the family of adjoint L-functions in view of the identity $L(s, \mathrm{Ad}\phi) = \frac{L(s,\,\phi\times\bar{\phi})}{\zeta(s)}$, and the fact that $\zeta(s)$ has no low-lying zeros.

The case when $\rho(\phi) = \operatorname{sym}^2 \phi$:

It immediately follows from (9.2), (8.3), and (8.4), that

$$D(\operatorname{sym}^2 \phi;\, \psi) =$$
$$\widehat{\psi}(0) - \Sigma^1_{\operatorname{sym}^2 \phi} - \Sigma^2_{\operatorname{sym}^2 \phi} + 2r_{\operatorname{sym}^2 \phi}\, \psi\left(\frac{\log c_{\operatorname{sym}^2 \phi}}{4\pi i}\right) + \mathcal{O}\left(\frac{1}{\log c_{\operatorname{sym}^2 \phi}}\right),$$

where

$$\Sigma^1_{\operatorname{sym}^2 \phi} = -\sum_{p<\infty} \widehat{\psi}\left(\frac{\log p}{\log C_\rho}\right) \frac{\log p}{\log C_\rho}\, \frac{A_\phi(1,p^2) + A_\phi(p^2,1)}{p^{1/2}}$$

and

$$\Sigma^2_{\operatorname{sym}^2 \phi} = \sum_{p<\infty} \widehat{\psi}\left(\frac{2\log p}{\log C_\rho}\right) \frac{\log p}{\log C_\rho}$$
$$\times \left(\frac{A_\phi(p^2,1) + A_\phi(1,p^4) - A_\phi(p,p^2) - 2A_\phi(1,p^3) + A_\phi(1,p) - A_\phi(p,1)}{p}\right.$$
$$\left. + \frac{A_\phi(1,p^2) + A_\phi(p^4,1) - A_\phi(p^2,p) - 2A_\phi(p^3,1) + A_\phi(p,1) - A_\phi(1,p)}{p}\right).$$

In this case $r_{\operatorname{sym}^2 \phi} = 0$ unless ϕ is self dual, which happens only if ϕ is a symmetric square lift from $GL(2)$. This occurs for $\asymp T^2$ cases out of $\asymp T^6$, and hence contributes a negligible error term to the low-lying zeros sum. The method to estimate $\Sigma^1_{\operatorname{sym}^2 \phi}$ and $\Sigma^2_{\operatorname{sym}^2 \phi}$, using Theorem 1.1, is exactly the same as the method used above for the case of $\rho(\phi) = \phi \times \bar{\phi}$. Because of the presence of $A_\phi(1,p^2), A_\phi(p^2,1)$ in $\Sigma^1_{\operatorname{sym}^2 \phi}$ and $A_\phi(1,p^4), A_\phi(p^4,1)$ in $\Sigma^2_{\operatorname{sym}^2 \phi}$, we obtain the same value $\delta < \frac{2}{27}$ (which we obtained for the case $\rho(\phi) = \phi \times \bar{\phi}$) for the support of $\widehat{\psi}$. It follows that

$$\sum_j D(\operatorname{sym}^2 \phi;\, \psi)\, \frac{h_{T,R}(\nu^{(j)})}{\mathcal{L}_j} = \left[\widehat{\psi}(0) + O\left(\frac{\log\log T}{\log T}\right)\right] \cdot \sum_j \frac{h_{T,R}(\nu^{(j)})}{\mathcal{L}_j}.$$

The case when $\rho(\phi) = \phi$:

In this case, we have that the residue $r_\phi = 0$. It then follows from (9.2), (8.3), and (8.4), that

$$D(\phi;\, \psi) = \widehat{\psi}(0) - \Sigma^1_\phi - \Sigma^2_\phi + \mathcal{O}\left(\frac{1}{\log c_\phi}\right)$$

where

$$\Sigma_\phi^1 = -2\sum_{p<\infty} \widehat{\psi}\left(\frac{\log p}{\log C_\rho}\right)\frac{\log p}{\log C_\rho}\frac{A_\phi(1,p)+A_\phi(p,1)}{p^{1/2}}$$

and

$$\Sigma_\phi^2 = -\sum_{p<\infty} \widehat{\psi}\left(\frac{2\log p}{\log C_\rho}\right)\frac{2\log p}{\log C_\rho}\frac{A_\phi(1,p^2)-A_\phi(p,1)\ +\ A_\phi(p^2,1)-A_\phi(1,p)}{p}.$$

Assume the support of $\widehat{\psi}$ is in $(-\delta,\delta)$. As before, we have

$$\sum_j D(\phi_j;\psi)\frac{h_{T,R}(\nu^{(j)})}{\mathcal{L}_j} = \sum_j\left[\widehat{\psi}(0) - \Sigma_{\phi_j}^1 - \Sigma_{\phi_j}^2 + O\left(\frac{\log\log T}{\log T}\right)\right]\frac{h_{T,R}(\nu^{(j)})}{\mathcal{L}_j},$$

and invoking Theorem 1.1 again gives

$$\sum_j\left(-\Sigma_{\phi_j}^1 - \Sigma_{\phi_j}^2\right)\frac{h_{T,R}(\nu^{(j)})}{\mathcal{L}_j} =$$

$$\sum_j\left[\sum_{p\ll T^{3\delta}}\widehat{\psi}\left(\frac{\log p}{\log C_\rho}\right)\frac{\log p}{\log C_\rho}\frac{A_\phi(1,p)+A_\phi(p,1)}{p^{1/2}} + \sum_{p\ll T^{3\delta/2}}\widehat{\psi}\left(\frac{2\log p}{\log C_\rho}\right)\times\right.$$

$$\left.\frac{\log p}{\log C_\rho}\frac{A_\phi(1,p^2)-A_\phi(p,1)\ +\ A_\phi(p^2,1)-A_\phi(1,p)}{p}\right]\frac{h_{T,R}(\nu^{(j)})}{\mathcal{L}_j}$$

$$\ll T^{3+3R+\varepsilon}\left[\sum_{p\ll T^{3\delta}} p^{3/2} + \sum_{p\ll T^{3\delta/2}} p^3\right] \ll T^{3+3R+\varepsilon}\,T^{\frac{15}{2}\delta}.$$

So in this case, we need $\delta < \frac{4}{15}$.

This completes the proof of Theorem 1.7.

References

[AIL+11] N. Amersi, G. Iyer, O. Lazarev, S. Miller, and L. Zhang, *Low-lying zeros of cuspidal Maass forms, 2011*, Preprint arXiv:1111.6524v2.

[BF90] D. Bump and S. Friedberg, *The exterior square automorphic L-functions on* GL(n), In *Festschrift in honor of I. I. Piatetski-Shapiro on the occasion of his sixtieth birthday, Part II (Ramat Aviv, 1989)*, volume 3 of *Israel Math. Conf. Proc.*, pages 47–65. Weizmann, Jerusalem, 1990.

[BFG88] D. Bump, S. Friedberg, and D. Goldfeld, *Poincaré series and Kloosterman sums for* SL$(3, \mathbf{Z})$, Acta Arith. **50** (1) (1988), 31–89.

[BG92] D. Bump and D. Ginzburg, *Symmetric square L-functions on* $GL(r)$, Ann. of Math. (2) **136** (1) (1992), 137–205.

[Blo11] V. Blomer, *Applications of the Kuznetsov formula on* $GL(3)$, 2011, preprint.

[DM06] E. Dueñez and S. J. Miller, *The low lying zeros of a* GL(4) *and a* GL(6) *family of L-functions*, Compos. Math. **142** (2006) no. 6, 1403–1425.

[DM09] E. Dueñez and S. J. Miller, *The effect of convolving families of L-functions on the underlying group symmetries*, Proc. Lond. Math. Soc. (3), **99** (3) (2009), 787–820.

[GJ72] R. Godement and H. Jacquet, *Zeta functions of simple algebras*, Lecture Notes in Mathematics, Vol. 260. Springer-Verlag, Berlin, 1972.

[GK11] D. Goldfeld and A. Kontorovich. *On the determination of the Plancherel measure for Lebedev-Whittaker transforms on GL(n)*, 2011, to appear, Acta Arithmetica.

[GLS04] S. Gelbart, E. Lapid, and P. Sarnak, *A new method for lower bounds of L-functions*, C. R. Math. Acad. Sci. Paris, **339** (2) (2004), 91–94.

[Gol06] D. Goldfeld, *Automorphic forms and L-functions for the group* GL$(n, \mathbf{R})$, volume 99 of Cambridge Studies in Advanced Mathematics, Cambridge University Press, Cambridge, 2006, with an appendix by K.A. Broughan.

[Gul05] A. Güloğlu, *Low-lying zeroes of symmetric power L-functions*, Int. Math. Res. Not. (9), (2005), 517–550.

[HL94] J. Hoffstein and P. Lockhart, *Coefficients of Maass forms and the Siegel zero*, Ann. of Math. (2) **140** (1) (1994), 161–181. (With an appendix by D. Goldfeld, Hoffstein and D. Lieman).

[HM07] C. P. Hughes and S. J. Miller, *Low-lying zeros of L-functions with orthogonal symmetry*, Duke Math. J. **136** (1) (2007), 115–172.

[HR95] J. Hoffstein and D. Ramakrishnan, *Siegel zeros and cusp forms*, Internat. Math. Res. Not. **6** (1995), 279–308.

[IK04] H. Iwaniec and E. Kowalski, *Analytic number theory*, volume 53 of American Mathematical Society Colloquium Publications. American Mathematical Society, Providence, RI, 2004.

[ILS00] H. Iwaniec, W. Luo, and P. Sarnak, *Low lying zeros of families of L-functions*, Inst. Hautes Études Sci. Publ. Math. **91** 55–131 (2001), 2000.

[JPSS83] H. Jacquet, I. Piatetskii-Shapiro, and J. Shalika. *Rankin-Selberg convolutions*, Amer. J. Math, **105** (2) (1983), 367–464.

[JS90] H. Jacquet and J. Shalika, *Exterior square L-functions*, In Automorphic forms, Shimura varieties, and L-functions, Vol. II (Ann Arbor, MI, 1988), volume 11 of Perspect. Math., pages 143–226. Academic Press, Boston, MA, 1990.

[JS77] H. Jacquet and J. Shalika, *A non-vanishing theorem for zeta functions of $GL(n)$*, Invent. Math. **38** (1) (1976/77), 1–16.

[Kon10] A. V. Kontorovich, *The Dirichlet series for the exterior square L-function on* $\mathrm{GL}(n)$, Ramanujan J. **21** (3) (2010), 263–266.

[KS99] N. M. Katz and P. Sarnak, *Random matrices, Frobenius eigenvalues, and monodromy*, volume 45 of American Mathematical Society Colloquium Publications, American Mathematical Society, Providence, RI, 1999.

[Luo01] W. Luo, *Nonvanishing of L-values and the Weyl law*, Ann. of Math. (2) **154** (2) (2001), 477–502.

[Mil01] S. D. Miller, *On the existence and temperedness of cusp forms for* $\mathrm{SL}_3(\mathbb{Z})$. J. Reine Angew. Math. **533** (2001), 127–169.

[PŠ75] I. I. Pjateckij-Šapiro, *Euler subgroups*, In Lie groups and their representations (Proc. Summer School, Bolyai János Math. Soc., Budapest, 1971), pages 597–620. Halsted, New York, 1975.

[Roy01] E. Royer. *Petits zeros de fonctions L de formes modulaires*, Acta Arith. **99** (2) (2001), 147–172.

[Sar08] P. Sarnak, *Definition of families of L-functions*, 2008, Online note, available at http://publications.ias.edu/sarnak/paper/507.

[Sha73] J. A. Shalika, *On the multiplicity of the spectrum of the space of cusp forms of* GL_n, Bull. Amer. Math. Soc. **79** (1973), 454–461.

[ST12] S. Shin and N. Templier, *Sato-Tate theorem for families and low-lying zeroes of automorphic L-functions*, 2012, preprint, arXiv: 1208.1945.

[Sta01] E. Stade, *Mellin transforms of* $\mathrm{GL}(n,\mathbb{R})$ *Whittaker functions*, Amer. J. Math. **123** (1) (2001), 121–161.

[Sta02] E. Stade, *Archimedean L-factors on* $\mathrm{GL}(n)\times\mathrm{GL}(n)$ *and generalized Barnes integrals*, Israel J. Math. **127** (2002), 201–219.

[Wal92] N. R. Wallach. *Real reductive groups. II*, volume 132 of Pure and Applied Mathematics, Academic Press Inc., Boston, MA, 1992.

[You06] M. Young, *Low-lying zeros of families of elliptic curves*, J. Amer. Math. Soc, **19** (1) (2006), 205–250.

Dorian Goldfeld, Math Department, Columbia University, New York, NY 10027, USA.
E-mail: goldfeld@math.columbia.edu

Alex Kontorovich, Math Department, Yale University, New Haven, CT 06511, USA.
E-mail: alex.kontorovich@yale.edu

Automorphic Representations and L-Functions
Editors: D. Prasad, C.S. Rajan, A. Sankaranarayanan, J. Sengupta

Publisher: Hindustan Book Agency, New Delhi, India

Level Aspect Subconvexity for Rankin-Selberg L-functions

Roman Holowinsky and Ritabrata Munshi

Abstract

Let M be a square-free integer and let P be a prime not dividing M such that $P \sim M^{\eta}$ with $0 < \eta < 2/21$. We prove subconvexity bounds for $L(\frac{1}{2}, f \otimes g)$ when f and g are two primitive holomorphic cusp forms of levels P and M. These bounds are achieved through an unamplified second moment method.

1 Introduction and Statement of Results

Several authors have recently been successful in implementing the amplification method in order to establish level aspect subconvexity results for Rankin-Selberg convolutions of two GL(2) forms when one form is fixed and the other form is varying. For example, if f is a Hecke cusp form of fixed level and g is a Hecke cusp form of varying level M, then various bounds of the form

$$L(\tfrac{1}{2}, f \otimes g) \ll_f M^{1/2-\delta}$$

for some absolute positive constant δ have been shown by Kowalski-Michel-VanderKam [22], Michel [24] and Harcos-Michel [12]. Furthermore, results for the Rankin-Selberg convolution of two independently varying forms have been established in the works of Michel-Ramakrishnan [25], Feigon-Whitehouse [9] and Nelson [26] in situations where positivity of the central L-values is known. Of particular interest, yet seemingly out of reach by means of current technology, are level aspect subconvexity results for the Rankin-Selberg convolution of two GL(2) forms of same level (or similarly, spectral aspect subconvexity results for forms of the same weight or Laplace eigenvalue). These L-values appear naturally in many areas of number theory and in particular, have important connections with quantum chaos and equidistribution problems.

Subconvexity bounds for an individual L-function are often the result of sufficient bounds for a weighted average over an appropriate family of

L-functions. In this note, we consider the subconvexity problem for the Rankin-Selberg convolution of two varying GL(2) forms with co-prime levels through the use of a second moment method. With the L-function here being constructed from data associated with two independently varying forms, one has a large collection of natural families to choose from.

The ideas presented here may be applied to other Rankin-Selberg convolutions constructed out of multiple independently varying forms. This is the first installment of recent work by the two authors related to the subconvexity problem and its purpose is to demonstrate the existence of situations in which subconvexity may be established through a second moment average without amplification.

Acknowledgements We thank IAS Princeton for the wonderful working conditions in which many of the ideas for this collaboration were initially conceived. We also thank MSRI Berkeley, MF Oberwolfach and TIFR Mumbai for providing the opportunity for further discussions. The first author is supported by the Sloan fellowship BR2011-083 and the NSF grant DMS-1068043.

1.1 Holomorphic cusp forms

Let $N > 0$ be an integer and $k > 0$ be an even integer. We denote by $\mathcal{S}_k(N)$ the linear space of holomorphic cusp forms of weight k, level N and trivial nebentypus. Such forms are holomorphic functions on the upper half-plane $f : \mathbb{H} \to \mathbb{C}$ satisfying

$$f(\gamma z) = (cz + d)^k f(z)$$

for every $\gamma = \left(\begin{smallmatrix} a & b \\ c & d \end{smallmatrix}\right) \in \Gamma_0(N)$ and which vanish at every cusp. Any form $f \in \mathcal{S}_k(N)$ has a Fourier series expansion

$$f(z) = \sum_{n \geqslant 1} \psi_f(n) n^{\frac{k-1}{2}} e(nz)$$

with coefficients $\psi_f(n)$ satisfying

$$\psi_f(n) \ll_f \tau(n)$$

as proven by Deligne [3].

The space $\mathcal{S}_k(N)$ is a finite dimensional Hilbert space with respect to the Petersson inner product

$$\langle f_1, f_2 \rangle = \int_{\Gamma_0(N)\backslash\mathbb{H}} y^k f_1(z) \bar{f}_2(z) \frac{dxdy}{y^2}.$$

We can choose an orthogonal basis $\mathcal{H}_k(N)$ for $\mathcal{S}_k(N)$ which consists of common eigenfunctions of all the Hecke operators T_n with $(n, N) = 1$. That is, each $f \in \mathcal{H}_k(N)$ satisfies

$$(T_n f)(z) = \frac{1}{\sqrt{n}} \sum_{\substack{ad=n \\ (a,N)=1}} \left(\frac{a}{d}\right)^{k/2} \sum_{b \,(\mathrm{mod}\, d)} f\left(\frac{az+b}{d}\right) = \lambda_f(n) f(z)$$

for all $(n, N) = 1$. Such f are called *Hecke eigen cusp forms*. The Hecke operators are multiplicative and one has that

$$\psi_f(m)\lambda_f(n) = \sum_{d|(m,n)} \psi_f\left(\frac{mn}{d^2}\right)$$

for any $m, n \geqslant 1$ with $(n, N) = 1$. In particular, $\psi_f(1)\lambda_f(n) = \psi_f(n)$ if $(n, N) = 1$. Therefore,

$$\lambda_f(m)\lambda_f(n) = \sum_{d|(m,n)} \lambda_f\left(\frac{mn}{d^2}\right) \tag{1.1}$$

if $(nm, N) = 1$. The Hecke eigenbasis $\mathcal{H}_k(N)$ also contains a subset of *newforms* $\mathcal{H}_k^*(N)$, those forms which are simultaneous eigenfunctions of all the Hecke operators T_n for any $n \geqslant 1$ and normalized to have first Fourier coefficient $\psi_f(1) = 1$. For $f \in \mathcal{H}_k^*(N)$, the Hecke relations (1.1) hold for all integers $n, m \geqslant 1$ and it is also known (see [18]) that

$$|\lambda_f(p)| = p^{-1/2} \quad \text{for any} \quad p|N. \tag{1.2}$$

1.2 Rankin-Selberg convolutions of forms with co-prime levels

Let N and M be two positive square-free co-prime integers and let k and κ be two fixed positive even integers. Given two newforms $f \in \mathcal{H}_k^*(N)$ and $g \in \mathcal{H}_\kappa^*(M)$, we consider the associated Rankin-Selberg convolution L-function (see [12])

$$L(s, f \otimes g) = \prod_p \prod_{i=1}^{2} \prod_{j=1}^{2} \left(1 - \frac{\alpha_{f,i}(p)\alpha_{g,j}(p)}{p^s}\right)^{-1} = \zeta^{(NM)}(2s) \sum_{n \geqslant 1} \lambda_f(n)\lambda_g(n) n^{-s}$$

where the $\{\alpha_{f,i}\}$ and $\{\alpha_{g,j}\}$ are the local parameters of the L-functions associated to f and g respectively and $\zeta^{(NM)}(2s)$ is the partial Riemann zeta function with the local factors at primes dividing NM removed. The local parameters satisfy the relations $\alpha_{f,1}(p)+\alpha_{f,2}(p) = \lambda_f(p)$ and $\alpha_{f,1}(p)\alpha_{f,2}(p) = \chi_0(p)$ with χ_0 the principal character of modulus N and similarly for the local parameters associated with g. The completed L-function is then defined as

$$\Lambda(s, f \otimes g) := \mathcal{Q}^{s/2} L_\infty(s, f \otimes g, s) L(s, f \otimes g)$$

where the conductor (see [24]) is given by $\mathcal{Q} := \mathcal{Q}(f \otimes g) = (NM)^2$ and the local factor at infinity (see [17]) is a product of gamma factors

$$L_\infty(s, f \otimes g) :=$$
$$\pi^{-2s}\Gamma\Big(\frac{s + \frac{|k-\kappa|}{2}}{2}\Big)\Gamma\Big(\frac{s + \frac{k+\kappa}{2}}{2}\Big)\Gamma\Big(\frac{s + \frac{|k-\kappa|}{2} + 1}{2}\Big)\Gamma\Big(\frac{s + \frac{k+\kappa}{2} - 1}{2}\Big).$$

The completed L-function satisfies the functional equation

$$\Lambda(s, f \otimes g) = \Lambda(1 - s, f \otimes g).$$

Remark We have restricted our discussion to the case of trivial nebentypus in the Rankin-Selberg convolution as this case provides additional difficulties (see [12], [22], [24]). If one of the forms has non-trivial nebentypus, then there is additional oscillation which one may take advantage of. We have taken $(N, M) = 1$ to ensure that the conductor is as large as possible. For general N and M we have that $(NM)^2/(N, M)^4 \leqslant \mathcal{Q}(f \otimes g) \leqslant (NM)^2/(N, M)$ (see [12]).

The convexity bound for $L(s, f \otimes g)$ at the point $s = 1/2$ is

$$L(\tfrac{1}{2}, f \otimes g) \ll_\varepsilon \mathcal{Q}^{1/4+\varepsilon}$$

for any $\varepsilon > 0$ and may be established in this case simply by the approximate functional equation and Deligne's bound. It has recently been shown by Heath-Brown [14], in the general setting of Selberg class L-functions using Jensen's formula for strips, that the ε in the above bound may be removed

$$L(\tfrac{1}{2}, f \otimes g) \ll \mathcal{Q}^{1/4}.$$

Furthermore, the general results of Soundararajan [30] provide a "weak-subconvexity" bound of the form

$$L(\tfrac{1}{2}, f \otimes g) \ll \frac{\mathcal{Q}^{1/4}}{(\log \mathcal{Q})^{1-\varepsilon}}$$

for any $\varepsilon > 0$.

1.3 Main results

Our purpose here is to provide level aspect subconvexity bounds for the Rankin-Selberg convolution of two newforms of varying levels N and M in situations where both forms are varying at different rates, say $N \sim M^\eta$ for some $0 < \eta < 1$. The main point we wish to stress, is that we take advantage of the size of the smaller level N. The method we present here does not produce subconvexity bounds when $N = 1$ nor when N is the same size as M. Both levels must contribute to the complexity of the problem and they must do so in a manner which is sufficiently distinguishable for the method to work. We restrict to the case of $N = P$ prime to simplify our presentation. Recall that our conductor in this case is of size $\mathcal{Q} = (PM)^2$.

We start by reducing our L-function to a smooth sum over Hecke eigenvalues by a standard approximate functional equation argument, see for example [17], [19], [24]. Since we are working with newforms of trivial nebentypus, we have

$$L(\tfrac{1}{2}, f \otimes g) = 2 \sum_{n=1}^{\infty} \frac{\lambda_f(n)\lambda_g(n)}{\sqrt{n}} W\left(\frac{n}{\sqrt{\mathcal{Q}}}\right)$$

where

$$W(y) = \frac{1}{2\pi i} \int_{(3)} G(u) \frac{L_\infty(\frac{1}{2}+u, f \otimes g)}{L_\infty(\frac{1}{2}, f \otimes g)} \zeta^{(NM)}(1+2u) y^{-u} \frac{du}{u}$$

and

$$G(u) = \left(\cos \frac{\pi u}{4A}\right)^{-16A}$$

for any positive integer A. The derivatives of $W(y)$ satisfy

$$y^j W^{(j)}(y) \ll_{k,\kappa} \mathcal{Q}^\varepsilon (1+y)^{-A} \log(2 + y^{-1})$$

for any $\varepsilon > 0$. Applying a smooth partition of unity one may derive that (see e.g. [19])

$$L(\tfrac{1}{2}, f \otimes g) \ll_{k,\kappa} \mathcal{Q}^\varepsilon \sum_X \frac{|L_{f\otimes g}(X)|}{\sqrt{X}} \left(1 + \frac{X}{\sqrt{\mathcal{Q}}}\right)^{-A}$$

where

$$L_{f\otimes g}(X) = \sum_n \lambda_f(n)\lambda_g(n) h\left(\frac{n}{X}\right)$$

and h is a smooth function, compactly supported on $[\frac{1}{2}, \frac{5}{2}]$ with bounded derivatives and X runs over values 2^ν with $\nu = -1, 0, 1, 2, \ldots$.

Since $L_{f\otimes g}(X)$ is trivially bounded by $X^{1+\varepsilon}$ for any $\varepsilon > 0$, the contribution from those $X > \mathcal{Q}^{1/2+\varepsilon}$ is made negligible by choosing A above to be sufficiently large. Likewise, if $X < \mathcal{Q}^{1/2-\delta}$ for some $\delta > 0$, then $L_{f\otimes g}(X)X^{-1/2} \ll \mathcal{Q}^{1/4-\delta/2}$. Therefore, we are left with

$$L(\tfrac{1}{2}, f\otimes g) \ll_\varepsilon \mathcal{Q}^\varepsilon \left\{ \mathcal{Q}^{1/4-\delta/2} + \max_{\mathcal{Q}^{1/2-\delta}\leqslant X\leqslant \mathcal{Q}^{1/2+\varepsilon}} \frac{|L_{f\otimes g}(X)|}{\sqrt{X}} \right\}$$

for any $\delta > 0$. Subconvexity bounds will now follow if one is able to sufficiently bound $L_{f\otimes g}(X)$ in the remaining range for X. We shall do so by averaging over an orthogonal basis $\mathcal{B}_\kappa(M)$ for $\mathcal{S}_\kappa(M)$ which contains the above newform g.

Theorem 1.1 *Let M be a positive square-free integer and let P be a prime not dividing M. Let k and κ be two fixed positive even integers and let $\mathcal{B}_\kappa(M)$ be an orthogonal basis for $\mathcal{S}_\kappa(M)$. Set $\mathcal{Q} = (PM)^2$. Let $\varepsilon, \delta > 0$ and choose any $\mathcal{Q}^{1/2-\delta} \leqslant X \leqslant \mathcal{Q}^{1/2+\varepsilon}$. For any newform $f \in \mathcal{H}_k^*(P)$ we have*

$$\sum_{g\in\mathcal{B}_\kappa(M)} \omega_g^{-1} \left| \sum_n \psi_f(n)\psi_g(n) h\left(\frac{n}{X}\right)\right|^2 \ll_{\varepsilon,\delta}$$

$$XP\mathcal{Q}^\varepsilon \left(\frac{1}{P} + \frac{1}{\mathcal{Q}^\delta} + \mathcal{Q}^{\frac{5}{4}\delta}\frac{P^{\frac{21}{8}}}{M^{\frac{1}{4}}} + \mathcal{Q}^{3\delta}\frac{P^2}{M^{\frac{1}{4}}} \right)$$

where the spectral weights are given as $\omega_g := \frac{(4\pi)^{\kappa-1}}{\Gamma(\kappa-1)}\langle g, g\rangle$.

Note that a second moment bound of the form

$$\sum_{g\in\mathcal{B}_\kappa(M)} \omega_g^{-1} \left| \sum_n \psi_f(n)\psi_g(n) h\left(\frac{n}{X}\right)\right|^2 \ll XP\mathcal{Q}^\varepsilon \tag{1.3}$$

for all $X \leqslant \mathcal{Q}^{1/2+\varepsilon}$ and any $\varepsilon > 0$ would produce the convexity bound for any individual $L(\frac{1}{2}, f\otimes g)$ with f and g both newforms since then $\psi = \lambda$ and $\omega_g \ll_\kappa M^{1+\varepsilon}$ (see [18] Lemma 2.5). Therefore, the bound in Theorem 1.1 produces a subconvexity bound when $P \sim M^\eta$ with $0 < \eta < 2/21$.

Corollary 1.2 *Let M be a positive square-free integer and let P be a prime not dividing M. Let $\eta = \frac{\log P}{\log M}$. Let k and κ be two fixed positive even integers. For two newforms $f \in \mathcal{H}_k^*(P)$ and $g \in \mathcal{H}_\kappa^*(M)$ we have*

$$L(\tfrac{1}{2}, f\otimes g) \ll \mathcal{Q}^{\frac{1}{4}+\varepsilon}\left(\frac{1}{\mathcal{Q}^{\frac{\eta}{2(1+\eta)}}} + \frac{1}{\mathcal{Q}^{\frac{2-21\eta}{64(1+\eta)}}} \right).$$

Proof Soften the bound in Theorem 1.1 to

$$\sum_{g\in\mathcal{B}_\kappa(M)} \omega_g^{-1}\left|\sum_n \psi_f(n)\psi_g(n)h\left(\frac{n}{X}\right)\right|^2 \ll_{\varepsilon,\delta} XPQ^\varepsilon\left(\frac{1}{P}+\frac{1}{Q^\delta}+Q^{3\delta}\frac{P^{\frac{21}{8}}}{M^{\frac{1}{4}}}\right)$$

and equate the second and third terms on the right hand side above while replacing all occurrences of P by M^η.

□

The estimates that we have obtained in Theorem 1.1 and Corollary 1.2 are the result of analysis of the shifted convolution sum problem through the δ-method ([5], [13]) with explicit dependence on the level P of the form f. It is possible to push our arguments further to improve these estimates by considering the shifted convolution sum problem on average over shifts while again maintaining explicit dependence on the level P of f and we shall do so in a later work. For our purposes here, we prove the following theorem for a fixed non-zero shift.

Theorem 1.3 *Let ℓ be a non-zero integer and let $X, Y \geqslant 1$. Let F be a smooth function supported on $[1/2, 5/2]\times[1/2, 5/2]$ with partial derivatives satisfying*

$$x^i y^j \frac{\partial^i}{\partial x^i}\frac{\partial^j}{\partial y^j}F\left(\frac{x}{X},\frac{y}{Y}\right) \ll Z{Z_x}^i{Z_y}^j$$

for some $Z>0$ and $Z_x, Z_y \geqslant 1$. For any newforms $f_1, f_2 \in \mathcal{H}_k^*(P)$ we have

$$\sum\sum_{m=nP+\ell} \lambda_{f_1}(n)\lambda_{f_2}(m)F\left(\frac{n}{X},\frac{m}{Y}\right) \ll_\varepsilon$$

$$P\max\{XP,Y\}^{3/4}Z\sqrt{Z_xZ_y}\max\{Z_x,Z_y\}^{5/4}(XYP)^\varepsilon.$$

For other works involving estimates of shifted sums see [1], [2], [5], [8], [12], [15], [16], [20], [21], [23], [27], [28], [29] and [11] for dependence on the level of the forms. The above bound in Theorem 1.3 does not follow easily from any of the above works. The main advantage here is uniformity with respect to the shift ℓ and the coefficient P. Furthermore, we note that if $\ell\equiv 0 \bmod P$ then one also has the trivial bound $ZX/\sqrt{P}$ by using (1.2).

2 Preliminaries

2.1 Bessel functions

We record here some standard facts about the J-Bessel functions as can be seen in [31] as well as several estimates for integrals involving Bessel functions which will be required for our application. One may write the J-Bessel functions as

$$J_k(x) = e^{ix}W_k(x) + e^{-ix}\overline{W}_k(x) \tag{2.1}$$

where

$$W_k(x) = \frac{e^{i(\frac{\pi}{2}k-\frac{\pi}{4})}}{\Gamma(k+\frac{1}{2})}\sqrt{\frac{2}{\pi x}}\int_0^\infty e^{-y}(y(1+\frac{iy}{2x}))^{k-\frac{1}{2}}dy \tag{2.2}$$

which, when k is a positive integer, one has that

$$x^j W_k^{(j)}(x) \ll \frac{x}{(1+x)^{3/2}}. \tag{2.3}$$

Using the above facts leads us to the following results.

Lemma 2.1 *Let $k, \kappa \geqslant 2$ be integers and let $a, b, x, y > 0$. Define*

$$I(x,y) := \int_0^\infty h(\xi)\, J_{\kappa-1}\left(4\pi a\sqrt{x\xi}\right) J_{k-1}\left(4\pi b\sqrt{y\xi}\right) d\xi$$

where h is a smooth function compactly supported on $\left[\frac{1}{2}, \frac{5}{2}\right]$ with bounded derivatives. We have

$$I(x,y) \ll_j |a\sqrt{x} - b\sqrt{y}|^{-j}$$

for any $j \geqslant 0$.

Proof A change of variables, $\xi = w^2$, gives

$$I(x,y) = 2\int_0^\infty h(w^2)\, w\, J_{\kappa-1}\left(4\pi a\sqrt{x}w\right) J_{k-1}\left(4\pi b\sqrt{y}w\right) dw.$$

Therefore, we see from (2.1) that $I(x,y)$ may be written as the sum of four similar terms, one of them being

$$\int_0^\infty e\left(2w(a\sqrt{x} - b\sqrt{y})\right) h(w^2)\, w\, W_{\kappa-1}\left(4\pi a\sqrt{x}w\right) \overline{W}_{k-1}\left(4\pi b\sqrt{y}w\right) dw.$$

Repeated integration by parts gives the desired result. □

Lemma 2.2 *For $I(x,y)$ as in Lemma 2.1, we have*

$$x^i y^j \frac{\partial^i}{\partial x^i}\frac{\partial^j}{\partial y^j} I(x,y) \ll_{i,j} \frac{a\sqrt{x}}{(1+a\sqrt{x})^{3/2}}\frac{b\sqrt{y}}{(1+b\sqrt{y})^{3/2}}\left(1+a\sqrt{x}\right)^i\left(1+b\sqrt{y}\right)^j.$$

Proof Differentiate and use the bound in (2.3).

□

Lemma 2.3 *Let k, P, q be positive integers with $k \geqslant 2$ and let ℓ be a non-zero integer. Take $Q > 1$ and $X, Y \geqslant 1$. For any $a, b > 0$, define*

$$J(a,b) :=$$
$$\int_0^\infty \int_0^\infty F\left(\frac{x}{X}, \frac{y}{Y}\right) h\left(\frac{q}{Q}, \frac{xP+\ell-y}{Q^2}\right) J_{k-1}\left(4\pi a\sqrt{x}\right) J_{k-1}\left(4\pi b\sqrt{y}\right) dx\, dy$$

where $h\left(\frac{q}{Q}, \frac{xP+\ell-y}{Q^2}\right)$ is the function from Lemma 2.7 in §2.2 and F is a smooth function supported on $[1/2, 5/2] \times [1/2, 5/2]$ with partial derivatives satisfying

$$x^i y^j \frac{\partial^i}{\partial x^i} \frac{\partial^j}{\partial y^j} F\left(\frac{x}{X}, \frac{y}{Y}\right) \ll Z {Z_x}^i {Z_y}^j$$

for some $Z > 0$ and $Z_x, Z_y \geqslant 1$. We have that $J(a,b)$ is of size at most

$$\ll ZXY \frac{Q}{q} \frac{a\sqrt{X}}{(1+a\sqrt{X})^{3/2}} \frac{b\sqrt{Y}}{(1+b\sqrt{Y})^{3/2}} \left[\frac{1}{a\sqrt{X}}\left\{Z_x + \frac{XP}{qQ}\right\}\right]^i \times \left[\frac{1}{b\sqrt{Y}}\left\{Z_y + \frac{Y}{qQ}\right\}\right]^j \quad (2.4)$$

for any non-negative integers i and j. Furthermore,

$$J(a,b) \ll \frac{ZXY}{(1+a\sqrt{X})^{3/2}\,(1+b\sqrt{Y})^{3/2}} \frac{Q}{q} \min\{Z_x\, b\sqrt{Y}, Z_y\, a\sqrt{X}\}\, Q^{\varepsilon}. \quad (2.5)$$

Proof A change of variables, integrating by parts once in x and applying the given bounds for the functions F, h and the Bessel functions gives

$$J(a,b) \ll ZXY \frac{Q}{q} \frac{a\sqrt{X}}{(1+a\sqrt{X})^{3/2}} \frac{b\sqrt{Y}}{(1+b\sqrt{Y})^{3/2}} \left[\frac{1}{a\sqrt{X}}\left\{Z_x + XPI\right\}\right]$$

with

$$I := \int_{1/2}^{5/2} \int_{1/2}^{5/2}{}_{2|xXP+\ell-yY|>qQ} \frac{1}{|xXP+\ell-yY|}\, dx\, dy.$$

Trivially, $I \ll (qQ)^{-1}$ and this is how one arrives at (2.4) with $i = 1$ and $j = 0$. Repeated integration by parts would then establish (2.4) for all i

and j. Otherwise, replace x by $u = xXP + \ell - yY$ so that $dx = (XP)^{-1}du$ and

$$I \ll (XP)^{-1} \int_{1/2}^{5/2} \int_{qQ/2}^{(XP+Y+|\ell|)Q^{\varepsilon}} \frac{1}{u}\, du\, dy \ll (XP)^{-1}Q^{\varepsilon}.$$

Repeating the argument, for y instead of x, gives the bound (2.5).

□

2.2 Summation Formulae, Large Sieve and the δ-method

Let $k \geqslant 2$ be an integer. For any $n, m, c \in \mathbb{N}$, let $S(n, m; c)$ denote the Kloosterman sum

$$S(n, m; c) = \sum_{\alpha(c)}^{*} e\left(\frac{n\alpha + m\overline{\alpha}}{c}\right).$$

The Kloosterman sums satisfy the Weil bound

$$|S(n, m; c)| \leqslant (n, m, c)^{1/2} c^{1/2} \tau(c)$$

where $\tau(c)$ is the number of divisors of c. This bound is best possible for an individual Kloosterman sum. Sums of Kloosterman sums appear in the following spectral average (see §14.2 in [17] for a derivation).

Lemma 2.4 *Let $N \geqslant 1$ be an integer. Let $\mathcal{B}_k(N)$ be any orthogonal basis for $\mathcal{S}_k(N)$. For any $n, m \geqslant 1$, we have*

$$\sum_{f \in \mathcal{B}_k(N)} \omega_f^{-1} \psi_f(n) \overline{\psi_f(m)} = \delta(n, m) + 2\pi i^{-k} \sum_{\substack{c>0 \\ c \equiv 0(N)}} \frac{1}{c} S(n, m; c) J_{k-1}\left(\frac{4\pi\sqrt{nm}}{c}\right)$$

where the spectral weights ω_f are given by

$$\omega_f := \frac{(4\pi)^{k-1}}{\Gamma(k-1)} \langle f, f \rangle$$

and $\delta(n, m) = 1$ if $n = m$ and $\delta(n, m) = 0$ otherwise.

One also has the following large sieve estimate ([22] Prop. 5.1, [6], [7], [4]).

Lemma 2.5 *Let η be a smooth function supported on $[C/2, 5C/2]$ such that $\eta^{(j)} \ll_j C^{-j}$ for all $j \geqslant 0$. For any sequences of complex numbers x_n, y_m we have*

$$\sum_{n \leqslant X} \sum_{m \leqslant Y} x_n y_m \sum_{\substack{c>0 \\ c \equiv 0(N)}} \frac{\eta(c)}{c} S(n, m; c) J_{k-1}\left(\frac{4\pi\sqrt{nm}}{c}\right)$$
$$\ll_{\varepsilon,k} C^{\varepsilon}\left(\frac{\sqrt{XY}}{C}\right)^{k-3/2}\left(1+\frac{X}{N}\right)^{1/2}\left(1+\frac{Y}{N}\right)^{1/2}\|x\|_2\|y\|_2$$

with any $\varepsilon > 0$. Moreover the exponent $k - 3/2$ may be replaced by $1/2$.

The above estimate will be useful in controlling the size of Kloosterman sum moduli. For all remaining moduli we will apply the following analogue to Poisson summation ([22], Theorem A.4).

Lemma 2.6 *Let $(a, q) = 1$ and let h be a smooth function, compactly supported in $(0, \infty)$. Let f be a holomorphic newform of level N and weight k. Set $N_2 := N/(N, q)$. Then there exists a complex number η of modulus 1 (depending on a, q and f) and a newform f^* of the same level N and the same weight k such that*

$$\sum_n \lambda_f(n) e\left(n\frac{a}{q}\right) h(n) =$$
$$\frac{2\pi\eta}{q\sqrt{N_2}} \sum_n \lambda_{f^*}(n) e\left(-n\frac{\overline{aN_2}}{q}\right) \int_0^\infty h(\xi) J_{k-1}\left(\frac{4\pi\sqrt{n\xi}}{q\sqrt{N_2}}\right) d\xi$$

where $\overline{x}$ denotes the multiplicative inverse of x.

We will now briefly recall a version of the circle method introduced in [5] and [13]. The starting point is a smooth approximation of the δ-symbol. We will follow the exposition of Heath-Brown in [13].

Lemma 2.7 *For any $Q > 1$ there is a positive constant c_Q, and a smooth function $h(x, y)$ defined on $(0, \infty) \times \mathbb{R}$, such that*

$$\delta(n, 0) = \frac{c_Q}{Q^2} \sum_{q=1}^{\infty} \sideset{}{^\star}\sum_{a \bmod q} e\left(\frac{an}{q}\right) h\left(\frac{q}{Q}, \frac{n}{Q^2}\right). \tag{2.6}$$

The constant c_Q satisfies $c_Q = 1 + O_A(Q^{-A})$ for any $A > 0$. Moreover $h(x, y) \ll x^{-1}$ for all y, and $h(x, y)$ is non-zero only for $x \leq \max\{1, 2|y|\}$.

In practice, to detect the equation $n = 0$ for a sequence of integers in the range $[-X, X]$, it is logical to choose $Q = X^{1/2}$. The smooth function $h(x, y)$ satisfies (see [13])

$$x^i \frac{\partial^i}{\partial x^i} h(x,y) \ll_i x^{-1} \quad \text{and} \quad \frac{\partial}{\partial y} h(x,y) = 0 \tag{2.7}$$

for $x \leq 1$ and $|y| \leq x/2$. Also for $|y| > x/2$, we have

$$x^i y^j \frac{\partial^i}{\partial x^i} \frac{\partial^j}{\partial y^j} h(x,y) \ll_{i,j} x^{-1}. \tag{2.8}$$

3 Initial Reduction of the Second Moment

Let M be a positive square-free integer and let P be a prime not dividing M. Let k and κ be two positive even fixed integers. Fix a newform $f \in \mathcal{H}_k^*(P)$ and choose an orthogonal basis $\mathcal{B}_\kappa(M)$ for $\mathcal{S}_\kappa(M)$. Set $\mathcal{Q} := (PM)^2$. Let $\varepsilon, \delta > 0$ and choose any $\mathcal{Q}^{1/2-\delta} \leqslant X \leqslant \mathcal{Q}^{1/2+\varepsilon}$. As seen in the statement of Theorem 1.1, we are interested in obtaining upper bounds for the sum

$$S_f(X) := \sum_{g \in \mathcal{B}_\kappa(M)} \omega_g^{-1} \left| \sum_n \psi_f(n) \psi_g(n) h\left(\frac{n}{X}\right) \right|^2 \tag{3.1}$$

where $\omega_g = \frac{(4\pi)^{\kappa-1}}{\Gamma(\kappa-1)} \langle g, g \rangle$ and h is smooth, compactly supported on $[1/2, 5/2]$ with bounded derivatives. We start by opening the square and applying the Petersson trace formula in g. Since f is a newform, we have $\psi_f(n) = \lambda_f(n)$ and so

$$S_f(X) = \sum_n \lambda_f(n)^2 h\left(\frac{n}{X}\right)^2$$
$$+2\pi i^{-\kappa} \sum_n \sum_m \lambda_f(n) h\left(\frac{n}{X}\right) \lambda_f(m) h\left(\frac{m}{X}\right) \sum_{\substack{d>0 \\ d \equiv 0(M)}} \frac{S(n,m;d)}{d} J_{\kappa-1}\left(\frac{4\pi\sqrt{nm}}{d}\right).$$

The "diagonal term" satisfies

$$\sum_n \lambda_f(n)^2 h\left(\frac{n}{X}\right)^2 \ll X \mathcal{Q}^\varepsilon$$

for any $\varepsilon > 0$. This is the first term seen in the bound in Theorem 1.1. We are now left with the "off-diagonal" terms

$$\sum_n \sum_m \lambda_f(n) h\left(\frac{n}{X}\right) \lambda_f(m) h\left(\frac{m}{X}\right) \sum_{\substack{d>0 \\ d \equiv 0(M)}} \frac{S(n,m;d)}{d} J_{\kappa-1}\left(\frac{4\pi\sqrt{nm}}{d}\right).$$

We start by truncating the sum over d. By the Weil bound for individual Kloosterman sums and bounds for the Bessel functions in §2.1, there exist positive values A and B such that the sum over d may be truncated to those $d \leqslant X^A$ up to an error term of size at most $X^{-B}M^{-1}$. For the remaining sum over $d \leqslant X^A$, we introduce another smooth partition of unity and break the sum into dyadic segments of size D, as we did with our n-sum above, so that we are left with sums of type

$$R_{f,D}(X) :=$$
$$\sum_n \sum_m \lambda_f(n) h\left(\frac{n}{X}\right) \lambda_f(m) h\left(\frac{m}{X}\right) \sum_{\substack{d>0 \\ d\equiv 0(M)}} \frac{S(n,m;d)}{d} J_{\kappa-1}\left(\frac{4\pi\sqrt{nm}}{d}\right) \eta_D(d) \tag{3.2}$$

where η_D is a smooth function supported on $[D/2, 5D/2]$. Note that D must be of size at least M by the congruence condition. Furthermore, an application of Lemma 2.5 shows that

$$R_{f,D}(X) \ll \left(\frac{X}{D}\right)^{k-3/2} \left(1 + \frac{X}{M}\right) X Q^{\varepsilon}$$

which is smaller than the bound in Theorem 1.1 as soon as $D > XQ^{2\delta}$. Therefore, bounding the second moment in (3.1) has reduced to the following statement.

Lemma 3.1 *Let $\delta > 0$. For any $Q^{1/2-\delta} \leqslant X \leqslant Q^{1/2+\varepsilon}$ we have*

$$S_f(X) \ll_{\varepsilon,\delta} Q^{\varepsilon} \left(X + PXQ^{-\delta} + \sum_{M \leqslant D \leqslant XQ^{2\delta}} R_{f,D}(X) \right) \tag{3.3}$$

where $R_{f,D}(X)$ is given by (3.2) above and D runs over dyadic values.

Remark With additional work, one might also eliminate all $D < XQ^{-\theta}$, for some $\theta > 0$ depending on δ, in order to improve the final range of sizes P relative to M for which subconvexity is achieved. To keep our presentation short, we shall only show how one may remove $D < \sqrt{P}MQ^{-\delta}$ (see Lemma 5.1).

We emphasize here the significance of the level P in our problem. Note that the first term XQ^{ε} in the above bound (3.3), which came from the diagonal term after applying the Petersson trace formula in g, beats the convexity bound for $S_f(X)$ by P. If P were fixed, then Lemma 3.1 would already be insufficient for subconvexity.

4 Reduction to Shifted Convolution Sums

Let $\delta > 0$. We now proceed with the analysis of $R_{f,D}(X)$, as defined by (3.2) above, when $M \leqslant D \leqslant X\mathcal{Q}^{2\delta}$ with $\mathcal{Q}^{1/2-\delta} \leqslant X \leqslant \mathcal{Q}^{1/2+\varepsilon}$. Opening the Kloosterman sums and changing the order of summation, one is left to study

$$\sum_{\substack{d>0\\ d\equiv 0(M)}} \frac{\eta_D(d)}{d} \cdot \sum_{\beta(d)}^{*} \sum_n \lambda_f(n) e\left(n\frac{\beta}{d}\right) h\left(\frac{n}{X}\right)$$
$$\sum_m \lambda_f(m) e\left(m\frac{\overline{\beta}}{d}\right) h\left(\frac{m}{X}\right) J_{\kappa-1}\left(\frac{4\pi\sqrt{nm}}{d}\right). \quad (4.1)$$

As in the works [12], [22] and [24], an application of Voronoi summation in m and the evaluation of the resulting Ramanujan sums will lead to a collection of shifted convolution sums. Switching from Kloosterman sums to Ramanujan sums in such a manner was already seen in the work of Goldfeld [10]. Since the application of Voronoi summation will be for a newform f of level P and therefore depends on the divisibility of d by P, we first break apart our d sum as

$$\sum_{LR=P} \sum_{\substack{d>0\\ (d,L)=1\\ d\equiv 0(RM)}} \frac{\eta_D(d)}{d} \cdot \sum_{\beta(d)}^{*} \sum_n \lambda_f(n) e\left(n\frac{\beta}{d}\right) h\left(\frac{n}{X}\right)$$
$$\sum_m \lambda_f(m) e\left(m\frac{\overline{\beta}}{d}\right) h\left(\frac{m}{X}\right) J_{\kappa-1}\left(\frac{4\pi\sqrt{nm}}{d}\right).$$

Voronoi summation in m then gives that the inner sum, up to a constant, is equal to

$$\frac{1}{d\sqrt{L}} \sum_m \lambda_{f^*}(m) e\left(-m\frac{\beta\overline{L}}{d}\right) \int_0^\infty h\left(\frac{\xi}{X}\right) J_{\kappa-1}\left(\frac{4\pi\sqrt{n\xi}}{d}\right) J_{k-1}\left(\frac{4\pi\sqrt{m\xi}}{d\sqrt{L}}\right) d\xi.$$

This produces a Ramanujan sum over β for each modulus d, which we write as

$$\sum_{\beta(d)}^{*} e\left(\frac{\beta(nL-m)}{d}\right) = \sum_{bc=d} \mu(b) \sum_{\beta(c)} e\left(\frac{\beta(nL-m)}{c}\right).$$

Summing over β will now produce a congruence condition between n and m modulo c. Thus, we have reduced (4.1) to the following.

Lemma 4.1 *Let $\delta > 0$ and let $R_{f,D}(X)$ be as in (3.2) with $\mathcal{Q}^{1/2-\delta} \leqslant X \leqslant \mathcal{Q}^{1/2+\varepsilon}$. For any $M \leqslant D \leqslant X\mathcal{Q}^{2\delta}$ we have*

$$R_{f,D}(X) \ll \sum_{LR=P} \frac{1}{\sqrt{L}} \sum_{\substack{d>0 \\ (d,L)=1 \\ d\equiv 0(RM)}} \frac{\eta_D(d)}{d} \sum_{bc=d} \frac{1}{b} |\Sigma_d(L;c)|$$

with shifted convolution sums

$$\Sigma_d(L;c) = \sum_n \sum_{m\equiv nL(c)} \lambda_f(n)\lambda_{f^*}(m) I_d(n,m)$$

where

$$I_d(n,m) = h\left(\frac{n}{X}\right) \int_0^\infty h\left(\frac{\xi}{X}\right) J_{\kappa-1}\left(\frac{4\pi\sqrt{n\xi}}{d}\right) J_{k-1}\left(\frac{4\pi\sqrt{m\xi}}{d\sqrt{L}}\right) d\xi.$$

In the above, $I_d(n,m)$ determines the main contribution in the sum over n and m which occurs when $n \sim X$ and $m = nL + O(dL(1+d/X)\mathcal{Q}^\varepsilon)$. The other ranges of summation are negligible as can be seen by Lemma 2.1.

5 Proof of Theorem 1.1

Theorem 1.1 will follow after an appropriate treatment of the shifted convolution sums $\Sigma_d(L;c)$ in Lemma 4.1. We break this apart into cases according to the value of L.

5.1 Treatment of the shifted sums $\Sigma_d(1;c)$

Since we are dealing with forms of level P prime, we only have two types of shifted convolution sums to consider, those with $L = P$ and those with $L = 1$. In the latter case, the moduli d must be of size at least PM by the congruence condition. Applying Lemma 2.1 and the bound $I_d(n,m) \ll X\min\{1, X/d\}$ obtained from Lemma 2.2 one has that

$$\Sigma_d(1;c) \ll \mathcal{Q}^\varepsilon \frac{X^2}{d} \sum_{\substack{n\sim X \\ m=n+O(\frac{d^2}{X}\mathcal{Q}^\varepsilon) \\ m\equiv n(c)}}\sum 1 \ll \frac{X^3}{d}\left(1+\frac{d^2}{Xc}\right)\mathcal{Q}^\varepsilon,$$

so that this contribution to bounding $R_{f,D}(X)$ is

$$\sum_{\substack{d>0 \\ d\equiv 0(PM)}} \frac{\eta_D(d)}{d} \sum_{bc=d} \frac{1}{b} |\Sigma_d(1;c)| \ll_\varepsilon \frac{X^2}{PM}\mathcal{Q}^\varepsilon \ll_\varepsilon X\mathcal{Q}^\varepsilon, \tag{5.1}$$

which matches the first term in (3.3).

5.2 Treatment of the zero shift in $\Sigma_d(P;c)$

We now examine the case of $L = P$ and the contribution of the sums

$$\frac{1}{\sqrt{P}} \sum_{\substack{d>0\\(d,P)=1\\d\equiv 0(M)}} \frac{\eta_D(d)}{d} \sum_{bc=d} \frac{1}{b} |\Sigma_d(P;c)|$$

to $R_{f,D}(X)$. We first treat the "zero shift" in the shifted sums $\Sigma_d(P;c)$, i.e. when $m = nP$. One has

$$\frac{1}{\sqrt{P}} \sum_{\substack{d>0\\(d,P)=1\\d\equiv 0(M)}} \frac{\eta_D(d)}{d} \sum_{bc=d} \frac{1}{b} \Big|\sum_n \lambda_f(n)\lambda_{f^*}(nP) I_d(n,nP)\Big| \ll_\varepsilon \frac{X^2}{PM}\mathcal{Q}^\varepsilon \ll_\varepsilon X\mathcal{Q}^\varepsilon \tag{5.2}$$

by using the fact that $|\lambda_{f^*}(nP)| = |\lambda_{f^*}(n)\lambda_{f^*}(P)| = |\lambda_{f^*}(n)|P^{-1/2}$ (using (1.2)) and again the bound $I_d(n,nP) \ll X \min\{1, X/d\}$. This also matches the first term in (3.3). In fact, for the same reasons, one may also show that

$$\frac{1}{\sqrt{P}} \sum_{\substack{d>0\\(d,P)=1\\d\equiv 0(M)}} \frac{\eta_D(d)}{d} \sum_{bc=d} \frac{1}{b} \Big|\sum_n \sum_{\substack{m\equiv nP(c)\\m\equiv 0(P)}} \lambda_f(n)\lambda_{f^*}(m) I_d(n,m)\Big| \ll_\varepsilon X\mathcal{Q}^\varepsilon.$$

However, we will not use this fact in what follows.

5.3 Treatment of the non-zero shifts in $\Sigma_d(P;c)$

Finally, we are left with the non-zero shifts

$$\sum_n \sum_{\substack{m\equiv nP(c)\\m\neq nP}} \lambda_f(n)\lambda_{f^*}(m) I_d(n,m).$$

By Lemma 2.1, we need only consider those $m \equiv nP(c)$ with $n \sim X$ and $m = nP + O(dP(1 + d/X)\mathcal{Q}^\varepsilon)$. Therefore, the congruence in the inner sums may be rewritten as an equation

$$\sum_{0\neq|r|\ll\frac{dP}{c}(1+\frac{d}{X})\mathcal{Q}^\varepsilon} \sum\sum_{m=nP+cr} \lambda_f(n)\lambda_{f^*}(m) I_d(n,m). \tag{5.3}$$

We proceed by taking a smooth partition of unity for the sum over m writing

$$I_d(n,m) =: X \sum_Y F\left(\frac{n}{X}, \frac{m}{Y}\right)$$

where Y runs over values 2^v with $v = -1, 0, 1, \ldots$ such that $m = nP + cr$ is soluble when $m \sim Y$ and F is supported on $[1/2, 5/2] \times [1/2, 5/2]$. Furthermore, by Lemma 2.2 and the support of F, one has that

$$x^i y^j \frac{\partial^i}{\partial x^i} \frac{\partial^j}{\partial y^j} F\left(\frac{x}{X}, \frac{y}{Y}\right)$$
$$\ll \frac{X/d}{(1+X/d)^{3/2}} \frac{\sqrt{XY/(d^2P)}}{(1+\sqrt{XY/(d^2P)})^{3/2}} \left(1+\frac{X}{d}\right)^i \left(1+\frac{\sqrt{XY}}{d\sqrt{P}}\right)^j \tag{5.4}$$

for any non-negative integers i and j. Therefore, we may split apart the sums in (5.3) as

$$X \sum_{Y} \sum_{0 \neq |r| \ll \frac{dP}{c}(1+\frac{d}{X})Q^{\varepsilon}} \sum\sum_{m=nP+cr} \lambda_f(n)\lambda_{f^*}(m) F\left(\frac{n}{X}, \frac{m}{Y}\right)$$

which is bounded by

$$X^2 \sum_{Y} \frac{X/d}{(1+X/d)^{3/2}} \frac{\sqrt{XY/(d^2P)}}{(1+\sqrt{XY/(d^2P)})^{3/2}} \frac{dP(1+d/X)}{c} Q^{\varepsilon} \tag{5.5}$$

through an application of (5.4) with $i = j = 0$. For general X and d, this may be bounded by

$$X^2 \frac{dP}{c} Q^{\varepsilon}.$$

However, in the case of $d \ll XQ^{-\delta}$, one has that $m = nP + cr$ is soluble only when $Y \sim XP$ so that (5.5) then satisfies the stronger bound

$$X^2 \frac{X/d}{(1+X/d)^2} \frac{dP}{c} Q^{\varepsilon} \ll X \frac{d^2 P}{c} Q^{\varepsilon}.$$

Therefore, one has the following Lemma.

Lemma 5.1 *Let $\delta > 0$. For any $M \leqslant D \leqslant XQ^{2\delta}$ we have*

$$R_{f,D}(X) \ll XP^{3/2} Q^{\varepsilon}. \tag{5.6}$$

Furthermore, if $D \ll XQ^{-\delta}$ then

$$R_{f,D}(X) \ll XPQ^{\varepsilon} \left(\frac{D}{\sqrt{PM}}\right). \tag{5.7}$$

Since the bound for $R_{f,D}(X)$ in (5.7) is better than the convexity bound in (1.3) when $D < \sqrt{P}M\mathcal{Q}^{-\delta}$, we may restrict now to the case of $\sqrt{P}M\mathcal{Q}^{-\delta} \leqslant D \leqslant X\mathcal{Q}^{2\delta}$. The remaining task is to show that one can improve on the bound (5.6) by more than $\sqrt{P}$ when D is of that size.

For such values of D, an application of Theorem 1.3 to the shifted convolution sums

$$S_{X,Y}(cr) := \sum_{m=nP+cr}\sum \lambda_f(n)\lambda_{f^*}(m)F\left(\frac{n}{X},\frac{m}{Y}\right)$$

gives

$$S_{X,Y}(cr) \ll \mathcal{Q}^{\varepsilon} P \max\{XP,Y\}^{3/4} Z\sqrt{Z_x Z_y}\max\{Z_x,Z_y\}^{5/4},$$

where

$$Z = \frac{X/d}{(1+X/d)^{3/2}}\frac{\sqrt{XY/(d^2P)}}{(1+\sqrt{XY/(d^2P)})^{3/2}},$$

$$Z_x = \left(1+\frac{X}{d}\right) \quad \text{and}$$

$$Z_y = \left(1+\frac{\sqrt{XY}}{d\sqrt{P}}\right).$$

Hence the contribution of these non-zero shifts to Lemma 4.1 is bounded by

$$\mathcal{Q}^{\varepsilon}P^{3/2}\sum_{\substack{d>0\\(d,P)=1\\d\equiv 0(M)}}\eta_D(d)\left(1+\frac{X}{d}\right)\sum_{Y}\max\{XP,Y\}^{3/4}Z\sqrt{Z_xZ_y}\max\{Z_x,Z_y\}^{5/4}. \tag{5.8}$$

First consider $\sqrt{P}M\mathcal{Q}^{-\delta} \leqslant D < X$. In this case, we have that $Y \ll XP\mathcal{Q}^{\varepsilon}$ and (5.8) reduces to

$$\mathcal{Q}^{\varepsilon}XP^{3/2}(XP)^{3/4}\sum_{\substack{d>0\\(d,P)=1\\d\equiv 0(M)}}\frac{\eta_D(d)}{d}\left(\frac{X}{d}\right)^{5/4} \ll \mathcal{Q}^{\varepsilon}(XP)\frac{X^2P^{\frac{5}{4}}}{MD^{\frac{5}{4}}}$$

$$\ll \mathcal{Q}^{\frac{5}{4}\delta+\varepsilon}(XP)\frac{X^2P^{\frac{5}{8}}}{M^{\frac{9}{4}}} \ll \mathcal{Q}^{\frac{5}{4}\delta+\varepsilon}(XP)\frac{P^{\frac{21}{8}}}{M^{\frac{1}{4}}}. \tag{5.9}$$

Next consider $X \leqslant D \leqslant XQ^{2\delta}$. In this case, we have that $Y \ll D^2PQ^{\varepsilon}/X$ and (5.8) reduces to

$$Q^{\varepsilon}XP^{3/2}\left(\frac{D^2P}{X}\right)^{3/4}\sum_{\substack{d>0\\(d,P)=1\\d\equiv 0(M)}}\frac{\eta_D(d)}{d} \ll Q^{\varepsilon}(XP)\frac{D^{\frac{3}{2}}P^{\frac{5}{4}}}{MX^{\frac{3}{4}}}$$

$$\ll Q^{3\delta+\varepsilon}(XP)\frac{X^{\frac{3}{4}}P^{\frac{5}{4}}}{M} \ll Q^{3\delta+\varepsilon}(XP)\frac{P^2}{M^{\frac{1}{4}}}. \quad (5.10)$$

Combining (5.9) and (5.10) with (5.1) and (5.2) in §5.1, §5.2 and inserting these bounds into Lemma 3.1, completes the proof of Theorem 1.1.

6 Proof of Theorem 1.3

Let $X, Y \geqslant 1$ and let F be a smooth function supported on $[1/2, 5/2] \times [1/2, 5/2]$ with partial derivatives bounded by

$$x^i y^j \frac{\partial^i}{\partial x^i}\frac{\partial^j}{\partial y^j}F\left(\frac{x}{X},\frac{y}{Y}\right) \ll Z{Z_x}^i{Z_y}^j \quad (6.1)$$

for some $Z > 0$ and $Z_x, Z_y \geqslant 1$. Let P be a prime, and let k be a fixed positive even integer. For any $f_1, f_2 \in \mathcal{H}_k^*(P)$ we consider the shifted convolution sums

$$S_{X,Y}(\ell) := \sum\sum_{m=nP+\ell} \lambda_{f_1}(n)\lambda_{f_2}(m)F\left(\frac{n}{X},\frac{m}{Y}\right) \quad (6.2)$$

with ℓ a fixed non-zero integer satisfying $|\ell| \leqslant 10(XP+Y)$ such that the sum is non-trivial. Detecting the equation $m = nP + \ell$ in (6.2) through an application of the δ-method gives

$$S_{X,Y}(\ell) = \frac{1}{Q^2}\sum_{q=1}^{\infty}\sum_{a(q)}^{\star} e\left(\frac{a\ell}{q}\right)\sum_n \lambda_{f_1}(n)e\left(\frac{anP}{q}\right)$$

$$\times \sum_m \lambda_{f_2}(m)e\left(\frac{-am}{q}\right)F\left(\frac{n}{X},\frac{m}{Y}\right)h\left(\frac{q}{Q},\frac{nP+\ell-m}{Q^2}\right) \quad (6.3)$$

up to a negligible error term with the function h as in Lemma 2.7. As mentioned in §2.1 one expects to take Q to be roughly of size $\max\{\sqrt{XP}, \sqrt{Y}\}$.

Remark Consider the case of $X \sim PM$ and $Y \sim P^2M$. Such is the situation in our subconvexity application if one initially takes $X \sim Q^{1/2}$

and Kloosterman sum moduli of size $D \sim PM$ in order to focus on the transition range of the Bessel function. Taking moduli q of size up to $Q = P\sqrt{M}$ may therefore be regarded as a reduction of size M to the conductor of the n and m sums. One then returns to Kloosterman sums, of moduli q rather than d, by further applications of Voronoi summation.

6.1 Voronoi summation in m

We are now set to treat $S_{X,Y}(\ell)$ in the form seen in display (6.3). Since we will be applying Voronoi summation to our sums in n and m, the resulting sums will depend on the divisibility of the moduli q by powers of P. Indeed, an application of Voronoi summation to the m-sum gives, up to a constant factor,

$$\frac{1}{q\sqrt{P_q}}\sum_m \lambda_{f_2^*}(m)e\left(\frac{\overline{aP_q}m}{q}\right)\int_0^\infty F\left(\frac{n}{X},\frac{y}{Y}\right)h\left(\frac{q}{Q},\frac{nP+\ell-y}{Q^2}\right)\times J_{k-1}\left(\frac{4\pi\sqrt{my}}{q\sqrt{P_q}}\right)dy$$

where $P_q = P/(P,q)$. Therefore, (6.3) reduces to

$$\frac{1}{Q^2}\sum_{q=1}^\infty \frac{1}{q\sqrt{P_q}}\sum_n\sum_m \lambda_{f_1}(n)\lambda_{f_2^*}(m)S(\ell+nP,m\overline{P_q};q) \times\int_0^\infty F\left(\frac{n}{X},\frac{y}{Y}\right)h\left(\frac{q}{Q},\frac{nP+\ell-y}{Q^2}\right)J_{k-1}\left(\frac{4\pi\sqrt{my}}{q\sqrt{P_q}}\right)dy. \tag{6.4}$$

Although we have gained the Kloosterman sum structure, an application of the Weil bound here would still be insufficient for our goal.

6.2 Voronoi summation in n

Define

$$J_\alpha(n,m;q) := J\left(\frac{\sqrt{nP^\alpha}}{q\sqrt{P_q}},\frac{\sqrt{m}}{q\sqrt{P_q}}\right) \tag{6.5}$$

where $P^\alpha = (q,P^2)$, $P_q = P/(P,q)$ and $J\left(\frac{\sqrt{nP^\alpha}}{q\sqrt{P_q}},\frac{\sqrt{m}}{q\sqrt{P_q}}\right)$ is the function in Lemma 2.3. Opening the Kloosterman sum in (6.4) and applying Voronoi summation to the n-sum gives, up to a constant factor,

$$\frac{1}{Q^2}\sum_q \frac{\sqrt{P^\alpha}}{q^2P_q}\sum_n\sum_m \lambda_{f_1^*}(n)\lambda_{f_2^*}(m)S_\alpha(n,m,\ell;q)J_\alpha(n,m;q) \tag{6.6}$$

where

$$S_\alpha(n,m,\ell;q) = \begin{cases} S(\ell,(mP-n)\overline{P^2};q) & \text{if } \alpha=0,\\ S(\ell\overline{P},(m-n)\overline{P};q/P)S(\ell\overline{q/P},m\overline{q/P};P) & \text{if } \alpha=1,\\ S(\ell,m-nP;q) & \text{if } \alpha=2. \end{cases}$$

6.3 Application of Weil bound

We now break apart the sums in (6.6) according to the size of q. First, we note that the bound (2.4) in Lemma 2.3 allows one to truncate the n and m sums to be of size

$$n \leqslant T_1 := \frac{q^2P_q}{P^\alpha X}\left(Z_x + \frac{XP}{qQ}\right)^2 (XYP)^\varepsilon$$

$$\text{and} \quad m \leqslant T_2 := \frac{q^2P_q}{Y}\left(Z_y + \frac{Y}{qQ}\right)^2 (XYP)^\varepsilon. \tag{6.7}$$

When the parameters are such that either $T_1 < 1$ or $T_2 < 1$ in (6.7), then one has arbitrary saving in these situations. Otherwise, we apply the bound (2.5) from Lemma 2.3 to $J_\alpha(n,m;q)$ and the Weil bound for Kloosterman sums in order to bound (6.6) by

$$\frac{ZXY}{Q}\sum_{q\leqslant Q}\frac{\sqrt{P^\alpha}}{q^3P_q}\sum_{n\leqslant T_1}\sum_{m\leqslant T_2}(\ell,q)^{1/2}q^{1/2}\left(\frac{q^2P_q}{\sqrt{nmP^\alpha XY}}\right)^{3/2}\times$$
$$\frac{\min\{Z_y\sqrt{nP^\alpha X}, Z_x\sqrt{mY}\}}{q\sqrt{P_q}}(XYP)^\varepsilon$$

which is bounded by

$$\frac{Z(XY)^{1/4}}{Q}\sum_{\delta|\ell}\delta^{1/2}\sum_{\substack{q\leqslant Q\\(q,\ell)=\delta}}\frac{(T_1T_2)^{1/4}}{q^{1/2}P^{\alpha/4}}\min\{Z_y\sqrt{T_1P^\alpha X}, Z_x\sqrt{T_2Y}\}(XYP)^\varepsilon.$$

Bounding the minimum by the geometric mean, and using (6.7), we get the bound

$$\frac{ZP}{Q}\sqrt{Z_xZ_y}\sum_{\delta|\ell}\delta^2\sum_{\substack{q\leqslant \frac{Q}{\delta}\\(q,\ell)=1}}q^{3/2}\left(Z_x+\frac{XP}{q\delta Q}\right)\left(Z_y+\frac{Y}{q\delta Q}\right)(XYP)^\varepsilon,$$

which is dominated by

$$ZP\sqrt{Z_xZ_y}Q^{3/2}\left(Z_x+\frac{XP}{Q^2}\right)\left(Z_y+\frac{Y}{Q^2}\right)(XYP)^\varepsilon.$$

We bound the last expression by

$$ZP\sqrt{Z_xZ_y}Q^{3/2}\left(\max\{Z_x, Z_y\} + \frac{\max\{XP, Y\}}{Q^2}\right)^2 (XYP)^\varepsilon \tag{6.8}$$

Choosing $Q = \left(\frac{\max\{XP,Y\}}{\max\{Z_x,Z_y\}}\right)^{1/2}$ in (6.8) produces the final bound

$$S_{X,Y}(\ell) \ll (XYP)^\varepsilon ZP\sqrt{Z_xZ_y}\max\{XP, Y\}^{3/4}\max\{Z_x, Z_y\}^{5/4}.$$

References

[1] Valentin Blomer, *Shifted convolution sums and subconvexity bounds for automorphic L-functions*, Int. Math. Res. Not. **(73)** (2004), 3905–3926.

[2] Valentin Blomer and Gergely Harcos, *The spectral decomposition of shifted convolution sums*, Duke Math. J. **144** (2) (2008), 321–339.

[3] Pierre Deligne, *La conjecture de Weil. I*, Inst. Hautes Études Sci. Publ. Math. **(43)** (1974), 273–307.

[4] J.-M. Deshouillers and H. Iwaniec, *Kloosterman sums and Fourier coefficients of cusp forms*, Invent. Math. **70** (2) (1982/83), 219–288.

[5] W. Duke, J. Friedlander, and H. Iwaniec, *Bounds for automorphic L-functions*, Invent. Math. **112** (1) (1993), 1–8.

[6] W. Duke, J. B. Friedlander, and H. Iwaniec, *Bounds for automorphic L-functions II*, Invent. Math. **115** (2) (1994), 219–239.

[7] W. Duke, J. Friedlander, and H. Iwaniec, *Erratum: "Bounds for automorphic L-functions. II"* [Invent. Math. **115** (1994), no. 2, 219–239; MR1258904 (95a:11044)]. Invent. Math. **140** (1) (2000), 227–242.

[8] W. Duke, J. B. Friedlander, and H. Iwaniec, *A quadratic divisor problem*, Invent. Math. **115** (2) (1994), 209–217.

[9] Brooke Feigon and David Whitehouse, *Averages of central L-values of Hilbert modular forms with an application to subconvexity*, Duke Math. J. **149** (2) (2009), 347–410.

[10] Dorian Goldfeld, *Analytic and arithmetic theory of Poincaré series*, In *Journées Arithmétiques de Luminy (Colloq. Internat. CNRS, Centre Univ. Luminy, Luminy, 1978)*, volume 61 of Astérisque, pg. 95–107. Soc. Math. France, Paris, 1979.

[11] Gergely Harcos, *An additive problem in the Fourier coefficients of cusp forms*, Math. Ann. **326** (2) (2003), 347–365.

[12] Gergely Harcos and Philippe Michel, *The subconvexity problem for Rankin-Selberg L-functions and equidistribution of Heegner points. II*, Invent. Math. **163** (3) (2006), 581–655.

[13] D. R. Heath-Brown, *A new form of the circle method, and its application to quadratic forms*, J. Reine Angew. Math. **481** (1996), 149–206.

[14] D. R. Heath-Brown, *Convexity bounds for L-functions*, Acta Arith. **136** (4) (2009), 391–395.

[15] Roman Holowinsky, *A sieve method for shifted convolution sums*, Duke Math. J. **146** (3) (2009), 401–448.

[16] Roman Holowinsky, *Sieving for mass equidistribution*, Ann. of Math. (2) **172** (2) (2010), 1499–1516.

[17] H. Iwaniec and E. Kowalski, *Analytic number theory*, American Mathematical Society Colloquium Publications **53** American Mathematical Society, Providence, RI, 2004.

[18] Henryk Iwaniec, Wenzhi Luo, and Peter Sarnak, *Low lying zeros of families of L-functions*, Inst. Hautes Études Sci. Publ. Math. **91** (2000), 55–131.

[19] H. Iwaniec and P. Michel, *The second moment of the symmetric square L-functions*, Ann. Acad. Sci. Fenn. Math. **26** (2) (2001), 465–482.

[20] Matti Jutila, *The additive divisor problem and its analogs for Fourier coefficients of cusp forms. I*, Math. Z. **223** (3) (1996), 435–461.

[21] Matti Jutila, *The additive divisor problem and its analogs for Fourier coefficients of cusp forms. II*, Math. Z. **225** (4) (1997), 625–637.

[22] E. Kowalski, P. Michel, and J. VanderKam, *Rankin-Selberg L-functions in the level aspect*, Duke Math. J. **114** (1) (2002), 123–191.

[23] Wenzhi Luo and Peter Sarnak, *Mass equidistribution for Hecke eigenforms*, Comm. Pure Appl. Math. **56** (7) (2003), 874–891. Dedicated to the memory of Jürgen K. Moser.

[24] Philippe Michel, *The subconvexity problem for Rankin-Selberg L-functions and equidistribution of Heegner points*, Ann. of Math. (2) **160** (1) (2004), 185–236.

[25] P. Michel and D. Ramakrishnan, *Consequences of the Gross/Zagier formulae: stability of average L-values, subconvexity, and non-vanishing mod p*, http://arxiv.org/abs/0709.4668v1, 2007.

[26] Paul Nelson, *Stable averages of central values of Rankin-Selberg L-functions: Some new variants*, http://arxiv.org/abs/1202.6313v1, 2010.

[27] Nigel Pitt, *On an analogue of Titchmarsh's divisor problem for holomorphic cusp forms*, submitted, 2011.

[28] G. Ricotta, *Real zeros and size of Rankin-Selberg L-functions in the level aspect*, Duke Math. J. **131** (2) (2006), 291–350.

[29] Peter Sarnak, *Estimates for Rankin-Selberg L-functions and quantum unique ergodicity*, J. Funct. Anal. **184** (2) (2001), 419–453.

[30] Kannan Soundararajan, *Weak subconvexity for central values of L-functions*, Ann. of Math. (2), **172** (2) (2010), 1469–1498.

[31] G. N. Watson, *A treatise on the theory of Bessel functions*, Cambridge Mathematical Library. Cambridge University Press, Cambridge, 1995. Reprint of the second (1944) edition.

Roman Holowinsky, Department of Mathematics, The Ohio State University, 100 Math Tower, 231 West 18th Avenue, Columbus, OH 43210-1174, USA.
E-mail: holowinsky.1@osu.edu

Ritabrata Munshi, School of Mathematics, Tata Institute of Fundamental Research, 1 Dr. Homi Bhabha Road, Mumbai 400005, India.
E-mail: rmunshi@math.tifr.res.in

Automorphic Representations and L-Functions
Editors: D. Prasad, C.S. Rajan, A. Sankaranarayanan, J. Sengupta

Publisher: Hindustan Book Agency, New Delhi, India

On the Harish-Chandra Schwartz Space of $G(F)\backslash G(\mathbb{A})$

Erez Lapid[1]
With an appendix by Farrell Brumley

Abstract

We study the Harish-Chandra Schwartz space of an adelic quotient $G(F)\backslash G(\mathbb{A})$. We state a conjectural spectral decomposition of it in terms of parabolic induction. We verify a cuspidal version of this conjecture under additional hypotheses on the group G, which are known to be satisfied for $G = \mathrm{GL}_n$.

1 Introduction

In the harmonic analysis of reductive groups over a local field, an important companion of the Plancherel formula is the Paley-Wiener Theorem for the Harish-Chandra Schwartz space (cf. [Art75, Ber88, Wal03]). In this paper we will be interested in the Harish-Chandra Schwartz space of an adelic quotient. It was defined in this context by Bernstein [Ber88] (see also §2 below). We will state a conjectural Paley-Wiener Theorem for this space. Unfortunately, in the global setup, we can formulate (let alone prove) a reasonable statement only assuming a conjectural nontrivial analytic condition on the intertwining operators, which is known completely only for the groups GL_n. This condition has both a global aspect, (UT), and a local aspect, (WR). The property (UT) (uniform temperedness) is pertaining to the global normalization factor of the intertwining operator while the property (WR) (weak Ramanujan) is pertaining to the poles of the local normalized intertwining operators. They are explicated in §3. Assuming these properties, we prove a *cuspidal* version of the Paley-Wiener Theorem (Theorem 4.6). The main step is a majorization of cuspidal Eisenstein series which is uniform in the spectral and the group variable (Proposition 5.1).

[1]Partially supported by the Israel Science Foundation Center of Excellence grant 1691/10

The property (WR) had been considered earlier (in an equivalent form) in the work of Werner Müller on the spectral side of Arthur's trace formula. Specifically, it was shown in [Mül02] that (WR) implies the absolute convergence of the spectral side of Arthur's trace formula. For GL_n, (WR) directly follows from the results of Luo-Rudnick-Sarnak [LRS99] (see [MS04]). Subsequently, a different approach for the absolute convergence of the spectral side of the trace formula which avoids (WR) was given in [FLM11, FL11]. On the other hand, for the analysis of the Harish-Chandra Schwartz space it seems that the properties (UT) and (WR), at least in some form, are indispensable. We also mention that both (UT) and (WR) were encountered in the analytic study of Jacquet's relative trace formula in [Lap06], where a majorization of cuspidal Eisenstein series (in a slightly weaker form) was also considered.

We mention that in a different context, an approach for the (conjectural) Paley-Wiener Theorem for the (much smaller) space of rapidly decreasing functions was set forth by Casselman – cf. [Cas04, Cas89]. The analysis of the two problems turns out to be quite different. However, they both have a common goal, namely to study the cohomology (in various contexts) of arithmetic groups (see [Cas84] for such an approach). We will not go into any further detail here but refer the reader to [Fra98, FS98] for a related theme.

The paper is organized as follows. After introducing notation we recall the definition and the basic properties of the Harish-Chandra Schwartz space (§2). We also recall some standard facts from reduction theory. Then, in §3 we discuss the analytic properties which are required for the main result. We prove them for GL_n and conjecture that they hold (even in a stronger form) in general. In §4 we give the conjectural Paley-Wiener statement in this context. We also state our main result, Theorem 4.6, which is the cuspidal version of this conjecture, conditional on the analytic properties discussed above. The technical heart of the proof is the majorization of the Eisenstein series and their derivatives (in both the group and the spectral variables) near the imaginary axis. This is carried out in §5 using the Maass-Selberg relations. Finally, we prove the main result in §6 by a simple induction, together with the standard principle of approximation by the constant term.

I would like to thank MPI, Bonn for generous hospitality during the summer of 2011. I am indebted to Bill Casselman for useful discussions which spurred me to write this note and for drawing my attention to [Cas84] and [Fra98]. I also thank Joseph Bernstein and Patrick Delorme for their interest in this work. I am very grateful to Farrell Brumley for providing an appendix, sharpening his results on lower bounds of Rankin-Selberg L-functions at the edge of the critical strip. I thank Jean-Pierre Labesse who drew my attention to the preprint [LW13]. Finally, I would like to thank Guy Henniart, Xiannan Li, Phillip Michel and Peter Sarnak for helpful correspondence.

1.1 Notation

Let G be a reductive group over a number field F. We fix throughout a maximal F-split torus T_0. Any F-parabolic subgroup P of G containing T_0 admits a unique Levi decomposition $P = M \ltimes U$ with Levi part M containing T_0. We refer to the M's arising this way simply as the Levi subgroups of G containing T_0 and denote this set by $\mathcal{L}$.

For any algebraic group Y over F we write $\mathfrak{a}_Y^* = X^*(Y) \otimes \mathbb{R}$ where $X^*(Y)$ is the lattice of F-rational characters of Y; let $\mathfrak{a}_Y$ be the dual space of $\mathfrak{a}_Y^*$. We also set

$$Y(\mathbb{A})^1 = \bigcap_{\chi \in X^*(Y)} \operatorname{Ker} |\chi|_{\mathbb{A}^*}$$

where we extend any $\chi \in X^*(Y)$ to a homomorphism $\chi : Y(\mathbb{A}) \to \mathbb{A}^*$ and take the standard norm on the ideles. We have a surjective homomorphism

$$H_Y : Y(\mathbb{A}) \to \mathfrak{a}_Y$$

given by $e^{\langle \chi, H_Y(y) \rangle} = |\chi(y)|$, $\chi \in X^*(Y)$, $y \in Y(\mathbb{A})$. The kernel of H_Y is $Y(\mathbb{A})^1$. Denote by $r(Y)$ the dimension of $\mathfrak{a}_Y$. Finally, we write δ_Y for the modulus function of $Y(\mathbb{A})$.

Fix a minimal F-parabolic subgroup $P_0 = M_0 \ltimes U_0$ containing T_0. Its Levi part M_0 is the centralizer of T_0. We set $\delta_0 = \delta_{P_0}$.

For any $M \in \mathcal{L}$, the set of Levi subgroups containing M will be denoted by $\mathcal{L}(M)$ and the set of parabolic subgroups whose Levi part equals (resp. contains) M will be denoted by $\mathcal{P}(M)$ (resp. $\mathcal{F}(M)$). (For simplicity we write $\mathcal{F} = \mathcal{F}(M_0)$ for the set of semistandard parabolic subgroups.) The parabolic subgroup opposite to P containing M is denoted by $\overline{P}$. We have canonically $\mathfrak{a}_P = \mathfrak{a}_M = \mathfrak{a}_{T_M}$ for any $P \in \mathcal{P}(M)$ where T_M is the split part of the center of M. We write $R(T_M, G)$ for the set of reduced roots of T_M on $\operatorname{Lie} G$. For any $P = M \ltimes U \in \mathcal{P}(M)$ we write $\Sigma_P = R(T_M, U) \subseteq \mathfrak{a}_M^*$ for the set of reduced roots of T_M on $\operatorname{Lie} U$ and Δ_P for the subset of simple roots. Similarly $\Delta_P^\vee \subseteq \mathfrak{a}_M$ will denote the set of simple co-roots.

We set $\mathfrak{a}_0 = \mathfrak{a}_{P_0} = \mathfrak{a}_{M_0} = \mathfrak{a}_{T_0}$ and for any $M \in \mathcal{L}$ we view $\mathfrak{a}_M$ canonically as a subspace of $\mathfrak{a}_0$ with a canonical projection $\mathfrak{a}_0 \to \mathfrak{a}_M$. Similarly for $\mathfrak{a}_M^*$. We denote by Δ_0 the simple roots of T_0 on $\operatorname{Lie}(U_0)$. We endow $\mathfrak{a}_0$ with a W-invariant inner product where $W = W^G = N_{G(F)}(T_0)/M_0$ is the Weyl group of G. We write the corresponding norms on $\mathfrak{a}_0$ and $\mathfrak{a}_0^*$ by $\|\cdot\|$. We write $\mathfrak{a}_{0,+}$ for the positive Weyl chamber of $\mathfrak{a}_0$, i.e.

$$\mathfrak{a}_{0,+} = \{X \in \mathfrak{a}_0 : \langle \alpha, X \rangle > 0 \text{ for all } \alpha \in \Delta_0\}.$$

Note that $\mathfrak{a}_{0,+}$ is invariant under translation by $\mathfrak{a}_G$.

For any $M, L \in \mathcal{L}$ we write $W(M, L)$ for the set of right W^M-cosets of elements of W such that $wMw^{-1} = L$.

Fix a maximal compact subgroup $\mathbf{K}$ of $G(\mathbb{A})$ in good position with respect to P_0. Let $\mathbf{K}_\infty = \mathbf{K} \cap G(F_\infty)$ where $F_\infty = F \otimes_{\mathbb{Q}} \mathbb{R}$.

For any $P = M \ltimes U \in \mathcal{F}$ the homomorphism $H_M : M(\mathbb{A}) \to \mathfrak{a}_M$ extends to a left-$U(\mathbb{A})$ right-$\mathbf{K}$-invariant map $H_P : G(\mathbb{A}) \to \mathfrak{a}_M$. In particular, we write $H = H_{P_0}$. Note that if P is standard then H_P is the composition of H with the projection $\mathfrak{a}_0 \to \mathfrak{a}_M$.

Let A_0 be the identity component (in the usual Hausdorff topology) of $T_0(\mathbb{R}) \subseteq T_0(\mathbb{A})$ where we view $\mathbb{R} \hookrightarrow \mathbb{A}$ via $\mathbb{R} \hookrightarrow \mathbb{A}_{\mathbb{Q}} \hookrightarrow \mathbb{A} = \mathbb{A}_{\mathbb{Q}} \otimes_{\mathbb{Q}} F$. Then H restricts to an isomorphism $H : A_0 \to \mathfrak{a}_0$. We write $X \mapsto e^X$ for the inverse isomorphism. Similarly for any $M \in \mathcal{L}$ let $A_M = A_0 \cap T_M$, so that $H_M : A_M \to \mathfrak{a}_M$ is an isomorphism.

We will use the following notational convention. Suppose that X and Y are certain quantities (depending on parameters). We write $X \ll Y$ if there exists a constant $c > 0$ such that $|X| \le cY$. If the constant c depends on additional data, say D, we write $X \ll_D Y$.

For instance, for any $g, x \in G(\mathbb{A})$ we have

$$\|H(gx) - H(g)\| \le \sup_{k \in \mathbf{K}} \|H(kx)\|.$$

Therefore, for any compact subset $C \subseteq G(\mathbb{A})$ we have

$$\|H(gx) - H(g)\| \ll_C 1 \text{ for all } g \in G(\mathbb{A}),\ x \in C. \tag{1.1}$$

Fix $T_1 \in \mathfrak{a}_0$ throughout such that $\langle \alpha, T_1 \rangle$ is sufficiently small for all $\alpha \in \Delta_0$. Let

$$\begin{aligned}\mathfrak{s} &= \{g \in G(\mathbb{A}) : H(g) \in T_1 + \mathfrak{a}_{0,+}\} \\ &= \{pe^X k : p \in P_0(\mathbb{A})^1, X \in T_1 + \mathfrak{a}_{0,+}, k \in \mathbf{K}\}\end{aligned}$$

and $\mathfrak{s}^1 = \mathfrak{s} \cap G(\mathbb{A})^1$, so that $\mathfrak{s} = A_G \mathfrak{s}^1$. Clearly, $\mathfrak{s}$ is an open left $P_0(F)$-invariant subset of $G(\mathbb{A})$. By reduction theory we have $G(F)\mathfrak{s} = G(\mathbb{A})$ and $G(F)\mathfrak{s}^1 = G(\mathbb{A})^1$.

In particular, for any non-negative measurable function f on $G(F)\backslash G(\mathbb{A})$ we have

$$\begin{aligned}\int_{G(F)\backslash G(\mathbb{A})} f(g)\, dg &\le \int_{P_0(F)\backslash \mathfrak{s}} f(g)\, dg \\ &= \int_{P_0(F)\backslash P_0(\mathbb{A})^1} \int_{T_1 + \mathfrak{a}_{0,+}} \int_{\mathbf{K}} \delta_0(e^X)^{-1} f(pe^X k)\, dp\, dX\, dk. \end{aligned}\tag{1.2}$$

Another basic fact is that there exists a compact subset Ω of $G(\mathbb{A})$ (which can be taken in the form $\Omega_0\mathbf{K}$ where $\Omega_0 \subseteq P_0(\mathbb{A})^1$) such that

$$\mathfrak{s} \subseteq P_0(F)\{e^X : X \in T_1 + \mathfrak{a}_{0,+}\}\Omega. \tag{1.3}$$

We write $C(X)$ for the space of complex valued continuous functions on a topological space X. If X is a smooth manifold, we denote by $C^m(X)$, $m \in \mathbb{N}$ (resp., C^∞) the subspace of m-times continuously differentiable (resp., smooth) functions in $C(X)$. Similarly we use the notation $C_c(X)$, $C_c^m(X)$, $C_c^\infty(X)$ for the subspaces of compactly supported functions in $C(X)$, $C^m(X)$ and $C^\infty(X)$ respectively.

The space $G(\mathbb{A})$ is not a smooth manifold, but for any compact open subgroup $K \subseteq G(\mathbb{A}_{\text{fin}})$, $G(\mathbb{A})/K$ is a smooth manifold (namely, countably many copies of $G(F_\infty)$). We define $C^\infty(G(\mathbb{A}))$ to be the union over all compact open subgroups $K \subseteq G(\mathbb{A}_{\text{fin}})$ of $C^\infty(G(\mathbb{A})/K)$. Similarly for any closed subgroup H of $G(\mathbb{A})$ let $C^\infty(H\backslash G(\mathbb{A}))$ denote the space of smooth left H-invariant functions on $G(\mathbb{A})$ which are right K-invariant for some compact open subgroup K of $G(\mathbb{A}_{\text{fin}})$. For brevity we will refer to a compact open subgroup K of $G(\mathbb{A}_{\text{fin}})$ simply as a "level" of G.

We denote by $\mathfrak{g}$ the complexified Lie algebra of $G(F_\infty)$. Its universal algebra will be denoted by $\mathcal{U}(\mathfrak{g})$ and the center of the latter by $\mathfrak{z}$.

Let $(V, \|\cdot\|)$ be a finite-dimensional normed vector space over $\mathbb{R}$. Denote by $\mathcal{D}(V)$ the graded ring of translation invariant differential operators on V (isomorphic to the symmetric algebra of V). Occasionally, we will also view elements of $\mathcal{D}(V)$ as holomorphic differential operators on $V_\mathbb{C}$ with constant coefficients. Given $D \in \mathcal{D}(V)$ we sometimes write $D_v f(v)$ for $(Df)(v)$ in order to emphasize the variable of differentiation. Given a Fréchet space $\mathcal{F}$ we denote by $\mathcal{S}(V;\mathcal{F})$ the Fréchet space of smooth functions $f : V \to \mathcal{F}$ such that

$$\sup_{v\in V}(1 + \|v\|)^n \mu(D_v f(v)) < \infty$$

for any $n \in \mathbb{N}$, $D \in \mathcal{D}(V)$ and a continuous seminorm μ on $\mathcal{F}$.

For an inductive limit $U = \cup\mathcal{F}_n$ of Fréchet spaces we write $\mathcal{S}(V;U) = \cup\mathcal{S}(V;\mathcal{F}_n)$ with the inductive limit topology.

2 Schwartz Space of Automorphic Forms

In this section we recall the definition of the Schwartz space in this context. For a more flexible and elaborate setup we refer the reader to [Fra98].

Define the function Ξ on $\mathfrak{s}$ by

$$\Xi(pk) = \delta_{P_0}(p)^{\frac{1}{2}},\ p \in P_0(\mathbb{A}), H(p) \in T_1 + \mathfrak{a}_{0,+}, k \in \mathbf{K}.$$

This function plays the role of the Harish-Chandra standard spherical function (which is usually denoted by the same letter) in the local case (cf. [Wal03]).

For any $T \in \mathfrak{a}_0$ define

$$\mathfrak{s}_{>T} = \{g \in G(\mathbb{A}) : H(g) \in T + \mathfrak{a}_{0,+}\}.$$

Thus, $\mathfrak{s} = \mathfrak{s}_{>T_1}$. We will also set

$$\mathfrak{s}_D = \{g \in G(\mathbb{A}) : H(g) \in D\}$$

for any subset $D \subseteq \mathfrak{a}_0$.

The following Lemma is a standard result in reduction theory (see e.g. [LW13, §3.5] or [Fra98, §2.1]). For convenience we provide a proof.

Lemma 2.1 *For any $T_1' \in \mathfrak{a}_0$ we have*

$$\|H(x)\| \ll_{T_1'} 1 + \|H(\gamma x)\| \tag{2.1}$$

for any $x \in \mathfrak{s}_{>T_1'}$ and $\gamma \in G(F)$, and if moreover $\gamma x \in \mathfrak{s}_{>T_1'}$ then

$$\|H(\gamma x) - H(x)\| \ll_{T_1'} 1. \tag{2.2}$$

In addition,

$$\left|\{\gamma \in P_0(F)\backslash G(F) : \gamma x \in \mathfrak{s}_{>T_1'}\}\right| \ll_{T_1'} 1 \tag{2.3}$$

for any $x \in G(\mathbb{A})$. In particular, if $x \in \mathfrak{s}_{>T_1'}$ then

$$\mathfrak{s} \cap G(F)x \subseteq \mathfrak{s}_{H(x)+B} \tag{2.4}$$

where B is a bounded set of $\mathfrak{a}_0$ depending only on T_1'. If $H(x) \in \mathfrak{a}_{0,+}$ and sufficiently regular then in fact $G(F)x \cap \mathfrak{s} = P_0(F)x$.

Proof Suppose $x \in \mathfrak{s}_{>T_1'}$ and $\gamma \in G(F)$. Let $X = H(x)$ and $Y = H(\gamma x)$. By assumption $X \in T_1' + \mathfrak{a}_{0,+}$. Fix representatives $n_w \in N_G(T_0)$, $w \in W$. Multiplying γ on the left by an element of $P_0(F)$ and using the Bruhat decomposition, we may assume without loss of generality that $\gamma = n_w u$ where $w \in W$ and $u \in U_w^- = U_0 \cap w^{-1}\overline{U_0}w$. Let $x = ptk$ where $p \in P_0(\mathbb{A})^1$, $t \in A_0$ and $k \in \mathbf{K}$. Then $X = H(t)$ and

$$Y = wX + H(n_w u')$$

where $u' = t^{-1}p^{-1}upt \in U_0(\mathbb{A})$. It is well known (e.g., [LW13, Lemme 3.3.2]) that

$$H(n_w u') = \sum_{\beta \in \Phi_w} d_\beta \beta^\vee$$

where

$$\Phi_w = \{\beta \in \Phi_+ : w^{-1}\beta < 0\}$$

and $d_\beta \leq d$ where d is a constant. Thus,

$$w^{-1}H(n_w u') - H(n_w u') = \sum_{\beta \in \Phi_w} d_\beta (w^{-1}\beta^\vee - \beta^\vee).$$

We conclude that $w^{-1}H(n_w u') - H(n_w u')$ is a sum of simple co-roots with coefficients bounded from below and that

$$\|H(n_w u')\| \ll \sum |d_\beta| \ll 1 + \|w^{-1}H(n_w u') - H(n_w u')\|.$$

Since $X \in T_1' + \mathfrak{a}_{0,+}$, we have $wX - X = \sum_{\beta \in \Phi_w} c_\beta \beta^\vee$ with $c_\beta \leq c$ where c is a constant depending only on T_1'. From the relation

$$\begin{aligned} w^{-1}Y - Y &= X - wX - H(n_w u') + w^{-1}H(n_w u') \\ &= \sum_{\beta \in \Phi_w} (d_\beta w^{-1}\beta^\vee - (c_\beta + d_\beta)\beta^\vee) \end{aligned} \tag{2.5}$$

we conclude that

$$\|wX - X\| \ll \sum |c_\beta| \ll_{T_1'} 1 + \|w^{-1}Y - Y\|$$

and

$$\|H(n_w u')\| \ll 1 + \|w^{-1}H(n_w u') - H(n_w u')\| \ll_{T_1'} 1 + \|w^{-1}Y - Y\|.$$

Thus,

$$\|Y - X\| = \|wX - X + H(n_w u')\| \ll_{T_1'} 1 + \|w^{-1}Y - Y\| \leq 1 + 2\|Y\|$$

and hence,

$$\|X\| \leq \|Y - X\| + \|Y\| \ll_{T_1'} 1 + \|Y\|$$

which is the relation (2.1).

Moreover, if $\gamma x \in \mathfrak{s}_{>T_1'}$, that is, if $Y \in T_1' + \mathfrak{a}_{0,+}$, then we also have $w^{-1}Y - Y = \sum_{\beta \in \Phi_{w^{-1}}} c'_\beta \beta^\vee$ with $c'_\beta \leq c$. (Note that $w^{-1}\Phi_w = -\Phi_{w^{-1}}$.) The relation (2.5) immediately implies that $|c_\beta|$, $\left|c'_\beta\right|$, $|d_\beta|$, $\beta \in \Phi_w$ are bounded in terms of T_1'. Therefore, the same holds for $wX - X$, $H(n_w u')$ and $Y - X = wX - X + H(n_w u')$. This proves (2.2).

In addition, since $wX - X$ is bounded and $X \in T_1' + \mathfrak{a}_{0,+}$, we infer that X lies in a compact translate of $\mathfrak{a}_L$ where L is the smallest standard Levi subgroup containing n_w. Also, since $H(n_w u')$ is bounded, u' lies in a compact subset of $U_w^-(\mathbb{A})$. Write $t = t_1 t_2$ where $t_1 \in A_L$ and t_2 is in a compact set. Let $p' = t_1^{-1} p t_1 \in P_0(\mathbb{A})^1$ and write $p' = p_1 p_2$ where $p_1 \in P_0(F)$ and p_2 lies in a fixed compact set of $P_0(\mathbb{A})^1$. Since $L \supseteq U_w^-$, t_1 commutes with u and we have $p_2 t_2 u' t_2^{-1} p_2^{-1} = p_1^{-1} u p_1$. Therefore, $p_1^{-1} u p_1$ lies in a fixed compact set C of $U_0(\mathbb{A})$. However, $\left|\{v \in U_0(F) : p_1^{-1} v p_1 \in C\}\right| = |U_0(F) \cap C|$ is finite and bounded. This gives (2.3).

Finally, the last statement follows from the fact that if $wX - X$ is bounded and $X \in \mathfrak{a}_{0,+}$ is sufficiently regular then $w = 1$.

□

For $g \in G(\mathbb{A})$ let $\sigma(g) = 1 + \min_{x \in \mathfrak{s} \cap G(F)g} \|H(x)\|$. The following properties are clear.

1. σ is left $G(F)$-invariant.
2. If $x, x' \in \mathfrak{s}$ and $x' \in G(F)x$ then $1 + \|H(x')\| \ll 1 + \|H(x)\|$. (Lemma 2.1, or alternatively, [MW95, I.2.2].) Therefore,

$$1 + \max_{x \in \mathfrak{s} \cap G(F)g} \|H(x)\| \ll \sigma(g)$$

for all $g \in G(\mathbb{A})$ and

$$\sigma(x) \leq 1 + \|H(x)\| \ll \sigma(x)$$

for all $x \in \mathfrak{s}$.
3. $\log \Xi(x) \ll \sigma(x) \ll \max(1, \log \Xi(x))$ for any $x \in \mathfrak{s}^1$.
4. For any compact set $C \subseteq G(\mathbb{A})$ we have

$$\sigma(xy) \ll_C \sigma(x) \quad \text{for all } x \in G(\mathbb{A}), y \in C.$$

5. For $n \gg 1$ (in fact, for any $n > r/2$ where r is the F-rank of G) we have

$$\Xi\sigma^{-n} \text{ (or equivalently } \Xi(1 + \|H\|)^{-n}) \in L^2(P_0(F)\backslash\mathfrak{s}). \tag{2.6}$$

For any $f \in C_c(G(\mathbb{A}))$ let K_f be the automorphic kernel

$$K_f(x, y) = \sum_{\gamma \in G(F)} f(x^{-1}\gamma y), \quad x, y \in G(F)\backslash G(\mathbb{A}).$$

Similarly, let

$$\tilde{K}_f(x, y) = \int_{A_G} \sum_{\gamma \in G(F)} f(ax^{-1}\gamma y)\, da, \quad x, y \in A_G G(F)\backslash G(\mathbb{A}).$$

The following Lemma is also proved in [LW13, §12.2].

Lemma 2.2 *Let* $C \subseteq G(\mathbb{A})$ *be compact and* $n = 0, 1, 2, \ldots$. *Then*

1. $\mathrm{vol}(G(F)\backslash G(F)xC) \ll_C \Xi(x)^{-2}$ *for all* $x \in \mathfrak{s}$. *(The volume is taken in* $G(F)\backslash G(\mathbb{A})$.*)*
2. *For any* $x \in \mathfrak{s}$ *and* $y \in G(\mathbb{A})$ *we have*

$$\sum_{\gamma \in G(F)} 1_C(x^{-1}\gamma y)\sigma(x)^n \sigma(y)^{-n} \ll_{C,n} 1_{G(F)xC}(y)\Xi(x)^2. \qquad (2.7)$$

Consequently, for any $f \in C_c(G(\mathbb{A}))$ *we have*

1. $K_f(x, y)\sigma(x)^n\sigma(y)^{-n} \ll_{f,n} \Xi(x)^2$ *for any* $x \in \mathfrak{s}, y \in G(\mathbb{A})$.
2. $\|K_f(x,\cdot)\sigma^{-n}\|_{L^2(G\backslash G(\mathbb{A}))} \ll_{f,n} \sigma(x)^{-n}\Xi(x)$ *for all* $x \in \mathfrak{s}$.

Similarly, we have

$$\|\tilde{K}_f(x,\cdot)\sigma^{-n}\|_{L^2(G\backslash G(\mathbb{A})^1)} \ll_{f,n} \sigma(x)^{-n}\Xi(x), \quad x \in \mathfrak{s}^1.$$

Proof It follows from (1.1) and (2.4) that $G(F)xC \cap \mathfrak{s} \subseteq \mathfrak{s}_{H(x)+B}$ where B is a bounded subset of $\mathfrak{a}_0$ depending only on C. The first part follows from (1.2) since

$$\mathrm{vol}(P_0(F)\backslash \mathfrak{s}_{H(x)+B}) \ll_C \Xi(x)^{-2}.$$

To prove (2.7), we first observe that the left-hand side of (2.7) is supported (with respect to y) in $G(F)xC$. Therefore, $\sigma(x)^n\sigma(y)^{-n}$ is bounded in terms of C and n only, and we can ignore it from the estimation. It remains to bound the cardinality of $G(F) \cap xCy^{-1}$. Clearly, if $\gamma_1, \gamma_2 \in G(F) \cap xCy^{-1}$ then $\gamma_1\gamma_2^{-1} \in G(F) \cap xCC^{-1}x^{-1}$. Therefore, by passing to a larger C, it suffices to bound the size of $G(F) \cap xCx^{-1}$. Once again, using (1.3), upon enlarging C and multiplying x on the left by an element of $P_0(F)$, we may assume that $x = t = e^X \in A_0$ with $X \in T_1 + \mathfrak{a}_{0,+}$. Suppose that $\gamma \in G(F) \cap tCt^{-1}$. Using the Bruhat decomposition, write $\gamma = u_1 a n_w u_2$ where $w \in W$, $a \in M_0(F)$, $u_1 \in U_0(F)$, $u_2 \in U_w^-(F)$, and we recall that $U_w^- = U_0 \cap w^{-1}\overline{U_0}w$. Since $t^{-1}\gamma t \in C$, $H(t^{-1}\gamma t)$ is bounded. On the other hand, we may write $t^{-1}\gamma t = u_1' t^{-1} a n_w t u_2'$ where $u_1' = t^{-1}u_1 t \in U_0(\mathbb{A})$ and $u_2' = t^{-1}u_2 t \in U_w^-(\mathbb{A})$. Therefore

$$H(t^{-1}\gamma t) = wX - X + H(n_w u_2').$$

As in the proof of Lemma 2.1, we have $wX - X = \sum_{\beta \in \Phi_w} c_\beta \beta^\vee$ and $H(n_w u_2') = \sum_{\beta \in \Phi_w} d_\beta \beta^\vee$ with $c_\beta, d_\beta \le c$ (depending only on T_1) and $\Phi_w = \{\beta \in \Phi_+ : w^{-1}\beta < 0\}$. It follows that both $wX - X$ and $H(n_w u_2')$ are bounded. Let L be the smallest standard Levi subgroup containing n_w. The boundedness of $wX - X$ implies that t is in a compact translate of A_L, while the boundedness of $H(n_w u_2')$ implies that u_2' lies in a compact subset of $U_w^-(\mathbb{A})$. Thus, upon enlarging C, we may assume that $t \in A_L$. In this case t commutes with u_2, so that $u_2' = u_2$. Hence, u_2 is confined to a compact set (hence, a finite set) depending only on C. Also, since $t^{-1}\gamma t = u_1' a n_w u_2$, it follows that a is confined to a finite set and u_1' is confined to a compact set D of $U_0(\mathbb{A})$, both depending only on C. Finally, we can bound the number of possible u_1's since $t^{-1}u_1 t = u_1'$ and

$$\left|\{v \in U_0(F) : t^{-1}vt \in D\}\right| = \left|\{Y \in \operatorname{Lie} U_0(F) : \operatorname{Ad}(t)^{-1}Y \in \log(D)\}\right| \ll \delta_0(t).$$

The second part follows.

The next two statements of the Lemma are an immediate consequence. Finally, the last statement is proved by a similar argument.

□

Recall the regular representation R of $G(\mathbb{A})$ on $L^2(G(F)\backslash G(\mathbb{A}))$. Thus,

$$R(f)\varphi(x) = \int_{G(\mathbb{A})} f(g)\varphi(xg)\, dg = \int_{G(F)\backslash G(\mathbb{A})} K_f(x,y)\varphi(y)\, dy.$$

From the Cauchy-Schwarz inequality we get

Corollary 2.3 *Let $f \in C_c(G(\mathbb{A}))$ and $n = 0, 1, 2, \ldots$. Then for any function ϕ on $G(F)\backslash G(\mathbb{A})$ such that $\sigma^n\phi \in L^2(G(F)\backslash G(\mathbb{A}))$ we have*

$$\sup_{x\in\mathfrak{s}} \Xi(x)^{-1}\sigma(x)^n\left|R(f)\phi(x)\right| \ll_f \|\sigma^n\phi\|_{L^2(G(F)\backslash G(\mathbb{A}))}.$$

For future use we will need a variant of this Corollary. Recall Arthur's truncation operator Λ^T (with $T \in \mathfrak{a}_0$) on locally bounded functions on $G(F)\backslash G(\mathbb{A})^1$ [Art80]. We will always implicitly assume that $T \in \mathfrak{a}_{0,+}$ is sufficiently regular.

Corollary 2.4 *Let $f \in C_c(G(\mathbb{A}))$. Then for any locally bounded measurable function ϕ on $G(F)\backslash G(\mathbb{A})$ which is (A_G,χ)-equivariant with respect to a unitary character χ of A_G we have*

$$R(f)\phi(x) \ll_f \Xi(x)\|\Lambda^T\phi\|_{L^2(G(F)\backslash G(\mathbb{A})^1)}$$

for any $x \in \mathfrak{s}^1$ where T is a fixed translate of $H(x)$ (depending on the support of f).

Proof Indeed, let $f^\chi(g) = \int_{A_G} f(ag)\chi(a)\,da$. By [Lap06, Lemma 6.2] and the assumption of T we may write

$$R(f)\phi(x) = \int_{G(\mathbb{A})^1} f^\chi(y)\phi(xy)\,dy = \int_{G(\mathbb{A})^1} f^\chi(y)\Lambda^T\phi(xy)\,dy$$
$$= \int_{G(\mathbb{A})^1} f^\chi(x^{-1}y)\Lambda^T\phi(y)\,dy.$$

Hence,

$$|R(f)\phi(x)| \le \int_{G\backslash G(\mathbb{A})^1} \tilde{K}_{|f|}(x,\cdot)\left|\Lambda^T\phi(y)\right|\,dy.$$

The corollary follows from the last part of Lemma 2.2 and the Cauchy-Schwarz inequality.

□

Recall the following standard result (cf. [Art78, Lemma 4.1]).

Lemma 2.5 *Let G' be a real reductive group. Then for any $m \in \mathbb{N}$ there exist $f_1, f_2 \in C_c^m(G')$ and $X \in \mathcal{U}(\mathfrak{g}')$ such that $f_1 * X + f_2$ is the Dirac distribution at the identity of G'. Consequently, if π is a Banach representation of G' and V is the Fréchet space of smooth vectors then the following conditions are equivalent for a seminorm μ on V.*

1. *μ is continuous.*
2. *There exists $m \in \mathbb{N}$ such that for any $f \in C_c^m(G')$ the seminorm $v \mapsto \mu(\pi(f)v)$ is continuous.*
3. *There exists $m \in \mathbb{N}$ such that for any $f \in C_c^m(G')$ we have $\mu(\pi(f)v) \ll_f \|v\|$.*

For any irreducible representation π of $G(F_\infty)$ we will write

$$\Lambda_\pi = \sqrt{\lambda_\pi^2 + \lambda_\tau^2} \tag{2.8}$$

where τ is a minimal $\mathbf{K}_\infty$-type of π and λ_π (resp. λ_τ) is the Casimir eigenvalue of π (resp. τ). (λ_τ does not depend on the choice of τ.) We will also write $\|\tau\|$ for the norm of the highest weight of τ, that is $\lambda_\tau = \|\tau\|^2$.

Corollary 2.6
The following two conditions are equivalent for $\phi \in C^\infty(G(F)\backslash G(\mathbb{A}))$.

1. *For any $X \in \mathcal{U}(\mathfrak{g})$ and $n \in \mathbb{N}$ we have $\sigma^n \cdot R(X)\phi \in L^2(G(F)\backslash G(\mathbb{A}))$.*
2. *For any $X \in \mathcal{U}(\mathfrak{g})$ and $n \in \mathbb{N}$ we have*

$$\sup_{x \in \mathfrak{S}} \Xi(x)^{-1}\sigma(x)^n\,|R(X)\phi(x)| < \infty. \tag{2.9}$$

Denote by $\mathcal{S}(G(F)\backslash G(\mathbb{A}))$ the space of functions satisfying the conditions above. This space is invariant under translation by $G(\mathbb{A})$. Given a level K of G, the two sets of seminorms $\|\sigma^n \cdot R(X)\phi\|_{L^2(G(F)\backslash G(\mathbb{A}))}$, $X \in \mathcal{U}(\mathfrak{g}), n \in \mathbb{N}$ and (2.9) give rise to the same Fréchet space $\mathcal{S}(G(F)\backslash G(\mathbb{A}))^K$ of right-K-invariant functions in $\mathcal{S}(G(F)\backslash G(\mathbb{A}))$.

In one direction this follows from (2.6). In the other direction this follows from Corollary 2.3 and Lemma 2.5, applied, for any $n \in \mathbb{N}$, to the space of right K-invariant functions φ on $G(F)\backslash G(\mathbb{A})$ such that $\sigma^n\varphi \in L^2(G(F)\backslash G(\mathbb{A}))$.

We call $\mathcal{S}(G(F)\backslash G(\mathbb{A})) = \cup_K \mathcal{S}(G(F)\backslash G(\mathbb{A}))^K$, equipped with the inductive limit topology, the Harish-Chandra Schwartz space of $G(F)\backslash G(\mathbb{A})$.

Similarly, let $P = M \ltimes U$ be a standard parabolic subgroup of G. Define

$$\mathfrak{s}^P = \{g \in G(\mathbb{A}) : \langle \alpha, H(g) - T_1 \rangle > 0 \text{ for all } \alpha \in \Delta_0^M\}.$$

We have

1. $G(\mathbb{A}) = P(F)\mathfrak{s}^P$.
2. There exists C such that $\left|P_0(F)\backslash(\mathfrak{s}^P \cap P(F)x)\right| \le C$ and the diameter of $H(P_0(F)\mathfrak{s}^P \cap P(F)x)$ is bounded by C for any $x \in G(\mathbb{A})$.

Define the function Ξ^P on $\mathfrak{s}^P$ by

$$\Xi^P(pk) = \delta_0(p)^{\frac{1}{2}}, \quad p \in P_0(\mathbb{A}),\ k \in \mathbf{K},\ \langle \alpha, H(p) - T_1 \rangle > 0 \text{ for all } \alpha \in \Delta_0^M.$$

For $g \in G(\mathbb{A})$ let $\sigma^P(g) = 1 + \min_{x \in \mathfrak{s}^P \cap P(F)g} \|H(x)\|$. Clearly, σ is left $U(\mathbb{A})P(F)$-invariant.

We will define the Harish-Chandra Schwartz space of $P(F)U(\mathbb{A})\backslash G(\mathbb{A})$ (denoted by $\mathcal{S}(P(F)U(\mathbb{A})\backslash G(\mathbb{A}))$) to be the inductive limit over K of the spaces of right-K-invariant functions $\phi \in C^\infty(P(F)U(\mathbb{A})\backslash G(\mathbb{A}))$ such that the seminorms

$$\|(\sigma^P)^n \cdot R(X)\phi\|_{L^2(P(F)U(\mathbb{A})\backslash G(\mathbb{A}))}, \quad X \in \mathcal{U}(\mathfrak{g}), n \in \mathbb{N},$$

are finite. Alternatively, we could use the equivalent sequence of seminorms

$$\sup_{x \in \mathfrak{s}^P} \Xi^P(x)^{-1}\sigma^P(x)^n \left|R(X)\phi(x)\right|, \quad X \in \mathcal{U}(\mathfrak{g}), n \in \mathbb{N}.$$

It is also easy to see that the map $\phi \mapsto (k \mapsto \delta_P^{-\frac{1}{2}}\phi(\cdot k))$ defines a topological isomorphism between $\mathcal{S}(P(F)U(\mathbb{A})\backslash G(\mathbb{A}))$ and the space of smooth functions $f : \mathbf{K} \to \mathcal{S}(M(F)\backslash M(\mathbb{A}))$ which are right-K-invariant for some level K and such that $f(umk) = f(k)[\cdot m]$ for any $u \in U(\mathbb{A}) \cap \mathbf{K}$, $m \in M(\mathbb{A}) \cap \mathbf{K}$, $k \in \mathbf{K}$.

Note that

$$\sigma\big|_{\mathfrak{s}} \ll \sigma^P\big|_{\mathfrak{s}} \ll \sigma\big|_{\mathfrak{s}} \quad \text{and} \quad \Xi^P\big|_{\mathfrak{s}} = \Xi. \tag{2.10}$$

3 Growth Conditions

3.1 (G, M)-families

Following Arthur ([Art81]), a (G, M)-family is a collection of smooth functions $c_P \in C^\infty(i\mathfrak{a}_M^*)$, $P \in \mathcal{P}(M)$, satisfying the compatibility relations

$$c_P \equiv c_{P'} \text{ on the hyperplane } \langle \Lambda, \alpha^\vee \rangle = 0$$

whenever P, P' are adjacent along the root α, i.e. when $\Sigma_{\overline{P}} \cap \Sigma_{P'} = \{\alpha\}$. For any such a (G, M)-family one defines the function

$$c_M = \sum_{P \in \mathcal{P}(M)} \frac{c_P}{\theta_P}$$

where

$$\theta_P(\Lambda) = v_M^{-1} \prod_{\alpha \in \Delta_P} \langle \Lambda, \alpha^\vee \rangle$$

and v_M is the co-volume of the lattice spanned by the co-roots in $\mathfrak{a}_M$. The basic result [Art81, Lemma 6.2] is that $c_M \in C^\infty(i\mathfrak{a}_M^*)$.

Occasionally, we will also consider (G, M)-families of meromorphic functions on $\mathfrak{a}_{M,\mathbb{C}}^*$ which are holomorphic on $i\mathfrak{a}_M^*$. Then c_M will also be a meromorphic function on $\mathfrak{a}_{M,\mathbb{C}}^*$, holomorphic on $i\mathfrak{a}_M^*$

If $(c_P)_{P \in \mathcal{P}(M)}$ is a (G, M)-family and $Q = L \ltimes V \in \mathcal{F}(M)$ we may consider the (L, M)-family

$$c_R^Q = c_{R \ltimes V} \quad R \in \mathcal{P}^L(M).$$

Note that $R \ltimes V$ is the unique parabolic subgroup of G contained in Q which intersects L in R. Correspondingly $c_M^Q \in C^\infty(i\mathfrak{a}_M^*)$ is defined.

If $(c_P)_{P \in \mathcal{P}(M)}$ and $(d_P)_{P \in \mathcal{P}(M)}$ are (G, M)-families, then so is their product $(c_P d_P)_{P \in \mathcal{P}(M)}$. To compute $(cd)_M$ we use a formula of Arthur ([Art88, Proposition 7.1 and Corollary 7.4])[2]: there exist constants α_{Q_1,Q_2} for all pairs $Q_1, Q_2 \in \mathcal{F}(M)$, with α_{Q_1,Q_2} nonzero only if $\mathfrak{a}_M^G = \mathfrak{a}_M^{L_1} \oplus \mathfrak{a}_M^{L_2} = \mathfrak{a}_{L_1}^G \oplus \mathfrak{a}_{L_2}^G$, such that

$$(cd)_M = \sum_{Q_1, Q_2 \in \mathcal{F}(M)} \alpha_{Q_1,Q_2} c_M^{Q_1} d_M^{Q_2}. \tag{3.1}$$

Note that the constants α_{Q_1,Q_2} are not uniquely determined – they depend on certain auxiliary choices. For our purposes the exact value of α_{Q_1,Q_2} is immaterial.

[2] This is only stated for the value at 0 but the argument is valid for any Λ.

Consider now a (G, M)-family of a special form as in [Art82, §7] (cf. [Lap06, §4]). Namely, suppose that for any reduced root $\beta \in R(T_M, G)$ we are given $c_\beta \in C^\infty(\mathrm{i}\mathbb{R})$ with $c_\beta(0) = 1$. Let

$$c_P(\Lambda) = \prod_{\beta \in \Sigma_P} c_\beta(\langle \Lambda, \beta^\vee \rangle), \quad P \in \mathcal{P}(M). \tag{3.2}$$

This is clearly a (G, M)-family. By [Lap06, (4.4)] for any $Q = L \ltimes V$ with $L \in \mathcal{L}(M)$ we have

$$c_M^Q(\Lambda) = \sum_{\mathfrak{B}_1, \mathfrak{B}_2} \alpha_{\mathfrak{B}_1, \mathfrak{B}_2} \prod_{\beta \in \mathfrak{B}_1} \frac{c_\beta(\langle \Lambda, \beta^\vee \rangle) - 1}{\langle \Lambda, \beta^\vee \rangle} \prod_{\beta \in \mathfrak{B}_2} c_\beta(\langle \Lambda, \beta^\vee \rangle), \tag{3.3}$$

where the sum is over disjoint subsets $\mathfrak{B}_1, \mathfrak{B}_2 \subseteq R(T_M, Q)$ such that $\mathfrak{B}_2 \supseteq R(T_M, V)$ and $\mathfrak{B}_1$ forms a basis for $(\mathfrak{a}_M^L)^*$, and $\alpha_{\mathfrak{B}_1, \mathfrak{B}_2}$ are certain constant (whose values are unimportant for us).

We will use one more elementary fact about (G, M)-families.

Lemma 3.1 *Let $\delta > 0$. Suppose that $(c_P(\lambda))_{P \in \mathcal{P}(M)}$ is a (G, M)-family of functions which are holomorphic and bounded on the strip $\|\mathrm{Re}\,\lambda\| < \delta$. Then for any $\epsilon > 0$, c_M is bounded on $\|\mathrm{Re}\,\lambda\| < \delta - \epsilon$ and in this region*

$$c_M(\lambda) \ll_{\delta,\epsilon} \sum_{P \in \mathcal{P}(M)} \sup_{\|\mathrm{Re}\,\lambda\| < \delta} |c_P(\lambda)| \,.$$

Proof Fix $\lambda_0 \in \mathfrak{a}_M^*$ in general position such that $\|\lambda_0\| = 1$. Let

$$c_0 = \min_{\alpha \in R(T_M, G)} |\langle \lambda_0, \alpha^\vee \rangle|$$

and assume that $c_0 > 0$. Let $\lambda \in \mathfrak{a}_{M,\mathbb{C}}^*$. By the pigeonhole principle there exists $0 \le r \le \epsilon$ such that

$$\Big| |\langle \lambda, \alpha^\vee \rangle| - r \, |\langle \lambda_0, \alpha^\vee \rangle| \Big| > c_0 \frac{\epsilon}{2\,|R(T_M, G)| + 1}$$

for all $\alpha \in R(T_M, G)$. For such r we have

$$\theta_P(\lambda + z\lambda_0)^{-1} \ll_{\delta,\epsilon} 1$$

for all $z \in \mathbb{C}$ with $|z| = r$. Hence, if $\|\mathrm{Re}\,\lambda\| < \delta - \epsilon$ then

$$|c_M(\lambda)| \le \sup_{|z|=r} |c_M(\lambda + z\lambda_0)| \ll_{\delta,\epsilon} \sum_{P \in \mathcal{P}(M)} \sup_{\|\mathrm{Re}\,\lambda\| < \delta} |c_P(\lambda)| \,.$$

□

3.2 Intertwining operators

Let M be a Levi subgroup. Consider the discrete part $L^2_{\mathrm{disc}}(A_M M(F)\backslash M(\mathbb{A}))$ of $L^2(A_M M(F)\backslash M(\mathbb{A}))$, namely the closure of the sum of the irreducible subrepresentations of $L^2(A_M M(F)\backslash M(\mathbb{A}))$. We write

$$L^2_{\mathrm{disc}}(A_M M(F)\backslash M(\mathbb{A})) = \hat{\oplus}_{\pi\in\Pi_{\mathrm{disc}}(M(\mathbb{A}))} L^2_{\mathrm{disc}}(A_M M(F)\backslash M(\mathbb{A}))_\pi \quad (3.4)$$

where $L^2_{\mathrm{disc}}(A_M M(F)\backslash M(\mathbb{A}))_\pi$ is the π-isotypic component of $L^2_{\mathrm{disc}}(A_M M(F)\backslash M(\mathbb{A}))$. For any level K_M of M we write $\Pi_{\mathrm{disc}}(M(\mathbb{A}))^{K_M}$ for the subset of $\Pi_{\mathrm{disc}}(M(\mathbb{A}))$ consisting of the π's such that the K_M-fixed part is non-zero.

For any $P \in \mathcal{P}(M)$ we write L^2_P for $\mathrm{Ind}_{P(\mathbb{A})}^{G(\mathbb{A})} L^2_{\mathrm{disc}}(A_M M(F)\backslash M(\mathbb{A}))$ in the sense of L^2-induction. We can identify L^2_P with the Hilbert space $L^2(A_M M(F)U(\mathbb{A})\backslash G(\mathbb{A}), \delta_P^{\frac{1}{2}})$ of measurable functions $\varphi : M(F)U(\mathbb{A})\backslash G(\mathbb{A}) \to \mathbb{C}$ (up to functions which vanish almost everywhere) such that for almost all $g \in G(\mathbb{A})$ the function $m \mapsto \delta_P(m)^{-\frac{1}{2}}\varphi(mg)$ belongs to $L^2_{\mathrm{disc}}(A_M M(F)\backslash M(\mathbb{A}))$ and $\|\varphi\|_P^2 := \int_{M(F)A_M U(\mathbb{A})\backslash G(\mathbb{A})} |\varphi(g)|^2 \, dg < \infty$. Corresponding to (3.4) we write $L^2_P = \hat{\oplus}_\pi L^2_{P,\pi}$. We denote the regular representation of $G(\mathbb{A})$ on L^2_P by I_P, or simply by I if P is clear from the context. For any level K of G we write $(L^2_P)^K$ for the K-fixed part of L^2_P. Moreover, for any $\tau \in \widehat{\mathbf{K}_\infty}$ we denote by $(L^2_{P,\pi})^{\tau,K}$ the $(\mathbf{K}_\infty, \tau)$-isotypic part of $(L^2_{P,\pi})^K$. The space $(L^2_{P,\pi})^{\tau,K}$ is finite-dimensional.

We also have the representations $I_P(\lambda) = I(\lambda)$ on L^2_P, $\lambda \in \mathfrak{a}^*_{M,\mathbb{C}}$ given by $(I_P(g,\lambda)\varphi)_\lambda(x) = \varphi_\lambda(xg)$, $x, g \in G(\mathbb{A})$ where $\varphi_\lambda(x) = \varphi(x)e^{\langle\lambda, H(x)\rangle}$.

For any $P = M \ltimes U, Q = L \ltimes V \in \mathcal{F}$ and $w \in W(M, L)$ let $M_{Q|P}(w, \lambda) : L^2_P \to L^2_Q$, $\lambda \in i\mathfrak{a}^*_M$ be the unitary intertwining operators defined in [Art82, §1]. If $M = L$ and $w = 1$ then we simply write $M_{Q|P}$. In general, if $P' = w^{-1}Qw$ then up to an immaterial constant

$$M_{Q|P}(w, \lambda) \text{ is the composition of left translation by } w^{-1} \text{ with } M_{P'|P}(\lambda) \quad (3.5)$$

[ibid., (1.4)].

For every $\pi \in \Pi_{\mathrm{disc}}(M(\mathbb{A}))$ we denote the restriction of $M_{Q|P}(\lambda)$ to $L^2_{P,\pi}$ by $M_{Q|P}(\pi, \lambda)$. We normalize $M_{Q|P}(\pi, \lambda)$ as in [Art82, §6] (cf. [Art89, Theorem 2.1]). Thus, we write

$$M_{Q|P}(\pi, \lambda) = n_{Q|P}(\pi, \lambda) N_{Q|P}(\pi, \lambda)$$

where $N_{Q|P}$ are the normalized intertwining operators satisfying the properties [Art82, (6.3)–(6.6)] and $n_{Q|P}$ are the normalizing factors. In particular, $N_{Q|P}(\pi, \lambda) = \prod_v N_{Q|P}(\pi_v, \lambda)$ where $N_{Q|P}(\pi_v, \lambda)$, $\lambda \in i\mathfrak{a}^*_M$ are unitary operators. The normalizing factors are given by

$$n_{Q|P}(\lambda) = \prod_{\beta \in \Sigma_Q \cap \Sigma_{\overline{P}}} n_\beta(\pi, \langle \lambda, \beta^\vee \rangle) = \prod_{\beta \in \Sigma_{\overline{Q}} \cap \Sigma_P} n_\beta(\pi, \langle \lambda, \beta^\vee \rangle)^{-1} \tag{3.6}$$

where $n_\beta(\pi, z)$ are certain meromorphic functions on $\mathbb{C}$ which are holomorphic on $\mathrm{i}\mathbb{R}$. We have the functional equations

$$n_{-\beta}(\pi, z) n_\beta(\pi, -z) = 1,$$
$$\overline{n_\beta(\pi, z)} = n_{-\beta}(\pi, \bar{z}).$$

Thus,

$$n_\beta(\pi, z)^{-1} = \overline{n_\beta(\pi, -\bar{z})}, \tag{3.7}$$

so that $|n_\beta(\pi, \mathrm{i}t)| = 1$ for $t \in \mathbb{R}$.

3.3

We will now state the first analytic condition which is crucial for our analysis. Recall the notation (2.8).

Definition 3.2 We say that G satisfies uniform temperedness (UT) if there exist $k, l > 0$ such that for any maximal parabolic subgroup $P = M \ltimes U$ of G and level K_M of M, there exists a constant $c > 0$ such that for any $\pi \in \Pi_{\mathrm{disc}}(M(\mathbb{A}))^{K_M}$ we have

$$n_\alpha(\pi, z) \ll_{K_M} (1 + \Lambda_{\pi_\infty} + |z|)^k \tag{3.8}$$

in the region $|\mathrm{Re}\, z| < c(1 + \Lambda_{\pi_\infty} + |\mathrm{Im}\, z|)^{-l}$, where $\Sigma_P = \{\alpha\}$.

This growth condition was introduced (in a slightly different form) in [Lap06]. Of course, if we are not interested in optimizing the exponents we could take $k = l$.

If G satisfies (UT) then by Cauchy's formula we get a bound similar to (3.8) (with a larger k and with c replaced by $c/2$) for any derivative of $n_\alpha(\pi, z)$. By the mean value theorem we also get

$$\frac{n_\alpha(\pi, z) - n_\alpha(\pi, z')}{z - z'} \ll_{K_M} (1 + \Lambda_{\pi_\infty} + |z| + |z'|)^k \tag{3.9}$$

provided that $|\mathrm{Re}\, z| < c(1 + \Lambda_{\pi_\infty} + |\mathrm{Im}\, z|)^{-l}$ and $|\mathrm{Re}\, z'| < c(1 + \Lambda_{\pi_\infty} + |\mathrm{Im}\, z'|)^{-l}$.

Lemma 3.3 *Suppose that G and its Levi subgroups satisfy (UT) and let $M \in \mathcal{L}$. Then there exists $l > 0$ and*

- *for any $D \in \mathcal{D}(\mathfrak{a}_M^*)$ there exists $k > 0$*
- *for any level K_M of M there exists $c > 0$*

such that for any $\pi \in \Pi_{\mathrm{disc}}(M(\mathbb{A}))^{K_M}$ we have

$$D_\lambda n_{P'|P}(\pi, \lambda) \ll_{D,K_M} (1 + \|\lambda\| + \Lambda_{\pi_\infty})^k$$

in the region

$$\mathcal{R}_{\pi,c,l} = \{\lambda \in \mathfrak{a}_{M,\mathbb{C}}^* : \|\mathrm{Re}\,\lambda\| < c(1 + \Lambda_{\pi_\infty} + \|\mathrm{Im}\,\lambda\|)^{-l}\}.$$

Similarly, for any $P \in \mathcal{P}(M)$ and $\lambda \in \mathfrak{a}_{M,\mathbb{C}}^*$ let $\nu_Q(P, \pi, \lambda, \Lambda)$, $Q \in \mathcal{P}(M)$ be the (G, M)-family (in Λ) given by

$$\nu_Q(P, \pi, \lambda, \Lambda) = n_{Q|P}(\pi, \lambda)^{-1} n_{Q|P}(\pi, \lambda + \Lambda).$$

Then there exists $k > 0$ such that for any π and $R \in \mathcal{F}(M)$ we have

$$\nu_M^R(P, \pi, \lambda, \Lambda) \ll_{K_M} (1 + \|\lambda\| + \|\Lambda\| + \Lambda_{\pi_\infty})^k$$

provided that $\lambda, \lambda + \Lambda \in \mathcal{R}_{\pi,c,k}$.

Proof The first part follows from (3.6), (3.8) and the Leibniz rule. To prove the second part we write the (G, M)-family $\nu_Q(P, \pi, \lambda, \Lambda)$ in the form (3.2) with

$$c_\beta(z) = \begin{cases} n_\beta(\pi, \langle\lambda, \beta^\vee\rangle)^{-1} n_\beta(\pi, \langle\lambda, \beta^\vee\rangle + z) & \text{if } \beta \in \Sigma_{\overline{P}}, \\ 1 & \text{otherwise} \end{cases}$$

(cf. [Art82, p. 1323]). The required estimate follows from (3.3), (3.7) and (3.9).

□

It is reasonable to conjecture that any reductive group satisfies (UT). In fact, it is conceivable that we can take *any* $k, l \in \mathbb{R}^{>0}$ in the definition of (UT) (and consequently, in Lemma 3.3). However, at this stage we can only prove (UT) for the general linear group.

3.4

Proposition 3.4 *The group $G = \mathrm{GL}_r$ satisfies (UT).*

We will prove the Proposition below. The proof is based on known analytic properties of the Rankin-Selberg L-function which we now recall. Let π_i be cuspidal representations of $\mathrm{GL}_{n_i}(\mathbb{A})$, $i = 1, 2$, whose central character is trivial on the scalar matrices with (the same) positive real scalar in the archimedean places. (The integers n_1, n_2 as well as the field F will be fixed in the estimates below.) Let $L(s, \pi_1 \times \tilde{\pi}_2)$ and $\epsilon(s, \pi_1 \times \tilde{\pi}_2)$ be the completed Rankin-Selberg L-function and epsilon factors defined in [JPSS83]. We have the functional equation

$$L(s, \pi_1 \times \tilde{\pi}_2) = \epsilon(s, \pi_1 \times \tilde{\pi}_2) L(1 - s, \tilde{\pi}_1 \times \pi_2)$$

where $\epsilon(s, \pi_1 \times \tilde{\pi}_2) = \epsilon_0 q^{\frac{1}{2} - s}$ for some ϵ_0 of modulus 1 and $q \in \mathbb{N}$ (the arithmetic conductor of $\pi_1 \times \tilde{\pi}_2$). The function $[s(1-s)]^{\delta_{\pi_1, \pi_2}} L(s, \pi_1 \times \tilde{\pi}_2)$ is an entire function of order one where $\delta_{\pi_1, \pi_2} = 1$ if $\pi_2 = \pi_1$ and 0 otherwise (see e.g. [RS96]).

We write

$$L(s, \pi_1 \times \tilde{\pi}_2) = L_\infty(s, \pi_{1,\infty} \times \tilde{\pi}_{2,\infty}) L^\infty(s, \pi_1 \times \tilde{\pi}_2)$$

where $L_\infty(s, \pi_{1,\infty} \times \tilde{\pi}_{2,\infty})$ is the Archimedean part of the Rankin-Selberg L-function. We write the latter as $\prod_{j=1}^m \Gamma_{\mathbb{R}}(s - \alpha_j)$ for certain complex parameters $\alpha_1, \dots, \alpha_m$ where $m = n_1 n_2 [F : \mathbb{Q}]$. Here $\Gamma_{\mathbb{R}}(s) = \pi^{-s/2} \Gamma(s/2)$ (for the usual Γ function). By the Jacquet-Shalika bounds, $L_\infty(s, \pi_{1,\infty} \times \tilde{\pi}_{2,\infty})$ is holomorphic for $\operatorname{Re} s \geq 1$, so that $\operatorname{Re} \alpha_j < 1$ for all j. In fact, by [MS04, Proposition 3.3] (which is based on [LRS99]) we have

$$1 - \operatorname{Re} \alpha_j > \frac{1}{n_1^2 + 1} + \frac{1}{n_2^2 + 1}. \tag{3.10}$$

Following [IS00] we define the analytic conductor $\mathfrak{c}(\pi_1 \times \tilde{\pi}_2, s) = q \prod_{j=1}^m (1 + |s - \alpha_j|)$ and $\mathfrak{c}(\pi_1 \times \tilde{\pi}_2) = \mathfrak{c}(\pi_1 \times \tilde{\pi}_2, 0)$. Similarly, we define $\mathfrak{c}(\pi)$ for any cuspidal representation of $\mathrm{GL}_n(\mathbb{A})$.

We have

$$\mathfrak{c}(\pi_1 \times \tilde{\pi}_2) \leq \mathfrak{c}(\pi_1)^{n_2} \mathfrak{c}(\pi_2)^{n_1}. \tag{3.11}$$

The p-adic aspect follows from [BH97] and the Archimedean aspect boils down to an easy GL_2 computation (cf. [RS96, Appendix]).

A basic fact which is proved using the Phragmén-Lindelöf principle is polynomial growth on vertical strips for $L^\infty(s, \pi_1 \times \tilde{\pi}_2)$. For our purposes it suffices to know that there exist $k > 0$ and $c > 0$ (independent of π_1, π_2) such that

$$\left(\frac{s-1}{s}\right)^{\delta_{\pi_1, \pi_2}} L^\infty(s, \pi_1 \times \tilde{\pi}_2) \ll \mathfrak{c}(\pi_1 \times \tilde{\pi}_2, s)^k, \tag{3.12}$$

in the region $\operatorname{Re} s \geq 1 - c\mathfrak{c}(\pi_1 \times \tilde{\pi}_2, s)^{-k}$. As before, this implies bounds of a similar nature for all derivatives. In fact it follows from the results of [Li10] that we can take *any* $k \in \mathbb{R}^{>0}$ but we will not need to use this fact.[3] This is still far off from the "correct" estimates which can be obtained assuming both the generalized Riemann Hypothesis and the Ramanujan Hypothesis, e.g.

$$\left(\frac{s-1}{s}\right)^{\delta_{\pi_1,\pi_2}} L^{\infty}(s, \pi_1 \times \tilde{\pi}_2) \ll (\log\log \mathfrak{c}(\pi_1 \times \tilde{\pi}_2, s))^m, \quad \operatorname{Re} s = 1.$$

Another property that we will use is the following result due to Brumley ([Bru06], see also Appendix A below) giving coarse lower bounds on Rankin-Selberg L-functions at the edge of the critical strip.

Proposition 3.5 (Brumley) *There exist $k > 0$ and $c > 0$ (independent of π_1, π_2) such that*

$$\left(\frac{s}{s-1}\right)^{\delta_{\pi_1,\pi_2}} L^{\infty}(s, \pi_1 \times \tilde{\pi}_2)^{-1} \ll (\mathfrak{c}(\pi_1) + \mathfrak{c}(\pi_2) + |s|)^k \tag{3.13}$$

in the region $\operatorname{Re} s \geq 1 - c\mathfrak{c}(\pi_1 \times \tilde{\pi}_2, s)^{-k}$.

Remark 3.6 The bound (3.13) can certainly be improved in many cases. For instance, it is well known that

$$\zeta(s)^{-1} \ll \log(1 + |s|), \quad \operatorname{Re} s = 1,$$

and under the Riemann Hypothesis we have

$$\zeta(s)^{-1} \ll \log\log(2 + |s|), \quad \operatorname{Re} s = 1,$$

which is (up to determining the constant) best possible. More generally, as in the case of the upper bounds, if we assume both the generalized Ramanujan Hypothesis and the generalized Riemann Hypothesis then we would have the "correct" bound

$$\left(\frac{s}{s-1}\right)^{\delta_{\pi_1,\pi_2}} L^{\infty}(s, \pi_1 \times \tilde{\pi}_2)^{-1} \ll (\log\log \mathfrak{c}(\pi_1 \times \tilde{\pi}_2, s))^m, \quad \operatorname{Re} s = 1.$$

For our purposes the bound (3.13) suffices.

Note that it follows from Stirling's formula that for any $c > 0$ there exists $k > 0$ such that

$$\frac{\Gamma_{\mathbb{R}}(-s + \overline{\beta})}{\Gamma_{\mathbb{R}}(s + \beta)} \ll_c 1 + |s + \beta|^k \tag{3.14}$$

uniformly for $\operatorname{Re} s \in [-c/2, c/2]$ and $\operatorname{Re} \beta \geq c$.

[3] The case $\pi_1 = \pi_2$ is not formally covered by [ibid.] but it follows from the technique.

Proof (of Proposition 3.4) Assume that $\pi = \pi_1 \otimes \pi_2 \in \Pi_{\mathrm{disc}}(M(\mathbb{A}))$. Writing $\Sigma_{\overline{P}} = \{\beta\}$ we have

$$\xi(s) := n_\beta(\pi, -s) = \frac{L(s, \pi_1 \times \tilde{\pi}_2)}{\epsilon(s, \pi_1 \times \tilde{\pi}_2) L(s+1, \pi_1 \times \tilde{\pi}_2)} = \frac{L(1-s, \tilde{\pi}_1 \times \pi_2)}{L(1+s, \pi_1 \times \tilde{\pi}_2)}.$$

Assume first that π is cuspidal. We write ξ as the product of the following three factors

$$A(s) = \left(\frac{s}{1-s}\right)^{\delta_{\pi_1,\pi_2}} L^\infty(1-s, \tilde{\pi}_1 \times \pi_2),$$
$$B(s) = \prod_{j=1}^{m} \frac{\Gamma_{\mathbb{R}}(1 - s - \overline{\alpha_j})}{\Gamma_{\mathbb{R}}(1 + s - \alpha_j)},$$
$$C(s) = \left(\frac{1-s}{s}\right)^{\delta_{\pi_1,\pi_2}} L^\infty(s+1, \pi_1 \times \tilde{\pi}_2)^{-1}.$$

For $A(s)$ we apply (3.12). For $B(s)$ we apply (3.14) with $\beta = 1 - \alpha_j$, $j = 1, \ldots, m$ and $c = 1/(n_1^2+1) + 1/(n_2^2+1)$ which is applicable by (3.10). For $C(s)$ we apply (3.13). To finish the proof in the cuspidal case, it remains to use (3.11) and to note that $\mathfrak{c}(\pi) \ll_{K_M} \Lambda_{\pi_\infty}$ since the arithmetic conductor is bounded in terms of K_M.

Now, we drop the assumption that π_1 and π_2 are cuspidal. By [MW89] there exist factorizations $n_i = (2d_i + 1)m_i$ (d_i half-integers) and cuspidal representations σ_i on $\mathrm{GL}_{m_i}(\mathbb{A})$, $i = 1, 2$ such that π_i is obtained by taking residues of Eisenstein series induced from $\sigma_i \left|\det \cdot\right|^{d_i} \otimes \cdots \otimes \sigma_i \left|\det \cdot\right|^{-d_i}$. Assume without loss of generality that $d_1 \le d_2$. Then

$$\xi(s) = \prod_{j=-d_1}^{d_1} \frac{L(d_2 + j + 1 - s, \tilde{\sigma}_1 \times \sigma_2)}{L(d_2 + j + 1 + s, \sigma_1 \times \tilde{\sigma}_2)}.$$

We can apply the same argument as above for each factor.

□

3.5

We turn to the local aspect of the analytic condition on G. We formulate a property about the local components of an irreducible representation occurring in the discrete spectrum of a Levi subgroup of G.

Definition 3.7 We say that G satisfies the weak Ramanujan property (WR) with constant $c > 0$ if for any

- maximal parabolic subgroup $P = MU$ with corresponding fundamental weight ϖ,
- $\pi \in \Pi_{\mathrm{disc}}(M(\mathbb{A}))$,
- place v of F,
- level K of G,
- $\tau \in \widehat{\mathbf{K}_\infty}$,

the normalized intertwining operators $N_{\overline{P}|P}(\pi_v, s\varpi)$ on $(L^2_{P,\pi})^{\tau,K}$ is holomorphic in the strip $|\mathrm{Re}\, s| < c$.

A closely related property was considered in [Mül02]. (See Lemma 3.11 below.)

For our purposes we may allow c to depend on K, and even inverse polynomially on Λ_{π_∞}. However, this weaker property doesn't seem to be any easier to prove.

For GL_n the property (WR) was established in [MS04] as a consequence of the results of Luo-Rudnick-Sarnak [LRS99] and the properties of the local intertwining operators. More precisely, we have

Theorem 3.8 *[MS04, Proposition 4.2] The group $G = \mathrm{GL}_n$ (and hence any Levi subgroup of G) satisfies (WR) with constant $c = 2/(n^2+1)$.*

Remark 3.9 Following the argument of [MS04] it is easy to see that the Ramanujan Hypothesis for GL_n for all n is equivalent to (WR) for all GL_n's with constant $c = 1$. More generally, we expect (WR) to hold for any G with $c = \frac{1}{2}$ (which is best possible for the symplectic group of rank two).

Lemma 3.10 *Let V be a normed vector space and let $z_1, \dots, z_m \notin \mathbf{S^1}$. Suppose that $A : \mathbb{C} \setminus \{z_1, \dots, z_m\} \to V$ is such that $(z - z_1) \dots (z - z_m)A(z)$ is a polynomial on $\mathbb{C}$ of degree $\le n$ with coefficients in V. Assume that $\|A(z)\| \le 1$ for all $z \in \mathbf{S^1}$ and that there exist $r < 1 < R$ such that $|z_i| \notin [r, R]$ for all i. Then for any $0 < \epsilon < (R-r)/2$ and $k = 0, 1, 2 \dots$ we have*

$$\sup_{r+\epsilon<|z|<R-\epsilon} \|A^{(k)}(z)\| \ll_{k,m,n,R,\epsilon} 1. \tag{3.15}$$

Proof By Cauchy's formula it is enough to prove this for $k = 0$. Consider

$$B(z) = \Big(\prod_{|z_i|<1} \frac{z - z_i}{1 - \overline{z_i} z} \Big) A(z).$$

Then $B(z)$ is holomorphic for $|z| \le 1$ and by the maximum modulus principle we have $\|B(z)\| \le 1$ for $|z| \le 1$. We infer that

$$\|A(z)\| \le \left(\frac{2}{\epsilon}\right)^m$$

for $1 \ge |z| \ge r + \epsilon$. Similarly,

$$z^{n-m}\left(\prod_{|z_i|>1} \frac{z - z_i^{-1}}{1 - \overline{z_i}^{-1} z}\right) A(1/z)$$

is holomorphic and bounded in absolute value by 1 for $|z| \le 1$ and therefore

$$\|A(z)\| \le \left(\frac{2R}{\epsilon}\right)^m R^{\max(m,n)}$$

for $1 \le |z| \le R - \epsilon$.

□

For Hilbert spaces $\mathfrak{H}_1, \mathfrak{H}_2$ we will write $\mathcal{B}(\mathfrak{H}_1, \mathfrak{H}_2)$ for the Banach space of bounded operators from $\mathfrak{H}_1$ to $\mathfrak{H}_2$ with the standard operator norm. If $\mathfrak{H}_2 = \mathfrak{H}_1$ we simply write $\mathcal{B}(\mathfrak{H}_1)$.

Lemma 3.11 *Suppose that G and its Levi subgroups satisfy (WR) with constant c. Then there exists $c' > 0$ such that for any $M \in \mathcal{L}$ and $D \in \mathcal{D}(\mathfrak{a}_M^*)$ there exists $k > 0$ such that for any level K of G, $P, P' \in \mathcal{P}(M)$ and $\pi \in \Pi_{\mathrm{disc}}(M(\mathbb{A}))$ we have*

$$\|D_\lambda N_{P'|P}(\pi_v, \lambda)\|_{\mathcal{B}((L^2_{P,\pi})^K, (L^2_{P',\pi})^K)} \ll_{D,K} 1$$

for any finite place v and

$$\|D_\lambda N_{P'|P}(\pi_v, \lambda)\|_{\mathcal{B}((L^2_{P,\pi})^\tau, (L^2_{P',\pi})^\tau)} \ll_D (1 + \|\tau\|)^k$$

for any $v|\infty$ and $\tau \in \widehat{\mathbf{K}_v}$. Both estimates are valid for $\|\mathrm{Re}\,\lambda\| < c'$.

Proof By factoring the intertwining operator and passing to a smaller Levi subgroup if necessary, we can reduce to the case where P is maximal and $P' = \overline{P}$. The archimedean case follows from [MS04, Corollary A.3] (which is stated only for $\lambda \in i\mathfrak{a}_M^*$ but is valid in the larger region) and the assumption on G. In the non-archimedean case we can write $N_{P'|P}(\pi_v, s\varpi)\big|_{(L^2_{P,\pi})^K} = A_v(q_v^{-s})$ where A_v is a rational function whose degree is bounded in terms of K only. By the assumption on G we can apply Lemma 3.10 with $R = q_v^c$ and $r = q_v^{-c}$. We used of course the fact that only finitely many v's need to be considered (depending on K).

□

Let $M \in \mathcal{L}$, $\pi \in \Pi_{\mathrm{disc}}(M(\mathbb{A}))$, $P \in \mathcal{P}(M)$ and $\lambda \in \mathfrak{a}^*_{M,\mathbb{C}}$. Define $(\mathfrak{N}_Q(P,\pi,\lambda,\Lambda))_{Q\in\mathcal{P}(M)}$ to be the operator valued (G,M)-family (in Λ) given by

$$\mathfrak{N}_Q(P,\pi,\lambda,\Lambda) = N_{Q|P}(\pi,\lambda)^{-1}N_{Q|P}(\pi,\lambda+\Lambda) = N_{P|Q}(\pi,\lambda)N_{Q|P}(\pi,\lambda+\Lambda). \tag{3.16}$$

From Lemma 3.11 (with $D = \mathrm{Id}$) and Lemma 3.1 we infer

Corollary 3.12 *Suppose that G and its Levi subgroups satisfy (WR). Then there exists $c' > 0$ and $k > 0$ such that for any level K of G we have*

$$\|\mathfrak{N}^S_M(P,\pi,\lambda,\Lambda)\|_{\mathcal{B}((L^2_{P,\pi})^{\tau,K})} \ll_K (1+\|\tau\|)^k$$

for any $S \in \mathcal{F}(P)$, $\pi \in \Pi_{\mathrm{disc}}(M(\mathbb{A}))$ and $\tau \in \widehat{\mathbf{K}_\infty}$ provided that $\|\mathrm{Re}\,\lambda\|$, $\|\mathrm{Re}\,\Lambda\| < c'$.

3.6

Finally, we combine the properties (UT) and (WR). Namely, we say that G satisfies property (HP) (hereditary property) if G and its Levi subgroups satisfy (UT) and (WR). By the above, the group $G = \mathrm{GL}_n$ satisfies (HP), and we expect that any reductive group satisfies (HP).

From Lemmas 3.3 and 3.11 and (3.5) we conclude

Corollary 3.13 *Suppose that G satisfies (HP) and let $M \in \mathcal{L}$. Then there exists $l > 0$ and*

- *for any $D \in \mathcal{D}(\mathfrak{a}^*_M)$ there exists $k > 0$,*
- *for any level K there exists $c > 0$,*

such that for any $P' \in \mathcal{P}(M')$, $w \in W(M,M')$, $\pi \in \Pi_{\mathrm{disc}}(M(\mathbb{A}))$ and $\tau \in \widehat{\mathbf{K}_\infty}$ we have

$$\|D_\lambda M_{P'|P}(w,\lambda)|_{(L^2_{P,\pi})^{\tau,K}}\|_{\mathcal{B}((L^2_{P,\pi})^{\tau,K},(L^2_{P',w\pi})^{\tau,K})} \ll_K (1+\|\lambda\|+\Lambda_{\pi_\infty}+\|\tau\|)^k$$

for $\lambda \in \mathcal{R}_{\pi,c,l}$.

Let $(\mathfrak{M}_Q(P,\lambda,\Lambda))_{Q\in\mathcal{P}(M)}$ be the operator valued (G,M)-family

$$\mathfrak{M}_Q(P,\lambda,\Lambda) = M_{Q|P}(\lambda)^{-1}M_{Q|P}(\lambda+\Lambda)|_{(L^2_P)^{\mathbf{K}\text{-fin},\mathfrak{z}\text{-fin}}} \tag{3.17}$$

where

$$(L^2_P)^{\mathbf{K}\text{-fin},\mathfrak{z}\text{-fin}} = \sum_{\pi\in\Pi_{\mathrm{disc}}(M(\mathbb{A})),\tau\in\widehat{\mathbf{K}_\infty},K} (L^2_{P,\pi})^{\tau,K}$$

is the $\mathfrak{z}$-finite, $\mathbf{K}$-finite part of L_P^2 (where K ranges over the compact open subgroups of $G(\mathbb{A}_{\text{fin}})$). We write $\mathfrak{M}_Q(P,\pi,\lambda,\Lambda)$ for the restriction of $\mathfrak{M}_Q(P,\lambda,\Lambda)$ to $L_{P,\pi}^2$.

Corollary 3.14 *Assume that G satisfies (HP). Then there exists $k>0$ and for any level K there exists $c>0$ such that for any $M\in\mathcal{L}$, $R\in\mathcal{F}(M)$, $\pi\in\Pi_{\text{disc}}(M(\mathbb{A}))$ and $\tau\in\widehat{\mathbf{K}_\infty}$ we have*

$$\|\mathfrak{M}_M^R(P,\pi,\lambda,\Lambda)\|_{\mathcal{B}((L_{P,\pi}^2)^{\tau,K})} \ll_K (1+\|\lambda\|+\|\Lambda\|+\|\tau\|+\Lambda_{\pi_\infty})^k$$

provided that $\lambda,\lambda+\Lambda\in\mathcal{R}_{\pi,c,k}$.

Proof Using the normalization of the intertwining operators we may write

$$\mathfrak{M}_Q(P,\pi,\lambda,\Lambda)=\nu_Q(P,\pi,\lambda,\Lambda)\mathfrak{N}_Q(P,\pi,\lambda,\Lambda)$$

where the (G,M)-family $\nu_Q(P,\pi,\lambda,\Lambda)$ is given by

$$\nu_Q(P,\pi,\lambda,\Lambda)=n_{Q|P}(\pi,\lambda)^{-1}n_{Q|P}(\pi,\lambda+\Lambda)$$

and $\mathfrak{N}_Q$ was defined in (3.16). Applying the product formula (3.1) to

$$\mathfrak{M}_Q^R(P,\pi,\lambda,\Lambda)=\nu_Q^R(P,\pi,\lambda,\Lambda)\mathfrak{N}_Q^R(P,\pi,\lambda,\Lambda),$$

it remains to invoke Lemma 3.3 and Corollary 3.12.

□

4 Conjectural Spectral Decomposition of the Harish-Chandra Schwartz Space

Recall the space L_P^2 defined in the previous section. We denote its smooth part by $\mathcal{A}_P$. Thus, $\mathcal{A}_P$ is the space of $\varphi\in C^\infty(M(F)U(\mathbb{A})\backslash G(\mathbb{A}))$ such that for all $X\in\mathcal{U}(\mathfrak{g})$ and $g\in G(\mathbb{A})$ the function $m\mapsto\delta_P(m)^{-\frac12}X\varphi(mg)$ belongs to $L^2_{\text{disc}}(A_M M(F)\backslash M(\mathbb{A}))$. We endow $\mathcal{A}_P$ with the inductive limit of $\mathcal{A}_P^K=\mathcal{A}_P\cap(L_P^2)^K$ topologized by the seminorms $\|X\varphi\|_P$, $X\in\mathcal{U}(\mathfrak{g})$. We similarly write $\mathcal{A}_{P,\pi}$.

Let $\mathcal{A}_{P,\mathfrak{z}\text{-fin}}$ be the subspace of $\mathcal{A}_P$ consisting of $\mathfrak{z}$-finite functions. It is the algebraic direct sum of $\mathcal{A}_{P,\pi}$, $\pi\in\Pi_{\text{disc}}(M(\mathbb{A}))$.

The following results are standard.

Lemma 4.1 *For any $X\in\mathcal{U}(\mathfrak{g})$ there exist a seminorm μ on $\mathcal{A}_P$ and $m\in\mathbb{N}$ such that $\|I(X,\lambda)\varphi\|_P\le(1+\|\lambda\|)^m\mu(\varphi)$ for all $\varphi\in\mathcal{A}_P$, $\lambda\in\mathfrak{a}_{M,\mathbb{C}}^*$.*

Lemma 4.2 *The topology on $\mathcal{A}_P^K$ is given by the seminorms*

$$\left(\sum_{\pi\in\Pi_{\mathrm{disc}}(M(\mathbb{A}))^K,\tau\in\widehat{\mathbf{K}_\infty}} (1+\Lambda_{\pi_\infty}+\|\tau\|)^k\|p_{\pi,\tau}^K\varphi\|_P^2\right)^{\frac{1}{2}}, \quad k\in\mathbb{N},$$

where $p_{\pi,\tau}^K : (L_P^2)^K \to (L_{P,\pi}^2)^{\tau,K}$ is the orthogonal projection.

Recall that $\mathcal{S}(\mathrm{i}\mathfrak{a}_M^*;\mathcal{A}_P)$ is the union over K of the Fréchet spaces $\mathcal{S}(\mathrm{i}\mathfrak{a}_M^*;\mathcal{A}_P^K)$ of smooth functions $\varphi : \mathrm{i}\mathfrak{a}_M^* \to \mathcal{A}_P^K$ such that the seminorms

$$\sup_{\lambda\in\mathrm{i}\mathfrak{a}_M^*}(1+\|\lambda\|)^n\|I(X)(D_\lambda\varphi(\lambda))\|_P$$

are finite for any $X\in\mathcal{U}(\mathfrak{g})$ and $D\in\mathcal{D}(\mathrm{i}\mathfrak{a}_M^*)$. We endow $\mathcal{S}(\mathrm{i}\mathfrak{a}_M^*;\mathcal{A}_P)$ with the inductive limit topology. By Lemma 4.2, we have

Lemma 4.3 *The topology on $\mathcal{S}(\mathrm{i}\mathfrak{a}_M^*;\mathcal{A}_P^K)$ is given by the seminorms*

$$\sup_{\lambda\in\mathrm{i}\mathfrak{a}_M^*}\left(\sum_{\pi\in\Pi_{\mathrm{disc}}(M(\mathbb{A}))^K,\tau\in\widehat{\mathbf{K}_\infty}} (1+\Lambda_{\pi_\infty}+\|\tau\|+\|\lambda\|)^k\|p_{\pi,\tau}^K D_\lambda\varphi(\lambda)\|_P^2\right)^{\frac{1}{2}},$$

$D\in\mathcal{D}(\mathrm{i}\mathfrak{a}_M^*)$, $k\in\mathbb{N}$.

Let $P = M\ltimes U$ and $P' = M'\ltimes U'$ be standard parabolic subgroups and let $w\in W(M,M')$. For any $\varphi\in\mathcal{S}(\mathrm{i}\mathfrak{a}_M^*;\mathcal{A}_P)$ define $\mathcal{I}_w\varphi(w\lambda) = M_{P'|P}(w,\lambda)\varphi(\lambda)$. By Lemma 4.3 and Corollary 3.13 we get

Lemma 4.4 *Suppose that G satisfies the property (HP). Then the map $\varphi\mapsto\mathcal{I}_w\varphi$ defines an isomorphism of topological vector spaces $\mathcal{S}(\mathrm{i}\mathfrak{a}_M^*;\mathcal{A}_P)\xrightarrow{\sim}\mathcal{S}(\mathrm{i}\mathfrak{a}_{M'}^*;\mathcal{A}_{P'})$.*

Denote by $\mathcal{M}$ the set of standard Levi subgroups.

Define

$$(\oplus\mathcal{S}(\mathrm{i}\mathfrak{a}_M^*;\mathcal{A}_P))^W = \{(f_M)_{M\in\mathcal{M}} : f_M\in\mathcal{S}(\mathrm{i}\mathfrak{a}_M^*;\mathcal{A}_P), f_{M'}\equiv\mathcal{I}_w(f_M), \forall w\in W(M,M')\}.$$

Similarly define $C_c^\infty(\mathrm{i}\mathfrak{a}_M^*;\mathcal{A}_P)$ to be the space of compactly supported smooth functions $\varphi : \mathrm{i}\mathfrak{a}_M^*\to\mathcal{A}_P$ with image in $\mathcal{A}_P^K$ for some K, and let

$$(\oplus C_c^\infty(\mathrm{i}\mathfrak{a}_M^*;\mathcal{A}_P))^W = \{(\varphi_M)_{M\in\mathcal{M}}\in(\oplus\mathcal{S}(\mathrm{i}\mathfrak{a}_M^*;\mathcal{A}_P))^W : \varphi_M\in C_c^\infty(\mathrm{i}\mathfrak{a}_M^*;\mathcal{A}_P)\}.$$

We also define $L^2(i\mathfrak{a}_M^*; L_P^2)$ to be the space of measurable functions $\varphi : i\mathfrak{a}_M^* \to L_P^2$ such that $\int_{i\mathfrak{a}_M^*} \|\varphi(\lambda)\|_P^2 \, d\lambda < \infty$ (up to functions which are zero almost everywhere) and

$$(\oplus L^2(i\mathfrak{a}_M^*; L_P^2))^W = \{(\varphi_M)_{M \in \mathcal{M}} : \varphi_M \in L^2(i\mathfrak{a}_M^*; L_P^2),$$
$$\varphi_{M'}(w\lambda) = M_{P'|P}(w, \lambda)\varphi_M(\lambda) \text{ for almost all } \lambda \in i\mathfrak{a}_M^*, \forall w \in W(M, M')\},$$

with the inner product

$$\|(\varphi_M)\|^2 = \sum_{M \in \mathcal{M}} n_M \int_{i\mathfrak{a}_M^*} \|\varphi_M(\lambda)\|_P^2 \, d\lambda$$

where $n_M = \sum_{M' \in \mathcal{M}} |W(M, M')|$.

We write $C_c^\infty(i\mathfrak{a}_M^*; \mathcal{A}_{P,\mathfrak{z}\text{-fin}}) = \oplus_{\pi \in \Pi_{\mathrm{disc}}(M(\mathbb{A}))} C_c^\infty(i\mathfrak{a}_M^*; \mathcal{A}_{P,\pi})$.

Recall the Eisenstein series $E(\varphi, \lambda)$. The Eisenstein transform

$$\mathcal{E} : C_c^\infty(i\mathfrak{a}_M^*; \mathcal{A}_{P,\mathfrak{z}\text{-fin}}) \to C^\infty(G(F)\backslash G(\mathbb{A}))$$

is given by

$$\mathcal{E}(\varphi) = \int_{i\mathfrak{a}_M^*} E(\varphi(\lambda), \lambda) \, d\lambda.$$

By Langlands, $\mathcal{E}$ extends to a continuous linear map of Hilbert spaces

$$\tilde{\mathcal{E}} : L^2(i\mathfrak{a}_M^*; L_P^2) \to L^2(G(F)\backslash G(\mathbb{A}))$$

which induces an isomorphism of Hilbert spaces

$$\tilde{\mathcal{E}}^W : (\oplus L^2(i\mathfrak{a}_M^*; L_P^2))^W \xrightarrow{\sim} L^2(G(F)\backslash G(\mathbb{A})).$$

The inverse of $\tilde{\mathcal{E}}$ is given by $\phi \mapsto (\varphi_M)_M$ where

$$(\varphi_M(\lambda), \psi)_P = \frac{1}{n_M}(\phi, E(g, \psi, \lambda))_{G(F)\backslash G(\mathbb{A})}, \quad \lambda \in i\mathfrak{a}_M^*, \ \psi \in \mathcal{A}_P.$$

Following Harish-Chandra in the local case (cf. [Art75, Wal92, Wal03]) it is natural to make the following conjecture.

Conjecture 4.5 *The map $\mathcal{E}$ uniquely extends to a continuous linear map*

$$\mathcal{S}(i\mathfrak{a}_M^*; \mathcal{A}_P) \to \mathcal{S}(G(F)\backslash G(\mathbb{A}))$$

which induces an isomorphism of topological vector spaces

$$(\oplus \mathcal{S}(i\mathfrak{a}_M^*; \mathcal{A}_P))^W \xrightarrow{\sim} \mathcal{S}(G(F)\backslash G(\mathbb{A})).$$

Note that a special case of the conjecture is that $\mathcal{A}_G \subseteq \mathcal{S}(G(F)\backslash G(\mathbb{A}))$ if $A_G = 1$. This assertion already seems nontrivial. (The local analogue is the finiteness of discrete series representations with a given K-type.) On the other hand, it is well known that the cuspidal part of $\mathcal{A}_G$ is contained in $\mathcal{S}(G(F)\backslash G(\mathbb{A}))$ (if $A_G = 1$). Our modest goal in this paper is to prove a cuspidal version of Conjecture 4.5 for groups satisfying (HP). More precisely, let $\mathcal{A}_{P,\mathrm{cusp}}$ be the cuspidal part of $\mathcal{A}_P$ and denote by $\mathcal{E}_{\mathrm{cusp}}$ the restriction of $\mathcal{E}$ to $C_c^\infty(\mathrm{i}\mathfrak{a}_M^*; \mathcal{A}_{P,\mathrm{cusp},\mathfrak{z}\text{-fin}})$. Similarly for $\tilde{\mathcal{E}}_{\mathrm{cusp}}$. Let $L_c^2(G(F)\backslash G(\mathbb{A}))$ be the image of $\oplus L^2(\mathrm{i}\mathfrak{a}_M^*; L^2_{P,\mathrm{cusp}})$ under $\tilde{\mathcal{E}}_{\mathrm{cusp}}$. It is a closed subspace of $L^2(G(F)\backslash G(\mathbb{A}))$. Let $\mathcal{S}_c(G(F)\backslash G(\mathbb{A})) = L_c^2(G(F)\backslash G(\mathbb{A})) \cap \mathcal{S}(G(F)\backslash G(\mathbb{A}))$, a closed subspace of $\mathcal{S}(G(F)\backslash G(\mathbb{A}))$. Then we have

Theorem 4.6 *Assume that G satisfies (HP). Then the map $\mathcal{E}_{\mathrm{cusp}}$ uniquely extends to a continuous linear map*

$$\mathcal{S}(\mathrm{i}\mathfrak{a}_M^*; \mathcal{A}_{P,\mathrm{cusp}}) \to \mathcal{S}_c(G(F)\backslash G(\mathbb{A}))$$

which gives rise to an isomorphism of topological vector spaces

$$(\oplus \mathcal{S}(\mathrm{i}\mathfrak{a}_M^*; \mathcal{A}_{P,\mathrm{cusp}}))^W \xrightarrow{\sim} \mathcal{S}_c(G(F)\backslash G(\mathbb{A})).$$

5 Majorization of Cuspidal Eisenstein Series

The key step in the proof of Theorem 4.6, which is of independent interest, will be the following majorization of Eisenstein series in the spirit of [Lap06, Proposition 6.1].

Throughout this section we assume that G satisfies (HP).

Proposition 5.1 *For any $X \in \mathcal{U}(\mathfrak{g})$ there exists $k > 0$ such that*

$$\varphi \mapsto \sup_{\lambda \in \mathrm{i}\mathfrak{a}_M^*} (1 + \|\lambda\|)^{-k} \sup_{x \in \mathfrak{s}^1} \Xi(x)^{-1} \sigma(x)^{(r(G)-r(P))/2} |XE(x, \varphi, \lambda)|$$

is a continuous seminorm on $\mathcal{A}_{P,\mathrm{cusp}}$. Moreover, for any $N > 1$ there exists $m \in \mathbb{N}$ such that for any $f \in C_c^m(G(F_\infty))$ and level K of G we have

$$[f * E(\varphi, \lambda)](x) = E(x, I(f, \lambda)\varphi, \lambda) \ll_{f,N,K} \Xi(x)\sigma(x)^{(r(P)-r(G))/2}(1 + \|\lambda\|)^{-N} \|\varphi\|_P \quad (5.1)$$

for any $\varphi \in \mathcal{A}^K_{P,\mathrm{cusp}}$, $\lambda \in \mathrm{i}\mathfrak{a}_M^*$, $x \in \mathfrak{s}^1$.

Proof Note that by Lemma 2.5 and Lemma 4.1 the first part of the proposition follows from the second part. In fact, the first part is equivalent to the following statement: there exists $k > 0$ such that for any level K of G there exists $m \in \mathbb{N}$ such that for any $f \in C_c^m(G(F_\infty))$ we have

$$E(x, I(f,\lambda)\varphi, \lambda) \ll_{f,K} \Xi(x)\sigma(x)^{(r(P)-r(G))/2}(1 + \|\lambda\|)^k \|\varphi\|_P$$

for any $\varphi \in \mathcal{A}^K_{P,\mathrm{cusp}}$, $\lambda \in i\mathfrak{a}^*_M$, $x \in \mathfrak{s}^1$.

To prove the bound (5.1) we may assume that $f = f_1 * f_2$ for sufficiently smooth $f_1, f_2 \in C_c(G(F_\infty))$, since any $f \in C_c^m(G(F_\infty))$ is a linear combination of these. Corollary 2.4 applied to $\phi = f_2 * E(\varphi,\lambda)$ and $f = f_1$ will reduce the Proposition to the following Lemma.

□

Lemma 5.2 *For any $N > 1$ there exists $m \in \mathbb{N}$ such that for any $f \in C_c^m(G(F_\infty))$ and level K of G we have*

$$\|\Lambda^T E(I(f,\lambda)\varphi, \lambda)\|_{L^2(G(F)\backslash G(\mathbb{A})^1)} \ll_{f,N,K} \|T\|^{(r(P)-r(G))/2}(1+\|\lambda\|)^{-N}\|\varphi\|_P$$

for any $\varphi \in \mathcal{A}^K_{P,\mathrm{cusp}}$, $\lambda \in i\mathfrak{a}^*_M$, and $T \in \mathfrak{a}_{0,+}$ sufficiently regular.

Proof We recall the Maass-Selberg relations worked out in [Lan76] and [Art80]. We will follow the discussion in [Art82]. As in [ibid., p. 1310] consider the (G,M)-families (in $\Lambda \in i\mathfrak{a}^*_M$)

$$c_Q(T,\Lambda) = e^{\langle \Lambda, Y_Q(T)\rangle},$$
$$\mathfrak{M}^T_Q(P,\lambda,\Lambda) = c_Q(T,\Lambda)\mathfrak{M}_Q(P,\lambda,\Lambda),$$

where the Y_Q's are certain affine functions which we don't need to know explicitly and $\mathfrak{M}_Q$ was defined in (3.17). Then

$$\|\Lambda^T E(\varphi,\lambda)\|^2_{L^2(G(F)\backslash G(\mathbb{A})^1)} = \sum_{s \in W(M,M)} (\mathfrak{M}^T_M(P,\lambda,s\lambda-\lambda)M_{P|P}(s,\lambda)\varphi,\varphi)_P. \quad (5.2)$$

We use the product formula (3.1). It is clear from the formula [ibid., (3.1)] that for any $Q_1 \in \mathcal{F}(M)$ we have $c_M^{Q_1}(T,\Lambda) \ll \|T\|^{r(P)-r(Q_1)}$. By Lemma 4.1 it remains to show that any $N > 1$, $Q_2 \in \mathcal{F}(M)$ and $s \in W(M,M)$ there exists $m \in \mathbb{N}$ such that for any $f \in C_c^m(G(F_\infty))$ and level K we have

$$\sup_{\lambda \in i\mathfrak{a}^*_M} (1 + \|\lambda\|)^N \|\mathfrak{M}^{Q_2}_M(P,\lambda,s\lambda-\lambda)M_{P|P}(s,\lambda)I(f,\lambda)\varphi\|_P \ll_{f,N,K} \|\varphi\|_P$$

for all $\varphi \in \mathcal{A}_P^K$. Since $M_{P|P}(s,\lambda)$ is unitary and $M_{P|P}(s,\lambda)I(f,\lambda) = I(f,s\lambda)M_{P|P}(s,\lambda)$, it suffices to prove that

$$\sup_{\lambda\in i\mathfrak{a}_M^*}(1+\|\lambda\|)^N\|\mathfrak{M}_M^{Q_2}(P,\lambda,s\lambda-\lambda)I_P(f,s\lambda)\|_{\mathcal{B}((L_P^2)^K)} < \infty,$$

or equivalently, that

$$\sup_{\lambda\in i\mathfrak{a}_M^*}(1+\|\lambda\|)^N \sup_{\pi\in\Pi_{\mathrm{disc}}(M(\mathbb{A}))}\|\mathfrak{M}_M^{Q_2}(P,\pi,\lambda,s\lambda-\lambda)I_P(f,\pi,s\lambda)\|_{\mathcal{B}((L_{P,\pi}^2)^K)} < \infty.$$

Expand $f = \sum_{\tau\in\widehat{\mathbf{K}_\infty}} f_\tau$ according to the left action of $\mathbf{K}_\infty$. Then

$$\begin{aligned}&\|\mathfrak{M}_M^{Q_2}(P,\pi,\lambda,s\lambda-\lambda)I_P(f,\pi,s\lambda)\|_{\mathcal{B}((L_{P,\pi}^2)^K)}\\ &\quad\le \sum_{\tau\in\widehat{\mathbf{K}_\infty}}\|\mathfrak{M}_M^{Q_2}(P,\pi,\lambda,s\lambda-\lambda)\|_{\mathcal{B}((L_{P,\pi}^2)^{\tau,K})}\|I_P(f_\tau,\pi,s\lambda)\|_{\mathcal{B}((L_{P,\pi}^2)^K)}.\end{aligned}$$

To prove the lemma it therefore remains to apply Corollary 3.14 together with the fact that for any $N > 1$ there exists $m \in \mathbb{N}$ such that for any $f \in C_c^m(G(\mathbb{A}))$ we have

$$\|I_P(\pi,f_\tau,s\lambda)\| \ll_{f,K,N} (1+\Lambda_{\pi_\infty}+\|\tau\|+\|\lambda\|)^{-N}.$$

□

Remark 5.3 It is likely that with a finer analysis the factor $\sigma(x)^{(r(P)-r(G))/2}$ can be eliminated from the statement of Proposition 5.1. However, we will not address this issue here.

Remark 5.4 Consider the standard Eisenstein series for $\mathrm{SL}_2(\mathbb{Z})$, namely, the analytic continuation in $s \in \mathbb{C}$ of

$$E(z,s) = \sum_{(m,n)=1}\frac{y^{s+\frac12}}{|mz+n|^{2s+1}}, \qquad z = x+\mathrm{i}y, y>0.$$

Fix $c > 0$. The argument of [Sar04] together with the bounds $\zeta(1+s)$, $\zeta(1+s)^{-1} \ll \log(1+|s|)$, $s\in \mathrm{i}\mathbb{R}$ show that for any $\epsilon > 0$,

$$E(z,s) \ll_{c,\epsilon} y^{\frac12} + |s|^{\frac12+\epsilon}y^{-\frac12}$$

uniformly for $y > c$ and $s \in \mathrm{i}\mathbb{R}$. This bound is essentially sharp for $y > |s|^{\frac12}$ but at least if y is confined to a compact set, the argument of [IS95] gives a better exponent in $|s|$. It is an extremely difficult problem to determine precisely the growth of $E(z,s)$. For instance, the estimate $E(\mathrm{i},s) \ll_\epsilon (1+|s|)^\epsilon$, $s \in \mathrm{i}\mathbb{R}$ is equivalent to the Lindelöf hypothesis for the Dedekind zeta function $\zeta_{\mathbb{Q}(\mathrm{i})}$.

Next, we will give a consequence of Proposition 5.1. For $\kappa \geq 0$ and any level K of G, denote by $A^\kappa_{\mathrm{mg}}(G(F)\backslash G(\mathbb{A})^1)^K$ the Fréchet space of right-K-invariant functions $\phi \in C^\infty(G(F)\backslash G(\mathbb{A})^1)$ such that

$$\sup_{x \in \mathfrak{s}^1} \Xi(x)^{-1}\sigma(x)^{-\kappa}\,|R(X)\phi(x)| < \infty$$

for all $X \in \mathcal{U}(\mathfrak{g}^1)$. We also denote by $A^\kappa_{\mathrm{mg}}(G(F)\backslash G(\mathbb{A}))^K$ the Fréchet space of right-K-invariant functions $\phi \in C^\infty(G(F)\backslash G(\mathbb{A}))$ such that

$$\sup_{x \in \mathfrak{s}^1, a \in A_G} \Xi(x)^{-1}\sigma(a)^n\sigma(x)^{-\kappa}\,|R(X)\phi(ax)| < \infty$$

for all $X \in \mathcal{U}(\mathfrak{g})$ and $n \in \mathbb{N}$. (Note that we require ϕ to be a Schwartz function in A_G.) We write $A^\kappa_{\mathrm{mg}}(G(F)\backslash G(\mathbb{A})^1)$ for the union over K of $A^\kappa_{\mathrm{mg}}(G(F)\backslash G(\mathbb{A}))^K$ with the inductive limit topology. Similarly for $A^\kappa_{\mathrm{mg}}(G(F)\backslash G(\mathbb{A}))$. Note that we can identify $A^\kappa_{\mathrm{mg}}(G(F)\backslash G(\mathbb{A}))$ with $\mathcal{S}(A_G; A^\kappa_{\mathrm{mg}}(G(F)\backslash G(\mathbb{A})^1))$ by $f \mapsto (a \mapsto f(a\cdot))$.

Corollary 5.5 *Let κ be half the F-rank of G. Then the map $\varphi \mapsto \mathcal{E}(\varphi)$ extends to a continuous linear map from $\mathcal{S}(\mathrm{i}\mathfrak{a}_M^*; \mathcal{A}_{P,\mathrm{cusp}})$ to $A^\kappa_{\mathrm{mg}}(G(F)\backslash G(\mathbb{A}))$.*

Proof It easily follows from Proposition 5.1 that the map

$$\varphi \in \mathcal{S}(\mathrm{i}\mathfrak{a}_M^*; \mathcal{A}_{P,\mathrm{cusp}}) \mapsto \int_{\mathrm{i}(\mathfrak{a}_M^G)^*} E(\varphi(\lambda+\cdot),\lambda)\,d\lambda \in \mathcal{S}(\mathrm{i}\mathfrak{a}_G^*; A^\kappa_{\mathrm{mg}}(G(F)\backslash G(\mathbb{A})^1))$$

is continuous. Composing this map with the Fourier transform on $\mathrm{i}\mathfrak{a}_G^*$ we get $\mathcal{E}_{\mathrm{cusp}}$.

□

For the second part of Theorem 4.6 we will need to extend Proposition 5.1 to the derivatives of $E(\varphi,\lambda)$ in λ. To that end we first extend Proposition 5.1 for $\mathfrak{z}$-finite φ to a larger domain of λ's.

Proposition 5.6 *There exists $l > 0$ and*

- *for any level K of G there exists $c > 0$,*
- *for any $N > 1$ there exists $m \in \mathbb{N}$,*

such that for any $f \in C_c^m(G(F_\infty))$ and $\pi \in \Pi_{\text{cusp}}(M(\mathbb{A}))^{K_M}$ we have

$$f * E(x, \varphi, \lambda) \ll_{f,N,K} \Xi(x) e^{\|\operatorname{Re}\lambda\|\sigma(x)} \sigma(x)^{(r(P)-r(G))/2} (1 + \Lambda_{\pi_\infty} + \|\lambda\|)^{-N} \|\varphi\|_P$$

for any $\varphi \in \mathcal{A}_{P,\text{cusp},\pi}^K$, $x \in \mathfrak{s}^1$ and $\lambda \in \mathcal{R}_{\pi,c,l}$.

Proof We proceed as in Proposition 5.1 and Lemma 5.2 and use the notation of the latter. The difference is that now we have to use the Maass-Selberg relations for general $\lambda \in \mathfrak{a}_{M,\mathbb{C}}^*$. Thus, instead of (5.2) we have

$$\|\Lambda^T E(\varphi, \lambda)\|^2_{L^2(G(F)\backslash G(\mathbb{A})^1)} = \sum_{s \in W(M,M)} (\mathfrak{M}_M^T(P, -\overline{\lambda}, s\lambda + \overline{\lambda}) M_{P|P}(s, \lambda)\varphi, \varphi)_P$$

(cf. the proof of [Lap06, Proposition 6.1]). Since $c_M^{Q_1}(T, \Lambda)$ is the Fourier transform of a certain polytope (cf. [Art82, §3]) we get

$$c_M^{Q_1}(T, \Lambda) \ll e^{\|\operatorname{Re}\Lambda\| \|T\|} \|T\|^{r(P)-r(G)}.$$

The rest of the proof follows that of Lemma 5.2 except that we have to bound $\mathfrak{M}_M^T(P, \pi, -\overline{\lambda}, \Lambda)$ and $M_{P|P}(s, \lambda)$ off the unitary axis. To that end we apply Corollaries 3.13 (with $D = \text{Id}$) and 3.14.

□

Corollary 5.7 *There exists $l > 0$ and*

- *for any level K of G there exists $c' > 0$,*
- *for any $D \in \mathcal{D}(\mathfrak{a}_M^*)$ there exists $k > 0$,*
- *for any $N > 1$ there exists $m \in \mathbb{N}$,*

such that for any $f \in C_c^m(G(F_\infty))$ and $\pi \in \Pi_{\text{cusp}}(M(\mathbb{A}))^{K_M}$ we have

$$D_\lambda(f * E(x, \varphi, \lambda)) \ll_{f,D,N,K} \Xi(x) e^{\|\operatorname{Re}\lambda\|\sigma(x)} \sigma(x)^k (1 + \Lambda_{\pi_\infty} + \|\lambda\|)^{-N} \|\varphi\|_P$$

for any $\varphi \in \mathcal{A}_{P,\text{cusp},\pi}^K$, $x \in \mathfrak{s}^1$ and $\lambda \in \mathcal{R}_{\pi,c',l}$.

This follows from Proposition 5.6 by taking $c' = c/2$ and using Cauchy's formula for a circle of radius $\min(\sigma(x)^{-1}, c'(1 + \Lambda_{\pi_\infty} + \|\operatorname{Im}\lambda\|)^{-l})$.

Corollary 5.8 *For any $D \in \mathcal{D}(i\mathfrak{a}_M^*)$ there exists $k > 0$ and for any $N > 1$ there exists $m \in \mathbb{N}$ such that for any $f \in C_c^m(G(F_\infty))$ and level K of G we have*

$$D_\lambda(f * E(x, \varphi, \lambda)) \ll_{f,D,N,K} \Xi(x)\sigma(x)^k (1 + \|\lambda\|)^{-N} \|\varphi\|_P$$

for any $\varphi \in \mathcal{A}_{P,\text{cusp}}^K$, $x \in \mathfrak{s}^1$ and $\lambda \in i\mathfrak{a}_M^*$.

This follows from Corollary 5.7 together with the fact that there exists $N > 1$ such that for any K we have

$$\sum_{\pi \in \Pi_{\mathrm{cusp}}(M(\mathbb{A}))^{K_M}} m_{\mathrm{cusp}}(\pi)(1+\Lambda_{\pi_\infty})^{-N} < \infty$$

where $m_{\mathrm{cusp}}(\pi) = \dim \mathrm{Hom}(\pi, L^2_{\mathrm{cusp}}(A_M M(F)\backslash M(\mathbb{A})))$ (e.g., [Don82]).

6 Proof of Main Theorem

We will apply induction and approximation by the constant term. For convenience we introduce an auxiliary space of functions on $P_0(F)\backslash\mathfrak{s}$, analogous to $\mathcal{S}(G(F)\backslash G(\mathbb{A}))$. Namely, for any level K of G let $A_{\mathrm{sch}}(P_0(F)\backslash\mathfrak{s})^K$ be the Fréchet space of smooth right-K-invariant functions $\phi : P_0(F)\backslash\mathfrak{s} \to \mathbb{C}$ such that for all $n \in \mathbb{N}$ and $X \in \mathcal{U}(\mathfrak{g})$ we have

$$\sup_{x\in\mathfrak{s}} \Xi(x)^{-1}\sigma(x)^n \left|X\phi(x)\right| < \infty. \tag{6.1}$$

We write $A_{\mathrm{sch}}(P_0(F)\backslash\mathfrak{s})$ for the union over K of $A_{\mathrm{sch}}(P_0(F)\backslash\mathfrak{s})^K$ with the inductive limit topology.

Note that we can identify $\mathcal{S}(G(F)\backslash G(\mathbb{A}))$ with a closed subspace of $A_{\mathrm{sch}}(P_0(F)\backslash\mathfrak{s})$.

For any measurable locally bounded function φ on $U_0(F)\backslash G(\mathbb{A})$ and a standard parabolic subgroup $P = M \ltimes U$ let φ_P be the constant term

$$\varphi_P(g) = \int_{U(F)\backslash U(\mathbb{A})} \varphi(ug)\, du, \quad g \in G(\mathbb{A}).$$

The map $\phi \mapsto \mathfrak{d}\phi = \sum_{P\supseteq P_0}(-1)^{r(P)-r(G)}\phi_P$ defines a linear operator on the space of measurable locally bounded functions on $P_0(F)\backslash\mathfrak{s}$.

The following version of Langlands's Lemma is proved exactly as in [MW95, Lemma I.2.10 and Corollary I.2.11].

Proposition 6.1 *Fix $\lambda_0 \in \mathfrak{a}_0^*$. Then $\mathfrak{d}$ defines a continuous map from the space of functions on $P_0(F)\backslash\mathfrak{s}$ defined by the seminorms*

$$\sup_{x\in\mathfrak{s}^1, a\in A_G} \Xi(x)^{-1}\sigma(a)^n e^{-\langle\lambda_0, H(x)\rangle}\left|X\phi(ax)\right|, \quad X \in \mathcal{U}(\mathfrak{g}), n \in \mathbb{N}$$

to the space of functions on $P_0(F)\backslash\mathfrak{s}$ defined by the seminorms

$$\sup_{x\in\mathfrak{s}^1, a\in A_G} \Xi(x)^{-1}\sigma(a)^n e^{\langle\lambda, H(x)\rangle}\left|X\phi(ax)\right|, \quad X \in \mathcal{U}(\mathfrak{g}), n \in \mathbb{N}, \lambda \in \mathfrak{a}_0^*.$$

Remark 6.2 The factor $\Xi(x)^{-1}$ can be eliminated from the statement of the proposition without changing its contents. We include it in order to conform to the previous notation.

For our purposes we will only need the following much weaker consequence.

Corollary 6.3 *Fix* $\kappa \geq 0$. *Then* $\mathfrak{d}$ *defines a continuous map from* $A^{\kappa}_{\mathrm{mg}}(G(F)\backslash G(\mathbb{A}))$ *to* $A_{\mathrm{sch}}(P_0(F)\backslash\mathfrak{s})$.

More generally, for any standard parabolic subgroup $Q = LV$, we can define $A_{\mathrm{sch}}(P_0(F)\backslash\mathfrak{s}^Q)$ in a similar way (with $\mathfrak{s}$, Ξ and σ replaced by $\mathfrak{s}^Q$, σ^Q and Ξ^Q respectively in (6.1)), and we can identify $\mathcal{S}(L(F)V(\mathbb{A})\backslash G(\mathbb{A}))$ with a closed subspace of $A_{\mathrm{sch}}(P_0(F)\backslash\mathfrak{s}^Q)$.

Note that by (2.10), the restriction map

$$A_{\mathrm{sch}}(P_0(F)\backslash\mathfrak{s}^Q) \to A_{\mathrm{sch}}(P_0(F)\backslash\mathfrak{s})$$

is continuous. Therefore,

$$\text{the restriction map } \mathcal{S}(L(F)V(\mathbb{A})\backslash G(\mathbb{A})) \to A_{\mathrm{sch}}(P_0(F)\backslash\mathfrak{s}) \text{ is continuous} \tag{6.2}$$

(not necessarily an embedding).

Proof (of Theorem 4.6) The proof will be by induction on the semisimple rank of G. The case where G is anisotropic modulo its center is trivial. For the induction step, we can assume by induction hypothesis that the theorem holds for any proper Levi subgroup of G. The constant term of the cuspidal Eisenstein transform is given by

$$\mathcal{E}(\varphi)_Q = \sum_{M'} \sum_{w \in W(M,M')} \mathcal{E}^Q(\mathcal{I}_w\varphi),$$

where M' ranges over the standard Levi subgroups of L and $\mathcal{E}^Q$ is the relative Eisenstein transform. By the induction hypothesis and Lemma 4.4, we conclude that for any proper Q the map $\mathcal{S}(\mathrm{i}\mathfrak{a}^*_M;\mathcal{A}_{P,\mathrm{cusp}}) \to \mathcal{S}(L(F)V(\mathbb{A})\backslash G(\mathbb{A}))$ given by $\varphi \mapsto \mathcal{E}(\varphi)_Q$ is continuous. Thus, by (6.2), for all $Q \subsetneq G$

$$\text{the map } \varphi \mapsto \mathcal{E}(\varphi)_Q\big|_{P_0(F)\backslash\mathfrak{s}} \text{ from } \mathcal{S}(\mathrm{i}\mathfrak{a}^*_M;\mathcal{A}_{P,\mathrm{cusp}}) \text{ to } A_{\mathrm{sch}}(P_0(F)\backslash\mathfrak{s}) \text{ is continuous.} \tag{6.3}$$

On the other hand, it follows from Corollaries 5.5 and 6.3 that for κ equals half the F-rank of G, the map $\varphi \mapsto \mathfrak{d}(\mathcal{E}(\varphi))$ defines a continuous linear map from $\mathcal{S}(\mathrm{i}\mathfrak{a}^*_M;\mathcal{A}_{P,\mathrm{cusp}})$ to $A_{\mathrm{sch}}(P_0(F)\backslash\mathfrak{s})$. From the definition of the map $\mathfrak{d}$ and (6.3) we conclude that the map $\varphi \mapsto \mathcal{E}(\varphi)\big|_{P_0(F)\backslash\mathfrak{s}}$ from $\mathcal{S}(\mathrm{i}\mathfrak{a}_M^*;\mathcal{A}_{P,\mathrm{cusp}})$ to $A_{\mathrm{sch}}(P_0(F)\backslash\mathfrak{s})$ is continuous. Equivalently, the map $\varphi \mapsto \mathcal{E}(\varphi)$ to $\mathcal{S}(G(F)\backslash G(\mathbb{A}))$ is continuous.

To show the second part of Theorem 4.6 we will construct the inverse

$$\iota : \mathcal{S}_c(G(F)\backslash G(\mathbb{A})) \to (\oplus \mathcal{S}(\mathrm{i}\mathfrak{a}_M^*;\mathcal{A}_P))^W$$

to $\mathcal{E}_{\mathrm{cusp}}$ (which will be automatically continuous by the open mapping theorem). Let $\phi \in \mathcal{S}_c(G(F)\backslash G(\mathbb{A}))$. We first claim that for any $\lambda \in \mathrm{i}\mathfrak{a}_M^*$ the linear form $\varphi \mapsto (\phi, E(\varphi,\lambda))_{G(F)\backslash G(\mathbb{A})}$ extends to $L^2_{P,\mathrm{cusp}}$. Indeed, we may assume that ϕ is of the form $R(f_0)\phi_0$ for some $f_0 \in C_c^m(G(F_\infty))$ (with $m \in \mathbb{N}$ arbitrary) and then the claim follows from Proposition 5.1 and the relation

$$(\phi, E(\varphi,\lambda))_{G(F)\backslash G(\mathbb{A})} = (\phi_0, E(I(f_0^*,\lambda)\varphi,\lambda))_{G(F)\backslash G(\mathbb{A})}$$

where $f_0^*(g) = \overline{f_0(g^{-1})}$. Let $\iota_P(\phi)(\lambda) \in L^2_{P,\mathrm{cusp}}$ be such that

$$(\iota_P(\phi)(\lambda),\varphi)_P = \frac{1}{n_M}(\phi, E(\varphi,\lambda))_{G(F)\backslash G(\mathbb{A})}.$$

By a similar reasoning, using Proposition 5.1 once again, we have $\iota_P(\phi)(\lambda) \in \mathcal{A}_{P,\mathrm{cusp}}$, i.e. $\iota_P(\phi)(\lambda)$ is a smooth vector in $L^2_{P,\mathrm{cusp}}$. By the functional equation of the Eisenstein series we have $\iota_{P'}(\phi)(w\lambda) = M_{P'|P}(w,\lambda)(\iota_P(\phi)(\lambda))$ for all $w \in W(M,M')$ and $\lambda \in \mathrm{i}\mathfrak{a}_M^*$. It remains therefore to show that $\iota_P(\phi) \in \mathcal{S}(\mathrm{i}\mathfrak{a}_M^*;\mathcal{A}_{P,\mathrm{cusp}})$. This follows easily from Corollary 5.8.

□

It would be desirable to extend Theorem 4.6, as well as Proposition 5.1, to the non-cuspidal case. We will not discuss this problem here but we mention that a first step in this direction, namely the Maass-Selberg relations for general Eisenstein series, was carried out in [LO12].

A Lower Bounds on Rankin-Selberg L-functions, by Farrell Brumley[4]

The purpose of this appendix is to update and improve the statement of [Bru06, Theorem 5] on the lower bounds of Rankin-Selberg L-functions at the edge of the critical strip. We use this opportunity to correct some errors that appeared in [ibid.]. (See footnotes below.)

[4]Partially supported by the ANR grant ArShiFO ANR-BLANC-114-2010.

Fix a number field F and integers n_1, n_2. Write $d = n_1 + n_2$. Let π_1 and π_2 be automorphic cuspidal representations of $\mathrm{GL}_{n_1}(\mathbb{A}_F)$ and $\mathrm{GL}_{n_2}(\mathbb{A}_F)$, respectively. We assume that their central characters are unitary and normalized so that viewed as Hecke characters on $F^*\backslash\mathbb{A}_F^*$, they are trivial on $\mathbb{R}_{>0}$ embedded diagonally.

In the following all constants depend implicitly on F, n_1, n_2. As usual we write $X \ll_\varepsilon Y$ to mean that $|X| \leq cY$ for some constant c depending on ε (as well as on F, n_1, n_2).

Theorem A.1 *For every $\varepsilon > 0$ there exists a constant $c_1 > 0$ such that for any two distinct π_1 and π_2 as above and all*

$$\mathrm{Re}\, s \geq 1 - c_1 \mathfrak{c}(\Pi \times \tilde{\Pi})^{-(\frac{1}{2} - \frac{1}{2d} + \varepsilon)}, \tag{A.1}$$

we have

$$L^\infty(s, \pi_1 \times \tilde{\pi}_2)^{-1} \ll_\varepsilon \mathfrak{c}(\Pi \times \tilde{\Pi})^{\frac{1}{2} - \frac{1}{2d} + \varepsilon}, \tag{A.2}$$

where, setting $t = \mathrm{Im}\, s$,

$$\mathfrak{c}(\Pi \times \tilde{\Pi}) = \mathfrak{c}(\pi_1 \times \tilde{\pi}_1)\mathfrak{c}(\pi_2 \times \tilde{\pi}_2)\mathfrak{c}(\pi_1 \times \tilde{\pi}_2, 1 + it)^2. \tag{A.3}$$

Similarly, for every $\varepsilon > 0$ and n there exists a constant $c_2 > 0$ such that for any cuspidal representation π of $\mathrm{GL}_n(\mathbb{A}_F)$ we have

$$\frac{s}{s-1} L^\infty(s, \pi \times \tilde{\pi})^{-1} \ll_\varepsilon \mathfrak{c}(\Pi \times \tilde{\Pi})^{\frac{7}{8} - \frac{5}{8n} + \varepsilon}, \tag{A.4}$$

for $\mathrm{Re}\, s \geq 1 - c_2 \mathfrak{c}(\Pi \times \tilde{\Pi})^{-(\frac{7}{8} - \frac{5}{8n} + \varepsilon)}$ *where $\mathfrak{c}(\Pi \times \tilde{\Pi})$ is as before (with $\pi_2 = \pi_1$).*

Remark A.2 Note that this makes the following improvements to the original statement of Theorem 5 of [Bru06].

1. We have explicated the exponent. With Xiannan Li's recent convexity bound [Li10] on Rankin-Selberg L-functions, this seemed like a good time to put into print an explicit power.
2. We have included the case where $\pi_1 = \pi_2$, which was not considered in [Bru06].
3. We have extended the range of s from the 1-line, as it was originally stated, to the wider region that extends slightly within the critical strip.

We would like to thank Erez Lapid for bringing these improvements to our attention and for suggesting the necessary modifications to the proof to handle the points (2) and (3).

Remark A.3 Besides the narrow (or coarse) zero free regions and lower bounds at the edge of the critical strip for Rankin-Selberg L-functions, we also established in [Bru06, Theorem 7] an effective multiplicity one result for cusp forms on GL_n. The more recent papers [Wan08, LW09] explicate the exponents of this other result, using a more efficient argument that doesn't pass by zero-free regions.

Proof (of Theorem A.1) It suffices to prove (A.2) for $\mathrm{Re}\, s = 1$. This bound can then be extended to all $\mathrm{Re}\, s \geq 1$ in the following way. On $\mathrm{Re}\, s \geq 2$ one bounds $|L^\infty(s, \pi_1 \times \tilde{\pi}_2)|^{-1}$ by an absolute constant by inserting the Jacquet-Shalika bounds. One then interpolates between the two bounds within the strip $1 \leq \mathrm{Re}\, s \leq 2$ by the Phragmén-Lindelöf principle. Moreover the same bounds can be extended to the small region within the critical strip (A.1) by applying the convexity bound of Li [Li10] on the derivative of Rankin-Selberg L-functions.

To proceed, we first assume that $\pi_1 \neq \pi_2$. For $t \in \mathbb{R}$ put[5]

$$\Pi = (\pi_1 \otimes |\det|^{\mathrm{i}t/2}) \boxplus (\pi_2 \otimes |\det|^{-\mathrm{i}t/2}). \tag{A.5}$$

The associated Rankin-Selberg L-function $L(s, \Pi \times \tilde{\Pi})$ factorizes as

$$L(s, \pi_1 \times \tilde{\pi}_1) L(s, \pi_2 \times \tilde{\pi}_2) L(s + \mathrm{i}t, \pi_1 \times \tilde{\pi}_2) L(s - \mathrm{i}t, \tilde{\pi}_1 \times \pi_2). \tag{A.6}$$

On the right half-plane $\mathrm{Re}\, s > 1$, $L(s, \Pi \times \tilde{\Pi})$ is an absolutely convergent Euler product of degree $d^2[F : \mathbb{Q}]$. Let

$$D(s) = L^\infty(s, \Pi \times \tilde{\Pi})$$

be the finite part L-function. When expanded into a Dirichlet series

$$D(s) = \sum_{\mathfrak{n}} b(\mathfrak{n})\, \mathrm{N}_{F/\mathbb{Q}}(\mathfrak{n})^{-s},$$

the coefficients $b(\mathfrak{n})$ are non-negative. Moreover $D(s)$ has a double pole at $s = 1$ and nowhere else.[6]

We write $D(s)$ in its Laurent series expansion about $s = 1$ as

$$D(s) = \sum_{j=-2}^{\infty} r_j (s-1)^j. \tag{A.7}$$

---

[5] In [Bru06] there is a missing minus sign in the exponent of the second isobaric factor of Π.

[6] In [Bru06] we make a big deal of the possibility that $D(s)$ might have a pole at $s = 0$. This in fact never happens because for any cuspidal representation π, $L(s, \pi_v \times \tilde{\pi}_v)$ has a pole at $s = 0$ for all places v, in particular the archimedean ones. (This holds for any generic representation, regardless of its temperedness.)

Theorem 3 of [Bru06] bounds the polar part from below by

$$\mathfrak{c}(\Pi \times \tilde{\Pi})^{-\frac{1}{2}+\frac{1}{2d}-\varepsilon} \ll_\varepsilon |r_{-1}| + |r_{-2}| . \tag{A.8}$$

Let us recall the proof of (A.8). Fix a non-negative $\psi \in C_c^\infty(0,\infty)$ such that $\psi\big|_{[1,2]} \equiv 1$. For a parameter $Y \geq 1$ we form the sum

$$S(Y) = \sum_{\mathfrak{n}} b(\mathfrak{n})\psi(\mathrm{N}_{F/\mathbb{Q}}(\mathfrak{n})/Y).$$

We have

$$S(Y) = \frac{1}{2\pi \mathrm{i}} \int_{\mathrm{Re}\, s=2} D(s)\hat{\psi}(s)Y^s ds.$$

Shifting contours far to the left to $\mathrm{Re}\, s = \sigma$ we pick up no other poles and hence

$$S(Y) = \operatorname*{Res}_{s=1} \hat{\psi}(s)D(s)Y^s + \frac{1}{2\pi \mathrm{i}} \int_{\mathrm{Re}\, s=\sigma} D(s)\hat{\psi}(s)Y^s ds. \tag{A.9}$$

The second term above is

$$O_{\varepsilon,\sigma}(\mathfrak{c}(\Pi \times \tilde{\Pi})^{\frac{1-\sigma}{2}+\varepsilon}Y^\sigma), \tag{A.10}$$

by Li's convexity bound [Li10]. On the other hand, by [Bru06, Lemma 1]

$$S(Y) \geq \#\{\mathfrak{n} : Y \leq \mathrm{N}_{F/\mathbb{Q}}\, \mathfrak{n}^d \leq 2Y\} \gg Y^{1/d}. \tag{A.11}$$

(This is an improvement and simplification of [Bru06, Lemma 2]. Thanks to Erez Lapid for pointing this out.) Picking $Y = \mathfrak{c}(\Pi \times \tilde{\Pi})^{B(\sigma)+\varepsilon}$ where

$$B(\sigma) = \frac{1}{2} \cdot \frac{\sigma - 1}{\sigma - 1/d},$$

the second term in (A.9) is smaller than the lower bound (A.11). For this value of Y we thus obtain

$$Y^{1/d} \ll \operatorname*{Res}_{s=1} \hat{\psi}(s)D(s)Y^s.$$

Calculating this residue, we find

$$\frac{1}{Y^{1-1/d}} \ll (|r_{-1}| + |r_{-2}|) \log Y.$$

Inserting the value of Y and noting that $B(\sigma)$ tends to $1/2$ (from below) as σ tends to $-\infty$ we obtain (A.8).

Although we used it in the above reasoning, the bound (A.8) does not require Li's convexity result as an input. Indeed, if we use only the pre-convex bound, the error term in (A.10) is $O_{\varepsilon,\sigma}(\mathfrak{c}(\Pi\times\tilde{\Pi})^{\frac{c-\sigma}{2}+\varepsilon}Y^{\sigma})$ for some potentially large (fixed) constant $c \geq 1$. The exponent $B(\sigma)$ would then have a $\sigma - c$ in the numerator rather than a $\sigma - 1$. Nevertheless the limit of $B(\sigma)$ as $\sigma \to -\infty$ would still be $1/2$ as before.[7] Li's convexity result will be used more crucially in the next paragraph (to optimize exponents).

To complete the argument, the idea is that we can factor out the L-value we're interested in from the polar part $r_{-1} + r_{-2}$. We first let

$$L^{\infty}(s,\pi_1\times\tilde{\pi}_1) = \sum_{j=-1}^{\infty} A_j(s-1)^j, \quad L^{\infty}(s,\pi_2\times\tilde{\pi}_2) = \sum_{j=-1}^{\infty} B_j(s-1)^j$$

be the Laurent series expansions about $s = 1$. Then

$$r_{-2} = A_{-1}B_{-1}\left|L^{\infty}(1+it,\pi_1\times\tilde{\pi}_2)\right|^2$$

and

$$r_{-1} = (A_{-1}B_0 + A_0B_{-1})\left|L^{\infty}(1+it,\pi_1\times\tilde{\pi}_2)\right|^2 \\ + 2A_{-1}B_{-1}\operatorname{Re}(L^{\infty\prime}(1+it,\pi_1\times\tilde{\pi}_2)\overline{L^{\infty}(1+it,\pi_1\times\tilde{\pi}_2)}).$$

From Li's convexity bound we have

$$A_j \ll_{\varepsilon,j} \mathfrak{c}(\pi_1\times\tilde{\pi}_1)^{\varepsilon}, \qquad B_j \ll_{\varepsilon,j} \mathfrak{c}(\pi_2\times\tilde{\pi}_2)^{\varepsilon},$$

and

$$L^{\infty(k)}(s,\pi_1\times\tilde{\pi}_2) \ll_{\varepsilon,k} \mathfrak{c}(\pi_1\times\tilde{\pi}_2,s)^{\varepsilon},$$

along $\operatorname{Re} s = 1$. This gives

$$r_{-2}, r_{-1} \ll_{\varepsilon} \mathfrak{c}(\Pi\times\tilde{\Pi})^{\varepsilon}\left|L^{\infty}(1+it,\pi_1\times\tilde{\pi}_2)\right|. \tag{A.12}$$

Putting (A.8) and (A.12) together yields (A.2) on the 1-line, as desired.

Now we tackle the more delicate case where $\pi_1 = \pi_2 (= \pi)$ and $n_1 = n_2 (= n)$. Here the proof breaks up into two parts according to whether s is close to 1 or not.

If $s = 1$ then we already know by [Bru06, Theorem 3] applied to $L^{\infty}(s,\pi\times\tilde{\pi})$ (as explained above) that

$$A_{-1} = \operatorname*{Res}_{s=1} L^{\infty}(s,\pi\times\tilde{\pi})$$

---

[7] We point out that in displays (3) and (6) of [Bru06] the absolute values on the parameters $\mu_{\pi}(v,i)$ and $\mu_{\pi\times\pi'}(v,i,j)$ should not appear. This misprint does not affect the reasoning leading to display (9) of [ibid.].

is bounded below by

$$\mathfrak{c}(\pi \times \tilde{\pi})^{-\frac{1}{2}+\frac{1}{2n}-\varepsilon}.$$

As before, Li's upper bounds on the derivatives of $L^\infty(s, \pi \times \tilde{\pi})$ allow us to extend this to

$$((s-1)L^\infty(s, \pi \times \tilde{\pi}))^{-1} \ll_\varepsilon \mathfrak{c}(\pi \times \tilde{\pi}, s)^{\frac{1}{2}-\frac{1}{2n}+\varepsilon},$$

for all $s = 1 + it$ with $|t| \leq \mathfrak{c}(\pi \times \tilde{\pi})^{-\frac{1}{2}+\frac{1}{2n}-2\varepsilon}$.

It remains to treat the range $|t| \geq \mathfrak{c}(\pi \times \tilde{\pi})^{-\frac{1}{2}+\frac{1}{2n}-2\varepsilon}$. Define Π as in (A.5) so that

$$L(s, \Pi \times \tilde{\Pi}) = L(s, \pi \times \tilde{\pi})^2 L(s + it, \pi \times \tilde{\pi}) L(s - it, \pi \times \tilde{\pi}).$$

Again write $D(s) = L^\infty(s, \Pi \times \tilde{\Pi})$ for the finite part L-function. This time $D(s)$ has a double pole at $s = 1$ and two additional poles, both simple, at $s = 1 \pm it$. We continue to use the notation r_j, introduced in (A.7), for the Laurent series coefficients of $D(s)$ at $s = 1$. Let $r_{-1}^{\pm}$ be the residues of $D(s)$ at $s = 1 \pm it$. The proof of (A.8) given above gives

$$\mathfrak{c}(\Pi \times \tilde{\Pi})^{-\frac{1}{2}+\frac{1}{4n}-\varepsilon} \ll_\varepsilon |r_{-1}| + |r_{-2}| + \left|r_{-1}^{+}\right| + \left|r_{-1}^{-}\right|. \qquad \text{(A.13)}$$

Indeed, the contour shift picks out all of these poles. We calculate

$$r_{-1}^{\pm} = A_{-1} L^\infty(1 \pm it, \pi \times \tilde{\pi})^2 L^\infty(1 \pm 2it, \pi \times \tilde{\pi}).$$

If $|t| \geq 1$ then we can apply Li's convexity result simultaneously to $L^\infty(1 + it, \pi \times \tilde{\pi})$ and $L^\infty(1 + 2it, \pi \times \tilde{\pi})$ to obtain (A.12) and

$$r_{-1}^{\pm} \ll_\varepsilon \mathfrak{c}(\Pi \times \tilde{\Pi})^\varepsilon \left|L^\infty(1 \pm it, \pi \times \tilde{\pi})\right|. \qquad \text{(A.14)}$$

On the other hand, for $|t| \leq 1$ we claim that

$$r_{-1}, r_{-2}, r_{-1}^{\pm} \ll_\varepsilon |tL^\infty(1 + it, \pi \times \tilde{\pi})| \, |t|^{-3} \, \mathfrak{c}(\pi \times \tilde{\pi}, s)^\varepsilon.$$

This follows from the convexity bounds

$$(s-1)L^\infty(s, \pi \times \tilde{\pi}) \ll_\varepsilon \mathfrak{c}(\pi \times \tilde{\pi}, s)^\varepsilon$$

(applied to $s = 1 \pm it$ and $s = 1 \pm 2it$) and

$$(s-1)^2 (L^\infty)'(s, \pi \times \tilde{\pi}) \ll_\varepsilon \mathfrak{c}(\pi \times \tilde{\pi}, s)^\varepsilon$$

which are valid for $\operatorname{Re} s = 1$, $s - 1 = O(1)$. We conclude that for $\mathfrak{c}(\pi \times \tilde{\pi})^{-\frac{1}{2}+\frac{1}{2n}-2\varepsilon} \leq |t| \leq 1$ we have

$$r_{-1}, r_{-2}, r_{-1}^{\pm} \ll_\varepsilon |tL^\infty(1 + it, \pi \times \tilde{\pi})| \, \mathfrak{c}(\Pi \times \tilde{\Pi})^{\frac{3}{4}(\frac{1}{2}-\frac{1}{2n})+\varepsilon},$$

where we used that $\mathfrak{c}(\Pi \times \tilde{\Pi})$ and $\mathfrak{c}(\pi \times \tilde{\pi})^4$ are within a constant apart. Combining the three regimes we get the required lower bound from (A.13).

□

Remark A.4 Note that except in the range $\mathfrak{c}(\pi \times \tilde{\pi})^{-\frac{1}{2}+\frac{1}{2n}-2\varepsilon} \le |t| \le 1$ we can improve the bound (A.4). It would be nice to improve the bound in the above regime as well.

Remark A.5 As in the case of the upper bounds, it would be more natural to replace $\mathfrak{c}(\Pi \times \tilde{\Pi})$ by $\mathfrak{c}(\pi_1 \times \tilde{\pi}_2, s)$ in the formulation of Theorem A.1. This is possible since as explained to us by Guy Henniart (work in progress), one has

$$\mathfrak{c}(\pi_1 \times \tilde{\pi}_1)\mathfrak{c}(\pi_2 \times \tilde{\pi}_2) \le \mathfrak{c}(\pi_1 \times \tilde{\pi}_2, s)^2.$$

This is a purely local question. In the supercuspidal case we have in fact the inequality

$$\max(\mathfrak{c}(\pi_1 \times \tilde{\pi}_1), \mathfrak{c}(\pi_2 \times \tilde{\pi}_2)) \le \mathfrak{c}(\pi_1 \times \tilde{\pi}_2, s)$$

for the local conductors – see [BH03, Theorem C]. We take this opportunity to thank Guy Henniart for providing us this information.

References

[Art75] J. Arthur, *A theorem on the Schwartz space of a reductive Lie group*, Proc. Nat. Acad. Sci. U.S.A. **72** (1975), no. 12, 4718–4719. MR 0460539 (57 #532)

[Art78] ———, *A trace formula for reductive groups. I. Terms associated to classes in $G(\mathbf{Q})$*, Duke Math. J. **45** (1978), no. 4, 911–952. MR 518111 (80d:10043)

[Art80] ———, *A trace formula for reductive groups. II. Applications of a truncation operator*, Compositio Math. **40** (1980), no. 1, 87–121. MR 558260 (81b:22018)

[Art81] ———, *The trace formula in invariant form*, Ann. of Math. (2) **114** (1981), no. 1, 1–74. MR 625344 (84a:10031)

[Art82] ———, *On a family of distributions obtained from Eisenstein series. II. Explicit formulas*, Amer. J. Math. **104** (1982), no. 6, 1289–1336. MR 681738 (85d:22033)

[Art88] ———, *The invariant trace formula. I. Local theory*, J. Amer. Math. Soc. **1** (1988), no. 2, 323–383. MR 928262 (89e:22029)

[Art89] ———, *Intertwining operators and residues. I. Weighted characters*, J. Funct. Anal. **84** (1989), no. 1, 19–84. MR 999488 (90j:22018)

[Ber88] J. N. Bernstein, *On the support of Plancherel measure*, J. Geom. Phys. **5** (1988), no. 4, 663–710 (1989). MR 1075727 (91k:22027)

[Bru06] F. Brumley, *Effective multiplicity one on* GL_N *and narrow zero-free regions for Rankin-Selberg L-functions*, Amer. J. Math. **128** (2006), no. 6, 1455–1474. MR 2275908 (2007h:11062)

[BH97] C. J. Bushnell and G. Henniart, *An upper bound on conductors for pairs*, J. Number Theory **65** (1997), no. 2, 183–196. MR 1462836 (98h:11153)

[BH03] ———, *Local tame lifting for* $\mathrm{GL}(n)$. *IV. Simple characters and base change*, Proc. London Math. Soc. (3) **87** (2003), no. 2, 337–362. MR 1990931 (2004f:22017)

[Cas84] W. Casselman, *Automorphic forms and a Hodge theory for congruence subgroups of* $\mathrm{SL}_2(\mathbf{Z})$, Lie group representations, II (College Park, Md., 1982/1983), Lecture Notes in Math., vol. 1041, Springer, Berlin, 1984, pp. 103–140. MR 748506 (86f:22012)

[Cas89] ———, *Introduction to the Schwartz space of* $\Gamma\backslash G$, Canad. J. Math. **41** (1989), no. 2, 285–320. MR 1001613 (90e:22014)

[Cas04] ———, *Harmonic analysis of the Schwartz space of* $\Gamma\backslash\mathrm{SL}_2(\mathbb{R})$, Contributions to automorphic forms, geometry, and number theory, Johns Hopkins Univ. Press, Baltimore, MD, 2004, pp. 163–192. MR 2058608 (2005b:22014)

[Don82] H. Donnelly, *On the cuspidal spectrum for finite volume symmetric spaces*, J. Differential Geom. **17** (1982), no. 2, 239–253. MR MR664496 (83m:58079)

[FL11] T. Finis and E. Lapid, *On the spectral side of Arthur's trace formula—combinatorial setup*, Ann. of Math. (2) **174** (2011), no. 1, 197–223. MR 2811598

[FLM11] T. Finis, E. Lapid and W. Müller, *On the spectral side of Arthur's trace formula—absolute convergence*, Ann. of Math. (2) **174** (2011), no. 1, 173–195. MR 2811597

[Fra98] J. Franke, *Harmonic analysis in weighted* L_2*-spaces*, Ann. Sci. École Norm. Sup. (4) **31** (1998), no. 2, 181–279. MR 1603257 (2000f:11065)

[FS98] J. Franke and J. Schwermer, *A decomposition of spaces of automorphic forms, and the Eisenstein cohomology of arithmetic groups*, Math. Ann. **311** (1998), no. 4, 765–790. MR 1637980 (99k:11077)

[IS95] H. Iwaniec and P. Sarnak, *L^∞ norms of eigenfunctions of arithmetic surfaces*, Ann. of Math. (2) **141** (1995), no. 2, 301–320. MR MR1324136 (96d:11060)

[IS00] ______, *Perspectives on the analytic theory of L-functions*, Geom. Funct. Anal. (2000), no. Special Volume, Part II, 705–741, GAFA 2000 (Tel Aviv, 1999). MR MR1826269 (2002b:11117)

[JPSS83] H. Jacquet, I. I. Piatetskii-Shapiro, and J. A. Shalika, *Rankin-Selberg convolutions*, Amer. J. Math. **105** (1983), no. 2, 367–464. MR 701565 (85g:11044)

[LW13] J.-P. Labesse and J.-L. Waldspurger, *La formule des traces tordue d'aprs le Friday Morning Seminar*, CRM Monograph Series, vol. 31, American Mathematical Society, Providence, RI, 2013.

[Lan76] R. P. Langlands, *On the functional equations satisfied by Eisenstein series*, Springer-Verlag, Berlin, 1976, Lecture Notes in Mathematics, Vol. 544. MR MR0579181 (58 #28319)

[Lap06] E. M. Lapid, *On the fine spectral expansion of Jacquet's relative trace formula*, J. Inst. Math. Jussieu **5** (2006), no. 2, 263–308. MR 2225043 (2007d:11059)

[LO12] E. Lapid and K. Ouellette, *Truncation of Eisenstein series*, Pacific J. Math. **260** (2012), no. 2, 665-685.

[Li10] X. Li, *Upper bounds on L-functions at the edge of the critical strip*, Int. Math. Res. Not. IMRN (2010), no. 4, 727–755. MR 2595006 (2011a:11160)

[LW09] J. Liu and Y. Wang, *A theorem on analytic strong multiplicity one*, J. Number Theory **129** (2009), no. 8, 1874–1882. MR 2522710 (2010i:11070)

[LRS99] W. Luo, Z. Rudnick, and P. Sarnak, *On the generalized Ramanujan conjecture for* GL(n), Automorphic forms, automorphic representations, and arithmetic (Fort Worth, TX, 1996), Proc. Sympos. Pure Math., vol. 66, Amer. Math. Soc., Providence, RI, 1999, pp. 301–310. MR MR1703764 (2000e:11072)

[MW89] C. Mœglin and J.-L. Waldspurger, *Le spectre résiduel de* $\mathrm{GL}(n)$, Ann. Sci. École Norm. Sup. (4) **22** (1989), no. 4, 605–674. MR 1026752 (91b:22028)

[MW95] ______, *Spectral decomposition and Eisenstein series*, Cambridge Tracts in Mathematics, vol. 113, Cambridge University Press, Cambridge, 1995, Une paraphrase de l'Écriture [A paraphrase of Scripture]. MR 1361168 (97d:11083)

[Mül02] W. Müller, *On the spectral side of the Arthur trace formula*, Geom. Funct. Anal. **12** (2002), no. 4, 669–722. MR 1935546 (2003k:11086)

[MS04] W. Müller and B. Speh, *Absolute convergence of the spectral side of the Arthur trace formula for* GL_n, Geom. Funct. Anal. **14** (2004), no. 1, 58–93, With an appendix by E. M. Lapid. MR 2053600 (2005m:22021)

[RS96] Z. Rudnick and P. Sarnak, *Zeros of principal L-functions and random matrix theory*, Duke Math. J. **81** (1996), no. 2, 269–322, A celebration of John F. Nash, Jr. MR MR1395406 (97f:11074)

[Sar04] P. Sarnak, *A letter to Cathleen Morawetz*, Available at `http://www.math.princeton.edu/sarnak`.

[Wal03] J.-L. Waldspurger, *La formule de Plancherel pour les groupes p-adiques (d'après Harish-Chandra)*, J. Inst. Math. Jussieu **2** (2003), no. 2, 235–333. MR 1989693 (2004d:22009)

[Wal92] N. R. Wallach, *Real reductive groups. II*, Pure and Applied Mathematics, vol. 132, Academic Press Inc., Boston, MA, 1992. MR 1170566 (93m:22018)

[Wan08] Y. Wang, *The analytic strong multiplicity one theorem for* $\mathrm{GL}_m(\mathbb{A}_K)$, J. Number Theory **128** (2008), no. 5, 1116–1126. MR 2406482 (2009g:11062)

Erez Lapid, Institute of Mathematics, The Hebrew University of Jerusalem, Jerusalem 91904, Israel.
Faculty of Mathematics and Computer Science, The Weizmann Institute of Science, Rehovot 76100, Israel.
E-mail: `erez.m.lapid@gmail.com`

Farrell Brumley, Laboratoire Analyse, Géométrie & Applications, UMR 7539, Institut Galilé, Université Paris 13, 99 avenue J.-B. Clément, 93430 Villetaneuse, France.
E-mail: `brumley@math.univ-paris13.fr`

Automorphic Representations and L-Functions
Editors: D. Prasad, C.S. Rajan, A. Sankaranarayanan, J. Sengupta

Publisher: Hindustan Book Agency, New Delhi, India

Degenerate Principal Series of Metaplectic Groups and Howe Correspondence

Soo Teck Lee and Chen-Bo Zhu[1]

Abstract

The main purpose of this article is to supplement the authors' results on degenerate principal series representations of real symplectic groups with the analogous results for metaplectic groups. The basic theme, as in the previous case, is that their structures are anticipated by certain natural subrepresentations constructed from Howe correspondence. This supplement is necessary as these representations play a key role in understanding the basic structure of Howe correspondence (and its complications in the archimedean case), and their global counterparts play an equally essential part in the proof of Siegel-Weil formula and its generalizations (work of Kudla-Rallis). The full results in the metaplectic case also shed light on the seeming peculiarities, when the results in the symplectic case are viewed in their isolation.

1 Introduction: Classical Invariant Theory and Its Transcendental Analog

Let the complex orthogonal group O_m act on the space of complex matrices $M_{m,n}$ by matrix multiplication on the left:

$$\mathrm{O}_m \curvearrowright M_{m,n}.$$

The First Fundamental Theorem (FFT) of classical invariant theory asserts that the ring of O_m-invariant polynomials are generated by the fundamental invariants of degree 2:

$$P(M_{m,n})^{\mathrm{O}_m} =< r_{ij} | \ 1 \leq i, j \leq n >,$$

where $r_{ij}(X) = \sum_{k=1}^{m} x_{ki}x_{kj}$, for $X = (x_{ij}) \in M_{m,n}$.

[1]Both authors are supported by the MOE2010-T2-2-113.

Let $S^2(\mathbb{C}^n)$ denote the space of complex $n \times n$ symmetric matrices and define

$$Q: \; M_{m,n} \to S^2(\mathbb{C}^n), \qquad X \mapsto X^t X.$$

Then FFT asserts that the pull-back map

$$Q^*: \; P(S^2(\mathbb{C}^n)) \to P(M_{m,n})^{\mathrm{O}_m}$$

is surjective. Alternatively, we have an affine embedding:

$$\psi: \; M_{m,n}//\mathrm{O}_m \hookrightarrow S^2(\mathbb{C}^n),$$

where $M_{m,n}//\mathrm{O}_m$ is the affine quotient of $M_{m,n}$ by O_m. The Second Fundamental Theorem (SFT) of classical invariant theory is then a statement on the image of ψ as an affine subvariety of $S^2(\mathbb{C}^n)$. Note that both $M_{m,n}$ and $S^2(\mathbb{C}^n)$ carry natural actions of $\mathrm{GL}(n, \mathbb{C})$, and all maps (Q, Q^*, ψ) are $\mathrm{GL}(n, \mathbb{C})$-equivariant.

We now state results of Kudla-Rallis [KR1] and Lee-Zhu [LZ2], which may be viewed as transcendental analogs of FFT and SFT alluded to above.

Let

$$\mathrm{O}(p, q) \curvearrowright M_{p+q,n}(\mathbb{R}),$$

again by matrix multiplication on the left. Denote by $G = \widetilde{\mathrm{Sp}}(2n, \mathbb{R})$ the real metaplectic group of rank n and P its Siegel parabolic subgroup. The rest of notation will be explained in Section 3.

Theorem 1.1 (Kudla-Rallis) *There is a natural topological embedding with closed image:*

$$\psi_{p,q} : \mathcal{S}(M_{p+q,n}(\mathbb{R}))_{\mathrm{O}(p,q)} \hookrightarrow I^{\alpha}(\sigma)(:= \mathrm{Ind}_P^G(\chi_\sigma^\alpha)),$$

as G-representations.

Theorem 1.2 (Lee-Zhu) *describes the image of $\psi_{p,q}$.*

In [LZ2], the authors describe the image $\Omega^{p,q}$ of $\psi_{p,q}$ when $p+q$ is even, in which case the representations concerned factor through the linear group $\mathrm{Sp}(2n, \mathbb{R})$. In fact only the subcase of $p - q \equiv 0 \pmod 4$ was treated in detail and the other subcase of $p - q \equiv 2 \pmod 4$ was left to the reader.

The aim of the current article is to complete the description of $\Omega^{p,q}$ (without any restriction on p and q). This is necessary and useful as the representations involved (the coinvariants) play a key role in understanding the basic structure of Howe correspondence (through the doubling method

[H1, Ra, Ku]) and their global counterparts feature prominently in the proof of Siegel-Weil formula and its generalizations [KR3]. For a recent application of these results to first occurrence conjecture of Kudla-Rallis, see [SZ]. As an additional benefit, this full description enables us to organize the statements in a more coherent way so that their structures emerge clearer to the reader. The basic idea is very simple: since both $\Omega^{p,q}$ and $I^{\alpha}(\sigma)$ are K-multiplicity free (K being a maximal compact subgroup of G), and since the structure of $I^{\alpha}(\sigma)$ is known (Section 5; [L2] for the linear group), one can identify the image $\Omega^{p,q}$ as a G-representation by knowing its K-types. The fascinating point is that from this description, one concludes that the reducibilities of $I^{\alpha}(\sigma)$ are completely accounted for by the possible embeddings of $\Omega^{p,q}$'s (in a precise way). We summarize this assertion in Theorem 6.1, which should be viewed as analogous to SFT (proverbially stated) as "all relations among the fundamental invariants are generated by the obvious ones".

Here are some words on the organization of this article. In Section 2, we review the basics of Howe duality correspondence. In Section 3, we introduce the space of coinvariants, as a special case of Howe's construction of maximal quotients, and its embedding into the degenerate principal series $I^{\alpha}(\sigma)$ of the metaplectic group. In Section 4, we describe the transition coefficients of $I^{\alpha}(\sigma)$ and give its immediate consequences for irreducibility and complementary series. In Section 5, we give the detailed structure of $I^{\alpha}(\sigma)$ at points of reducibility, which are again arrived by analyzing the transition coefficients. In Section 6, we complete our description of the image of $\psi_{p,q}$ in $I^{\alpha}(\sigma)$. As proofs of (any new) results follow a similar line as those of [L2] or [LZ2], we shall omit them.

Finally we mention the following works which are closely related to the theme of this article: [KR1, LZ1, LZ2, LZ3, Ya] (for real groups) and [KR2, KS, Ya] (for p-adic groups).

2 Howe Duality Correspondence

In this section, we review briefly Howe's theory of dual pair correspondence, for the case at hand.

Let $H = \mathrm{O}(p,q)$, which acts on $M_{p+q,n}(\mathbb{R})$ by matrix multiplication on the left. We have the dualised action of H on $\mathcal{Y} = L^2(M_{p+q,n}(\mathbb{R}))$, the space of square integrable functions on $M_{p+q,n}(\mathbb{R})$. As is well-known, there is a (unitary) representation ω of a metaplectic group $\widetilde{\mathrm{Sp}}(2N,\mathbb{R})$ ($N = (p+q)n$) on $\mathcal{Y}$, called the Schrodinger model of an oscillator representation. Here and after, for any real symplectic group $\mathrm{Sp}(2N,\mathbb{R})$, $\widetilde{\mathrm{Sp}}(2N,\mathbb{R})$ denotes its metaplectic two fold cover. The oscillator representation depends on a choice of a non-trivial unitary character of $\mathbb{R}$, which we fix once for all.

The main feature of this set-up is the following: there is a reductive dual pair

$$(H, G) = (\mathrm{O}(p,q), \mathrm{Sp}(2n,\mathbb{R})) \subseteq \mathrm{Sp}(2N,\mathbb{R}),$$

namely a pair of reductive subgroups in $\mathrm{Sp}(2N,\mathbb{R})$ which are mutual centralizers.

Let ω^∞ be the smooth representation of ω realized in $\mathcal{Y}^\infty$. For the case at hand, $\mathcal{Y}^\infty = \mathcal{S}(M_{p+q,n}(\mathbb{R}))$, the Schwartz space of rapidly decreasing functions on $M_{p+q,n}(\mathbb{R})$.

For any subgroup E of $\mathrm{Sp}(2N,\mathbb{R})$, denote by $\widetilde{E}$ the preimage of E under the covering map $\widetilde{\mathrm{Sp}}(2N,\mathbb{R}) \to \mathrm{Sp}(2N,\mathbb{R})$. If E is also reductive, denote by $\mathcal{R}(\widetilde{E},\omega)$ the set of infinitesimal equivalent classes of irreducible admissible representations of $\widetilde{E}$ which are realizable as quotients by $\omega^\infty(\widetilde{E})$-invariant closed subspaces of $\mathcal{Y}^\infty$.

Howe Duality Theorem ([H2]): $\mathcal{R}(\widetilde{H}\cdot\widetilde{G},\omega)$ is the graph of a bijection between $\mathcal{R}(\widetilde{H},\omega)$ and $\mathcal{R}(\widetilde{G},\omega)$.

This is the Howe quotient correspondence. Let us make it more concrete. As usual, we may twist the oscillator representation (by a character of $\widetilde{H}$) so that it will factor through the standard linear action of H on $\mathcal{Y}$. We will assume that this has been done (but retain the same notation) and will be concerned with H, rather than $\widetilde{H}$. Let ρ be an irreducible admissible representation of H, which is realizable as a quotient by a H-invariant closed subspace of $\mathcal{Y}^\infty$. Define $\Omega(\rho)$ to be the maximal quotient of $\mathcal{Y}^\infty$ on which H acts by a representation of class ρ. We have (isomorphism class to mean infinitesimal equivalent class):

$$\Omega(\rho) \cong \rho \otimes \Theta(\rho), \tag{2.1}$$

where $\Theta(\rho)$ is a $\widetilde{G}$-module.

Howe duality theorem asserts that $\Theta(\rho)$ is a finitely generated admissible quasisimple representation of $\widetilde{G}$, and has a unique irreducible $\widetilde{G}$-quotient, denoted by $\theta(\rho)$:

$$\Theta(\rho) \twoheadrightarrow \theta(\rho).$$

The correspondence

$$\rho \mapsto \theta(\rho)$$

is the Howe quotient correspondence.

Remark If we take ρ to be in the Casselman-Wallach class [Wal, Chapter 11], then we will have the isomorphism of Casselman-Wallach representations $\Omega(\rho) \cong \rho\widehat{\otimes}\Theta(\rho)$, where "$\widehat{\otimes}$" stands for the completed projective tensor product.

3 The Coinvariants and the Embedding

For the dual pair $(H, G) = (\mathrm{O}(p,q), \mathrm{Sp}(2n, \mathbb{R}))$, let

$$\begin{aligned}\Omega^{p,q} &= \text{Howe's maximal quotient corresponding to} \\ &\quad\ \text{the trivial representation } \mathbb{1} \text{ of } \mathrm{O}(p,q) \\ &= \mathcal{S}(M_{p+q,n}(\mathbb{R}))_{\mathrm{O}(p,q)} \quad \text{(the space of coinvariants).}\end{aligned}$$

This is a representation of $\widetilde{G}$.

Remark The continuous dual of $\Omega^{p,q}$ is $\mathcal{S}^*(M_{p+q,n}(\mathbb{R}))^{\mathrm{O}(p,q)}$, the space of $\mathrm{O}(p,q)$-invariant tempered distributions on $M_{p+q,n}(\mathbb{R})$. This is investigated in [KR1] and [Zh].

From now on, we shift notation and will denote $G = \widetilde{\mathrm{Sp}}(2n, \mathbb{R})$. The case of $n = 1$ differs from the general case slightly, but since it is straightforward, we omit it and will assume that $n \geq 2$ throughout this paper. We shall identify $\widetilde{\mathrm{Sp}}(2n, \mathbb{R})$ as a set with

$$\mathrm{Sp}(2n, \mathbb{R}) \times \mathbb{Z}_2 = \{(g, \varepsilon) :\ g \in \mathrm{Sp}(2n, \mathbb{R}),\ \varepsilon = \pm 1\}.$$

For $a \in \mathrm{GL}(n, \mathbb{R})$ and $b \in M_n(\mathbb{R})$ such that $b = b^t$, we let

$$\begin{aligned} m_a &= \begin{pmatrix} a & 0 \\ 0 & (a^{-1})^t \end{pmatrix}, \\ n_b &= \begin{pmatrix} I_n & b \\ 0 & I_n \end{pmatrix}. \end{aligned}$$

Let

$$M = \{(m_a, \varepsilon) :\ a \in \mathrm{GL}(n, \mathbb{R}),\ \varepsilon = \pm 1\}$$

and

$$N = \{(n_b, 1) :\ b \in M_n(\mathbb{R}),\ b = b^t\}.$$

Then $P = MN$ is a maximal parabolic subgroup of G, called the Siegel parabolic.

Let $\chi : M \longrightarrow \mathbb{C}^\times$ be given by

$$\chi(m_a, \varepsilon) = \varepsilon \cdot \begin{cases} i & \text{if } \det a < 0, \\ 1 & \text{if } \det a > 0. \end{cases}$$

This is a character of M and it is of order 4. For $\alpha = 0, 1, 2, 3$ and $\sigma \in \mathbb{C}$, let χ_σ^α be the character of P given by

$$\chi_\sigma^\alpha[(m_a, \varepsilon)(n_b, 1)] = |\det a|^\sigma \chi(m_a, \varepsilon)^\alpha. \tag{3.1}$$

Let $I^\alpha(\sigma)$ be the normalized induced representation:

$$I^\alpha(\sigma) = \mathrm{Ind}_P^G \chi_\sigma^\alpha.$$

The representation space of $I^\alpha(\sigma)$ is

$$\{f \in C^\infty(G) : \ f(pg) = \delta^{\frac{1}{2}}(p)\chi_\sigma^\alpha(p) f(g), \ \forall g \in G, p \in P\},$$

and G acts by right translation:

$$g \cdot f(h) = f(hg), \qquad (g, h \in G).$$

Here δ denotes the modular function of P, and is given by

$$\delta[(m_a, \varepsilon)(n_b, 1)] = |\det a|^{2\rho_n}$$

and

$$\rho_n = \frac{n+1}{2}.$$

When $\alpha = 0$ or 2, the representation $I^\alpha(\sigma)$ descends to a representation of the linear group $\mathrm{Sp}(2n, \mathbb{R})$.

Define the map $\psi_{p,q} : \mathcal{S}(M_{p+q,n}(\mathbb{R})) \mapsto C^\infty(G)$ by

$$\psi_{p,q}(f)(g) = (\omega(g)f)(0), \quad f \in \mathcal{S}(M_{p+q,n}(\mathbb{R})), \ g \in G.$$

From the well-known formula of the oscillator representation ω, we see that

$$\psi_{p,q} : \mathcal{S}(M_{p+q,n}(\mathbb{R})) \mapsto I^\alpha(\sigma),$$

where

$$\sigma = \frac{p+q}{2} - \rho_n, \quad \text{and} \quad \alpha \equiv p - q \ (\mathrm{mod}\ 4). \tag{3.2}$$

The following is the fundamental result of Kudla-Rallis.

Theorem 3.1 ([KR1]) *The map $\psi_{p,q}$ induces a topological embedding with closed image:*

$$\Omega^{p,q} \hookrightarrow I^\alpha(\sigma)$$

Remark Analogous results hold for other classical groups. See [Zh].

4 Degenerate Principal Series: The Transition Coefficients

Fix a maximal compact subgroup $K_1 \simeq \mathrm{U}(n)$ of $\mathrm{Sp}(2n,\mathbb{R})$ and let K be the inverse image of K_1 in $\widetilde{\mathrm{Sp}}(2n,\mathbb{R})$, thus a maximal compact subgroup of $\widetilde{\mathrm{Sp}}(2n,\mathbb{R})$. The Lie algebra of $\mathrm{Sp}(2n,\mathbb{R})$ has a Cartan decomposition

$$\mathfrak{sp}(2n,\mathbb{R}) = \mathfrak{k} \oplus \mathfrak{p}$$

where $\mathfrak{k}$ is the Lie algebra of K_1 and is isomorphic to $\mathfrak{u}(n)$.

Let

$$\Lambda_n^+ = \{(\lambda_1,\ldots,\lambda_n) \in \mathbb{Z}^n : \lambda_1 \geq \lambda_2 \geq \cdots \geq \lambda_n\},$$

and

$$\mathbf{1} = (1,\ldots,1) \in \Lambda_n^+.$$

Also let $e_j = (\overbrace{0,\ldots,0,1}^{j},0,\ldots,0)$, for $1 \leq j \leq n$.

We have the K-type decomposition:

$$I^\alpha(\sigma)|_K = \oplus_{\lambda\in\Lambda_n^+} V_{2\lambda+\frac{\alpha}{2}\mathbf{1}}, \tag{4.1}$$

where $V_{2\lambda+\frac{\alpha}{2}\mathbf{1}}$ is an irreducible K-module with highest weight $2\lambda + \frac{\alpha}{2}\mathbf{1}$, for $\lambda \in \Lambda_n^+$.

For $\mu = 2\lambda + \frac{\alpha}{2}\mathbf{1}$, let V_μ be a K-type in $I^\alpha(\sigma)$ and fix γ_μ to be the unique (up to a multiple) K-highest weight vector in V_μ. We consider the tensor product $\mathfrak{p}_\mathbb{C} \otimes V_\mu$ of $\mathfrak{k}_\mathbb{C}$ modules, and the $\mathfrak{k}_\mathbb{C}$ map

$$m : \mathfrak{p}_\mathbb{C} \otimes V_\mu \longrightarrow I^\alpha(\sigma)_K$$

$$m(p \otimes v) = p \cdot v.$$

For each $1 \leq j \leq n$, there exists an element X_j in $\mathcal{U}(\mathfrak{sp}(2n,\mathbb{C}))$ (the universal enveloping algebra of the complexified Lie algebra) with the property that $X_j \cdot \gamma_\mu$ is the image of the unique $\mathfrak{k}_\mathbb{C}$ highest weight vector of weight $\mu + 2e_j$ in $\mathfrak{p}_\mathbb{C} \otimes V_\mu$, valid for all μ. C.f. [L2, Section 3]. Thus $X_j \cdot \gamma_\mu$ is a multiple of the unique $\mathfrak{k}_\mathbb{C}$ highest weight vector $\gamma_{\mu+2e_j}$ in $V_{\mu+2e_j}$. We call this multiple the transition coefficient from V_μ to $V_{\mu+2e_j}$. Likewise there is a transition coefficient from V_μ to $V_{\mu-2e_j}$. Note that the transition coefficients depend on the choice of γ_μ's.

The main use of the transition coefficients is as follows: if the transition coefficient from V_μ to $V_{\mu+2e_j}$ is nonzero, then $V_{\mu+2e_j}$ is in the submodule generated by V_μ, and vice versa. We shall indicate by the symbol: $V_\mu \to V_{\mu+2e_j}$; If this transition coefficient is zero, then $\sum_{\mu'_j \leq \mu_j} V_{\mu'}$ is a submodule of $I^\alpha(\sigma)$ and we say that we have a barrier to block jth rightward movement $V_\mu \to V_{\mu+2e_j}$. Likewise for the transition coefficient from V_μ to $V_{\mu-2e_j}$. To summarize, the collection of transition coefficients allows us to determine the lattice of submodules in $I^\alpha(\sigma)$, explicitly.

By constructing an appropriate choice of $\mathfrak{k}_{\mathbb{C}}$ highest weight vector γ_μ for all μ (c.f. [L2, Section3]), the transition coefficients $A_j^+(\mu)$ from V_μ to $V_{\mu+2e_j}$, and the transition coefficients $A_j^-(\mu)$ from V_μ to $V_{\mu-2e_j}$, are computed to be

$$A_j^+(\mu) = B_j^+ - 2\lambda_j, \quad A_j^-(\mu) = 2\lambda_j - B_j^-, \tag{4.2}$$

where

$$B_j^+ = -\sigma - \rho_n - \tfrac{\alpha}{2} + j - 1, \quad B_j^- = \sigma - \rho_n - \tfrac{\alpha}{2} + j + 1. \tag{4.3}$$

Corollary 4.1 ([KR1]) *$I^\alpha(\sigma)$ is irreducible if and only if*

$$\sigma + \rho_n + \tfrac{\alpha}{2} \notin \mathbb{Z}.$$

The explicit form of transition coefficients also allows us to determine the complementary series. We will follow the method used in [L2]. Each K-type $V_{2\lambda+\frac{\alpha}{2}\mathbf{1}}$ of $I^\alpha(\sigma)$ has a K-invariant inner product given by

$$\langle f_1, f_2\rangle_\lambda = \int_K f_1(k)\overline{f_2(k)}dk.$$

Since $V_{2\lambda+\frac{\alpha}{2}\mathbf{1}}$ is an irreducible K-module, any K-invariant inner product on $V_{2\lambda+\frac{\alpha}{2}\mathbf{1}}$ is a multiple of $\langle .,.\rangle_\lambda$. Thus if $\langle .,.\rangle$ is a $\widetilde{\mathrm{Sp}}(2n,\mathbb{R})$-invariant inner product on $I^\alpha(\sigma)$ then there exists positive constants $\{c_\lambda\}_{\lambda\in\Lambda_n^+}$ such that

$$\langle f_1, f_2\rangle = c_\lambda\langle f_1, f_2\rangle_\lambda, \qquad \forall f_1, f_2 \in V_{2\lambda+\frac{\alpha}{2}\mathbf{1}}.$$

Since the K-types of $I^\alpha(\sigma)$ are mutually orthogonal with respect $\langle .,.\rangle$, $\langle .,.\rangle$ is completely determined by the constants $\{c_\lambda\}$. Using similar arguments as the $U(n,n)$ case (see [L1, section 9]), we obtain the following:

Lemma 4.2 *The inner product on $I^\alpha(\sigma)$ defined by the constants $\{c_\lambda\}_{\lambda\in\Lambda_n^+}$ is $\widetilde{\mathrm{Sp}}(2n,\mathbb{R})$-invariant if and only if*

$$(-\sigma - \rho_n - \tfrac{\alpha}{2} - 2\lambda_j + j - 1)c_{\lambda+e_j} + (-\overline{\sigma} + \rho_n + \tfrac{\alpha}{2} + 2\lambda_j - j + 1)c_\lambda = 0$$

for all $\lambda \in \Lambda_n^+$ and all $1 \le j \le n$.

We let

$$N_{\lambda,j} = \frac{-\sigma - \rho_n - \frac{\alpha}{2} - 2\lambda_j + j - 1}{-\overline{\sigma} + \rho_n + \frac{\alpha}{2} + 2\lambda_j - j + 1} = -\frac{c_\lambda}{c_{\lambda+e_j}}.$$

Then $I^\alpha(\sigma)$ is unitarizable if and only if $N_{\lambda,j} < 0$ for all $\lambda \in \Lambda_n^+$ and for all j.

Note that

$$N_{\lambda,j} = \frac{-\sigma - (\rho_n + \frac{\alpha}{2} + 2\lambda_j - j + 1)}{-\overline{\sigma} + (\rho_n + \frac{\alpha}{2} + 2\lambda_j - j + 1)}.$$

We write $\xi = \rho_n + \frac{\alpha}{2} + 2\lambda_j - j + 1$. Then $N_{\lambda,j} = (-\sigma - \xi)/(-\overline{\sigma} + \xi)$. Thus $N_{\lambda,j}$ is real for all λ and for j if and only if either $\mathrm{Re}(\sigma) = 0$ or σ is real. The case $\mathrm{Re}(\sigma) = 0$ corresponds to the unitary axis. If σ is real, then

$$N_{\lambda,j} = \frac{-\sigma - \xi}{-\overline{\sigma} + \xi} < 0 \Longleftrightarrow |\sigma| < |\xi|.$$

The minimum value of $|\xi|$ is $\frac{1}{2}$ if $n + \alpha$ is even, and 0 if $n + \alpha$ is odd. This leads to the following

Theorem 4.3 *If $n + \alpha$ is even, then $I^\alpha(\sigma)$ is unitarizable for $|\sigma| < \frac{1}{2}$.*

5 Subquotients of $I^\alpha(\sigma)$

In this section, we shall give a detailed description of the module structure of $I^\alpha(\sigma)$ when it is reducible. We shall describe all the irreducible subquotients of $I^\alpha(\sigma)$ and determine which of them are unitarizable, i.e., possess a G-invariant positive definite inner product. We also describe the socle series and module diagram of $I^\alpha(\sigma)$.

Let

$$\tilde{\sigma} = \sigma + \rho_n + \frac{\alpha}{2} = \sigma + \frac{n+1+\alpha}{2}. \tag{5.1}$$

By Corollary 4.1, $I^\alpha(\sigma)$ is irreducible if and only if $\tilde{\sigma} \notin \mathbb{Z}$. Thus we shall assume that $\tilde{\sigma} \in \mathbb{Z}$ throughout this section.

Let $\mu = 2\lambda + (\alpha/2)\mathbf{1}_n$, as before. Then by the results on transition coefficients in Section 4, we have

$$V_\mu \to V_{\mu+2e_j} \Longleftrightarrow 2\lambda_j \neq B_j^+ \quad \text{and} \quad V_\mu \to V_{\mu-2e_j} \Longleftrightarrow 2\lambda_j \neq B_j^-$$

where

$$B_j^+ = B_j^+(\alpha,\sigma) = -\sigma - \frac{n+\alpha+1}{2} + j - 1 = -\tilde{\sigma} + j - 1$$

and

$$B_j^- = B_j^-(\alpha, \sigma) = \sigma - \frac{n+\alpha+1}{2} + j + 1 = \tilde{\sigma} - (n+\alpha) + j.$$

As discussed in Section 4, we may determine the lattice of submodules of $I^\alpha(\sigma)$ from the transition coefficients $\{B_j^\pm\}_{1\le j\le n}$. Schematically it goes as follows:

1. Represent V_μ by the point $2\lambda \in \mathbb{R}^n$ with standard coordinates $(x_1, \ldots, x_n)$.
2. Barrier to block j-th rightward movement $V_\mu \to V_{\mu+2e_j}$ is at hyperplane $\ell_j^+ : x_j = B_j^+$.
3. Barrier to block j-th leftward movement $V_\mu \to V_{\mu-2e_j}$ is at hyperplane $\ell_j^- : x_j = B_j^-$.
4. Analyze barriers systematically (Subsections 5.1 and 5.2).

Note that the barrier $\ell_j^+ : x_j = B_j^+$ is effective if and only if it cuts at an "even" point, i.e. $B_j^+ \in 2\mathbb{Z}$. The parity of B_j^+ depends on the parity of $\tilde{\sigma} = \sigma + \frac{n+\alpha+1}{2}$ and the parity of j. Similarly for the barrier ℓ_j^-, but the parity of B_j^- also depends on the parity of $n+\alpha$. So our analysis will be divided into 4 cases.

$\tilde{\sigma}\backslash n+\alpha$	odd	even
odd	Case 1a	Case 2a
even	Case 1b	Case 2b

We also define the "gap" between ℓ_j^+ and ℓ_j^- by

$$\text{gap} = B_j^+ - B_j^- = -2\sigma - 2. \tag{5.2}$$

From the lattice of submodules of $I^\alpha(\sigma)$, we may give its module structure a diagrammatic representation, called the module diagram. It is a directed simple graph $\mathcal{G}$, defined as follows: The vertex set of $\mathcal{G}$ is the set of all irreducible subquotients in $I^\alpha(\sigma)$, to be identified with its collection of K-types. There is a directed edge from the node R_1 to the node R_2 if and only if there are submodules U and V of $I^\alpha(\sigma)$ such that $V \subseteq U$ and there is a nonsplit exact sequence of infinitesimal $\widetilde{\text{Sp}}(2n,\mathbb{R})$-modules $0 \to R_2 \to U/V \to R_1 \to 0$. We shall also arrange the nodes in $\mathcal{G}$ in such a way that all the edges are directed downward. Then one can recover the lattice of submodules of $I^\alpha(\sigma)$ from the graph $\mathcal{G}$. For a more detailed explanation on module diagram, see Section 7 of [L1].

5.1 Subquotients of $I^\alpha(\sigma)$: $n+\alpha$ odd

Case 1: $n+\alpha$ odd.

Since $\tilde{\sigma} \in \mathbb{Z}$ and $\frac{n+\alpha+1}{2} \in \mathbb{Z}$, we have $\sigma \in \mathbb{Z}$. It follows from this and equation (5.2) that the gap $B_j^+ - B_j^-$ is even, i.e. $B_j^+ \equiv B_j^- (\mathrm{mod}\ 2)$. This means that for each j, either both barriers ℓ_j^+ and ℓ_j^- are effective, or both are not effective.

We consider two subcases:

Case 1a: $\tilde{\sigma} = \sigma + \frac{n+\alpha+1}{2}$ is odd   *Case 1b:* $\tilde{\sigma} = \sigma + \frac{n+\alpha+1}{2}$ is even.

For Case 1a, We have

$$B_j^+ = -(\sigma + \frac{n+\alpha+1}{2}) + j - 1 \in 2\mathbb{Z} \Longleftrightarrow j \text{ is even.}$$

Hence the effective barriers are ℓ_{2i}^+ and ℓ_{2i}^- for all $1 \leq i \leq n_0/2$, where

$$n_0 = 2\left[\frac{n}{2}\right] = \text{largest even integer less than or equal to } n. \tag{5.3}$$

But for Case 1b, We have

$$B_j^+ = -(\sigma + \frac{n+\alpha+1}{2}) + j - 1 \in 2\mathbb{Z} \Longleftrightarrow j \text{ is odd.}$$

So the effective barriers are ℓ_{2i-1}^+ and ℓ_{2i-1}^- for all $1 \leq i \leq (n_1+1)/2$, where

$$n_1 = 2\left[\frac{n+1}{2}\right] - 1 = \text{largest odd integer less than or equal to } n. \tag{5.4}$$

Case 1a: $\tilde{\sigma} = \sigma + \frac{n+\alpha+1}{2}$ is odd

Let i and j be such that $0 \leq i + j \leq n_0/2$. We will define $R_{ij}(n, \sigma, \alpha)$ as follows:

(i) For $\sigma \leq -1$, let $R_{ij}(n, \sigma, \alpha)$ be the set of 2λ such that

$$2\lambda_{2i} \geq B_{2i+2}^+ \geq 2\lambda_{2i+2} \tag{5.5}$$

$$2\lambda_{n_0-2j} \geq B_{n_0-2j}^- \geq 2\lambda_{n_0-2j+2} \tag{5.6}$$

where

$$B_{2i+2}^+ = -\sigma - \frac{n+\alpha+1}{2} + 2i + 1 = -\sigma - \frac{n+\alpha-1}{2} + 2i,$$

$$B_{n_0-2j}^- = \sigma - \frac{n+\alpha+1}{2} + n_0 - 2j + 1 = \sigma - \frac{n+\alpha-1}{2} + n_0 - 2j.$$

(ii) For $\sigma \geq 0$, let $R_{ij}(n, \sigma, \alpha)$ be the set of 2λ such that

$$2\lambda_{2i} \geq B_{2i}^{-} \geq 2\lambda_{2i+2} \tag{5.7}$$

$$2\lambda_{n_0-2j} \geq B_{n_0-2j+2}^{+} \geq 2\lambda_{n_0-2j+2} \tag{5.8}$$

where

$$B_{2i}^{-} = \sigma - \frac{n+\alpha+1}{2} + 2i + 1 = \sigma - \frac{n+\alpha-1}{2} + 2i,$$

$$B_{n_0-2j+2}^{+} = -\sigma - \frac{n+\alpha+1}{2} + n_0 - 2j + 1 = -\sigma - \frac{n+\alpha-1}{2} + n_0 - 2j.$$

Note that the set $R_{ij}(n, \sigma, \alpha)$ defined above may be empty. When it is nonempty, it will be identified with the direct sum of all the K-representations $V_{2\lambda+\frac{\alpha}{2}\mathbf{1}}$ in $I^\alpha(\sigma)$ such that $2\lambda \in R_{ij}(n, \sigma, \alpha)$.

Remark When $\sigma = 0$ and $i+j = n_0/2$, the two conditions (5.7) and (5.8) coincide. More precisely,

$$R_{ij}(n, 0, \alpha) = \{2\lambda : \ 2\lambda_{2i} \geq B_{2i}^{-} \geq 2\lambda_{2i+2}\}.$$

In what follows, $\oplus$ denotes sum in the Grothendieck group of G-modules. If we have a direct sum of submodules, we will state it explicitly.

Theorem 5.1 (Case 1a) *Assume that $n+\alpha$ is odd, σ is an integer and $\tilde{\sigma} = \sigma + \frac{n+\alpha+1}{2}$ is odd.*

(a) *If $\sigma \leq -1$, then*

$$I^\alpha(\sigma) = \bigoplus \left\{ R_{ij}(n, \sigma, \alpha) : \ r_1 \leq i+j \leq \frac{n_0}{2} \right\},$$

where

$$r_1 = \max\left(\frac{n_0}{2} + \sigma, 0\right).$$

In this case, the module diagram of $I^\alpha(\sigma)$ can be obtained from Figure 1 by removing those $R_{ab}(n, \sigma, \alpha)$ which are empty. In particular, the socle series of $I^\alpha(\sigma)$ is given by

$$\mathrm{Soc}^l(I^\alpha(\sigma)) = \begin{cases} \bigoplus_{r_1 \leq i+j \leq r_1+l-1} R_{ij}(n, \sigma, \alpha) & 1 \leq l \leq \frac{n_0}{2} - r_1, \\ I^\alpha(\sigma) & l \geq \frac{n_0}{2} - r_1 + 1. \end{cases}$$

An irreducible constituent $R_{ij}(n, \sigma, \alpha)$ of $I^\alpha(\sigma)$ is unitarizable if and only if $-\frac{n_0}{2} \leq \sigma \leq -1$ and $i+j = r_1$.

(b) *If* $\sigma = 0$, *then*

$$I^\alpha(0) = \bigoplus_{i+j=n_0/2} R_{ij}(n,0,\alpha)$$

is a direct sum of irreducible unitary submodules.

(c) *If* $\sigma \geq 1$, *then*

$$I^\alpha(\sigma) = \bigoplus \left\{ R_{ij}(n,\sigma,\alpha) : \ r_2 \leq i+j \leq \frac{n_0}{2} \right\},$$

where

$$r_2 = \max\left(\frac{n_0}{2} - \sigma, 0\right).$$

In this case, the module diagram of $I^\alpha(\sigma)$ *can be obtained from Figure 2 by removing those* $R_{ab}(n,\sigma,\alpha)$ *which are empty. In particular, the socle series of* $I^\alpha(\sigma)$ *is given by*

$$\mathrm{Soc}^l(I^\alpha(\sigma)) = \begin{cases} \bigoplus_{\frac{n_0}{2}-l+1 \leq i+j \leq \frac{n_0}{2}} R_{ij}(n,\sigma,\alpha) & 1 \leq l \leq \frac{n_0}{2} - r_2, \\ I^\alpha(\sigma) & l \geq \frac{n_0}{2} - r_2 + 1. \end{cases}$$

An irreducible constituent $R_{ij}(n,\sigma,\alpha)$ *of* $I^\alpha(\sigma)$ *is unitarizable if and only if* $1 \leq \sigma \leq \frac{n_0}{2}$ *and* $i + j = r_2$.

Case 1b: $\tilde{\sigma} = \sigma + \frac{n+\alpha+1}{2}$ is even

Let i and j be such that $0 \leq i+j \leq (n_1+1)/2$. We will define $R_{ij}(n,\sigma,\alpha)$ as follows:

(i) For $\sigma \leq -1$, let $R_{ij}(n,\sigma,\alpha)$ be the set of 2λ such that

$$2\lambda_{2i-1} \geq B^+_{2i+1} \geq 2\lambda_{2i+1} \tag{5.9}$$

$$2\lambda_{n_1-2j} \geq B^-_{n_1-2j} \geq 2\lambda_{n_1-2q+2} \tag{5.10}$$

where

$$B^+_{2i+1} = -\sigma - \frac{n+\alpha+1}{2} + 2i,$$

$$B^-_{n_1-2j} = \sigma - \frac{n+\alpha+1}{2} + n_1 - 2j + 1 = \sigma - \frac{n+\alpha-1}{2} + n_1 - 2j.$$

(ii) For $\sigma \geq 0$, let $R_{ij}(n,\sigma,\alpha)$ be the set of 2λ such that

$$2\lambda_{2i-1} \geq B^-_{2i-1} \geq 2\lambda_{2i+1} \tag{5.11}$$

$$2\lambda_{n_1-2j} \geq B^+_{n_1-2j+2} \geq 2\lambda_{n_1-2j+2} \tag{5.12}$$

where

$$B^-_{2i-1} = \sigma - \frac{n+\alpha+1}{2} + 2i,$$

$$B^+_{n_1-2j+2} = -\sigma - \frac{n+\alpha+1}{2} + n_1 - 2j + 1 = -\sigma - \frac{n+\alpha-1}{2} + n_1 - 2j.$$

Remark When $\sigma = 0$ and $i + j = (n_1+1)/2$, the two conditions (5.9) and (5.10) coincide, that is, $2\lambda_{n_1-2j} = 2\lambda_{(2i+2j-1)-2j} = 2\lambda_{2i-1}$,

$$B^-_{n_1-2j+2} = -\frac{n+\alpha-1}{2} + (2i+2j-1) - 2j = -\frac{n+\alpha+1}{2} + 2i = B^-_{2i-1},$$

and $2\lambda_{n_1-2j+2} = 2\lambda_{(2i+2j-1)-2j+2} = 2\lambda_{2j+1}$. Precisely,

$$R_{ij}(n,0,\alpha) = \{2\lambda : \ 2\lambda_{2i-1} \geq B^-_{2i-1} \geq 2\lambda_{2i+1}\}.$$

Theorem 5.2 (Case 1b) *Assume that $n+\alpha$ is odd, σ is an integer and $\tilde{\sigma} = \sigma + \frac{n+\alpha+1}{2}$ is even.*

(a) *If $\sigma \leq -1$, then*

$$I^\alpha(\sigma) = \bigoplus \left\{ R_{ij}(n,\sigma,\alpha) : \ r_1 \leq i+j \leq \frac{n_1+1}{2} \right\},$$

where

$$r_1 = \max\left(\frac{n_1+1}{2} + \sigma, 0\right).$$

In this case, the module diagram of $I^\alpha(\sigma)$ can be obtained from Figure 1 by removing those $R_{ab}(n,\sigma,\alpha)$ which are empty. In particular, the socle series of $I^\alpha(\sigma)$ is given by

$$\mathrm{Soc}^l(I^\alpha(\sigma)) = \begin{cases} \bigoplus_{r_1 \leq i+j \leq r_1+l-1} R_{ij}(n,\sigma,\alpha) & 1 \leq l \leq \frac{n_1+1}{2} - r_1, \\ I^\alpha(\sigma) & l \geq \frac{n_1+1}{2} - r_1 + 1. \end{cases}$$

An irreducible constituent $R_{ij}(n,\sigma,\alpha)$ of $I^\alpha(\sigma)$ is unitarizable if and only if

(i) $-\frac{n_1+1}{2} \leq \sigma \leq -1$ *and* $i+j = r_1$; *or*
(ii) n *is odd,* $\alpha \in \{0,2\}$ *and* $(i,j) \in \left\{ \left(\frac{n+1}{2}, 0\right), \left(0, \frac{n+1}{2}\right) \right\}$.

(b) *If $\sigma = 0$, then*

$$I^\alpha(0) = \bigoplus_{i+j=\frac{n_1+1}{2}} R_{ij}(n,0,\alpha)$$

is a direct sum of irreducible unitary submodules.

(c) *If $\sigma \geq 1$, then*

$$I^\alpha(\sigma) = \bigoplus \left\{ R_{ij}(n,\sigma,\alpha) : \; r_2 \leq i+j \leq \frac{n_1+1}{2} \right\},$$

where

$$r_2 = \max\left(\frac{n_1+1}{2} - \sigma, 0 \right).$$

In this case, the module diagram of $I^\alpha(\sigma)$ can be obtained from Figure 2 by removing those $R_{ab}(n,\sigma,\alpha)$ which are empty. In particular, the socle series of $I^\alpha(\sigma)$ is given by

$$\mathrm{Soc}^l(I^\alpha(\sigma)) = \begin{cases} \bigoplus_{\frac{n_1+1}{2}-l+1 \leq i+j \leq \frac{n_1+1}{2}} R_{ij}(n,\sigma,\alpha) & 1 \leq l \leq \frac{n_1+1}{2} - r_2, \\ I^\alpha(\sigma) & l \geq \frac{n_1+1}{2} - r_2 + 1. \end{cases}$$

An irreducible constituent $R_{ij}(n,\sigma,\alpha)$ of $I^\alpha(\sigma)$ is unitarizable if and only if

(i) $1 \leq \sigma \leq \frac{n_1+1}{2}$ *and* $i+j = r_2$; *or*
(ii) n *is odd,* $\alpha \in \{0,2\}$ *and* $(i,j) \in \left\{ \left(\frac{n+1}{2}, 0\right), \left(0, \frac{n+1}{2}\right) \right\}$.

The module diagram of $I^\alpha(\sigma)$ for $\sigma \leq -1$ and $\sigma \geq 1$ are given in Figure 1 and Figure 2 below. Here

$$k = \begin{cases} \frac{n_0}{2} = [\frac{n}{2}] & \text{if } \tilde{\sigma} \text{ is odd,} \\ \frac{n_1+1}{2} = [\frac{n+1}{2}] & \text{if } \tilde{\sigma} \text{ is even.} \end{cases} \tag{5.13}$$

and we write a constituent $R_{ij}(n,\sigma,\alpha)$ simply as R_{ij}.

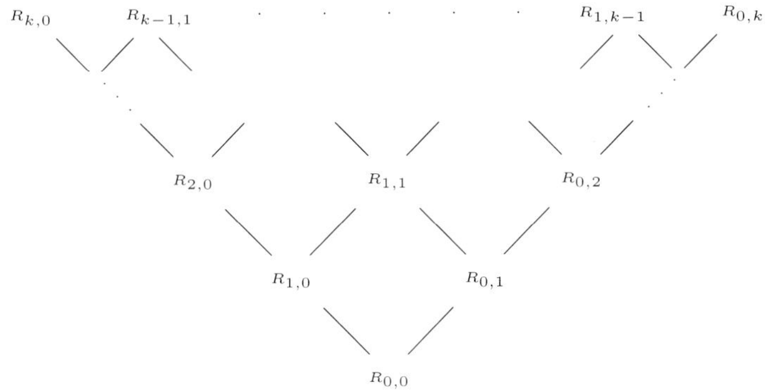

Figure 1: Module diagram for $I^\alpha(\sigma)$ $(\sigma \leq -1)$

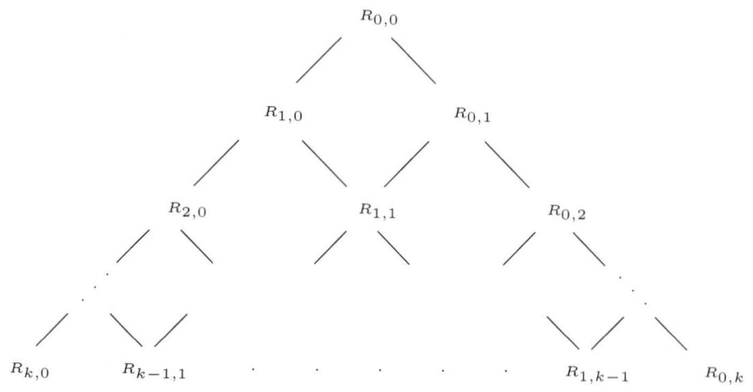

Figure 2: Module diagram for $I^{\alpha}(\sigma)$ $(\sigma \geq 1)$

5.2 Subquotients of $I^{\alpha}(\sigma)$: $n+\alpha$ even

Case 2: $n+\alpha$ **even.**

Recall that $\tilde{\sigma} = \sigma + \frac{n+\alpha+1}{2} \in \mathbb{Z}$. So in this case, $\sigma + \frac{1}{2} \in \mathbb{Z}$, and consequently the gap between the barriers ℓ_j^+ and ℓ_j^- given by

$$\text{gap} = -2\sigma - 2$$

is odd. It follows that along each coordinate axis, exactly one barrier is effective. More precisely, for each $1 \leq j \leq n$, either ℓ_j^+ is effective or ℓ_j^- is effective, but not both. Again, we consider two subcases:

Case 2a: $\tilde{\sigma} = \sigma + \frac{n+\alpha+1}{2}$ is odd Case 2b: $\tilde{\sigma} = \sigma + \frac{n+\alpha+1}{2}$ is even.

For Case 2a, ℓ_j^+ is effective when j is even, and ℓ_j^- is effective when j is odd. Case 2b is the other way round.

Case 2a: $\tilde{\sigma} = \sigma + \frac{n+\alpha+1}{2}$ is odd

Since exactly one barrier is effective along each coordinate axis, the barrier partitions the K-types into 2 subsets. For $1 \leq r \leq \frac{n_1+1}{2} + 1$, let

$$\begin{aligned} X_1^r &= \{2\lambda : 2\lambda_{2r-1} < B_{2r-1}^-\}, \\ X_2^r &= \{2\lambda : 2\lambda_{2r-1} \geq B_{2r-1}^-\}, \end{aligned}$$

and for $1 \leq s \leq \frac{n_0}{2} + 1$,

$$\begin{aligned} Y_1^s &= \{2\lambda : 2\lambda_{2s} \leq B_{2s}^+\}, \\ Y_2^s &= \{2\lambda : 2\lambda_{2s} > B_{2s}^+\}. \end{aligned}$$

Now we let

$$S(n) = \{(i,j):\ 0 \le i \le \frac{n_1+1}{2},\ 0 \le j \le \frac{n_0}{2}\}. \tag{5.14}$$

Then for $(i,j) \in S(n)$, we form the intersection

$$\begin{aligned} L_{i,j}(n,\sigma,\alpha) = \quad & (X_2^1 \cap \cdots \cap X_2^i \cap X_1^{i+1} \cap \cdots \cap X_1^{\frac{n_1+1}{2}+1}) \cap \\ & (Y_2^1 \cap \cdots \cap Y_2^j \cap Y_1^{j+1} \cap \cdots \cap Y_1^{\frac{n_0}{2}+1}). \end{aligned}$$

More precisely, $2\lambda \in L_{i,j}(n,\sigma,\alpha)$ if and only if

$$2\lambda_{2i-1} \ge B^-_{2i-1} \ge 2\lambda_{2i+1} \quad \text{and} \quad 2\lambda_{2j} \ge B^+_{2j+2} \ge 2\lambda_{2j+2}, \tag{5.15}$$

where

$$B^-_{2i-1} = \sigma - \frac{n+\alpha+1}{2} + 2i - 1 + 1 = \sigma - \frac{n+\alpha}{2} + 2i - \frac{1}{2}$$

and

$$B^+_{2j+2} = -\sigma - \frac{n+\alpha+1}{2} + 2j + 2 - 1 = -\sigma - \frac{n+\alpha}{2} + 2j + \frac{1}{2}.$$

Note that $L_{a,b}(n,\sigma,\alpha)$ may be empty. If it is nonempty, then we shall identify it with the direct sum of all the K-representations $V_{2\lambda+\frac{\alpha}{2}\mathbf{1}}$ with $2\lambda \in L_{a,b}(n,\sigma,\alpha)$.

Remark Let $k = [n/2]$. Then

$$S(n) = \begin{cases} \{(i,j):\ 0 \le i \le k+1,\ 0 \le j \le k\} & n \text{ odd}, \\ \{(i,j):\ 0 \le i,j \le k\} & n \text{ even}. \end{cases} \tag{5.16}$$

The following combinatorial result is elementary.

Lemma 5.3 (i) *Assume that $(i,j) \in S(n)$ and $i-j \le -1$. Then $L_{i,j} \ne \emptyset$ if and only if $i - j \ge -\sigma + \frac{1}{2}$. In particular, if $\sigma \le -1/2$, then all such $L_{i,j}$ is empty.*

(ii) *For $0 \le i \le k$, $L_{i,i}$ is always nonempty.*

(iii) *If $(j+1,j) \in S(n)$, then $L_{j+1,j}$ is always nonempty.*

(iv) *Assume that $(i,j) \in S(n)$ and $i - j \ge 2$. Then $L_{i,j} \ne \emptyset$ if and only if $i - j \le -\sigma + \frac{1}{2}$. In particular, if $\sigma \ge 1/2$, then all such $L_{i,j}$ is empty.*

Theorem 5.4 (Case 2a) *Assume that $n+\alpha$ is even, $\sigma+\frac{1}{2}\in\mathbb{Z}$, and $\tilde{\sigma}=\sigma+\frac{n+\alpha+1}{2}$ is odd.*

(i) *If $\sigma\geq 1/2$, then*

$$I^\alpha(\sigma)=\bigoplus\{L_{i,j}(n,\sigma,\alpha):\ (i,j)\in S(n),\ -1\leq j-i\leq r_1\},$$

where

$$r_1=\min\left(\sigma-\tfrac{1}{2},\left[\tfrac{n}{2}\right]\right),$$

and each $L_{i,j}(n,\sigma,\alpha)$ which appears in the sum forms an irreducible constituent of $I^\alpha(\sigma)$. The module diagram of $I^\alpha(\sigma)$ can be obtained from Figure 3 by removing those $L_{a,b}(n,\sigma,\alpha)$ which are empty. In particular, the socle series of $I^\alpha(\sigma)$ is given by

$$\mathrm{Soc}^l(I^\alpha(\sigma)=\begin{cases}\bigoplus_{(i,j)\in S(n),\ -1\leq j-i\leq l-2}L_{i,j}(n,\sigma,\alpha) & 1\leq l\leq r_1+1,\\ I^\alpha(\sigma) & l\geq r_1+2.\end{cases}$$

An irreducible constituent $L_{i,j}(n,\sigma,\alpha)$ of $I^\alpha(\sigma)$ is unitarizable if and only if $i=j+1$ or $\frac{1}{2}\leq\sigma\leq[\frac{n}{2}]+\frac{1}{2}$ and $j-i=r_1$.

(ii) *If $\sigma\leq -1/2$, then*

$$I^\alpha(\sigma)=\bigoplus\{L_{i,j}(n,\sigma,\alpha):\ (i,j)\in S(n),\ 0\leq i-j\leq r_2\},$$

where

$$r_2=\min(-\sigma+\tfrac{1}{2},\left[\tfrac{n+1}{2}\right]),$$

and each $L_{i,j}(n,\sigma,\alpha)$ which appears in the sum forms an irreducible constituent of $I^\alpha(\sigma)$. The module diagram of $I^\alpha(\sigma)$ can be obtained from Figure 3 by removing those $L_{a,b}(n,\sigma,\alpha)$ which are empty. In particular, the socle series of $I^\alpha(\sigma)$ is given by

$$\mathrm{Soc}^l(I^\alpha(\sigma))=\begin{cases}\bigoplus_{(i,j)\in S(n),\ r_2-l+1\leq i-j\leq r_2}L_{i,j}(n,\sigma,\alpha) & 1\leq l\leq r_2,\\ I^\alpha(\sigma) & l\geq r_2+1.\end{cases}$$

An irreducible constituent $L_{i,j}(n,\sigma,\alpha)$ of $I^\alpha(\sigma)$ is unitarizable if and only if $i=j$ or $-\left[\frac{n+1}{2}\right]+\frac{1}{2}\leq\sigma\leq-\frac{1}{2}$ and $i-j=r_2$.

The module diagram of $I^\alpha(\sigma)$ in the case when $n+\alpha$ is even and $\sigma+\frac{n+\alpha+1}{2}$ is odd can be obtained from the following rectangle (Figure 3) by removing those $L_{a,b}(n,\sigma,\alpha)$ which are empty. Here $k=[n/2]$ and we write a constituent $L_{a,b}(n,\sigma,\alpha)$ simply as $L_{a,b}$. Note also that when n is even, the spaces $L_{k+1,j}$ $(0\leq j\leq k)$ are not defined. In this case, the rectangle below reduces to a square.

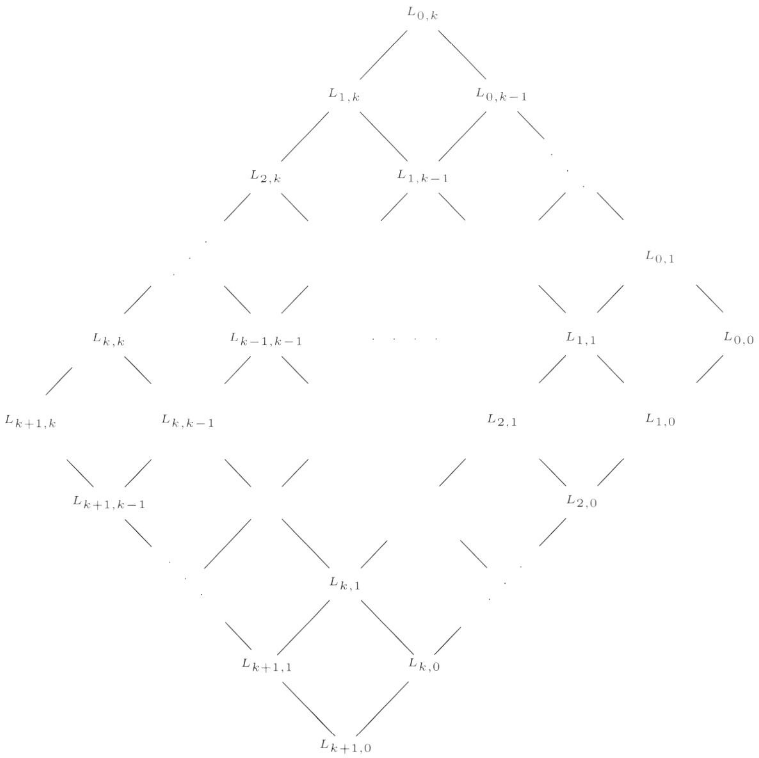

Figure 3: Diagram for $I^{\alpha}(\sigma)$ when $n + \alpha$ is even and $\sigma + \frac{n+\alpha+1}{2}$ is odd

Case 2b: $\sigma + \frac{n+\alpha+1}{2}$ is even

As in Case 2a, exactly one barrier is effective along each coordinate axis. But now ℓ_j^+ is effective when j is odd, and ℓ_j^- is effective when j is even. So we have to define $L_{a,b}(n, \sigma, \alpha)$ differently. For $1 \leq r \leq \frac{n_1+1}{2} + 1$, let

$$\begin{aligned} X_1^r &= \{2\lambda : 2\lambda_{2r-1} \leq B_{2r-1}^+\}, \\ X_2^r &= \{2\lambda : 2\lambda_{2r-1} > B_{2r-1}^-\}, \end{aligned}$$

and for $1 \leq s \leq \frac{n_0}{2} + 1$,

$$\begin{aligned} Y_1^s &= \{2\lambda : 2\lambda_{2s} < B_{2s}^-\}, \\ Y_2^s &= \{2\lambda : 2\lambda_{2s} \geq B_{2s}^-\}. \end{aligned}$$

We define the set $S(n)$, as in Case 2a in equation (5.14). Then for $(i,j) \in S(n)$, we form the intersection

$$\begin{aligned} L_{i,j}(n,\sigma,\alpha) = \quad & (X_2^1 \cap \cdots \cap X_2^i \cap X_1^{i+1} \cap \cdots \cap X_1^{\frac{n_1+1}{2}+1}) \cap \\ & (Y_2^1 \cap \cdots \cap Y_2^j \cap Y_1^{j+1} \cap \cdots \cap Y_1^{\frac{n_0}{2}+1}). \end{aligned}$$

More precisely, $2\lambda \in L_{i,j}(\sigma,\alpha)$ if and only if

$$2\lambda_{2i-1} \geq B_{2i+1}^{+} \geq 2\lambda_{2i+1} \quad \text{and} \quad 2\lambda_{2j} \geq B_{2j}^{-} \geq 2\lambda_{2j+2}. \tag{5.17}$$

where

$$B_{2i+1}^{+} = -\sigma - \frac{n+\alpha+1}{2} + 2i + 1 - 1 = -\sigma - \frac{n+\alpha}{2} + 2i - \frac{1}{2}$$

and

$$B_{2j}^{-} = \sigma - \frac{n+\alpha+1}{2} + 2j + 1 = \sigma - \frac{n+\alpha}{2} + 2j + \frac{1}{2}.$$

Theorem 5.5 (Case 2b) *Assume that $n+\alpha$ is even, $\sigma + \frac{1}{2} \in \mathbb{Z}$, and $\tilde{\sigma} = \sigma + \frac{n+\alpha+1}{2}$ is even.*

(i) *If $\sigma \geq 1/2$, then*

$$I^{\alpha}(\sigma) = \bigoplus \{L_{i,j}(n,\sigma,\alpha) : \ (i,j) \in S(n), \ 0 \leq i-j \leq r_2\},$$

where

$$r_2 = \min(\sigma + \tfrac{1}{2}, \left[\tfrac{n+1}{2}\right]),$$

and each $L_{i,j}(n,\sigma,\alpha)$ which appears in the sum forms an irreducible constituent of $I^{\alpha}(\sigma)$. The module diagram of $I^{\alpha}(\sigma)$ can be obtained from Figure 4 by removing those $L_{a,b}(n,\sigma,\alpha)$ which are empty. In particular, the socle series of $I^{\alpha}(\sigma)$ is given by

$$\mathrm{Soc}^{l}(I^{\alpha}(\sigma) = \begin{cases} \bigoplus_{(i,j)\in S(n), 0\leq i-j\leq l-1} L_{i,j}(n,\sigma,\alpha) & 1 \leq l \leq r_2, \\ I^{\alpha}(\sigma) & l \geq r_2 + 1. \end{cases}$$

An irreducible constituent $L_{i,j}(n,\sigma,\alpha)$ of $I^{\alpha}(\sigma)$ is unitarizable if and only if $i = j$ or $\frac{1}{2} \leq \sigma \leq \left[\frac{n+1}{2}\right] - \frac{1}{2}$ and $i - j = r_2$.

(ii) *If $\sigma \leq -1/2$, then*

$$I^{\alpha}(\sigma) = \bigoplus \{L_{i,j}(n,\sigma,\alpha) : \ (i,j) \in S(n), \ -1 \leq j-i \leq r_1\},$$

where

$$r_1 = \min\left(-\sigma - \tfrac{1}{2}, \left[\tfrac{n}{2}\right]\right),$$

and each $L_{i,j}(n,\sigma,\alpha)$ which appears in the sum forms an irreducible constituent of $I^{\alpha}(\sigma)$. The module diagram of $I^{\alpha}(\sigma)$ can be obtained from Figure 4 by removing those $L_{a,b}(n,\sigma,\alpha)$ which are empty. In particular, the socle series of $I^{\alpha}(\sigma)$ is given by

$$\mathrm{Soc}^l(I^{\alpha}(\sigma)) = \begin{cases} \bigoplus_{(i,j)\in S(n),\ r_1-l+1\le j-i\le r_1} L_{i,j}(n,\sigma,\alpha) & 1 \le l \le r_1+1, \\ I^{\alpha}(\sigma) & l \ge r_1+2. \end{cases}$$

An irreducible constituent $L_{i,j}(n,\sigma,\alpha)$ of $I^{\alpha}(\sigma)$ is unitarizable if and only if $i = j+1$ or $-[\frac{n}{2}] - \frac{1}{2} \le \sigma \le -\frac{1}{2}$ and $j - i = r_1$.

Finally, we describe the module diagram of $I^{\alpha}(\sigma)$ in the case when both $n+\alpha$ and $\sigma + \frac{n+\alpha+1}{2}$ are even. It can be obtained from the following rectangle (Figure 4) by removing those $L_{a,b}(n,\sigma,\alpha)$ which are empty. As before, $k = [n/2]$, and we write a constituent $L_{a,b}(n,\sigma,\alpha)$ simply as $L_{a,b}$. When n is even, $L_{k+1,j}$ $(0 \le j \le k)$ is empty so that this rectangle reduces to a square.

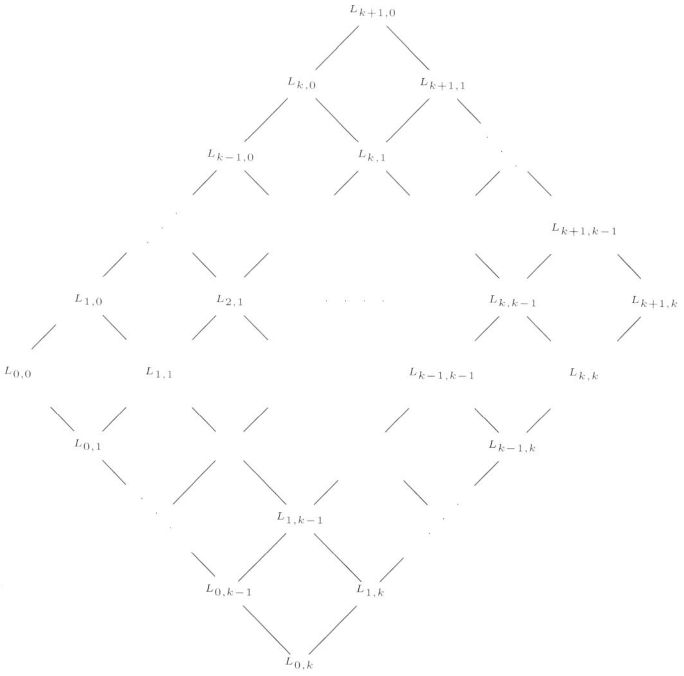

Figure 4: Diagram for $I^{\alpha}(\sigma)$ when both $n+\alpha$ and $\sigma + \frac{n+\alpha+1}{2}$ are even

6 The Image of $\psi_{p,q}$ in $I^\alpha(\sigma)$

In this section, we will describe the image of $\psi_{p,q}$ in $I^\alpha(\sigma)$:

$$\psi_{p,q}: \quad \Omega^{p,q} \hookrightarrow I^\alpha(\sigma),$$

where

$$\sigma = \frac{p+q}{2} - \frac{n+1}{2}, \quad \text{and} \quad \alpha \equiv p-q \pmod 4.$$

We allow $(p,q) = (0,0)$, in which case we understand $\Omega^{0,0}$ to be the trivial representation of $G = \widetilde{\mathrm{Sp}}(2n,\mathbb{R})$.

Denote $m = p+q$. Observe that

$$n+\alpha \equiv n+m \pmod 2,$$

and

$$\tilde{\sigma} = \sigma + \frac{n+\alpha+1}{2} = \frac{m+\alpha}{2} \equiv p \pmod 2.$$

Recall that our analysis of $I^\alpha(\sigma)$ is divided into 4 cases:

$\tilde{\sigma} \backslash n+\alpha$	odd	even
odd	Case 1a	Case 2a
even	Case 1b	Case 2b

Correspondingly, our analysis of $\Omega^{p,q}$ will also be divided into 4 cases:

$p \backslash n+m$	odd	even
odd	Case 1a	Case 2a
even	Case 1b	Case 2b

Throughout this section, we assume that we are at a point of reducibility, namely

$$\sigma + \frac{n+\alpha+1}{2} \in \mathbb{Z}.$$

If $\frac{p+q-(n+1)}{2} = \sigma$ and $p-q \equiv \alpha \pmod 4$, we say that we have a possible embedding of $\Omega^{p,q}$ (into $I^\alpha(\sigma)$).

We shall give the detailed description of $\Omega^{p,q}$ in $I^\alpha(\sigma)$ in the following two subsections. We end this section with a corollary which gives conceptual underpinning to our results.

Theorem 6.1 *The relationship between $\Omega^{p,q}$'s and $I^\alpha(\sigma)$ is as follows.*

(a) $-\rho_n \leq \sigma < 0$: *The irreducible submodules of $I^\alpha(\sigma)$ are given by the possible embeddings of $\Omega^{p,q}$'s, and all of them are unitary.*

(b) *Unitary axis ($\sigma = 0$) (when $n + \alpha$ is odd):*

$$I^{\alpha}(0) = \bigoplus_{\substack{p+q=n+1 \\ p-q\equiv\alpha \pmod 4}} \Omega^{p,q}.$$

(c) $\sigma > 0$: *The reducibilities of $I^{\alpha}(\sigma)$ are completely accounted for by the possible embeddings of $\Omega^{p,q}$'s.*

- *See Part b) of Theorems 6.2, 6.3, 6.4 and 6.5 for the precise statements.*

6.1 The image of $\psi_{p,q}$ in $I^{\alpha}(\sigma)$: m and n different parity

Case 1: $n + m$ odd.

We consider two subcases:

Case 1a: p is odd   *Case 1b:* p is even.

Case 1a: we write

$$p = 2i + 1, \quad q = 2j + \epsilon,$$

so that $p + q = m$. Here

$$\epsilon = \begin{cases} 1 & m \text{ even}, \\ 0 & m \text{ odd}. \end{cases}$$

Note that

$$i + j = \frac{m - 1 - \epsilon}{2} = \frac{m + n_0 - (n+1)}{2} = \frac{n_0}{2} + \sigma.$$

Case 1b: we write

$$p = 2i, \quad q = 2j + \epsilon',$$

so that $p + q = m$. Here

$$\epsilon' = \begin{cases} 0 & m \text{ even}, \\ 1 & m \text{ odd}. \end{cases}$$

Note that

$$i + j = \frac{m - \epsilon'}{2} = \frac{m + (n_1 + 1) - (n+1)}{2} = \frac{n_1 + 1}{2} + \sigma.$$

In accordance with 5.13, we let

$$k = \begin{cases} \frac{n_0}{2} = [\frac{n}{2}] & \text{if } p \text{ is odd,} \\ \frac{n_1+1}{2} = [\frac{n+1}{2}] & \text{if } p \text{ is even.} \end{cases}$$

Recall that we have the following module diagrams:

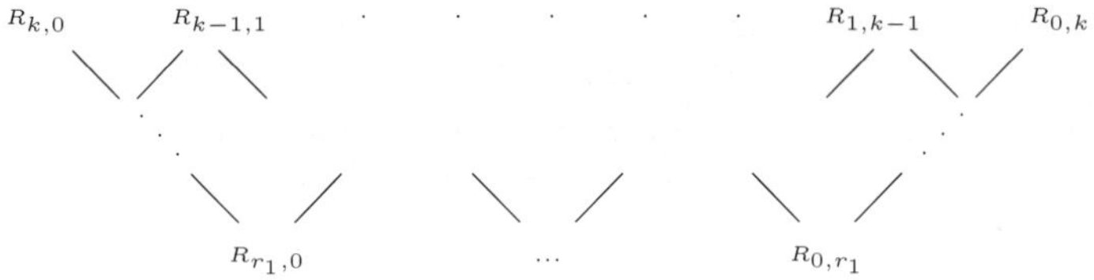

Case 1: Module diagram for $I^\alpha(\sigma)$ $(-k \le \sigma \le 0)$; $r_1 = k + \sigma$

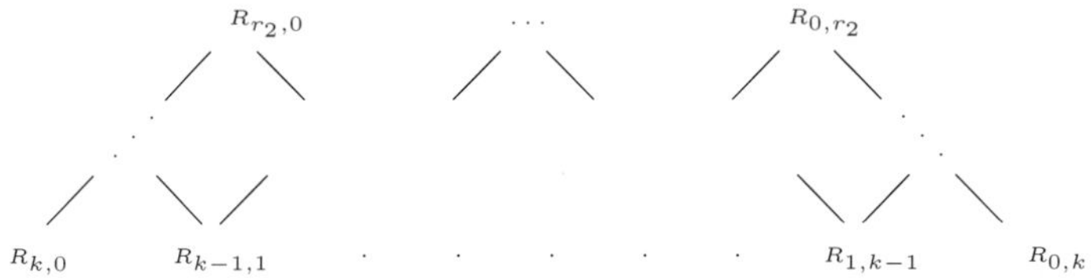

Case 1: Module diagram for $I^\alpha(\sigma)$ $(1 \le \sigma \le k)$; $r_2 = k - \sigma$

We introduce one notation. For an irreducible subquotient $R_{s,t}$ of $I^\alpha(\sigma)$ where $\sigma > 0$, denote by $\prec R_{s,t} \succ$ the submodule of $I^\alpha(\sigma)$ generated by $R_{s,t}$:

$$\prec R_{s,t} \succ = \oplus\{R_{i,j} : i \ge s, j \ge t\}. \tag{6.1}$$

Recall that

$$\sigma = \frac{m-(n+1)}{2}, \quad \text{and} \quad p - q \equiv \alpha \ (\text{mod } 4).$$

Theorem 6.2 (Case 1a) *Assume that $n+m$ is odd, and p is odd.*

(a) *If $\sigma \le 0$, then*

$$\Omega^{p,q} = R_{\frac{p-1}{2},\frac{q-\epsilon}{2}}.$$

(b) *If* $\sigma \geq 1$, *then*

$$\Omega^{p,q} = \prec R_{s,t} \succ,$$

where $s = \max\left(0, \frac{n-q}{2}\right)$, *and* $t = \max\left(0, \frac{n+1-\epsilon-p}{2}\right)$.

Remarks 1

(a) For $(i,j) = (\frac{p-1}{2}, \frac{q-\epsilon}{2})$, we have noted that $i+j = \frac{n_0}{2} + \sigma = r_1$. Thus $R_{i,j}$ is at the bottom layer of the module diagram of $I^\alpha(\sigma)$, namely is an irreducible submodule. The collection of $R_{i,j}$'s exhausts the set of all irreducible submodules of $I^\alpha(\sigma)$.

(b) If $q \leq n$, and $p \leq n+1-\epsilon$, then $s = \max\left(0, \frac{n-q}{2}\right) = \frac{n-q}{2}$, $t = \max\left(0, \frac{n+1-\epsilon-p}{2}\right) = \frac{n+1-\epsilon-p}{2}$, and $s+t = \frac{n-\epsilon}{2} - \sigma = \frac{n_0}{2} - \sigma = r_2$. Such $R_{s,t}$'s are exactly those at the top layer of the module diagram of $I^\alpha(\sigma)$, namely irreducible quotients. The rest of $R_{s,t}$'s are those on the "left boundary" $(s=0)$ or the "right boundary" $(t=0)$.

Theorem 6.3 (Case 1b) *Assume that* $n+m$ *is odd, and* p *is even.*

(a) *If* $\sigma \leq 0$, *then*

$$\Omega^{p,q} = R_{\frac{p}{2}, \frac{q-\epsilon'}{2}}.$$

(b) *If* $\sigma \geq 1$, *then*

$$\Omega^{p,q} = \prec R_{s,t} \succ,$$

where $s = \max\left(0, \frac{n+1-q}{2}\right)$, *and* $t = \max\left(0, \frac{n+1-\epsilon'-p}{2}\right)$.

Remarks 2

(a) For $(i,j) = (\frac{p}{2}, \frac{q-\epsilon'}{2})$, we have noted that $i+j = \frac{n_1+1}{2} + \sigma = r_1$. Thus $R_{i,j}$ is at the bottom layer of the module diagram of $I^\alpha(\sigma)$, namely is an irreducible submodule. The collection of $R_{i,j}$'s exhausts the set of all irreducible submodules of $I^\alpha(\sigma)$.

(b) If $q \leq n+1$, and $p \leq n+1-\epsilon'$, then $s = \max\left(0, \frac{n+1-q}{2}\right) = \frac{n+1-q}{2}$, $t = \max\left(0, \frac{n+1-\epsilon'-p}{2}\right) = \frac{n+1-\epsilon'-p}{2}$, and $s+t = \frac{n+1-\epsilon'}{2} - \sigma = \frac{n_1+1}{2} - \sigma = r_2$. Such $R_{s,t}$'s are exactly those at the top layer of the module diagram of $I^\alpha(\sigma)$, namely irreducible quotients. The rest of $R_{s,t}$'s are those on the "left boundary" $(s=0)$ or the "right boundary" $(t=0)$.

6.2 The image of $\psi_{p,q}$ in $I^\alpha(\sigma)$: m and n same parity

Case 2: $n+m$ even.

We consider two subcases:

Case 2a: p is odd *Case 2b:* p is even.

We introduce one similar notation in this case. For an irreducible subquotient $L_{s,t}$ of $I^\alpha(\sigma)$ where $\sigma > 0$, denote by $\prec L_{s,t} \succ$ the submodule of $I^\alpha(\sigma)$ generated by $L_{s,t}$:

$$\prec L_{s,t} \succ = \begin{cases} \oplus\{L_{i,j} : i \geq s, j \leq t\}, & \text{(Case 2a)}, \\ \oplus\{L_{i,j} : i \leq s, j \geq t\}, & \text{(Case 2b)}. \end{cases} \tag{6.2}$$

As in Subsection 5.2, we let $k = [n/2]$. First recall the module diagram of $I^\alpha(\sigma)$ for Case 2a:

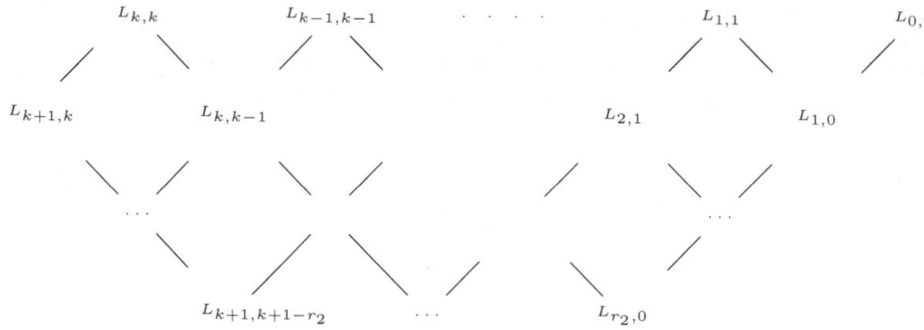

Case 2a: Diagram for $I^\alpha(\sigma)$ when $n+m$ is even and p is odd; $\sigma \leq -\frac{1}{2}$; $r_2 = -\sigma + \frac{1}{2} \leq [\frac{n+1}{2}]$

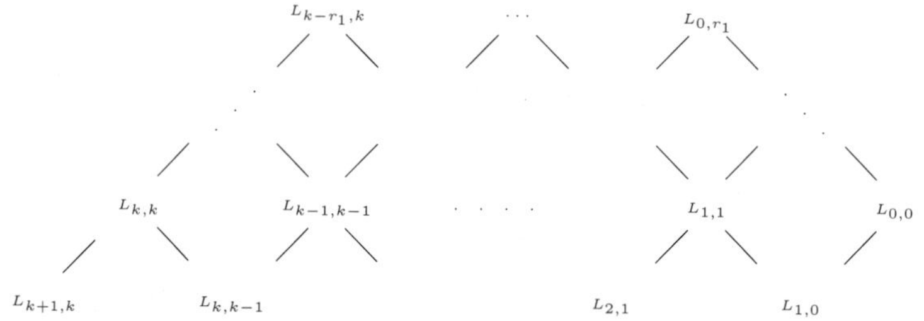

Case 2a: Diagram for $I^\alpha(\sigma)$ when $n+m$ is even and p is odd; $\sigma \geq \frac{1}{2}$; $r_1 = \sigma - \frac{1}{2} \leq [\frac{n}{2}]$

Theorem 6.4 (Case 2a) *Assume that $n+m$ is even, and p is odd.*

(a) *If $\sigma \leq -\frac{1}{2}$, then*

$$\Omega^{p,q} = L_{\frac{n+1-q}{2}, \frac{p-1}{2}}.$$

(b) *If $\sigma \geq \frac{1}{2}$, then*

$$\Omega^{p,q} = \prec L_{s,t} \succ,$$

where $s = \max\left(0, \frac{n+1-q}{2}\right)$, *and* $t = \min\left([\frac{n}{2}], \frac{p-1}{2}\right)$.

Remarks 3

(a) For $(i,j) = (\frac{n+1-q}{2}, \frac{p-1}{2})$, we have $i - j = \frac{n+2-m}{2} = -\sigma + \frac{1}{2} = r_2$. Thus $L_{i,j}$ is at the bottom layer of the module diagram of $I^\alpha(\sigma)$, namely is an irreducible submodule. The collection of $L_{i,j}$'s exhausts the set of all irreducible submodules of $I^\alpha(\sigma)$.

(b) If $q \leq n+1$, and $\frac{p-1}{2} \leq [\frac{n}{2}]$, then $s = \max\left(0, \frac{n+1-q}{2}\right) = \frac{n+1-q}{2}$, $t = \min\left([\frac{n}{2}], \frac{p-1}{2}\right) = \frac{p-1}{2}$, and $t - s = \frac{m-(n+2)}{2} = \sigma - \frac{1}{2} = r_1$. Such $L_{s,t}$'s are exactly those at the top layer of the module diagram of $I^\alpha(\sigma)$, namely irreducible quotients. The rest of $L_{s,t}$'s are those on the "left boundary" ($t = k = [\frac{n}{2}]$) or the "right boundary" ($s = 0$). Note that when n is even, the submodule $L_{k+1,k}$ is empty.

Next recall the module diagram of $I^\alpha(\sigma)$ for Case 2b:

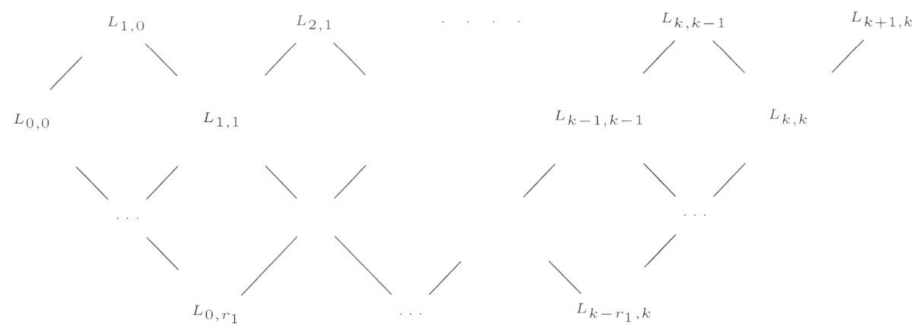

Case 2b: Diagram for $I^\alpha(\sigma)$ when $n+m$ and p are even; $\sigma \leq -\frac{1}{2}$; $r_1 = -\sigma - \frac{1}{2} \leq [\frac{n}{2}]$

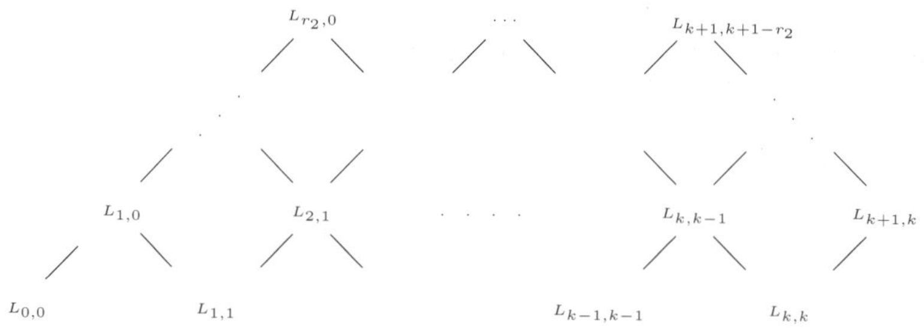

Case 2b: Diagram for $I^{\alpha}(\sigma)$ when both $n+m$ and p are even; $\sigma \geq \frac{1}{2}$; $r_2 = \sigma + \frac{1}{2} \leq [\frac{n+1}{2}]$

Theorem 6.5 (Case 2b) *Assume that $n+m$ is even, and p is even.*

(a) *If $\sigma \leq -\frac{1}{2}$, then*

$$\Omega^{p,q} = L_{\frac{p}{2}, \frac{n-q}{2}}.$$

(b) *If $\sigma \geq \frac{1}{2}$, then*

$$\Omega^{p,q} = \prec L_{s,t} \succ,$$

where $s = \min\left([\frac{n+1}{2}], \frac{p}{2}\right)$, and $t = \max\left(0, \frac{n-q}{2}\right)$.

Remarks 4

(a) For $(i,j) = (\frac{p}{2}, \frac{n-q}{2})$, we have $j - i = \frac{n-m}{2} = -\sigma - \frac{1}{2} = r_1$. Thus $L_{i,j}$ is at the bottom layer of the module diagram of $I^{\alpha}(\sigma)$, namely is an irreducible submodule. The collection of $L_{i,j}$'s exhausts the set of all irreducible submodules of $I^{\alpha}(\sigma)$.

(b) If $\frac{p}{2} \leq [\frac{n+1}{2}]$ and $q \leq n$, then $s = \min\left([\frac{n+1}{2}], \frac{p}{2}\right) = \frac{p}{2}$, $t = \max\left(0, \frac{n-q}{2}\right) = \frac{n-q}{2}$, and $s - t = \frac{m-n}{2} = \sigma + \frac{1}{2} = r_2$. Such $L_{s,t}$'s are exactly those at the top layer of the module diagram of $I^{\alpha}(\sigma)$, namely irreducible quotients. The rest of $L_{s,t}$'s are those on the "left boundary" $(t = 0)$ or the "right boundary" $(s = [\frac{n+1}{2}])$. Note that when n is even, the subquotient $L_{k+1,j}$ is empty.

References

[H1] R. Howe, *θ-series and invariant theory*, in Automorphic Forms, Representations and L-functions, Proc. Symp. Pure Math. **33** (1979), 275–285.

[H2] R. Howe, *Transcending classical invariant theory*, J. Amer. Math. Soc. **2** (1989), 535–552.

[Ku] S. S. Kudla, *Notes on the local theta correspondence*, Lecture notes from the European School of Group Theory, 1996. http://www.math.toronto.edu/~skudla/ssk.research.html.

[KR1] S. S. Kudla and S. Rallis, *Degenerate principal series and invariant distributions*, Israel J. Math. **69** (1990), 25–45.

[KR2] S. S. Kudla and S. Rallis, *Ramified degenerate principal series for* $Sp(n)$, Israel J. Math. **78** (1992), 209–256.

[KR3] S. S. Kudla and S. Rallis, *A regularized Siegel-Weil formula: The first term identity*, Annals Math. **140** (1994), 1–80.

[KS] S. S. Kudla and W. J. Sweet, Jr., *Degenerate principal series representations for* $U(n,n)$, Israel J. Math. **98** (1997), 253–306.

[L1] S. T. Lee, *On some degenerate principal series representations for* $U(n,n)$, J. Funct. Anal. **126** (1994), 305–366.

[L2] S. T. Lee, *Degenerate principal series representations for* $Sp(2n,\mathbb{R})$, Compositio Math. **103** (1996), 123–151.

[LZ1] S. T. Lee and C.-B. Zhu, *Degenerate principal series and local theta correspondence*, Trans. Amer. Math. Soc. **350** (1998), 5017–5046.

[LZ2] S. T. Lee and C.-B. Zhu, *Degenerate principal series and local theta correspondence II*, Israel J. Math. **100** (1997), 29–59.

[LZ3] S. T. Lee and C.-B. Zhu, *Degenerate principal series and local theta correspondence III: the case of complex groups*, J. Algebra **319** (2008), 336–359.

[Ra] S. Rallis, *On the Howe duality conjecture*, Compositio Math. **51** (1984), 333–399.

[SZ] B. Sun and C.-B. Zhu, *Conservation relations for local theta correspondence*, arXiv:1204.2969.

[Wal] N. Wallach, *Real Reductive Groups II*, Academic Press, San Diego, 1992.

[Ya] S. Yamana, *Degenerate principal series representations for quaternionic unitary groups*, Israel J. Math. **185** (2011), 77–124.

[Zh] C.-B. Zhu, *Invariant distributions of classical groups*, Duke Math. J. **65** (1992), 85–119.

Soo Teck Lee, Department of Mathematics, National University of Singapore, 10 Lower Kent Ridge Road, Singapore 119076.
E-mail: matleest@nus.edu.sg

Chen-Bo Zhu, Department of Mathematics, National University of Singapore, 10 Lower Kent Ridge Road, Singapore 119076.
E-mail: matzhucb@nus.edu.sg

Automorphic Representations and L-Functions
Editors: D. Prasad, C.S. Rajan, A. Sankaranarayanan, J. Sengupta

Publisher: Hindustan Book Agency, New Delhi, India

The Fibonacci Zeta Function

M. Ram Murty[1]

Abstract

We consider the lacunary Dirichlet series obtained by taking the reciprocals of the s-th powers of the Fibonacci numbers. This series admits an analytic continuation to the entire complex plane. Its special values at integral arguments are then studied. If the argument is a negative integer, the value is algebraic. If the argument is a positive even integer, the value is transcendental by Nesterenko's work. This is a result of Duverney, Ke. Nishioka, Ku. Nishioka and Shiokawa. We present a simplified proof of their result. If the argument is 1, the value has been shown to be irrational by André-Jeannin. We present a slight modification of Duverney's proof of this fact. At the same time, we highlight the "modular connection" of these questions as well as signal some new results in the theory of special values of q-analogues of classical Dirichlet L-functions.

1 Introduction

The sequence of Fibonacci numbers is defined by the recurrence relation

$$f_{n+1} = f_n + f_{n-1}, \quad n \geq 1$$

with initial values $f_0 = 0$ and $f_1 = 1$. The Fibonacci zeta function is the series

$$F(s) := \sum_{n=1}^{\infty} f_n^{-s}.$$

Since the n-th Fibonacci number has exponential growth, it is easy to see that the series converges for $\Re(s) > 0$. Its analytic continuation is easily derived using a technique the author and Sinha [12] used to derive the analytic continuation for the Riemann, Hurwitz and multiple Hurwitz zeta functions. Indeed, if $\alpha = \frac{1+\sqrt{5}}{2}$ and $\beta = -1/\alpha$, then,

$$f_n = \frac{\alpha^n - \beta^n}{\alpha - \beta}.$$

[1]Research of the author was partially supported by an NSERC Discovery grant.

Thus, using the binomial theorem,

$$\begin{aligned}(\alpha-\beta)^{-s}\sum_{n=1}^{\infty}1/f_n^s &= \sum_{n=1}^{\infty}\alpha^{-ns}\left(1-(\beta/\alpha)^n\right)^{-s}\\ &= \sum_{n=1}^{\infty}\alpha^{-ns}\sum_{j=0}^{\infty}\binom{-s}{j}(-1)^j(\beta/\alpha)^{nj}\\ &= \sum_{j=0}^{\infty}\binom{-s}{j}(-1)^j\frac{\beta^j/\alpha^{s+j}}{1-(\beta^j/\alpha^{s+j})}.\end{aligned}$$

The right hand side is easily seen to converge for all complex values of s. From this derivation, we immediately see that the special values of $F(s)$ are all algebraic (and in fact lying in the quadratic field $\mathbb{Q}(\sqrt{5})$) when s is either zero or a negative integer, since in these cases the sum is a finite sum.

In this paper, we will be interested in the special values $F(k)$ when k is a natural number. What kind of numbers are these? Are they transcendental? Are they algebraic? It turns out that much like the case with the Riemann zeta function, the special values $F(2k)$ are all transcendental and the nature of $F(2k+1)$ is still unknown. Analogous questions can be asked for any second order recurrence sequence. The discussion below extends to some of these sequences but not all. There are some subtle issues that enter into the derivations that seem at present to be insurmountable. We discuss these issues in the final section of the paper. For the sake of elegance of exposition, we focus on the Fibonacci sequence first.

The problem of evaluating $F(1)$ seems to go back to Laisant in 1899 (see [11]). Later that year, Edmund Landau [11] addressed the question and was unable to give a definite answer. However, he noted that the sum

$$\sum_{n=1}^{\infty}\frac{1}{f_{2n+1}}$$

can be expressed as special values of classical theta functions, giving us the first hint of a "modular connection." More precisely, Landau proved that

$$\sum_{n=1}^{\infty}\frac{1}{f_{2n+1}}=\frac{\sqrt{5}}{4}\theta_2^2\left(\frac{3-\sqrt{5}}{2}\right),$$

where

$$\theta_2(q)=\sum_{n=-\infty}^{\infty}q^{(n+1/2)^2}.$$

Beyond this, Landau was unable to say if this number is transcendental but this is now known due to recent advances in transcendental number theory.

Our current state of knowledge on this problem is recorded in the following two theorems:

Theorem 1.1 (André-Jeannin, 1989) *$F(1)$ is irrational.*

Theorem 1.2 (Duverney, Ke. Nishioka, Ku. Nishioka, Shiokawa, 1998, [6]) *$F(2k)$ is transcendental for all $k \geq 1$.*

As will be explained below, the essential ingredient in the proof of Theorem 1.2 is a deep theorem of Nesterenko regarding the transcendence of special values of Eisenstein series, proved in 1996. We will attempt to explain why the case of even exponents can be solved and where the difficulty lies for the odd exponents.

The proof of Theorem 1.2 is somewhat "easier" than Theorem 1.1. But both proofs require our entry into the world of q-series. Theorem 1.1 requires the q-exponential and q-logarithm functions and the identities seem to suggest a connection to Ramanujan's "mock-theta" world.

For the most part, this paper is a survey of known results. However, the presentation and arrangement of ideas is new. In particular, the emphasis on the use of the q-exponential and q-logarithm in the proof of Theorem 1.1 following [5] is with a view to give conceptual unity to these isolated results. We also introduce q-analogues of Dirichlet L-series $L(s,\chi)$ and report on the transcendence of some of their special values. These results will appear in a forthcoming paper [3].

2 Nesterenko's Theorem

To facilitate the proof of Theorem 1.2, we review Nesterenko's theorem regarding transcendental values of Eisenstein series. Recall that the Eisenstein series of weight 2, 4, and 6 for the full modular group are given by

$$E_2(q) = 1 - 24\sum_{n=1}^{\infty}\sigma_1(n)q^n,$$

$$E_4(q) = 1 + 240\sum_{n=1}^{\infty}\sigma_3(n)q^n,$$

$$E_6(q) = 1 - 504\sum_{n=1}^{\infty}\sigma_5(n)q^n,$$

where $\sigma_j(n) = \sum_{d|n} d^j$. We will make fundamental use of:

Theorem 2.1 (Nesterenko, 1996) *For any q with $|q| < 1$, the transcendence degree of the field*

$$\mathbb{Q}(q, E_2(q), E_4(q), E_6(q))$$

is at least 3. Thus, for q algebraic, $E_2(q), E_4(q)$, and $E_6(q)$ are algebraically independent.

Let us also recall that the general Eisenstein series

$$E_{2k}(q) = 1 - \frac{4k}{B_{2k}} \sum_{n=1}^{\infty} \sigma_{2k-1}(n) q^n,$$

where B_{2k} is the $2k$-th Bernoulli number, is a polynomial in E_4 and E_6 (see for example, [15]).

We also record here the following theorems.

Theorem 2.2 *For $k \geq 1$ and $m \geq 1$, we have that $E_{2k}(q^m)$ is algebraic over the field generated by E_2, E_4, E_6.*

Theorem 2.3 *Let K be the field generated by E_2, E_4, E_6 over the rational numbers. Suppose that f is a non-constant function which is algebraic over the function field K. If α is an algebraic number with $0 < |\alpha| < 1$ and $f(\alpha)$ is defined, then $f(\alpha)$ is transcendental.*

For proofs, we refer the reader to Lemmas 2 and 3 of [6].

3 Proof of Theorem 1.2

Let $\alpha = \frac{1+\sqrt{5}}{2}$ and $\beta = -1/\alpha$. Then,

$$f_n = \frac{\alpha^n - \beta^n}{\alpha - \beta},$$

as is easily verified by induction. Thus, the study of the series $F(k)$ reduces to the study of the series

$$\sum_{n=1}^{\infty} \frac{1}{(\alpha^n - \beta^n)^k},$$

which is

$$(-1)^k \sum_{n=1}^{\infty} \frac{1}{(\beta^n - (-1)^n \beta^{-n})^k}. \tag{3.1}$$

Breaking up the sum into n even and n odd leads us to consider the two series

$$A_k(q) = \sum_{n=1}^{\infty} \frac{1}{(q^n - q^{-n})^k},$$

and

$$B_k(q) = \sum_{n=1}^{\infty} \frac{1}{(q^n + q^{-n})^k}.$$

For then, we can re-write (3.1) as

$$(-1)^k \left(A_k(\beta^2) + B_k(\beta) - B_k(\beta^2)\right),$$

We now look at $A_k(q)$ and $B_k(q)$ separately. Clearly,

$$(-1)^k A_k(q) = \sum_{n=1}^{\infty} \frac{q^{nk}}{(1 - q^{2n})^k}. \tag{3.2}$$

Now for $|q| < 1$, we have

$$\frac{1}{1-q} = \sum_{m=0}^{\infty} q^m.$$

Differentiating this $(k-1)$ times with respect to q leads to

$$\frac{(k-1)!}{(1-q)^k} = \sum_{m=k-1}^{\infty} m(m-1)\cdots(m-(k-2))q^{m-k+1}.$$

Thus,

$$(-1)^k A_k(q)(k-1)! =$$
$$\sum_{n=1}^{\infty} q^{nk} \sum_{m=k-1}^{\infty} m(m-1)\cdots(m-(k-2))q^{2n(m-k+1)}$$

which is equal to

$$\sum_{n\geq 1; m\geq k-1} q^{n(2m-k+2)} m(m-1)\cdots(m-(k-2)).$$

Now suppose $k = 2j$ is even. Then,

$$A_{2j}(q)(2j-1)! = \sum_{n\geq 1; m\geq 2j-1} q^{2n(m-j+1)} m(m-1)\cdots(m-2j+2).$$

For $j = 1$, we have

$$A_2(q) = \sum_{n,m\geq 1} q^{2nm} m = \sum_{r=1}^{\infty} \sigma_1(r) q^{2r}.$$

We immediately recognize that this is, apart from the constant term, the weight 2 Eisenstein series, $E_2(q^2)$ (upto a constant). For $j = 2$, we have

$$A_4(q)3! = \sum_{n\geq 1; m\geq 3} q^{2n(m-1)} m(m-1)(m-2).$$

Putting $m-1$ as d, we see that $m(m-1)(m-2)$ is $(d+1)d(d-1)$ so that setting $r = nd$, the above can be re-written as

$$\sum_{r=1}^{\infty} q^{2r} \left(\sum_{d|r} d(d^2-1) \right)$$

and again we see that this is a linear combination of $E_2(q^2)$ and $E_4(q^2)$ (minus the constant terms). A similar calculation shows that $A_6(q)$ is a linear combination of $E_2(q^2)$, $E_4(q^2)$ and $E_6(q^2)$ (minus the constant terms). Since every Eisenstein series $E_{2j}(q)$ with $j \geq 2$ is a polynomial is $E_4(q)$ and $E_6(q)$, we see that $A_{2j}(q)$ is a non-zero polynomial in $E_2(q^2), E_4(q^2)$ and $E_6(q^2)$ with rational coefficients. We now invoke Theorem 2.1 to deduce that $E_2(\beta^2), E_4(\beta^2)$ and $E_6(\beta^2)$ are algebraically independent. Hence, $A_{2j}(\beta^2)$ is transcendental. (From (3.2), it is clear that $A_{2j}(\beta^2) \neq 0$.)

Now, what about $B_k(q)$? Here again, we need only observe that

$$\frac{(k-1)!}{(1+q)^k} = \sum_{m=k-1}^{\infty} m(m-1)\cdots(m-(k-2))q^{m-k+1}(-1)^{m-k+1},$$

so that

$$B_k(q) = \sum_{n=1}^{\infty} \frac{q^{nk}}{(1+q^{2n})^k},$$

and

$$B_k(q)(k-1)! = \sum_{n\geq 1; m\geq k-1} q^{n(2m-k+2)}(-1)^{m-k+1} m(m-1)\cdots(m-(k-2)).$$

For $k = 2j$, we have

$$B_{2j}(q)(2j-1)! = \sum_{n\geq 1; m\geq 2j-1} q^{2n(m-j+1)}(-1)^{m-1} m(m-1)\cdots(m-(2j-2)).$$

For $j = 1$, we have

$$-B_2(q) = \sum_{n,m \geq 1} q^{2nm}(-1)^m m = \sum_{r=1}^{\infty} q^{2r} \left(\sum_{d|r} (-1)^d d \right).$$

Now let us note that

$$-B_2(q) + A_2(q) = 2 \sum_{r=1}^{\infty} q^{2r} \left(\sum_{d|r, d \text{ even}} d \right).$$

Writing $r = 2r_1$ and $d = 2d_1$, we may rewrite this as

$$-B_2(q) + A_2(q) = 2 \sum_{r_1=1}^{\infty} q^{4r_1} \left(\sum_{d_1|r_1} 2d_1 \right).$$

This is equal to

$$4 \sum_{r_1=1}^{\infty} q^{4r_1} \sigma_1(r_1) = 4A_2(q^2).$$

It follows that

$$B_2(q) = A_2(q) - 4A_2(q^2).$$

Since the sum in question is (3.1) which is equal to

$$(-1)^k \left(A_k(\beta^2) + B_k(\beta) - B_k(\beta^2) \right),$$

we deduce from the above calculations that for $k = 2$, the series $F(2)$ can be expressed as a polynomial in $E_2(\beta^2), E_2(\beta^4)$ and $E_2(\beta^8)$. Now, the function

$$f(q) := A_2(q^2) + B_2(q) - B_2(q^2)$$

is a non-constant polynomial in $E_2(q^2), E_2(q^4)$ and $E_2(q^8)$. By Theorem 2.2, $E_2(q^m)$ is algebraic over K. Thus, $f(q)$ is algebraic over K and by Theorem 2.3, the specialization $f(\beta)$ is transcendental. Proceeding inductively, we see that in the general case, the non-constant function

$$\sum_{n=1}^{\infty} \frac{1}{(q^n - (-1)^n q^{-n})^k}$$

is algebraic over the function field K. By Theorem 2.3, the specialization (3.1) is transcendental. This completes the proof of Theorem 1.2.

The method can be used to treat the case of the series

$$\sum_{n=1}^{\infty} \frac{1}{f_{2n+1}^k}.$$

We need only make a few elementary observations and reduce the calculation to the above. First, we observe that the sum in question is essentially (apart from an algebraic factor) $B_k(\beta) - B_k(\beta^2)$. If k is even, we are done by the above argument. If k is odd, the argument is more delicate. We make some remarks on this point at the end of the paper.

4 The Case of Odd Arguments

The situation of the special values of the Fibonacci zeta function at odd arguments is somewhat similar to the case of the Riemann zeta function and there are some striking analogies between the two. In later sections, we will outline the proof of André Jeannin that $F(1)$ is irrational. This can be viewed as the "Fibonacci analogue" of Apéry's theorem that $\zeta(3)$ is irrational. As for $\zeta(2k+1)$, we know from the work of Ball and Rivoal [2] that there is a positive constant $c > 0$ such that

$$\dim_{\mathbb{Q}} \mathbb{Q}(\zeta(3), \zeta(5), ..., \zeta(2a+1)) \geq c \log a.$$

In particular, there are infinitely many $\zeta(2k+1)$ that are irrational. There is a "Fibonacci analogue" of this result too. To state it, we need to introduce the following q-analogues of the Riemann zeta function, namely,

$$\zeta_q(s) = \sum_{n=1}^{\infty} q^n \left(\sum_{d|n} d^{s-1} \right). \tag{4.1}$$

We will not go into a discussion of why this can be considered as the q-analogue of the Riemann zeta function here except to say that some hint will be given in the next section when we review q-series. The problem of transcendence of $F(s)$ at $s = k$ leads to the study of $\zeta_q(s)$ at $s = k$ for $q = (\sqrt{5} - 1)/2$ and other algebraic numbers. When $s = 2k$, then $\zeta_q(2k)$ is the classical Eisenstein series (without the constant term) $E_{2k}(q)$. Since these are polynomials in $E_4(q)$ and $E_6(q)$ (or equal to $E_2(q)$ when $k = 1$) with algebraic coefficients, the theory of Nesterenko enters into the discussion in a fundamental way in the even case. In the odd case, nothing is known about $\zeta_q(2k+1)$ and these are not modular forms. Their nature is a mystery. Motivated by the work on $\zeta(2k+1)$, Krattenthaler, Rivoal and Zudilin [10] proved:

Theorem 4.1 (2006) *Fix q such that $1/q$ is an integer different from ± 1. Then, infinitely many $\zeta_q(2k+1)$ are irrational. More precisely,*

$$\dim_{\mathbb{Q}} \mathbb{Q}(\zeta_q(3), \ldots, \zeta_q(2a+1)) \geq c\sqrt{a}.$$

5 Introduction to q-series

It is becoming increasingly clear that the theory of q-series, often relegated to the theory of combinatorics, actually will come to play a dominant conceptual role unifying large tracts of the mathematical world. Often called "quantum analogues", the q-analogues provide not only an appealing domain of investigation but also give rise to greater conceptual understanding of the world of "natural" numbers.

In its simplest form, the analogy is best understood by replacing the natural number n by $q^n - 1$ to obtain the "q-analogue." Until recently, mathematicians preferred

$$\frac{q^n - 1}{q - 1}$$

instead of $q^n - 1$, simply because

$$\lim_{q \to 1} \frac{q^n - 1}{q - 1} = n.$$

But the "Birch-Swinnerton-Dyer philosophy" has taken over in which we divide by the "obvious" factor eliminating the zero at $q = 1$ and study the constant term. Thus, the exponential function

$$e^x = 1 + \sum_{n=1}^{\infty} \frac{x^n}{n!}$$

has the q-analogue:

$$E_q(x) = 1 + \sum_{n=1}^{\infty} \frac{x^n}{(q^n - 1)(q^{n-1} - 1) \cdots (q - 1)}.$$

The logarithm function

$$-\log(1 - x) = \sum_{n=1}^{\infty} \frac{x^n}{n}$$

would have the q-analogue

$$L_q(x) = \sum_{n=1}^{\infty} \frac{x^n}{q^n - 1}.$$

Let us quickly return to our original motivating question and study the number

$$\theta = \sum_{n=1}^{\infty} 1/f_n.$$

Since

$$f_n = \frac{\alpha^n - \beta^n}{\alpha - \beta}, \quad \alpha = \frac{1+\sqrt{5}}{2}, \quad \beta = \frac{1-\sqrt{5}}{2} = -\frac{1}{\alpha},$$

we see that

$$\theta = (\alpha - \beta)\sum_{n=1}^{\infty} \frac{1}{\alpha^n - \beta^n} = (\alpha - \beta)\sum_{n=1}^{\infty} \frac{1}{\alpha^n - (-1/\alpha)^n} =$$
$$(\alpha - \beta)\sum_{n=1}^{\infty} \frac{(-\alpha)^n}{(-\alpha^2)^n - 1} = (\alpha - \beta)L_q(-\alpha),$$

with $q = -\alpha^2$. Thus, the original question of the algebraic nature of $F(1)$ reduces to the study of a special value of the q-logarithm.

Returning to the q-analogue of the Riemann zeta function, it would seem logical, in view of the discussion above, to define it as

$$Z_q(s) := \sum_{n=1}^{\infty} \frac{1}{(q^n - 1)^s}.$$

That this is related to $\zeta_q(s)$ becomes evident when we write $\zeta_q(s)$ given by (4.1) as a Lambert series:

$$\zeta_q(s) = \sum_{d=1}^{\infty}\sum_{e=1}^{\infty} d^{s-1} q^{de} = \sum_{d=1}^{\infty} \frac{d^{s-1} q^d}{1 - q^d}.$$

Indeed,

$$Z_q(s) = (-1)^s \sum_{n=1}^{\infty} \frac{1}{(1 - q^n)^s}$$

and when $s = k$,

$$\frac{1}{(1-q^n)^k} = \frac{1}{(k-1)!} \sum_{m=k-1}^{\infty} m(m-1)\cdots(m-(k-2))q^{n(m-k+1)}.$$

Putting $d = m - k + 2$ and writing $m(m-1)\cdots(m-(k-2))$ as $d(d+1)\cdots(d+k-2)$ we see that

$$\sum_{n=1}^{\infty} \frac{1}{(1-q^n)^k}$$

is a linear combination of series of the form

$$\sum_{d,n=1}^{\infty} d^j q^{dn} = \sum_{d=1}^{\infty} \frac{d^j q^d}{1-q^d}$$

and these are the values $\zeta_q(j+1)$. So this gives some clue as to why the q-Riemann zeta function is defined as (4.1). For other perspectives on this definition, we refer the reader to [9].

6 The q-exponential and q-logarithm

Now

$$E_q(x) - E_q(x/q) = \sum_{n=1}^{\infty} \frac{x^n - (x/q)^n}{(q^n-1)\cdots(q-1)} = \sum_{n=1}^{\infty} \frac{(x/q)^n}{(q^{n-1}-1)\cdots(q-1)} = \frac{x}{q} E_q(x/q).$$

Thus, we get the functional equation

$$E_q(x) = (1 + x/q)E_q(x/q).$$

Iterating this, we get

$$E_q(x) = \left(1 + \frac{x}{q}\right)\left(1 + \frac{x}{q^2}\right)\cdots$$

so that we deduce:

Theorem 6.1 *For* $|q| > 1$, *we have*

$$E_q(x) = \prod_{n=1}^{\infty}\left(1 + \frac{x}{q^n}\right).$$

We recognize immediately that the q-exponential function has some resemblance to the Dedekind η-function and thus we see the emergence of a "modular" link.

Let us look at the q-logarithm. We have

$$L_q(x) - L_q(x/q) = \sum_{n=1}^{\infty} \frac{x^n}{q^n-1} - \sum_{n=1}^{\infty} \frac{(x/q)^n}{q^n-1} = \sum_{n=1}^{\infty} \frac{q^n x^n - x^n}{q^n(q^n-1)} = \sum_{n=1}^{\infty} (x/q)^n = \frac{x/q}{1-x/q} = \frac{x}{q-x},$$

provided $|x| < |q|$. Iterating this, we obtain

$$L_q(x) = \sum_{n=1}^{\infty} \frac{x}{q^n - x}.$$

This leads to

Theorem 6.2 *For* $|q| > \max(1, |x|)$, *we have*

$$L_q(x) = \sum_{n=1}^{\infty} \frac{x}{q^n - x}.$$

Taking the logarithmic derivative of $E_q(x)$, we deduce immediately that

Theorem 6.3 *For* $|q| > \max(1, |x|)$, *we have*

$$L_q(x) = x\frac{E_q'(-x)}{E_q(-x)},$$

where the derivative is with respect to x.

7 Proof of Theorem 1.1.

Now suppose that θ is rational and equal to $-A/B$ (say), with A, B coprime integers. Recall that we showed

$$\theta = \sqrt{5}L_q(-\alpha), \quad q = -\alpha^2, \quad \alpha = \frac{1+\sqrt{5}}{2}.$$

Then,

$$B\sqrt{5}L_q(-\alpha) + A = 0.$$

But

$$L_q(x) = x\frac{E_q'(-x)}{E_q(-x)}.$$

Thus, we may re-write this as

$$-B\sqrt{5}\alpha E_q'(\alpha) + AE_q(\alpha) = 0.$$

Using the q-series for $E_q(x)$, we obtain

$$A + \sum_{n=1}^{\infty} \frac{A - Bn\sqrt{5}}{(1+\alpha^2)(1-\alpha^4)\cdots(1-(-\alpha^2)^n)}(-\alpha)^n = 0.$$

We split this as

$$A + \sum_{n=1}^{N} \frac{A - Bn\sqrt{5}}{(1+\alpha^2)(1-\alpha^4)\cdots(1-(-\alpha^2)^n)}(-\alpha)^n =$$
$$- \sum_{n=N+1}^{\infty} \frac{A - Bn\sqrt{5}}{(1+\alpha^2)(1-\alpha^4)\cdots(1-(-\alpha^2)^n)}(-\alpha)^n.$$

We clear the denominators and estimate:

$$|A(1+\alpha^2)\cdots(1-(-\alpha^2)^N) + \sum_{n=1}^{N}(A - Bn\sqrt{5})(-\alpha)^n(\cdots)|$$
$$\leq \sum_{n=N+1}^{\infty} |A + Bn\sqrt{5}| \frac{\alpha^n}{(\alpha^{2N+2}-1)\cdots(\alpha^{2n}-1)}.$$

To estimate this tail, we proceed as follows. The summand in the tail is easily seen to be

$$\ll \frac{n}{\alpha^n}$$

so that the tail is estimated by

$$\ll \sum_{n=N+1}^{\infty} \frac{n}{\alpha^n} \ll \frac{N}{\alpha^N}.$$

The number X_N (say) given by

$$X_N := A(1+\alpha^2)\cdots(1-(-\alpha^2)^N) + \sum_{n=1}^{N}(A - Bn\sqrt{5})(-\alpha)^n(\cdots)$$

lies in $\mathbb{Q}(\sqrt{5})$ and is an algebraic integer. We take the algebraic conjugate $\widetilde{X_N}$. Since the conjugate β of α satisfies $|\beta| < 1$, one finds

$$|\widetilde{X_N}| \ll N^2.$$

Thus,

$$|X_N\widetilde{X_N}| \ll N^3\alpha^{-N},$$

which tends to zero as N tends to infinity. Now, $X_N\widetilde{X_N}$ is an integer, and so we conclude that either X_N or $\widetilde{X_N}$ is zero for N large. But since X_N and $\widetilde{X_N}$ are conjugates of each other, we deduce that both X_N and $\widetilde{X_N}$ are zero for N sufficiently large. But then, this implies that $\sqrt{5}$ is rational, a contradiction. This completes the proof.

8 q-analogues of Dirichlet L-series

In analogy with our definition of the q-Riemann zeta function (4.1), it seems natural to define for each Dirichlet character $\chi \pmod N$, the q-Dirichlet L-function by

$$L_q(s,\chi) := \sum_{n=1}^{\infty} q^n \left(\sum_{d|n} \chi(d) d^{s-1} \right).$$

As before, this can be written as a Lambert series of the form:

$$\sum_{d=1}^{\infty} \frac{\chi(d) d^{s-1} q^d}{1-q^d}.$$

For certain values of k and suitable characters χ, this function coincides with a classical Eisenstein series of level N. More precisely, if $\chi(-1) = (-1)^k$, and $k \geq 3$, then, $L_q(k,\chi)$ coincides (apart from a rational scalar factor) to an Eisenstein series relative to to the congruence subgroup $\Gamma_1(N)$ (see Theorem 4.5.1 of [4]). One has similar results for $k = 1$ and $k = 2$ also. In each of these cases, with q algebraic, one can deduce the transcendence of $L_q(k,\chi)$. The essential idea is to use the modularity of $L_q(k,\chi)$ and observe that if Δ is Ramanujan's cusp form of weight 12, then $L_q(k,\chi)^{12}/\Delta^k$ is a modular function of level N. Since the field of modular functions of level N is algebraic over the field $\overline{\mathbb{Q}}(j)$, where j denotes the j-function, we derive a contradiction to Nesterenko's therem if $L_q(k,\chi)$ is algebraic. The details of this result along with a discussion of the cases not covered by modularity will appear in [3].

9 Concluding Remarks

A natural question that arises is to what extent these results can be generalized for other second-order recurrence sequences. In [6], the authors show the following: let α, β be algebraic numbers with $\alpha \neq \beta$ and $|\beta| < 1$. Put

$$U_n = \frac{\alpha^n - \beta^n}{\alpha - \beta}, \quad V_n = \alpha^n + \beta^n.$$

If $\alpha\beta = \pm 1$, then the numbers

$$\sum_{n=1}^{\infty} \frac{1}{U_n^{2s}}, \quad \sum_{n=1}^{\infty} \frac{1}{V_n^{2s}}$$

are transcendental for any positive integer s. If $\alpha\beta = 1$, then the number

$$\sum_{n=1}^{\infty} \frac{1}{V_n^s}$$

is transcendental for any positive integer s and if $\alpha\beta = -1$, then

$$\sum_{n=1}^{\infty} \frac{1}{U_{2n+1}^s}$$

is transcendental for any positive integer s. It would be interesting to investigate the cases not covered by this theorem.

If k is odd, it is possible to show that $B_k(q)$ is algebraic over the field generated by E_2, E_4 and E_6. But this is not a totally trivial deduction (see [6]). However, it is possible to deduce the transcendence result for the sum

$$\sum_{n=1}^{\infty} \frac{1}{f_{2n+1}}.$$

Indeed,

$$B_1(q) = \sum_{n=1}^{\infty} \frac{q^n}{(1+q^{2n})}.$$

We recall the classical Jacobi identity:

$$\theta(q)^2 = 1 + 4\sum_{n=1}^{\infty} \frac{q^n}{1+q^{2n}},$$

where

$$\theta(q) = \sum_{n=-\infty}^{\infty} q^{n^2},$$

is one of the basic theta functions. Now, $\theta(q)^{24}/\Delta$ is a modular function of level 4. Thus, it is algebraic over the field $\overline{\mathbb{Q}}(j)$ where j denotes the j-function. Since the j function can be expressed as a rational function involving E_4 and E_6, we immediately deduce the desired transcendence result. The discussion for

$$\sum_{n=1}^{\infty} \frac{1}{f_{2n+1}^k}$$

proceeds along similar lines.

There are still many open problems concerning reciprocal sums of Fibonacci numbers. For instance, is $F(3)$ irrational. Recently, the authors in [8] showed that $F(2s_1)$, $F(2s_2)$ and $F(2s_3)$ are algebraically independent if and only if the three integers s_1, s_2, s_3 are all distinct and at least one of them is even thus settling an old problem of whether $F(2)$, $F(4)$ and $F(6)$ are algebraically independent posed in [7]. No doubt, there are still more fascinating facts yet to be discovered about modularity and non-modulairty of the Fibonacci zeta function $F(s)$.

Acknowledgements I thank Tapas Chatterjee, Michael Dewar and Sanoli Gun for their helpful comments on an earlier version of this paper. I also thank the referees for their useful remarks that improved the readability of this paper.

References

[1] R. André-Jeannin, *Irrationalité de la somme des inverses de certaines séries récurrentes*, C.R. Acad. Sci. Paris **308** Sér. I, (1989), no. 19, 539–541.

[2] K. Ball and T. Rivoal, *Irrationalité d'une infinité de valeurs de la fonction zeta aux entiers impairs*, Inventiones Math. **146** (1) (2001), 193–207.

[3] M. Dewar and M. Ram Murty, *Special values of q-analogues of Dirichlet L-series*, in preparation.

[4] F. Diamond and J. Shurman, *A first course in modular forms*, Springer, 2005.

[5] D. Duverney, *Irrationalité de la somme des inverses de la suite de Fibonacci*, Elemente der Mathematik, **52** (1997), 31–36.

[6] D. Duverney, Ke. Nishioka, Ku. Nishioka, I. Shiokawa, *Transcendence of Jacobi's theta series and related results*, in Number Theory (Eger, 1996), 157–168, de Gruyter, Berlin, 1998.

[7] D. Duverney, Ke. Nishioka, Ku. Nishioka, I. Shiokawa, *Transcendence of Rogers-Ramanujan continued fraction and reciprocal sums of Fibonacci numbers*, Surikaisekikenkyusho Kokyuroku No. 1060 (1998), 91–100.

[8] C. Elsner, S. Shimomura, I. Shiokawa, *Algebraic independence results for reciprocal sums of Fibonacci numbers*, Acta Arith. **148** (2011), 205–223.

[9] M. Kaneko, N. Kurokawa, and M. Wakayama, *A variation of Euler's approach to values of the Riemann zeta function*, Kyushu J. Math. **57** (2003), no. 1, 175–192.

[10] C. Krattenthaler, T. Rivoal and W. Zudilin, *Séries hypergeometriques basiques, q-analogues des valeurs de la fonction zéta et séries d'Eisenstein*, J. Inst. Math. Jussieu **5** (1) (2006), 53–79.

[11] E. Landau, *Sur la série des inverse de nombres de Fibonacci*, Bull. Soc. Math. France **27** (1899), 298–300.

[12] M. Ram Murty and K. Sinha, *Multiple Hurwitz zeta functions*, in Multiple Dirichlet Series, Automorphic Forms and Analytic Number Theory, 135–156, Proc. Sympos. Pure Math. **75**, Amer. Math. Society, Providence, RI, 2006.

[13] Yu.V. Nesterenko, *Modular functions and transcendence problems*, C.R. Acad. Sci. Paris Sér. I **322** (1996), 909–914.

[14] Yu.V. Nesterenko, *Modular functions and transcendence problems*, Mat. Sb. **187** (9) (1996), 65–96, (Russian). English translation, Sb. Math. **187** (9-10) (1996), 1319-1348.

[15] J.-P. Serre, *A course in arithmetic*, Springer-Verlag, New York - Heidelberg, 1973.

Department of Mathematics, Queen's University, Kingston, Ontario, K7L 3N6, Canada.
E-mail: murty@mast.queensu.ca

Automorphic Representations and L-Functions
Editors: D. Prasad, C.S. Rajan, A. Sankaranarayanan, J. Sengupta

Publisher: Hindustan Book Agency, New Delhi, India

Decomposition and Parity of Galois Representations Attached to GL(4)

Dinakar Ramakrishnan[1]

Introduction

Let F be a number field, and π an isobaric ([La]), algebraic ([Cℓ1]) automorphic representation of $\mathrm{GL}_n(\mathbb{A}_F)$. We will call π *quasi-regular* iff at every archimedean place v of F, the associated n-dimensional representation σ_v of the Weil group W_{F_v} (defined by the archimedean local correspondence) is *multiplicity free*. For example, when $n = 2$ and $F = \mathbb{Q}$, a cuspidal π is quasi-regular exactly when it is generated by a holomorphic newform f of weight ≥ 1. Recall that π is *regular* iff at each archimedean v the *restriction* of σ_v to $\mathbb{C}^*$ is multiplicity free; hence any regular π is quasi-regular, but not conversely.

When F is totally real, an algebraic automorphic representation π of $\mathrm{GL}_n(\mathbb{A}_F)$ is said to be *totally odd* iff it is odd at each archimedean place v, i.e., iff the difference in the multiplicities of 1 and -1 as eigenvalues of complex conjugation in σ_v is at most 1; in particular if n is even, these two eigenvalues occur with the same multiplicity. One readily sees that any quasi-regular π is totally odd, but not conversely. When $n = 2$, π is said to be *even* (or that it has even parity) at an archimedean place v (of F) iff if it is not odd at v.

Still with F totally real, let c be one of the $[F : \mathbb{Q}]$ complex conjugations in the absolute Galois group $\mathfrak{G}_F$ =Gal$(\overline{F}/F)$. Recall that an n-dimensional $\overline{\mathbb{Q}}_p$-representation ρ of $\mathfrak{G}_F$ is *odd relative to* c if the trace of $\rho(c)$ lies in $\{1, 0, -1\}$. It is *odd* if it is so relative to every c. When $n = 2$, ρ is said to be *even relative to* c if it is not odd relative to c, which is the same as the determinant of $\rho(c)$ being 1.

One says that an isobaric, algebraic π on $\mathrm{GL}(n)/F$ has a fixed *archimedean weight* iff there is an integer w such that at every archimedean place v of F, the restriction to $\mathbb{C}^*$ of $\sigma_v \otimes |\cdot|^{(n-1)/2}$ is a direct sum of characters

[1]Partially supported by a grant from the NSF

$z \to z^{p_j}\bar{z}^{q_j}$ with $p_j + q_j = w$ $(\forall j)$. Every *cuspidal* algebraic π has such a weight ([Cℓ1]). The first object here is to provide a proof of the following assertion, which was established for the regular (algebraic), cuspidal case in an earlier preprint [Ra1] which has remained unpublished:

Theorem A *Let F be a totally real number field, $n \leq 4$, p a prime, π a quasi-regular algebraic, isobaric automorphic representation of $GL_n(\mathbb{A}_F)$ of a fixed archimedean weight, and ρ an associated n-dimensional, Hodge-Tate $\overline{\mathbb{Q}_p}$-representation of $\mathfrak{G}_F$ whose local L-factors agree with those of π (up to a shift) at almost all primes P of F. Then the semisimplification $\rho^{\rm ss}$ (of ρ) does not contain any irreducible 2-dimensional Galois representation which is even relative to some complex conjugation c.*

Since π is *algebraic*, it is expected by a conjecture of Clozel ([Cℓ1] that there is an associated Galois representation ρ (see also [BuzG]). When π is *regular and selfdual* and F totally real, the existence of ρ has been well known for some time for GL(n) ([Cℓ2], [PL], [CℓHLN]). Now it is also known for π *non-selfdual* and regular, by the important recent work of Harris, Lan, Taylor and Thorne ([HLTT]). When π is quasi-regular, the algebraicity is already in [BHR] (for F totally real), where one finds the equivalent notion of a *semi-regular, motivically odd* cusp form, and much progress has been made in the construction of ρ in the selfdual case in the thesis of Goldring ([Go1, Go2]), when the base change of π to a suitable CM extension K with $K^+ = F$ is known to descend to a holomorphic form on a suitable unitary group associated to K/F.

Here are some remarks on the proof of Theorem A. When $\rho^{\rm ss}$ is a direct sum $\eta_1 \oplus \eta_2$ with each η_j two-dimensional, one sees easily that the summands must have the same parity. But then we are left with the subtle task of ruling out both of them being *even*. We first appeal to the exterior square construction of Kim ([K]), which supplies (using the Langlands-Shahidi method) an isobaric automorphic form $\Pi = \Lambda^2(\pi)$ on GL$(6)/F$, and then, more importantly, we make use of the fact that one has some control, thanks to the works of Shahidi ([Sh1, Sh2]) and Ginzburg-Rallis ([GR]) concerning the analytic properties of the exterior cube L-function of Π, which has degree 20. Luckily, this L-function is also related to the square of a twist of the symmetric square L-function of π, and we exploit this. We also appeal to base change for GL(n) ([AC]). The proof given here is a strengthened form of the one in [Ra1], and we avoid making use of either regularity or cuspidality (of π). This article is completely self-contained, however, and one does not need to refer to [Ra1].

When the semisimplification of an p-adic representation τ decomposes as $\oplus_{j=1}^{r} tau_j$ with each τ_j irreducible of dimension n_j, we will say that τ

has decomposition type $(n_1, n_2, \ldots, n_r)$. Similarly, if π is an isobaric automorphic representation of $\mathrm{GL}_n(\mathbb{A}_F)$ of the form $\boxplus_{j=1}^r \pi_j$, with each π_j cuspidal of $\mathrm{GL}_{n_j}(\mathbb{A}_F)$, we will say that π has isobaric type $(n_1, n_2, \ldots, n_r)$.

One motivation for the assertion of Theorem A (in [Ra1]) was to help establish the *irreducibility of ρ for π cuspidal and regular,algebraic* on $\mathrm{GL}(4)/F$, at least *for large enough p*, generalizing the classical results of Ribet for $n = 2$, and Blasius-Rogawski ([BRo2]) for $n = 3$ *and* π essentially self-dual:

Theorem B *Let F be totally real, $n \leq 4$, and π an isobaric, algebraic automorphic representation of $GL_n(\mathbb{A}_F)$ which is regular at every archimedean place, with associated Hodge-Tate representation ρ of $\mathfrak{G}_F$. Assume that ρ is crystalline and $p - 1$ is greater than the twice the largest difference in the Hodge-Tate weights. Then the (isobaric) type of π determines the (reducibility) type of ρ^{ss}. In particular, if π is cuspidal, then ρ is irreducible.*

It should be remarked that recently, such an irreducibility result for essentially selfdual, cuspidal π has been established by F. Calegari and T. Gee ([CaG]) *with no condition on p* (and even for $n = 5$), and the route they take appeals to, and further extends, recent modularity results, as well as [BCh] on signs of selfdual representations and the results on even Galois representations in [Ca]; see also [DV] and all other related references in [CaG]. In the non-selfdual case, the authors of [CaG] give a condensed form of a simple argument of [Ra1], which we give in its entirety in section 3 below, both to correct an identity slightly and also to check that the isobaric type of π agrees with the decomposition type of ρ.

Our proof of Theorem B makes use of Theorem A in conjunction with a theorem of Richard Taylor establishing the potential automorphy (over a totally real extension) of a class of 2-dimensional p-adic representations, as well as some tricks involving L-functions. Our original approach was to require potential automorphy over a common totally real extension *simultaneously* for two 2-dimensional odd representations of the type considered in [Ta2, Ta3]. The proof given here is simpler and does not require this more stringent condition. We also make use of the recent work [HLTT] on non-selfdual representations. In general, the arguments here are a bit more involved than in [Ra1] so as to deal (for Theorem A) also with the quasi-regular, non-regular case. If $F = \mathbb{Q}$, in many cases one can also use [Ki] for the proof of the $(2, 2)$-case of Theorem B (instead of [Ta2, Ta3]), once the even summands are ruled out (as in Theorem A).

As a companion to Theorem B, we will also establish the following

Theorem B *Let F be a totally real number field, p a rational prime. and σ, σ' 2-dimensional semisimple, odd, crystalline p-adic representations of*

$\mathfrak{G}_F$ of the same weight w, such that each has distinct Hodge-Tate weights in an interval of length at most $(p-1)/2$. Assume moreover that $\sigma \oplus \sigma'$ is associated to an isobaric automorphic representation π of $GL_4(\mathbb{A}_F)$. Then there are a finite set S of places of $\mathbb{Q}$ such that $L^S(1+w/2,\sigma) \neq 0$. Moreover,

$$-\mathrm{ord}_{s=1} L^S(s, \sigma^\vee \otimes \sigma') = \dim(\sigma^\vee \otimes \sigma')^{\mathfrak{G}_F}.$$

This assertion is as predicted by the Tate conjecture when σ, σ' occur in the p-adic étale cohomology of smooth projective varieties. It is essentially a consequence of Theorem B if we also assume that π is regular and algebraic (in which case the oddness of σ, σ' would in fact be a consequence, thanks to Theorem A). Furthermore, this result is stated (for $F = \mathbb{Q}$) in [Ra1], and referred to by Skinner and Urban in [SU] (Thm. 3.2.4), without even assuming that $\sigma \oplus \sigma'$ is automorphic, and this stronger version will be the subject of a sequel to this paper. However, Theorem C as stated above appears to be sufficient for the application in [SU], section 3.2.6. To elaborate, in [SU], $\sigma \oplus \sigma'$ is modular, corresponding to a cusp form Π on GSp(4)/$\mathbb{Q}$. By Arthur's recent work ([A]), one can transfer Π to an isobaric automorphic form π on GL(4)/$\mathbb{Q}$ such that the degree 4 L-functions of π and of Π coincide at all but a finite number of primes. (One can say more but it is not needed for this application.) Now Theorem C above applies with this π associated to (σ, σ'). It will be interesting to extend this result without making the regularity assumption on the Hodge-Tate weights, possibly by making use of recent works of Calegari and Geraghty.

We would like to thank Don Blasius, Dipendra Prasad, Freydoon Shahidi, Chris Skinner and Eric Urban for their interest and encouragement, Clozel for some remarks, and Richard Taylor for helpful comments on an earlier ($\sim$ 2004) version of the preprint [Ra1], which had clearly been inspired by [Ta2, Ta3]. Thanks are also due to the referee for carefully reading the article and making helpful comments which improved the presentation. Finally, we happily acknowledge partial support from the NSF through the grant DMS-1001916.

1 Reductions

From here on, let π be as in Theorem A, with associated ρ. Before deriving some lemmas, let us make some simple observations. To begin, note that if π is an isobaric sum of idele class characters, then ρ^{ss} must be a direct sum of one-dimensional p-adic representations. Indeed, each idele class character ν appearing as an isobaric summand of π is algebraic and corresponds to an p-adic character ν by [Se], and the fact that ρ^{ss} is a direct sum of such characters follows by Tchebotarev. So we may take $2 \leq n \leq 4$; of course the case $n = 1$ is classical.

Next suppose there is a cuspidal automorphic representation β of $\mathrm{GL}_2(\mathbb{A}_F)$ occurring as an isobaric summand of π. Then by our hypothesis (in Theorem A), β is a totally odd, algebraic cuspidal automorphic representation of $\mathrm{GL}_2(\mathbb{A}_F)$, which corresponds to a holomorphic Hilbert modular newform with multi-weight $(k_1, k_2, \ldots, k_m)$, $m = [F : \mathbb{Q}]$, such that the $k_j \geq 1$ are all of the same parity. By Blasius-Rogawski ([BRo1]), Taylor ([Ta1]) in the regular case ($k_j \geq 2$), and by Jarvis ([Jar]) when one of the k_j is 1, we know the existence of an irreducible 2-dimensional p-adic representation τ of the absolute Galois group of F attached to β, i.e., with coincident local factors almost everywhere. Then τ is totally odd, and there is nothing else to say if $n = 2$; therefore we assume in the rest of this section that n is 3 or 4. If we write $\pi = \beta \boxplus \beta'$, with β' is an algebraic, isobaric automorphic representation of $\mathrm{GL}_{n-2}(\mathbb{A}_F)$, then β' also corresponds to an $(n-2)$-dimensional p-adic representation τ' which is totally odd if $n = 4$.

Suppose $n = 3$ and π is a cuspidal. Then $\pi' := \pi \boxplus 1$ is an isobaric automorphic representation of $\mathrm{GL}_4(\mathbb{A}_F)$ satisfying the same hypotheses of Theorem A, with corresponding Galois representation $\rho' = \rho \oplus 1$. The proof of Theorem A for π' will imply the same for π.

In view of the above remarks, we may assume from here on that $n = 4$ and that either π is cuspidal or isobaric of type $(3, 1)$. Let $\Lambda^2(\pi)$ denote the isobaric automorphic representation of $\mathrm{GL}_6(\mathbb{A}_F)$ defined by Kim ([K]), which corresponds to $\Lambda^2(\rho)$.

Assume moreover that

$$\rho^{\mathrm{ss}} \simeq \tau_1 \oplus \tau_2, \text{ with } \dim(\tau_j) = 2, \ j = 1, 2. \tag{1.1}$$

Lemma 1.1 *The decomposition* (1.1) *of* ρ^{ss} *precludes the possibility of* π *being of type* $(3, 1)$, *i.e., we cannot have an isobaric sum decomposition of the form*

$$(*) \qquad \pi \simeq \eta \boxplus \nu,$$

with η *a cuspidal automorphic representation of* $GL_3(\mathbb{A}_F)$ *and* ν *an idele class character. In fact, when* π *is of type* $(3, 1)$, ρ^{ss} *must be of the same type, i.e.,* ρ *must contain a sub or a quotient which is irreducible of dimension* 3.

Proof Let ω, resp. ω_η, be the central character of π, resp. ν, so that $\omega_\nu = \omega\nu^{-1}$. Note that $(*)$ implies

$$\Lambda^2(\pi) \simeq (\eta^\vee \otimes \omega_\pi \nu^{-1}) \boxplus (\eta \otimes \nu),$$

which can be seen by checking the unramified local factors and applying the strong multiplicity one theorem ([JS1]). It follows that

$$\Lambda^2(\pi) \boxplus \nu^2 \boxplus \omega\nu^{-1} \simeq (\pi^\vee \otimes \omega\nu^{-1}) \boxplus (\pi \otimes \nu). \qquad (1.2)$$

The algebraicity of π implies the same for η, ν and ω, and the latter pair defines (by [Se]) p-adic characters, again denoted by ν, ω, of the absolute Galois group of F. We get (by Tchebotarev)

$$\Lambda^2(\rho^{\rm ss}) \oplus \nu^2 \oplus \omega\nu^{-1} \simeq (\rho^{\rm ss})^\vee \otimes \omega\nu^{-1} \boxplus (\rho^{\rm ss} \otimes \nu). \qquad (1.3)$$

On the other hand, applying the exterior square operation to both sides of the decomposition (1.1) yields

$$\Lambda^2(\rho^{\rm ss}) \oplus \nu^2 \oplus \omega\nu^{-1} \simeq (\tau_1 \otimes \tau_2) \oplus \omega_1 \oplus \omega_2 \oplus \nu^2 \oplus \omega\nu^{-1}, \qquad (1.4)$$

where ω_j is the determinant of τ_j. Comparing (1.3) and (1.4), and using (1.1) and the isomorphism $\tau_j^\vee \simeq \tau_j \otimes \omega_j^{-1}$, we see that τ_1 and τ_2 must both be reducible. If we write

$$\tau_j \simeq \mu_j \oplus \mu_j', \ \dim(\mu_j) = \dim(\mu_j') = 1,$$

then the characters on the right are both Hodge-Tate, since ρ has that property (by assumption), hence are locally algebraic and correspond to idele class characters of F (cf. [Se]). This then forces the isobaric decomposition of the form

$$\pi \simeq \mu_1 \boxplus \mu_1' \boxplus \mu_2 \boxplus \mu_2',$$

which contradicts the cuspidality of η.

To finish the proof of the Lemma, we still need to check that ρ cannot be irreducible when $(*)$ holds. For this, note that (1.3) still holds because its proof did not use (1.1). But then, since the left hand side contains 1-dimensionals, ρ cannot be irreducible.

□

Lemma 1.2 *Let π be cuspidal and satisfy the hypotheses of Theorem A, together with the decomposition* (1.1) *of the semisimplification of the associated ρ. Let K be a totally real number field which is solvable and Galois over F. If the base change π_K of π to $GL(4)/K$ is Eisensteinian, then ρ must be irreducible.*

Proof If π_K is Eisensteinian, the cuspidality of π implies that $\mathrm{Gal}(K/F)$ acts transitively on the set of cuspidal automorphic representations occurring in the isobaric sum decomposition of π_K (cf. [AC]). This forces a decomposition $\pi_K \simeq \beta_1 \boxplus \beta_2$, where each β_j is an isobaric automorphic representation of $\mathrm{GL}(2, \mathbb{A}_K)$. It follows that there are intermediate fields L, E such that $F \subset E \subset L \subset K$ with $[L : E] = 2$, and a cuspidal automorphic representation β of $\mathrm{GL}(2, \mathbb{A}_L)$ such that π_E is cuspidal and is the automorphic induction $I_L^E(\beta)$. Now since π_E is the base change of π, at every archimedean place w of E, the eigenvalue set of complex conjugation in the associated 4-dimensional representation σ_w of $W_{E_w} \simeq W_{\mathbb{R}}$ must remain $\{1, -1, 1, -1\}$. As L is totally real, w splits in it, say into $\{\tilde{w}_1, \tilde{w}_2\}$, and so $\beta_{\tilde{w}_j}$ is forced to be odd for each j. In other words, β is a totally odd, algebraic cusp form on $\mathrm{GL}(2)/L$, and by [BRo1], [Ta1] and [Jar], it corresponds to a 2-dimensional, irreducible, odd Galois representation r. Then the induction R, say, of r to the absolute Galois group $\mathfrak{G}_E$ of E corresponds by functoriality to π_E. Moreover, the cuspidality of π implies that β, and hence r, is not θ-invariant, where θ denotes the non-trivial automorphism of L/E. Then R is irreducible by Mackey, and must be the restriction of ρ to $\mathfrak{G}_E$; this can be checked, for example, at almost all places and then deduced by Tchebotarev. In any case, it implies that ρ itself must be irreducible. □

Note that in this section we did not need π to be quasi-regular, only that it is totally odd.

2 Proof of Theorem A

Fix a totally real number field F. In view of the Lemmas of the previous section, we may assume that $n = 4$ with π satisfying the hypotheses of Theorem A with associated ρ, and that π remains cuspidal upon base change to any finite solvable Galois extension K which is totally real.

Suppose we have a decomposition of the form (1.1), i.e., with ρ^{ss} being the direct sum of two 2-dimensional p-adic representations τ_1, τ_2 of $\mathfrak{G}_F$, with $\omega_j = \det(\tau_j)$, which we will also view (by class field theory) as idele class characters ω_j of F. The Hodge-Tate hypothesis on ρ implies that as a p-adic character, ω_j is locally algebraic for $j = 1, 2$, and is thus associated to an algebraic Hecke character, again denoted ω_j, of the idele class group C_F of (the totally real) F, which must be an integral power of the norm character times a finite order character ν_j. It follows that as a p-adic character,

$$\omega_j = \chi^{a_j} \nu_j,$$

where χ is the p-adic cyclotomic character. Clearly, $\omega = \det(\rho) = \omega_1 \omega_2$. Note that τ_j is even relative to some c iff a_j and ν_j have the same parity at an archimedean place.

Lemma 2.1 *We have $a_1 = a_2$. Also, ν_1 and ν_2 have the same parity.*

Proof We may, and we will, assume that $a_1 \geq a_2$. A twist of π by $|\cdot|^t$ is, for some $t \in \mathbb{R}$, unitary; call this representation π^u. Since π is algebraic, t lies in $\frac{1}{2}\mathbb{Z}$. Then at any finite place v where π is unramified, we know by the Rankin-Selberg theory that the inverse roots $\alpha_{j,v}$, $1 \leq j \leq 4$, defining the local factor of π^u, satisfy $|\alpha_{j,v}| < (Nv)^{1/2}$. Consequently, any inverse root of $\Lambda^2(\pi_v^u)$ is strictly bounded in absolute value by Nv. Now,

$$\Lambda^2(\pi^u) \simeq \Lambda^2(\pi \otimes |\cdot|^t) \simeq \Lambda^2(\pi) \otimes |\cdot|^{2t}.$$

So $|a_j - 2t| < 1$ for $j = 1, 2$, and since $a_j, 2t \in \mathbb{Z}$, $a_1 = 2t = a_2$ (as desired).

Consequently, the central character of π, which is associated to $\omega = \det(\rho)$, equals $\nu_1\nu_2|\cdot|^{2a}$, where $a = a_1 = a_2$. Since ω is even, ν_1 and ν_2 have the same parity.

□

Assume the following:

For $j = 1, 2$, ν_j has the same parity as a at some archimedean place v_0. (2.1)

We will obtain a contradiction below, leading to Theorem A.

Lemma 2.2 *Let K be a finite abelian extension of F in which v_0 splits. Then $\omega_{j,K}$ $(= \omega_j \circ N_{K/F})$ cannot occur, for either j, in the isobaric sum decomposition of $\Lambda^2(\pi_K)$. In particular, $L(s, \Lambda^2(\pi) \otimes \omega_{j,K}^{-1})$ has no pole at $s = 1$.*

It should be noted that this crucial Lemma will be false if ω_j were to be totally odd; in that case $L(s, \pi; \mathrm{sym}^2 \otimes \omega_{j,K}^{-1})$ will be regular at $s = 1$.

Proof First consider when π is regular at the archimedean place v_0, so that the associated 4-dimensional representation of $W_{\mathbb{R}}$ is of the form

$$\sigma_{v_0} \simeq I(\xi) \oplus I(\xi'),$$

for distinct, non-conjugate characters $\xi : z \mapsto z^p\overline{z}^{w-p}$ and $\xi' : z \mapsto z^r\overline{z}^{w-r}$ of $\mathbb{C}^*$, where I denotes the induction from $\mathbb{C}^*$ to $W_{\mathbb{R}}$. Then $I(\xi) \otimes I(\xi')$ decomposes as $I(\xi\xi') \oplus I(\xi\overline{\xi}')$. It follows that

$$\Lambda^2(\sigma_\infty) \simeq I(z^{p+r}\overline{z}^{2w-p-r}) \oplus I(z^{p+w-r}\overline{z}^{w-p+r}) \oplus sgn^{w+1}|\cdot|^w \oplus sgn^{w+1}|\cdot|^w, \tag{2.2}$$

where w is the archimedean weight of π (with $w - a_j \in \mathbb{Z}$). Since K/F is abelian, the base change π_K (of π) makes sense ([AC]), and for any archimedean place $\tilde{v}_0$ of K above v_0, $K_{\tilde{v}_0} \simeq \mathbb{R}$ (since v_0 splits in K by assumption) and $\sigma_{\tilde{v}_0} \simeq \sigma_{v_0}$ is the parameter of $\pi_{K,\tilde{v}_0}$ if $\tilde{v}_0 \mid v_0$. So the occurrence of $\omega_{j,K}$ (for either j) in the isobaric decomposition of $\Lambda^2(\pi_K)$ implies that $\omega_{j,K,\tilde{v}_0}$ will occur in $\Lambda^2(\sigma_\infty)$, and so must equal $sgn^{w+1}|\cdot|^w$. Now since ω_{j,v_0} is of the form $\omega_{j,v_0}|\cdot|^a$, we see that $w = a$ and that the parity of ω_j at v_0 is the opposite of that of a, a contradiction!

Next look at when π_{v_0} is quasi-regular, but not regular. Then we must have

$$\sigma_{v_0} \simeq I(\xi) \oplus (1 + \mathrm{sgn})|\cdot|^{w/2},$$

where again ξ is given as above. In this case, since $I(\xi)$ is twist invariant under sgn, the sign character, we obtain

$$\Lambda^2(\sigma_\infty) \simeq (I(\xi) \oplus \mathrm{sgn}|\cdot|^w)^{\oplus 2}. \tag{2.3}$$

Again, we see that $\omega_{j,\tilde{v}_0}$ cannot be a summand on the right because of parity.

□

As the proof shows, a is necessarily the archimedean weight of π. Let

$$\alpha = \nu_1/\nu_2, \ \nu = \nu_1\nu_2, \quad \text{and} \quad E = F(\alpha, \nu),$$

the compositum of the cyclic extensions of F cut out by α and ν. Then by Lemma 2.1, $\alpha = \omega_1/\omega_2$, and $\omega = \nu|\cdot|^{2a}$. Moreover, the total evenness of α and ν implies that E is totally real. By construction, $\nu_{1,E} = \nu_{2,E}$ and $\nu_E = 1$, so if we put

$$\mu := \nu_{1,E}|\cdot|_E^a,$$

then μ^2 is $|\cdot|_E^{2a}$ on C_E.

The reason for considering E is that π becomes essentially selfdual over E:

Lemma 2.3 *We have*

(a) $\pi_E^\vee \simeq \pi_E \otimes \mu^{-1}$;

(b) *The incomplete L-function* $L^T(s, \pi_E; \mathrm{sym}^2 \otimes \mu^{-1}))$ *has a pole at* $s = 1$ *of order* 1*, where* T *is a finite set of places of* E *containing the archimedean and ramified primes;*

(c) $\Lambda^2(\pi_E \otimes |\cdot|_E^{-a})$ *is selfdual.*

Proof (a) By assumption, ρ^{ss} is $\tau_1 \oplus \tau_2$, and for either j, $\tau_j^\vee \simeq \tau_j \otimes \omega_j^{-1}$. So with μ being the restriction of ω_j to E, then

$$\rho_E^{\mathrm{ss}\vee} \simeq \rho_E^{\mathrm{ss}} \otimes \mu^{-1}.$$

Comparing L-functions, we get

$$L^T(s, \pi_E^\vee) = L^T(s, \pi_E \otimes \mu^{-1}).$$

It follows by the strong multiplicity one theorem for isobaric automorphic representations ([JS1]) that $\pi_E^\vee$ is isomorphic to $\pi_E \otimes \mu^{-1}$.

(b) We have

$$L^T(s, \pi_E \times \pi_E \otimes \mu^{-1}) = L^T(s, \pi_E, \Lambda^2 \otimes \mu^{-1}))L^T(s, \pi_E, \mathrm{sym}^2 \otimes \mu^{-1})).$$

One knows by the Rankin-Selberg theory that the L-function on the left has a pole at $s = 1$ of order ≥ 1, and so the assertion follows in view of Lemma 2.2 (with $K = E$).

(c) This follows from (a).

□

For any isobaric automorphic representation Π of $\mathrm{GL}(6)/F$ and a character ξ of F, let $L^T(s, \Pi; \Lambda^3 \otimes \xi)$ denote, for a finite set T of places, the incomplete ξ-twisted *exterior cube* L-function of Π of degree 20. One knows — see [Sh1], Corollary 6.8, that this L-function admits a meromorphic continuation to the whole s-plane and satisfies a standard functional equation.

Proposition 2.4 *Let $\Pi = \Lambda^2(\pi)$, and T a sufficiently large finite set of places of E containing the archimedean and ramified places as well as those dividing p. Then for any character ξ of E,*

(a) $L^T(s, \Pi_E; \Lambda^3 \otimes \xi) = L^T(s, \pi_E; \mathrm{sym}^2 \otimes \xi)^2$;
(b) $L^T(s, \Pi_E; \Lambda^2 \otimes \xi)L^T(s, \xi) = L^T(s, \pi_E \times \pi_E \otimes \xi\mu)$.

Corollary 2.5

(a) $-\mathrm{ord}_{s=1} L^T(s, \Pi_E; \Lambda^3 \otimes \mu) \geq 2$;
(b) $\mathrm{ord}_{s=1} L^T(s, \Pi_E; \Lambda^2 \otimes \mu^{-2}) = 0$.

Clearly, part (a) (resp. (b)) of the Corollary follows from part (b) (resp. (a)) of Lemma 2.3 and part (a) (resp. (b) of Proposition 2.4.

To prove the Proposition we need the following

Lemma 2.6 *Let $\sigma = \sigma_1 \oplus \sigma_2$ be a representation of a group G, with each σ_j being of dimension 2 with determinant α_j. Then* $\mathrm{sym}^2(\sigma)$ *is* $\sigma_1 \otimes \sigma_2 \oplus \mathrm{sym}^2(\sigma_1) \oplus \mathrm{sym}^2(\sigma_2)$. *Moreover,*

$$\Lambda^3(\Lambda^2(\sigma)) \simeq$$
$$(\sigma_1 \otimes \sigma_2 \otimes \alpha_1\alpha_2)^{\oplus 2} \oplus \mathrm{sym}^2(\sigma_1) \otimes (\alpha_1\alpha_2 \oplus \alpha_2^2) \oplus \mathrm{sym}^2(\sigma_2) \otimes (\alpha_1\alpha_2 \oplus \alpha_1^2),$$
$$\sigma^{\otimes 2} \simeq \mathrm{sym}^2(\sigma_1) \otimes \alpha_1 \oplus \mathrm{sym}^2(\sigma_2) \otimes \alpha_2 \oplus (\sigma_1 \otimes \sigma_2)^{\oplus 2},$$

and

$$\Lambda^2(\Lambda^2(\sigma)) \simeq (\sigma_1 \otimes \sigma_2)^{\oplus 2} \oplus \mathrm{sym}^2(\sigma_1) \otimes \alpha_2 \oplus \mathrm{sym}^2(\sigma_2) \oplus \alpha_1.$$

In particular, if $\alpha_1 = \alpha_2 = \alpha$, we have

(a) $\Lambda^3(\Lambda^2(\sigma)) \otimes \alpha^{-2} \simeq \mathrm{sym}^2(\sigma)^{\oplus 2}$

and

(b) $\Lambda^2(\Lambda^2(\sigma)) \oplus \alpha^2 \simeq \sigma \otimes \sigma \otimes \alpha.$

Proof (of Lemma 2.6) The first identity (involving the symmetric square) is evident. For the second, observe that since $\Lambda^2(\sigma)$ is $\sigma_1 \otimes \sigma_2 \oplus \alpha_1 \oplus \alpha_2$,

(i) $\Lambda^3(\Lambda^2(\sigma)) \simeq \Lambda^3(\sigma_1 \otimes \sigma_2) \oplus \Lambda^2(\sigma_1 \otimes \sigma_2) \otimes (\alpha_1 \oplus \alpha_2) \oplus \sigma_1 \otimes \sigma_2 \otimes \alpha_1\alpha_2.$

We have

(ii) $\Lambda^2(\sigma_1 \otimes \sigma_2) \simeq \mathrm{sym}^2(\sigma_1) \otimes \Lambda^2(\sigma_2) \oplus \Lambda^2(\sigma_1) \otimes \mathrm{sym}^2(\sigma_2).$

Moreover, the non-degenerate G-pairing

$$\sigma_1 \otimes \sigma_2 \times \Lambda^3(\sigma_1 \otimes \sigma_2) \to \Lambda^4(\sigma_1 \otimes \sigma_2) = \det(\sigma_1 \otimes \sigma_2) = \alpha_1^2\alpha_2^2$$

identifies $\Lambda^3(\sigma_1 \otimes \sigma_2)$ with $(\sigma_1 \otimes \sigma_2)^\vee \otimes \alpha_1^2\alpha_2^2$, which is isomorphic to $\sigma_1 \otimes \sigma_2 \otimes \alpha_1\alpha_2$. The second identity and part (a) of the Lemma now follow by putting these together.

Moreover,

$$\Lambda^2(\Lambda^2(\sigma)) \simeq \Lambda^2(\sigma_1 \otimes \sigma_2) \oplus \alpha_1\alpha_2 \oplus (\sigma_1 \otimes \sigma_2 \otimes (\alpha_1 \oplus \alpha_2)).$$

Applying (ii), we also get the third identity and part (b).

□

Proof (of Proposition 2.4) Since ρ is associated to π, there is a finite set S of places of F containing the ramified places and ∞, such that for all $u \notin S$,

$$L(s, \pi_u) \;=\; L(s, \rho_u^{\rm ss}),$$

which is just an equality of 4-tuples of inverse roots defining the respective L-factors. It follows using (1.3) that for any such u,

$$L(s, \Lambda^2(\pi)_u) \;=\; L(s, \tau_{1,u} \otimes \tau_{2,u}) L(s, \omega_{1,u}) L(s, \omega_{2,u}).$$

Since by construction, ω_1 and ω_2 become the same over E, it follows that at any place v of E outside the inverse image T of S, we can write

$$\sigma_v(\pi_E) \;\simeq\; \sigma_{1,v} \oplus \sigma_{2,v},$$

with $\sigma_{1,v}$, $\sigma_{2,v}$ both being 2-dimensional of the same determinant. Here, $\sigma_v(\pi_E)$ is the 4-dimensional representation of the Weil group of $F_{1,v}$ attached to $\pi_{E,v}$. Now we are done by appealing to Lemma 2.6.

□

Proof (of Theorem A (contd.))

As noted at the beginning of this section, we have already reduced to the case when π remains cuspidal over any solvable totally real extension of F, in particular over E. Recall that π_E is essentially selfdual relative to the character μ, and we may replace π by its twist by a power of $|\cdot|$ so that Π_E is selfdual.

Suppose $\Pi_E = \Lambda^2(\pi_E)$ is cuspidal. Then by a result of Ginzburg and Rallis (cf. [GR], Theorem 3.2) we know that $L^T(s, \Pi_E; \Lambda^3 \otimes \xi)$, for any character ξ, will have at most a simple pole at $s = 1$, for any finite set T of places of E containing the archimedean and ramified places. (If Π_E has a supercuspidal component, this also follows from [KSh], where one finds such a result even for the complete L-function.) But this contradicts Corollary 2.5, part (a).

So Π_E is not cuspidal, and we may write

$$\Pi_E \;\simeq\; \boxplus_{j=1}^m \beta_j,\; m > 1, \tag{2.4}$$

where each β_j is a cuspidal automorphic representation of $\mathrm{GL}(n_j, \mathbb{A}_E)$ with $\sum_j n_j = 6$ and $n_i \leq n_j$ if $i \leq j$. Note that each β_j will necessarily be algebraic.

It remains to get a contradiction when Π_E is Eisenteinian. Note that since π_E has trivial central character, $\Lambda^2(\pi_E)$ is selfdual.

Lemma 2.7 *$n_1 \geq 2$ and $m \leq 3$.*

Proof (of Lemma 2.7) First note that if some $n_j = 1$, then twisting by the inverse of the corresponding character β_j, we get a pole at $s = 1$ of the L-function

$$L^T(s, \pi_E \times \pi_E \otimes \beta_j^{-1}) = L^T(s, \Pi_E \otimes \beta_j^{-1})L^T(s, \pi_E, \mathrm{sym}^2 \otimes \beta_j^{-1}),$$

because $\prod_{j\geq 2} L^T(s, \beta_j)$ and $L(s, \pi_E; \mathrm{sym}^2)$ has no zero at $s = 1$ (cf. [Sh1]). Consequently, by the cuspidality of π_E and [JS1], we obtain

$$\pi_E^\vee \simeq \pi_E \otimes \beta_j^{-1} \ \ (\text{if } \ n_j = 1).$$

Moreover, at any archimedean place v of E lying above above u (of F), the expressions (2.2) and (2.3) (for the possible shapes of) $\Lambda^2(\sigma_u(\pi))$ imply, since u splits in E, that *at most two of the* n_j *could be* 1, and the corresponding β_j must be totally odd. Suppose $n_1, n_2 = 1$. Since π is cuspidal,

$$\gamma := \beta_1\beta_2^{-1} \neq 1,$$

and π admits a self-twist by this totally even character, π must be induced from the totally real quadratic extension $L = E(\gamma)$ of E; so π_L is not cuspidal. But this cannot happen, as we have earlier reduced (see the first paragraph of this section) to the situation where the base change of π to any finite solvable Galois, totally real extension remains cuspidal. Hence $n_2 \geq 2$ and $m \leq 3$.

If $n_1 = 1$, then there are then two possibilities for the type of Π_E, namely $(1, 5)$ and $(1, 2, 3)$. In the former case,

$$L^T(s, \Pi_E; \Lambda^3) = L^T(s, \beta_2; \Lambda^2 \otimes \beta_1)L^T(s, \beta_2; \Lambda^3).$$

The second L-function can be identified with $L^T(s, \Lambda^2(\beta_2)^\vee \otimes \lambda)$, where λ is the central character of β_2. So each factor on the right of (5.13) is an abelian twist of the exterior square L-function of a cusp form on GL(5)/F, and by a theorem of Jacquet and Shalika ([JS2]), it admits no pole at $s = 1$ (because 5 is odd). This contradicts part (a) of Corollary 2.5, and so the case $(1, 5)$ cannot happen.

Suppose $(n_1, n_2, n_3) = (1, 2, 3)$. Here each β_j must be selfdual. In particular, the square of the idele class character β_1 is 1; similarly for the square of the central characters δ_j of β_j for $j = 2, 3$. But β_1 must be non-trivial, because the L-function of Π_E has no pole at $s = 1$. Similarly, $L^T(s, \delta_2)$ divides $L^T(s, \Pi_E; \Lambda^2)$, which (by part (b) of Corollary 2.5 is invertible at $s = 1$, implying that δ_2 is non-trivial. Consequently, $\beta_2 \simeq \beta_2^\vee = \beta_2 \otimes \delta_2$ and so β_2 is dihedral. Moreover, the selfduality of the GL(3)-cusp form β_3 implies (cf. [Ra5], for example) that it is of the form $\mathrm{sym}^2(\eta) \otimes \xi$ for a cusp form η on $\mathrm{GL}(2)/E$ and a character ξ. A direct computation yields (using $\Lambda^2(\beta_3) \simeq \beta_3$ by selfduality):

$$L^T(s, \Pi_E; \Lambda^3) = \\ L^T(s, \beta_1 \times \delta_2)L^T(s, \beta_1 \otimes \beta_2 \times \mathrm{sym}^2(\eta) \otimes \xi)L^T(s, (\beta_1\delta_3^{-1}) \otimes \beta_3) \times \\ L^T(s, \delta_2 \otimes \beta_3)L(s, \xi_3).$$

Since the left hand side equals $L^T(s, \pi_E; \mathrm{sym}^2)^2$, the incomplete L-function $L^T(s, \beta_1\xi \otimes \beta_2 \times \mathrm{sym}^2(\eta))$ must be the square of a degree 3 automorphic L-function $L(s)$. This forces η to be dihedral, which is not possible since $\mathrm{sym}^2(\eta)$ is not cuspidal. Hence the $(1, 2, 3)$ case is not possible either. □

End of proof of Theorem A

To recap, we have already established Theorem A when Π_E is cuspidal, and we may assume by Lemma 2.7 that Π_E is an isobaric sum as in (2.4) with $n_1 \geq 2$ and $m \leq 3$. As above, we may also replace π by its twist by a power of $|\cdot|$ to assume that Π_E is selfdual.

Suppose Π_E is of *type* $(2, 2, 2)$, i.e., an isobaric sum of three algebraic cusp forms $\beta_1, \beta_2, \beta_3$ on $\mathrm{GL}(2)/E$, with α_j being the central character of β_j, we get (for suitable finite set T of places containing the archimedean ones):

$$L^T(s, \Pi_E; \Lambda^3) = L^T(s, (\beta_1 \boxtimes \beta_2) \times \beta_3) \prod_{i \neq j} L^T(s, \beta_i \otimes \alpha_j). \qquad (2.5)$$

We know by part (b) of Corollary 2.5 that this exterior cube L-function of Π_E has a pole of order ≥ 2 at $s = 1$. Since each $L^T(s, \beta_i \otimes \alpha_j)$ is invertible at $s = 1$ (as β_i is cuspidal), we deduce from (2.5) that $L^T(s, (\beta_1 \boxtimes \beta_2) \times \beta_3)$ must have at least a double pole, which implies ([Ra3]) that $\beta_1 \boxtimes \beta_2$ must be of type $(2, 2)$ and contain $\beta_2^\vee$ as an isobaric summand with multiplicity 2. This forces (see [PRa]) each β_i to be dihedral, in fact corresponding to a character χ_i of a common quadratic extension E/E. Then each β_i is associated to an irreducible 2-dimensional representation τ_i of $\mathfrak{G}_E$. It follows that the restriction of $\Lambda^2(\rho^{\mathrm{ss}})$ to $\mathfrak{G}_E$ must be $\tau_1 \oplus \tau_2 \oplus \tau_3$, which contradicts the fact that the (supposed) decomposition (1.1) results in $\Lambda^2(\rho^{\mathrm{ss}})$ having one-dimensional summands. Hence Π_E cannot be of type $(2, 2, 2)$.

Next suppose Π_E is of *type* $(2, 4)$, i.e., with β_1, resp. β_2, being a selfdual cusp form on $\mathrm{GL}(2)/E$, resp. $\mathrm{GL}(4)/E$. Let ξ_j be the central character of β_j whose square is 1. Then $L^T(s, \xi_1)$ divides $L^T(s, \Pi_E, \Lambda^2)$, which has no pole at $s = 1$; thus $\xi_1 \neq 1$, implying that the selfdual β_1 must be dihedral, say of the form $I_L^E(\chi)$ for a character χ of a quadratic extension L of E. We obtain

$$L^T(s, \Pi_E; \Lambda^3) = L^T(s, \beta_1 \times \Lambda^2(\beta_2)) L^T(s, \xi_1 \otimes \beta_2)^2. \tag{2.6}$$

The occurrence of at least a double pole then forces the isobaric decomposition

$$\Lambda^2(\beta_2) \simeq (\beta_1 \otimes \xi_1)^{\boxplus 2} \boxplus I_L^E(\chi'), \tag{2.7}$$

with $I_L^E(\chi\chi') \simeq I_L^E(\chi^\theta\chi')$, where θ is the non-trivial automorphism of L/E. This forces χ' to be θ-invariant, forcing $I_L^E(\chi')$ to be Eisensteinian and of the form $\lambda \boxplus \lambda\delta$, where λ is the restriction of χ' to C_E and δ is the quadratic character attached to L. Since β_2 is selfdual, it follows that it admits a self-twist by δ and is hence induced from L; say $\beta_2 = I_L^E(\eta)$, for a cusp form η on GL(2)$/L$ with central character γ. Then we have

$$\Lambda^2(\beta_2) \simeq \mathrm{As}_{L/E}(\eta) \otimes \delta \boxplus I_L^E(\gamma), \tag{2.8}$$

where $\mathrm{As}_{L/E}(\eta)$ is the Asai transfer of η to a form on GL(4)$/E$ ([Ra4]). Comparing (2.7) and (2.8) we see that this Asai representation is not cuspidal and of type (2, 2). Then by Theorem B of [PRa], we see that η must be dihedral, induced by a character λ, say, of a biquadratic extension N of E containing L. Thus

$$\beta_2 \simeq I_L^E(\eta) \simeq I_N^E(\lambda),$$

with λ algebraic. So we may attach to β_2 an irreducible 4-dimensional representation σ obtained by Galois induction from the p-adic Galois character λ associated to λ in [Se]. Since $\Lambda^2(\pi_E) = \beta_1 \boxplus \beta_2$, we obtain (by Tchebotarev) an isomorphism of $\mathfrak{G}_E$-modules

$$\Lambda^2(\rho^{\mathrm{ss}}) \simeq \mathrm{Ind}_L^E(\chi) \oplus \mathrm{Ind}_N^E(\lambda), \tag{2.9}$$

with the two summands on the right being irreducible. This contradicts the fact that (1.1) implies that the left hand side of (2.9) admits one-dimensional summands, already over F. Hence the type (2, 4) is not possible in our case.

Finally, suppose Π_E is of *type* (3, 3), and write

$$\Lambda^2(\pi_E) \simeq \beta \boxplus \beta',$$

with β, β' cusp forms on GL(3)$/E$. Since E is a totally real extension of F, the archimedean type is the same as over F. It follows from (2.2), (2.3) that β, β' must both be *regular* algebraic. Either they are both selfdual or duals of each other. In the former situation, one has known since [Pic] how to attach irreducible 3-dimensional p-adic representations σ, σ' to them; see also [Cℓ2], [PL]. In the non-selfdual case, one still has such a construction in the recent work [HLTT]. Combining this with the exterior square of (1.1), we get (by Tchebotarev) the following as $\mathfrak{G}_E$-modules:

$$\tau \otimes \tau' \oplus \omega \oplus \omega' \simeq \Lambda^2(\rho^{\rm ss}) \simeq \sigma \oplus \sigma',$$

which gives the needed contradiction.

□

3 The $(3,1)$-case

The following provides a step towards establishing Theorem B:

Proposition D *Let π be an isobaric, algebraic representation of $GL_4(\mathbb{A}_F)$ which is not of type $(3,1)$, with associated 4-dimensional Hodge-Tate, $\overline{\mathbb{Q}}_p$-representation ρ of $\mathfrak{G}_F$. Then $\rho^{\rm ss}$ is not of type $(3,1)$. Conversely, assuming in addition that π is quasi-regular, if the decomposition type of ρ is not $(3,1)$ for a sufficiently large p, then the isobaric type of π is not $(3,1)$ either.*

Proof Now suppose we have a decomposition as $\mathfrak{G}_F$-modules over $\overline{\mathbb{Q}}_p$:

$$\rho^{\rm ss} \simeq \tau \oplus \chi, \tag{3.1}$$

with τ (resp. χ) of dimension 3 (resp. 1). Since every subrepresentation of a crystalline, resp. Hodge-Tate, representation is crystalline, resp. Hodge-Tate, χ is crystalline. It follows that χ is locally algebraic and by [Se], it is defined by an algebraic Hecke character χ. Put

$$\nu = \det(\tau). \tag{3.2}$$

Then ν is also Hodge-Tate and corresponds to an algebraic Hecke character ν.

Taking the contragredients of both sides of (3.1), then twisting by $\nu\chi^{-1}$, and noting that these processes commute with taking semisimplification, we obtain

$$(\rho^\vee \otimes \nu\chi^{-1})^{\rm ss} \simeq (\tau^\vee \otimes \nu\chi^{-1}) \oplus \nu\chi^{-2}. \tag{3.3}$$

Appealing to hypothesis (b) of Theorem A, we see that outside a finite set S of places, the decompositions (3.1) and (3.3) imply the following identity of L-functions:

$$L^S(s,\pi)L^S(s,\pi^\vee \otimes \nu\chi^{-1}) = L^S(s,\tau)L^S(s,\tau^\vee \otimes \nu\chi^{-1})L^S(s,\chi)L^S(s,\nu\chi^{-2}). \tag{3.4}$$

Lemma 3.1 *Assuming* (3.1), *we have*

$$\Lambda^2(\rho^{ss}) \simeq (\tau^\vee \otimes \nu) \oplus (\tau \otimes \chi).$$

Proof Using (3.1) and the fact that χ is one-dimensional and hence has trivial exterior square, we get

$$\Lambda^2(\rho^{\rm ss}) \simeq \Lambda^2(\tau) \oplus \tau \otimes \chi. \tag{3.5}$$

So the Lemma will be proved if we verify that

$$\Lambda^2(\tau) \simeq \tau^\vee \otimes \nu. \tag{3.6}$$

By the Tchebotarev density theorem, it suffices to check this easy identity at the primes where all the representations are unramified. Let P such a prime, and let $\{\alpha_P, \beta_P, \gamma_P\}$ denote the inverse roots of ϕ_P, the Frobenius at P, acting on τ. Then the inverse roots on $\Lambda^2(\tau)$ are given by the set

$$\{\alpha_P\beta_P, \alpha_P\gamma_P, \beta_P\gamma_P\} = \nu(\phi_P)\{\alpha_P^{-1}, \beta_P^{-1}, \gamma_P^{-1}\}.$$

Here we have used the fact that ν is the determinant of τ. The identity (3.6) now follows because the inverse roots of ϕ_P on $\tau^\vee$ are given by the inverses of those on τ. Done.

□

Proposition 3.2 *Let* ρ, π *satisfy the hypotheses of Theorem A. Then the decomposition* (3.1) *cannot hold.*

Proof Combining Lemma 3.1 with the identity (3.4), we get

$$L^S(s,\pi)L^S(s,\pi^\vee \otimes \nu\chi^{-1}) = L^S(s,\Lambda^2(\rho^{\rm ss}) \otimes \chi^{-1})L^S(s,\chi)L^S(s,\nu\chi^{-2}). \tag{3.7}$$

Now we appeal to a beautiful recent theorem of H. Kim ([K]) which establishes the automorphy in GL(6) of the exterior square of any cusp form on GL(4). Applying this to our form π, we get an isobaric automorphic representation $\Lambda^2(\pi)$ which is functorial at all the unramified primes. Using this information in (3.7), and twisting by χ^{-1}, we get

$$L^S(s,\pi \otimes \chi^{-1})L^S(s,\pi^\vee \otimes \nu\chi^{-2}) = L^S(s,\Lambda^2(\pi) \otimes \chi^{-1})\zeta^S(s)L^S(s,\nu\chi^{-3}). \tag{3.8}$$

(This identity slightly corrects (7.10) in [Ra1], where χ^{-2} in the second L-function on the left turns up mistakenly as χ^{-1}, which however does not affect the proof.)

One knows by Jacquet and Shalika ([JS1]) that for any isobaric automorphic representation Π_E of $\mathrm{GL}(n)/\mathbb{Q}$ for any $n \geq 1$, the incomplete

L-function $L^S(s, \Pi_E)$ has no zero at $s = 1$. Consequently, $L^S(s, \Lambda^2(\pi) \otimes \chi^{-1})L^S(s, \nu\chi^{-3})$ is non-vanishing at $s = 1$. Since $\zeta^S(s)$ has a pole at $s = 1$, we see then that the right hand side, and hence the left hand side, of (3.8) admits a pole at $s = 1$. On the other hand, since π and $\pi^\vee$ are cusp forms on GL(4)/$\mathbb{Q}$, the left hand side of (3.8) does not have a pole at $s = 1$. So we get a contradiction. The only possibility is that the decomposition of $\rho^{\rm ss}$ given by (3.1) cannot hold.

□

Now let us prove the converse direction in Proposition D. Suppose we have an isobaric decomposition

$$\pi \simeq \eta \boxplus \nu, \tag{3.9}$$

where η is a cuspidal automorphic representation of $\mathrm{GL}_3(\mathbb{A}_F)$ and ν an idele class character of F, necessarily with η, ν algebraic. Moreover, the quasi-regularity hypothesis on π implies that η is regular. Hence by the recent powerful theorem of Harris, Lan, Thorne and Taylor ([HLTT]), one may associate to η a semisimple, 3-dimensional $\overline{\mathbb{Q}}_p$-representation β. Since ν is algebraic, it also corresponds to an abelian p-adic representation ν of $\mathfrak{G}_F$. On the other hand, ρ is associated to π. It follows by Tchebotarev that

$$\rho^{\rm ss} \simeq \beta \oplus \nu \tag{3.10}$$

Hence the Proposition will be proved if (the semisimple representation) β is irreducible. Suppose not. If β is a direct sum of three p-adic character, each of which is necessarily Hodge-Tate, we will deduce, by the strong multiplicity one theorem, that η is isomorphic to an isobaric sum of three idele class characters of F, contradicting the cuspidality of η. So we may assume that we have

$$\beta \simeq \sigma \oplus \mu, \tag{3.11}$$

with σ (resp. μ) irreducible of dimension 2 (resp. 1). Let μ be the Hecke character attached to the (Hodge-Tate) p-adic character μ.

Now the regularity of η implies the same about the Hodge-Tate type of β. It follows that σ ha distinct Hodge-Tate weights and by Taylor ([Ta2, Ta3]), there is a finite solvable Galois extension E of F such that the restriction σ_E of σ to $\mathfrak{G}_E$ is modular, i.e., is associated to a cusp form π_0 of $\mathrm{GL}_2(\mathbb{A}_E)$, which is regular algebraic. One can, by well known arguments via cyclic layers, descend π_0 to any subfield M of E with E/M Galois and solvable and obtain a cusp form π_0^M on GL(2)/M which base changes to π_0 over E and in fact corresponds to the restriction σ_M of σ to $\mathfrak{G}_M$.

As in [Ta2], we now appeal to Brauer's theorem. By the inductive nature of L-functions, we have

$$L^{S^E}(s, \sigma_E \otimes \mu_E^{-1}) = L^S(s, \sigma \otimes \mu^{-1} \otimes \mathrm{Ind}_E^F(1_E)), \tag{3.12}$$

where $\mathrm{Ind}_E^F(1_E))$ is the representation of $\mathrm{Gal}(\overline{F}/F)$ induced by the trivial representation 1_E of $\mathrm{Gal}(\overline{F}/E)$, S is a finite set of places of F containing the archimedean and ramified places as well as the ones above p, and S^E the finite set of places of E above S. We can write

$$\mathrm{Ind}_E^F(1_E)) \simeq 1_F \oplus a_{E/F}, \tag{3.13}$$

for a unique representation $a_{E/F}$ of $\mathrm{Gal}(\overline{F}/F)$ called the *augmentation representation*. Using Brauer's theorem we can write it as a virtual sum of monomial representations. More precisely,

$$a_{E/F} \simeq \oplus_{j=1}^r n_j \mathrm{Ind}_{E_j}^F(\alpha_j), \tag{3.14}$$

where for each j, n_j is in $\mathbb{Z}$, E_j a subfield of E with E/E_j cyclic, and α_j a finite order character of $\mathrm{Gal}(\overline{F}/E_j)$ with trivial restriction to $\mathrm{Gal}(\overline{F}/E)$. Consequently, using (3.13), (3.14), the inductive nature and additivity,

$$L^S(s, \sigma \otimes \mu^{-1} \otimes \mathrm{Ind}_E^F(1_E)) = L^S(s, \sigma \otimes \mu^{-1}) \prod_{j=1}^r L^{S_j}(s, \sigma_{E_j} \otimes \mu_{E_j}^{-1} \otimes \alpha_j)^{n_j}, \tag{3.15}$$

where S_j is the set of places of E_j above S. Now we can appeal to (3.12) and the modularity of σ_{E_j} $(\forall j)$ to obtain

$$L^S(s, \sigma \otimes \mu^{-1}) = \prod_{j=0}^r L^{S_j}(s, \pi_0^{E_j} \otimes \mu_{E_j}^{-1}\alpha_j)^{-n_j}, \tag{3.16}$$

where $n_0 = -1$, $E_0 = E$ and $\alpha_0 = 1$.

Since $\pi_0^{E_j}$ is cuspidal, $L^{S_j}(s, \pi_0^{E_j} \otimes \mu_{E_j}^{-1}\alpha_j)$ has no zero or pole at the right edge $s = 1$. It then follows by (3.16) that

$$\mathrm{ord}_{s=1} L^S(s, \sigma \otimes \mu^{-1}) = 0. \tag{3.17}$$

Now applying (3.11) and the fact that β is associated to η, we get

$$L^S(s, \eta \otimes \mu^{-1}) = L^S(s, \sigma \otimes \mu^{-1})\zeta_F^S(s). \tag{3.18}$$

In view of (3.17), the right hand side of (3.18) has a pole at $s = 1$, which yields a contradiction as the left hand side (of (3.18)) is entire, η being a cusp form on $\mathrm{GL}(3)/F$.

This finishes the proof of Proposition D.

□

4 The $(2,2)$-case; End of Proof of Theorem B

We have just proved that π is of type $(3,1)$ iff ρ is also of the same type, under the running hypothesis that ρ is crystalline and p sufficiently large, i.e., with each of the Hodge-Tate weights of ρ being $< 2(p-1)$. By a similar argument we can also prove that for π Eisensteinian, $\rho^{\rm ss}$ has the same type as π. So we may *assume from here on that π is cuspidal*, which is the key case. We already know that $\rho^{\rm ss}$ is not of type $(3,1)$, and it is easy to see that it cannot also be of type $(1,1,1,1)$ as then π would, by the strong multiplicity one theorem, an isobaric sum of four Hecke characters.

Hence we will be done that it is impossible for $\rho^{\rm ss}$ to be of type $(2,2)$ or $(2,1,1)$ when π is cuspidal. Suppose not. Then we have

$$\rho^{\rm ss} \simeq \sigma \oplus \sigma', \tag{4.1}$$

with $\dim(\sigma) = \dim(\sigma') = 2$ and σ irreducible. Thanks to Theorem A, both σ and σ' are odd. The Hodge-Tate types of σ and σ' are regular and they are crystalline as ρ is by hypothesis. By Taylor ([Ta2, Ta3]), we know that there is a finite Galois, totally real extension E of F such that σ is modular over E, i.e., associated to a cusp form π_0 on $\mathrm{GL}(2)/E$.

Suppose σ' is reducible, i.e., of the form $\nu \oplus \nu'$ with $\dim(\nu) = \dim(\nu') = 1$, then there are associated Hecke characters ν, ν' of F, and by the argument at the end of the previous section, $L^S(s, \sigma \otimes \nu^{-1})$ is invertible at $s = 1$. Consequently, the cuspidal L-function $L^S(s, \pi \otimes \nu^{-1})$, which equals $L^S(s, \sigma \otimes \nu^{-1})\zeta_F^S(s) L^S(s, \nu'/\nu)$, has a pole at $s = 1$, which is a contradiction.

Hence we may assume that σ, σ' are both irreducible and crystalline with p sufficiently large. Since π evidently satisfies the hypotheses of Theorem A as well, we may apply Lemma 1.2 and assume that π remains cuspidal when base changed to any finite solvable normal extension which is totally real. Moreover, since the determinants ν, ν' of σ, σ' respectively are both odd by section 2, we may replace F by an abelian, totally real extension over which $\nu = \nu'$, still with π cuspidal. Appealing to Taylor's theorem for σ' as well, one gets a finite Galois, totally real extension E' of F and a cuspidal automorphic representation π_0' of $\mathrm{GL}_2(\mathbb{A}_{E'})$ which is associated to $\sigma'_{E'}$. By Gelbart-Jacquet ([GJ]), there is an isobaric automorphic representation $\mathrm{sym}^2(\pi_0)$, resp. $\mathrm{sym}^2(\pi_0')$, of $\mathrm{GL}_3(\mathbb{A}_E)$, resp. $\mathrm{GL}_3(\mathbb{A}_{E'})$ whose standard L-function equals $L^S(s, \pi_0; \mathrm{sym}^2)$, resp. $L^S(s, \pi_0'; \mathrm{sym}^2)$. Applying Brauer's theorem as in the previous section, and the modularity of σ, resp. σ', over any sub $M \subset E$, resp. $M' \subset E'$, with $[E : M]$, resp. $[E' : M']$, solvable and normal, we get the analogue of (3.17):

$$\mathrm{ord}_{s=1}\, L^S(s, \mathrm{Ad}(\tau)) = 0, \quad \text{for } \tau \in \{\sigma, \sigma'\}, \tag{4.2}$$

where $\mathrm{Ad}(\tau) = \mathrm{sym}^2(\tau) \otimes \nu^{-1}$. In addition, $L^S(s, \mathrm{Ad}(\tau))$ is meromorphic on the s-plane.

Applying the exterior square to (4.1) and twisting by ν^{-1}, we obtain

$$L^S(s, \Lambda^2(\pi) \otimes \nu^{-1}) = L^S(\sigma \otimes \sigma' \otimes \nu^{-1})\zeta_F^S(s)^2, \tag{4.3}$$

which also gives the meromorphic continuation for $L^S(\sigma \otimes \sigma' \otimes \nu^{-1})$. Moreover, since π is cuspidal, the left hand side of (4.3) can have at most a simple pole at $s = 1$. So it follows that

$$\mathrm{ord}_{s=1}\, L^S(\sigma \otimes \sigma' \otimes \nu^{-1}) = 1. \tag{4.4}$$

On the other hand, applying the symmetric square to (4.1) and twisting, we get

$$L^S(s, \pi, \mathrm{sym}^2 \otimes \nu^{-1}) = L^S(\sigma \otimes \sigma' \otimes \nu^{-1})L^S(s, \mathrm{Ad}(\sigma))L^S(s, \mathrm{Ad}(\sigma')), \tag{4.5}$$

where the properties of the left hand side are given in [BuG] and [Sh1]. Moreover, by [BuG], $L^S(s, \pi, \mathrm{sym}^2 \otimes \nu^{-1})$ has no zero at $s = 1$. On the other hand, the right hand side of (4.5) must have a zero at $s = 1$ by the conjunction of (4.2) and (4.4). This gives the requisite contradiction to the decomposition (4.1).

We have now proved that ρ is irreducible when π is cuspidal satisfying the hypotheses of Theorem B.

□

5 Proof of Theorem C

Let σ, σ' be 2-dimensional odd, semisimple, crystalline representations as in Theorem C, with $\sigma \oplus \sigma'$ associated to an isobaric automorphic form π on $\mathrm{GL}(4)/F$. If σ, σ' are both reducible, they correspond to sums of algebraic Hecke characters, and the assertion of Theorem C follows by Hecke. More generally, if σ, σ' become reducible over a finite Galois extension, then they are both of Artin-Hecke type, and the claim is well known (see [De]). So we may, and we will, assume that at least one of them, say σ, remains irreducible over any finite extension of F.

Lemma 5.1 *Under the hypothesis of irreducibility of σ, we must have*

$$\pi \simeq \eta \boxplus \eta',$$

where η, η' are isobaric automorphic forms on $GL(2)/F$, with η cuspidal. Moreover, η' is cuspidal iff σ' is irreducible.

Before proving this Lemma, let us note that if we knew π to be algebraic and quasi-regular, we can conclude by the results in the earlier sections that π cannot be of type $(3,1)$, $(2,1,1)$ or $(1,1,1,1)$. Similarly, if π were regular and algebraic, we can also rule out π being cuspidal.

Proof (of Lemma 5.1) First suppose σ' is reducible. Then evidently, since σ is irreducible, $\sigma^\vee \otimes \sigma'$ has no Galois invariants. Moreover, applying the potential automorphy result of Taylor ([Ta1, Ta2]) to the crystalline, odd 2-dimensional σ, whose Hodge-Tate type satisfies his hypothesis relative to p, we deduce by Brauer (as it was done earlier) the following for any character μ of F:

$$\mathrm{ord}_{s=e} L^S(s, \sigma \otimes \mu) = 0, \tag{5.1}$$

where e denotes the right edge ("Tate point"), which is $1 + w/2$ if μ is of finite order. For suitable finite set S of places of F, $L^S(s, \pi \otimes \mu)$ is a product of two abelian twists of the L^S-function of σ, which implies, by (5.1), that π must be an isobaric sum of the form $\eta \boxplus \mu \boxplus \mu'$, with η a form on $\mathrm{GL}(2)/F$ and μ, μ' idele class characters of F. Moreover, η must be cuspidal, for otherwise π will be an isobaric sum of four characters, yielding an associated abelian 4-dimensional representation of the Weil group of F, contradicting the irreducibility of σ. Putting $\eta' = \mu \boxplus \mu'$, we get, as asserted, $\pi \simeq \eta \boxplus \eta'$, with η cuspidal and η' not cuspidal.

We may now assume that σ and σ' are both irreducible, still with σ remaining irreducible upon restriction to any open subgroup. Since they have the same parity and weight, we may consider the cyclic, totally real extension E of F such that their determinants are the same over E, say $\chi^w \nu$ with ν of finite order. Then the unitarily normalized π_E, the base change of π, has central character $\omega_E = \nu^2$ and satisfies $\pi_E^\vee \simeq \pi_E \otimes \nu^{-1}$. Comparing L-functions we get, for a finite set T of places of E,

$$L^T(s, \Lambda^2(\pi_E) \otimes \nu^{-1}) = L^T(s, \sigma_E^\vee \otimes \sigma')\zeta_E^T(s)^2. \tag{5.2}$$

Suppose π_E were cuspidal. Then the left hand side of (5.1) will have at most a simple pole at $s = 1$, implying that $L^T(s, \sigma_E^\vee \otimes \sigma')$ must have a zero at $s = 1$. On the other hand, we have the identity (4.5) over E, which implies that $L^T(s, \mathrm{Ad}(\sigma_E))$ or $L^T(s, \mathrm{Ad}(\sigma'))$ must have a pole at $s = 1$, since $L^T(s, \pi_E, \mathrm{sym}^2 \otimes \nu^{-1})$ does not have a zero at $s = 1$ by Shahidi ([Sh1, Sh2]). Now we apply Taylor's theorem ([Ta1, Ta2]) to conclude as before that σ, σ' are potentially automorphic over finite Galois, totally real extensions M, M' respectively of F, and we may deduce as before that for any character μ of F, and for any finite set S of places of F,

$$\mathrm{ord}_{s=1} L^S(s, \mathrm{Ad}(\tau) \otimes \mu) = 0, \text{ for } \tau \in \{\sigma, \sigma'\}. \tag{5.3}$$

More precisely, we see that as $L^T(s, \sigma^\vee \otimes \sigma' \otimes \mu)$ has no zero at $s = 1$,

$$-\mathrm{ord}_{s=1} L^T(s, \Lambda^2(\pi_E) \otimes \nu^{-1}) \;=\; -\mathrm{ord}_{s=1} L^T(s, \sigma_E^\vee \otimes \sigma'_E) + 2 \;\geq\; 2. \quad (5.4)$$

Similarly,

$$-\mathrm{ord}_{s=1} L^T(s, \pi_E; \mathrm{sym}^2 \otimes \nu^{-1}) \;=\; -\mathrm{ord}_{s=1} L^T(s, \sigma_E^\vee \otimes \sigma'_E) \;\geq\; 0, \quad (5.5)$$

implying

$$-\mathrm{ord}_{s=1} L^T(s, \pi_E^\vee \times \pi_E) \;=\; -2\mathrm{ord}_{s=1} L^T(s, \sigma_E^\vee \otimes \sigma'_E) + 2 \;\geq\; 2. \quad (5.6)$$

In particular, π_E is not cuspidal. We claim that π itself is not cuspidal. Suppose not. Then there exists a quadratic extension M/L with $F \subset L \subset M \subset E$ such that π_L is cuspidal, but π_M is not, implying that π_L admits a self twist by the quadratic character δ, say, of L associated to M. Then $\sigma_L \oplus \sigma'_L$ must also be isomorphic to its twist by δ. Since σ remains irreducible under restriction to any open subgroup, σ remain irreducible, which forces an isomorphism $\sigma'_L \simeq \sigma_L \otimes \delta$. Then for non-trivial automorphism α of L/F, if any, $\sigma_L \simeq \sigma_L \otimes (\delta/\delta^\alpha)$, which implies that δ must be α-invariant (since otherwise σ_L would become reducible over $L(\delta/\delta^\alpha)$). It follows that already over F, $\sigma' \simeq \sigma \otimes \delta_0$, where δ_0 is a descent of δ to F. Then σ and σ' have the same determinant, say $\nu\chi^w$, and (the assumption of cuspidality of π implies)

$$\sigma^\vee \otimes \sigma' \;\simeq\; (\mathrm{Ad}(\sigma) \otimes \delta) \;\oplus\; \delta \simeq \sigma'^\vee \otimes \sigma.$$

This implies by (5.3) that $L^S(s, \sigma^\vee \otimes \sigma')$ does not vanish at $s = 1$. Consequently, $L^S(\pi^\vee \times \pi)$, which equals $L^S(s, (\sigma \oplus \sigma')^\vee \otimes (\sigma \oplus \sigma'))$, has a pole of order > 1 at $s = 1$, which contradicts (by [JS1]) the cuspidality assumption on π. Thus π is not cuspidal. Moreover, if π is of type $(3, 1)$, then no abelian twist of its exterior square L-function can admit any pole at $s = 1$. Thus π must be of type $(2, 2), (2, 1, 1)$ or $(1, 1, 1, 1)$. As above we can eliminate $(1, 1, 1, 1)$, and we can eliminate $(2, 1, 1)$ as well, for otherwise $L^S(s, \sigma \otimes \mu) L^S(s, \sigma' \otimes \mu)$ will have a pole at $s = e$ for a suitable character μ, contradicting (5.1) (which holds for σ' as well). Hence the assertion of the Lemma holds when σ' is irreducible as well, with both η, η' cuspidal.

□

Proof of Theorem C (contd.) Comparing the exterior square L-functions of $\pi = \eta \boxplus \eta'$ and of $\sigma \oplus \sigma'$, we see that for any character μ of F,

$$L^S(s, \sigma^\vee \otimes \sigma' \otimes \mu) \;=\; L^S(s, \eta^\vee \boxtimes \eta' \otimes \mu), \quad (5.7)$$

where $\eta^\vee \boxtimes \eta'$ is the isobaric automorphic form on $\mathrm{GL}(4)/F$ associated to the pair $(\eta^\vee, \eta')$ in [Ra3].

Suppose σ' is reducible. Then $\sigma \otimes \sigma'$ has no Galois invariants, since σ is irreducible. On the other hand, as the right hand side of (5.7) is a product of two abelian twists of the standard L-function of the cusp form η, the order of $L^S(s, \sigma^\vee \otimes \sigma)$ at $s = 1$ is 0. So Theorem C is proved in this case.

It remains to consider when both σ and σ' are irreducible, in fact upon restriction to any open subgroup. Then $\sigma^\vee \otimes \sigma'$ has Galois invariants iff $\sigma \simeq \sigma'$, in which case the dimension of $\mathfrak{G}_F$-invariants is 1. On the other hand, by Lemma 5.1, η and η' are both cuspidal on GL(2)/F. Then $L^S(s, \eta^\vee \boxtimes \eta')$ has a pole at $s = 1$ iff η is isomorphic to η', in which case the pole is of order 1. We get the same assertion for $L^S(s, \sigma^\vee \otimes \sigma'$ by (5.7). So we have only to prove the *claim* that $\sigma \simeq \sigma'$ iff $\eta \simeq \eta'$. Arithmetically normalizing both sides, an isomorphism on either side furnishes an equality (at good primes v of norm q_v) of the form:

$$\{\alpha_v, q_v^w \alpha_v^{-1}, \beta_v, q_v^w \beta_v^{-1}\} = \{\gamma_v, q_v^w \gamma_v^{-1}, \gamma_v, q_v^w \gamma_v^{-1}\}.$$

Up to renaming α_v as β_v and switching γ_v with $q_v^w \gamma_v^{-1}$, we may assume that $\alpha_v = \gamma_v$. Then $q_v^w \alpha_v^{-1}$ equals $q_v^w \gamma_v^{-1}$, resulting in the equality $\{\beta_v, q_v^w \beta_v^{-1}\} = \{\gamma_v, q_v^w \gamma_v^{-1}\}$. The claim follows.

□

References

[A] J. Arthur, *The endoscopic classification of representations: orthogonal and symplectic groups*, to appear as a Colloquium Publication of the American Mathematical Society (2012).

[AC] J. Arthur and L. Clozel, *Simple Algebras, Base Change and the Advanced Theory of the Trace Formula*, Ann. Math. Studies **120** (1989), Princeton, NJ.

[BCh] J. Bellaïche, G. Chenevier, *The sign of Galois representations attached to automorphic forms for unitary groups*, Compos. Math. **147** no. 5 (2011), 1337–1352

[BHR] D. Blasius, M. Harris and D. Ramakrishnan, Coherent cohomology, limits of discrete series, and Galois conjugation. Duke Math. J. **73** no. 3 (1994), 647–685.

[BRo1] D. Blasius and J. Rogawski, *Galois representations for Hilbert modular forms*, Bull. Amer. Math. Soc. (N.S.) **21** no. 1 (1989), 65–69.

[BRo2] D. Blasius and J. Rogawski, *Tate Classes and Arithmetic Quotients of the Two-Ball*, The Zeta functions of Picard modular surfaces, CRM Publications, 421–444, 1992.

[BuG] D. Bump and D. Ginzburg, *Symmetric square L-functions on* GL(r), Ann. of Math. (2) **136** no. 1 (1992), 137–205.

[BuzG] K. Buzzard and T. Gee, *The conjectural connections between automorphic representations and Galois representations*, preprint, arXiv:1009.0785v2.

[Ca] F. Calegari, *Even Galois Representations and the Fontaine-Mazur Conjecture*, Invent. Math. **185** (2011), 1–16.

[CaG] F. Calegari and T. Gee, *Irreducibility of automorphic Galois representations of* $GL(n)$, n *at most* 5, To appear in Annales de l'Institut Fourier.

[Cℓ1] L. Clozel, *Motifs et formes automorphes*, in *Automorphic Forms, Shimura varieties, and L-functions*, vol. I, 77–159, Perspectives in Math. **10** (1990).

[Cℓ2] L. Clozel, *Représentations galoisiennes associées aux représentations automorphes autoduales de GL(n)*, Inst. Hautes Études Sci. Publ. Math. **73** (1991), 97–145.

[CℓHLN] L. Clozel, M. Harris, J.-P. Labesse and B.-C. Ngo, *On the Stabilization of the Trace Formula*, International Press (2011).

[De] P. Deligne, Les constantes des équations fonctionnelles des fonctions L, Lecture Notes in Math. **349** (Springer-Verlag 1973) pp. 501–597.

[DV] L. Dieulefait and N. Vila, *Geometric families of 4-dimensional Galois representations with generically large images*, Math. Z. **259** no. 4(2008), 879–893.

[Fo] J.-M. Fontaine, *Arithmétique des représentations Galoisiennes p-adique*, prépublication d'Orsay **24** (March 2000).

[GJ] S. Gelbart and H. Jacquet, *A relation between automorphic representations of* $GL(2)$ *and* $GL(3)$, Ann. Scient. Éc. Norm. Sup. (4) **11** (1979), 471–542.

[GR] D. Ginzburg and S. Rallis, *On the adjoint L-function of* $GL(4)$, J. Reine Angew. Math. **505**, (1988), 119–172.

[Go1] W. Goldring, *Galois Representations Associated to Holomorphic Limits of Discrete Series*, Thesis, Harvard University (2011).

[Go2] W. Goldring, *Galois representations associated to holomorphic limits of discrete series I: Unitary groups, with an appendix by Sug Woo Shin*, preprint (2012), available at http://www.math.hardvard.edu/~wushi.

[HaT] M. Harris and R. Taylor, *On the geometry and cohomology of some simple Shimura varieties*, preprint (2000), to appear in the Annals of Math. Studies, Princeton.

[HLTT] M. Harris, Kai-Wen Lan, R. Taylor and J. Thorne, *On the rigid cohomology of certain Shimura varieties*, in preparation (2012).

[He] G. Henniart, *Une preuve simple des conjectures de Langlands pour* GL(n) *sur un corps p-adique*, Invent. Math. **139** no. 2 (2000), 439–455.

[JS1] H. Jacquet and J.A. Shalika, *Euler products and the classification of automorphic forms* I & II, Amer. J. Math. **103** (1981), 499–558 & 777–815.

[JS2] H. Jacquet and J.A. Shalika, *Exterior square L-functions*, in *Automorphic forms, Shimura varieties, and L-functions*, Vol. II, 143–226, Perspectives in Math. **11** (1990), Academic Press, Boston, MA.

[Jar] F. Jarvis, *On Galois representations associated to Hilbert modular forms*, J. Reine Angew. Math. **491** (1997), 199–216.

[Kh] C. Khare, *Serre's conjecture and its consequences*, Japanese J. Math. **5** (2010), no. 1, 103–125.

[K] H. Kim, *Functoriality for the exterior square of GL_4 and the symmetric fourth of $GL(2)$*, J. Amer. Math. Soc. **16** no. 1 (2003), 139–183.

[KSh] H. Kim and F. Shahidi, *On the holomorphy of certain L-functions*, in *Contributions to Automorphic Forms, Geometry and Number theory*, 561–572, Johns Hopkins Univ. Press (2004).

[Ki] M. Kisin, *The Fontaine-Mazur conjecture for GL_2*, J. Amer. Math. Soc. **22** no. 3 (2009), 641–690.

[La] R.P. Langlands, *On the notion of an automorphic representation. A supplement*, in *Automorphic forms, Representations and L-functions*, ed. by A. Borel and W. Casselman, Proc. Symp. Pure Math **33**, part 1, 203–207, Amer. Math. Soc. Providence (1979).

[Pic] *The Zeta functions of Picard modular surfaces*, CRM Publications, ed. by R.P. Langlands and D. Ramakrishnan (1992).

[PL] Projet de Livre, edited by M. Harris, et al, http://fa.institut.math.jussieu.fr/node/29.

[PRa] D. Prasad and D. Ramakrishnan, *On the cuspidality criterion for the Asai transfer to GL(4)*, preprint (2011), Appendix to "Determination of cusp forms on GL(2) by coefficients restricted to quadratic subfields" by M. Krishnamurthy, J. Number Theory **132** (2012), 1376–1383.

[Ra1] D. Ramakrishnan, *Irreducibility of ℓ-adic representations associated to regular cusp forms on $GL(4)/\mathbb{Q}$*, preprint (2009).

[Ra2] D. Ramakrishnan, *Irreducibility and Cuspidality*, Progress in Math. **255**, 1–27, Birkhuser Boston, Boston, MA (2008).

[Ra3] D. Ramakrishnan, *Modularity of the Rankin-Selberg L-series, and Multiplicity one for $SL(2)$*, Ann. of Math. **152** (2000), 45–111.

[Ra4] D. Ramakrishnan, Modularity of solvable Artin representations of GO(4)-type, Int. Math. Res. Not. **2002**, No. **1** (2002), 1–54.

[Ra5] D. Ramakrishnan, *On the Selfdual Cusp Forms on $GL(3)$*, preprint (2009), 1–8, Indian J. Pure and App. Math. Ramanujan 125th birth centenary issue, to appear (2012/13).

[Ri] K. Ribet, *Galois representations attached to eigenforms with Nebentypus*, in "Modular functions of one variable V" (1977), 17–51, Lecture Notes in Math. **601**, Springer-Verlag, Berlin-NY.

[Se] J.-P. Serre, *Abelian ℓ-adic representations*, With the collaboration of Willem Kuyk and John Labute, Revised reprint of the 1968 original, Res. Notes in Math. **7**, A.K. Peters Ltd., Wellesley, MA (1998).

[Sh1] F. Shahidi, *On the Ramanujan conjecture and the finiteness of poles for certain L-functions*, Ann. of Math. (2) **127** (1988), 547–584.

[Sh2] F. Shahidi, *On non-vanishing of twisted symmetric and exterior square L-function for GL(n)*, Olga Taussky-Todd: In Memoriam, Pacific J. Math. 1997, Special Issue, 311–322.

[SU] C. Skinner and E. Urban, *Sur les déformations p-adiques de certaines représentations automorphes*, J. Inst. Math. Jussieu **5** (0) (2006), 1–70.

[SW1] C. Skinner and A. Wiles, *Residually reducible representations and modular forms*, Publ. Math. Inst. Hautes Études Sci. **89** (2000), 5–126.

[SW2] C. Skinner and A. Wiles, *Base change and a problem of Serre*, Duke Math. J. **107** (2001), no. 1, 15–25.

[T] J. Tate, *Les conjectures de Stark sur les fonctions L d'Artin en* $s = 0$, Lecture notes edited by D. Bernardi and N. Schappacher, Progress in Mathematics **47** (1984), Birkhauser, Boston, MA.

[Ta1] R.L. Taylor, *On Galois representations associated to Hilbert modular forms*, Invent. Math. **98** (1989), no. 2, 265–280.

[Ta2] R.L. Taylor, *Remarks on a conjecture of Fontaine and Mazur*, J. Inst. Math. Jussieu **1** (2002), 1–19.

[Ta3] R.L. Taylor, *On the meromorphic continuation of degree two L-functions*, Doc. Math. 2006, Extra Vol., 729–779 (electronic).

[U] E. Urban, *Selmer groups and the Eisenstein-Klingen Ideal*, Duke Math. J. **106**, No. 3 (2001), 485–525.

[Zeta] *Zeta functions of Picard modular surfaces*, CRM Publications, Montréal, 1992.

Dinakar Ramakrishnan, 253-37 Caltech Pasadena, CA 91125, USA.

E-mail: `dinakar@caltech.edu`

Automorphic Representations and L-Functions
Editors: D. Prasad, C.S. Rajan, A. Sankaranarayanan, J. Sengupta

Publisher: Hindustan Book Agency, New Delhi, India

Self-dual Artin Representations

David E. Rohrlich

Functional equations in number theory are relations between an L-function and some sort of dual L-function, and in general, the L-function and its dual need not coincide. For example, if χ is a primitive Dirichlet character then the functional equation relates $L(s,\chi)$ to $L(1-s,\overline{\chi})$, and $L(s,\overline{\chi}) = L(s,\chi)$ if and only if $\chi^2 = 1$. Or if f is a primitive cusp form of weight two for $\Gamma_1(N)$ and $f^\vee$ is the complex-conjugate form then the functional equation relates $L(s,f)$ to $L(2-s,f^\vee)$, and $L(s,f^\vee) = L(s,f)$ if and only if f is a cusp form for $\Gamma_0(N)$ with trivial character. Let us call an L-function *self-dual* if its functional equation is a relation between the L-function and itself. While self-dual L-functions are often of special interest, the preceding examples suggest that they may also be rare. Indeed the number of Dirichlet characters modulo N is the quantity

$$\varphi(N) = N\prod_{p|N}(1-p^{-1})$$

and is therefore $\gg N^{1-\varepsilon}$ for every $\varepsilon > 0$, but the number of quadratic Dirichlet characters modulo N is $\ll N^{\varepsilon}$. Similarly, if $N \geqslant 5$ then the dimension of the space $\mathrm{S}_2(\Gamma_1(N))$ of cusp forms of weight two for $\Gamma_1(N)$ is given by

$$\dim \mathrm{S}_2(\Gamma_1(N)) = 1 + \frac{N^2}{24}\prod_{p|N}(1-p^{-2}) - \frac{1}{4}\sum_{N_1N_2=N}\varphi(N_1)\varphi(N_2)$$

and is therefore $\gg N^2$, but the dimension of the space of cusp forms of weight two for $\Gamma_0(N)$ is $\ll N^{1+\varepsilon}$. Is it perhaps the case that self-dual L-functions are of density zero among all L-functions?

It is tidier, although not *a priori* equivalent, to replace the L-functions by the objects underlying them. If the L-functions are motivic then the underlying objects are motives, and one can ask whether "essentially self-dual motives" (in other words, pure motives which are self-dual up to Tate twist) have density zero among all pure motives of a given rank and weight. However if we insist on full generality then the preceding question is not yet amenable to a precise formulation, because the set of isomorpism classes of pure motives of a given rank and weight over a given number field with conductor below a given bound is not known to be finite. So instead we shall focus on motives of weight zero. By an *Artin representation* of a number field F we mean as usual a continuous representation ρ of $\mathrm{Gal}(\overline{F}/F)$ on a finite-dimensional complex vector space. Such a representation always factors through the quotient of $\mathrm{Gal}(\overline{F}/F)$ by an open normal subgroup and so will be regarded as a representation of $\mathrm{Gal}(L/F)$ for some finite Galois extension L of F. The conductor of ρ is an integral ideal $\mathfrak{q}(\rho)$ of F, the absolute norm of which will be denoted $q(\rho)$. According to a theorem of Ralph Greenberg (unpublished) and of Anderson, Blasius, Coleman, and Zettler [1] (who consider more generally the case of representations of the global Weil group of F), if we fix F and n then the set of isomorphism classes of n-dimensional Artin representations ρ of F with $q(\rho) \leqslant x$ is finite. Write $\vartheta_{F,n}(x)$ for the number of such isomorphism classes and $\vartheta_{F,n}^{\mathrm{sd}}(x)$ for the number of classes such that ρ is self-dual. Dropping the subscripts F and n for simplicity, we ask whether $\lim_{x\to\infty} \vartheta^{\mathrm{sd}}(x)/\vartheta(x) = 0$.

If $F = \mathbb{Q}$ and $n = 1$ then an affirmative answer is implicit already in our remarks about Dirichlet characters, and it is easy to see that in fact $\vartheta^{\mathrm{sd}}(x)/\vartheta(x) \sim \pi^2/(3x)$ in this case. Using the work of Bhargava [3], [4] and of Bhargava, Cojocaru, and Thorne [5], we shall prove that the answer is also affirmative for $F = \mathbb{Q}$ and $n = 2$. For $F = \mathbb{Q}$ and $n = 3$ we show at least that an affirmative answer would follow from a conjecture of Malle [26] on the distribution of Galois groups, but for $n \geqslant 4$ we are unable to derive an affirmative answer even conditionally, and if F is an arbitrary number field then we are able to confirm that $\lim_{x\to\infty} \vartheta^{\mathrm{sd}}(x)/\vartheta(x) = 0$ only for $n = 1$, when the assertion follows from a theorem of M. J. Taylor [36].

Before describing the contents of the paper in more detail we introduce some refinements of $\vartheta_{F,n}(x)$. Recall that a finite-dimensional complex representation of a finite group G is *abelian* if it is a direct sum of one-dimensional characters of G, *reducible* if it is a direct sum of two proper subrepresentations, *irreducible* if it is of positive dimension but not reducible, *monomial* if it is induced by a one-dimensional character of a subgroup of G, and *primitive* if it is not induced from any proper subgroup of G. We use the superscripts "ab," "irr," "im," and "ip" to refer to abelian, irreducible, irreducible monomial, and irreducible primitive representations respectively. For example, $\vartheta_{F,n}^{\mathrm{ab}}(x)$ is the number of isomorphism classes of n-dimensional abelian Artin representations ρ of F with $q(\rho) \leqslant x$, and $\vartheta_{F,n}^{\mathrm{ab,sd}}(x)$ is the number of such isomorphism classes that are self-dual. The notation is illustrated by the self-evident assertions

$$\vartheta_{\mathbb{Q},2}^{\mathrm{sd}}(x) = \vartheta_{\mathbb{Q},2}^{\mathrm{ab,sd}} + \vartheta_{\mathbb{Q},2}^{\mathrm{im,sd}}(x) + \vartheta_{\mathbb{Q},2}^{\mathrm{ip,sd}}(x) \tag{0.8}$$

and

$$\vartheta^{\mathrm{sd}}_{\mathbb{Q},3}(x) = \vartheta^{\mathrm{ab,sd}}_{\mathbb{Q},3} + \vartheta^{1+2,\mathrm{sd}}_{\mathbb{Q},3}(x) + \vartheta^{\mathrm{irr,sd}}_{\mathbb{Q},3}(x), \tag{0.9}$$

where $\vartheta^{1+2,\mathrm{sd}}_{\mathbb{Q},3}(x)$ is the number of isomorphism classes of self-dual Artin representations of $\mathbb{Q}$ of the form $\rho \cong \rho' \oplus \rho''$ with ρ' one-dimensional, ρ'' irreducible and two-dimensional, and $q(\rho')q(\rho'') \leqslant x$. Of course (0.8) and (0.9) remain valid without the superscript "sd" and with $\mathbb{Q}$ replaced by any number field F.

In addition to $\vartheta_{F,n}(x)$ and its refinements, we need two functions which count discriminants rather than conductors. Given a finite extension K of F, write $\mathfrak{d}_{K/F}$ for the relative discriminant ideal of K over F and $d_{K/F}$ for the absolute norm of $\mathfrak{d}_{K/F}$. If $F = \mathbb{Q}$ then we write simply $\mathfrak{d}_K$ and d_K. Now fix an integer $m \geqslant 2$. We write $\eta_{F,m}(x)$ for the number of extensions K of F inside our fixed algebraic closure $\overline{F}$ such that $[K : F] = m$ and $d_{K/F} \leqslant x$. Also, if G is a transitive subgroup of the symmetric group S_m, then $\eta^{G}_{F,m}(x)$ denotes the number of such extensions K for which $\mathrm{Gal}(L/F) \cong G$ as permutation groups, where L is a normal closure of K over F and $\mathrm{Gal}(L/F)$ is viewed as a permutation group via its action on the set of conjugates $\alpha_1, \alpha_2, \ldots, \alpha_m$ of a primitive element of K over F. The requirement that $\mathrm{Gal}(L/F)$ and G be isomorphic as permutation groups means of course that there is a bijection of $\{\alpha_1, \alpha_2, \ldots, \alpha_m\}$ onto $\{1, 2, \ldots, m\}$ such that the resulting map $\mathrm{Gal}(L/F) \hookrightarrow S_m$ has image G.

With these notations in hand let us now describe the contents of the paper section by section. We have included a considerable amount of expository material throughout, because our aim is in part pedagogical.

The first four sections are devoted to the abelian case. The tauberian method, recalled in Section 1, leads to asymptotic formulas for $\vartheta_{\mathbb{Q},1}(x)$ and $\vartheta^{\mathrm{sd}}_{\mathbb{Q},1}(x)$ in Section 2 and for $\vartheta^{\mathrm{ab}}_{\mathbb{Q},n}(x)$ and $\vartheta^{\mathrm{ab,sd}}_{\mathbb{Q},n}(x)$ in Section 3. Our dicussion of the abelian case is completed in Section 4, where we attempt to replace $\mathbb{Q}$ by an arbitrary number field F. If F is neither $\mathbb{Q}$ nor an imaginary quadratic field then the asymptotic behavior of $\vartheta_{F,1}(x)$ appears to be unknown, and we argue that what is needed is a horizontal analogue of Leopoldt's conjecture.

In the next two sections we bound $\vartheta^{\mathrm{im,sd}}_{\mathbb{Q},2}(x)$. Whether monomial or not, an irreducible self-dual Artin representation is either *orthogonal* or *symplectic* – in other words, relative to an appropriate choice of basis, its image is contained in either the real orthogonal group $\mathrm{O}_n(\mathbb{R})$ or the complex symplectic group $\mathrm{Sp}_{2n}(\mathbb{C})$ – and hence in particular $\vartheta^{\mathrm{im,sd}}_{\mathbb{Q},2}(x)$ is the sum of an orthogonal term and a symplectic term. These terms are bounded in Sections 5 and 6 respectively. The orthogonal term is bounded by a reduction to the asymptotic formulas of Siegel [35], and then the symplectic term is bounded by a reduction to the orthogonal term.

Our treatment of the primitive case begins in Section 7 with some background on Schur covers. In Section 8 we bound $\vartheta^{\mathrm{ip,sd}}_{\mathbb{Q},2}(x)$ in terms of $\eta_{\mathbb{Q},4}(x)$ and $\eta^{A_5}_{\mathbb{Q},5}(x)$, to which we then apply the results of Bhargava [3] and Bhargava, Cojocaru, and Thorne [5] (the latter work being itself an application of Bhargava's asymptotics for quintic fields [4]). In principle we could have adopted a different strategy, in the spirit of Serre's paper [31]: bound the dimension of spaces of holomorphic cusp forms of weight one and spaces of Maass forms of eigenvalue 1/4, and then appeal to the Langlands correspondence to deduce a bound for $\vartheta^{\mathrm{ip,sd}}_{\mathbb{Q},2}(x)$. In fact the relevant bounds on spaces of automorphic forms can simply be quoted from the work of Michel and Venkatesh [28], who vastly generalize the original breakthrough (in the case of holomorphic cusp forms of weight one, prime level, and character the Legendre symbol) of Duke [11]. However, in spite of the enormous progress of recent years, the Langlands correspondence for two-dimensional Artin representations of $\mathbb{Q}$ of icosahedral type and *even* determinant remains conjectural, and for the sake of an unconditional result and a uniform treatment our argument will be carried out on the Galois side of the correspondence.

By the end of Section 8 we will have assembled upper bounds for each of the terms on the right-hand side of (0.8). The upshot will be that

$$\vartheta^{\mathrm{sd}}_{\mathbb{Q},2}(x) = O(x^{2-\gamma}) \tag{0.10}$$

for every $\gamma < 1/60$. On the other hand, from our asymptotic formula for $\vartheta^{\mathrm{ab}}_{\mathbb{Q},n}(x)$ we will also have

$$\vartheta^{\mathrm{ab}}_{\mathbb{Q},2}(x) \gg x^2 \log x. \tag{0.11}$$

Since $\vartheta_{\mathbb{Q},2}(x) \geqslant \vartheta^{\mathrm{ab}}_{\mathbb{Q},2}(x)$, it follows from (0.10) and (0.11) that $\lim_{x\to\infty} \vartheta^{\mathrm{sd}}(x)/\vartheta(x)$ is indeed 0 for $F = \mathbb{Q}$ and $n = 2$.

Perhaps it is disappointing to arrive at this conclusion by comparing the totality of self-dual representations with the abelian representations only. Thus in Section 9 we go on to show that $\lim_{x\to\infty} \vartheta^{\mathrm{irr,sd}}(x)/\vartheta^{\mathrm{irr}}(x) = 0$ for $F = \mathbb{Q}$ and $n = 2$. But even the latter assertion rests on the trivial inequalities $\vartheta^{\mathrm{irr,sd}}(x) \leqslant \vartheta^{\mathrm{sd}}(x)$ and $\vartheta^{\mathrm{irr}}(x) \geqslant \vartheta^{\mathrm{im}}(x)$. Unfortunately, a direct comparison between, say, $\vartheta^{\mathrm{ip,sd}}(x)$ and $\vartheta^{\mathrm{ip}}(x)$ seems to be out of our reach.

Apart from a short appendix, the remainder of the paper is devoted to Malle's conjecture and two of its consequences. One consequence, derived in Sections 10 and 11, is an upper bound for $\vartheta^{\mathrm{ip,sd}}(x)$ valid for arbitrary F and $n \geqslant 2$. The other consequence, a variant of the first, is a bound for the term $\vartheta^{\mathrm{irr,sd}}_{\mathbb{Q},3}(x)$ in (0.9). Using this bound we prove in Section 12 that under Malle's conjecture we have $\lim_{x\to\infty} \vartheta^{\mathrm{sd}}(x)/\vartheta(x) = 0$ for $F = \mathbb{Q}$ and $n = 3$.

The many questions left open by this paper are so glaringly obvious that it would be superfluous to enumerate them. But it may be worthwhile to point out a parallel line of inquiry in the domain of automorphic forms: Do lifts from orthogonal and symplectic groups have density zero among all cuspidal automorphic representations of $\mathrm{GL}(n)$? The question seems amenable to a precise formulation, and perhaps also to a solution.

I am deeply grateful to Manjul Bhargava for providing me with a preprint of [5] before publication. I would also like to thank Josh Zelinsky for drawing my attention to the paper of Collins [7]. Finally, I thank the referee for a careful reading of the text and Tata Institute and the organizers of the International Colloquium on Automorphic Representations and L-Functions for their warm hospitality.

1 A Tauberian Theorem

The tauberian theorem that will be needed in this paper is a special case of Theorem 7.7 on p. 154 of the book [2] by Bateman and Diamond. Let $\psi(1), \psi(2), \psi(3), \dots$ be a sequence of nonnegative real numbers, and let

$$D(s) = \sum_{q\geqslant 1} \psi(q)q^{-s}$$

be the associated Dirichlet series and

$$\vartheta(x) = \sum_{q\leqslant x} \psi(q)$$

the associated summatory function. We assume that there are positive real numbers a and a' with $a' < a$ together with an integer $b \geqslant 1$ such that the following conditions are satisfied:

(i) The series $\sum_{q\geqslant 1} \psi(q)q^{-s}$ converges for $\Re(s) > a$ and thus defines $D(s)$ as a holomorphic function in this region.

(ii) $D(s)$ extends to a meromorphic function in the region $\Re(s) > a'$.

(iii) $D(s)$ has a pole of order b at $s = a$ and is otherwise holomorphic for $\Re(s) > a'$.

Let κ be the residue of $(s-a)^{b-1}D(s)$ at $s = a$, and put $c = \kappa/(a\cdot(b-1)!)$. It follows from the hypotheses that $\kappa > 0$ and hence that $c > 0$.

Proposition 1.1 $\vartheta(x) \sim cx^a(\log x)^{b-1}$.

To deduce Proposition 1.1 from Theorem 7.7 of [2], note the definition of $\widehat{F}$ given on p. 109 of [2], the special case of the definition embodied in the displayed equation at the top of p. 110, and the definition of $\sigma_c(\widehat{F})$ on p. 119, and keep in mind that our a, a', and b correspond to the constants α, β, and γ of [2].

2 Dirichlet Characters

Given a positive integer q, write $\psi(q)$ for the number of primitive Dirichlet characters of conductor q. We consider the Dirichlet series

$$D(s) = \sum_{q \geqslant 1} \psi(q) q^{-s},$$

convergent for $\Re(s) > 2$.

Proposition 2.1 $D(s) = \zeta(s-1)/\zeta(s)^2$.

Proof Assertions of this sort are antique (cf. [14], p. 268, Theorem 330), but we include a proof nonetheless. Let μ and φ denote as usual the Möbius and Euler functions, and put $C(s) = \sum_{q \geqslant 1} \varphi(q) q^{-s}$. Since $\psi(q) = \sum_{q'|q} \mu(q/q')\varphi(q')$ we have

$$D(s) = C(s)/\zeta(s). \tag{2.1}$$

Now φ is multiplicative, so

$$C(s) = \prod_p \Big(\sum_{\nu \geqslant 0} \varphi(p^\nu) p^{-\nu s}\Big).$$

Write $C_p(s)$ for the Euler factor on the right-hand side. Since $\varphi(1) = 1$ and $\varphi(p^\nu) = (p-1)p^{\nu-1}$ for $\nu \geqslant 1$, we have

$$C_p(s) = 1 + \sum_{\nu \geqslant 1} (p-1)p^{-1}p^{\nu(1-s)} = 1 + (p-1)p^{-s}/(1-p^{1-s})$$

and consequently

$$C_p(s) = 1 + \frac{p^{1-s} - p^{-s}}{1 - p^{1-s}} = \frac{1 - p^{-s}}{1 - p^{1-s}}.$$

Hence $C(s) = \zeta(s-1)/\zeta(s)$. The proposition now follows from (2.1).

□

Identifying one-dimensional characters of $\mathrm{Gal}(\overline{\mathbb{Q}}/\mathbb{Q})$ with primitive Dirichlet characters in the usual way, we see that

$$\vartheta_{\mathbb{Q},1}(x) = \sum_{q \leqslant x} \psi(q).$$

In other words $\vartheta_{\mathbb{Q},1}(x)$ is the summatory function corresponding to $D(s)$. On the other hand, it follows from Proposition 2.1 that $D(s)$ is holomorpic for $\Re(s) > 1$ apart from a simple pole at $s = 2$ with residue $36/\pi^4$. Hence Proposition 1.1 gives:

Corollary 2.2 $\vartheta_{\mathbb{Q},1}(x) \sim 18x^2/\pi^4$.

Next consider the Dirichlet series

$$D^{\mathrm{sd}}(s) = \sum_{q \geqslant 1} \psi^{\mathrm{sd}}(q) q^{-s},$$

where $\psi^{\mathrm{sd}}(q)$ is the number of primitive Dirichlet characters χ of conductor q such that $\chi^2 = 1$.

Proposition 2.3 $D^{\mathrm{sd}}(s) = (1 + 4^{-s} + 2 \cdot 8^{-s}) \dfrac{\zeta(s)(1 - 2^{-s})}{\zeta(2s)(1 - 2^{-2s})}$.

Proof The conductor of a primitive quadratic Dirichlet character can be written $2^\nu r$, where $\nu = 0$, 2, or 3 and r is a square-free odd positive integer. Conversely, every number of this form is the conductor of exactly one (if $\nu = 0$ or 2) or exactly two (if $\nu = 3$) primitive Dirichlet characters χ with $\chi^2 = 1$. It follows that

$$D^{\mathrm{sd}}(s) = (1 + 4^{-s} + 2 \cdot 8^{-s}) R(s), \tag{2.2}$$

where $R(s)$ is the Dirichlet series $\sum r^{-s}$, the sum being taken over square-free odd positive integers r. Now if the sum were taken over all square-free positive integers then the resulting Dirichlet series would be $\zeta(s)/\zeta(2s)$, so to deduce a formula for $R(s)$ we remove the Euler factor at 2 in $\zeta(s)/\zeta(2s)$. Substitution in (2.2) yields the stated formula.

□

Another appeal to Proposition 1.1 gives:

Corollary 2.4 $\vartheta^{\mathrm{sd}}_{\mathbb{Q},1}(x) \sim 6x/\pi^2$.

Comparing this corollary with the previous one, we see that

$$\vartheta^{\mathrm{sd}}_{\mathbb{Q},1}(x)/\vartheta_{\mathbb{Q},1}(x) \sim \pi^2/(3x), \tag{2.3}$$

as mentioned in the introduction.

3 Abelian Representations

Given positive integers n and q, let $\psi_n(q)$ be the number of isomorphism classes of n-dimensional abelian Artin representations of $\mathbb{Q}$ of conductor q. We put

$$D_n(s) = \sum_{q \geqslant 1} \psi_n(q) q^{-s}.$$

In the notation of Section 2 we have $\psi_1(q) = \psi(q)$ and hence $D_1(s) = D(s)$.

Proposition 3.1 *For $n \geqslant 1$,*

$$D_n(s) = \sum_{k=1}^{n} \frac{1}{k!} \sum_{\nu_1+\nu_2+\cdots+\nu_k=n} \frac{D(\nu_1 s) D(\nu_2 s) \cdots D(\nu_k s)}{\nu_1 \nu_2 \cdots \nu_k},$$

where the inner sum on the right runs over k-tuples $(\nu_1, \nu_2, \cdots, \nu_k)$ of positive integers summing to n.

Proof Given a one-dimensional character χ of $\mathrm{Gal}(\overline{\mathbb{Q}}/\mathbb{Q})$, let us write $\chi^{\oplus \nu}$ for the direct sum of ν copies of χ. If

$$\rho \cong \chi_1^{\oplus n_1} \oplus \chi_2^{\oplus n_2} \oplus \cdots \oplus \chi_k^{\oplus n_k}$$

with one-dimensional characters $\chi_1, \chi_2, \ldots, \chi_k$ of $\mathrm{Gal}(\overline{\mathbb{Q}}/\mathbb{Q})$ and positive integers $n_1, n_2, \ldots, n_k$ then

$$q(\rho) = q(\chi_1)^{n_1} q(\chi_2)^{n_2} \cdots q(\chi_k)^{n_k}.$$

Thus we have the following identity of formal power series in x with coefficients in the ring of formal Dirichlet series:

$$\sum_{\rho} q(\rho)^{-s} x^{\dim(\rho)} = \prod_{\chi} (1 - q(\chi)^{-s} x)^{-1},$$

where ρ runs over a set of representatives for the distinct isomorphism classes of abelian Artin representations of $\mathbb{Q}$ and χ runs over one-dimensional characters of $\mathrm{Gal}(\overline{\mathbb{Q}}/\mathbb{Q})$. Equivalently,

$$1 + \sum_{n \geqslant 1} \sum_{q \geqslant 1} \psi_n(q) q^{-s} x^n = \prod_{q \geqslant 1} (1 - q^{-s} x)^{-\psi(q)}.$$

Summing over q on the left-hand side while expressing the right-hand side as the exponential of its logarithm, we obtain

$$1 + \sum_{n \geqslant 1} D_n(s) x^n = \exp\left(\sum_{\nu \geqslant 1} D(\nu s) \frac{x^\nu}{\nu} \right).$$

The proposition follows on comparing the coefficient of x^n on both sides. $\square$

Proposition 3.2 *$D_n(s)$ is holomorphic for $\Re(s) > 1$ except for a pole of order n at $s = 2$. Furthermore, the residue of $(s-2)^{n-1}D_n(s)$ at $s = 2$ is $(1/n!)(36/\pi^4)^n$.*

Proof Rewrite Proposition 3.1 in the form

$$D_n(s) = \frac{D(s)^n}{n!} + \sum_{k=1}^{n-1} \frac{1}{k!} \sum_{\nu_1+\nu_2+\cdots+\nu_k=n} \frac{D(\nu_1 s)D(\nu_2 s)\cdots D(\nu_k s)}{\nu_1\nu_2\cdots\nu_k}. \quad (3.1)$$

From Proposition 2.1 we know that $D(s)$ is holomorphic for $\Re(s) > 1$ except for a simple pole at $s = 2$ with residue $36/\pi^4$. Thus $D(s)^n/n!$ has the properties claimed for $D_n(s)$. To deduce that $D_n(s)$ itself has these properties it suffices to observe that for $k \leqslant n-1$ the term $D(\nu_1 s)D(\nu_2 s)\cdots D(\nu_k s)/(\nu_1\nu_2\cdots\nu_k)$ on the right-hand side of (3.1) has at most $n-2$ factors $D(\nu_i s)$ with $\nu_i = 1$. Hence the pole (if any) of such a term at $s = 2$ is of order at most $n-2$.

□

As $\vartheta^{\text{ab}}_{\mathbb{Q},n}(x)$ is the summatory function of $D_n(s)$, Proposition 1.1 gives:

Theorem 3.3 $\vartheta^{\text{ab}}_{\mathbb{Q},n}(x) \sim (1/2)(1/(n-1)!)(1/n!)(36/\pi^4)^n \cdot x^2(\log x)^{n-1}$.

A similar argument can be applied in the self-dual case. Write $\psi^{\text{sd}}_n(q)$ for the number of isomorphism classes of n-dimensional self-dual abelian Artin representations of $\mathbb{Q}$ of conductor q, and put

$$D^{\text{sd}}_n(s) = \sum_{q\geqslant 1} \psi^{\text{sd}}_n(q)q^{-s}.$$

Then $\psi^{\text{sd}}_1 = \psi^{\text{sd}}$ and $D^{\text{sd}}_1 = D^{\text{sd}}$ in the notation of Section 2. Given a positive integer ν, it is also convenient to set

$$D[\nu](s) = \begin{cases} D^{\text{sd}}(\nu s) & \text{if } \nu \text{ is odd} \\ D(\nu s) & \text{if } \nu \text{ is even.} \end{cases}$$

Note in particular that $D[1] = D^{\text{sd}}$.

Proposition 3.4 *For $n \geqslant 1$,*

$$D^{\text{sd}}_n(s) = \sum_{k=1}^{n} \frac{1}{k!} \sum_{\nu_1+\nu_2+\cdots+\nu_k=n} \frac{D[\nu_1](s)D[\nu_2](s)\cdots D[\nu_k](s)}{\nu_1\nu_2\cdots\nu_k},$$

where the inner sum on the right runs over k-tuples $(\nu_1, \nu_2, \cdots, \nu_k)$ of positive integers summing to n.

Proof An abelian Artin representation ρ of $\mathbb{Q}$ is self-dual if and only if it has the form

$$\rho \cong \Big(\bigoplus\nolimits_{\chi^2=1} \chi^{\oplus\nu(\chi)}\Big) \oplus \Big(\bigoplus\nolimits'_{\chi^2\neq 1} (\chi \oplus \chi^{-1})^{\oplus\nu(\chi)}\Big),$$

where the direct sum inside the first set of parentheses runs over one-dimensional characters χ of $\mathrm{Gal}(\overline{\mathbb{Q}}/\mathbb{Q})$ of order $\leqslant 2$, the direct sum inside the second set of parentheses runs over *pairs* $\{\chi, \chi^{-1}\}$ of complex conjugate characters (this is the significance of the prime) of order $\geqslant 3$, and $\nu(\chi) = 0$ for all but finitely many χ. As $q(\chi \oplus \chi^{-1}) = q(\chi)^2$, it follows that

$$\begin{aligned} 1 + \sum_{n\geqslant 1} D_n^{\mathrm{sd}}(s)x^n &= \prod\nolimits_{\chi^2=1}(1 - q(\chi)^{-s}x)^{-1} \cdot \prod\nolimits'_{\chi^2\neq 1}(1 - q(\chi)^{-2s}x^2)^{-1} \\ &= \prod_{q\geqslant 1}(1 - q^{-s}x)^{-\psi^{\mathrm{sd}}(q)} \cdot \prod_{q\geqslant 1}(1 - q^{-2s}x^2)^{-\psi^*(q)} \end{aligned} \tag{3.2}$$

with $\psi^*(q) = (\psi(q) - \psi^{\mathrm{sd}}(q))/2$. Set $D^*(s) = \sum_{q\geqslant 1}\psi^*(q)q^{-s}$. Then $D^*(s) = (D(s) - D^{\mathrm{sd}}(s))/2$. Writing the two products in the last expression in (3.2) as the exponentials of their logarithms, we obtain

$$\begin{aligned} 1 + \sum_{n\geqslant 1} D_n^{\mathrm{sd}}(s)x^n &= \exp\Big(\sum_{\nu\geqslant 1} D^{\mathrm{sd}}(\nu s)x^\nu/\nu\Big) \cdot \exp\Big(\sum_{\mu\geqslant 1} D^*(2\mu s)x^{2\mu}/\mu\Big) \\ &= \exp\Big(\sum_{\nu\geqslant 1} D[\nu](s)x^\nu/\nu\Big). \end{aligned}$$

The proposition follows on inspecting the coefficient of x^n in this last expression.

□

Proposition 3.5 *$D_n^{\mathrm{sd}}(s)$ is holomorphic for $\Re(s) > 1/2$ except for a pole of order n at $s = 1$. Furthermore, the residue of $(s-1)^{n-1}D_n^{\mathrm{sd}}(s)$ at $s = 1$ is $(1/n!)(6/\pi^2)^n$.*

Proof We observe first of all that if ν is a positive integer then $D[\nu](s)$ is holomorphic for $\Re(s) > 1/2$ except possibly for a simple pole at $s = 1$. Indeed if ν is odd then $D[\nu](s) = D^{\mathrm{sd}}(\nu s)$ and our assertion follows from Proposition 2.3, while if ν is even then $D[\nu](s) = D(\nu s)$ with $\nu \geqslant 2$ and $D(s) = \zeta(s-1)/\zeta(s)^2$ (Proposition 2.1). Now Proposition 3.4 gives

$$D_n^{\mathrm{sd}}(s) = \frac{1}{n!}D^{\mathrm{sd}}(s)^n + \sum_{k=1}^{n-1}\frac{1}{k!}\sum_{\nu_1+\nu_2+\cdots+\nu_k=n}\frac{D[\nu_1](s)D[\nu_2](s)\cdots D[\nu_k](s)}{\nu_1\nu_2\cdots\nu_k}, \tag{3.3}$$

and by Proposition 2.3 we know that $D^{\mathrm{sd}}(s)$ is holomorphic for $\Re(s) > 1/2$ except for a simple pole at $s = 1$ with residue $6/\pi^2$. Thus $D^{\mathrm{sd}}(s)^n/n!$ has the properties claimed for $D_n^{\mathrm{sd}}(s)$. These properties are inherited by $D_n^{\mathrm{sd}}(s)$ itself, because for $k < n$ the term $D[\nu_1](s)D[\nu_2](s)\cdots D[\nu_k](s)/(\nu_1\nu_2\cdots\nu_k)$ on the right-hand side of (3.3) has at most $n-1$ factors of the form $D[\nu](s)$, and thus its pole (if any) at $s = 1$ is of order at most $n - 1$. Of course each such factor and hence their product is holomorphic elsewhere in the region $\Re(s) > 1/2$.

□

Once again we appeal to Proposition 1.1, obtaining:

Theorem 3.6 $\vartheta_{\mathbb{Q},n}^{\mathrm{ab,sd}}(x) \sim (1/(n-1)!)(1/n!)(6/\pi^2)^n \cdot x(\log x)^{n-1}$.

Combining Theorems 3.3 and 3.6, we see that

$$\vartheta_{\mathbb{Q},n}^{\mathrm{ab,sd}}(x)/\vartheta_{\mathbb{Q},n}^{\mathrm{ab}}(x) \sim 2 \cdot \pi^{2n}/(6^n x), \tag{3.4}$$

a straightforward generalization of (2.3).

4 Does Leopoldt's Conjecture Have a Horizontal Analogue?

When $\mathbb{Q}$ is replaced by an arbitrary number field F no asymptotic relationship comparable to (2.3) seems to be known, but thanks to a theorem of M. J. Taylor ([36], Theorem 1) we can assert that at least

$$\lim_{x\to\infty} \vartheta_{F,1}^{\mathrm{sd}}(x)/\vartheta_{F,1}(x) = 0. \tag{4.1}$$

Indeed let m be a positive integer which is relatively prime to the discriminant of F and not divisible by 4, and let $\vartheta_{F,1}^{(m)}(x)$ be the number of characters of $\mathrm{Gal}(\overline{F}/F)$ of order m and absolute conductor $\leqslant x$. Then Taylor proves that

$$\vartheta_{F,1}^{(m)}(x) \sim cx(\log x)^{\tau(m)-2}, \tag{4.2}$$

where c is a positive constant depending on F and m, and $\tau(m)$ is the number of positive divisors of m. (For the sake of simplicity we are not stating Taylor's result in full generality.) Taking $m = 2$ gives

$$\vartheta_{F,1}^{\mathrm{sd}}(x) \sim cx, \tag{4.3}$$

and taking $m = p^2$ with an odd prime p not dividing the discriminant of F gives

$$\vartheta_{F,1}(x) \gg x(\log x). \tag{4.4}$$

Equation (4.1) is an immediate consequence of (4.3) and (4.4).

Of course by making different choices of m we can replace the right-hand side of (4.4) by $x(\log x)^\nu$ for arbitrarily large ν. But this lower bound is far from the trivial upper bound, so the asymptotic behavior of $\vartheta_{F,1}(x)$ remains a mystery:

Proposition 4.1 *$\vartheta_{F,1}(x) = O(x^2)$, where the implied constant depends on F.*

In the case where F has units of infinite order, Josh Zelinsky has proved the stronger assertion that $\vartheta_{F,1}(x) = o(x^2)$. But let us prove Proposition 4.1 as it stands: First of all, we identify one-dimensional characters of $\mathrm{Gal}(\overline{F}/F)$ with idele class characters of F of finite order, or equivalently with primitive ray class characters of F. Given a nonzero integral ideal $\mathfrak{q}$ of F, write $h_F^{\mathrm{nar}}(\mathfrak{q})$ for the order of the narrow ray class group of F to the modulus $\mathfrak{q}$. Then

$$\vartheta_{F,1}(x) \leqslant \sum_{\mathbf{N}\mathfrak{q}\leqslant x} h_F^{\mathrm{nar}}(\mathfrak{q}), \tag{4.5}$$

because $h_F^{\mathrm{nar}}(\mathfrak{q})$ is equal to the number of primitive ray class characters of F of conductor dividing $\mathfrak{q}$ and is thus an upper bound for the number of such characters of conductor exactly $\mathfrak{q}$.

On the other hand, let $\mathcal{O}_F$ be the ring of integers of F and $\mathcal{O}_F^\times$ its unit group. It is convenient to put $U_F = \mathcal{O}_F^\times$ and to write $U_F(\mathfrak{q})$ for the subgroup of U_F consisting of units congruent to 1 modulo $\mathfrak{q}$. We also write $U_F^+(\mathfrak{q})$ for the subgroup of totally positive units in $U_F(\mathfrak{q})$. Finally, let h_F be the class number and $r_1(F)$ and $2r_2(F)$ the number of real and complex embeddings of F. According to a classic formula (cf. [23], p. 127, Theorem 1),

$$h_F^{\mathrm{nar}}(\mathfrak{q}) = 2^{r_1(F)} \cdot h_F \cdot \varphi_F(\mathfrak{q})/[U_F : U_F^+(\mathfrak{q})], \tag{4.6}$$

where $\varphi_F(\mathfrak{q}) = |(\mathcal{O}_F/\mathfrak{q})^\times|$. As $\varphi_F(\mathfrak{q}) \leqslant \mathbf{N}\mathfrak{q}$ and $[U_F : U_F^+(\mathfrak{q})] \geqslant 1$, we see on returning to (4.5) that $\vartheta_{F,1}(x)$ is bounded by a constant times $\sum_{\mathbf{N}\mathfrak{q}\leqslant x} \mathbf{N}(\mathfrak{q})$. The latter expression is the summatory function associated to $\zeta_F(s-1)$, where $\zeta_F(s)$ is the Dedekind zeta function of F, so Proposition 4.1 now follows from Proposition 1.1.

Question 4.2 Is it the case that $\vartheta_{F,1}(x) \sim c \cdot x^a$ with constants $c > 0$ and $a > 1$ depending on F?

The underlying issue here is the average size of $[U_F : U_F^+(\mathfrak{q})]$, about which little seems to be known. Of some relevance, perhaps, is the literature on analogues of Artin's primitive root conjecture for units of number fields (see for example [9], [18], [19], [20], [25], [29], and [30]). In any case,

$[U_F : U_F^+(\mathfrak{q})]$ differs by a factor dividing $2^{r_1(F)}$ from the order of the image of the natural map from U_F to $(\mathcal{O}_F/\mathfrak{q})^\times$, so the problem is to understand the image of the global units in an approximation to a group of local units. This formulation is reminiscent of Leopoldt's conjecture, which we now revisit for the sake of the analogy.

Fix a prime number p and let θ_n be the number of one-dimensional characters of $\mathrm{Gal}(\overline{F}/F)$ of conductor dividing $p^n\mathcal{O}_F$. We think of θ_n as a vertical analogue of $\vartheta_{F,1}(x)$. To simplify the notation, write $U_F^+(p^n\mathcal{O}_F)$ as $U_F^+(p^n)$, and put

$$E_n = U_F^+(p^n) \tag{4.7}$$

for $n \geqslant 2$. Also put $E = E_2$. Via the map $u \mapsto u \otimes 1$ we may view E as a subset of $\mathcal{O}_F \otimes_{\mathbb{Z}} \mathbb{Z}_p$ and more precisely as a subgroup of $(\mathcal{O}_F \otimes_{\mathbb{Z}} \mathbb{Z}_p)^\times$ and indeed of $1 + p^2(\mathcal{O}_F \otimes_{\mathbb{Z}} \mathbb{Z}_p)$. We denote the p-adic closure of a subset S of $\mathcal{O}_F \otimes_{\mathbb{Z}} \mathbb{Z}_p$ by $\overline{S}$, and we write $r_1(F)$ and $r_2(F)$ simply as r_1 and r_2. Leopoldt's conjecture is usually stated as (i) or (ii) below.

Proposition 4.3 *The following statements are equivalent:*

(i) $\mathrm{rk}_{\mathbb{Z}_p}\mathrm{Hom}(\mathrm{Gal}(\overline{F}/F), \mathbb{Z}_p) = r_2 + 1.$

(ii) $\mathrm{rk}_{\mathbb{Z}_p}\overline{E} = r_1 + r_2 - 1.$

(iii) $\log \theta_n \sim (r_2 + 1)\log p \cdot n.$

Thus (iii) *is another formulation of Leopoldt's conjecture.*

Proof The equivalence of (i) and (ii) is well known, cf. [37], p. 265, Theorem 13.4. (Strictly speaking, the unit group E_1 in [37] is not quite the same as our E, but our E is a subgroup of finite index in E_1 and so the p-adic closures have the same $\mathbb{Z}_p$-rank.) For the sake of completeness we will verify that (ii) is equivalent to (iii), although the argument is in principle the same as in [37].

Put $s = \mathrm{rk}_{\mathbb{Z}_p}\overline{E}$ and $t = [F : \mathbb{Q}] - \mathrm{rk}_{\mathbb{Z}_p}\overline{E}$, so that

$$s + t = r_1 + 2r_2. \tag{4.8}$$

It suffices to see that there is a constant $c > 0$ such that

$$\theta_n = cp^{tn} \tag{4.9}$$

for n sufficiently large. Indeed (4.9) implies that $\log \theta_n \sim (t \log p)\cdot n$, whence (iii) becomes equivalent to $t = r_2 + 1$; but (ii) is equivalent to $s = r_1 + r_2 - 1$, and the equations $t = r_2 + 1$ and $s = r_1 + r_2 - 1$ are equivalent by (4.8).

To derive (4.9) we use the fact that $\theta_n = h_F^{\text{nar}}(p^n\mathcal{O}_F)$. It is readily verified that $\varphi_F(p^n\mathcal{O}_F) = p^{n[F:\mathbb{Q}]}\prod_{\mathfrak{p}|p}(1-(\mathbf{N}\mathfrak{p})^{-1})$, so (4.6) gives

$$\theta_n = c_1 \cdot p^{n[F:\mathbb{Q}]}/[U_F : U_F^+(p^n)] \tag{4.10}$$

with $c_1 = 2^{r_1} \cdot h_F \cdot \prod_{\mathfrak{p}|p}(1-(\mathbf{N}\mathfrak{p})^{-1})$.

On the other hand, recalling the notation (4.7), we can write

$$[U_F : U_F^+(p^n)] = [U_F : E][E : E_n]$$

for $n \geqslant 2$. As the natural map $E/E_n \to \overline{E}/\overline{E_n}$ is an isomorphism, it follows that

$$[U_F : U_F^+(p^n)] = c_2[\overline{E} : \overline{E_n}] \tag{4.11}$$

with $c_2 = [U_F : E]$. Now the p-adic logarithm $\log_p$ gives an isomorphism

$$\overline{E}/\overline{E_n} \cong (\log_p \overline{E})/((\log_p \overline{E}) \cap p^n\mathcal{O}_F),$$

so we have

$$[\overline{E} : \overline{E_n}] = [L : L \cap (p^n\mathcal{O}_F)] \tag{4.12}$$

with $L = \log_p \overline{E}$. Put $m = [F : \mathbb{Q}_p]$. As $\mathcal{O}_F \otimes_{\mathbb{Z}} \mathbb{Z}_p$ is a free $\mathbb{Z}_p$-module of rank m and L is a $\mathbb{Z}_p$-submodule of rank s, there exists a basis $e_1, \ldots, e_m$ for $\mathcal{O}_F \otimes_{\mathbb{Z}} \mathbb{Z}_p$ together with integers $\nu_1, \ldots \nu_s \geqslant 0$ such that $p^{\nu_1}e_1, \ldots, p^{\nu_s}e_s$ is a basis for L. Returning to (4.12), we see that if $n \geqslant \max(\nu_1, \ldots, \nu_s)$ then

$$[\overline{E} : \overline{E_n}] = c_3 p^{ns} \tag{4.13}$$

with $c_3 = p^{-(\nu_1+\nu_2+\cdots+\nu_s)}$. Finally, combining (4.13) with (4.10) and (4.11), and setting $c = c_1/(c_2c_3)$, we obtain (4.9) for n sufficiently large.

□

5 Dihedral Representations

A finite subgroup G of $\mathrm{GL}_n(\mathbb{C})$ is *irreducible* if the tautological representation $\iota : G \hookrightarrow \mathrm{GL}_n(\mathbb{C})$ is irreducible. Similarly, G is *monomial* if ι is monomial, and G is *self-dual* if ι is self-dual. Let D_{2m} denote the dihedral group of order $2m$ ($m \geqslant 3$) and Q_{4m} the quaternion group of order $4m$ ($m \geqslant 2$). The term "quaternion group" is used here as in [27], p. 72, but since it is often reserved for the case $m = 2$, let us recall the standard presentations: D_{2m} has generators a, b with $a^m = 1 = b^2$ and $bab^{-1} = a^{-1}$, while Q_{4m} has generators a, b with $a^{2m} = 1$, $a^m = b^2$, and $bab^{-1} = a^{-1}$. These are the only groups that figure in $\vartheta_{\mathbb{Q},2}^{\text{im,sd}}(x)$:

Proposition 5.1 *Let G be a finite subgroup of* $\mathrm{GL}_2(\mathbb{C})$*. If G is irreducible, monomial, and self-dual then either $G \cong D_{2m}$ with $m \geqslant 3$ or $G \cong Q_{4m}$ with $m \geqslant 2$. In the former case G is conjugate to a subgroup of* $\mathrm{O}_2(\mathbb{R})$ *and in the latter case* $G \subset \mathrm{SL}_2(\mathbb{C})$.

A two-dimensional irreducible monomial self-dual Artin representation ρ will be called *dihedral* or *quaternionic* according as the image of ρ is isomorphic to D_{2m} ($m \geqslant 3$) or to Q_{4m} ($m \geqslant 2$). We also put $m(\rho) = m$. Since $\mathrm{SL}_2(\mathbb{C})$ and $\mathrm{Sp}_2(\mathbb{C})$ coincide, we see that the orthogonal and symplectic terms in the decomposition

$$\vartheta_{\mathbb{Q},2}^{\mathrm{im,sd}}(x) = \vartheta_{\mathbb{Q},2}^{\mathrm{im,orth}}(x) + \vartheta_{\mathbb{Q},2}^{\mathrm{im,symp}}(x) \tag{5.1}$$

count dihedral and quaternionic Artin representations of $\mathbb{Q}$ respectively. In this section we bound the dihedral term $\vartheta_{\mathbb{Q},2}^{\mathrm{im,orth}}(x)$.

Proposition 5.1 is a standard remark, as are Propositions 5.2 and 5.3 below, but for want of a suitable reference we supply proofs of all three assertions in an appendix (Section 13). Given a group G, a normal subgroup H, a one-dimensional character χ of H, and an element $g \in G$, write χ^g for the character $h \mapsto \chi(ghg^{-1})$ of H.

Proposition 5.2 *Let G be a finite group and ρ a faithful irreducible monomial self-dual representation of G of dimension two. Then G has a cyclic subgroup of index two, and if H is any such subgroup then ρ is induced by a faithful one-dimensional character ξ of H of order $\geqslant 3$ satisfying $\xi^g = \xi^{-1}$ for $g \in G \smallsetminus H$. Furthermore, ξ and ξ^{-1} are the only two characters of H inducing ρ.*

Given a finite group G and a subgroup H, write G^{ab} and H^{ab} for their maximal abelian quotients and $\mathrm{tran}_H^G : G^{\mathrm{ab}} \to H^{\mathrm{ab}}$ for the transfer. If ξ is a one-dimensional character of H then ξ factors through H^{ab}, whence we can form the composition $\xi \circ \mathrm{tran}_H^G$ and view it as a one-dimensional character of G. We write sign_H^G for the sign of the permutation representation of G on the left cosets of H in G, and we write 1 for the trivial one-dimensional character of any group. Finally, if λ is a representation of H then $\mathrm{ind}_H^G \lambda$ denotes the representation of G induced by λ. Part (a) of the following proposition is a converse to Proposition 5.2 and part (b) is a refinement of it.

Proposition 5.3 *Let G be a finite group and H a subgroup of index two, and let ξ be a faithful one-dimensional character of H of order $\geqslant 3$. Put $\rho = \mathrm{ind}_H^G \xi$.*

(a) *If $\xi^g = \xi^{-1}$ for $g \in G \smallsetminus H$ then ρ is faithful, irreducible, and self-dual.*

(b) *The hypothesis of* (a) *holds if and only if* $\xi \circ \mathrm{tran}_H^G$ *is either* 1 *or* sign_H^G, *and these two alternatives imply respectively that* ρ *is orthogonal or symplectic.*

Now let F be a number field. Given a finite extension K of F (always understood to be contained in some fixed algebraic closure $\overline{F}$ of F) and an Artin representation λ of K, we write $\mathrm{ind}_{K/F}\lambda$ for the Artin representation of F induced by λ. We may think of $\mathrm{ind}_{K/F}$ either as induction from $\mathrm{Gal}(\overline{F}/K)$ to $\mathrm{Gal}(\overline{F}/F)$ or as induction from $\mathrm{Gal}(L/K)$ to $\mathrm{Gal}(L/F)$, where L is any finite Galois extension of F containing K such that λ factors through $\mathrm{Gal}(L/K)$. Similarly, $\mathrm{tran}_{K/F}$ denotes the transfer from $\mathrm{Gal}(\overline{F}/F)^{\mathrm{ab}}$ to $\mathrm{Gal}(\overline{F}/K)^{\mathrm{ab}}$ or alternatively the transfer from $\mathrm{Gal}(L/F)^{\mathrm{ab}}$ to $\mathrm{Gal}(L/K)^{\mathrm{ab}}$, where L is any finite Galois extension of F containing K. Of course in the case of a topological group like $\mathrm{Gal}(\overline{F}/F)$ the notation G^{ab} refers to the quotient of G by the *closure* of its commutator subgroup.

Proposition 5.4 *Consider pairs* (K,ξ) *with* $[K : F] = 2$ *and* ξ *a one-dimensional character of* $\mathrm{Gal}(\overline{F}/K)$ *of order* $m \geqslant 3$ *such that* $\xi \circ \mathrm{tran}_{K/F} = 1$. *The formula* $\rho = \mathrm{ind}_{K/F}\xi$ *defines a two-to-one map from the set of such* (K,ξ) *onto the set of isomorphism classes of dihedral Artin representations* ρ *of* F *with* $m(\rho) = m$, *the other preimage of the isomorphism class of* ρ *being the pair* (K,ξ^{-1}).

Proof Given a dihedral Artin representation ρ of F, let L be the fixed field of $\mathrm{Ker}\,\rho$ and put $G = \mathrm{Gal}(L/F)$. By Proposition 5.2 and part (b) of Proposition 5.3, $\rho = \mathrm{ind}_{K/F}\xi$ for some (K,ξ) as above. Conversely, given (K,ξ), we have $\xi^g = \xi^{-1}$ for $g \in \mathrm{Gal}(\overline{F}/F) \smallsetminus \mathrm{Gal}(\overline{F}/K)$ by part (b) of Proposition 5.3. Hence the fixed field L of $\mathrm{Ker}\,\xi$ is Galois over F, and part (a) of Proposition 5.3 shows that the representation $\rho = \mathrm{ind}_{K/F}\xi$ is irreducible and orthogonal, as well as faithful as a representation of $\mathrm{Gal}(L/F)$. Hence it follows from Proposition 5.1 that ρ has image D_{2m}. Since a cyclic subgroup of index two in D_{2m} is unique, ρ determines K uniquely, and the last assertion of Proposition 5.2 then implies that ξ is unique up to replacement by ξ^{-1}.

□

If $\rho = \mathrm{ind}_{K/F}\xi$ then $\mathfrak{q}(\rho) = \mathfrak{d}_{K/F}\,\mathfrak{q}(\xi)$ by the conductor-discriminant formula (cf. [32], p. 104, Proposition 6), whence $q(\rho) = d_{K/F}\,q(\xi)$ on taking absolute norms. Thus Proposition 5.4 gives

$$\vartheta_{F,2}^{\mathrm{im,orth}}(x) = \frac{1}{2} \sum_{d_{K/F}q(\xi) \leqslant x} 1, \tag{5.2}$$

where the sum runs over ordered pairs (K, ξ) satisfying the stated inequality.

We now rewrite (5.2) using class field theory: A one-dimensional character ξ of $\mathrm{Gal}(\overline{F}/K)$ becomes an idele class character of K of finite order, and the condition $\xi \circ \mathrm{tran}_{K/F} = 1$ becomes $\xi|\mathbb{A}_F^\times = 1$, where $\mathbb{A}_F^\times$ is the idele group of F.

Lemma 5.5 *There is an ideal $\mathfrak{q}$ of $\mathcal{O}_F$ such that $\mathfrak{q}(\xi) = \mathfrak{q}\mathcal{O}_K$.*

Proof This is a straightforward deduction from the fact that $\xi|\mathbb{A}_F^\times = 1$. Only one point deserves comment: If v is a finite place of F which ramifies in K and w is the place of K above v, then the local component ξ_w of ξ has *even* conductor-exponent $a(\xi_w)$. To see this, let $\mathcal{O}_{F,v}$ and $\mathcal{O}_{K,w}$ be the completions of $\mathcal{O}_F$ and $\mathcal{O}_K$, and let π_w be a uniformizer of $\mathcal{O}_{K,w}$. If $a = a(\xi_w)$ is odd then the cosets of $1+\pi_w^a\mathcal{O}_{K,w}$ in $1+\pi_w^{a-1}\mathcal{O}_{K,w}$ (or in $\mathcal{O}_{K,w}^\times$, if $a = 1$) are represented by elements of $\mathcal{O}_{F,v}^\times$, whence the nontriviality of ξ_w on the quotient contradicts the triviality of ξ on $\mathbb{A}_F^\times$.

□

Given a nonzero integral ideal $\mathfrak{q}$ of F, let $g_{K/F}(\mathfrak{q})$ be the number of idele class characters of K of finite order $\geqslant 3$ which are trivial on $\mathbb{A}_F^\times$ and of conductor $\mathfrak{q}\mathcal{O}_K$. Returning to (5.2), we see that $\vartheta_{F,2}^{\mathrm{im,orth}}(x) = 1/2 \sum g_{K/F}(\mathfrak{q})$, where the sum runs over pairs $(K, \mathfrak{q})$ with $d_{K/F}(\mathbf{N}\mathfrak{q})^2 \leqslant x$. It follows in particular that

$$\vartheta_{F,2}^{\mathrm{im,orth}}(x) \leqslant \frac{1}{2} \sum_{d_{K/F}(\mathbf{N}\mathfrak{q})^2 \leqslant x} h_{K/F}^{\mathrm{nar}}(\mathfrak{q}), \tag{5.3}$$

where $h_{K/F}^{\mathrm{nar}}(\mathfrak{q})$ is the number of idele class characters of K of *arbitrary* finite order which are trivial on $\mathbb{A}_F^\times$ and of conductor *dividing* $\mathfrak{q}\mathcal{O}_K$.

Now take $F = \mathbb{Q}$. We write $h_{K/F}^{\mathrm{nar}}(\mathfrak{q})$ simply as $h_{K/\mathbb{Q}}^{\mathrm{nar}}(q)$, where q is the positive integer such that $\mathfrak{q} = q\mathcal{O}_K$. If the quadratic field K is imaginary then $h_{K/\mathbb{Q}}^{\mathrm{nar}}(q)$ may be further abbreviated to $h_{K/\mathbb{Q}}(q)$. Thus (5.3) becomes

$$\vartheta_{\mathbb{Q},2}^{\mathrm{im,orth}}(x) \leqslant \frac{1}{2} \sum_{\substack{d_K q^2 \leqslant x \\ K \text{ imaginary}}} h_{K/\mathbb{Q}}(q) + \frac{1}{2} \sum_{\substack{d_K q^2 \leqslant x \\ K \text{ real}}} h_{K/\mathbb{Q}}^{\mathrm{nar}}(q). \tag{5.4}$$

Siegel [35] proved the asymptotic formulas

$$\sum_{\substack{d_K q^2 \leqslant x \\ K \text{ imaginary}}} h_{K/\mathbb{Q}}(q) \sim \pi x^{3/2}/(18\zeta(3)) \tag{5.5}$$

and

$$\sum_{\substack{d_K q^2 \leqslant x \\ K \text{ real}}} h^{\text{nar}}_{K/\mathbb{Q}}(q) \log \epsilon_{K,q} \sim \pi^2 x^{3/2}/(18\zeta(3)), \tag{5.6}$$

where $\epsilon_{K,q}$ is the fundamental totally positive unit of the order $\mathcal{O}_{K,q} = \mathbb{Z} + q\mathcal{O}_K$: In other words, $\epsilon_{K,q}$ is the unique generator > 1 of the group $U^+_{K,q} = U^+_K \cap U_{K,q}$, where $U_{K,q} = \mathcal{O}^\times_{K,q}$. Since $\log \epsilon_{K,q} \gg 1$ (indeed $\epsilon_{K,q} > q\sqrt{d}/2 \geqslant \sqrt{5}/2$) we deduce the following bound from (5.4), (5.5), and (5.6).

Proposition 5.6 $\vartheta^{\text{im,orth}}_{\mathbb{Q},2}(x) = O(x^{3/2})$.

One point deserves clarification. Put $d = \pm d_K q^2$, choosing the sign so that $\pm d_K$ is the discriminant of K. The quantity $h^{\text{nar}}_{K/\mathbb{Q}}(q)$ as we have defined it is *the narrow ring class number of* K *to the modulus* q, whereas the results which we have quoted from [35] pertain to *the narrow class number of primitive binary quadratic forms of discriminant* d. The equality of these two quantities is of course classical and can be established conceptually, but we will take the shortcut of recalling a standard formula for $h^{\text{nar}}_{K/\mathbb{Q}}(q)$, which upon comparison with formulas (10) and (19) of [35] (and an application of Dirichlet's class number formula) will assure us that Siegel's h_d coincides with our $h^{\text{nar}}_{K/\mathbb{Q}}(q)$. Let χ_K be the primitive quadratic Dirichlet character corresponding to K. We write h^{nar}_K for the narrow ideal class number of K (equal to h_K if K is imaginary).

Proposition 5.7 $h^{\text{nar}}_{K/\mathbb{Q}}(q) = \dfrac{h^{\text{nar}}_K}{[U^+_K : U^+_{K,q}]} \cdot q \prod_{p|q} (1 - \chi_K(p)/p)$.

Proof The argument is classical (see for example the references to Fueter and Weber on p. 95 of [24], where the analogous formula is proved for wide ring class numbers) but we recall it briefly nonetheless.

Suppose first that K is real. Write $C^{\text{nar}}_{\mathbb{Q}}(q)$ and $C^{\text{nar}}_K(q)$ for the narrow ray class groups of $\mathbb{Q}$ and K to the moduli $q\mathbb{Z}$ and $q\mathcal{O}_K$ respectively, and let ω be the natural map from $C^{\text{nar}}_{\mathbb{Q}}(q)$ to $C^{\text{nar}}_K(q)$. Then $h^{\text{nar}}_{K/\mathbb{Q}}(q)$ is the order of the cokernel of ω. Hence

$$h^{\text{nar}}_{K/\mathbb{Q}}(q) = \frac{h^{\text{nar}}_K(q)}{\varphi(q)} |\text{Ker}\ \omega|. \tag{5.7}$$

Let $U_{K/\mathbb{Q}}(q)$ be the subgroup of U_K consisting of units u for which there exists $a \in \mathbb{Z}$ with $au \equiv 1$ modulo $\mathfrak{q}\mathcal{O}_K$ and $au > 0$ at both real places of K. Also put $U^+_K(q) = U^+_K(q\mathcal{O}_K)$. One checks that the map sending the ray class of $a\mathbb{Z}$ to the coset of u modulo $\{\pm 1\}U_K^+(q)$ is an isomorphism from Ker ω onto $U_{K/\mathbb{Q}}(q)/\{\pm 1\}U_K^+(q)$. Hence (5.7) becomes

$$h_{K/\mathbb{Q}}^{\mathrm{nar}}(q) = \frac{h_K^{\mathrm{nar}}(q)}{\varphi(q)}[U_{K/\mathbb{Q}}(q) : \{\pm 1\}U_K^+(q)]. \tag{5.8}$$

Replacing F by K in (4.6) and inserting the result in (5.8), we deduce that

$$h_{K/\mathbb{Q}}^{\mathrm{nar}}(q) = h_K \cdot q \prod_{p|q}(1 - \chi_K(p)/p) \cdot \frac{2^2}{[U_K : U_{K/\mathbb{Q}}(q)][\{\pm 1\}U_K^+(q) : U_K^+(q)]}. \tag{5.9}$$

The stated formula follows from (5.9), because $[\{\pm 1\}U_K^+(q) : U_K^+(q)] = 2$ and $2h_K[U_K^+ : U_{K,q}^+] = h_K^{\mathrm{nar}}[U_K : U_{K/\mathbb{Q}}(q)]$. (To verify the latter equation, consider cases according as the fundamental unit of K does or does not have norm -1, and observe that the units in $U_{K/\mathbb{Q}}(q)$ all have norm 1.)

Next suppose that K is imaginary. We take ω to be the natural map of wide ray class groups $C_\mathbb{Q}(q) \to C_K(q)$. The order of $C_\mathbb{Q}(q)$ is $\varphi(q)/2$ or $\varphi(q)$ according as $q > 2$ or $q \leqslant 2$, hence it equals $\varphi(q)/[\{\pm 1\}U_K(q) : U_K(q)]$ in all cases. Thus putting $U_{K/\mathbb{Q}}(q) = \{\pm 1\}U_K(q)$, we have

$$h_{K/\mathbb{Q}}(q) = \frac{h_K(q)}{\varphi(q)}[U_{K/\mathbb{Q}}(q) : U_K(q)]. \tag{5.10}$$

in place of (5.8). Applying (4.6) as before, we obtain

$$h_{K/\mathbb{Q}}(q) = \frac{h_K}{[U_K : U_{K/\mathbb{Q}}(q)]} \cdot q \prod_{p|q}(1 - \chi_K(p)/p). \tag{5.11}$$

Now $[U_K : U_{K/\mathbb{Q}}(q)]$ is 1 if $q = 1$ and otherwise 1, 2, or 3 according as $d_K > 4$, $d_K = 4$, or $d_K = 3$. The same is true of $[U_K : U_{K,q}]$, so (5.11) is the stated formula.

□

6 Quaternionic Representations

Next we will prove an estimate for the quaternionic term in (5.1):

Proposition 6.1 $\vartheta_{\mathbb{Q},2}^{\mathrm{im,symp}}(x) = O(x^{3/2+\varepsilon})$ *for every* $\varepsilon > 0$, *where the implied constant depends on* ε.

Combining Propositions 6.1 and 5.6, we will have:

Proposition 6.2 $\vartheta^{\text{im,sd}}_{\mathbb{Q},2}(x) = O(x^{3/2+\varepsilon})$ *for every* $\varepsilon > 0$, *where the implied constant depends on* ε.

We begin with a general remark. Given Artin representations ρ and ρ' of a number field F, write $\mathrm{P}\rho$ and $\mathrm{P}\rho'$ for the projective representations of $\mathrm{Gal}(\overline{F}/F)$ determined by ρ and ρ', and call ρ and ρ' *projectively equivalent* if $\mathrm{P}\rho \cong \mathrm{P}\rho'$.

Proposition 6.3 *Suppose that* ρ *and* ρ' *are symplectic of dimension* n. *Then* ρ *and* ρ' *are projectively equivalent if and only if* $\rho' \cong \rho \otimes \chi$ *for some one-dimensional character* χ *of* $\mathrm{Gal}(\overline{F}/F)$ *with* $\chi^n = 1$.

Proof To say that $\mathrm{P}\rho \cong \mathrm{P}\rho'$ means precisely that $\rho' \cong \rho \otimes \chi$ for some one-dimensional character χ of $\mathrm{Gal}(\overline{F}/F)$. Taking determinants of both sides, we find that $\chi^n = 1$, because symplectic representations have trivial determinant.

□

Next we state an analogue for quaternionic Artin representations of an earlier assertion about dihedral Artin representations (Proposition 5.4). Given a quadratic extension K of F (understood to lie in some fixed algebraic closure $\overline{F}$ of F), write $\mathrm{sign}_{K/F}$ for the quadratic character of $\mathrm{Gal}(\overline{F}/F)$ with kernel $\mathrm{Gal}(\overline{F}/K)$.

Proposition 6.4 *Consider pairs* (K, ξ) *with* $[K : F] = 2$ *and* ξ *a one-dimensional character of* $\mathrm{Gal}(\overline{F}/K)$ *of even order* $2m \geqslant 6$ *such that* $\xi \circ \mathrm{tran}_{K/F} = \mathrm{sign}_{K/F}$. *The formula* $\rho = \mathrm{ind}_{K/F}\xi$ *defines a two-to-one map from the set of such* (K, ξ) *onto the set of isomorphism classes of quaternionic Artin representations* ρ *of* F *with* $m(\rho) = m$, *the other preimage of the isomorphism class of* ρ *being the pair* (K, ξ^{-1}).

This is simply Proposition 5.4 with three changes: the word "dihedral" is replaced by "quaternionic" and the conditions "order $m \geqslant 3$" and "$\xi \circ \mathrm{tran}_{K/F} = 1$" by "even order $2m \geqslant 6$" and "$\xi \circ \mathrm{tran}_{K/F} = \mathrm{sign}_{K/F}$." (Actually the requirement that ξ have *even* order is superfluous; it follows from the condition $\xi \circ \mathrm{tran}_{K/F} = \mathrm{sign}_{K/F}$). The proof of Proposition 6.4 is likewise identical to that of Proposition 5.4, apart from the obvious changes. Note in particular that in terms of the presentation of Q_{4m} given in Section 5, the elements $a^j b$ and $a^j b^3$ have order four, whence for $m \geqslant 3$ a cyclic subgroup of index two in Q_{4m} is unique, just as it is in D_{2m}. By contrast, Q_8 has three cyclic subgroups of index two, and as a result the analogue of Proposition 6.4 for $m(\rho) = 2$ is as follows:

Proposition 6.5 *Consider pairs (K, ξ) with $[K : F] = 2$ and ξ a one-dimensional character of $\mathrm{Gal}(\overline{F}/K)$ of order 4 such that $\xi \circ \mathrm{tran}_{K/F} = \mathrm{sign}_{K/F}$. The formula $\rho = \mathrm{ind}_{K/F}\xi$ defines a six-to-one map from the set of such (K, ξ) onto the set of isomorphism classes of quaternionic Artin representations ρ of F with $m(\rho) = 2$. If L is the fixed field of $\mathrm{Ker}\,\rho$ and K_1, K_2 and K_3 are the three quadratic extensions of F contained in L then the six preimages of the isomorphism class of ρ have the form $(K_j, \xi_j^{\pm 1})$ with $1 \leqslant j \leqslant 3$ and one-dimensional characters ξ_j of $\mathrm{Gal}(\overline{F}/K_j)$.*

Our strategy for bounding the quaternionic term in (5.1) rests on a simple remark: Given a quaternionic Artin representation ρ of F with $m(\rho) \geqslant 3$, we can define a dihedral Artin representation $\hat{\rho}$ of F by writing $\rho \cong \mathrm{ind}_{K/F}\xi$ as in Proposition 6.4 and setting $\hat{\rho} = \mathrm{ind}_{K/F}\xi^2$. That the isomorphism class of $\hat{\rho}$ is well defined follows from Proposition 5.4, which also gives

$$m(\rho) = m(\hat{\rho}). \tag{6.1}$$

Using Proposition 6.5, we can define $\hat{\rho}$ in the same way when $m(\rho) = 2$, but because of the nonuniqueness of K in Proposition 6.5 we must make an arbitrary but fixed choice of a quadratic extension K of F inside every biquadratic extension of F. Note that $\hat{\rho}$ is now reducible; in fact if L is the fixed field of $\mathrm{Ker}\,\rho$ then $\hat{\rho} \cong \chi \oplus \chi'$, where χ and χ' are the two quadratic characters of $\mathrm{Gal}(L/F)$ which do not factor through $\mathrm{Gal}(K/F)$. Thus $\hat{\rho}$ is no longer "dihedral," but we still set $m(\hat{\rho}) = 2$, so that (6.1) holds in all cases. Another formula which holds in all cases is

$$q(\rho) \geqslant q(\hat{\rho}), \tag{6.2}$$

because $q(\rho) = d_{K/F}\, q(\xi)$ and $q(\hat{\rho}) = d_{K/F}\, q(\xi^2)$ by the conductor-discriminant formula, and $q(\xi) \geqslant q(\xi^2)$. Finally, it follows from Proposition 5.1 that if ρ is a quaternionic Artin representation of F and χ is a one-dimensional character of $\mathrm{Gal}(\overline{F}/F)$ with $\chi^2 = 1$ then $\rho \otimes \chi$ is again a quaternionic Artin representation of F. Now if ρ is replaced by $\rho \otimes \chi$ then ξ is multiplied by $\mathrm{res}_{K/F}(\chi)$, the restriction of χ to $\mathrm{Gal}(\overline{F}/K)$. But as $\chi^2 = 1$ the character ξ^2 is unchanged, and hence $\hat{\rho}$ is unchanged up to isomorphism. Referring to Proposition 6.3, we deduce that the isomorphism class $\langle \hat{\rho} \rangle$ of $\hat{\rho}$ depends only on the projective equivalence class $[\rho]$ of ρ, so we obtain a map $[\rho] \mapsto \langle \hat{\rho} \rangle$.

Proposition 6.6 *The map $[\rho] \mapsto \langle \hat{\rho} \rangle$ is injective.*

Proof In view of (6.1), it suffices to verify injectivity on the subset of projective equivalence classes $[\rho]$ for which $m(\rho)$ has a fixed value m. To begin with we take $m \geqslant 3$. So suppose that we are given quaternionic

Artin representations ρ and ρ' of F with $m(\rho) = m(\rho') = m \geqslant 3$. Write $\rho \cong \mathrm{ind}_{K/F}\xi$ and $\rho' \cong \mathrm{ind}_{K'/F}\xi'$ with pairs (K, ξ) and (K', ξ') as in Proposition 6.4. We assume that

$$\mathrm{ind}_{K/F}\xi^2 \cong \mathrm{ind}_{K'/F}(\xi')^2 \tag{6.3}$$

and must deduce that $\mathrm{P}\rho \cong \mathrm{P}\rho'$.

Since $m \geqslant 3$, the representations $\mathrm{ind}_{K/F}\xi^2$ and $\mathrm{ind}_{K/F}(\xi')^2$ are dihedral. Hence in view of (6.3), we have $K = K'$ and $(\xi')^2 = \xi^{\pm 2}$ by Proposition 5.4. After replacing the pair (K, ξ) by (K, ξ^{-1}) if necessary, we may assume that $(\xi')^2 = \xi^2$, and then $\xi' = \xi\phi$ for some character ϕ of $\mathrm{Gal}(\overline{F}/K)$ with $\phi^2 = 1$. Since $\xi \circ \mathrm{tran}_{K/F}$ and $\xi' \circ \mathrm{tran}_{K/F}$ both coincide with $\mathrm{sign}_{K/F}$, it follows that $\phi \circ \mathrm{tran}_{K/F} = 1$. Let us now view ϕ as an idele class character of K. Then the condition $\phi \circ \mathrm{tran}_{K/F} = 1$ becomes $\phi | \mathbb{A}_F^\times = 1$. In particular, $\phi \circ N_{K/F} = 1$, where $N_{K/F}$ is the idelic norm from $\mathbb{A}_K^\times$ to $\mathbb{A}_F^\times$. Write σ for the nontrivial element of $\mathrm{Gal}(K/F)$, and view σ as an automorphism of $\mathbb{A}_K^\times$. Then $\phi(x^{\sigma+1}) = 1$ for all $x \in \mathbb{A}_K^\times$, and as $\phi^2 = 1$ we deduce that $\phi(x^{\sigma-1}) = 1$ also. Hilbert's Theorem 90 now implies that ϕ factors through $N_{K/F}$, so that $\phi = \chi \circ N_{K/F}$ for some one-dimensional character χ of $\mathbb{A}_F^\times$. Returning to the Galois setting, we see that $\phi = \mathrm{res}_{K/F}(\chi)$ when ϕ and χ are viewed as one-dimensional characters of $\mathrm{Gal}(\overline{F}/K)$ and $\mathrm{Gal}(\overline{F}/F)$ respectively. To recapitulate, we have $\rho \cong \mathrm{ind}_{K/F}\xi$, $\rho' \cong \mathrm{ind}_{K'/F}\xi'$, $K = K'$, $\xi' = \xi\phi$, and $\phi = \mathrm{res}_{K/F}(\chi)$. It follows that $\rho' \cong \rho \otimes \chi$, whence $\mathrm{P}\rho \cong \mathrm{P}\rho'$.

The case $m = 2$ is contained in Theorem 4 on p. 146 of [12], at least for $F = \mathbb{Q}$. However for the sake of completing the present argument, we first observe that if χ and χ' are distinct quadratic characters of $\mathrm{Gal}(\overline{F}/F)$, then there is a unique pair (K, ζ) consisting of a quadratic extension K of F and a quadratic character ζ of $\mathrm{Gal}(\overline{F}/K)$ such that $\mathrm{ind}_{K/F}(\zeta) \cong \chi \oplus \chi'$. Indeed if M and M' are the fixed fields of the kernels of χ and χ' respectively then K is the third quadratic extension of F contained in MM', and ζ is the unique quadratic character of $\mathrm{Gal}(\overline{F}/K)$ which factors through $\mathrm{Gal}(MM'/K)$. It follows that in the case $m = 2$, the isomorphism (6.3) still implies that $K = K'$ and $\xi^2 = (\xi')^2$. The proof is now completed as in the case $m \geqslant 3$. □

Now take $F = \mathbb{Q}$, and let X be the set of one-dimensional characters of $\mathrm{Gal}(\overline{\mathbb{Q}}/\mathbb{Q})$ satisfying $\chi^2 = 1$. The arguments to be given next will be needed again when we deal with primitive representations, so it is efficient to suspend our focus on quaternionic Artin representations in favor of a more general setting. Thus $\mathcal{A}$ will denote any class of two-dimensional Artin representations of $\mathbb{Q}$ which is *symplectic* and *closed under quadratic twists* in the sense that the following conditions hold:

- If $\rho \in \mathcal{A}$ then $\det \rho = 1$.
- If $\rho \in \mathcal{A}$ and $\chi \in X$ then $\rho \otimes \chi \in \mathcal{A}$.

- If $\rho \in \mathcal{A}$ and $\rho' \cong \rho$ then $\rho' \in \mathcal{A}$.

The third condition is an inessential nicety intended only to eliminate ambiguities. We write $\vartheta_{\mathcal{A}}(x)$ for the number of isomorphism classes of representations $\rho \in \mathcal{A}$ such that $q(\rho) \leqslant x$.

Let $\mathcal{E}$ denote the set of projective equivalence classes of $\mathcal{A}$, and write $[\rho]$ as before for the projective equivalence class of ρ. Proposition 6.3 implies that

$$\vartheta_{\mathcal{A}}(x) \leqslant \sum_{[\rho]\in\mathcal{E}} \sum_{\substack{\chi\in X \\ q(\rho\otimes\chi)\leqslant x}} 1, \tag{6.4}$$

the inner sum being independent of the choice of representative ρ of $[\rho]$. The reason for inequality rather than equality in (6.4) is that sometimes $\rho \otimes \chi \cong \rho$ with $\chi \neq 1$.

In order to bound the right-hand side of (6.4) it is convenient to introduce the notion of the "ρ-conductor" $q_\rho(\chi)$ of a character $\chi \in X$. Let ord_p denote the p-adic valuation of $\mathbb{Z}$. We define $q_\rho(\chi)$ by deleting from $q(\chi)$ the contributions of the primes dividing $q(\rho)$:

$$q_\rho(\chi) = \prod_{p\nmid q(\rho)} p^{\mathrm{ord}_p q(\chi)}. \tag{6.5}$$

Then we have the following elementary remark:

Proposition 6.7 *Each projective equivalence class $E \in \mathcal{E}$ has a representative ρ such that $q(\rho \otimes \chi) \geqslant q(\rho)q_\rho(\chi)^2$ for all $\chi \in X$.*

Proof Write $E = [\lambda]$ with $\lambda \in \mathcal{A}$. We must exhibit a character $\phi \in X$ such that the representation $\rho = \lambda \otimes \phi$ satisfies the stated inequality for all $\chi \in X$.

Given a prime p, let $X_p \subset X$ be the subset of characters $\chi \in X$ which are unramified outside p and infinity. Thus $|X_2| = 4$, and if p is odd then $|X_p| = 2$. In particular, X_p is finite, so for each p dividing $q(\lambda)$ we can choose $\phi_p \in X_p$ minimizing $\mathrm{ord}_p\, q(\lambda \otimes \phi_p)$. We put $\phi = \prod_{p|q(\lambda)} \phi_p$, and as already indicated, $\rho = \lambda \otimes \phi$. By construction, every prime p dividing $q(\rho)$ divides $q(\lambda)$, and for every such p and every $\chi \in X$ we have

$$\mathrm{ord}_p\, q(\rho \otimes \chi) \geqslant \mathrm{ord}_p\, q(\rho) \qquad (p|q(\rho)). \tag{6.6}$$

On the other hand, if $p \nmid q(\rho)$ then the restriction of ρ to an inertia subgroup at p is the two-dimensional trivial representation, whence the restriction of $\rho \otimes \chi$ coincides with that of $\chi \oplus \chi$. Therefore

$$\mathrm{ord}_p\, q(\rho \otimes \chi) = 2\, \mathrm{ord}_p\, q(\chi) \qquad (p \nmid q(\rho)). \tag{6.7}$$

The stated inequality follows from (6.6) and (6.7).

□

Henceforth we assume that in the sum in (6.4) over equivalence classes $[\rho] \in \mathcal{E}$, the representative ρ is chosen as in Proposition 6.7. Then

$$\vartheta_{\mathcal{A}}(x) \leqslant \sum_{[\rho]\in\mathcal{E}} \sum_{\substack{\chi\in X \\ q_\rho(\chi)\leqslant(x/q(\rho))^{1/2}}} 1, \tag{6.8}$$

because the summation in (6.4) runs over a subset of the set of summation in (6.8). The next step eliminates the inner sum in (6.8):

Proposition 6.8 *For every $\varepsilon > 0$,*

$$\vartheta_{\mathcal{A}}(x) \ll x^{1/2} \sum_{\substack{[\rho]\in\mathcal{E} \\ q(\rho)\leqslant x}} q(\rho)^{-1/2+\varepsilon},$$

where the implicit constant depends on ε.

Proof Given $[\rho] \in \mathcal{E}$, we define a map $\chi \mapsto \chi_\rho$ from X to itself as follows: Write $\chi = \prod_{p|q(\chi)} \chi_p$ with $\chi_p \in X$ and χ_p unramified outside p and infinity; then

$$\chi_\rho = \prod_{\substack{p|q(\chi) \\ p\nmid q(\rho)}} \chi_p.$$

Recalling the definition (6.5) of $q_\rho(\chi)$, we see that

$$q_\rho(\chi) = q(\chi_\rho). \tag{6.9}$$

for all $\chi \in X$. Furthermore an element $\lambda \in X$ has at most $2\tau(q(\rho))$ preimages under the map $\chi \mapsto \chi_\rho$, where $\tau(q)$ denotes the number of positive divisors of q. Hence on making the substitution (6.9) in (6.8) and setting $\lambda = \chi_\rho$, we obtain

$$\vartheta_{\mathcal{A}}(x) \leqslant 2 \sum_{[\rho]\in\mathcal{E}} \sum_{\substack{\lambda\in X \\ q(\lambda)\leqslant(x/q(\rho))^{1/2}}} \tau(q(\rho)). \tag{6.10}$$

The inner sum in (6.10) equals $\tau(q(\rho)) \cdot \vartheta^{\mathrm{sd}}_{\mathbb{Q},1}((x/q(\rho))^{1/2})$ if $q(\rho) \leqslant x$ and 0 otherwise. Furthermore $\tau(q) = O(q^\varepsilon)$ for every $\varepsilon > 0$. Hence the stated estimate for $\vartheta_{\mathcal{A}}(x)$ follows from the corollary to Proposition 2.3.

□

We now specialize to the case where $\mathcal{A}$ is the class of quaternionic Artin representations of $\mathbb{Q}$ and $\mathcal{E}$ is the set of projective equivalence classes of such representations. Combining (6.2) with Proposition 6.8, we find that

$$\vartheta^{\mathrm{im,symp}}_{\mathbb{Q},2}(x) \ll x^{1/2} \sum_{\substack{[\rho]\in\mathcal{E} \\ q(\hat{\rho})\leqslant x}} q(\hat{\rho})^{-1/2+\varepsilon} \tag{6.11}$$

provided $\varepsilon < 1/2$. In view of Proposition 6.6 we deduce that

$$\vartheta^{\mathrm{im,symp}}_{\mathbb{Q},2}(x) \ll x^{1/2} \sum_{\substack{\langle\varrho\rangle \text{ im, orth} \\ q(\varrho)\leqslant x}} q(\varrho)^{-1/2+\varepsilon} + x^{1/2} \sum_{\substack{\langle\varrho\rangle \text{ ab, sd} \\ q(\varrho)\leqslant x}} q(\varrho)^{-1/2+\varepsilon}, \tag{6.12}$$

where in the first sum $\langle\varrho\rangle$ denotes an *arbitrary* isomorphism class of dihedral Artin representations of $\mathbb{Q}$ (not just one of the form $\langle\hat{\rho}\rangle$) and in the second sum $\langle\varrho\rangle$ denotes an *arbitrary* isomorphism class of two-dimensional abelian self-dual Artin representations of $\mathbb{Q}$ (not just one of the form $\langle\chi\oplus\chi'\rangle$ with distinct quadratic characters χ and χ' of $\mathrm{Gal}(\overline{\mathbb{Q}}/\mathbb{Q})$). Next we apply Abel summation to the two sums in (6.12). However since similar appeals to Abel summation will occur later on, the referee has suggested that it would be efficient to formulate a statement that covers all cases.

Proposition 6.9 *Fix $\mu,\nu > 0$ with $\mu \neq \nu$, and let $n(1), n(2), n(3), \ldots$ be a fixed sequence of nonnegative integers. If $\sum_{q\leqslant y} n(q)$ is $O(y^\nu)$ then $\sum_{q\leqslant y} n(q)q^{-\mu}$ is $O(y^{\nu-\mu})$ or $O(1)$ according as $\nu > \mu$ or $\nu < \mu$.*

The proof is elementary. To apply the proposition to the first sum on the right-hand side of (6.12), take $n(q)$ to be the number of isomorphism classes of dihedral Artin representations ϱ of $\mathbb{Q}$ with $q(\varrho) = q$. Since $\vartheta^{\mathrm{im,orth}}_{\mathbb{Q},2}(x) = O(x^{3/2})$ by Proposition 5.6, we deduce that

$$\sum_{\substack{\langle\varrho\rangle \text{ im, orth} \\ q(\varrho)\leqslant x}} q(\varrho)^{-1/2+\varepsilon} = O(x^{1+\varepsilon}). \tag{6.13}$$

The second sum in (6.12) is handled similarly: Theorem 3.6 gives $\vartheta^{\mathrm{ab,sd}}_{\mathbb{Q},2}(x) = O(x\log x)$, so we find that

$$\sum_{\substack{\langle\varrho\rangle \text{ ab, sd} \\ q(\varrho)\leqslant x}} q(\varrho)^{-1/2+\varepsilon} = O(x^{1/2+2\varepsilon}). \tag{6.14}$$

Inserting (6.13) and (6.14) in (6.12), we obtain Proposition 6.1.

7 Schur Covers

As before, if G is a finite subgroup of $\mathrm{GL}_n(\mathbb{C})$ then we attribute properties of the tautological representation $\iota : G \hookrightarrow \mathrm{GL}_n(\mathbb{C})$ to G itself. Thus G is *irreducible* or *self-dual* or *primitive* if these adjectives are applicable to ι. We denote the image of G in $\mathrm{PGL}_n(\mathbb{C})$ by $\mathrm{P}G$, and we write S_n and A_n for the symmetric and alternating groups on n letters. The following result is classical (cf. [38], Section 68).

Proposition 7.1 *Let G be a finite subgroup of $\mathrm{GL}_2(\mathbb{C})$. If G is irreducible and primitive then $\mathrm{P}G \cong A_4$, S_4, or A_5.*

Note that it is $\mathrm{P}G$ and not G itself which is isomorphic to A_4, S_4, or A_5. In fact A_4, S_4, and A_5 do not have faithful two-dimensional representations over $\mathbb{C}$, so none of them is isomorphic to G. But if G is self-dual then there is an analogous tripartition for G itself (cf. [39], p. 131, Lemma 1). To state it, put $\widetilde{A}_4 = \mathrm{SL}_2(\mathbb{F}_3)$ and $\widetilde{A}_5 = \mathrm{SL}_2(\mathbb{F}_5)$, and let $\widetilde{S}_4$ be the subgroup of $\mathrm{SL}_2(\mathbb{F}_9)$ generated by $\mathrm{SL}_2(\mathbb{F}_3)$ and $i\eta$, where $i \in \mathbb{F}_9$ is a fixed square root of -1 and

$$\eta = \begin{pmatrix} -1 & 0 \\ 0 & 1 \end{pmatrix}.$$

Since η normalizes $\mathrm{SL}_2(\mathbb{F}_3)$ and $(i\eta)^2 = -1$ we have $\widetilde{S}_4 = \mathrm{SL}_2(\mathbb{F}_3) \cup (i\eta)\mathrm{SL}_2(\mathbb{F}_3)$. We denote the center of a group G by $Z(G)$.

Proposition 7.2 *Let G be a finite subgroup of $\mathrm{GL}_2(\mathbb{C})$. If G is irreducible, primitive, and self-dual then $G \cong \widetilde{A}_4$, $\widetilde{S}_4$, or $\widetilde{A}_5$. Furthermore $G \subset \mathrm{SL}_2(\mathbb{C})$ and $Z(G) = \{\pm 1\}$.*

Conversely, these three groups do all have faithful two-dimensional irreducible primitive self-dual representations over $\mathbb{C}$. In fact up to isomorphism $\widetilde{A}_4$ has exactly one such representation, while $\widetilde{S}_4$ and $\widetilde{A}_5$ have exactly two. These facts can all be read from a character table (see for example [15], p. 44 or [16], p. 89 in the case of $\widetilde{A}_4$; [15], p. 43 in the case of $\widetilde{S}_4$; and [16], p. 140 in the case of $\widetilde{A}_5$). On the other hand, to derive Proposition 7.2 from Proposition 7.1, we will use the theory of Schur covers, a few elements of which will now be recalled. All of the results about Schur covers to be quoted here can be found in [17], and some of them are also usefully summarized in [15]. Given a group G we denote its commutator subgroup by G', and we say that G is *perfect* if $G = G'$.

Let G and J be finite groups. We say that G is a *representation group* of J if there is a subgroup $C \subset Z(G) \cap G'$ such that $C \cong H^2(J, \mathbb{C}^\times)$ and

$G/C \cong J$. The group $H^2(J, \mathbb{C}^\times)$ is the *Schur multiplier* of J, and a representation group of J is also called a *Schur cover* of J. Every finite group has at least one Schur cover, and up to isomorphism it has only finitely many. Furthermore, if the orders of $H^2(J, \mathbb{C}^\times)$ and J/J' are relatively prime – in particular, if J is perfect — then the isomorphism class of a Schur cover of J is unique.

In keeping with tradition we have referred to G itself as a Schur cover of J, but it is also convenient to apply the term to any epimorphism $\varphi : G \to J$ with kernel C. In practice G and φ are largely interchangeable, for if $Z(J)$ is trivial (as it will be in the cases of primary interest to us) then G determines C: In fact $C = Z(G)$, because $Z(G)$ has trivial image in J and is therefore contained in C. Furthermore, the fundamental property of a "representation group" (and the property which explains the terminology itself) is that projective representations of J lift to genuine representations of G, and we claim that the validity of this property is unaffected by the choice of φ. To justify the claim, let us state the property at issue more precisely: If π is a projective representation of J then there exists a representation ρ of G such that $\mathrm{P}\rho \cong \pi \circ \varphi$, where $\mathrm{P}\rho$ denotes the projective representation determined by ρ. Now if $\psi : G \to J$ is another epimorphism with kernel $Z(G)$ then $\psi = \alpha \circ \varphi$ for some automorphism α of J, and as $\pi \circ \alpha$ is a projective representation of J there exists a representation ρ' of G such that $\mathrm{P}\rho' = (\pi \circ \alpha) \circ \varphi$. Then $\mathrm{P}\rho' = \pi \circ \psi$.

While the lifting property will be used in Section 8, our immediate concern is simply to identify the Schur covers of A_4, S_4, and A_5. It is actually more instructive to consider A_n and S_n for arbitrary n. First A_n: It is known that $H^2(A_n, \mathbb{C}^\times)$ is trivial if $n \leqslant 3$, cyclic of order six if $n = 6$ or 7, and cyclic of order two otherwise. Furthermore A_n is perfect for $n \geqslant 5$, while for $n = 4$ we have $|A_4/A_4'| = 3$. It follows that for all $n \geqslant 1$ the groups $H^2(A_n, \mathbb{C}^\times)$ and A_n/A_n' are of relatively prime order, whence a Schur cover of A_n is unique up to isomorphism. If $n \geqslant 4$ and $n \neq 6, 7$ then a Schur cover of A_n is typically denoted $\widetilde{A}_n$ or $\widehat{A}_n$. Granted, if $n = 4$ or 5 then $\widetilde{A}_n$ has already been assigned a meaning, but we will check in a moment that $\mathrm{SL}_2(\mathbb{F}_3)$ and $\mathrm{SL}_2(\mathbb{F}_5)$ are indeed Schur covers of A_4 and A_5.

The situation for S_n is as follows: $H^2(S_n, \mathbb{C}^\times)$ is trivial for $n \leqslant 3$ but cyclic of order two for all $n \geqslant 4$ without exception. Furthermore, if $n \geqslant 4$ and $n \neq 6$ then up to isomorphism there are exactly *two* Schur covers of S_n. In the literature, the two Schur covers are variously denoted $\widetilde{S}_n$ and $\widehat{S}_n$ (cf. [15], p. 23), or S_n^* and S_n^{**} (cf. [17], p. 523), or 2^+S_n and 2^-S_n (cf. [8], p. xxiii), the second member of each pair being characterized by the fact that the preimages of the transpositions of S_n have order two. (Warning: Although we follow [15] in distinguishing between $\widetilde{S}_n$ and $\widehat{S}_n$, the opposite convention is also in use; see e. g. the characterization of $\widetilde{S}_4$ in [22], p. 199

and of $\widetilde{S}_n$ in [34], p. 97.) If $n \geqslant 4$ and $n \neq 6, 7$ then the respective inverse images of A_n under $\widetilde{S}_n \to S_n$ and $\widehat{S}_n \to S_n$ are Schur covers of A_n and are therefore isomorphic, whence we obtain the notations $\widetilde{A}_n$ and $\widehat{A}_n$ already mentioned.

The next proposition will justify our original definition of $\widetilde{A}_4$, $\widetilde{S}_4$, and $\widetilde{A}_5$ and will show in addition that we may take $\widehat{S}_4 = \mathrm{GL}_2(\mathbb{F}_3)$. By an *involution* in a group we mean as usual an element of order two (which of course is central if unique).

Lemma 7.3 *Let G and J be finite groups. Assume:*

(i) *$H^2(J, \mathbb{C}^\times)$ has order two, and J' has even order.*

(ii) *G has a unique involution, and if C is the subgroup generated by the involution then $G/C \cong J$.*

Then G is a Schur cover of J.

Proof The only point to be checked is that $C \subset G'$. As C is the unique subgroup of order two in G it is contained in every subgroup of even order, and G' is of even order because its quotient J' is.

□

Proposition 7.4 *In each of the following cases, G is a Schur cover of J, and $J \cong G/C$ with $C = Z(G) = \{\pm 1\}$:*

- *$J = A_4$ and $G = \mathrm{SL}_2(\mathbb{F}_3)$.*
- *$J = A_5$ and $G = \mathrm{SL}_2(\mathbb{F}_5)$.*
- *$J = S_4$ and $G = \mathrm{SL}_2(\mathbb{F}_3) \cup (i\eta)\mathrm{SL}_2(\mathbb{F}_3)$.*
- *$J = S_4$ and $G = \mathrm{GL}_2(\mathbb{F}_3)$.*

Furthermore, -1 is the unique involution in G in the first three cases, but every transposition in S_4 lifts to an involution in $\mathrm{GL}_2(\mathbb{F}_3)$.

Proof Over any field F the scalar matrix -1 is the unique involution in $\mathrm{SL}_2(F)$, and we have identifications $A_4 \cong \mathrm{PSL}_2(\mathbb{F}_3)$ and $A_5 \cong \mathrm{SL}_2(\mathbb{F}_4)$ ($\cong \mathrm{PSL}_2(\mathbb{F}_5)$) by virtue of the transitive action of $\mathrm{PGL}_2(F)$ on the projective line $\mathbf{P}^1(F)$. The first two cases of the proposition now follow from the lemma.

To justify the fourth case we note that the identification $S_4 \cong \mathrm{PGL}_2(\mathbb{F}_3)$ is again a reflection of the action of $\mathrm{PGL}_2(F)$ on $\mathbf{P}^1(F)$. Since the subgroup $C = \{\pm 1\}$ of $\mathrm{GL}_2(\mathbb{F}_3)$ is both central and contained in $\mathrm{GL}_2(\mathbb{F}_3)' = \mathrm{SL}_2(\mathbb{F}_3)$, we conclude directly from the definition that $\mathrm{GL}_2(\mathbb{F}_3)$ is a Schur cover of S_4. Now when we identify $\mathrm{PGL}_2(\mathbb{F}_3)$ with S_4 via its action on $\mathbf{P}^1(F)$, the image of the matrix η in $\mathrm{PGL}_2(\mathbb{F}_3)$ maps to the transposition in S_4 interchanging the points $[1:1]$ and $[-1:1]$ of $\mathbf{P}^1(\mathbb{F}_3)$. Since the transpositions form a conjugacy class of S_4, we deduce that *every* transposition in S_4 lifts to an involution in $\mathrm{GL}_2(\mathbb{F}_3)$.

Finally, in the third case $G \subset \mathrm{SL}_2(\mathbb{F}_9)$, and consequently -1 is the unique involution in G. Hence to conclude from the lemma that G is a Schur cover of S_4 it suffice to see that $G/C \cong S_4$, or equivalently that $G/C \cong \mathrm{PGL}_2(\mathbb{F}_3)$. But $\mathrm{GL}_2(\mathbb{F}_3) = \mathrm{SL}(2, \mathbb{F}_3) \cup \eta \mathrm{SL}_2(\mathbb{F}_3)$, and η and $i\eta$ have the same image in $\mathrm{PGL}_2(\mathbb{F}_9)$. Thus the identity embedding of $\mathrm{PGL}_2(\mathbb{F}_3)$ into $\mathrm{PGL}_2(\mathbb{F}_9)$ is an isomorphism of $\mathrm{PGL}_2(\mathbb{F}_3)$ onto G/C.

□

Proof (of Proposition 7.2) By Proposition 7.1, PG is A_4, S_4, or A_5. As already noted, none of these groups has a faithful irreducible two-dimensional representation, so G intersects the group of scalar matrices in $\mathrm{GL}_2(\mathbb{C})$ nontrivally. On the other hand, G is self-dual, so the only scalar matrices which can belong to G are ± 1. It follows that $-1 \in G$, that the group $C = \{\pm 1\}$ coincides with $Z(G)$ (by Schur's lemma), and that $G/C \cong PG$. Now G is symplectic, for otherwise it is orthogonal, and a two-dimensional irreducible orthogonal representation is monomial (because $\mathrm{O}_2(\mathbb{R})$ contains the abelian subgroup $\mathrm{SO}_2(\mathbb{R})$ with index two). Thus $G \subset \mathrm{SL}_2(\mathbb{C})$. As -1 is the only involution in $\mathrm{SL}_2(\mathbb{C})$ and *a fortiori* the only involution in G, the lemma shows that G is a Schur cover of PG. Proposition 7.2 now follows from Proposition 7.4 and the fact that a Schur cover of A_4 or A_5 is unique up to isomorphism, as is a Schur cover of S_4 with only one involution.

□

8 Primitive Representations

As already mentioned, some of the arguments used to bound the quaternionic term in Section 6 will now find application in the primitive case. We take the class $\mathcal{A}$ of Section 6 to be the collection of two-dimensional irreducible self-dual primitive Artin representations of $\mathbb{Q}$. That $\mathcal{A}$ is symplectic and closed under quadratic twists follows from Proposition 7.2. Hence Proposition 6.8 gives

$$\vartheta^{\mathrm{ip,sd}}_{\mathbb{Q},2}(x) \ll x^{1/2} \sum_{\substack{[\rho]\in\mathcal{E} \\ q(\rho)\leqslant x}} q(\rho)^{-1/2+\varepsilon}, \tag{8.1}$$

where $\mathcal{E}$ is the set of projective equivalence classes of $\mathcal{A}$. Although the validity of (8.1) depends on the choice of a particular representative ρ for the equivalence class $[\rho]$, in the arguments that follow no further use will be made of this choice. Our goal is the following bound:

Proposition 8.1 *Fix $\gamma < 1/60$. Then $\vartheta^{\mathrm{ip,sd}}_{\mathbb{Q},2}(x) = O(x^{2-\gamma})$, where the implicit constant depends on γ.*

Our strategy for proving Proposition 8.1 is to replace conductors by discriminants in (8.1) and then to appeal to the results of Bhargava and of Bhargava, Cojocaru and Thorne. Consider the fixed field L of the kernel of $\mathrm{P}\rho$. By Proposition 7.1, $\mathrm{Gal}(L/\mathbb{Q})$ is isomorphic to one of A_4, S_4, and A_5, and we write m for the degree of the permutation group in question: thus $m = 4$ in the first two cases and $m = 5$ in the third. In the following proposition K is any subfield of L with $[K : \mathbb{Q}] = m$. While the choice of K is arbitrary, L is the normal closure of K over $\mathbb{Q}$ for every possible choice.

Proposition 8.2 *$d_K \leqslant cq(\rho)^{(m-1)/2}$ with an absolute constant $c > 1$.*

Proof A standard bound for wild ramification (cf. [33], p. 127, Proposition 2) gives

$$d_K \leqslant c \prod_{\substack{p \mid d_K \\ p > m}} p^{m-1}. \tag{8.2}$$

with $c = 2^{11}3^7$ if $m = 4$ and $c = 2^{14}3^9 5^9$ if $m = 5$. (Thus we may take $c = 2^{14}3^9 5^9$ in all cases.) On the other hand, let M be the fixed field of the kernel of ρ itself. Then Proposition 7.2 implies that $\mathrm{Gal}(M/\mathbb{Q})$ is $\widetilde{A}_4$, $\widetilde{S}_4$, or $\widetilde{A}_5$ according as $\mathrm{Gal}(L/\mathbb{Q})$ is A_4, S_4, or A_5. Thus if $p > m$ then p does not divide the order of the image of ρ, and consequently the restriction of ρ to an inertia group I at p factors through the tame quotient of I. Hence $\mathrm{ord}_p\, q(\rho)$ is $\dim(V/V^I)$, where V is the space of ρ and V^I the subspace of inertial invariants. Now if $p|q(\rho)$ then V/V^I has dimension 1 or 2, but if the dimension is 1 then V is the direct sum of a line on which I acts trivially and a line on which it acts nontrivially, contradicting the fact that $\det \rho = 1$ (Proposition 7.2 again). Therefore

$$q(\rho) \geqslant \prod_{\substack{p \mid q(\rho) \\ p > m}} p^2, \tag{8.3}$$

Since L is the normal closure of K, every prime dividing d_K divides $q(\rho)$, whence the proposition follows from (8.2) and (8.3).

□

Remarks 8.3

1) The inequalities (8.2) and (8.3) are both deduced from the fact that one side of the inequality is *divisible* by the other.
2) Using the fact that A_4 has no elements of order > 3, one finds that $d_K \leqslant cq(\rho)$ when $\mathrm{Gal}(L/\mathbb{Q}) \cong A_4$. However this improvement in Proposition 8.2 does not lead to an improvement in Proposition 8.1, because the latter combines all three cases.

Proposition 8.4 *Let L be a finite Galois extension of $\mathbb{Q}$ such that* $\mathrm{Gal}(L/\mathbb{Q})$ *is isomorphic to A_4, S_4, or A_5. Then the number of elements $[\rho] \in \mathcal{E}$ such that L is the fixed field of* Ker $(\mathrm{P}\rho)$ *is bounded by an absolute constant.*

Proof Put $J = \mathrm{Gal}(L/\mathbb{Q})$. We may assume that there is a quadratic extension M of L, Galois over $\mathbb{Q}$, such that the group $G = \mathrm{Gal}(M/\mathbb{Q})$ is isomorphic to $\widetilde{A}_4$, $\widetilde{S}_4$, or $\widetilde{A}_5$ according as J is isomorphic to A_4, S_4, or A_5. Indeed if there exists $[\rho] \in \mathcal{E}$ such that L is the fixed field of Ker $(\mathrm{P}\rho)$ then we may take M to be the fixed field of Ker (ρ), and if no such $[\rho]$ exists then there is nothing to prove. Now up to isomorphism, there are exactly three two-dimensional irreducible representations φ of G if $G \cong \widetilde{A}_4$ or $\widetilde{S}_4$ and exactly two if $G \cong \widetilde{A}_5$. (Note that we are not requiring φ to be faithful or self-dual or primitive.) Let us declare φ and φ' to be equivalent if $\varphi' \cong \varphi \otimes \chi$ for some one-dimensional character χ of G. Then there is exactly one equivalence class if $G \cong \widetilde{A}_4$ and there are exactly two if $G \cong \widetilde{S}_4$ or $\widetilde{A}_5$. So the proposition will follow (with the absolute constant equal to 2) if we define an injective map $[\rho] \mapsto [\varphi]$ from the set of $[\rho] \in \mathcal{E}$ such that L is the fixed field of Ker $(\mathrm{P}\rho)$ to the set of equivalence classes $[\varphi]$ as above.

Given $[\rho]$, view $P\rho$ as a projective representation of J. Since G is a Schur cover of J we can lift $P\rho$ to a genuine representation φ of G. It is immediately verified that the equivalence class $[\varphi]$ of φ is uniquely determined by $[\rho]$ and that the map $[\rho] \mapsto [\varphi]$ is injective.

□

Reviewing the preceding paragraphs, we see that we have defined a function $[\rho] \mapsto K$: Given $[\rho] \in \mathcal{E}$, we let L be the fixed field of Ker $(\mathrm{P}\rho)$ and then we choose a subfield $K \subset L$ with $[K : \mathbb{Q}] = m$. Since L is determined by K (indeed L is the normal closure of K) Proposition 8.4 shows that the number of preimages $[\rho]$ of K is bounded by an absolute constant. Thus Proposition 8.2 gives

$$\sum_{\substack{[\rho]\in\mathcal{E} \\ q(\rho)\leqslant x}} q(\rho)^{-1/2+\varepsilon} \ll \sum_{\substack{[K:\mathbb{Q}]=4 \\ d_K\leqslant cx^{3/2}}} d_K^{-1/3+2\varepsilon/3} + \sum_{\substack{[K:\mathbb{Q}]=5 \\ \mathrm{Gal}(L/\mathbb{Q})\cong A_5 \\ d_K\leqslant cx^2}} d_K^{-1/4+\varepsilon/2} \qquad (8.4)$$

for $0 < \varepsilon < 1/2$, where the first sum on the right-hand side runs over number fields K with $[K : \mathbb{Q}] = 4$ and $d_K \leqslant cx^{3/2}$, and the second sum runs over K with $[K : \mathbb{Q}] = 5$, $d_K \leqslant cx^2$, and $\mathrm{Gal}(L/\mathbb{Q}) \cong A_5$, L being the normal closure of K. Of course the first sum could be confined to K such that $\mathrm{Gal}(L/\mathbb{Q}) \cong A_4$ or $\mathrm{Gal}(L/\mathbb{Q}) \cong S_4$, but (8.4) will suffice as it stands.

We now apply Proposition 6.9 (i. e. Abel summation) to the first sum on the right-hand side of (8.4). Since $\eta_{\mathbb{Q},4}(x) = O(x)$ by [3], we deduce that

$$\sum_{\substack{[K:\mathbb{Q}]=4 \\ d_K \leqslant cx^{3/2}}} d_K^{-1/3+2\varepsilon/3} = O(x^{1+\varepsilon}). \tag{8.5}$$

The second sum on the right-hand side of (8.4) can be treated in the same way: By [5] we have $\eta_{\mathbb{Q},5}^{A_5}(x) = O(x^{1-\beta})$ for any $\beta < 1/120$, so we obtain

$$\sum_{\substack{[K:\mathbb{Q}]=5 \\ \mathrm{Gal}(L/\mathbb{Q})\cong A_5 \\ d_K \leqslant cx^2}} d_K^{-1/4+\varepsilon/2} \ll O(x^{3/2-2\beta+\varepsilon}). \tag{8.6}$$

Inserting (8.5) and (8.6) in (8.4) and then concatenating the result with (8.1), we obtain Proposition 8.1.

9 Monomial Representations Revisited

Assembling our estimates for the three terms on the right-hand side of (0.8), we see that Theorem 3.6, Proposition 6.2, and Proposition 8.1 together imply the upper bound for $\vartheta_{\mathbb{Q},2}^{\mathrm{sd}}(x)$ claimed in (0.10). On the other hand, Theorem 3.3 gives the lower bound for $\vartheta_{\mathbb{Q},2}^{\mathrm{ab}}(x)$ in (0.11), so we conclude that indeed $\lim_{x\to\infty} \vartheta^{\mathrm{sd}}(x)/\vartheta(x) = 0$ for $F = \mathbb{Q}$ and $n = 2$, as asserted in the introduction. We will now show that the limit of $\vartheta^{\mathrm{sd}}(x)/\vartheta^{\mathrm{irr}}(x)$ and *a fortiori* of $\vartheta^{\mathrm{irr,sd}}(x)/\vartheta^{\mathrm{irr}}(x)$ is 0 also. Since $\vartheta^{\mathrm{irr}}(x) \geqslant \vartheta^{\mathrm{im}}(x)$ it will suffice to show that

$$\vartheta_{\mathbb{Q},2}^{\mathrm{im}}(x) \gg x^2. \tag{9.1}$$

I do not know how to replace (9.1) by an asymptotic equality.

To prove (9.1), fix an imaginary quadratic field K, and let $\vartheta_{\mathbb{Q},2}^{K}(x)$ be the number of isomorphism classes of two-dimensional monomial Artin representations of $\mathbb{Q}$ which are induced from K and of absolute conductor $\leqslant x$. Write $\vartheta_{\mathbb{Q},2}^{\mathrm{im},K}(x)$ and $\vartheta_{\mathbb{Q},2}^{\mathrm{ab},K}(x)$ for the number of such classes of irreducible representations and abelian representations respectively. Then

$$\vartheta_{\mathbb{Q},2}^{\mathrm{im},K}(x) = \vartheta_{\mathbb{Q},2}^{K}(x) - \vartheta_{\mathbb{Q},2}^{\mathrm{ab},K}(x). \tag{9.2}$$

We shall prove that

$$\vartheta^{\mathrm{ab},K}_{\mathbb{Q},2}(x) \sim 18x/(d_K\pi^4) \tag{9.3}$$

and then deduce that

$$\vartheta^{\mathrm{im},K}_{\mathbb{Q},2}(x) \sim cx^2 \tag{9.4}$$

with a constant $c > 0$ depending on K. Since $\vartheta^{\mathrm{im}}_{\mathbb{Q},2}(x) \geqslant \vartheta^{\mathrm{im},K}_{\mathbb{Q},2}(x)$ the lower bound (9.1) will then follow. In principle we would get a better result in (9.1) if instead of fixing K we were to sum (9.4) over all K, taking account of any duplications. However even after summing over K we would not be able to replace (9.1) by an asymptotic formula, because we do not have the analogue of (9.4) for real quadratic fields.

To prove (9.3), we observe that the two-dimensional abelian Artin representations of $\mathbb{Q}$ induced from K are precisely the representations $\rho \cong \chi \oplus \chi \cdot \mathrm{sign}_{K/\mathbb{Q}}$, where χ is an arbitrary one-dimensional character of $\mathrm{Gal}(\overline{\mathbb{Q}}/\mathbb{Q})$ and $\mathrm{sign}_{K/\mathbb{Q}}$ is the character with kernel $\mathrm{Gal}(\overline{\mathbb{Q}}/K)$. As $q(\rho) = q(\chi)^2 d_K$ we have

$$\vartheta^{\mathrm{ab},K}_{Q,2}(x) = \sum_{q(\chi)^2 \leqslant x/d_K} 1,$$

where the sum runs over all χ satisfying the stated inequality. Recognizing this sum as $\vartheta_{\mathbb{Q},1}(\sqrt{x/d_K})$, we obtain (9.3) from the corollary to Proposition 2.1.

It remains to prove (9.4). The Artin representations of $\mathbb{Q}$ counted by $\vartheta^{K}_{\mathbb{Q},2}(x)$ are precisely the representations of the form $\rho \cong \mathrm{ind}_{K/\mathbb{Q}}\xi$, where ξ runs over one-dimensional characters of $\mathrm{Gal}(\overline{\mathbb{Q}}/K)$ such that $d_K q(\xi) \leqslant x$. Furthermore, if ρ is irreducible then there are precisely two characters ξ such that $\mathrm{ind}_{K/\mathbb{Q}}\xi \cong \rho$, while if ρ is abelian then ρ uniquely determines ξ. Therefore (9.2) becomes

$$\vartheta^{\mathrm{im},K}_{\mathbb{Q},2}(x) = (1/2)\vartheta_{K,1}(x/d_K) - (1/2)\vartheta^{\mathrm{ab},K}_{\mathbb{Q},2}(x). \tag{9.5}$$

Now put

$$c = (\pi/(2d_K^{5/2})) \cdot (h_K/(w_K\zeta_K(2)))^2, \tag{9.6}$$

where $\zeta_K(s)$, h_K, and w_K are as usual the Dedekind zeta function, class number, and number of roots of unity in K. We obtain (9.4) with c as in (9.6) by combining (9.5) and (9.3) with the following assertion:

Theorem 9.1 $\vartheta_{K,1}(x) \sim (\pi/\sqrt{d_K})(h_K x/(w_K\zeta_K(2)))^2$.

Proof Given a nonzero integral ideal $\mathfrak{q}$ of K, put $\varphi_K(\mathfrak{q}) = |(\mathcal{O}_K/\mathfrak{q})^\times|$ as before, and set $\mu_K(\mathfrak{q}) = (-1)^t$ if $\mathfrak{q}$ is the product of exactly t distinct prime ideals of K and $\mu_K(\mathfrak{q}) = 0$ otherwise. Also write $h_K^*(\mathfrak{q})$ for the number of primitive ray class characters of K of conductor $\mathfrak{q}$, so that

$$\vartheta_{K,1}(x) = \sum_{\mathbf{N}\mathfrak{q} \leqslant x} h_K^*(\mathfrak{q}) \tag{9.7}$$

and

$$h_K^*(\mathfrak{q}) = \sum_{\mathfrak{q}'|\mathfrak{q}} \mu_K(\mathfrak{q}/\mathfrak{q}')h_K(\mathfrak{q}'). \tag{9.8}$$

Let $w_K(\mathfrak{q})$ the number of roots of unity in K which are congruent to 1 modulo $\mathfrak{q}$. Since K has no real embeddings, the narrow ray class number $h_K^{\mathrm{nar}}(\mathfrak{q})$ is indistinguishable from the wide ray class number $h_K(\mathfrak{q})$, and consequently

$$h_K(\mathfrak{q}) = h_K \cdot \varphi_K(\mathfrak{q}) \cdot (w_K(\mathfrak{q})/w_K) \tag{9.9}$$

by (4.6). Combining (9.8) and (9.9), we have

$$h_K^*(\mathfrak{q}) = (h_K/w_K) \sum_{\mathfrak{q}'|\mathfrak{q}} \mu_K(\mathfrak{q}/\mathfrak{q}')\varphi_K(\mathfrak{q}')w_K(\mathfrak{q}'). \tag{9.10}$$

Put $\psi_K(\mathfrak{q}) = \sum_{\mathfrak{q}'|\mathfrak{q}} \mu_K(\mathfrak{q}/\mathfrak{q}')\varphi_K(\mathfrak{q}')$. It is convenient to rewrite (9.10) in the form

$$h_K^*(\mathfrak{q}) = (h_K/w_K)\psi_K(\mathfrak{q}) + O(1) \tag{9.11}$$

with $O(1) = (h_K/w_K) \sum_{\mathfrak{q}'|\mathfrak{q}} \mu_K(\mathfrak{q}/\mathfrak{q}')\varphi_K(\mathfrak{q}')(w_K(\mathfrak{q}') - 1)$.

The expression which we have denoted $O(1)$ is indeed bounded by a constant depending only on K, because $w_K(\mathfrak{q}') = 1$ unless $\mathfrak{q}'|6\mathcal{O}_K$. Hence by substituting (9.11) in (9.7) we obtain

$$\vartheta_{K,1}(x) = (h_K/w_K) \sum_{\mathbf{N}\mathfrak{q} \leqslant x} \psi_K(\mathfrak{q}) + O(\sum_{\mathbf{N}\mathfrak{q} \leqslant x} 1). \tag{9.12}$$

Denote the first and second sums on the right-hand side of (9.12) by Σ_1 and Σ_2:

$$\vartheta_{K,1}(x) = (h_K/w_K)\Sigma_1 + O(\Sigma_2). \tag{9.13}$$

Then Σ_2 is the summatory function of $\zeta_K(s)$, and consequently Proposition 1.1 gives $\Sigma_2 \sim \lambda_K x$, where λ_K is the residue of $\zeta_K(s)$ at $s = 1$. In particular, $\Sigma_2 = O(x)$. On the other hand, if we redo the proof of Proposition 2.1 with φ and ψ replaced by φ_K and ψ_K and with the rational prime p replaced by a prime ideal $\mathfrak{p}$ of K or by $\mathbf{N}\mathfrak{p}$, as appropriate, then we find that Σ_1 is the summatory function of $\zeta_K(s-1)/\zeta_K(s)^2$. Hence another appeal to Proposition 1.1 gives

$$\Sigma_1 \sim \lambda_K/(2\zeta_K(2)^2) \cdot x^2. \tag{9.14}$$

Since $\Sigma_2 = O(x)$, it follows from (9.12) and (9.14) that

$$\vartheta_{K,1}(x) \sim \lambda_K h_k x^2/(2w_K\zeta_K(2)^2).$$

Substituting $\lambda_K = (2\pi)h_K/(w_K\sqrt{d_K})$, we obtain the stated asymptotic formula.

□

10 Malle's Conjecture

Only a weak form of Malle's conjecture will be needed here, but for the sake of completeness we first state the conjecture in its original form: Given a number field F, an integer $m \geqslant 2$, and a transitive subgroup G of S_m, there are constants a, b, and c satisfying $0 < a \leqslant 1$, $b \geqslant 1$, and $c > 0$ such that

$$\eta^G_{F,m} \sim cx^a(\log x)^{b-1}. \tag{10.1}$$

What distinguishes Malle's conjecture from previous hypotheses of this type (cf. Cohen [6]) is that explicit values are proposed for a and b, as we now describe.

The value of a depends only on G, not on F; Malle denotes it $a(G)$. To define $a(G)$ we recall that the *index* of an element $g \in G$ is the quantity

$$\operatorname{ind}(g) = m - \operatorname{cyc}(g),$$

where $\operatorname{cyc}(g)$ is the number of cycles in the exhaustive disjoint cycle decomposition of g. Here "exhaustive" means that cycles of length 1 are included; for example if $g = 1$ then we write $g = (1)(2)\cdots(m)$ and find that $\operatorname{cyc}(g) = m$ and $\operatorname{ind}(g) = 0$, while if $g \neq 1$ then $\operatorname{ind}(g) > 0$. We put

$$\operatorname{ind}(G) = \min_{\substack{g\in G\\ g\neq 1}} \operatorname{ind}(g)$$

and $a(G) = \operatorname{ind}(G)^{-1}$.

The quantity b depends on F as well as G. The function $g \mapsto \operatorname{ind}(g)$ is constant on conjugacy classes of G, so we can speak of the index of a conjugacy class, and we let $\mathcal{C}$ be the set consisting of all conjugacy classes C such that $\operatorname{ind}(C) = \operatorname{ind}(G)$. We define an action of $\operatorname{Gal}(\overline{F}/F)$ on $\mathcal{C}$ by setting $\sigma \cdot C = C^{\omega(\sigma)}$ for $\sigma \in \operatorname{Gal}(\overline{F}/F)$ and $C \in \mathcal{C}$, where $\omega : \operatorname{Gal}(\overline{F}/F) \to \widehat{\mathbb{Z}}^\times$ is the cyclotomic character ($\widehat{\mathbb{Z}}$ being the ring of adelic integers) and $C^{\omega(\sigma)}$ is the conjugacy class consisting of the elements $g^{\omega(\sigma)}$ with $g \in C$. If one prefers one can take ω to be the mod-e cyclotomic character $\operatorname{Gal}(\overline{F}/F) \to (\mathbb{Z}/e\mathbb{Z})^\times$ for any positive integer e divisible by the order of every element of G. In any case, b is the number of orbits of $\mathrm{Gal}(\overline{F}/F)$ on $\mathcal{C}$.

A counterexample of Klüners [21] shows that with these definitions Malle's original conjecture (10.1) is false: If $F = \mathbb{Q}$, $n = 6$, and

$$G = ((\mathbb{Z}/3\mathbb{Z}) \times (\mathbb{Z}/3\mathbb{Z})) \rtimes (\mathbb{Z}/2\mathbb{Z})$$

(embedded in S_6 by identifying the first factor of $\mathbb{Z}/3\mathbb{Z}$ with $\langle(123)\rangle$, the second with $\langle(456)\rangle$, and $\mathbb{Z}/2\mathbb{Z}$ with $\langle(14)(25)(36)\rangle$) then $a = 1/2$ and $b = 1$, but Klüners shows that the left-hand side of (10.1) is $\gg x^{1/2} \log x$. However if we state Malle's conjecture in the weaker form

$$\eta^G_{F,m}(x) \ll x^{a(G)+\varepsilon} \tag{10.2}$$

for all $\varepsilon > 0$, where the implicit constant depends on F, G, and ε, then the conjecture has so far proved unassailable, and henceforth it is (10.2) to which reference will be made. We shall call (10.2) *the weak form of Malle's conjecture.*

At this juncture we change perspective slightly by viewing G as an *abstract* group of order $m \geqslant 2$. If we wish to regard G as a permutation group then we do so via the regular representation, so that the associated embedding $G \hookrightarrow S_m$ is uniquely determined up to conjugacy in S_m. Now fix an integer $n \geqslant 2$ and let $\vartheta^G_{F,n}(x)$ be the number of isomorphism classes of n-dimensional irreducible Artin representations ρ of F with image isomorphic to G and $q(\rho) \leqslant x$. We would like to compare $\vartheta^G_{F,n}(x)$ with $\eta^G_{F,m}(x)$. In Section 11 we will prove the inequality

$$d_{L/F} \leqslant q(\rho)^{|G|-n(n-1)}, \tag{10.3}$$

where L is the fixed field of the kernel of ρ and ρ is as before an n-dimensional irreducible Artin representation of F with image isomorphic to G. Granting (10.3), and making the trivial remark that if $q(\rho) \leqslant x$ then $q(\rho)^{|G|-n(n-1)} \leqslant x^{|G|-n(n-1)}$, we see that

$$\vartheta^G_{F,n}(x) \leqslant \mathrm{i}_n(G) \cdot \eta^G_{F,m}(x^{|G|-n(n-1)}), \tag{10.4}$$

where $\mathrm{i}_n(G)$ is the number of isomorphism classes of faithful n-dimensional irreducible complex representations of G.

Proposition 10.1 *Let p be the smallest prime divisor of $|G|$, and fix*

$$\gamma < pn(n-1)/((p-1)|G|).$$

If the weak form of Malle's conjecture holds then

$$\vartheta^G_{F,n}(x) \ll x^{p/(p-1)\,-\,\gamma},$$

where the implied constant depends on F, G, n, and γ.

Proof Since G is a permutation group via the regular representation, we have $\mathrm{cyc}(g) = |G|/|g|$ for $g \in G$, where $|g|$ is the order of g. Thus $\mathrm{ind}(G) = |G| - |G|/p$ and $a(G) = p/((p-1)|G|)$. Inserting this value in (10.2) and then combining (10.2) with (10.4) gives the stated estimate. □

Next we recall a theorem of Jordan: If G is a finite subgroup of $\mathrm{GL}_n(\mathbb{C})$ then G has an abelian normal subgroup of index bounded by a constant depending only on n. We denote the optimal choice of this constant $j(n)$. The value $j(2) = 60$ is classical, and the value of $j(n)$ for arbitrary n was determined by Collins [7]. For example if $n \geqslant 71$ then $j(n) = (n+1)!$.

Proposition 10.2 *Let G be a finite irreducible self-dual subgroup of* $\mathrm{GL}_n(\mathbb{C})$. *If G is primitive then $|G| \leqslant 2j(n)$.*

Proof Let $\iota : G \hookrightarrow \mathrm{GL}_n(\mathbb{C})$ be the tautological representation and A an abelian normal subgroup of G of index $\leqslant j(n)$. Then $\iota|A$ is a direct sum of one-dimensional characters of A. Let χ be a one-dimensional character of A occurring in $\iota|A$. If the multiplicity of χ in $\iota|A$ is $< n$ then the subgroup of G stabilizing χ is a proper subgroup from which ι is induced, contradicting the primitivity of ι. Hence $\iota|A = \chi^{\oplus n}$, and therefore

$$|G| = [G : A]|\mathrm{Ker}\ \chi|.$$

But ι is self-dual, hence so is $\iota|A$. Thus $\chi^2 = 1$ and consequently $|\mathrm{Ker}\ \chi| \leqslant 2$. □

Finally we deduce a conditional bound on $\vartheta^{\mathrm{ip,sd}}_{F,n}(x)$:

Proposition 10.3 *Fix $\gamma < n(n-1)/j(n)$. If the weak form of Malle's conjecture holds then*

$$\vartheta^{\mathrm{ip,sd}}_{F,n}(x) \ll x^{2-\gamma},$$

where the implied constant depends on F, n, and γ.

Proof The only irreducible self-dual representation of a group of odd order is the one-dimensional trivial representation. Hence it follows from Proposition 10.2 that

$$\vartheta^{\mathrm{ip,sd}}_{F,n}(x) \leqslant \sum_{\substack{|G| \leqslant 2j(n) \\ |G|\ \text{even}}} \vartheta^{G}_{F,n}(x), \tag{10.5}$$

where the sum on the right-hand side runs over a set of representatives for the distinct isomorphism classes of groups of even order $\leqslant 2j(n)$. Taking $p = 2$ in Proposition 10.1, we obtain the stated estimate.

□

11 Lower Bounds for the Conductor

We must still prove (10.3), the inequality between conductors and discriminants. Fix a number field F and a finite Galois extension L of F, and put $G = \mathrm{Gal}(L/F)$.

Lemma 11.1 *Let ρ and λ be finite-dimensional complex representations of G, with ρ faithful. Then $\mathfrak{q}(\lambda)$ divides $\mathfrak{q}(\rho)^{\dim(\lambda)}$.*

Proof Fix a prime ideal $\mathfrak{p}$ of F, and let $a(\rho)$ and $a(\lambda)$ be the exponent of $\mathfrak{p}$ in $\mathfrak{q}(\rho)$ and $\mathfrak{q}(\lambda)$ respectively. It suffices to see that

$$a(\lambda) \leqslant \dim(\lambda)a(\rho). \tag{11.1}$$

Let $I \subset G$ be the inertia subgroup of some fixed prime ideal of L above $\mathfrak{p}$. If $I = \{1\}$ then both sides of (11.1) are 0 and there is nothing to prove. Hence we may assume that $I \neq \{1\}$.

Let $G_0 = I \supseteq G_1 \supseteq G_2 \supseteq \ldots$ be the higher ramification subgroups of I in the lower numbering (cf. [32], p. 62). Since I is nontrivial there exists an integer $n \geqslant 0$ such that $G_i \neq \{1\}$ for $0 \leqslant i \leqslant n$ and $G_i = \{1\}$ for $i \geqslant n+1$. Writing V for the space of ρ and V^{G_i} for the subspace of vectors fixed by G_i, we have

$$a(\rho) = \sum_{i=0}^{n} \frac{|G_i|}{|G_0|} \dim(V/V^{G_i}) \tag{11.2}$$

(cf. [32], p. 100). Similarly,

$$a(\lambda) = \sum_{i=0}^{n} \frac{|G_i|}{|G_0|} \dim(W/W^{G_i}), \tag{11.3}$$

where W is the space of λ. Now as ρ is faithful we have $V^{G_i} \neq V$ for $0 \leqslant i \leqslant n$ and hence $\dim(V/V^{G_i}) \geqslant 1$. Thus

$$\dim(W/W^{G_i}) \leqslant \dim(W) = \dim(\lambda) \leqslant \dim(\lambda)\dim(V/V^{G_i}).$$

Substituting this inequality in (11.3) and comparing the result with (11.2), we obtain (11.1).

□

The inequality (10.3) is an immediate consequence of the following proposition:

Proposition 11.2 *Let ρ be a faithful irreducible complex representation of G. Then $\mathfrak{d}_{L/F}$ divides $\mathfrak{q}(\rho)^{|G|-(n^2-n)}$, where $n = \dim(\rho)$.*

Proof We apply the lemma to a set of representatives λ for the distinct isomorphism classes of irreducible representations of G. Raising both ideals in the divisibility of the lemma to the power $\dim\lambda$ and then taking the product over $\lambda \not\cong \rho$, we see that

$$\prod_{\lambda \not\cong \rho} \mathfrak{q}(\lambda)^{\dim\lambda} \text{ divides } \mathfrak{q}(\rho)^{\sum_{\lambda \not\cong \rho}(\dim\lambda)^2}. \tag{11.4}$$

Let reg_G denote the regular representation of G, and multiply the divisor and dividend in (11.4) by the same ideal $\mathfrak{q}(\rho)^n$. Since $\mathrm{reg}_G \cong \oplus_\lambda \lambda^{\oplus \dim\lambda}$, we obtain

$$\mathfrak{q}(\mathrm{reg}_G) \text{ divides } \mathfrak{q}(\rho)^{(\sum_\lambda (\dim\lambda)^2)-n^2+n}.$$

Now $\mathfrak{q}(\mathrm{reg}_G) = \mathfrak{d}_{L/F}$ by Artin's conductor-discriminant formula (cf. [32], p. 104). Since $|G| = \sum_\lambda (\dim\lambda)^2$, the proposition follows.

□

12 A Conditional Result in Dimension Three

To evaluate $\lim_{x\to\infty} \vartheta^{\mathrm{sd}}(x)/\vartheta(x)$ conditionally when $F = \mathbb{Q}$ and $n = 3$, we must bound each of the three terms on the right-hand side of (0.9). The first term is easily dealt with:

$$\vartheta^{\mathrm{ab,sd}}_{\mathbb{Q},3}(x) = O(x(\log x)^2). \tag{12.1}$$

by Theorem 3.6.

To bound $\vartheta^{1+2,\mathrm{sd}}_{\mathbb{Q},3}(x)$, we observe that if a self-dual representation is a direct sum of a one-dimensional and an irreducible two-dimensional representation then the one-dimensional and two-dimensional representations are self-dual. Thus

$$\vartheta^{1+2,\mathrm{sd}}_{\mathbb{Q},3}(x) = \sum_{q \leqslant x} \psi^{\mathrm{sd}}(q)\vartheta^{\mathrm{irr,sd}}_{\mathbb{Q},2}(x/q), \tag{12.2}$$

where $\psi^{\mathrm{sd}}(q)$ is the number of primitive Dirichlet characters χ of conductor q satisfying $\chi^2 = 1$, as in Section 2. But (0.10) gives

$$\vartheta^{\mathrm{irr,sd}}_{\mathbb{Q},2}(x/q) \leqslant \vartheta^{\mathrm{sd}}_{\mathbb{Q},2}(x/q) \ll (x/q)^{2-\varepsilon}$$

for any $\varepsilon < 1/60$, so (12.2) becomes

$$\vartheta_{\mathbb{Q},3}^{1+2,\mathrm{sd}}(x) \ll x^{2-\varepsilon} \sum_{q \leqslant x} \frac{\psi^{\mathrm{sd}}(q)}{q^{2-\varepsilon}}. \tag{12.3}$$

Applying Proposition 6.9 with $n(q) = \psi^{\mathrm{sd}}(q)$, we find that

$$\sum_{q \leqslant x} \frac{\psi^{\mathrm{sd}}(q)}{q^{2-\varepsilon}} = O(1), \tag{12.4}$$

because $\vartheta_{\mathbb{Q},1}^{\mathrm{sd}}(t) \ll t$ by the corollary to Proposition 2.3. In view of (12.4) we have

$$\vartheta_{\mathbb{Q},3}^{1+2,\mathrm{sd}}(x) \ll x^{2-\varepsilon} \tag{12.5}$$

after substitution in (12.3).

It remains to bound $\vartheta_{\mathbb{Q},3}^{\mathrm{irr},\mathrm{sd}}(x)$. We will use a variant of Proposition 10.2:

Proposition 12.1 *Let G be a finite irreducible self-dual subgroup of $\mathrm{GL}_n(\mathbb{C})$. If n is odd then $|G| \leqslant 2^n j(n)$.*

Proof The proof is similar to the proof of Proposition 10.2. If A is an abelian normal subgroup of G of index $\leqslant j(n)$ and $\iota : G \hookrightarrow \mathrm{GL}_n(\mathbb{C})$ is the tautological representation then $\iota|A$ is a direct sum of one-dimensional characters of A, and it follows from the self-duality of $\iota|A$ that the multiplicity of any character χ occurring in $\iota|A$ equals the multiplicity of χ^{-1}. Furthermore, since A is normal in G and ι is irreducible, all of the one-dimensional characters χ of A occurring in $\iota|A$ are conjugate under the action of G and thus have the same order w. If $w \geqslant 3$ then $\chi \neq \chi^{-1}$, whence $\iota|A$ is a direct sum of two-dimensional representations of the form $\chi \oplus \chi^{-1}$, contradicting the assumption that n is odd. Thus $w = 2$. Since A is abelian we may assume after a conjugation in $\mathrm{GL}_n(\mathbb{C})$ that A is contained in the group of diagonal matrices, hence in the group of diagonal matrices of order $\leqslant 2$. Thus $|A| \leqslant 2^n$ and $|G| = [G : A]|A| \leqslant j(n)2^n$.

□

We apply the proposition with $n = 3$. Using the value $j(3) = 360$ [7], and recalling once again that a group of odd order does not have nontrivial irreducible self-dual representations, we see that

$$\vartheta_{\mathbb{Q},3}^{\mathrm{irr},\mathrm{sd}}(x) \leqslant \sum_{\substack{|G| \leqslant 2880 \\ |G| \text{ even}}} \vartheta_{\mathbb{Q},3}^{G}(x), \tag{12.6}$$

where the sum on the right-hand side runs over a set of representatives for the distinct isomorphism classes of groups of even order $\leqslant 2880$. Applying Proposition 10.1 with $p = 2$, we obtain:

Proposition 12.2 *Fix $\gamma < 1/240$. If the weak form of Malle's conjecture holds then*

$$\vartheta_{\mathbb{Q},3}^{\mathrm{irr,sd}}(x) \ll x^{2-\gamma},$$

where the implied constant depends on γ.

Using the proposition together with (12.1) and (12.5) on the right-hand side of (0.9), we obtain the conditional bound $\vartheta_{\mathbb{Q},3}^{\mathrm{sd}}(x) = O(x^{2-\gamma})$ for every $\gamma < 1/240$. Since $\vartheta_{\mathbb{Q},3}^{\mathrm{ab}}(x) \gg (x \log x)^2$ by Theorem 3.3, we conclude under Malle's conjecture that $\lim_{x\to\infty} \vartheta^{\mathrm{sd}}(x)/\vartheta(x) = 0$ for $F = \mathbb{Q}$ and $n = 3$.

13 Appendix: Proof of Propositions 5.1, 5.2, and 5.3

We shall prove the propositions in reverse order.

Proof (of Proposition 5.3) (a) The irreducibility of ρ follows from Mackey's criterion, because the assumption that χ has order $\geqslant 3$ means that $\chi \neq \chi^{-1}$ and hence that $\chi \neq \chi^g$ for $g \in G \smallsetminus H$. The self-duality of ρ follows from the calculation

$$\rho^\vee = (\mathrm{ind}_H^G \chi)^\vee \cong \mathrm{ind}_H^G \chi^{-1} \cong \mathrm{ind}_H^G \chi^g \cong \rho.$$

Finally, induction preserves faithfulness.

(b) A straightforward calculation shows that if $g \in G \smallsetminus H$ and $h \in H$ then $\mathrm{tran}_H^G(h) = hghg^{-1}$. Consequently $\chi \circ \mathrm{tran}_H^G|H = \chi\chi^g$, whence $\chi^g = \chi^{-1}$ if and only if $\chi \circ \mathrm{tran}_H^G|H = 1$. Since sign_H^G and 1 are precisely the characters of G trivial on H we obtain the first half of (b). Now an irreducible self-dual representation is either orthogonal or symplectic, and we have just observed that if $\chi^g = \chi^{-1}$ then $\chi \circ \mathrm{tran}_H^G$ is either 1 or sign_H^G. Thus to prove the second half of (b) it suffices to see that $\chi \circ \mathrm{tran}_H^G = \mathrm{sign}_H^G$ if and only if ρ is symplectic, or equivalently (since $\mathrm{Sp}_2(\mathbb{C}) = \mathrm{SL}_2(\mathbb{C})$) if and only if $\det \rho = 1$. We now appeal to the formula for the determinant of an induced representation (cf. [13] or [10], p. 508, Proposition 1.2), which takes the form $\det \rho = (\mathrm{sign}_H^G)(\chi \circ \mathrm{tran}_H^G)$ in the case at hand.

□

Proof (of Proposition 5.2) Since ρ is monomial, there exists a subgroup H of index two in G and a one-dimensional character χ of H such that ρ is induced by χ. Since ρ is irreducible, $\chi \neq \chi^g$ for $g \in G \smallsetminus H$, and $\rho|H = \chi \oplus \chi^g$. Thus by Frobenius reciprocity χ and χ^g are precisely the two characters of H inducing ρ. But ρ is self-dual, so χ^{-1} also induces ρ.

Hence either $\chi^{-1} = \chi^g$ and χ is of order $\geqslant 3$ or else $\chi^{-1} = \chi$ and χ is quadratic. In the latter case ρ is realizable over $\mathbb{R}$, hence orthogonal. Viewing G as a subgroup of $\mathrm{O}_2(\mathbb{R})$, we can replace H by $\mathrm{SO}_2(\mathbb{R}) \cap G$ to get a *cyclic* subgroup of index two in G. On the other hand, if $\chi^{-1} = \chi^g$ then $\rho|H \cong \chi \oplus \chi^{-1}$. Since ρ is faithful, so is χ, whence H is cyclic.

Thus G has a cyclic subgroup of index two. If H is any such subgroup then $\rho|H \cong \chi \oplus \chi'$ with one-dimensional characters χ and χ' of H, and $\chi \neq \chi'$ because ρ is irreducible (if H is central then G is abelian). The irreducibility also gives $\chi' = \chi^g$ for $g \in G \smallsetminus H$, whence $\rho \cong \mathrm{ind}_H^G \chi$. We are now in the situation of the previous paragraph, but this time H is cyclic and so has at most one quadratic character. Thus if χ is quadratic then χ^g, which is consequently also quadratic, coincides with χ, a contradiction. Hence χ has order $\geqslant 3$ and χ^{-1}, which induces ρ and thus coincides with one of χ and χ^g, coincides with χ^g.

□

Proof (of Proposition 5.1) Applying Proposition 5.2 to the tautological representation $\iota : G \to \mathrm{GL}_2(\mathbb{C})$, we see that $\iota = \mathrm{ind}_H^G \chi$ for some cyclic subgroup H of index two in G and some character χ of H as in the proposition. Let a be a generator of H and choose $b \in G \smallsetminus H$. Then $\chi^b = \chi^{-1}$, and since χ is faithful we get $bab^{-1} = a^{-1}$. Also $b^2 \in H$ as $[G : H] = 2$. If $b^2 = 1$ then $G \cong D_{2m}$ with $m \geqslant 3$. Otherwise b^2 is a nontrivial element of the center of G, whence b^2 $(= \iota(b^2))$ is a scalar $\neq 1$ (Schur's lemma). Since ι is self-dual we get $b^2 = -1$. But $b^2 \in H$, so H has even order. Write $|H| = 2m$; then $a^m = b^2$ and $G \cong H_{4m}$.

The second assertion of the proposition follows from Proposition 5.3, because $\mathrm{tran}_H^G(b) = 1$ or -1 according as $G \cong D_{2m}$ or H_{4m}.

□

References

[1] G. Anderson, D. Blasius, R. Coleman, and G. Zettler, *On representations of the Weil group with bounded conductor*, Forum Math. **6** (1994), 537–545.

[2] P. T. Bateman and H. G. Diamond, *Analytic Number Theory: An Introductory Course*, World Scientific, 2004.

[3] M. Bhargava, *The density of discriminants of quartic rings and fields*, Ann. Math. **162** (2005), 1031–1063.

[4] M. Bhargava, *The density of discriminants of quintic rings and fields*, Ann. Math. **172** (2010), 1559–1591.

[5] M. Bhargava, A. Cojocaru, and F. Thorne, *The square sieve and the number of A_5-quintic extensions of bounded discriminant*, to appear.

[6] H. Cohen, *Advanced topics in computational number theory*, Grad. Texts in Math. **193**, Springer (2000).

[7] M. J. Collins, *On Jordan's theorem for complex linear groups*, J. Group Theory **10** (2007), 411–423.

[8] J. H. Conway, R. T. Curtis, S. P. Norton, R. A. Parker, and R. A. Wilson. *Atlas of finite groups: maximal subgroups and ordinary characters for simple groups*, Clarendon Press, 1985.

[9] G. Cooke and P. J. Weinberger, *On the construction of division chains in algebraic number fields, with applications to SL_2*, Commun. Algebra **3** (1975), 481–524.

[10] P. Deligne, *Les constantes des équations fonctionelles des fonctions L*, In: *Modular Functions of One Variable, II*, Lect. Notes in Math. **349** Springer (1973), 501–595.

[11] W. Duke, *The dimension of the space of cusp forms of weight one*, Internat. Math. Res. Notices (1995), 99–109.

[12] A. Fröhlich, *Artin root numbers and normal integral bases for quaternion fields*, Invent. Math. **17** (1972), 143–166.

[13] P. X. Gallagher, *Determinants of representations of finite groups*, Abh. Math. Sem. Univ. Hamburg **28** (1965), 162–167.

[14] G. H. Hardy and E. M. Wright, *An Introduction to the Theory of Numbers, 5th ed.*, Clarendon Press, Oxford (1979).

[15] P. N. Hoffman and J. F. Humphreys, *Projective Representations of the Symmetric Groups: Q-Functions and Shifted Tableaux*, Oxford Mathematical Monographs, Clarendon Press, Oxford (1992).

[16] B. Huppert, *Character Theory of Finite Groups*, de Gruyter (1998).

[17] G. Karpilovsky, *Group Representations* vol. 2, North-Holland Mathematics Studies **177**, (1993).

[18] N. Kataoka, *The distribution of prime ideals in a real quadratic field with units having a given index in the residue class field*, J. Number Theory **101** (2003), 349–375.

[19] Y. Kitaoka, *Distribution of units of a cubic field with negative discriminant*, J. Number Theory **91** (2001), 318–355.

[20] Y. Kitaoka, *Distribution of units of an algebraic number field.* In: *Galois Theory and Modular Forms,* edited by K. Hashimoto, K. Miyake, and H. Nakamura, Kluwer Academic Publishers (2003).

[21] J. Klüners, *A counter example to Malle's conjecture on the asymptotics of discriminants*, C. R. Acad. Sci. Paris, Ser. I **340** (2005), 411–414.

[22] S. Lang, *Introduction to Modular Forms*, Springer, Grundlehren der math. Wissen. **222**, (1976). Appendix by W. Feit: *Exceptional subgroups of* GL_2.

[23] S. Lang, *Algebraic Number Theory*, 2nd. ed., Graduate Texts in Mathematics **110** Springer, 1994.

[24] S. Lang, *Elliptic Functions*, Springer, Graduate Texts in Mathematics **112**, Springer, 1987.

[25] H. W. Lenstra, Jr., *On Artin's conjecture and Euclid's algorithm in global fields*, Inventiones Math. **42** (1977), 201–224.

[26] G. Malle, *On the distribution of Galois groups, II*, Experimental Math. **13** (2004), 129–135.

[27] J. Martinet, *Character theory and Artin L-functions*, In: *Algebraic Number Fields, Proceedings of the Durham Symposium*, A. Fröhlich ed. Academic Press (1977), 1–87

[28] P. Michel and A. Venkatesh, *On the dimension of the space of cusp forms associated to 2-dimensional complex Galois representations*, Internat. Math. Res. Notices (2002), 2021–2027.

[29] M. R. Murty, *Artin's conjecture for primitive roots*, Math. Intelligencer **10** (1988), 59–67.

[30] H. Roskam, *A quadratic analogue of Artin's conjecture on primitve roots*, J. Number Theory **81** (2000), 93–109.

[31] J.-P. Serre, *Modular forms of weight one and Galois representations* In: *Algebraic Number Fields, Proceedings of the Durham Symposium*, A. Fröhlich ed. Academic Press (1977), 193 – 268. (=*Oeuvres* vol. III, no. 110.)

[32] J.-P. Serre, *Local Fields*, translated from the French by M. J. Greenberg, Graduate Texts in Mathematics **67**, Springer, 1979.

[33] J.-P. Serre, *Quelques applications du théorème de densité de Chebotarev*, Inst. Hautes Études Sci. Publ. Math. **54** (1981), 123–201. (=*Oeuvres* vol. III, no. 125.)

[34] J.-P. Serre, *Topics in Galois Theory*, Notes written by H. Darmon, Research Notes in Mathematics **1**, Jones and Bartlett Publishers, 1992.

[35] C. L. Siegel, *The average measure of quadratic forms with given determinant and signature*, Ann. of Math. **45** (1944), 667–685.

[36] M. J. Taylor, *On the equidistribution of Frobenius in cyclic extensions of a number field*, J. London Math. Soc. **29** (1984), 211–213.

[37] L. C. Washington, *Introduction to cyclotomic fields*, Graduate Texts in Mathematics **83**, Springer, 1982.

[38] H. Weber,*Lehrbuch der Algebra, Bd. II (zweite Auflage)*, Braunschweig (1899).

[39] S. Wong, *Automorphic forms on GL(2) and the rank of class groups*, J. Reine Angew. Math. **515** (1999), 125–153.

Department of Mathematics and Statistics, Boston University, Boston, MA 02215, USA.
E-mail: `rohrlich@math.bu.edu`

Automorphic Representations and L-Functions
Editors: D. Prasad, C.S. Rajan, A. Sankaranarayanan, J. Sengupta

Publisher: Hindustan Book Agency, New Delhi, India

Determination of Modular Forms by Fundamental Fourier Coefficients

Abhishek Saha

Abstract

It is an interesting question when a natural subset of the Fourier coefficients is sufficient to uniquely determine a modular form. This article deals with this question for two kinds of modular forms: a) classical modular forms of half-integral weight, and b) Siegel modular forms of genus 2 and integral weight. These two apparently different scenarios turn out to be closely related. Our results have several applications to automorphic L-functions and Bessel models.

1 Introduction

Let V be a fixed set consisting of some "modular forms".[1] Let $\mathcal{S}$ be an indexing set for the "Fourier coefficients" of elements of V. This means that for all $\Phi \in V$, we have an expansion[2]

$$\Phi(z) = \sum_{n \in \mathcal{S}} \Phi_n(z).$$

Let $\mathcal{D}$ be an *interesting* subset of $\mathcal{S}$. We are interested in situations where the following implication is true for all $\Phi \in V$: $\Phi_n = 0 \ \forall \ n \in \mathcal{D} \Rightarrow \Phi = 0$; equivalently: $\Phi \neq 0 \Rightarrow$ there exists $n \in \mathcal{D}$ such that $\Phi_n \neq 0$. Another way of phrasing this problem is: *When does an interesting subset of Fourier coefficients determine a modular form?*

In this article, we will consider a special case of this question for the following two types of modular forms:

1. Classical modular forms of half-integral weight (automorphic forms on $\widetilde{\mathrm{SL}_2}$)

[1]We are being deliberately vague here in order to include a wide variety of objects that generalize or are similar to classical modular forms.

[2]The model to keep in mind here is the well-known Fourier expansion of classical modular forms.

2. Siegel modular forms of degree 2 and trivial central character (automorphic forms on PGSp_4)

For both types, the interesting subset $\mathcal{D}$ will be related to $\mathcal{S}$ in a similar manner as the set of squarefree integers is related to the set of all integers. Moreover, as we will see, the proof of our main result for Siegel modular forms hinges crucially on the corresponding result for classical modular forms of half-integral weight.

This article in based on the talk the author gave at the *International Colloquium on Automorphic Representations and L-Functions, 2012,* held at the Tata Institute of Fundamental Research. Some of the results of this article were obtained jointly with Ralf Schmidt. Most of the results are contained either in the author's paper [23] or in his joint paper with Schmidt [25]. The presentation here will therefore be less formal and of a more descriptive nature; our goal is to tie several related results into a coherent framework and emphasize the underlying motivations and ideas. The reader is invited to consult [23] and [25] for the technical details of the proofs that are sketched or omitted from this article.

We briefly summarize the structure of this article. In Section 2 we deal with the case of half-integral weight modular forms. In Section 3, we deal with the case of Siegel cusp forms of degree 2. The proofs of Section 3 need the results of Section 2 in an essential manner. Finally, in Section 4, we give some important applications of our results to global Bessel models for GSp_4 and simultaneous non-vanishing of dihedral twists of modular L-functions.

2 The Case of Half-integral Weight Modular Forms

2.1 Notations and preliminaries

The group $\mathrm{SL}_2(\mathbb{R})$ acts on the upper half-plane $\mathbb{H}$ via

$$\gamma z := \frac{az+b}{cz+d}$$

for $\gamma = \begin{pmatrix} a & b \\ c & d \end{pmatrix} \in \mathrm{SL}_2(\mathbb{R})$ and $z = x + iy \in \mathbb{H}$. For a positive integer N, let $\Gamma_0(N)$ denote the congruence subgroup consisting of matrices $\begin{pmatrix} a & b \\ c & d \end{pmatrix}$ in $\mathrm{SL}_2(\mathbb{Z})$ such that N divides c. For a complex number z, let $e(z)$ denote $e^{2\pi iz}$.

Let $\theta(z) = \sum_{n=-\infty}^{\infty} e(n^2 z)$ be the standard theta function on $\mathbb{H}$. If $A = \begin{pmatrix} a & b \\ c & d \end{pmatrix} \in \Gamma_0(4)$, we have $\theta(Az) = j(A,z)\theta(z)$, where $j(A,z)$ is the so called θ-multiplier. For an explicit formula for $j(A,z)$, see [27] or [26].

Let k be an integer. For a positive integer N divisible by 4, let $S_{k+\frac{1}{2}}(N)$ denote the space of holomorphic cusp forms of weight $k + \frac{1}{2}$ for the group $\Gamma_0(N)$. Precisely, a function $f : \mathbb{H} \to \mathbb{C}$ belongs to $S_{k+\frac{1}{2}}(N)$ if

1. $f(Az) = j(A,z)^{2k+1} f(z)$ for every $A = \begin{pmatrix} a & b \\ c & d \end{pmatrix} \in \Gamma_0(N)$,
2. f is holomorphic,
3. f vanishes at the cusps.

Any $f \in S_{k+\frac{1}{2}}(N)$ has the Fourier expansion

$$f(z) = \sum_{n>0} a(f,n) e(nz).$$

We let $\tilde{a}(f,n)$ denote the "normalized" Fourier coefficients, defined by

$$\tilde{a}(f,n) = a(f,n) n^{\frac{1}{4}-\frac{k}{2}}.$$

The Kohnen plus-space $S^+_{k+\frac{1}{2}}(N)$ is defined to be the subspace of $S_{k+\frac{1}{2}}(N)$ consisting of forms f for which $a(f,n) = 0$ whenever $n \equiv (-1)^{k+1}$ or 2 mod 4. Kohnen [8] developed a theory of *newforms* for the space $S^+_{k+\frac{1}{2}}(N)$ in the case that $N/4$ is odd and squarefree. He also proved [9] a precise version of Waldspurger's theorem in this setting. If $f \in S^+_{k+\frac{1}{2}}(N)$ is a newform and $(-1)^k d$ is a *fundamental discriminant*,[3] then Kohnen's formula implies that $|\tilde{a}(f,d)|^2$ is essentially equal to $L(\frac{1}{2}, \pi_f \times \chi_{(-1)^k d})$; here π_f is the automorphic representation of PGL_2 attached to f by the Shimura correspondence and $\chi_{(-1)^k d}$ is the quadratic Dirichlet character associated via class field theory to the field $\mathbb{Q}(\sqrt{(-1)^k d})$.

2.2 The problem

Let $\mathcal{S}$ equal the set of integers n such that $(-1)^k n$ is a discriminant, i.e., $n \equiv 0$ or 1 mod 4. If $f \in S^+_{k+\frac{1}{2}}(N)$, the Fourier coefficients $a(f,n)$ are indexed by $n \in \mathcal{S}$. A natural subset of $\mathcal{S}$ is the set $\mathcal{D}$ consisting of integers

[3]Recall that an integer n is a fundamental discriminant if *either* n is a squarefree integer congruent to 1 modulo 4 *or* $n = 4m$ where m is a squarefree integer congruent to 2 or 3 modulo 4.

n such that $(-1)^k n$ is a fundamental discriminant. It is clear from the Kohnen–Waldspurger formula that if $d \in \mathcal{D}$ and $f \in S^+_{k+\frac{1}{2}}(N)$ is a newform (so in particular, a Hecke eigenform at all places), then $a(f,d)$ has deep arithmetic significance. So, it is a natural question to ask whether elements of $S^+_{k+\frac{1}{2}}(N)$ are determined by the Fourier coefficients $a(f,d)$ with $d \in \mathcal{D}$.

In the language of the introduction, we want to prove that given any non-zero modular form Φ in V, there exists $n \in \mathcal{D}$ such that $a(f,n) \neq 0$. Here V is a suitable subset of $S^+_{k+\frac{1}{2}}(N)$. The simplest version of the problem is when V is a finite Hecke basis consisting of newforms. Here, we may assume that $N = 4M$ where M is odd and squarefree[4]. In this case, the problem was solved by Kohnen. The proof of Kohnen's result, stated immediately below, is straightforward and uses only some combinatorics with Hecke operators at the prime 2.

Theorem 2.1 (Kohnen, p.70 of [8]) *Suppose that M is odd and squarefree and $0 \neq f \in S^+_{k+\frac{1}{2}}(4M)$ is a newform. Then there exists a d such that $(-1)^k d$ is a fundamental discriminant and $a(f,d) \neq 0$.*

The next simplest version of the question is obtained by taking V to be set of all elements of the form $a_1 g_1 + a_2 g_2$ where g_1 and g_2 lie in a fixed Hecke basis consisting of newforms in $S^+_{k+\frac{1}{2}}(N)$. This question was raised explicitly by Kohnen [10] in 1992 and solved by Luo–Ramakrishnan [11] in 1997.

Theorem 2.2 (Luo-Ramakrishnan [11]) *Suppose that M is odd and squarefree. Let g_1 and g_2 be multiples of newforms in $S^+_{k+\frac{1}{2}}(4M)$. Assume further that $a(g_1,d) = a(g_2,d)$ for all d such that $(-1)^k d$ is a fundamental discriminant. Then $g_1 = g_2$.*

The above theorem clearly implies the result of Kohnen stated earlier (simply take $g_2 = 0$). The proof of the theorem is quite involved and uses the Kohnen–Waldspurger formula in a crucial way. Indeed, the squares of Fourier coefficients of half-integral weight newforms are related to twisted L-values of integral weight newforms; Luo–Ramakrishnan were able to prove that integral weight newforms are uniquely determined by the central L-values of their twists with quadratic characters.

Finally, one may consider the hardest (and most general) version of the question, where we take the set V to be the entire set $S^+_{k+\frac{1}{2}}(4M)$. Thus the problem becomes:

[4]These are the assumptions under which Kohnen developed his newform theory.

Problem 2.3 Let $f \in S^+_{k+\frac{1}{2}}(4M)$, M squarefree, $f \neq 0$. Does there exist d such that $(-1)^k d$ is a fundamental discriminant and $a(f,d) \neq 0$?

Note that a solution to the above problem will automatically imply the result of Luo–Ramakrishnan (take $f = g_1 - g_2$). However, because f is no longer assumed to be a Hecke eigenform or a difference of two eigenforms, it appears impossible to reduce the problem to one about central L-values via Waldspurger's formula. This is in sharp contrast to the results stated earlier in this subsection.

2.3 The main result for half-integral forms

The next theorem gives an affirmative answer, in a strong quantitative form, to the problem posed in the previous subsection, whenever $k \geq 2$. Note that the statement of the result does not involve the Kohnen plus-space $S^+_{k+\frac{1}{2}}(4M)$ at all, but only the larger space $S_{k+\frac{1}{2}}(4M)$.

Theorem 2.4 *Let $k \geq 2$ and let M be a squarefree integer. Suppose that $f \in S_{k+\frac{1}{2}}(4M)$ is non-zero. Then, one has the lower bound*

$$\#\{0 < d < X : d \text{ squarefree}, \ a(f,d) \neq 0\} \gg_{f,\delta} X^{\delta},$$

where $\delta > 0$ is an absolute constant (any value of $\delta < 5/8$ is admissible). In particular, there are infinitely many squarefree integers d such that $a(f,d) \neq 0$.

The above theorem is not true if $k < 2$; counter-examples can be obtained by considering suitable theta-series. Note also that if $k \geq 2$, it gives a positive answer to Problem 2.3. Indeed, if $f \in S^+_{k+\frac{1}{2}}(4M)$, then $a(f,n)$ is supported on the integers n for which $(-1)^k n$ is a discriminant. If such a n is squarefree, it is automatically a fundamental discriminant.

Theorem 2.4 is a special case of a more general result (involving not necessarily squarefree levels and possibly nontrivial nebentypus) that was proven in [23] and further generalized in [25]. So, we will only briefly sketch the proof here; the reader is invited to consult [23] for further details.

Sketch of Proof For any positive integer M, define

$$S(M,X;f) := \sum_{\substack{d \text{ squarefree} \\ (d,M)=1}} |\tilde{a}(f,d)|^2 e^{-d/X}.$$

Suppose that we can show there exists M such that $S(M,X;f) \gg_f X$. The result then follows immediately from any bound of the form

$$|\tilde{a}(f,d)|^2 \ll_{f,\delta} d^{1-\delta}.$$

The first non-trivial bound of this form is due to Iwaniec [7] who showed that one can take any $\delta < 4/7$. The best current bound is due to Bykovskiĭ [3] who improved this to 5/8; see also Blomer–Harcos [1]. The (unproven) Lindelöf hypothesis for twisted L-functions of f implies that one take any $\delta < 1$.

So the key point is to prove that $S(M,X;f) \gg_f X$ for some M. For this, we re-write $S(M,X;f)$ using a "squarefree sieve", as follows.

$$S(M,X;f) = \sum_{\substack{r \text{ squarefree} \\ (r,M)=1}} \mu(r) \sum_{\substack{n>0 \\ (n,M)=1}} |\tilde{a}(f,nr^2)|^2 e^{-r^2n/X},$$

where $\mu(n)$ denotes the Mobius function. We can bound the leading term (corresponding to $r = 1$) from below by using a theorem of Duke and Iwaniec [4]. The other terms can be bounded from above by using the theory of Hecke operators and the Deligne bound in the Hecke eigenvalues (see [23] for details). Putting things together, one sees that for large enough M, we have $S(M,X;f) \gg_f X$. As noted earlier, this completes the proof.

□

3 The Case of Siegel Cusp Forms of Degree 2

3.1 Notations and preliminaries

For any commutative ring R and positive integer n, let $M_n(R)$ denote the ring of n by n matrices with entries in R and $\mathrm{GL}_n(R)$ denote the group of invertible matrices. If $A \in M_n(R)$, we let ${}^tA$ denote its transpose. We say that a symmetric matrix in $M_n(\mathbb{Z})$ is semi-integral if it has integral diagonal entries and half-integral off-diagonal ones. Denote by J the 4 by 4 matrix given by

$$J = \begin{pmatrix} 0 & I_2 \\ -I_2 & 0 \end{pmatrix}.$$

where I_2 is the identity matrix of size 2.

Define the algebraic group Sp_4 over $\mathbb{Z}$ by

$$\mathrm{Sp}_4(R) = \{g \in \mathrm{GL}_4(R) \mid {}^tgJg = J\}$$

for any commutative ring R.

The Siegel upper-half space of degree 2 is defined by

$$\mathbb{H}_2 = \{Z \in M_2(\mathbb{C}) \mid Z = {}^tZ,\ \mathrm{Im}(Z) \text{ is positive definite}\}.$$

The group $\mathrm{Sp}_4(\mathbb{R})$ acts on $\mathbb{H}_2$ via

$$gZ := (AZ+B)(CZ+D)^{-1} \qquad \text{for } g = \begin{pmatrix} A & B \\ C & D \end{pmatrix} \in \mathrm{Sp}_4(\mathbb{R}),\ Z \in \mathbb{H}_2.$$

We let $J(g,Z) = CZ + D$. For any positive integer N, define

$$\Gamma_0^{(2)}(N) := \left\{ \begin{pmatrix} A & B \\ C & D \end{pmatrix} \in \mathrm{Sp}_4(\mathbb{Z}) \mid C \equiv 0 \pmod{N} \right\}. \tag{3.1}$$

Let $S_k^{(2)}(N)$ denote the space of holomorphic functions F on $\mathbb{H}_2$ which satisfy the relation

$$F(\gamma Z) = \det(J(\gamma, Z))^k F(Z) \tag{3.2}$$

for $\gamma \in \Gamma_0^{(2)}(N)$, $Z \in \mathbb{H}_2$, and vanish at all the cusps. Elements of $S_k^{(2)}(N)$ are often referred to as Siegel cusp forms of degree (genus) 2, weight k and level N. The Siegel cusp forms of degree 2 can be viewed as the simplest generalization of the classical cusp forms, and they arise naturally in number theory and representation theory. They also have applications to coding theory and conformal field theory.

Any $F \in S_k^{(2)}(N)$ has a Fourier expansion

$$F(Z) = \sum_{T>0} a(F,T) e(\mathrm{Tr}(TZ)),$$

where T runs through all symmetric, semi-integral, positive-definite matrices of size two, or equivalently, all positive, integral, binary quadratic forms. In fact, because $\begin{pmatrix} A & 0 \\ 0 & {}^tA^{-1} \end{pmatrix} \in \Gamma_0^{(2)}(N)$ for all $A \in \mathrm{SL}_2(\mathbb{Z})$, we have, using (3.2), that

$$a(F, {}^tATA) = a(F,T) \tag{3.3}$$

for all $A \in \mathrm{SL}_2(\mathbb{Z})$, thus showing that $a(F,T)$ only depends on the $\mathrm{SL}_2(\mathbb{Z})$-equivalence class of T, i.e., only on the proper equivalence class of the associated binary quadratic form.

We denote by $S_k^{(2),\mathrm{O}}(N)$ the linear subspace of $S_k^{(2)}(N)$ spanned by the set

$$\{F(dZ) : F \in S_k^{(2)}(M),\ dM|N,\ M \neq N\}.$$

Note that if $N = 1$, then $S_k^{(2),\mathrm{O}}(N) = \{0\}$.

3.2 The problem

If $F \in S_k^{(2)}(N)$, the Fourier coefficients $a(F, S)$ are indexed by $S \in \mathcal{S}$, where $\mathcal{S}$ equals the set of matrices of the form

$$S = \begin{pmatrix} a & b/2 \\ b/2 & c \end{pmatrix}, \qquad a, b, c \in \mathbb{Z}, \qquad a > 0, \qquad \mathrm{disc}(S) := b^2 - 4ac < 0.$$

If $\gcd(a, b, c) = 1$, then S is called *primitive*. If $\mathrm{disc}(S)$ is a fundamental discriminant, then S is called *fundamental*. Observe that if S is fundamental, then it is automatically primitive. Observe also that the following conditions are equivalent: a) $\mathrm{disc}(S)$ is squarefree, b) S is fundamental with $\mathrm{disc}(S)$ odd.

One possible choice of the interesting subset $\mathcal{D}$ of $\mathcal{S}$ consists of all the primitive matrices. In this context, the problem stated in the introduction was solved in the special case $N = 1$ by Zagier [29, p. 387]. This result was generalized by Yamana [28] to cusp forms with level and of higher degree. Using Yamana's result, Ibukiyama and Katsurada proved the following theorem.[5]

Theorem 3.1 (Ibukiyama–Katsurada [6]) *Let $F \in S_k^{(2)}(N)$ belong to the orthogonal complement of $S_k^{(2),O}(N)$ and suppose that $a(F, S) = 0$ for all the primitive matrices S. Then $F = 0$.*

In the language of the introduction, Theorem 3.1 addresses the case when V equals the orthogonal complement of $S_k^{(2),\mathrm{O}}(N)$ and $\mathcal{D}$ equals the subset of $\mathcal{S}$ consisting of the primitive matrices. Note that it is not possible to extend Theorem 3.1 so that V equals the full set $S_k^{(2)}(N)$. This is because forms such as $F(NZ)$ with $F \in S_k^{(2)}(1)$ are only supported on non-primitive coefficients.

From the point of view of representation theory, it is the smaller set of fundamental Fourier coefficients that are more interesting than the primitive ones. In the language of the introduction, let $\mathcal{S}$ be as before and let $\mathcal{D}$ consist of the fundamental matrices. Then, the simplest question in this direction is obtained by taking V to be a finite set consisting of eigenforms at all places.

Problem 3.2 Let $F \in S_k^{(2)}(N)$ be a non-zero element in the orthogonal complement of $S_k^{(2),\mathrm{O}}(N)$ and an eigenfunction for the local Hecke algebras at all primes. Show that there exists a fundamental matrix S such that $a(F, S) \neq 0$.

[5]The Theorem of Ibukiyama–Katsurada is actually for Siegel cusp forms of any degree but we state it only for degree 2 here for convenience.

The above problem is a neat analogue of Theorem 2.1, and turns out to have deep consequences to the theory of Bessel models for automorphic forms, as well as to non-vanishing of periods and simultaneous non-vanishing of dihedral twists of modular L-functions. These applications will form the heart of Section 4. It is perhaps surprising (given how easy the proof of Theorem 2.1 is) that Problem 3.2 does not appear to be amenable to any attacks relying on the theory of Hecke operators or local representation theory (even in the simplest case $N = 1$).

One can be much more ambitious, of course, and take V as in Theorem 3.1. This leads to the following significant generalization of Problem 3.2.

Problem 3.3 Let $F \in S_k^{(2)}(N)$ be a non-zero element in the orthogonal complement of $S_k^{(2),\mathrm{O}}(N)$. Show that there exists a fundamental matrix S such that $a(F,S) \neq 0$.

3.3 The main result for Siegel cusp forms

The next theorem, which is a mild variation of [25, Thm. 2] gives an affirmative answer, in a strong quantitative form, to Problem 3.3, whenever N is squarefree and $k > 2$ is even.

Theorem 3.4 *Let $k > 2$ and N be a squarefree integer. Moreover, if $N > 1$, assume that k is even. Let $0 \neq F \in S_k^{(2)}(N)$ belong to the orthogonal complement of $S_k^{(2),O}(N)$. Let $\delta > 0$ be an absolute constant as in Theorem 2.4. Then, one has the lower bound*

$$|\{0 < d < X,\ d \text{ squarefree },\ a(F,S) \neq 0 \text{ for some } S \text{ with } d = -\mathrm{disc}(S)\}| \gg_{F,\delta} X^{\delta}.$$

In particular, there are infinitely many fundamental matrices S such that $a(F,S) \neq 0$.

Note that if $N = 1$, the conditions $k > 2$ and F belonging to the orthogonal complement of $S_k^{(2),\mathrm{O}}(N)$ are both automatic.

Sketch of Proof By Theorem 3.1, there exists a *primitive* matrix T' such that $a(F,T') \neq 0$. Suppose that $T' = \begin{pmatrix} a & b/2 \\ b/2 & c \end{pmatrix}$. It is a classical result (going back at least to Weber) that the primitive quadratic form $ax^2 + bxy + cy^2$ represents infinitely many primes. So, let x_0, y_0 be such that $ax_0^2 + bx_0y_0 + cy_0^2$ is an odd prime not dividing N. Since this implies

$\gcd(x_0, y_0) = 1$, we can find integers x_1, y_1 such that $A = \begin{pmatrix} y_1 & y_0 \\ x_1 & x_0 \end{pmatrix} \in \mathrm{SL}_2(\mathbb{Z})$. Then $T = {}^tAT'A$ has the property that $a(F,T) \neq 0$ and T is of the form $\begin{pmatrix} a_0 & b_0/2 \\ b_0/2 & p \end{pmatrix}$ where p is an odd prime not dividing N.

Note that so far, we have only used the $\mathrm{SL}_2(\mathbb{Z})$-invariance of the Fourier coefficients and the theorem of Ibukiyama and Katsurada. To complete the proof, we are going to cook up a classical half-integral weight form and appeal to Theorem 2.4. For simplicity, assume henceforth that k is even. For all integers n, r with $4np > r^2$, let us denote

$$c(n,r) = a\left(F, \begin{pmatrix} n & r/2 \\ r/2 & p \end{pmatrix}\right).$$

Now, let

$$h(\tau) = \sum_{m=1}^{\infty} c(m)e(m\tau).$$

where

$$c(m) = \sum_{\substack{0 \le \mu \le 2p-1 \\ \mu^2 \equiv -m \pmod{4p}}} c\left((m+\mu^2)/4p,\ \mu\right).$$

By Theorem 4.8 of [12], we know that $h \in S_{k-\frac{1}{2}}(4pN)$.

It is easy to see that h is not identically equal to 0. Indeed put $d_0 = 4a_0p - b_0^2$. Then $c(d_0)$ equals

$$a\left(F, \begin{pmatrix} a_0 & b_0/2 \\ b_0/2 & p \end{pmatrix}\right) + a\left(F, \begin{pmatrix} a_0 + p - b_0 & p - b_0/2 \\ p - b_0/2 & p \end{pmatrix}\right),$$

which is simply $2a\left(F, \begin{pmatrix} a_0 & b_0/2 \\ b_0/2 & p \end{pmatrix}\right)$ by (3.3) and hence non-zero.

Now, by Theorem 2.4, it follows that

$$|\{0 < d < X,\ d \text{ squarefree },\ c(d) \neq 0\}| \gg_{h,\delta} X^{\delta}.$$

For any of these d, there exists a μ such that

$$c\left(\frac{d+\mu^2}{4p}, \mu\right) = a\left(F, \begin{pmatrix} \frac{d+\mu^2}{4p} & \mu/2 \\ \mu/2 & p \end{pmatrix}\right)$$

is not equal to zero. This completes the proof. □

Remark 3.5 The construction of a half-integral weight form h from the Fourier coefficients of F in the proof above is best understood as arising from the isomorphism between the space of Jacobi forms and the space of modular forms of half-integral weight. In the case $N = 1$, this isomorphism was first investigated in great detail by Skoruppa in his thesis.

Remark 3.6 In the applications, we will only need Theorem 3.4 when F is a Hecke eigenform. However, even in that case, the half-integral form h constructed above need not be a Hecke eigenform! That is why, even if we wished to prove Theorem 3.4 only for eigenforms, we would still need to prove Theorem 2.4 in general.

3.4 A mild restatement

In this subsection, we will restate Theorem 3.4 in a form that will be useful for applications. Let $-d < 0$ be a fundamental discriminant and put $K = \mathbb{Q}(\sqrt{-d})$. Let Cl_K denote the ideal class group of K. It is a fact going back to Gauss that the $\mathrm{SL}_2(\mathbb{Z})$−equivalence classes of binary quadratic forms of discriminant $-d$ are in natural bijective correspondence with the elements of Cl_K. In view of (3.3) and the comments immediately after, it follows that for any $c \in \mathrm{Cl}_K$ the notation $a(F, c)$ makes sense.

In particular, for any $F \in S_k^{(2)}(N)$, any imaginary quadratic field K (with discriminant equal to $-d$) and any character Λ of the finite group Cl_K, we can make the definition

$$R(F, K, \Lambda) = \sum_{c \in \mathrm{Cl}_K} a(F, c)\Lambda^{-1}(c). \tag{3.4}$$

An immediate corollary of Theorem 3.4 is the following proposition.

Proposition 3.7 *Let $k > 2$ and N be a squarefree integer. Moreover, if $N > 1$, assume that k is even. Let $F \in S_k^{(2)}(N)$ be a non-zero form belonging to the orthogonal complement of $S_k^{(2),O}(N)$. Then there are infinitely many pairs (K, Λ) with K an imaginary quadratic field and Λ an ideal class character of Cl_K, such that $R(F, K, \Lambda) \neq 0$.*

Proof Take $K = \mathbb{Q}(\sqrt{-d})$ to be any field for which there exists a semi-integral matrix S with $-d = \mathrm{disc}(S)$ and $a(F, S) \neq 0$. The proposition is now clear from Theorem 3.4 by Fourier inversion on the finite group Cl_K. □

4 Applications

4.1 Existence of nice Bessel models for Siegel cusp forms

The goal of this subsection is to explain a certain application of Theorem 3.4 that was the original motivation for the author's work in this direction.

When working with automorphic representations on some reductive group, it is useful to have a *model* for the representation consisting of a space of nicely transforming functions on the group. For automorphic representations on GL_n, such a model is provided by the space of adelic Whittaker functions. Indeed, every cuspidal automorphic representation on $\mathrm{GL}_n(\mathbb{A})$ has a Whittaker model, that is to say, the representation space consisting of Whittaker functions on $\mathrm{GL}_n(\mathbb{A})$ contains (with multiplicity one!) each cuspidal automorphic representation of $\mathrm{GL}_n(\mathbb{A})$. For automorphic representations on other classical groups, a Whittaker model does not necessarily exist. In particular, Siegel cusp forms of degree 2 correspond to automorphic representations on $\mathrm{GSp}_4(\mathbb{A})$ which *never* have Whittaker models. A useful substitute for the missing Whittaker model is the so-called Bessel model.

In contrast to Whittaker models, which are essentially canonical, Bessel models depend on some arithmetic data. In the case of cuspidal automorphic representations on $\mathrm{GSp}_4(\mathbb{A})$, there are two ingredients that go into the data defining a Bessel model[6]. One of them is a two by two non-degenerate symmetric matrix $S \in M_2(\mathbb{Q})$ such that $-d = -4\det(S)$ belongs to $\mathbb{Q}^\times - (\mathbb{Q}^\times)^2$. The other ingredient is a Hecke character Λ of $K^\times \backslash \mathbb{A}_K^\times$ where $K = \mathbb{Q}(\sqrt{-d})$. For any such S and Λ, it makes sense to ask whether a cuspidal automorphic representation π of $\mathrm{GSp}_4(\mathbb{A})$ has a global Bessel model of type (S, Λ). The interested reader should consult [15, 5, 20] for further details.

We say that the type (S, Λ) is *fundamental* if each of the following conditions is satisfied:

1. $S = \begin{pmatrix} a & b/2 \\ b/2 & 1 \end{pmatrix}$, $\quad a, b \in \mathbb{Z}, \quad -d := b^2 - 4a < 0$ is a fundamental discriminant.
2. The Hecke character Λ is unramified at all finite places of $K = \mathbb{Q}(\sqrt{-d})$ and trivial at infinity, and hence is a character on the ideal class group of K.

The automorphic representations we are interested in come from Siegel cusp forms of degree 2. Indeed, let $F \in S_k^{(2)}(N)$ be an eigenform for

[6] Here we are suppressing a third ingredient, which is the choice of additive character.

the local Hecke algebras at all places. Then F gives rise to an irreducible cuspidal representation π_F of $\mathrm{GSp}_4(\mathbb{A})$; see [14]. It can be shown that there exist infinitely many pairs (S, Λ) such that π_F has a global Bessel model of type (S, Λ). What is highly desirable, however, is that at least one of these pairs be fundamental, i.e., π_F have at least one global Bessel model of fundamental type. Indeed, working with a fundamental global Bessel model means that all the associated local Bessel data is unramified, which makes it much easier to compute local zeta integrals and deduce various properties of global L-functions.

In a pioneering work, Furusawa [5] used Bessel models to prove an integral representation for the $\mathrm{GSp}_4 \times \mathrm{GL}_2$ L-function of the twist of an eigenform F in $S_k^{(2)}(1)$ with an elliptic (classical) eigenform g of the same weight. He used this to deduce special value results in the spirit of Deligne's conjecture. However, his results were all obtained under the assumption that π_F has a global Bessel model of fundamental type. Subsequent works by various people built upon Furusawa's results and proved various analytic, special value and functoriality properties for π_F and its various associated twisted L-functions. We refer the reader to the papers [17, 21, 22, 19] for details. Since all these works depended upon Furusawa's unramified calculation, they were also only valid for Siegel cusp forms which have a global Bessel model of fundamental type. And till recently, it was completely unknown how restrictive this assumption about existence of fundamental Bessel model is.

The link between global Bessel models and the theme of this article is provided by the following lemma.

Lemma 4.1 *Let N be squarefree and $F \in S_k^{(2)}(N)$ be an eigenform for the local Hecke algebras at all places. Let π_F be the irreducible cuspidal representation attached to F. The following are equivalent:*

1. *π_F has a global Bessel model of fundamental type (S, Λ).*
2. *$R(F, K, \Lambda) \neq 0$ where $K = \mathbb{Q}(\sqrt{\mathrm{disc}(S)})$ is the imaginary quadratic field associated to S as above.*

In other words, π_F has a global Bessel model of fundamental type , if and only if F has a non-zero fundamental Fourier coefficient!

Proof This follows from [25, Prop. 4.3] using recent work of Pitale and Schmidt [18] that tells us that a local eigenvector fixed by the Siegel parahoric is a test vector for the local Bessel functional.

□

Together, Proposition 3.7 and Lemma 4.1 immediately imply the following theorem.

Theorem 4.2 *Let N be squarefree and $F \in S_k^{(2)}(N)$ lie in the orthogonal complement of $S_k^{(2),O}(N)$ and be an eigenform for the local Hecke algebras at all places. Let π_F be the irreducible cuspidal representation attached to F. Then π_F has a global Bessel model of fundamental type.*

Thus, our work on determination of Siegel cusp forms by fundamental Fourier coefficients has the very pleasant consequence of making *unconditional* all the results mentioned earlier due to Furusawa, Pitale, Schmidt and the author. It is worthwhile to remark here that most of those results are for $N = 1$; however, recent work by Pitale and Schmidt [18] should allow them to be extended to the case of arbitrary squarefree N.

4.2 Simultaneous non-vanishing of L-functions

In this subsection, we will describe an application of Proposition 3.7 to simultaneous non-vanishing of dihedral twists of modular L-functions. *This work was done jointly with Ralf Schmidt and will appear in more detailed form in [25].*

Let Λ be an ideal class group character of an imaginary quadratic field K of discriminant $-d$ and f be a classical holomorphic newform. One can then form the L-function $L(s, \pi_f \times \theta_\Lambda)$; this is the Rankin–Selberg convolution of the automorphic representation π_f attached to f and the θ-series

$$\theta_\Lambda(z) = \sum_{0 \neq \mathfrak{a} \subset O_K} \Lambda(\mathfrak{a}) e(N(\mathfrak{a})z).$$

Here, θ_Λ is a holomorphic modular form of weight 1 and nebentypus $\left(\frac{-d}{*}\right)$ on $\Gamma_0(d)$; it is a cusp form if and only if $\Lambda^2 \neq 1$.

The problem of studying the non-vanishing of the central values $L(\frac{1}{2}, \pi_f \times \theta_\Lambda)$ arises naturally in several contexts, and a considerable amount of work has been done in this direction. In this subsection, we will describe how our work so far leads to a *simultaneous* non-vanishing result for $L(\frac{1}{2}, \pi_f \times \theta_\Lambda)$, $L(\frac{1}{2}, \pi_g \times \theta_\Lambda)$ for two fixed forms f, g (but varying K and Λ) under certain hypotheses.

But first, back to Siegel cusp forms. Let $F \in S_k^{(2)}(N)$ be a Hecke eigenform and π_F be the irreducible cuspidal representation attached to F. Recall the definition of $R(F, K, \Lambda)$ from (3.4). A generalization of a famous conjecture of Böcherer [2] by several people (Furusawa, Shalika, Martin, Prasad, Takloo-Bighash), leads to the following very interesting Gross-Prasad type conjecture.

Conjecture 4.3 *Let $F \in S_k^{(2)}(N)$ be a Hecke eigenform, K an imaginary quadratic field and Λ an ideal class group character of K. Suppose that $R(F, K, \Lambda) \neq 0$. Then $L(\frac{1}{2}, \pi_F \times \theta_\Lambda) \neq 0$.*

While Conjecture 4.3 is open at the moment, it has been proved for certain special Siegel cusp forms known as Yoshida lifts.

The data required to define a Yoshida lift are as follows. Let N_1, N_2 be two squarefree integers that are *not* coprime and put $N = \mathrm{lcm}(N_1, N_2)$. Let f be a classical newform of weight 2 on $\Gamma_0(N_1)$ and g be a classical newform of weight $2k$ on $\Gamma_0(N_2)$. Finally, assume that for each prime p dividing $\gcd(N_1, N_2)$, f and g have the same Atkin-Lehner eigenvalue at p.

Existence of Yoshida lift: *Under the above assumptions, there exists a non-zero element $F \in S_{k+1}^{(2)}(N)$ that has the following properties:*

1. F lies in the orthogonal complement of $S_{k+1}^{(2),\mathrm{O}}(N)$.
2. F is an eigenform for the local Hecke algebras at all places.
3. For all automorphic representations σ on $\mathrm{GL}_n(\mathbb{A})$ for each n, we have $L(s, \pi_F \times \sigma) = L(s, \pi_f \times \sigma)L(s, \pi_g \times \sigma)$. Here and elsewhere, $L(s, \ )$ denotes the complete global Langlands L-function for the relevant representations.

The above lift was first investigated by Yoshida and has been studied extensively by Böcherer and Schulze-Pillot. In fact, the Yoshida lift is a certain case of Langlands functoriality; see Sect. [25, Sec. 3.2] for more details.

We briefly explain how the Yoshida lift F is constructed from the classical cusp forms f and g. First, we fix a definite quaternion algebra D which is unramified at all finite primes outside $\gcd(N_1, N_2)$. Via the Jacquet-Langlands correspondence, we transfer π_f, π_g to representations π'_f, π'_g on $D^\times(\mathbb{A})$. Using the isomorphism

$$(D^\times \times D^\times)/\mathbb{Q}^\times \cong GSO(4)$$

we obtain an automorphic representation $\pi'_{f,g}$ on $GSO(4, \mathbb{A})$. We use the theta lifting to transfer $\pi'_{f,g}$ to the automorphic representation π_F on $\mathrm{GSp}_4(\mathbb{A})$. Finally, the form F is constructed by picking a suitable vector in the space of π_F. For the details of this last step, as well as a more thorough explanation of Yoshida lifts from the representation theoretic point of view, we refer the reader to the author's paper with Schmidt [25, Sec. 3].

Now, recall the Conjecture 4.3 stated earlier. It turns out that this conjecture has been recently proved for Yoshida lifts by Prasad and Takloo-Bighash.

Theorem 4.4 (Prasad–Takloo-Bighash [20]) *Let f, g be classical newforms as above, and let $F \in S_{k+1}^{(2)}(N)$ be their Yoshida lift. Suppose that K is an imaginary quadratic field and Λ an ideal class group character of K such that $R(F, K, \Lambda) \neq 0$. Then $L(\frac{1}{2}, \pi_F \times \theta_\Lambda) \neq 0$.*

Putting together Proposition 3.7, Theorem 4.4 and the defining property of the Yoshida lift, we deduce the following result on simultaneous non-vanishing of L-functions.

Theorem 4.5 ([25]) *Let $k > 1$ be an odd integer. Let N_1, N_2 be two positive, squarefree integers such that $M = \gcd(N_1, N_2) > 1$. Let f be a holomorphic newform of weight 2 on $\Gamma_0(N_1)$ and g be a holomorphic newform of weight $2k$ on $\Gamma_0(N_2)$. Assume that for all primes p dividing M, the Atkin-Lehner eigenvalues of f and g coincide. Then there exists an imaginary quadratic field K and a character $\Lambda \in \widehat{\mathrm{Cl}_K}$ such that $L(\frac{1}{2}, \pi_f \times \theta_\Lambda) \neq 0$ and $L(\frac{1}{2}, \pi_g \times \theta_\Lambda) \neq 0$. In fact, if $D(f, g)$ is the set of d satisfying the following conditions:*

1. *$d > 0$ is an odd, squarefree integer and $-d$ is a fundamental discriminant,*
2. *There exists an ideal class group character Λ of $K = \mathbb{Q}(\sqrt{-d})$ such that $L(\frac{1}{2}, \pi_f \times \theta_\Lambda) \neq 0$ and $L(\frac{1}{2}, \pi_g \times \theta_\Lambda) \neq 0$,*

then, for an appropriate absolute constant $\delta > 0$, one has the lower bound

$$|\{0 < d < X,\ d \in D(f, g)\}| \gg_{f,g,\delta} X^{\delta}. \tag{4.1}$$

For related work on non-vanishing of L-functions, we refer the reader to the introductions of the papers [13] and [16].

Remark 4.6 By considering Yoshida lifts for more general (i.e. non squarefree) levels and more general congruence subgroups, it is possible to relax some of the conditions on f, g, N_1, N_2 above. For more details, we refer the reader to [24].

Remark 4.7 We say a few words about the restrictions on f and g in Theorem 4.5. The conditions that N_1, N_2 are squarefree and that the Atkin-Lehner eigenvalues of f and g coincide are needed to ensure that there exists a holomorphic Yoshida lift attached to (f, g) with respect to a

Siegel-type congruence subgroup $\Gamma_0^{(2)}(N) \subset \mathrm{Sp}_4(\mathbb{Z})$ of squarefree level N. Indeed, our key result (Theorem 3.4) on non-vanishing fundamental Fourier coefficients is only proved for Siegel cusp forms with respect to such congruence subgroups. However, even if these two conditions are removed, (f, g) will still have a Yoshida lift (possibly with respect to some other congruence subgroup) provided that there is a prime p dividing $\gcd(N_1, N_2)$ such that π_f, π_g are both discrete series at p.[7] So, an analogue of Theorem 3.4 for Siegel cusp forms with respect to more general congruence subgroups will allow us to remove some of the restrictions on f and g. This is currently work in progress by J. Marzec at the University of Bristol. To remove the restriction on the weight of g would require us to extend Theorem 3.4 to vector valued Siegel cusp forms, which seems possible at least in principle.

Acknowledgements I would like to thank Emmanuel Kowalski, Ameya Pitale and Ralf Schmidt for helpful comments on an earlier draft of this article. I would also like to thank Ramin Takloo-Bighash for suggesting the application of Section 4.2 to me.

References

[1] Valentin Blomer and Gergely Harcos, *Hybrid bounds for twisted L-functions*, J. Reine Angew. Math. **621** (2008), 53–79.

[2] Siegfried Böcherer, *Bemerkungen über die Dirichletreihen von Koecher und Maass*, Mathematica Gottingensis **68** (1986).

[3] V. A. Bykovskiĭ, *A trace formula for the scalar product of Hecke series and its applications*, Zap. Nauchn. Sem. S.-Peterburg. Otdel. Mat. Inst. Steklov. (POMI). **226** (1996), 14–36, 235–236.

[4] William Duke and Henryk Iwaniec, *Bilinear forms in the Fourier coefficients of half-integral weight cusp forms and sums over primes*, Math. Ann. **286** (1990), 783–802.

[5] Masaaki Furusawa, *On L-functions for* GSp(4) × GL(2) *and their special values*, J. Reine Angew. Math. **438** (1993), 187–218.

[7] The restriction that there is a prime dividing $\gcd(N_1, N_2)$ where π_f, π_g are both discrete series will probably be very difficult to remove by our method, because without this condition there are no Jacquet-Langlands transfers and hence no (holomorphic) Yoshida lifts. It is conceivable that one could still consider the "Fourier coefficients" of the non-holomorphic Yoshida lift in this setup and prove a non-vanishing result for those.

[6] T. Ibukiyama and H. Katsurada, *An Atkin-Lehner type theorem on Siegel modular forms and primitive Fourier coefficients*, Geometry and analysis of automorphic forms of several variables, Ser. Number Theory Appl. **7** (2012), 196–210.

[7] Henryk Iwaniec, *Fourier coefficients of modular forms of half-integral weight*, Invent. Math. **87** (1987), 385–401.

[8] Winfried Kohnen, *Newforms of half-integral weight*, J. Reine Angew. Math. **333** (1982), 32–72.

[9] Winfried Kohnen, *Fourier coefficients of modular forms of half-integral weight*, Math. Ann. **271** (1985), 237–268.

[10] Winfried Kohnen, *On Hecke eigenforms of half-integral weight*, Math. Ann. **293** (1992), 427–431.

[11] Wenzhi Luo and Dinakar Ramakrishnan, *Determination of modular forms by twists of critical L-values*, Invent. Math. **130** (1997), 371–398.

[12] M. Manickam and B. Ramakrishnan, *On Shimura, Shintani and Eichler-Zagier correspondences*, Trans. Amer. Math. Soc. **352** (2000), 2601–2617.

[13] Philippe Michel and Akshay Venkatesh, *Heegner points and non-vanishing of Rankin/Selberg L-functions*, Analytic number theory (Clay Math. Proc.), **7** (2007), 169–183.

[14] Hiro-aki Narita and Ameya Pitale and Ralf Schmidt, *Irreducibility criteria for local and global representations*, preprint.

[15] Mark E. Novodvorski and Ilya I. Piatetski-Shapiro, *Generalized Bessel models for the symplectic group of rank* 2, Mat. Sb. (N.S.) **90 (132)** (1973), 246–256.

[16] Ken Ono and Christopher Skinner, *Non-vanishing of quadratic twists of modular L-functions*, Invent. Math. **134** (1998), 651–660.

[17] Ameya Pitale and Ralf Schmidt, *Integral Representation for L-functions for* $\mathrm{GSp}_4 \times \mathrm{GL}_2$, J. Number Theory **129** (2009, 1272–1324.

[18] Ameya Pitale and Ralf Schmidt, *Bessel models for* GSp(4) *: Siegel vectors of square-free level*, preprint, 2012, available at http://www2.math.ou.edu/~rschmidt.

[19] Ameya Pitale, Abhishek Saha and Ralf Schmidt, *Transfer of Siegel cusp forms of degree 2*, to appear in Memoirs of Amer. Math. Soc..

[20] Dipendra Prasad and Ramin Takloo-Bighash, *Bessel models for* GSp(4), J. Reine Angew. Math. **655** (2011), 189–243.

[21] Abhishek Saha, *L-functions for holomorphic forms on* $\mathrm{GSp}(4)\times\mathrm{GL}(2)$ *and their special values*, Int. Math. Res. Not. IMRN **10** (2009), 1773–1837.

[22] Abhishek Saha, *Pullbacks of Eisenstein series from* $\mathrm{GU}(3,3)$ *and critical L-values for* $\mathrm{GSp}_4\times\mathrm{GL}_2$, Pacific J. Math. **246** (2010), 435–486.

[23] Abhishek Saha, *Siegel cusp forms of degree 2 are determined by their fundamental Fourier coefficients*, Math. Ann. (2012), DOI: 10.1007/s00208-012-0789-x.

[24] Abhishek Saha and Ralf Schmidt, *On the Petersson norms of Yoshida lifts*, forthcoming.

[25] Abhishek Saha and Ralf Schmidt, *Yoshida lifts and simultaneous non-vanishing of dihedral twists of modular L-functions*, J. London Math. Soc., to appear.

[26] Jean-Pierre Serre and Harold Stark, *Modular forms of weight* 1/2, Modular functions of one variable, VI (Proc. Second Internat. Conf., Univ. Bonn, Bonn, 1976). Lecture Notes in Math., **627**, 27–67.

[27] Goro Shimura, *On modular forms of half integral weight*, Ann. of Math. (2) **97** (1973), 440–481.

[28] Shunsuke Yamana, *Determination of holomorphic modular forms by primitive Fourier coefficients*, Math. Ann. **344** (2009), 853–862.

[29] Don Zagier, *Sur la conjecture de Saito-Kurokawa (d'après H. Maass)*, Seminar on Number Theory, Paris 1979–80. Progr. Math., **12**, 371–394.

Abhishek Saha, ETH Zürich – D-Math, Rämistrasse 101, 8092 Zürich, Switzerland.
E-mail: abhishek.saha@math.ethz.ch

Automorphic Representations and L-Functions
Editors: D. Prasad, C.S. Rajan, A. Sankaranarayanan, J. Sengupta

Publisher: Hindustan Book Agency, New Delhi, India

Beyond Endoscopy for the Relative Trace Formula I: Local Theory

Yiannis Sakellaridis

Abstract

For the group $G = \mathrm{PGL}_2$ we prove nonstandard matching and the fundamental lemma between two relative trace formulas: on one hand, the relative trace formula of Jacquet for the quotient $T\backslash G/T$, where T is a nontrivial torus; on the other, the Kuznetsov trace formula with *nonstandard* test functions. The matching is nonstandard in the sense that orbital integrals are related to each other not one-by-one, but via an explicit integral transform. These results will be used in [Saka] to compare the corresponding global trace formulas and reprove the celebrated result of Waldspurger [Wal85] on toric periods.

1 Introduction

1.1

With the present paper I launch an investigation of new ways to compare trace formulas in the field of automorphic forms, as a means of proving explicit relations between the spectra of two relative trace formulas (RTFs) — which is translated to relations between periods of automorphic forms. Such relations are predicted, in great generality, by the generalization of the Gross-Prasad-Ichino-Ikeda conjecture [II10] which is to appear in my joint work with Venkatesh [SV], and while the general conjecture is not yet very detailed, many special cases suggest the relevance of the RTF in formulating a detailed conjecture: the conjecture is not really about a *single* pair (G, H) consisting of a group and a subgroup, not even about the quotient space $X = H\backslash G$, but it involves certain *"pure inner forms"* of the pair (G, H) which can be understood in terms of the algebraic stack $H\backslash G/H$ (equivalently: $X \times X/G$). On the other hand, the relative trace formula, as currently being used following the paradigm of endoscopy for the Arthur-Selberg trace formula, seems to have no hope of proving relations in such generality, for reasons that will be explained below. Thus, we need new ways to compare trace formulas, which is what I am doing in the present paper, for two periods about which virtually everything is already known: the Whittaker period and the torus period, both for the group PGL_2. The continuation of the present paper in [Saka] will provide a new proof of the celebrated result of Waldspurger on toric periods of automorphic forms.

The topic of interest should not be seen as restricted to the study of periods and hence as something separate from the mainstream Langlands program. Indeed, the relative trace formula should be considered a potential generalization of the Arthur-Selberg trace formula (before stabilization), and the nonstandard comparison performed here is certainly within the spirit of the ambitious "beyond endoscopy" program proposed by Langlands [Lan04]. The difference lies in the level of ambition and difficulty: while Langlands wants to filter out only part of the spectrum of a trace formula by taking residues of L-functions, in order to detect the image of any chosen functorial lift, here I perform a full comparison of two trace formulas; thus, the spectral content is essentially dictated by the L-groups of the pertinant spherical varieties [SV]. On the other hand, many other features of the "beyond endoscopy" project are present. In particular, L-functions are inserted via nonstandard (not compactly supported) test functions; this is essential in view of the fact that, say, the Whittaker period and the torus period correspond to different L-functions; thus, a comparison of the corresponding RTFs using standard test functions would be impossible.

All known cases of "matching" between two RTFs — that is: direct comparison, up to scalar "transfer factors", of the functions obtained by (regular) orbital integrals — occur when the L-values associated to the pertinent periods agree. By "associated L-values" I mean the factors of the Euler products in the spectral expansion, assuming that the pertinent periods are Eulerian. In general, there is no reason to expect that for every Eulerian period that one wants to study there will be another "easier" one giving the same L-values, which is why it seems unlikely that the endoscopic paradigm of orbit-by-orbit comparisons will suffice to prove the period conjectures of [SV]. The comparison performed in the present paper is itself *nonstandard*: there is no matching of orbits such that the corresponding orbital integrals be preserved by the matching of functions; instead, matching is accomplished via a certain integral transform on the set of orbital integrals. While the existence of such a transform is more or less predicted by spectral matching, it is significant that for the example at hand *we are able to give an explicit formula for it*, in terms of Fourier transforms and birational maps. This is extremely important for the global story: since the comparison does not preserve orbital integrals, one will need to prove some kind of Poisson summation formula in order to obtain an identity between matching trace formulas; clearly, a Poisson summation formula for an arbitrary integral transform is no trivial thing — it might prove to be a reformulation of the functional equation of some intractable L-function. In my understanding, no one has yet proposed a conceptual reason why such a Poisson summation formula should be provable for the integral transforms that will appear in the "beyond endoscopy" project; thus, understanding instances where the transforms between spaces of orbital integrals are tractable is an important task.

1.2

Now I summarize the contents of the present paper. Throughout, G denotes the algebraic group PGL_2 over a local field F, or the F-points of G. The goal is to compare orbital integrals of test functions for the trace formula corresponding to the quotient $T\backslash G/T$, where T is a split or nonsplit torus, and orbital integrals of test functions for the Kuznetsov trace formula — this is the trace formula for the quotient $(N,\psi)\backslash G/(N,\psi)$, where N is a unipotent subgroup and ψ a nontrivial character of the F-points of N.

As explained before, such a comparison should not be possible with the usual test functions, for reasons related to different special values of L-functions appearing in global periods. Therefore, we need to modify the space of test functions — the standard choice is to take Schwartz functions on G, but now our test functions for the Kuznetsov trace formula will have nontrivial asymptotics at infinity. The best way to describe this is to think of the Kuznetsov orbital integrals not as distributions on G, but as G-invariant hermitian forms on $\mathcal{S}(N\backslash G,\psi)$. Then one should replace the standard Schwartz space $\mathcal{S}(N\backslash G,\psi)$ by sections which have a certain prescribed behavior at "infinity", where "infinity" means the partial compactification of $N\backslash G$ by $\mathbb{P}^1$. I describe these nonstandard sections in §4.5.

The affine quotient $N\backslash G /\!/ N$ is isomorphic to $\mathbb{A}^1$ (one-dimensional affine space), and so is the quotient $T\backslash G /\!/ T$. We will denote both by $\mathcal{B}$, the "base" of our quotient stacks. These quotients rougly parametrize orbits, and with suitable conventions that are explained in Sections 2, 3 and 4 we understand the orbital integrals as densely defined functions on $\mathcal{B}$. Here "densely" means that we only consider regular orbital integrals, where the stabilizers are trivial. Thus, we end up with two spaces of densely defined functions on $\mathcal{B}$, the space $\mathcal{S}(\mathcal{Z})$ (from orbital integrals for $T\backslash G/T$ equipped with standard test functions) and the space $\mathcal{S}(\mathcal{W})$ (from orbital integrals for the Kuznetsov quotient with nonstandard test functions).

Clearly (for anyone who has some experience with those trace formulas), these spaces of functions are not even closely related to each other as spaces of functions, that is: there is no orbit-by-orbit matching of the two trace formulas. However, the first main result (Theorem 5.1) is that there is an explicit integral transform which takes one to the other. To state it, let $\mathcal{F}$ denote usual Fourier transform in one variable (with respect to characters and measures that are described in the text), let η be the character of $F^\times$ corresponding to the splitting field of T, and let ι be the following operator on functions on $\mathcal{B}$:

$$\iota(f) = \frac{\eta(\bullet)}{|\bullet|} f\left(\frac{1}{\bullet}\right).$$

Define the operator:

$$\mathcal{G} = \mathcal{F} \circ \iota \circ \mathcal{F}.$$

Theorem ("Matching", 5.1) *The operator $|\bullet|\mathcal{G}$ is an isomorphism:*

$$\mathcal{S}(\mathcal{Z}) \to \mathcal{S}(\mathcal{W}).$$

The second main result, for F nonarchimedean and T unramified over $\mathbb{Q}_p$ or $\mathbb{F}_p((t))$ is a fundamental lemma for elements of the Hecke algebra. Here we need to specify (§5.3) "basic vectors" $f^0_{\mathcal{Z}}$, $f^0_{\mathcal{W}}$ for the spaces $\mathcal{S}(\mathcal{Z})$, $\mathcal{S}(\mathcal{W})$, which are obtained by the orbital integrals of certain $K := G(\mathfrak{o})$-invariant functions "upstairs". For the torus trace formula this will be the standard unramified test function, but for the Kuznetsov formula it has to be nonstandard, and tailored in order to produce the correct L-value $L(\pi, \frac{1}{2})L(\pi \otimes \eta, \frac{1}{2})$ on the spectral side of the trace formula.

Acting by an element h of the spherical Hecke algebra $\mathcal{H}(G, K)$ on those functions upstairs, we denote their orbital integrals by $h \star f^0_{\mathcal{Z}} \in \mathcal{S}(\mathcal{Z})$, $h \star f^0_{\mathcal{W}} \in \mathcal{S}(\mathcal{W})$. The fundamental lemma states that the above integral transform carries one to the other:

Theorem ("Fundamental Lemma", 5.4) *For any $h \in \mathcal{H}(G, K)$ the operator $|\bullet|\mathcal{G}$ carries $h \star f^0_{\mathcal{Z}}$ to $h \star f^0_{\mathcal{W}}$.*

1.3

The results of this paper will be used in the sequel [Saka] in order to reprove the theorem of Waldspurger [Wal85] on the Euler factorization of toric periods. This global application is far from a straightforward application of the local results; the first difficulty has to do with the fact that we do not have an orbit-by-orbit matching of trace formulas, and hence the matching of global trace formulas has to be proven by a quite nontrivial application of the Poisson summation formula. Other difficulties have to do with the fact that globally there are conconvergent Euler products, and we need to interpret them by analytic continuation; that is why in Section 6 we introduce variations of our spaces by a complex parameter $s \in \mathbb{C}$.

1.4

While I certainly hope that it will be possible to generalize such nonstandard comparisons to higher rank (hence proving new period relations that generalize the results of Waldspurger), I should add a word of caution: The relations that we get here can be seen as reflections at the level of orbital integrals of well-known results "upstairs": on one hand, the proof of Waldspurger's results by Jacquet [Jac86] relating orbital integrals for $T\backslash G/T$ to orbital integrals for the same quotient with a split torus; and on the other, the method of Hecke for calculating split torus periods via Fourier coefficients. I explain these relations in the proof of the fundamental lemma 5.4.

The point of the present paper is that while the work of Jacquet and Hecke does not generalize to all other periods one is interested in, the relations between orbital integrals might generalize. Of course, we will not be able to tell whether this is the case before studying many more examples and trying to find a pattern for nonstandard comparisons.

1.5 Acknowledgements

I would like to thank Akshay Venkatesh for pointing my attention to his thesis [Ven04] as a possible source of ideas for attacking the period conjectures of [SV]. I would also like to thank Joseph Bernstein, who taught me the correct way to think about several aspects of the relative trace formula. I am grateful to Bill Casselman for a careful reading and many helpful comments. Finally, it is my pleasure to acknowledge the support of the Institute for Advanced Study via NSF grant DMS-0635607 during the spring semester of 2011, when this work was initiated, and the support of NSF grant DMS-1101471.

1.6 Notation

Notation is mostly local, redefined in every section. For convenience of the reader, we give an overview of the symbols that are most frequently used (and will be defined in the text, if nonstandard):

- F is a local, locally compact field, E a quadratic etale extension of F. The quadratic character of $F^\times$ associated to E is denoted by $\eta = \eta_{E/F}$. If F is nonarchimedean, we denote by $\mathfrak{o}$ its ring of integers, by ϖ a uniformizer and by q the order of its residue field.
- We feel free to use the same symbol Y to denote the variety Y and its points over F, whenever this causes no confusion. For example, "functions on Y" means functions on the F-points of Y. When there is ambiguity, we will be denoting the latter by $Y(F)$.

- We fix a measure dx on F as explained in §2.3, and a unitary complex character ψ of F with respect to which dx is self-dual. The action of the multiplicative group $F^\times$ on functions on F is normalized in (2.14) in order to be unitary on L^2; this normalization makes Fourier transform on F anti-equivariant with respect to the action of $F^\times$.
- The symbols $\mathcal{X}, \mathcal{Z}, \mathcal{W}$ are reserved throughout the text for certain quotient stacks related, respectively, to the "baby case" of Section 2, the torus quotient of Section 3 or the Kuznetsov quotient of Section 4. In fact, the notion of "stack" is used in a rather symbolic way here and the reader only needs to understand the definitions of spaces $\mathcal{M}(\mathcal{X}), \mathcal{S}(\mathcal{X})$ etc. here, which are called the spaces of "measures" and "Schwartz functions" on those stacks but are actually defined in terms of coinvariants and orbital integrals on functions "upstairs" (on the homogeneous varieties). The only exception to this symbolic approach is a proof of certain isomorphisms of stacks in §3 and the subsequent derivation of isomorphisms between the pertinent Schwartz spaces; however, the arguments can easily be adapted to prove the isomorphisms of Schwartz spaces directly.

  Other than $\mathcal{X}, \mathcal{Z}, \mathcal{W}$, we define and redefine symbols locally, for instance the letter X usually denotes some homogeneous space, which is changing through the text.
- $\mathcal{B}$ denotes the "base" of our quotient stacks, i.e. the associated GIT quotient. In fact, throughout our examples we have an isomorphism: $\mathcal{B} \simeq \mathbb{G}_a$, which in general depends on some choices that we fix (cf. the remarks after Proposition 3.1). For each quotient space that we are considering, the "base" $\mathcal{B}$ has a regular set $\mathcal{B}^{\rm reg}$ (different in each case). We will be using this notation when it is clear which quotient space we are referring to, and the notation $\mathcal{B}^{\rm reg}_{\mathcal{X}}, \mathcal{B}^{\rm reg}_{\mathcal{Z}}$, etc. when we want to indicate the quotient space.
- For p-adic groups, the usual notion of "smooth" vectors and representations typically gives rise to inductive limits of Fréchet spaces. To achieve uniformity with the archimedean case, we describe in appendix A a notion of "almost smooth" vectors which gives rise to Fréchet space representations. For simplicity, we call these vectors "smooth" throughout the rest of the text and treat the archimedean and nonarchimedean cases together whenever possible, but the reader may ignore this and focus on smooth vectors in the traditional sense, replacing the Fréchet spaces that we consider with their corresponding limits of Fréchet spaces.
- The notion of Schwartz functions on an open semialgebraic set of a (smooth) real or p-adic manifold is defined in appendix A. When

"Schwartz function" appears without specifying on which set, we mean Schwartz function on F. Schwartz functions on a semialgebraic set X are denoted by $\mathcal{S}(X)$; we also use the letter $\mathcal{S}$ for the space of nonstandard Whittaker test functions that we define in §4.5, and for the "Schwartz space of a quotient stack", defined as the space of functions on $\mathcal{B}^{\text{reg}}$ obtained by taking regular orbital integrals. $\mathcal{M}$ denotes the space of Schwartz measures on the points of a real or p-adic variety, as well as on quotient stacks (defined as spaces of coinvariants).

- For spaces of smooth functions or smooth sections of line bundles, we use the word "stalk" as in §B.4 of the appendix, that is: an element of the stalk over a closed set is defined modulo Schwartz functions/sections on its complement — *not* modulo compactly supported functions/sections on its complement. In particular, the germ of a smooth function at a point is completely determined by its derivatives at that point.

- $O_\bullet$ is used to denote "regular" orbital integrals, while $\tilde{O}_\bullet$ is used to denote invariant distributions supported over the irregular points of the base. In particular, we fix the notation $\tilde{O}_0$ and $\tilde{O}_u$ for irregular orbital integrals defined for the split baby case in (2.9), (2.10), and the notation $\tilde{O}_{0+}$, $\tilde{O}_{0-}$ for the analogous orbital integrals in the nonsplit case, (2.22) and (2.23). For the other cases, we only fix the notation $\langle\ \rangle$ for a distinguished irregular orbital integral, called "inner product" and defined in §3.6 and §4.9.

- A basic integral transform used to compare spaces of orbital integrals is the transform $\mathcal{G}$ defined in (2.13).

2 A Baby Case

2.1

Before we work with non-commutative groups, we discuss the baby case of the relative trace formula for the variety $X := \mathbb{G}_a$ of the group $T := \mathbb{G}_m$, in order to examine certain integral transforms which will be useful in the sequel. We will also discuss a non-split form of the quotient $X \times X/T$, which although not under the general formalism of the relative trace formula, will provide us with some necessary integral transforms.

2.2 The split case

We let V denote an 1-dimensional vector space over F, and V^* its dual. In some of the calculations below, we will be identifying V and V^* with F under the pairing $\langle x, y\rangle = xy$. We let T be the group $\mathbb{G}_m$ acting diagonally on $V \times V^*$.

The stack-theoretic quotient $V \times V^*/\mathbb{G}_m$ will be denoted by $\mathcal{X}$; its "dual" quotient $V^* \times V/\mathbb{G}_m$ will be denoted by $\mathcal{X}^*$.

The pairing between V and V^* induces a canonical identification of the categorical quotient of affine spaces (the "bases"):

$$\mathcal{B} := V \times V^* /\!\!/ T \xrightarrow{\sim} \mathbb{G}_a \xleftarrow{\sim} V^* \times V /\!\!/ T. \tag{2.1}$$

We let $\mathcal{B}^{\mathrm{reg}}$, "the regular part of the base", denote the complement of zero. Notice that for every $\xi \in \mathcal{B}^{\mathrm{reg}}$ the fiber is a single orbit of T (both as a variety and in the sense of F-points).

2.3 Tamagawa measures

If instead of F we were talking about a global field k, we would be fixing the measure on its ring of adeles which comes from a globally defined differential form on $\mathrm{Res}_{k/\mathbb{Q}}\mathbb{G}_a$ and the usual measure on $\mathbb{A}_{\mathbb{Q}}$. We factorize this in the standard way locally [Tat67], namely:

If $F = \mathbb{R}$, the usual Lebesgue measure, if $F = \mathbb{C}$ the double of the usual Lebesgue measure and if F is nonarchimedean the Haar measure under which the ring of integers has measure equal to the inverse square root of the discriminant (hence, ≤ 1).

2.4 Measures and coinvariants

We let $\mathcal{S}(V \times V^*)$ denote the Fréchet space of Schwartz functions on $V \times V^*$. Notice that the usual notion of "Schwartz space" in the nonarchimedean case does not correspond to a Fréchet space but to a limit of Fréchet spaces. We explain in the appendix A how to obtain a Fréchet space completion thereof, consisting of functions which have the same decay at infinity as in the archimedean case (faster than the absolute value of any polynomial) and are *almost smooth* instead of smooth. For practical purposes, the difference between the two approaches is unimportant, and the reader can keep the traditional Schwartz space in their mind. The introduction of a Fréchet space just creates the convenience of treating the archimedean and nonarchimedean cases simultaneously. As explained in the introduction, we will be just using the word "smooth" for what should be "almost smooth"; also, any statement involving derivatives (other than the zeroth) should be considered as void in the nonarchimedean case.

We denote by $\mathcal{M}(V \times V^*)$ the corresponding Fréchet space of Schwartz measures on $V \times V^*$ (i.e. products of a Schwartz function with additive Haar measure). A choice of Haar measure on $V \times V^*$ defines an isomorphism: $\mathcal{S}(V \times V^*) \simeq \mathcal{M}(V \times V^*)$, but we will not need to fix such an isomorphism except as a convenience for certain calculations.

We define the *Schwartz space of measures on* $\mathcal{X}$ to be:

$$\mathcal{M}(\mathcal{X}) := \mathcal{M}(V \times V^*)_T, \tag{2.2}$$

the coinvariant space $\mathcal{M}(V \times V^*)_T$. By definition, the T-coinvariant space of a Fréchet representation W is the quotient by the closed subspace generated by vectors of the form $v - g \cdot v$, $v \in W, g \in T$, or equivalently the universal quotient: $W \to W_T$, with trivial T-action on the right, through which every continuous T-invariant functional factors. In particular, it is naturally a nuclear Fréchet space. The space $\mathcal{F}(\mathcal{X})$ of T-invariant, *tempered* generalized functions on $V \times V^*$ is, tautologically, the dual of $\mathcal{M}(\mathcal{X})$.

The following is standard:

Lemma 2.1 *$\mathcal{F}(\mathcal{X})$ is the weak-* closure of the space spanned by those invariant generalized functions which are each supported on a fiber of the map: $\mathcal{X} \to \mathcal{B}$.*

We will see a strengthening of it in Lemma 2.3.

2.5 Orbital integrals

For $\xi \in \mathcal{B}^{\text{reg}}$, $\Phi \in \mathcal{S}(V \times V^*)$ and a Haar measure dg on T we define the orbital integral:

$$O_\xi(\Phi) = \int_T \Phi(\tilde{\xi} \cdot g) dg. \tag{2.3}$$

Here $\tilde{\xi}$ is any lift of ξ to $V \times V^*$.

We define $\mathcal{S}(\mathcal{X})$ to be the space of functions on $\mathcal{B}^{\text{reg}}$ of the form $\xi \mapsto O_\xi(\Phi)$, for $\Phi \in \mathcal{S}(V \times V^*)$. Throughout the text, when we talk about "a lift of an element $f \in \mathcal{S}(\mathcal{X})$ to $\mathcal{S}(V \times V^*)$" we will implicitly mean a pair consisting of an element $\Phi \in \mathcal{S}(V \times V^*)$ and a Haar measure on T, so that f is obtained by the orbital integrals of Φ. Our first goal is to define (and normalize) a linear map: $\mathcal{M}(\mathcal{X}) \to \mathcal{S}(\mathcal{X})$. To do this, we start with the following integration formula:

Lemma 2.2 *For any $\Phi \in \mathcal{S}(V \times V^*)$ with image $f \in \mathcal{S}(\mathcal{X})$, and Haar measures on $V \times V^*$ and T, we have:*

$$\int_{V \times V^*} \Phi(v, v^*) dv dv^* = \int_{\mathcal{B}} f(\xi) d\xi, \tag{2.4}$$

where $d\xi$ is an additive Haar measure on $\mathcal{B} = F$. If we take on $V \times V^*$ the Haar measure corresponding to the differential form $dv \wedge dv^*$ and the standard Haar measure on F, where the coordinates v and v^* are defined using a dual basis, dg is the multiplicative Haar measure $|a|^{-1}da$ on $F^\times$, then $d\xi$ is the standard Haar measure on F (§2.3).

If we now *fix* the measure on $\mathcal{B}$ to be the standard measure on F discussed in §2.3, we get a map:

$$\mathcal{M}(\mathcal{X}) \to \mathcal{S}(\mathcal{X}), \tag{2.5}$$

as follows: a choice of compatible measures on $V \times V^*$ and T gives an isomorphism:

$$\mathcal{M}(V \times V^*) \simeq \mathcal{S}(V \times V^*)$$

and a map:

$$\mathcal{S}(V \times V^*) \to \mathcal{S}(\mathcal{X}),$$

and it is easy to see that the composition of the two depends only on the chosen measure on $\mathcal{B}$. For the purpose of calculations later in the chapter we will fix the Haar measures on $V \times V^*$ and T described in Lemma 2.2.[1]

The following strengthening of Lemma 2.1 will be a corollary of Proposition 2.5.

Lemma 2.3 *The functionals O_ξ, $\xi \in \mathcal{B}^{\text{reg}}$ span a weak-* dense subspace of $\mathcal{F}(\mathcal{X})$.*

This implies that *the map (2.5) is an isomorphism of vector spaces.* Therefore, we will not be distinguishing from now on between $\mathcal{M}(\mathcal{X})$ and $\mathcal{S}(\mathcal{X})$, and we will endow $\mathcal{S}(\mathcal{X})$ with the Fréchet topology induced from this identification.

In appendix B we introduce a notion of "Schwartz cosheaves". We point the reader there for definitions of restriction, stalks and other notions. By Corollary B.5.2, $\mathcal{S}(\mathcal{X})$ is the space of global sections of *a flabby* (i.e. extension maps are injective — in fact, closed embeddings) *Schwartz cosheaf* on $\mathcal{B}$, which for simplicity we will also be referring to by the symbol $\mathcal{S}(\mathcal{X})$ when there is no confusion. Via the regular orbital integrals, this cosheaf is identified with a cosheaf of functions on $\mathcal{B}^{\text{reg}}$; our purpose is to describe this cosheaf.

Lemma 2.4 *The restriction of the cosheaf $\mathcal{S}(\mathcal{X})$ to $\mathcal{B}^{\text{reg}}$ is equal to the cosheaf $\mathcal{S}(\mathcal{B}^{\text{reg}})$ (Schwartz functions in the usual sense).*

[1]Even fixing a measure on $\mathcal{B}$ locally is not important, since globally we always have canonical choices of measures (Tamagawa measures); nonetheless it will be helpful for calculations to fix the local maps (2.5).

Proof Over $\mathcal{B}^{\mathrm{reg}}$ we have a T-isomorphism of the variety $V \times V^*$ with $\mathcal{B}^{\mathrm{reg}} \times T$, and therefore $\mathcal{X}^{\mathrm{reg}} := \mathcal{X} \times_{\mathcal{B}} \mathcal{B}^{\mathrm{reg}}$ is isomorphic to $\mathcal{B}^{\mathrm{reg}}$.[2]

□

Now we focus our attention on the neighborhood of 0:

Proposition 2.5 *For $\Phi \in \mathcal{S}(V \times V^*)$ we have, for $\xi \in \mathcal{B}^{\mathrm{reg}}$ in a neighborhood of 0:*

$$O_\xi(\Phi) = -C_1(\xi) \cdot \ln|\xi| + C_2(\xi), \tag{2.6}$$

with C_1, C_2 (almost) smooth functions (which can be arbitrary).

Moreover, the distributions:

$$\tilde{O}_0(\Phi) := C_1(0) \tag{2.7}$$

and

$$\tilde{O}_u(\Phi) := C_2(0) \tag{2.8}$$

are a basis for the space of functionals on the fiber of $\mathcal{S}(\mathcal{X})$ over $0 \in \mathcal{B}$, and we have:

$$\tilde{O}_0(\Phi) = \mathrm{Vol}(T(F)_0)\Phi(0), \tag{2.9}$$

where $\mathrm{Vol}(T(F)_0)$ is the volume of the maximal compact subgroup of $T(F)$ described below, and:

$$\begin{aligned}\tilde{O}_u(\Phi) &= \lim_{s\to 0}\left(\zeta(\Phi|_{y=0}, s) + \zeta(\Phi|_{x=0}, -s)\right) = \\ &= \frac{d}{ds}\Big|_{s=0}\left(s\zeta(\Phi|_{y=0}, s) + s\zeta(\Phi|_{x=0}, s)\right),\end{aligned} \tag{2.10}$$

where ζ is the Tate integral of a function of one variable against unramified characters defined with the chosen measure on T, e.g. for $\zeta(\Phi|_{y=0}, s)$ we choose the isomorphism $T \simeq \mathbb{G}_m$ such that $\lim_{t\to 0}(1,0)\cdot t = (0,0)$ and have:

$$\zeta(\Phi|_{y=0}, s) = \int_T \Phi(t,0)|t|^s dt.$$

Remark 2.6 If we set $f(\xi) = O_\xi(\Phi)$ then the distributions $\tilde{O}_0, \tilde{O}_u$ have been defined in such a way that they depend only on $f \in \mathcal{S}(\mathcal{X})$ and not on Φ and the choice of Haar measure on T, i.e. if we modify Φ and the Haar measure simultaneously so that the orbital integrals of Φ continue to

---

[2] There is clearly some work to be done to establish that isomorphisms of stacks give rise to isomorphisms of their Schwartz spaces, but in each of the cases that we are considering in this paper this is easy to see explicitly.

give f, we get the same values of $\tilde{O}_0(\Phi), \tilde{O}_u(\Phi)$. It is therefore meaningful to write: $\tilde{O}_0(f), \tilde{O}_u(f)$.

The volume mentioned in the lemma is obtained as follows. Notice that the absolute value gives a canonical short exact sequence:

$$1 \to T(F)_0 \to T(F) \xrightarrow{|\bullet|} \mathrm{Hom}(\mathcal{X}^*(T)_F, |F^\times|) \to 1, \tag{2.11}$$

where $T(F)_0$ denotes the maximal compact subgroup of $T(F)$, $\mathcal{X}^*(T)_F \simeq \mathbb{Z}$ is the F-character group of T, and $|F^\times| \subset \mathbb{R}_+^\times$ denotes the group of absolute values of $F^\times$. The group $\mathrm{Hom}(\mathcal{X}(T)_F, |F^\times|)$ is canonically, up to inversion, a subgroup of $\mathbb{R}_+^\times$ (all of $\mathbb{R}_+^\times$ in the archimedean case, the group $q^{\mathbb{Z}}$ in the nonarchimedean case). *We endow it with a Haar measure $d|t|$ that is on average equal to the standard multiplicative measure $t^{-1}dt$ on $\mathbb{R}_+^\times$.* Hence, in the nonarchimedean case with residual degree q, $dt(\{1\}) = \ln q$. For a Haar measure μ on T, the disintegration of $\frac{d\mu}{d|t|}$ with respect to the map (2.11) is a Haar measure on $T(F)^0$, and we let:

$$\mathrm{Vol}(T(F)_0) = \frac{d\mu}{d|t|}(T(F)_0).$$

In particular, in the nonarchimedean case, this is $(\ln q)^{-1}$ times $\mu(T(F)_0)$.

Proof It is easy to see that all functions of the form $c_2 - c_1 \ln|\xi|$ can be obtained as orbital integrals in a neighborhood of zero, with $c_i = C_i(0)$ as claimed. It is then easy to see that the invariant distributions on the fiber of $\mathcal{S}(V \times V^*)$ over the preimage of 0 are given by (2.9) and (2.10). Thus, by Proposition B.4.1, the orbital integrals of all elements of $\mathcal{S}(V \times V^*)$ are of the form $C_2(\xi) - C_1(\xi)\ln|\xi|$ with C_i (almost) smooth functions.

This proves the lemma, and also Lemma 2.3.

□

Remark 2.7 In the archimedean case, a very novel and detailed analysis of orbital integrals, including most of the above theorem, is performed by Casselman and Tian in their preprint [CT].

2.6 Fourier transform

We choose a character $\psi : F \to \mathbb{C}^\times$ to identify the space $V^* \times V$ with the Pontryagin dual of $V \times V^*$; we choose it in such a way that the measure 2.3 on F is self-dual with respect to Fourier transform. Notice that, when working with a global field, adele class characters can be factorized as products of such characters.

In this paper "Fourier transform" stands for the usual, schoolbook Fourier transform from functions on F^n to functions on F^n, without any modifications to preserve equivariance, which in one variable $v \in V$ reads: $\hat{f}(v^*) = \int_F f(v)\psi^{-1}(\langle v, v^* \rangle)dv$, and in more variables is defined variable-by-variable. It will be denoted both by $\hat{}$ and also by the letter $\mathcal{F}$.

Hence we have:

$$\hat{\hat{\Phi}}(x, y) = \Phi(-x, -y).$$

Since Fourier transform in one variable satisfies:

$$\mathcal{F}(f(a\bullet))(y) = \frac{1}{|a|}\mathcal{F}(f)\left(\frac{y}{a}\right),$$

it is clear that Fourier transform on $V \times V^*$ is equivariant with respect to the action of T on $V \times V^*$ and on its dual, and therefore descends to an isomorphism:

$$\mathcal{S}(\mathcal{X}) \xrightarrow{\sim} \mathcal{S}(\mathcal{X}^*). \tag{2.12}$$

It is therefore natural to ask how it transforms orbital integrals, or in other words: For each $\Phi \in \mathcal{S}(V \times V^*)$, express the function $\mathcal{B}^{\text{reg}} \ni \xi \mapsto O_\xi(\Phi)$ in terms of the function $\xi \mapsto O_\xi(\hat{\Phi})$.

2.7 The integral transform $\mathcal{G}$.

Definition 2.8 We let $\mathcal{G}$ denote the transform which maps $f \in \mathcal{S}(\mathcal{X})$ to the Fourier transform of the (tempered) function $y \mapsto \frac{\eta(y)}{|y|}\hat{f}\left(\frac{1}{y}\right)$, that is:

$$\mathcal{G} = \mathcal{F} \circ \iota \circ \mathcal{F}, \tag{2.13}$$

where $\iota(f) = \frac{\eta(\bullet)}{|\bullet|} f\left(\frac{1}{\bullet}\right)$.

Recall that η is the quadratic character associated to the splitting field of T, so in the split case that we are currently discussing we have $\eta = 1$; however, this formula will be used in the nonsplit case, as well.

The following lemma shows, in particular, that the function $y \mapsto \frac{\eta(y)}{|y|}\hat{f}(\frac{1}{y})$ is in $L^2(\mathcal{B})$ and hence its Fourier transform makes sense as a function.

Lemma 2.9 *The Fourier transform of any $f \in \mathcal{S}(\mathcal{X})$ has the property that:*

$$\lim_{x \to \infty} |x|\hat{f}(x)$$

exists.

For the following proof, and later use, we normalize the action of $F^\times$ on functions on F in such a way that it is unitary (with respect to the $L^2(F)$-inner product):

$$(a \cdot f)(x) = |a|^{\frac{1}{2}} f(ax). \tag{2.14}$$

Hence, Fourier transform is anti-equivariant with respect to this action.

Proof The function $l : x \mapsto \ln|x|$ becomes smooth by application of the operator $(\mathrm{Id} - |a|^{-\frac{1}{2}} a\cdot)$, for all $a \in F^\times$. Therefore, its Fourier transform (considered as a tempered generalized function) will become rapidly decaying by application of the operator $(\mathrm{Id} - |a|^{-\frac{1}{2}} a^{-1}\cdot)$. Hence:

$$\hat{l}(x) = \frac{c}{|x|} + h_1(x)$$

in a neighborhood of infinity, for some constant c and some Schwartz function $h_1(x)$.

Thus, the Fourier transform of an element of $\mathcal{S}(\mathcal{X})$ will be of the form:

$$h(x) = h_2(x) \star \hat{l}(x) + h_3(x)$$

in a neighborhood of infinity, where h_i are Schwartz functions and $\star$ denotes convolution. It is easy to see that for such a function the limit: $\lim_{|x|\to\infty} |x| h(x)$ exists.

□

Hence, for $f \in \mathcal{S}(\mathcal{X})$ its Fourier transform $\hat{f}$ belongs to the space of continuous functions h on $\mathcal{B}$ with the property that $\lim_{|x|\to\infty} |x|h(x)$ exists. It is clear that ι is an involution on this space. The proof of the next Proposition will show that it preserves the image of $\mathcal{S}(\mathcal{X})$:

Proposition 2.10 *Let $\Phi \in \mathcal{S}(V \times V^*)$ and let $f(\xi) = O_\xi(\Phi)$. Then:*

$$O_\xi(\hat{\Phi}) = \mathcal{G}(f)(\xi). \tag{2.15}$$

Proof We denote by $\hat{\Phi}^1, \hat{\Phi}^2$ the partial Fourier transforms with respect to the first or second argument. We can treat f as a tempered distribution on $\mathcal{B}$; let h be a Schwartz function on $\mathcal{B}$, then according to the integration formula (2.2) we have:

$$\int_{\mathcal{B}} f(\xi)\overline{h(\xi)}d\xi = \iint \Phi(x,y)\overline{h(xy)}dxdy = \iint \hat{\Phi}^1(x,y)\overline{\hat{h}\left(\frac{x}{y}\right)}|y|^{-1}dxdy =$$
$$\iint \hat{\Phi}(x,y)\overline{\mathcal{G}(h)(xy)}dxdy == \int_{\mathcal{B}} O_\xi(\hat{\Phi})\overline{\mathcal{G}(h)}(\xi)d\xi.$$

It is easy to see that all the integrals above are absolutely convergent. Now, $\mathcal{G} = \mathcal{F} \circ \iota \circ \mathcal{F}$, and the operations $\mathcal{F}$ and ι preserve inner products, therefore:

$$\int_{\mathcal{B}} f(\xi)\overline{h(\xi)}d\xi = \int_{\mathcal{B}} \mathcal{G}(f)(\xi)\overline{\mathcal{G}(h)}(\xi)d\xi.$$

Hence, $O_\xi(\hat{\Phi}) = \mathcal{G}(f)(\xi)$.

□

It now follows that $\mathcal{G}f$ not only is a function on $\mathcal{B}^{\text{reg}}$, but it belongs to $\mathcal{S}(\mathcal{X})$. Indeed, since Fourier transform is a topological automorphism of $\mathcal{S}(V \times V^*)$, it follows that $\mathcal{G}$ is a topological automorphism of $\mathcal{S}(\mathcal{X})$, identified with the space of coinvariants. It also follows that the image of $\mathcal{S}(\mathcal{X})$ under Fourier transform is ι-stable:

Corollary 2.11 *The Fourier transform of $\mathcal{S}(\mathcal{X})$ is the space of those (almost) smooth functions on $\mathcal{B}$ which in a neighborhood of infinity are equal to $|x|^{-1}h\left(\frac{1}{x}\right)$, for some $h \in \mathcal{S}(\mathcal{B})$. Moreover, Fourier transform descends to a topological isomorphism between $\mathcal{S}(\mathcal{X})/\mathcal{S}(\mathcal{B})$ and the stalk[3] of functions of the form:*

$$|x|^{-1}h\left(\frac{1}{x}\right)$$

at ∞ (with the obvious topology, given by the derivatives of h at 0).

Recall that since $\mathcal{B}$ is a smooth variety, $\mathcal{S}(\mathcal{B})$ just denotes the usual space of Schwartz functions on $\mathcal{B}(F) = F$. The result is similar to, but not quite contained in, a special case of [Igu78, Theorem 2.1].

Proof It is clear that we have a short exact sequence:

$$0 \to \mathcal{S}(\mathcal{B}) \to \mathcal{F}\left(\mathcal{S}(\mathcal{X})\right) \to V \to 0,$$

arising as the Fourier transform of the sequence:

$$0 \to \mathcal{S}(\mathcal{B}) \to \mathcal{S}(\mathcal{X}) \to \mathcal{S}(\mathcal{X})/\mathcal{S}(\mathcal{B}) \to 0,$$

where $\mathcal{S}(\mathcal{B})$ is endowed with its usual topology, $\mathcal{F}\left(\mathcal{S}(\mathcal{X})\right)$ is endowed with the topology of $\mathcal{S}(\mathcal{X})$ and V is defined by this short exact sequence. Moreover, the first arrow is a closed embedding, and all elements of $\mathcal{F}\left(\mathcal{S}(\mathcal{X})\right)$ coincide with elements of $\mathcal{S}(\mathcal{B})$ on any compact subset of $\mathcal{B}$.

Since ι is a topological automorphism of $W := \mathcal{F}\left(\mathcal{S}(\mathcal{X})\right)$, this implies that W consists precisely of functions as in the statement of the corollary. Such functions are sections of a Schwartz cosheaf over $\overline{\mathcal{B}} = \mathbb{P}^1$ in an obvious

[3]Recall that the notion of "stalk" used for smooth functions is the one of appendix B; in particular, the germ of a smooth function at a point is determined by its derivatives.

way, and ι induces an isomorphism from the stalk at zero to the stalk at infinity — the latter being equal to the quotient V. From this it follows that the topology on V is given by the derivatives of h at 0 (where h appears in the expansion of a given element at ∞ as in the statement).

□

2.8 Mellin transform

Let us now view $\mathcal{B} \simeq \mathbb{G}_a$ as a vector space, and describe the integral operator $\mathcal{G}$ in terms of Mellin transforms with respect to the action of $\mathbb{G}_m$ on $\mathbb{G}_a$. We normalize the action of $F^\times$ on $L^2(\mathcal{B})$ as in (2.14).

By the asymptotic behavior of Lemma 2.9, any $f \in \mathcal{S}(\mathcal{X})$ satisfies the Mellin inversion formula:

$$f(\xi) = \int_{|\bullet|^\kappa \cdot \widehat{F^\times}} \check{f}(\chi)\chi(\xi)|\xi|^{-\frac{1}{2}} d\chi \tag{2.16}$$

for every $\kappa < -\frac{1}{2}$. Here $\check{f}(\chi)$ denotes the Mellin transform (not to be confused with the Fourier transform $\hat{f}$):

$$\check{f}(\chi) = \int_{F^\times} |\xi|^{\frac{1}{2}} f(\xi)\chi^{-1}(\xi) d^\times \xi. \tag{2.17}$$

We do not explicate the dual measures of the formulas above and below, because we will only be interested in gamma factors, which do not depend on the measures. For this reason, we ignore the fact that our normalization of multiplicative measures is different from the one of Tate's thesis.

We claim:

Lemma 2.12 *We have* $\widecheck{\mathcal{G}(f)}(\chi) = \gamma(\chi, \frac{1}{2}, \psi)^2 \check{f}(\chi^{-1})$, *where* $\gamma(\chi, s, \psi)$ *is the gamma factor of* χ *at* s *(cf. below).*

By the Mellin inversion formula (2.16), this completely characterizes the operator $\mathcal{G}$.

Proof By continuity of $\mathcal{G}$, it is enough to prove it for f in a dense subspace of $\mathcal{S}(\mathcal{X})$, so let us assume that $f(\xi) = O_\xi(\Phi)$ with $\Phi(x, y) = \Phi_1(x)\Phi_2(y)$. Then by the integration formula we can write:

$$\check{f}(\chi) = \iint \Phi_1(x)\Phi_2(y)\chi^{-1}(xy)|xy|^{\frac{1}{2}} d^\times x d^\times y =$$

$$\zeta(\Phi_1, \chi^{-1}, \frac{1}{2}) \cdot \zeta(\Phi_2, \chi^{-1}, \frac{1}{2}),$$

where $\zeta(\Phi_i, \chi, s) = \int_{F^\times} \Phi_i(x)\chi(x)|x|^s d^\times x$ denotes the *Tate integral* of Φ_i [Tat67].

Similarly, $\widetilde{\mathcal{G}(f)} = \zeta(\hat{\Phi}_1, \chi^{-1}, \frac{1}{2}) \cdot \zeta(\hat{\Phi}_2, \chi^{-1}, \frac{1}{2})$, and by the functional equation for Tate integrals we have, by definition:

$$\gamma(\chi, s, \psi)\zeta(\Phi_i, \chi, s) = \zeta(\hat{\Phi}_i, \chi^{-1}, 1-s).$$

This implies the claim of the lemma.

□

2.9 The non-split case

We discuss now the non-split version of the previous example, where $V \times V^*$ has been replaced by the space whose F-points are equal to the elements of a quadratic field extension E, under the action of the group of elements of norm 1. In other words, we take:

$$X = \mathrm{Res}_{E/F}\mathbb{G}_a,$$

the one-dimensional torus T over k defined by the short exact sequence:

$$1 \to T \to \mathrm{Res}_{E/F}\mathbb{G}_m \xrightarrow{N_F^E} \mathbb{G}_m \to 1,$$

and we will be interested in the quotient stack:

$$\mathcal{X} = X/T.$$

The quadratic character of $F^\times$ associated to the extension E will be denoted by $\eta_{E/F}$ or simply η.

Again we have a canonical isomorphism of categorical quotients: $\mathcal{B} := X /\!\!/ T \xrightarrow{\sim} \mathbb{G}_a$ given by the norm map.

The first thing to notice here is that the quotient stack has "points" corresponding to nontrivial torsors of T and which are, therefore, not accounted by F-points of X; this is already evident by the fact that the map: $X \twoheadrightarrow \mathcal{B}$ is not surjective at the level of F-points. We therefore propose the following two definitions of $\mathcal{M}(\mathcal{X})$, which can be seen to be equivalent (the first one was suggested to me by Joseph Bernstein):

1. We let $T \to \mathrm{GL}_2$ be an embedding, and let $\mathcal{M}(\mathcal{X})$ denote the $\mathrm{GL}_2(F)$-coinvariants of $\mathcal{M}\left((X \times^T \mathrm{GL}_2)(F)\right)$ (Schwartz measures).
2. We let:
$$\mathcal{M}(\mathcal{X}) = \oplus_\alpha \mathcal{M}(X^\alpha(F))_{T^\alpha(F)}, \tag{2.18}$$
the direct sum of coinvariant spaces, where α ranges over all isomorphism classes of T-torsors.

Here for a T-torsor R^α in the isomorphism class denoted by α we let $T^\alpha = \mathrm{Aut}(R^\alpha)^T$ and $X^\alpha = X \times^T R^\alpha$. (In terms of Galois cohomology, α can be regarded as denoting an element of $H^1(F,T)$, T^α is defined by its image in $H^1(F, \mathrm{Aut}(T))$ and X^α by its image in $H^1(F, \mathrm{Aut}(X))$.) Of course, in this case we have $T^\alpha \simeq T$ and $X^\alpha \simeq X$ (noncanonically) for all α, but these constructions make sense in a much more general setting — and explain inner forms appearing in the relative trace formula.

Although the first definition is more natural and geometric, the second one is more suitable for spectral expansions, and we will be working with that.

Notice that X^α has F-points if and only if R^α admits a T-equivariant morphism into X. Also, the definition $X^\alpha = X \times^T R^\alpha$ implies that as quotient stacks $X^\alpha/T^\alpha \simeq X/T$ canonically, and similarly for the GIT quotients we have: $X^\alpha /\!\!/ T^\alpha \simeq X /\!\!/ T = \mathcal{B}$. Since the GIT quotient is defined by the norm map from E to F, this means that the norm map extends to a (surjective) map: $\bigsqcup_\alpha X^\alpha(F) \to F$. Therefore, it makes sense to say that a T^α-orbit on X^α is "over" a point $\xi \in \mathcal{B}(F) \simeq F$; moreover, for every $\xi \in \mathcal{B}^{\mathrm{reg}}$ there is a unique such α and a unique such orbit, more precisely: if ξ belongs to the norms from E to F then α corresponds to the trivial torsor, while if ξ is not a norm then α corresponds to the nontrivial one.

2.10 Integration formula, Orbital integrals

The regular orbital integrals of a function $\Phi \in \oplus_\alpha \mathcal{S}(X^\alpha)$ are defined as:

$$O_\xi(\Phi) = \int_{T^\alpha} \Phi(\tilde{\xi} \cdot g) dg, \ \ \xi \in \mathcal{B}^{\mathrm{reg}}. \tag{2.19}$$

Here α is the torsor such that ξ has a lift $\tilde{\xi} \in X^\alpha(F)$. This definition depends on choosing measures on the tori T^α, which we are going to fix below. We define $\mathcal{S}(\mathcal{X})$ as the cosheaf over $\mathcal{B}$ of functions on $\mathcal{B}^{\mathrm{reg}}$ obtained as orbital integrals of elements of $\oplus_\alpha \mathcal{S}(X^\alpha)$ (this definition does not depend on choices of measures).

We endow $X(F) = E$, as we did with F, with the standard additive measure discussed in 2.3. When α denotes the (unique) isomorphism class of nontrivial torsors, we may fix an isomorphism $\iota : X \simeq X^\alpha$ which is equivariant with respect to some identification of T^α with T. If the image of $1 \in E$ is an element $e \in X^\alpha(F)$, we define a measure on $X^\alpha(F)$ as $|N_F^E(e)|$ times the push-forward of Haar measure on E. It is easy to see that this measure does not depend on the choice of isomorphism. Then:

Lemma 2.13 *There are is a compatible choices of Haar measures on the tori $T^\alpha(F)$ such that for any $\Phi \in \mathcal{S}(\bigsqcup_\alpha X^\alpha)$ with image $f \in \mathcal{S}(\mathcal{X})$ we have:*

$$\int_{\bigsqcup_\alpha X^\alpha(F)} \Phi(x)dx = \int_{\mathcal{B}} f(\xi)d\xi. \tag{2.20}$$

For the trivial torsor, this is the measure on $T(F)$ which disintegrates the multiplicative measures $|x|^{-1}dx$ on $E^\times$ and $F^\times$ (where $|\bullet|$ denotes the absolute value on each of them — notice that these do not coincide on $F^\times \subset E^\times$) with respect to the exact sequence:

$$1 \to T(F) \to E^\times \xrightarrow{N_F^E} F^\times.$$

Proposition 2.14 *The restriction of $\mathcal{S}(\mathcal{X})$ to $\mathcal{B}^{\text{reg}}$ is equal to $\mathcal{S}(\mathcal{B}^{\text{reg}})$. In a neighborhood of zero, the sections of $\mathcal{S}(\mathcal{X})$ are precisely those functions of the form:*

$$C_1(\xi) + C_2(\xi)\eta(\xi), \tag{2.21}$$

with C_1, C_2 (almost) smooth functions in one variable. Moreover, the values $C_1(0), C_2(0)$ are a basis for functionals on the fiber of $\mathcal{S}(\mathcal{X})$ over $\xi = 0$, and we have $C_1(0) = \tilde{O}_{0+}(\Phi)$, $C_2(0) = \tilde{O}_{0-}(\Phi)$, where:

$$\tilde{O}_{0+}(\Phi) = \frac{1}{2}\text{Vol}(T(F))\left(\Phi(0_X) + \Phi(0_{X^\alpha})\right), \tag{2.22}$$

$$\tilde{O}_{0-}(\Phi) = \frac{1}{2}\text{Vol}(T(F))\left(\Phi(0_X) - \Phi(0_{X^\alpha})\right). \tag{2.23}$$

Here α stands for the nontrivial class of torsors, and $0_X, 0_{X^\alpha}$ denote the "origins" (the points fixed by the tori) on X and X^α, respectively.

Proof It is very easy to see that if Φ is supported on $X(F)$ then, close to zero, $O_\xi(\Phi)$ is equal to 0 if ξ is not a norm from E and equal to a smooth function with value $\text{Vol}(T(F)) \cdot \Phi(0_X)$ at zero if ξ is a norm; similarly for the case that Φ is supported on $X^\alpha(F)$, but with ξ not a norm. The result follows.

□

2.11 Fourier transform

We define Fourier transform on E by identifying it with its Pontryagin dual via the pairing $(x, y) \mapsto \psi(\text{tr}(x\bar{y}))$, where ψ is as before and $\bar{y}$ denotes the Galois conjugate of y. The chosen measure on E is self-dual with respect to Fourier transform. Since we will be interested only in characters χ_E

of $E^\times$ which are base change of characters of $F^\times$, i.e. $\chi_E = \chi \circ N_F^E$ or, equivalently, $\bar{\chi}_E = \chi_E$, the functional equation of Tate integrals does not change under this alternative definition of duality, i.e.:

$$\gamma(\chi_E, s, \psi_E)\zeta_E(\Phi, \chi_E, s) = \zeta_E(\hat{\Phi}, \chi_E^{-1}, 1-s) \tag{2.24}$$

for such characters χ_E. The additive character ψ_E is the composition of the character ψ used previously on F with the trace map.

The correct way to define Fourier transform on $X^\alpha(F)$ (when α denotes the class of nontrivial torsors) is to notice that the hermitian map: $(x,y) \mapsto \mathrm{tr}(x\bar{y})$ extends naturally to $X^\alpha(F)$: if we choose any isomorphism $\iota : X \to X^\alpha$ which maps $1 \in E$ to the element $e \in X^\alpha(F)$ then we have:

$$\mathrm{tr}(\iota x \cdot \overline{\iota y}) := \mathrm{tr}(N_F^E(e)x\bar{y}),$$

and this definition clearly does not depend on ι. Then we have on X^α, as we had on X:

$$\widehat{\Phi^\alpha}(y) := \int_{X^\alpha} \Phi(x)\psi^{-1}\left(\mathrm{tr}(x \cdot \bar{y})\right) dx, \tag{2.25}$$

for the Haar measure defined previously, and this is self-dual, i.e.:

Lemma 2.15 *For $\Phi^\alpha \in C_c^\infty(X^\alpha)$,*

$$\widehat{\widehat{\Phi^\alpha}}(x) = \Phi^\alpha(-x). \tag{2.26}$$

Proof In what follows, it is important to distinguish between absolute values in F and in E, therefore we will be distinguishing them by an index. Choosing an isomorphism $\phi : X \to X^\alpha$ with $1 \mapsto e$ and $N_F^E(e) = a$, and denoting the pullback of Φ^α under this isomorphism by Φ^0, we have:

$$\phi^*\widehat{\Phi^\alpha}(y) = \int \Phi^\alpha(\phi x)\psi^{-1}(ax\bar{y})|a|_F dx = |a|_F\widehat{\Phi^0}(ay) \tag{2.27}$$

and, similarly,

$$\phi^*\widehat{\widehat{\Phi^\alpha}}(x) = |a|_F\widehat{\phi^*\widehat{\Phi^\alpha}}(ax) = |a|_F^2\mathcal{F}\left(\mathcal{F}(\Phi^0(a\cdot\bullet))\right)(ax) =$$

$$= |a|_F^2 \cdot \frac{1}{|a|_E}\widehat{\widehat{\Phi^0}}\left(\frac{ax}{a}\right) = \widehat{\widehat{\Phi^0}}(x) = \Phi^0(-x).$$

□

For $f \in \mathcal{S}(\mathcal{X})$ we now define the transform $\mathcal{G}$ as in (2.13), except that now η is nontrivial.

We claim:

Proposition 2.16 *Let $f \in \mathcal{S}(\mathcal{X})$ with lift $\Phi \in \mathcal{S}(X)$. Then:*

$$O_\xi(\hat{\Phi}) = \mathcal{G}(f)(\xi). \tag{2.28}$$

Proof Before we discuss the proof, let us extend the definition of Tate integrals to $X^\alpha(F)$ (where α stands for the nontrivial torsor). They will be defined as:

$$\zeta_E(\Phi^\alpha, \chi \circ N_F^E, s) = \int_{X^\alpha(F)} \Phi^\alpha(x)\chi(N_F^E x)|x|^{s-1} dx,$$

where $|x|$ denotes the absolute value, extended to $X^\alpha(F)$ via the norm map. The Tate integral is defined only for characters of the form $\chi \circ N_F^E$ in this case. If, now, Φ is a function supported on the union of X and X^α (with corresponding restrictions denoted by Φ^0 and Φ^α) we define:

$$\zeta_E(\Phi, \chi \circ N_F^E, s) = \zeta_E(\Phi^0, \chi \circ N_F^E, s) + \zeta_E(\Phi^\alpha, \chi \circ N_F^E, s).$$

It can be seen that, with these definitions, the local functional equation (2.24) still holds. Notice, moreover, that since $\chi_E = \chi \circ N_F^E$, we have:

$$\gamma(\chi_E, s, \psi_E) = \gamma(\chi, s, \psi) \cdot \gamma(\chi \otimes \eta_{E/F}, s, \psi) \tag{2.29}$$

It is now clear, as in the split case, that the Mellin transform of $f(\xi) = O_\xi(\Phi)$, $\Phi \in \mathcal{S}(\mathcal{X})$, can be written as:

$$\check{f}(\chi) = \zeta_E(\Phi, \chi^{-1} \circ N_F^E, \frac{1}{2}), \tag{2.30}$$

and similarly if $h_\xi = O_\xi(\hat{\Phi})$ we have:

$$\check{h}(\chi) = \zeta_E(\hat{\Phi}, \chi^{-1} \circ N_F^E, \frac{1}{2}).$$

Therefore:

$$\check{h}(\chi) = \gamma(\chi, \frac{1}{2}, \psi)\gamma(\chi \otimes \eta_{E/F}, \frac{1}{2}, \psi)\check{f}(\chi^{-1}).$$

Both h and $\mathcal{G}(f)$ satisfy the Mellin inversion formula 2.16, since they belong to $\mathcal{S}(\mathcal{X})$, therefore it suffices to check that their Mellin transforms coincide, which is immediate by the above calculation and an easy calculation of the Tate integrals of $G(f)$.

□

3 The Torus Quotient

3.1

From now on G will denote the group PGL_2 over F. By T we will be denoting a nontrivial torus in G, and E will be the quadratic etale extention of F such that $T = \ker(N_F^E)$. If T is split we have $E = F \oplus F$, and in that case we will sometimes be identifying T with some maximal torus inside of a chosen Borel subgroup, and will also be denoting it by A. As before, $\eta = \eta_{E/F}$ is the quadratic character associated to E.

The first main result of this section will be:

Proposition 3.1 *Let $\mathcal{Y}$ denote the stack $(\mathrm{Res}_{E/F}G_a)/T$ with the preimage of $\xi = -1 \in \mathcal{B} := (\mathrm{Res}_{E/F}G_a)/\!\!/T \simeq \mathbb{G}_a$ removed. Let $\mathcal{Y}^\times$ denote $\mathcal{Y}$ with the preimage of $\xi = 0$ removed; then the morphism to $\mathcal{B}$ defines an isomorphism between $\mathcal{Y}^\times$ and $\mathbb{A}^1 \smallsetminus \{-1, 0\}$.*

If $\mathcal{Y}_1, \mathcal{Y}_2$ denote two distinct copies of $\mathcal{Y}$, the stack $\mathcal{Z} := T\backslash G/T$ is isomorphic to the glueing of $\mathcal{Y}_1, \mathcal{Y}_2$ by the map:

$$\xi \mapsto -1 - \xi.$$

Among other things, this allows us to identify the GIT quotient $X \times X /\!\!/ G$ with $\mathcal{B} = \mathbb{G}_a$. We remark that this is a different parametrization from the one of Jacquet [Jac86]; we use it not only because it is more natural, but also because it turns out to be the one that makes the comparison of two RTFs via the integral transform $\mathcal{G}$ work. To preserve consistency of notation, the diagonal copy of X in $X \times X$ will have image $-1 \in \mathcal{B}$, and the open subset $\mathcal{Y}_1$ will be such that it contains the image of the diagonal copy (and hence the map $\mathcal{Y}_2 \to \mathcal{B}$ will be the one defined in the previous section, while the map $\mathcal{Y}_1 \to \mathcal{B}$ will be obtained from that by $\xi \mapsto -1 - \xi$). We let $\mathcal{B}_{\mathcal{Z}}^{\mathrm{reg}}$ be the complement of $\{0, -1\}$ in $\mathcal{B}$; as explained in §1.6, we will be denoting this by $\mathcal{B}^{\mathrm{reg}}$ when it is clear that we are referring to the torus quotient.

3.2 The open subset in the split case

We first consider the case $E = F \oplus F$, so we may assume that $T =$ the torus A of diagonal elements, and let B^+, B^- denote the Borel subgroups of upper and lower triangular matrices. Let $\mathcal{Z}_1 := A\backslash(B^-B^+ \cap B^+B^-)/A$, which is open in $\mathcal{Z}$.

Lemma 3.2 1. *The map:*[4]

$$F \oplus F \ni (x,y) \overset{\iota}{\mapsto} \begin{pmatrix} 1 & x \\ & 1 \end{pmatrix} \begin{pmatrix} 1 & \\ y & 1 \end{pmatrix} = \begin{pmatrix} 1+xy & x \\ y & 1 \end{pmatrix} \in G, \quad (3.1)$$

restricted to the set of x, y with $xy \neq -1$, descends to an isomorphism:

$$\mathcal{Y} \xrightarrow{\sim} \mathcal{Z}_1. \quad (3.2)$$

2. *Let w be an element in the F-points of the non-identity component of the normalizer of A, then the automorphism $g \mapsto {}^w g$ fixes the preimage of $\mathcal{Z}_1$ in G and has the property that:*

$$^w\iota(x,y) \sim \iota\left(y(1+xy), \frac{x}{1+xy}\right) \quad (3.3)$$

modulo the left action of A.

Proof Direct calculation. For the first statement, one easily sees that $A\backslash(B^-B^+ \cap B^+B^-)$ is isomorphic to the variety of matrices of the form (3.1) with $xy \neq -1$, and that, thinking of those matrices this way, the map ι is $\mathbb{G}_m$-equivariant with respect to the "baby case" action $(x,y)\cdot a = (ax, a^{-1}y)$ on F^2 and a suitable isomorphism $A \simeq \mathbb{G}_m$. The second statement is immediate.

□

3.3

Let $X = A\backslash G$ and let w denote the nontrivial G-automorphism of $X : Ag \mapsto Awg$, where we also use w to denote an element of the normalizer of A. We have a natural isomorphism of stacks:

$$X \times X/G \ni (x_1, x_2) \mapsto x_1 x_2^{-1} \in A\backslash G/A,$$

and applying the "w"-automorphism on the second copy of X induces an isomorphism of open substacks:

$$\mathcal{Z}_2 \xrightarrow{\sim} \mathcal{Z}_1,$$

where $\mathcal{Z}_2 = A\backslash(BwB \cap B^-wB^-)/A$.

Combined with the isomorphism of Lemma 3.2, this proves Proposition 3.1 in the split case.

---

[4]For notational clarity, we formulate in terms of F-points some statements which should, strictly speaking, be formulated in terms of schemes.

3.4 The non-split case

In the above setting, but with E now denoting a field extension, the cocycle which takes the nontrivial element $\sigma \in \mathrm{Gal}(E/F)$ to the inner automorphism of G by w (viewed also as an automorphism of A) defines a form of both PGL_2 and A over F. The form of A is $T \simeq \ker(N_F^E)$, while the (inner) form of PGL_2 could be split or non-split according as the cocycle chosen lifts to GL_2 or not. (This depends on the representative $w \in \mathcal{N}(A)(F)$ chosen, more precisely on whether the negative of the quotient of its eigenvalues is a norm from E or not.) This shows, in particular, that:

$$T\backslash G/T \simeq T\backslash G'/T, \tag{3.4}$$

as stacks where $G = \mathrm{PGL}_2$ and G' is an inner form of G which splits over E.

Notice that w preserves the open substacks $\mathcal{Z}_1, \mathcal{Z}_2$ of $\mathcal{Z}$ and hence defines forms of those.

At the same time, we have seen that the "w" automorphism on $A\backslash G$ corresponds, under the map ι of Lemma 3.2, to the automorphism:

$$\tau : (x, y) \mapsto \left(y(1 + xy), \frac{x}{1 + xy} \right)$$

of the subset of $(x, y) \in k \oplus k$ with $xy \neq -1$. This is the same as the composition of the automorphism: $(x, y) \mapsto (y, x)$ with the action of $(1 + xy) \in \mathbb{G}_m$, and therefore the form of the quotient stack defined by the cocycle $\sigma \mapsto \tau$ is isomorphic to:

$$\mathcal{Y}_1 \simeq \left(\mathrm{Res}_F^E(\mathbb{G}_a) \smallsetminus (N_F^E)^{-1}(-1) \right) /T.$$

Therefore, even in the non-split case we have: $\mathcal{Z}_1 \simeq \mathcal{Y}_1$ and, similarly, $\mathcal{Z}_2 \simeq \mathcal{Y}_2$, which completes the proof of Proposition 3.1.

3.5 Schwartz functions and orbital integrals

We define $\mathcal{M}(\mathcal{Z})$ to be the G-coinvariant space of $\mathcal{M}(X \times X)$ (the Fréchet space of Schwartz measures) in the split case. In the non-split case we must, as before, use one of the following equivalent definitions:

1. For some embedding of G into GL_n we let $\mathcal{M}(\mathcal{Z})$ be the space of $\mathrm{GL}_n(F)$-coinvariants of $\mathcal{M}\left(((X \times X) \times^G \mathrm{GL}_n)(F) \right)$.
2. We let:
$$\mathcal{M}(\mathcal{Z}) = \oplus_\alpha \mathcal{M}((X^\alpha \times X^\alpha)(F))_{G^\alpha(F)},$$
where α runs over all isomorphism classes of T-torsors R^α, $G^\alpha = \mathrm{Aut}_G(R^\alpha \times^T G)$, $X^\alpha = T\backslash G^\alpha$ and the index $_{G^\alpha(F)}$ denotes, as before, coinvariants. Equivalently, G^α ranges over inner forms of G which split over E.

By the above isomorphisms of stacks, we also have:

$$\mathcal{M}(\mathcal{Z}) \simeq (\mathcal{M}(\mathcal{Y}_1) \oplus \mathcal{M}(\mathcal{Y}_2)) / \mathcal{M}(\mathcal{Y}_1 \cap \mathcal{Y}_2). \tag{3.5}$$

We let $\mathcal{S}(\mathcal{Z})$ denote the cosheaf on $\mathcal{B}$ of functions on $\mathcal{B}^{\text{reg}}$ which are obtained as regular orbital integrals of Schwartz functions on $\bigsqcup_\alpha (X^\alpha \times X^\alpha)$.

$$\xi \mapsto O_\xi(\Phi) = \int_{G(F)} \Phi(\tilde{\xi} \cdot g), \tag{3.6}$$

where $\tilde{\xi}$ is a representative for the orbit parametrized by ξ.

The results of Section 2 immediately imply:

Proposition 3.3 *The restriction of $\mathcal{S}(\mathcal{Z})$ to $\mathcal{B}^{\text{reg}} = \mathbb{G}_a \smallsetminus \{0, -1\}$ is equal to the cosheaf of Schwartz functions $\mathcal{S}(\mathcal{B}^{\text{reg}})$. In neighborhoods of $\xi = 0$ and $\xi = -1$ they have the behavior of the germs of Propositions 2.5, 2.14 around zero.*

The choice of Haar measure on $G(F)$ does not matter for the definition of the sheaf $\mathcal{S}(\mathcal{Z})$, and again by a "lift of an element of $\mathcal{S}(\mathcal{Z})$ to $\mathcal{S}(X \times X)$" we will implicitly mean an element of $\mathcal{S}(X \times X)$ together with a choice of Haar measure on $G(F)$. However, we would now like to define a linear isomorphism:

$$\mathcal{M}(\mathcal{Z}) \xrightarrow{\sim} \mathcal{S}(\mathcal{Z}). \tag{3.7}$$

We do this locally on the open cover $\mathcal{Z}_1 \cup \mathcal{Z}_2$ by using the identification of $\mathcal{M}(\mathcal{Z}_i)$ with $\mathcal{M}(\mathcal{Y}_i)$, together with the identification (3.7) from the previous section: $\mathcal{M}(\mathcal{Y}_i) \simeq \mathcal{S}(\mathcal{Y}_i)$, which gives rise to an map to $\mathcal{S}(\mathcal{Z})$:

$$\mathcal{M}(\mathcal{Z}_i) \xrightarrow{\sim} \mathcal{M}(\mathcal{Y}_i) \xrightarrow{\sim} \mathcal{S}(\mathcal{Y}_i) \to \mathcal{S}(\mathcal{Z}).$$

Equivalently, this isomorphism arises from the standard additive measure on $\mathcal{B} = F$. Notice that the same integration formula as in the previous section (Propositions 2.2 and 2.13) follows from the local isomorphisms of stacks:

Lemma 3.4 *There are compatible choices of invariant measures on $\mathcal{B}$ (additive), G and $\bigsqcup_\alpha (X^\alpha \times X^\alpha)$ such that for any $\Phi \in \mathcal{S}(\bigsqcup_\alpha (X^\alpha \times X^\alpha))$ with image $f \in \mathcal{S}(\mathcal{Z})$ we have:*

$$\int_{\bigsqcup_\alpha (X^\alpha \times X^\alpha)} \Phi(x_1, x_2) dx_1 dx_2 = \int_{\mathcal{B}} f(\xi) d\xi. \tag{3.8}$$

3.6 Inner products

In order not to introduce excessive notation, we will not reserve any symbols for the irregular distributions on $\mathcal{Z}$ which are the analogs of $\tilde{O}_0, \tilde{O}_u, \tilde{O}_{0+}, \tilde{O}_{0-}$ of §2, with one exception:

Let α be a class of T-torsors. For $\Phi_1, \Phi_2 \in \mathcal{S}(X^\alpha)$, and an invariant measure dx on X^α (where "invariant" means, of course, invariant under the action of the pertinent inner form of G on each copy) we define the *inner product* of Φ_1, Φ_2 as:

$$\langle \Phi_1, \Phi_2 \rangle = \int_{X^\alpha} \Phi_1(x)\Phi_2(x)dx,$$

i.e. as a bilinear form. Clearly, this extends continuously to $\mathcal{S}(X^\alpha \times X^\alpha)$, and for an element Φ of the latter we will simply write $\langle \Phi \rangle$.

Now, given $f \in \mathcal{S}(\mathcal{Z})$, choose a pair $(\Phi, (dg_\alpha)_\alpha)$ consisting of an element $\Phi = \sum \Phi^\alpha \in \oplus_\alpha \mathcal{S}(X^\alpha \times X^\alpha)$ and a collection of Haar measures on the inner forms G^α such that f arises as the regular orbital integrals of Φ with respect to those measures. Let:

$$(-1)^\alpha = \begin{cases} 1, & \text{if } \alpha \text{ corresponds to the trivial torsor,} \\ -1 & \text{otherwise;} \end{cases} \tag{3.9}$$

that is, we are identifying $H^1(F,T)$ with $\mathbb{Z}/2$ in the nonsplit case.

Then we define the *"inner product"* of f as:

$$\langle f \rangle := (F^\times : N_F^E E^\times)^{-1} \cdot \sum_\alpha (-1)^\alpha \mathrm{Vol}(T(F)_0) \langle \Phi^\alpha \rangle, \tag{3.10}$$

where we have implicitly chosen a decomposition of dg as an invariant measure on $T(F)$ times a measure on $T\backslash G^\alpha(F)$ in order to define both the inner product on X^α and the volume of $T(F)_0$ according to the recipe of §2.5 (of course, in the non-split case $T(F)_0 = T(F)$ so no recipe is needed). Clearly, the definition does not depend on this choice, so we have a well-defined functional on $\mathcal{S}(\mathcal{Z})$. The following is easy to see by the results of the previous section:

Lemma 3.5 *In the split case $\langle f \rangle$ is equal to the distribution $\tilde{O}_0(f)$ of (2.7) when a neighborhood of $\xi = -1$ of $\mathcal{Z}$ is identified with a neighborhood of $\xi = 0$ of $\mathcal{X}$ according to Proposition 3.1. In the nonsplit case, $\langle f \rangle$ is equal to the distribution $\tilde{O}_{0-}(f)$ of (2.23) under the same identification.*

4 The Kuznetsov Quotient with Nonstandard Functions

4.1

The Kuznetsov trace formula is the relative trace formula for $\mathcal{S}(X, \mathcal{L}_\psi) \otimes \mathcal{S}(X, \mathcal{L}_\psi^{-1})$, where X is the quotient of PGL_2 by a nontrivial unipotent subgroup N and $\mathcal{L}_\psi$ is the complex G-line bundle on X defined by a character ψ of $N(F)$. Here, however, we will extend it to nonstandard sections of this line bundle, that is, sections which are not Schwartz, with prescribed asymptotic behavior at infinity. One can identify X with the quotient by $\{\pm 1\}$ of two-dimensional affine space, minus the origin, and "infinity" is precisely the partial compactification of this space by $\mathbb{P}^1$ at "infinity".

Let us start in a slightly different way: Let $G = \mathrm{Aut}(\mathbb{P}^1)$ and let $\bar{X}$ denote the total space of the line bundle $O(-2)$ over $\mathbb{P}^1$; it is G-linearizable, i.e. it carries an action of G which commutes with the natural action of $\mathbb{G}_m$. We denote by X the complement of the zero section — it is homogeneous under G, and stabilizers are unipotent subgroups. There is a unique, up to the action of $\mathbb{G}_m(F)$, $G(F)$-linear complex line bundle on the F-points of $\bar{X}$ on which the stabilizers of points on X act by a nontrivial unitary character; we fix such a line bundle, and denote it by $\mathcal{L}_\psi$. Over $\mathbb{P}^1$ this is G-isomorphic (non-canonically) to the trivial line bundle. For any subset S of $\bar{X} \times \bar{X}$ we will denote by S^+ the open subset of S lying over the open G-orbit on $\mathbb{P}^1 \times \mathbb{P}^1$. If S is stable under the diagonal action of G, then so is S^+.

4.2 Orbits

Let N be the subgroup of upper triangular unipotent matrices in PGL_2, N^- the subgroup of lower triangular matrices, both identified with the additive group $\mathbb{G}_a$ in the usual way, and A the torus of diagonal elements. Fix a nontrivial unitary character ψ of F, the same character that we used for Fourier transforms in previous sections. We claim that there is a canonical map from $(X \times X)^+$ to the open subset:

$$\mathbb{G}_m \simeq \left\{ N \begin{pmatrix} \xi & \\ & 1 \end{pmatrix} N^- \middle| \xi \in \mathbb{G}_m \right\}$$

of $N \backslash G / N^-$.

Indeed, let $(x, y) \in (X \times X)^+$, then there is a unique isomorphism of the triple (G, G_x, G_y) with (PGL_2, N, N^-) such that G_x acts on the fiber of $\mathcal{L}_\psi$ by the character ψ of $N(F) = \mathbb{G}_a(F)$ (standard isomorphism), and a unique isomorphism such that G_y acts on the fiber of $\mathcal{L}_\psi^{-1}$ by the character ψ^{-1} of $N^-(F) = \mathbb{G}_a(F)$. Then $\begin{pmatrix} \xi & \\ & 1 \end{pmatrix}$ is the unique element of A which conjugates one isomorphism to the other; our convention to distinguish between ξ and ξ^{-1} is that as (x,y) approach the complement of $(X\times X)^+$ in $X\times X$, ξ goes to infinity. This defines an isomorphism of quotient stacks (varieties):

$$(X\times X)^+/G \to \mathbb{G}_m. \tag{4.1}$$

We embed $\mathbb{G}_m \hookrightarrow \mathcal{B} := \mathbb{G}_a \hookrightarrow \overline{\mathcal{B}} := \mathbb{P}^1$, and set $\mathcal{B}^{\text{reg}} = \mathcal{B}^{\text{reg}}_{\mathcal{W}} = \mathcal{B} \smallsetminus \{0\}$. We will sometimes be denoting the regular set by $\mathcal{B}^\times$.

Lemma 4.1 *The map (4.1) extends to a rational map:*

$$\bar{X}\times X \to \overline{\mathcal{B}}, \tag{4.2}$$

which is regular away from the complement of $(\mathbb{P}^1\times X)^+$ *in* $\mathbb{P}^1\times X$, *that is: away from the set of points* $(p,x)\in\mathbb{P}^1\times X$ *such that* x *lies in the fiber over* p.

The reader should keep in mind that $\xi = 0$ *corresponds to points on* $(\mathbb{P}^1\times X)^+$, *while* $\xi=\infty$ *corresponds to the complement of* $(X\times X)^+$ *in* $X\times X$. This paradoxical way of parametrizing orbits plays a role when discussing Fourier transforms (where the vector space structure imposed on $\mathcal{B}$ is important).

There is a certain degree of arbitrariness in choosing the "standard" identifications of N, N^- with $\mathbb{G}_a$; therefore, there is nothing special about the orbit above $\xi = 1$. However, this choice should be compared to the choice that the "irregular" orbits of $\mathcal{Z}$ map to $\{0,-1\}\in\mathcal{B}$ in the discussion of Section 3; both are essentially choices of a generator of the ring of invariants, they are related and should be changed simultaneously.

4.3 Orbital integrals

In what follows, we try to make explicit the choices made in order to think of orbital integrals on the Kuznetsov trace formula as functions on $\mathcal{B}^{\text{reg}} \simeq F^\times$. The reader may wish to skip straight to (4.6).

Let Φ_1, Φ_2 be smooth sections of $\mathcal{L}_\psi$, resp. $\mathcal{L}_\psi^{-1}$, of compact support on X. Fix a Haar measure on $G = G(F)$. At a first stage, we define the (regular) orbital integrals of $\Phi_1\otimes\Phi_2$ to be the G-invariant section of $\mathcal{L}_\psi \boxtimes \mathcal{L}_\psi^{-1}$ on $(X\times X)^+$ obtained by integrating $\Phi_1\cdot\Phi_2$, that is:

$$O_{(x,y)}(\Phi_1\otimes\Phi_2) = \int_G g\cdot(\Phi_1,\Phi_2)(x,y)dg, \tag{4.3}$$

where $g\cdot$ denotes the right regular representation under the diagonal action of G.

At a second stage, we would like to represent these orbital integrals as functions on $\mathcal{B}^\times := \mathcal{B} \smallsetminus \{0\} = F^\times$. Let $(x_0, y_0) \in (X \times X)^+$, with image $1 \in \mathcal{B}$, let A denote the unique torus in G which normalizes both G_{x_0} and G_{y_0}, identified with $\mathbb{G}_m$ according to the image of $(x_0 a, y_0)$ in $\mathcal{B}$ $(a \in A)$. For $\xi \in \mathbb{G}_m$ corresponding to $a \in A$, let:

$$O_\xi(\Phi_1 \otimes \Phi_2) = O_{(x_0,y_0)}(a \cdot \Phi_1 \otimes \Phi_2), \tag{4.4}$$

where a acts on the section $O_{(x_0,y_0)}(\Phi_1 \otimes \Phi_2)$ by the regular representation on the first coordinate; hence, $O_\xi(\Phi_1 \otimes \Phi_2)$ is understood as an element in the fiber of $\mathcal{L}_\psi$ over (x_0, y_0). We remark that $a\cdot$ denotes the action of a as an element of G; the natural action of $\mathbb{G}_m$ on $O(-2)$ by dilations does not extend to the line bundle $\mathcal{L}_\psi$. When, later, we will replace $\mathcal{L}_\psi$ by the trivial line bundle, we will be denoting the action of $\mathbb{G}_m$ by dilations by $\mathscr{L}_a$ (where $\mathscr{L}$ is supposed to be reminiscent of "left action" in terms of the torus acting on $X \simeq N\backslash G$), in order to avoid confusion. Notice that the orbit map: $\mathcal{B}^\times \ni a \mapsto (x \cdot a, y) \in (X \times X)^+$ extends to:

$$\mathcal{B} \to (\bar{X} \times X)^+. \tag{4.5}$$

Finally, we choose an isomorphism of the fiber of $\mathcal{L}_\psi \boxtimes \mathcal{L}_\psi^{-1}$ over (x_0, y_0) with $\mathbb{C}$, in order to consider $O_\xi(\Phi_1 \otimes \Phi_2)$ as a complex-valued function on $F^\times$, as ξ varies. This last choice of isomorphism only affects the orbital integrals by a common scalar multiple, and will be reflected in our choice of unramified sections in the fundamental lemma (e.g. we will ask that some sections be equal to "1" at (x_0, y_0), which only makes sense after choosing this isomorphism).

Explicitly, if we identify the stabilizers of x_0, y_0 with N, N^- such that they act on $\mathcal{L}_\psi$, resp. $\mathcal{L}_\psi^{-1}$ by ψ, ψ^{-1}, respectively, and if we trivialize the fiber in order to think of $\Phi_1 \otimes \Phi_2$ as an element of $C^\infty(N\backslash G \times N^-\backslash G, \psi \otimes \psi^{-1})$, then:

$$O_\xi(\Phi_1 \otimes \Phi_2) = \int_G \Phi_1(\begin{pmatrix} \xi & \\ & 1 \end{pmatrix} g)\Phi_2(g)dg. \tag{4.6}$$

Given these choices — that is, the embedding (4.5) and the identification of the fiber over 1 with $\mathbb{C}$, we can easily see:

Lemma 4.2 1. *The action map:*

$$\mathcal{B} \times G \to (\bar{X} \times X)^+ \tag{4.7}$$

is an isomorphism.

2. *The chosen trivialization of the fiber over (x_0, y_0), the action of $\mathcal{B}^\times \times G$ (here $\mathcal{B}^\times$ is acting as the torus A on the first copy, as before), and the above isomorphism give rise to a G-equivariant isomorphism of:*

$$\mathcal{L}_\psi \boxtimes \mathcal{L}_\psi^{-1}\Big|_{(\bar{X}\times X)^+}$$

with the trivial line bundle over $\mathcal{B} \times G$.

4.4 Integration formula

Identifying sections $\Phi \in \mathcal{S}\left(\mathcal{L}_\psi \boxtimes \mathcal{L}_\psi^{-1}\Big|_{(\bar{X}\times X)^+} \right)$ with sections of the trivial line bundle according to Lemma 4.2, we have:

Lemma 4.3 *For suitable choices of a G-invariant measure on $X = X(F)$ and a Haar measure on $G = G(F)$, we have:*

$$\int_{(\bar{X}\times X)^+} \Phi(x,y)d(x,y) = \int_{\mathcal{B}} O_\xi(\Phi)|\xi|^{-2}d\xi, \tag{4.8}$$

where $d\xi$ is our fixed, standard additive measure on $\mathcal{B} \simeq F$.

4.5 Nonstandard sections

We let $\mathcal{M}(\bar{X}\times X, \mathcal{L}_\psi \boxtimes \mathcal{L}_\psi^{-1})$ (resp. $\mathcal{S}(\bar{X}\times X, \mathcal{L}_\psi \boxtimes \mathcal{L}_\psi^{-1})$) denote the Schwartz cosheaf over $\bar{X} \times X$ consisting of smooth measures (resp. functions) on $X \times X$, valued in $\mathcal{L}_\psi \boxtimes \mathcal{L}_\psi^{-1}$, with the following properties:

- the restriction of the cosheaf to $X \times X$ coincides with the standard cosheaf of Schwartz measures (resp. functions) valued in $\mathcal{L}_\psi \boxtimes \mathcal{L}_\psi^{-1}$;
- in a neighborhood of $\mathbb{P}^1 \times X$ they are finite sums of the form $\sum_i f_i F_i$, where:

    1. the f_i's are $\mathcal{L}_\psi \boxtimes \mathcal{L}_\psi^{-1}$-valued Schwartz functions on $\bar{X} \times X$;
    2. the F_i are scalar-valued measures (resp. functions) on $X \times X$ which are G-invariant in the second coordinate[5], and in the first coordinate are annihilated asymptotically by the operator:

$$\left(1 - \delta^{-\frac{1}{2}}(a)\mathscr{L}_a\right) \cdot \left(1 - \eta_{E/F}\delta^{-\frac{1}{2}}(a)\mathscr{L}_a\right). \tag{4.9}$$

[5]This is just one of many equivalent ways of describing our measures, and the reader should not get confused trying to figure out the purpose of invariance in the second coordinate; the point is that our resulting measures will be of Schwartz type in the second coordinate, which is taken care of by the f_i's, so we only need the F_i's in order to describe their asymptotic behavior in the first coordinate.

Our notation is hiding the details of the asymptotics, and is just replacing X by $\bar{X}$ to remind that these measures are not Schwartz on X; however, they cannot be considered as smooth measures on $\bar{X}$.

We explain what it means, for a *scalar-valued* function or measure on X, to be asymptotically annihilated by (4.9). Thinking of $\mathbb{G}_m$ as the dilation group of the bundle $O(-2)$ over $\mathbb{P}^1$, we have an L^2-isometric action of it on functions on X given by:

$$\mathscr{L}_a f(x) = \delta(a)^{-\frac{1}{2}} f(ax), \tag{4.10}$$

where $\delta(a)$ is the inverse of the character by which the non-normalized action of $\mathbb{G}_m$ transforms an invariant measure on X, written suggestively so that in an identification of X with $N\backslash\mathrm{PGL}_2$ it corresponds to:

$$\begin{pmatrix} a & \\ & 1 \end{pmatrix} \mapsto |a|.$$

Similarly, we have an L^2-isometric action of $\mathbb{G}_m$ on measures on X given by:

$$\mathscr{L}_a \mu(x) = \delta(a)^{\frac{1}{2}} \mu(ax), \tag{4.11}$$

and of course the map from functions to measures: $f \mapsto f dx$ is equivariant with respect to these actions.

Hence, "asymptotically annihilated" by (4.9) means that applying the operator (4.9) to the given function/measure produces a function/measure which is supported away from $\mathbb{P}^1$. Thus, these functions can be identified in a neighborhood of $\mathbb{P}^1$ with elements of a representation π of the form:

$$0 \to I(\delta^{\frac{1}{2}}) \to \pi \to I(\eta_{E/F}\delta^{\frac{1}{2}}) \to 0, \tag{4.12}$$

where the sequence is nonsplit if (and only if) $\eta_{E/F}$ is trivial. Here $I(\bullet)$ denotes the principal series representation obtained by normalized induction from the character $\bullet$. However, the isomorphisms with principal series are not canonical, and we prefer to think of π as the space of smooth functions on X which are annihilated by (4.9).

4.6 Coinvariants

We let $\mathcal{M}(\mathcal{W})$ denote the G-coinvariants of $\mathcal{M}(\bar{X} \times X, \mathcal{L}_\psi \boxtimes \mathcal{L}_\psi^{-1})$. Here the letter $\mathcal{W}$ is reminiscent of a stack, but for us it is just formal notation, because I do not know how to make sense of $\mathcal{L}_\psi \boxtimes \mathcal{L}_\psi^{-1}$ as a bundle on the stack. We denote by $\mathcal{M}(\mathcal{W}^+)$ the G-coinvarints of those measures which are almost supported on $(\bar{X} \times X)^+ \subset \bar{X} \times X$.

Using the trivializations of Lemma 4.2, we have a map from $\mathcal{M}(\mathcal{W})$ to smooth measures on $\mathcal{B}^\times$; as we shall see, this map is injective, so we feel free not to distinguish between an element of $\mathcal{M}(\mathcal{W})$ and the corresponding measure on $\mathcal{B}^\times$. We let $\mathcal{S}(\mathcal{W})$ be the space of functions on $\mathcal{B}^\times$ which are obtained as "regular" orbital integrals (understood as in §4.3), with respect to a Haar measure on G, of elements of $\mathcal{S}(\bar{X} \times X, \mathcal{L}_\psi \boxtimes \mathcal{L}_\psi^{-1})$. According to Lemma 4.3, the map $\mu \mapsto \frac{\mu}{|\xi|^{-2}d\xi}$ is a linear isomorphism:

$$\mathcal{M}(\mathcal{W}) \to \mathcal{S}(\mathcal{W}). \tag{4.13}$$

We would like to understand the spaces $\mathcal{M}(\mathcal{W}^+)$, $\mathcal{M}(\mathcal{W})$ as sections over $\mathcal{B}$, $\overline{\mathcal{B}}$ of the G-coinvariants of the push-forward to $\overline{\mathcal{B}}$ of the Schwartz cosheaf $\mathcal{M}(\bar{X} \times X, \mathcal{L}_\psi \boxtimes \mathcal{L}_\psi^{-1})$, in order to take advantage of the results of appendix B. This is not directly possible, as the map (4.2) is rational, not regular. We will see, however, in Proposition 4.10 that the stalk over the irregular locus S of (4.2) does not contribute at all to $\mathcal{M}(\mathcal{W})$; thus, we may indeed view $\mathcal{M}(\mathcal{W})$ as (sections of) the flabby cosheaf of coinvariants of the push-forward of $\mathcal{M}((\bar{X} \times X) \smallsetminus S, \mathcal{L}_\psi \boxtimes \mathcal{L}_\psi^{-1})$.

4.7 Limiting behavior at 0

Let $\mathcal{X}$ be the quotient stack of Section 2, that is: $\mathcal{X} = \mathrm{Res}_{E/F}\mathbb{G}_a/T$, where $T = \ker N_F^E$. We consider elements of $\mathcal{M}(\mathcal{X})$ as measures on $\mathcal{B} \smallsetminus \{0\} = F^\times$, as we do with elements of $\mathcal{M}(\mathcal{W})$. Recall that $\mathcal{M}(\mathcal{W}^+)$ denotes the sections of this cosheaf over $(\bar{X} \times X)^+$, and $\mathcal{S}(\mathcal{W}^+)$ their images under the operation of orbital integrals.

Proposition 4.4 *As spaces of measures on $\mathcal{F}^\times$, we have:*

$$\mathcal{M}(\mathcal{W}^+) = |\bullet|^{-1}\mathcal{M}(\mathcal{X}). \tag{4.14}$$

The symbol $|\bullet|^{-1}$ denotes multiplication of a measure $\mu(\xi)$ by $|\xi|^{-1}$.

Proof Using the isomorphisms of Lemma 4.2, the space $\mathcal{M}((\bar{X} \times X)^+, \mathcal{L}_\psi \boxtimes \mathcal{L}_\psi^{-1})$ can be identified with a space of scalar-valued measures on $\mathcal{B}^\times \times G$ with the following properties:

- away from a neighborhood of $\{0\} \times G$ they coincide with Schwartz measures;
- in a neighborhood of $\{0\} \times G$ they are equal to a Schwartz function on $\mathcal{B} \times G$ times a measure $\mu(b)dg$ with dg a Haar measure on G and:

$$\mu(b) - \mu(ab) - \eta_{E/F}(a)\mu(ab) + \eta_{E/F}(a)\mu(a^2b) = 0, \tag{4.15}$$

  for every $a \in F^\times$.

Their G-coinvariants coincide with their push-forwards to $\mathcal{B}^\times$, which are characterized by the analogous properties (i.e. Schwartz away from 0 and same condition on the measure μ). By the explicit description of $\mathcal{M}(\mathcal{X})$ in Propositions 2.5 and 2.14, the claim follows.

□

Corollary 4.5 *As spaces of functions on $\mathcal{B}^\times$, we have:*

$$\mathcal{S}(\mathcal{W}^+) = |\bullet|\mathcal{S}(\mathcal{X}). \tag{4.16}$$

This follows immediately from the integration formula of Lemma 4.3. In the next subsection we will identify the limiting behavior of an element of $\mathcal{S}(\mathcal{W}^+)$ as $\xi \to 0$ in terms of invariant distributions supported on the fiber over $\xi = 0$.

4.8 Explication

We would now like to explicate the "irregular" distributions that determine the limiting behavior at zero.

Let V be defined by the short exact sequence of Fréchet spaces:

$$0 \to \mathcal{S}(X \times X, \mathcal{L}_\psi \boxtimes \mathcal{L}_\psi^{-1}) \to \mathcal{S}(\bar{X} \times X, \mathcal{L}_\psi \boxtimes \mathcal{L}_\psi^{-1}) \to V \to 0.$$

(Recall that $\mathcal{S}(\bar{X} \times X, \mathcal{L}_\psi \boxtimes \mathcal{L}_\psi^{-1})$ denotes the nonstandard sections defined in 4.5 and not really sections of the line bundle over $\bar{X} \times X$.) That is, in the language of appendix B, V is the stalk over $\mathbb{P}^1 \times X$ of the Schwartz cosheaf whose global sections are $\mathcal{S}(\bar{X} \times X, \mathcal{L}_\psi \boxtimes \mathcal{L}_\psi^{-1})$.

Let $\mathscr{G}_\psi$ denote the Fréchet space of germs of smooth sections of $\mathcal{L}_\psi$ around $\mathbb{P}^1$. Since $\mathcal{L}_\psi$ is trivializable over $\mathbb{P}^1$, this space is isomorphic (non-canonically) to the space of germs of smooth functions around $\mathbb{P}^1$. Then we have an isomorphism of $G \times G$-representations:

$$V \simeq (\mathscr{G}_\psi \widehat{\otimes}_{C^\infty(\mathbb{P}^1)} \pi) \widehat{\otimes} \mathcal{S}(X), \tag{4.17}$$

(completed, projective tensor products), where π is as in (4.12).

Clearly, the *fiber* over $\mathbb{P}^1 \times X$ is obtained by evaluation of the element of $\mathscr{G}_\psi$ on $\mathbb{P}^1$, which gives a map:

$$\mathcal{S}(\bar{X} \times X, \mathcal{L}_\psi \boxtimes \mathcal{L}_\psi^{-1}) \to \left(C^\infty(\mathbb{P}^1, \mathcal{L}_\psi) \otimes_{C^\infty(\mathbb{P}^1)} \pi\right) \otimes \mathcal{S}(X, \mathcal{L}_\psi^{-1}). \tag{4.18}$$

The first tensor product is isomorphic to π, depending on a trivialization of $\mathcal{L}_\psi$ over $\mathbb{P}^1$; for simplicity, but to remember that this isomorphism is not canonical, we will be denoting it by π'. Since the elements of π can be thought of as sections of the trivial line bundle on X annihilated by the operator (4.9), elements of π' should be thought of as similar sections of the pullback of $\mathcal{L}_\psi$ under the projection map: $X \to \mathbb{P}^1$. We are going to encode the G-coinvariants of the fiber in two "irregular" orbital integrals.

For now we will define those irregular orbital integrals only for elements of $\mathcal{S}(\mathcal{W}^+)$. Let $f \in \mathcal{S}(\mathcal{W}^+)$ and (Φ, dg) be a pair consisting of an element $\Phi \in \mathcal{S}((\bar{X} \times X)^+, \mathcal{L}_\psi \boxtimes \mathcal{L}_\psi^{-1})$ and a Haar measure on G such that f is obtained by the orbital integrals of Φ. Under the isomorphisms of Lemma 4.2 (see also the proof of Proposition 4.4), Φ can be written as a function on $\mathcal{B} \times G$ of the form:

$$\begin{aligned} |\xi| \left(-h_1(\xi, g) \ln |\xi| + h_2(\xi, g)\right) &\quad \text{in the split case } (\eta = 1), \\ |\xi| \left(h_1(\xi, g) + \eta(\xi) h_2(\xi, g)\right) &\quad \text{in the non-split case } (\eta \neq 1) \end{aligned}$$

where $h_i(\xi, g)$ are Schwartz functions on $\mathcal{B} \times G$.

In the split case we set:

$$\tilde{O}_{0,\delta^{\frac{1}{2}}}(f) = \int_G h_1(0, g) dg, \tag{4.19}$$

$$\tilde{O}_{u,\delta^{\frac{1}{2}}}(f) = \int_G h_2(0, g) dg. \tag{4.20}$$

In the nonsplit case we set:

$$\tilde{O}_{0,\delta^{\frac{1}{2}}}(f) = \int_G h_1(0, g) dg, \tag{4.21}$$

$$\tilde{O}_{0,\eta\delta^{\frac{1}{2}}}(f) = \int_G h_2(0, g) dg. \tag{4.22}$$

Then it is easy to see:

Lemma 4.6 *For $f \in \mathcal{W}^+$, there are smooth functions C_1, C_2 so that in a neighborhood of zero:*

$$f(\xi) = |\xi| \left(-C_1(\xi) \ln |\xi| + C_2(\xi)\right) \quad \textit{in the split case,} \tag{4.23}$$

$$f(\xi) = |\xi| \left(C_1(\xi) + \eta(\xi) C_2(\xi)\right) \quad \textit{in the non-split case.} \tag{4.24}$$

Moreover, $C_1(0) = \tilde{O}_{0,\delta^{\frac{1}{2}}}(f)$ and $C_2(0) = \tilde{O}_{u,\delta^{\frac{1}{2}}}(f)$ in the split case, $C_1(0) = \tilde{O}_{0,\delta^{\frac{1}{2}}}(f)$ and $C_2(0) = \tilde{O}_{0,\eta\delta^{\frac{1}{2}}}(f)$ in the nonsplit case.

Thus, with the isomorphism of Corollary 4.5 and the distributions defined in Section 2, we have:

$$\begin{aligned} \tilde{O}_{0,\delta^{\frac{1}{2}}}(f) &= \tilde{O}_0(|\bullet|^{-1} f), \\ \tilde{O}_{u,\delta^{\frac{1}{2}}}(f) &= \tilde{O}_u(|\bullet|^{-1} f) \end{aligned}$$

in the split case, and:

$$\tilde{O}_{0,\delta^{\frac{1}{2}}}(f) = \tilde{O}_{0+}(|\bullet|^{-1}f),$$
$$\tilde{O}_{0,\eta\delta^{\frac{1}{2}}}(f) = \tilde{O}_{0-}(|\bullet|^{-1}f)$$

in the nonsplit case.

4.9 Inner product and limiting behavior at ∞

Let now Y be the complement of $(X \times X)^+$ in $X \times X$ (that is, the union of $\mathbb{G}_m$-translates of the diagonal copy of X), and denote by $\mathcal{S}(X \times X, \mathcal{L}_\psi \boxtimes \mathcal{L}_\psi^{-1})_Y$ the stalk of $\mathcal{S}(X \times X, \mathcal{L}_\psi \boxtimes \mathcal{L}_\psi^{-1})$ over Y. Notice that at this point we have restricted our attention to sections of our cosheaves over $X \times X$; that is, standard Schwartz sections of $\mathcal{L}_\psi \boxtimes \mathcal{L}_\psi^{-1}$ on $X \times X$. We denote by $\mathcal{S}(Y, \mathcal{L}_\psi \boxtimes \mathcal{L}_\psi^{-1})$ the fiber of $\mathcal{S}(X \times X, \mathcal{L}_\psi \boxtimes \mathcal{L}_\psi^{-1})$ over Y — it is the space of $\mathcal{L}_\psi \boxtimes \mathcal{L}_\psi^{-1}$-valued Schwartz functions on Y. Just for this subsection, we introduce the notation $\mathcal{S}(\mathcal{W}^0)$ for the G-coinvariants of $\mathcal{S}(X \times X, \mathcal{L}_\psi \boxtimes \mathcal{L}_\psi^{-1})$.

For $\Phi_1 \in \mathcal{S}(X, \mathcal{L}_\psi)$, $\Phi_2 \in \mathcal{S}(X, \mathcal{L}_\psi^{-1})$ and a measure dx on X, we define the inner product:

$$\langle \Phi_1, \Phi_2 \rangle = \int_X (\Phi_1 \cdot \Phi_2)(x) dx,$$

i.e. as a bilinear map. Clearly, it extends to a linear functional on $\mathcal{S}(X \times X, \mathcal{L}_\psi \boxtimes \mathcal{L}_\psi^{-1})$, and for Φ in this space we will be using the notation $\langle \Phi \rangle$.

Given $f \in \mathcal{S}(\mathcal{W}^0)$, choose a pair (Φ, dg) consisting of an element $\Phi \in \mathcal{S}(X \times X, \mathcal{L}_\psi \boxtimes \mathcal{L}_\psi^{-1})$ and a Haar measure on G so that f is obtained as the coinvariants of Φ with respect to this measure. The chosen measure on G induces a measure on X as follows: let $x \in X$ and $N = G_x$, the stabilizer of x; hence, $X = N \backslash G$. The group N acts by a character Ψ on the fiber of $\mathcal{L}_\psi$ over x, and we choose an identification of $N(F)$ with F such that the character Ψ becomes our fixed additive character ψ; we then let dn be the Haar measure on N corresponding to our fixed measure dx of §2.3, and we let dx be the measure on X corresponding to dg, dn. Clearly, it does not depend on the choice of point.

We then define the *"inner product"* of f to be the functional:

$$\langle f \rangle = \langle \Phi \rangle, \tag{4.25}$$

where the "inner product" of Φ is defined with respect to the measure described above.

The following is immediate:

Lemma 4.7 *The inner product spans the space of G-invariant functionals on $\mathcal{S}(Y, \mathcal{L}_\psi \boxtimes \mathcal{L}_\psi^{-1})$ (the fiber of $\mathcal{S}(X \times X, \mathcal{L}_\psi \boxtimes \mathcal{L}_\psi^{-1})$ over Y).*

Based on Proposition B.4.1 now, the stalk of $\mathcal{S}(\mathcal{W}^0)$ at $\xi = \infty$ is generated over the stalk of smooth functions by an element with nonzero "inner product".

Proposition 4.8 *The stalk of $\mathcal{S}(\mathcal{W}^0)$ at $\xi = \infty$ coincides with the set of germs of all functions f of the following form:*

- *in the nonarchimedean case:*
$$f(\xi) = C(\xi^{-1}) \cdot \int_{|x^2|=|\xi|} \psi(\xi x^{-1} - x)dx, \tag{4.26}$$
where C denotes a(n almost) smooth function[6] defined in a neighborhood of zero, with $C(0) = \langle f \rangle$.
- *in the archimedean case:*
$$f(\xi) = \int Z\left(\frac{\xi}{|x|^2+1}, x\right) \psi\left(\frac{\xi\bar{x}}{|x|^2+1} - x\right) dx, \tag{4.27}$$
where Z is a Schwartz function on $\mathbb{G}_m \times \mathbb{P}^1$, with $Z(-1, \infty) = \langle f \rangle$.
$$\tag{4.28}$$

Remark 4.9 It may not be clear at first from the above expressions, but it will become clear from the stationary phase analysis of the archimedean integrals in §5.2 that the stalks are generated over the stalk of smooth functions by a single element, as they should. Thus, we could also write them as $C(\xi^{-1})$ times the same integral with Z replaced by any preferred function such that $Z(-1, \infty) = 1$. (Jacquet has computed the integrals explicitly in [Jac05], and his work will be the basis for the analysis of §5.2.)

Proof We can easily see that the germ of f can be written as an integral of the form:
$$\int \Phi\left(\begin{pmatrix} \xi & \\ & 1 \end{pmatrix}\begin{pmatrix} 1 & \\ x & 1 \end{pmatrix}\right) \psi^{-1}(x)dx,$$
where $\Phi \in \mathcal{S}(N\backslash \mathrm{PGL}_2, \psi)$, with ψ here denoting the character $\begin{pmatrix} 1 & x \\ & 1 \end{pmatrix} \mapsto \psi(x)$, the measures being the standard ones, and $\Phi\left(\begin{pmatrix} & 1 \\ 1 & \end{pmatrix}\right) = \langle f \rangle$.

[6] Notice that this stalk is one dimensional; equivalently, the stalk is generated by such functions with C constant. This can be seen by direct computation, or by showing that the stalk is generated by the images of locally constant functions "upstairs".

Now we decompose in terms of the Iwasawa decomposition $G = NAK$, where $K = \mathrm{PGL}_2(\mathfrak{o})$ in the nonarchimedean case, and K is the "standard" SO(2) or SU(2) in the real and complex case, respectively.

We can easily see that in the nonarchimedean case, for:

$$\Phi(nak) = \begin{cases} \psi(n), & \text{if } a \in A(\mathfrak{o}) \\ 0, & \text{otherwise} \end{cases}$$

we get:

$$\int \psi(\xi x^{-1} - x) dx.$$

Since this particular Φ satisfies $\Phi\left(\begin{pmatrix} & 1 \\ 1 & \end{pmatrix}\right) = \langle f \rangle = 1 \neq 0$, it generates the fiber = stalk of $\mathcal{S}(\mathcal{W}^0)$ over ∞. It is easy to see that for large $|\xi|$ only the x with $|x^2| = |\xi|$ contribute (see the proof of Theorem 5.1), and this gives the desired claim.

In the archimedean case, the Cartan decomposition reads:

$$\begin{pmatrix} \xi & \\ & 1 \end{pmatrix} \begin{pmatrix} 1 & \\ x & 1 \end{pmatrix} =$$

$$\begin{pmatrix} 1 & \frac{\xi \bar{x}}{\sqrt{|x|^2+1}} \\ & 1 \end{pmatrix} \begin{pmatrix} \frac{\xi}{\sqrt{|x|^2+1}} & \\ & \sqrt{|x|^2+1} \end{pmatrix} \begin{pmatrix} \frac{1}{\sqrt{|x|^2+1}} & \frac{-\bar{x}}{\sqrt{|x|^2+1}} \\ \frac{x}{\sqrt{|x|^2+1}} & \frac{1}{\sqrt{|x|^2+1}} \end{pmatrix},$$

and the matrix $\begin{pmatrix} & 1 \\ 1 & \end{pmatrix}$ is obtained as the limit when $x \to \infty$, $\frac{\xi}{|x|^2+1} \to -1$, hence the claim.

□

4.10 Contribution of irregular locus and convergence of orbital integrals

Let $S \subset \bar{X} \times X$ be the irregular locus of the map (4.1), that is: the set of points $(\bar{x}, x)$ with $\bar{x} \in \mathbb{P}^1$ equal to the image of x under the natural map: $X \to \mathbb{P}^1$.

Proposition 4.10 *The embedding:*

$$\mathcal{S}((\bar{X} \times X) \smallsetminus S, \mathcal{L}_\psi \boxtimes \mathcal{L}_\psi^{-1}) \hookrightarrow \mathcal{S}(\bar{X} \times X, \mathcal{L}_\psi \boxtimes \mathcal{L}_\psi^{-1})$$

induces an isomorphism on G-coinvariants, that is:

$$\mathcal{S}(\mathcal{W}) = \mathcal{S}((\bar{X} \times X) \smallsetminus S, \mathcal{L}_\psi \boxtimes \mathcal{L}_\psi^{-1})_G.$$

This proposition already implies that all the invariant distributions that we have defined on $\mathcal{S}((\bar{X} \times X) \smallsetminus S, \mathcal{L}_\psi \boxtimes \mathcal{L}_\psi^{-1})$ (regular and irregular orbital integrals, including the inner product) extend to the whole space; for later use, we mention the following (which is easy to see):

Lemma 4.11 *In the nonarchimedean case, the regular orbital integrals of an element of* $\mathcal{S}(\bar{X} \times X, \mathcal{L}_\psi \boxtimes \mathcal{L}_\psi^{-1})$ *can be decomposed as:*

$$\int_{N\backslash G} \int_N^* ,$$

where $\int_N^*$ *is a stabilizing integral over large compact open subgroups of* N.

Proof (of Proposition 4.10) If we fix the stabilizer N^- of a point on X, and denote by ψ^{-1} the character by which it acts on the fiber of $\mathcal{L}_\psi^{-1}$, the problem is easily reduced to that of finding (N^-, ψ)-equivariant distributions on the stalk V of $\mathcal{S}(\bar{X}, \mathcal{L}_\psi)$ over the unique point y of $\mathbb{P}^1$ fixed by N^-. The notation $\mathcal{S}(\bar{X}, \mathcal{L}_\psi)$ means similar asymptotics as in §4.5, not sections of $\mathcal{L}_\psi$ over $\bar{X}$, but it is easy to see that as an N^--module this stalk has a filtration:

$$0 \to W \to V \to W \to 0,$$

where W is isomorphic to the stalk of smooth sections of $\mathcal{L}_\psi$ over y.

If $S' = \{y\} \subset \bar{X}$, in the notation of appendix B the stalk V of $\mathcal{S}(\bar{X}, \mathcal{L}_\psi)$ over S' has a separated decreasing filtration by $V_n := \overline{\mathcal{J}_{S'}^n V}$. Clearly, the group N^- acts trivially on $\mathcal{J}_{S'}/\mathcal{J}_{S'}^2$ = the cotangent space of y, and hence also on $\mathcal{J}_{S'}^n/\mathcal{J}_{S'}^{n+1}$ = the n-th symmetric power of the cotangent space. Moreover, recall that $\mathcal{L}_\psi$ is the trivial line bundle over $\mathbb{P}^1$. Therefore, there are no (N^-, ψ)-equivariant functionals on the n-th graded piece of this filtration, which is an image of (actually, isomorphic to):

$$\mathcal{J}_{S'}^n/\mathcal{J}_{S'}^{n+1} \otimes \mathcal{S}(\{y\}, \mathcal{L}_\psi).$$

□

5 Matching and the Fundamental Lemma

5.1 Matching

Theorem 5.1 *The operator* $|\bullet| \cdot \mathcal{G}$ *gives rise to a topological isomorphism:*

$$\mathcal{S}(\mathcal{Z}) \xrightarrow{\sim} \mathcal{S}(\mathcal{W}), \tag{5.1}$$

which satisfies:

$$\langle |\bullet|\mathcal{G}f\rangle = \gamma^*(\eta,0,\psi)\,\langle f\rangle\,, \tag{5.2}$$

where $\langle\ \rangle$ *denotes the inner products defined in §3.6, 4.9, and* $\gamma^*(\eta,0,\psi)$ *denotes the leading term in the Taylor expansion of the gamma factor* $\gamma(\eta,s,\psi)$ *around* $s=0$.

Proof We have short exact sequences:

$$0\to\mathcal{S}(\mathcal{X})\to\mathcal{S}(\mathcal{Z})\to\mathcal{S}(\mathcal{Z})/\mathcal{S}(X)\to 0 \tag{5.3}$$

and:

$$0\to|\bullet|\mathcal{S}(\mathcal{X})\to\mathcal{S}(\mathcal{W})\to\mathcal{S}(\mathcal{W})_\infty\to 0. \tag{5.4}$$

Recall that $\mathcal{S}(\mathcal{W})_\infty$ denotes the stalk of $\mathcal{S}(\mathcal{W})$ at $\xi=\infty$. The arrows on the left are closed embeddings and come from (3.5), where we restrict only to sections of $\mathcal{M}(\mathcal{Y}_1)$ with smooth orbital integrals — that is, we allow singularities only at $\xi=0$, not at $\xi=-1$; and from Corollary 4.5.

We have already seen that $\mathcal{G}$ is an automorphism of $\mathcal{S}(\mathcal{X})$, hence $|\bullet|\mathcal{G}$ is an automorphism between the leftmost terms of the above sequences. There remains to see that it induces isomorphisms of the quotients.

By Corollary 2.11 and standard properties of Fourier transform, the germs at $\xi=0$ of elements of $\iota\mathcal{F}(\mathcal{S}(\mathcal{Z}))$ are precisely the germs of functions of the form $f_1(\xi)+\psi\left(\frac{1}{\xi}\right)h(\xi)$ with f_1,h smooth. Moreover, we claim that for $\iota\mathcal{F}(f)\sim\psi\left(\frac{1}{\xi}\right)h(\xi)$ (where $\sim$ denotes equality of germs), we have:

$$h(0)=\gamma^*(\eta,0,\psi)\,\langle f\rangle\,. \tag{5.5}$$

It suffices to prove (5.5) for one element f for which $\langle f\rangle$ is nonzero. Recall that for (almost) every character χ of $F^\times$, considered as a tempered distribution on k by meromorphic continuation according to Tate's thesis, we have a relation:

$$\widehat{\chi(\bullet)}=\gamma(\chi^{-1},0,\psi)\cdot|\bullet|^{-1}\cdot\chi^{-1}(\bullet). \tag{5.6}$$

Indeed, this is just a reformulation of the functional equation for zeta integrals; in what follows, we denote the obvious *bilinear* (not hermitian) pairing by angular brackets, and use the exponent ψ when Fourier transform is taken with respect to the character ψ, instead of ψ^{-1} which is our standard convention. We denote Tate's zeta integral of a function $\phi\in\mathcal{S}(F)$ by $\zeta(\phi,\chi,s)$.

$$\langle\phi,\widehat{\chi}\rangle=\left\langle\widehat{\hat{\phi}}^{\psi},\widehat{\chi}\right\rangle=\left\langle\hat{\phi},\chi\right\rangle=\zeta(\hat{\phi},\chi,1)=$$
$$\gamma(\chi^{-1},0,\psi)\zeta(\phi,\chi^{-1},0)=\gamma(\chi^{-1},0,\psi)\left\langle\phi,\chi^{-1}(\bullet)\cdot|\bullet|^{-1}\right\rangle.$$

This implies that a function on F which is equal to $\chi(\xi)$ in a neighborhood of zero (and Schwartz elsewhere) has Fourier transform which is equal[7] to $\gamma(\chi^{-1},0,\psi)|\xi|^{-1}\chi^{-1}(\xi)$ in a neighborhood of infinity (and Schwartz elsewhere). In particular, (5.5) holds for the nonsplit case $\eta \neq 1$.

For the split case, we can obtain the function $1 - \ln|\xi|$ as the limit of:

$$\frac{1}{t} - \frac{|\xi|^t}{t}.$$

A function which is equal to this in a neighborhood of zero has Fourier transform which is equal to $-\frac{\gamma(1,-t,0)}{t}|\xi|^{-t-1}$ in a neighborhood of ∞, and in the limit $t \to 0$ we obtain $\gamma^*(1,-t,0)|\xi|^{-1}$.

There remains to show that Fourier transform gives a continuous surjection from the set of functions of the form $\psi\left(\frac{1}{\xi}\right)h(\xi)$ around $\xi = 0$ (and Schwartz otherwise) to $|\bullet|^{-1}$ times the germs of Kloosterman integrals described in Proposition 4.8. It will be an implicit byproduct of the proof that, if C is as in the remark following Proposition 4.8, then $C(0) = h(0)$, hence $\langle|\bullet|\mathcal{G}f\rangle = \gamma^*(\eta,0,\psi)\langle f\rangle$.

We perform this for the archimedean case in the next subsection. For the nonarchimedean case, let us say that $h = 1_{\mathfrak{o}}$. (Since the stalks are one-dimensional, it will be enough to check for one element.) Then:

$$\mathcal{F}\left(\psi\left(\frac{1}{\bullet}\right)h(\bullet)\right)(\xi) = \int_{\mathfrak{o}} \psi(x^{-1} - \xi x)dx.$$

For $|\xi|$ larger than $|\mathfrak{p}^{-2}\cdot\mathfrak{c}^2|$ (and larger than 1), where $\mathfrak{c}$ denotes the conductor of ψ, we claim that only the terms with $|x^2| = |\xi|^{-1}$ contribute. Indeed, set $u = x^{-1}$ and $v = \xi x$ and assume that $|u| > |v|$ (the case $|u| < |v|$ is identical). Then u has norm larger than $|\mathfrak{p}^{-1}\mathfrak{c}|$, and as it varies in a ball of radius $|\mathfrak{p}^{-1}\mathfrak{c}|$ around some point u_0, v varies in a ball of radius less or equal than $|\mathfrak{c}|$ around $v_0 = \xi u_0^{-1}$. Therefore:

$$\int_{u_0+\mathfrak{p}^{-1}\mathfrak{c}} \psi(u - \xi u^{-1})du = \int_{u_0+\mathfrak{p}^{-1}\mathfrak{c}} \psi(u)du = 0.$$

Hence,

$$\mathcal{F}\left(\psi\left(\frac{1}{\bullet}\right)1_{\mathfrak{o}}\right)(\xi) = \int_{|x|^2=|\xi|^{-1}} \psi(x^{-1} - \xi x)dx =$$

$$= |\xi|^{-1}\int_{|x|^2=|\xi|} \psi(\xi x^{-1} - x)dx.$$

□

[7] Asymptotically equal in the archimedean case, i.e. the quotient by the stated function tends to 1. This is proven by an easy argument multiplying the character by a smooth cutoff function.

5.2 Stationary phase

We complete the proof of matching in the archimedean case, and more precisely the identification of the Fourier transform of functions of the form $\psi\left(\frac{1}{\xi}\right)h(\xi)$ (in a neighborhood of 0) with the Kloosterman germ of (4.27) based on the arguments of [Jac05]. We only discuss the real case, as the complex case can be treated similarly.

Lemma 5.2 *Let $F = \mathbb{R}$ and let $\phi(u,\delta)$ be a Schwartz function in two variables. The integral:*

$$\int \phi(u, \frac{1}{\lambda})\psi(\lambda(u+u^{-1}))du$$

is equal to $f_1(\lambda) + |\lambda|^{-\frac{1}{2}}\psi(2\lambda)\theta_+\left(\frac{1}{\lambda}\right) + |\lambda|^{-\frac{1}{2}}\psi(-2\lambda)\theta_-\left(\frac{1}{\lambda}\right)$, where f_1 is a Schwartz function of λ, and $\theta_\pm$ are smooth functions (supported in a neighborhood of zero) whose derivatives at zero are polynomials, without constant terms, on the derivatives of $\phi(u,\delta)$ at $u=\pm 1$ (respectively), $\delta = 0$. In particular, $\theta_\pm(0)$ depends only on $\phi(\pm 1, 0)$ (respectively).

Moreover, in the special case that $\phi(u,\delta) = f(u\delta)$ for some smooth function f, each derivative of θ_+ at 0 depends on a finite number of derivatives of f at 0, and the germ of θ_+ at zero can be arbitrary. Similarly for θ_-.

Proof This is [Jac05][Proposition 1], except for the last statement.

It is proven in [Jac05] that, up to a certain nonzero constant:

$$\theta_+(\delta) = \int \phi_1(u,\delta)\psi\left(-\frac{u^2\delta}{4}\right)du,$$

where ϕ_1 is the partial Fourier transform in the variable $v = \frac{u-1}{\sqrt{u}}$ of the function:

$$\phi(u(v),\delta)\frac{du}{dv}.$$

(We assume without loss of generality that ϕ is supported close to $u=1$, so that the change of variables $v = \frac{u-1}{\sqrt{u}}$ is valid.)

Hence,

$$\theta_+^{(n)}(\delta) = \int \left[\left(\frac{\partial}{\partial\delta} - \frac{2\pi i u^2}{4}\right)^n \phi_1(u,\delta)\right]\psi\left(-\frac{u^2\delta}{4}\right)du.$$

(We assume without loss of generality that $\psi(x) = e^{2\pi i x}$.)

Therefore:

$$\theta_+^{(n)}(0) = \left(\frac{\partial}{\partial\delta} - \frac{1}{8\pi i}\frac{\partial}{\partial v}\right)^n \phi(u(v),\delta)\frac{du}{dv}\bigg|_{v=\delta=0}.$$

It is clear that, if $\phi = f(u\delta)$, this expression is bounded by a finite number of derivatives of f at 0, and that the evaluation of $\theta_+^{(n)}(0)$ involves higher derivatives of f at 0 than the evaluation of all $\theta_+^k(0)$, $k < n$. Therefore, the map $f \mapsto \theta_+$ is surjective onto the stalk of smooth functions at zero.

□

This allows us to complete the proof of Theorem 5.1 in the real case: Indeed, by the stationary phase method or the arguments of [Jac05] it is easy to see that (4.27) is a Schwartz function of ξ for $\xi > 0$ or Z supported away from -1 (in the first variable). For Z supported close to -1 and $\xi < 0$ we can make first the change of variables: $t = \frac{x^2+1}{x}$, and the integral (4.27) becomes:

$$\int Z_1\left(\frac{\xi}{t^2}, t\right)\psi(\xi t^{-1} - t)dt,$$

where Z_1 is another (arbitrary) smooth function on $\mathbb{G}_m \times \mathbb{P}^1$. Then we can make the change $u = -\sqrt{-\xi}^{-1}t$ to turn this into:

$$\sqrt{|\xi|}\int Z_1(-u, t)\psi(\sqrt{-\xi}(u^{-1} + u))du. \tag{5.7}$$

Similarly, for the Fourier transform of a function of the form $h(x)\psi\left(\frac{1}{x}\right)$ we have:

$$\int h(x)\psi(x^{-1} - \xi x)dx,$$

which again by the same arguments depends up to a Schwartz function of ξ only on the restriction of h in a neighborhood of zero, and only for $\xi < 0$. By the change of variables $u = -\sqrt{-\xi}x$ we get:

$$\sqrt{|\xi|}^{-1}\int h_1(-\sqrt{-\xi}^{-1}u)\psi(\sqrt{-\xi}(u + u^{-1}))du, \tag{5.8}$$

where h_1 is another (arbitrary) smooth function in a neighborhood of zero.

By the last statement of Lemma 5.2, the stalks at zero of (5.7), (5.8) coincide. This completes the proof of Theorem 5.1.

□

5.3 Basic vectors

From now on, until the end of this section, we assume that F is nonarchimedean, E (and hence F) is unramified over the base field $\mathbb{Q}_p$ or $\mathbb{F}_p((t))$, and endow the groups G, T, N, N^- with smooth group scheme structures over the ring of integers $\mathfrak{o}$. We set $K = G(\mathfrak{o})$, a hyperspecial maximal compact subgroup. The conductor of our fixed self-dual character ψ is equal to the ring of integers of F. We consider the $\mathfrak{o}$-schemes:

$$X_1 = T\backslash G, \ X_2 = N\backslash G,$$

where the latter is equipped with the line bundle $\mathcal{L}_\psi$ defined by ψ and an $\mathfrak{o}$-identification: $N \simeq \mathbb{G}_a$.

We endow the various groups with invariant volume forms defined over $\mathfrak{o}$, which are nonzero when reduced to the residue field. Based on our fixed measure on F of §2.3, this gives rise to invariant measures on their F-points, and the F-points of their quotients; these measures are canonical, as any two volume forms with these properties are multiples of each other by elements of $\mathfrak{o}^\times$.

We consider the spaces $\mathcal{S}(\mathcal{Z})$ and $\mathcal{S}(\mathcal{W})$ of coinvariants corresponding to X_1, resp. X_2, as defined previously. We will define distinguished vectors $f^0_{\mathcal{Z}}$, $f^0_{\mathcal{W}}$ on them, the *basic vectors.*

For $\mathcal{S}(\mathcal{Z})$ we define:

$$f^0_{\mathcal{Z}} := \text{ the image of } 1_{X_1(\mathfrak{o})} \otimes 1_{X_1(\mathfrak{o})} \text{ in } \mathcal{S}(\mathcal{Z}). \tag{5.9}$$

(Having fixed measures on the various groups, this image is a well-defined element of $\mathcal{S}(\mathcal{Z})$.)

The description for $f^0_{\mathcal{W}}$ will be more complicated, as it is a "nonstandard" test function, i.e. not compactly supported. Recall from 4.3 that, in order to define orbital integrals for the Kuznetsov trace formula as functions, we have chosen a point $(x_0, y_0) \in (X_2 \times X_2)^+$ with image $1 \in \mathcal{B}$, and have trivialized the fiber of $\mathcal{L}_\psi \boxtimes \mathcal{L}_\psi^{-1}$ over that point. We now assume that $(x_0, y_0) \in (X_2 \times X_2)^+(\mathfrak{o})$, hence after trivializing the fiber and choosing suitable $\mathfrak{o}$-isomorphisms of the stabilizers with N, N^- (and of the latter with $\mathbb{G}_a$) our sections become elements of $C^\infty(N\backslash G \times N^-\backslash G, \psi \otimes \psi^{-1})$. In fact, we may trivialize both the fibers of $\mathcal{L}_\psi$ over x_0 and $\mathcal{L}_\psi^{-1}$ over y_0, to consider smooth sections of $\mathcal{L}_\psi$ (resp. of $\mathcal{L}_\psi^{-1}$) as elements of $C^\infty(N\backslash G, \psi)$ (resp. $C^\infty(N^-\backslash G, \psi^{-1})$).

For $n \in \mathbb{N}$, we denote by $1_{x_n K}$ the section:

$$1_{x_n K}\left(u \begin{pmatrix} \varpi^m & \\ & 1 \end{pmatrix} k\right) \text{ (where } u \in N, k \in K) = \begin{cases} 0, & \text{if } m \neq n, \\ \psi(u), & \text{otherwise,} \end{cases} \tag{5.10}$$

of $\mathcal{L}_\psi$. As n varies in $\mathbb{N}$, these form a basis for the space of compactly supported, K-invariant sections of $\mathcal{L}_\psi$. We similarly define $1^-_{y_n K}$ for $\mathcal{L}_\psi^{-1}$.

For an algebraic representation V of the dual group $\check{G} = \mathrm{SL}_2$, denote by h_V the element of the spherical Hecke algebra $\mathcal{H}(G, K)$ corresponding under the Satake isomorphism:

$$\mathcal{H}(G,K) = \mathbb{C}[\mathrm{Rep}(\check{G})]$$

to the representation V. Here the monoid of dominant weights of $\check{G}$ is isomorphic to $\mathbb{N}$, and we will be writing h_n for h_{V_n}, where V_n is the n-th highest weight representation.

The Casselman-Shalika formula states that:

$$h_n \star 1_{x_0K} = q^{-\frac{n}{2}} 1_{x_nK}. \tag{5.11}$$

Let H_s be the formal series in the spherical Hecke algebra which *corresponds under the Satake isomorphism to the L-function*:

$$L(\pi, \frac{1}{2}+s)L(\pi\otimes\eta, \frac{1}{2}+s). \tag{5.12}$$

To understand what this means, we view an L-function $L(\pi,\rho,s)$ (where ρ is a representation of the dual group) as a formal series (in the parameter q^{-s}) of traces of representations:

$$L(\pi,\rho,s) = \sum_{i=0}^{\infty} q^{-is}\mathrm{tr}(S^i\rho(\hat{\pi})),$$

where $\hat{\pi}$ is the Satake parameter of π, hence the corresponding series in the Hecke algebra will be:

$$\sum_{i=0}^{\infty} q^{-is} h_{S^i\rho}.$$

We then define, for each s:

$$\Phi_s^0 := \Phi_{1,s}^0 \otimes \Phi_2^0 = (H_s \star 1_{x_0K}) \otimes 1^-_{y_0K} \in C^\infty(X_2\times X_2, \mathcal{L}_\psi \otimes \mathcal{L}_\psi^{-1})^{K\times K}. \tag{5.13}$$

To see that $H_s \star 1_{x_0K}$, a priori a formal series of elements of $C_c^\infty(X_2,\mathcal{L}_\psi)$, makes sense as a section of $\mathcal{L}_\psi$ when we fix s, write $H_s = h_{1,s}\star h_{2,s}$, where $h_{1,s}$ corresponds to the L-function $L(\pi,\frac{1}{2}+s)$ and $h_{2,s}$ corresponds to $L(\pi\otimes\eta,\frac{1}{2}+s)$. Recall that for a representation (ρ,V) of $\check{G}$, and $t\in\check{G}$, we have:

$$\overset{-1}{\det}(I - q^{-s}\rho(t)|_V) = \sum_{n\geq 0} q^{-ns}\mathrm{tr}\rho(t)|_{S^nV}. \tag{5.14}$$

Hence:

$$h_{1,s} = \sum_{n\geq 0} q^{-n(s+\frac{1}{2})} h_n,$$

$$h_{2,s} = \sum_{n\geq 0} q^{-n(s+\frac{1}{2})}\epsilon^n h_n,$$

where $\epsilon = \pm 1$, according as η is trivial or not.

Let V_n denote the highest weight representation of $\check{G}$ corresponding to the n-th dominant weight, then we have the Clebsch-Gordan formula:

$$V_m \otimes V_n = \sum_{l=0}^{\min(m,n)} V_{m+n-2l}. \tag{5.15}$$

We use it to compute the convolution of $h_{1,s}$ with $h_{2,s}$, i.e. to write the series:

$$\left(\sum_{m\geq 0} q^{-m(s+\frac{1}{2})}h_m\right) \star \left(\sum_{n\geq 0} q^{-n(s+\frac{1}{2})}\epsilon^n h_n\right) =$$

$$\sum_{n,m\geq 0} q^{-(n+m)(s+\frac{1}{2})}\epsilon^n h_m \star h_n = \sum_{n,m\geq 0}\sum_{l=0}^{\min(m,n)} q^{-(n+m)(s+\frac{1}{2})}\epsilon^n h_{m+n-2l}.$$

Let $k = m+n-2l$, then the restrictions between the different indices correspond to the system:

$$\begin{aligned} l \leq \min(m,n) &\leq l + \frac{k}{2} \\ m+n &= k+2l. \end{aligned}$$

To count all m, n for a given k, we add over all $l = 0, 1, \ldots$ and have two cases: either $m = \min(m,n)$, in which case m ranges over: $l \leq m \leq \lfloor \frac{k}{2} \rfloor$; or $m > \min(m,n)$ in which case n ranges over: $l \leq n \leq l + \lfloor \frac{k+1}{2} \rfloor$. Altogether, n ranges from l to $k+l$. Therefore, the coefficient for h_k will be:

$$\sum_{l=0}^{\infty} q^{-(k+2l)(s+\frac{1}{2})} \sum_{n=l}^{k+l} \epsilon^n = \sum_{l=0}^{\infty} \epsilon^l q^{-(k+2l)(s+\frac{1}{2})}. \begin{cases} k+1 & \text{if } \epsilon = 1, \\ 0 & \text{if } \epsilon = -1, k \text{is odd}, \\ 1 & \text{if } \epsilon = -1, k \text{is even}. \end{cases}$$

Hence,

$$\Phi^0_{1,s} = H_s \star 1_{x_0 K} = \sum_{n=0}^{\infty} \frac{q^{-n(s+1)}}{1-\epsilon q^{-2s-1}} \cdot 1_{x_n K} \cdot \begin{cases} k+1 & \text{if } \epsilon = 1, \\ 0 & \text{if } \epsilon = -1, k \text{is odd}, \\ 1 & \text{if } \epsilon = -1, k \text{is even}. \end{cases} \tag{5.16}$$

We deduce:

Lemma 5.3 *For each fixed s such that $1-\epsilon q^{-2s-1}\neq 0$, $\Phi^0_{1,s}$ makes sense as a smooth section of $\mathcal{L}_\psi$. Moreover, for $s=0$ we have: $\Phi^0_0\in\mathcal{S}(\bar{X}\times X, \mathcal{L}_\psi\boxtimes\mathcal{L}_\psi^{-1})$.*

Proof Only the last assertion remains to be proven. We denote by $F^0_{1,s}$ the K-invariant *function* on X_2 which, under the above trivializations, is equal to $\Phi^0_{1,s}$ on diagonal elements; that is, $F^0_{1,s}$ is given by the same series, but 1_{x_nK} is replaced by $1'_{x_nK}:=$ the characteristic function of the K-orbit represented by $\mathrm{diag}(\varpi^n,1)$. Then it is easy to see that $\Phi^0_{1,s}$ is the product of $F^0_{1,s}$ by a section of $\mathcal{L}_\psi$ which extends to $\mathbb{P}^1$. Therefore, it suffices to prove that $F^0_{1,0}$ satisfies, in the notation of §4.5:

$$\left(1-\delta^{-\frac{1}{2}}(a)\mathscr{L}_a\right)\cdot\left(1-\eta_{E/F}\delta^{-\frac{1}{2}}(a)\mathscr{L}_a\right)F^0_{1,0}=0.$$

This follows immediately from the fact that $\mathscr{L}_{\mathrm{diag}(\varpi^m,1)}1'_{x_nK}=q^{\frac{m}{2}}1'_{x_{n-m}K}$.

∎

Therefore, we may define the basic vector:

$$f^0_{\mathcal{W}}:=\text{ the image of }\Phi^0_0\text{ in }\mathcal{S}(\mathcal{W}). \tag{5.17}$$

5.4 Fundamental lemma

Finally, we arrive at the "fundamental lemma" for elements of the Hecke algebra. Notice that the Hecke algebra $\mathcal{H}(G,K)$ does not act on the quotients $\mathcal{S}(\mathcal{Z})$, $\mathcal{S}(\mathcal{W})$. However, the Bernstein center does, since these are quotients of $G\times G$ representations (and we accept the convention that it is the Bernstein center of the first copy of G which acts). The Bernstein center for the component of the spectrum corresponding to unramified principal series is isomorphic to $\mathcal{H}(G,K)$ under the natural map; therefore, we will abuse notation to write $h\star f$ for $h\in\mathcal{H}(G,K)$ and $f\in\mathcal{S}(\mathcal{Z})$ or $\mathcal{S}(\mathcal{W})$. Of course, this discussion serves only aesthetic purposes and is redundant otherwise, as we will only use such expressions for $f=$ the image of a $K\times K$-invariant function/section $\Phi_1\otimes\Phi_2$, and then $h\star f$ can be interpreted as the image of $(h\star\Phi_1)\otimes\Phi_2$.

Theorem 5.4 *For $f^0_{\mathcal{Z}}$, $f^0_{\mathcal{W}}$ the basic vectors defined in the previous subsection, and all $h\in\mathcal{H}(G,K)$, the integral transform $|\bullet|\mathcal{G}$ satisfies:*

$$|\bullet|\mathcal{G}\left(h\star f^0_{\mathcal{Z}}\right)=h\star f^0_{\mathcal{W}}. \tag{5.18}$$

This could be proven by explicit calculations as follows: On one hand, one can explicitly compute the orbital integrals of characteristic functions of K-orbits (or the "characteristic sections" $1_{x_n K} \otimes 1^-_{y_m K}$ of the previous subsection); some of those computations are exhibited in Section 6. On the other hand, one can use a Casselman-Shalika type formula (which in this case is one of the easiest cases of the general formula computed in [Sakb]) to explicitly describe the Hecke action in terms of those characteristic functions.

Since this is tedious and not particularly informative, but mainly in order to demonstrate how the contents of the present paper are simply reflections, at the level of orbital integrals, of certain transforms taking place "upstairs" at the level of G-spaces plus prior work of Jacquet on the results of Waldspurger, we follow a shortcut; it is important to realize, though, that nothing in the present paper depends on the existence of this shortcut, as it could be done directly.

Proof For the proof we will introduce intermediate "spaces" $\mathcal{Z}_1$ and $\mathcal{W}_1$ and we will prove "fundamental lemmas" for each step in the sequence:

$$\mathcal{S}(\mathcal{Z}) \leftrightarrow \mathcal{S}(\mathcal{Z}_1) \leftrightarrow \mathcal{S}(\mathcal{W}_1) \leftrightarrow \mathcal{S}(\mathcal{W}).$$

Just for this proof, we write P, P_1, Q_1, Q for the basic functions that have been defined, or will be defined, for the above spaces.

Symbolically, we have:

$$\mathcal{Z}_1 = A\backslash G/(A, \eta),$$

where A is the split torus of diagonal elements and $\eta(\mathrm{diag}(a, 1)) = \eta_{E/F}(a)$ — in particular, $\mathcal{Z}_1 = \mathcal{Z}$ in the split case; and:

$$\mathcal{W}_1 = (N, \psi)\backslash G/(N^-, \psi^{-1}),$$

but with different test functions than $\mathcal{W}$.

More precisely, now, we define $\mathcal{S}(\mathcal{Z}_1)$ as the space of functions on $\mathcal{B} \smallsetminus \{0, -1\}$ obtained by orbital integrals of elements of the space:

$$\mathcal{S}(A\backslash G \times A\backslash G, 1 \otimes \eta).$$

We use here the same parametrization for $A\backslash G/A$ as discussed in Section 3, but since there is a nontrivial character η we also need to specify representatives for the orbits which allow us to think of orbital integrals as functions on the regular set of $\mathcal{B}$. For $\Phi_1 \otimes \Phi_2 \in \mathcal{S}(A\backslash G \times A\backslash G, 1 \otimes \eta)$ we define:

$$O_\xi(\Phi_1 \otimes \Phi_2) = \int_G \Phi_1 \left(\begin{pmatrix} -\xi & 1+\xi \\ -1 & 1 \end{pmatrix} g \right) \Phi_2(g) dg. \tag{5.19}$$

Throughout we assume smooth $\mathfrak{o}$-models for our groups, and Haar measures arising from residually nontrivial integral volume forms. The "basic function" is, of course, the image of Φ_1 = the characteristic function of $(A\backslash G)(\mathfrak{o})$ and $\Phi_2(ak) = \eta(a)$ for $a \in A, k \in K = G(\mathfrak{o})$, $\Phi_2 = 0$ off AK.

Jacquet has shown in Proposition 5.1 of [Jac86] that there is a "fundamental lemma for the Hecke algebra" between $\mathcal{Z}$ and $\mathcal{Z}_1$, that is:

$$h \star P = h \star P_1 \tag{5.20}$$

for all $h \in \mathcal{H}(G, K)$ and $\xi \in \mathcal{B}_{\mathcal{Z}}^{\text{reg}} = \mathcal{B} \smallsetminus \{0, -1\}$. The parametrization of orbits is different in loc.cit., as are the volumes, but in the end there is no need to normalize by volume factors — as can easily be checked by taking $\xi \in \mathcal{B}_{\mathcal{Z}}^{\text{reg}}(\mathfrak{o})$.

Now we introduce the space $\mathcal{S}(\mathcal{W}_1)$, or rather just its basic vector Q_1. This space will consist of the orbital integrals of certain smooth — but not Schwartz — sections of $\mathcal{L}_\psi \boxtimes \mathcal{L}_\psi^{-1}$ over $X \times X$, where $X = N\backslash G$. The basic vector Q_1 will be obtained from the section $\Phi_1 \otimes \Phi_2$, where $\Phi_1 = H_1 \star 1_{x_0 K}$ and $\Phi_2 = H_2 \star 1^-_{y_0 K}$, in the notation of §5.3; Here H_1 and H_2 are the formal series in the Hecke algebra corresponding to the L-values:

$$L(\pi, \frac{1}{2})$$

and

$$L(\pi \otimes \eta, \frac{1}{2}),$$

respectively. How to make sense of Φ_1, Φ_2 as sections is completely analogous to the discussion of §5.3.

We claim that there is a "fundamental lemma for the Hecke algebra" between $\mathcal{S}(\mathcal{W}_1)$ and $\mathcal{S}(\mathcal{W})$, that is:

$$h \star Q_1 = h \star Q \tag{5.21}$$

for all $h \in \mathcal{H}(G, K)$ and $\xi \in \mathcal{B}_{\mathcal{W}}^{\text{reg}} = \mathcal{B}^\times$. It is convenient here to move to the domain of convergence by introducing a parameter s, i.e. functions H_i^s defined as before, with the L-values taken at $\frac{1}{2} + s$ instead of $\frac{1}{2}$; we let Q_1^s the corresponding function of orbital integrals. Then, writing $H_1^s = \sum_n c(n, s) h_n$, $H_2^s = \sum_n d(n, s) h_n$, where h_n is the Hecke element corresponding to the n-th dominant weight of the dual group, we have:

$$h \star Q_1^s(\xi) = \sum_{m,n} c(m, s) d(n, s) O_\xi(h \star h_m \star 1_{x_0 K}, h_n \star 1^-_{y_0 K})$$

for $\Re(s)$ large, by the fact that for such s the regular orbital integrals are actual, convergent, integrals.

For an element h in the full Hecke algebra of G, we denote by $h^\vee$ its linear dual: $h^\vee(g) = h(g^{-1})$. Elements of the spherical Hecke algebra of PGL_2 are all self-dual. Since orbital integrals are invariant by the diagonal action of G, we get:

$$\begin{aligned} h \star Q_1^s(\xi) &= \sum_{m,n} c(m,s) d(n,s) O_\xi(h_n \star h \star h_m \star 1_{x_0K}, 1^-_{y_0K}) \\ &= O_\xi(H_2^s \star h \star H_1^s \star 1_{x_0K}, 1^-_{y_0K}). \end{aligned}$$

Finally, using the commutativity of the spherical Hecke algebra, this is equal to: $O_\xi(h \star H_1^s \star H_2^s \star 1_{x_0K}, 1^-_{y_0K}) = h \star Q^s$. Hence, $h \star Q_1^s = h \star Q^s$ for $\Re(s)$ large.

Taking the limit (analytic continuation) as $s \to 0$ we obtain (5.21). Taking the limit is justified as follows: on one hand, the sections $H_1 \star 1_{x_0K}$ etc. are, by definition, pointwise limits of the sections $H_1^s \star 1_{x_0K}$ etc. On the other, for given ξ and sufficiently large m *or* sufficiently large n we have $O_\xi\left((h_m 1_{x_0K}) \otimes (h_n \star 1^-_{y_0K})\right) = 0$; this will be seen in §6.3.

We are left with showing the fundamental lemma for the passage $\mathcal{S}(\mathcal{Z}_1) \leftrightarrow \mathcal{S}(\mathcal{W}_1)$. To achieve that we will work on the level of spaces, and translate the "unfolding" method of Hecke to orbital integrals.

Let $f_1 \in \mathcal{S}(A\backslash G, \delta^s)$, $f_1' \in \mathcal{S}(A\backslash G, \eta\delta^s)$. Recall that $\delta^s(\mathrm{diag}(a,1)) = |a|^s$. We define:

$$f_2(g) = \int_N f_1(ng)\psi^{-1}(n)dn \in C^\infty(N\backslash G, \psi)$$

and:

$$f_2'(g) = \int_N f_1'(nwg)\psi^{-1}(n)dn \in C^\infty(N^-\backslash G, \psi^{-1}),$$

where $w = \begin{pmatrix} & 1 \\ -1 & \end{pmatrix}$.

We claim:

Lemma 5.5 *If f_1 is the basic function of $A\backslash G$, i.e. the function supported on AK with $f_1'(ak) = \delta^s(a)$, then $f_2 = L(\bullet, \frac{1}{2} + s) \star 1_{x_0K}$, where by the L-value we mean the corresponding element of the Hecke algebra $\mathcal{H}(G,K)$. If f_1' is the basic function of $(A\backslash G, \eta\delta^s)$, i.e. the function supported on AK with $f_1'(ak) = \eta\delta^{-s}(a)$, then $f_2' = L(\pi \otimes \eta, \frac{1}{2} + s) \star 1^-_{y_0K}$.*

Proof Assume that π is an irreducible unramified representation and W_π the spherical Whittaker function of π with respect to (N, ψ^{-1}), normalized so that $W_\pi(1) = 1$. Since by the Casselman-Shalika formula each $1_{x_n K}$ is up to a constant a multiple of $h_n \star 1_{x_0 K}$ we can write f_2 as a formal sum:

$$\sum_{n=0}^{\infty} c(n) h_n \star 1_{x_0 K},$$

so that if the integral:

$$\int_{N\backslash G} W_\pi(g) f_2(g) dg$$

is convergent, it is equal to:

$$\sum_n c(n) \mathrm{tr} V_n(\hat{\pi}) \int W_\pi(g) 1_{x_0}(g) dg = \mathrm{Vol}(N\backslash G(\mathfrak{o})) \sum_n c(n) \mathrm{tr} V_n(\hat{\pi}),$$

where V_n is the n-th irreducible representation of the dual group and $\hat{\pi}$ the Satake parameter of π. Therefore we just need to compute this integral.

We write:

$$\int_{N\backslash G} W_\pi(g) f_2(g) dg = \int_G W_\pi(g) f_1(g) dg =$$
$$\int_{A\backslash G} \int_A W_\pi(ag) \delta^s(a) da f_1(g) dg = \mathrm{Vol}(A\backslash G(\mathfrak{o})) \int_A W_\pi(ag) \eta \delta^s(a) da.$$

It is well-known (and follows easily from the Casselman-Shalika formula) that the last integral is absolutely convergent for $\Re(s) > -\frac{1}{2}$, and equal to $\mathrm{Vol}(A(\mathfrak{o})) L(\pi, \frac{1}{2} + s)$. This implies the claim, since $\frac{\mathrm{Vol}(A\backslash G(\mathfrak{o}))\mathrm{Vol}(A(\mathfrak{o}))}{\mathrm{Vol}(N\backslash G(\mathfrak{o}))} = \mathrm{Vol}(N(\mathfrak{o})) = 1$.

This proves the lemma for f_2, and the proof for f_2' is identical.

□

We continue with the proof of Theorem 5.4. By the previous lemma, when $s = 0$, we have $h \star P_1(\xi) = O_\xi(f_1 \otimes f_1')$ and $Q_1(\xi) = O_\xi(f_2 \otimes f_2')$ when $f_1 = h\star$(the basic function of $(A\backslash G, \delta^s)$) and $f_1' =$(the basic function of $(A\backslash G, \eta\delta^s)$). We want to investigate the relationship between orbital integrals for $f_1 \otimes f_1'$ and those for $f_2 \otimes f_2'$, when $s = 0$; as before, those will be the analytic continuation of the ones for $\Re(s) \gg 0$, where they are given by convergent integrals. We denote the Fourier transform of f_1 along N:

$$\hat{f}_1(y, g) = \int_N f_1(ng) \psi_y^{-1}(n) dn,$$

where $\psi_y\left(\begin{pmatrix} 1 & x \\ & 1 \end{pmatrix}\right) = \psi(yx)$, and similarly for $\hat{f}_1'(y, g)$.

Clearly, $f_2(g) = \hat{f}_1(1, g)$, $f'_2(g) = \hat{f}'_1(1, wg)$. Moreover, $\hat{f}_1(y, \mathrm{diag}(a, 1)g) = |a|^{s+1}\hat{f}_1(ay, g)$ and $\hat{f}'_1(y, w\mathrm{diag}(a, 1)(g)) = \eta(a)|a|^{-s-1}\hat{f}'_1(a^{-1}y, g)$. Hence we have:

$$O_\xi(f_2, f'_2) = \int_G \hat{f}_1(1, \mathrm{diag}(\xi, 1)g)\hat{f}'_1(1, wg)dg =$$
$$|\xi|^{s+1} \int_{A\backslash G}\int_{F^\times} \hat{f}_1(a\xi, g)\hat{f}'_1(a^{-1}, wg)\eta(a)dadg.$$

The function $\xi \mapsto \int_{F^\times} \hat{f}_1(a\xi, g)\hat{f}'_1(a^{-1}, wg)\eta(a)da$ can be seen as an orbital integral on ${\mathbb{G}_a}^2$ with respect to the action of the multiplicative group: $a \cdot (x, y) = (ax, a^{-1}y)$. Thus, we are in the split "baby case" of Section 2, except that we also have a character $\eta(a)$ in the orbital integrals. Moreover, we are applying those orbital integrals to the Fourier transform of a Schwartz function on ${\mathbb{G}_a}^2$ (indeed, the restrictions of f_1, f'_1 to unipotent orbits are Schwartz functions). We have then seen in 2 (for the case $\eta = 1$, but the case $\eta \neq 1$ is similar) that:

$$\int_{F^\times} \hat{f}_1(a\xi, g)\hat{f}'_1(a^{-1}, wg)\eta(a)da =$$
$$\mathcal{G}\left(c \mapsto \int_{F^\times} f_1\left(\begin{pmatrix}1 & ca\\ & 1\end{pmatrix} g\right) f'_1\left(\begin{pmatrix}1 & a^{-1}\\ & 1\end{pmatrix} wg\right)\eta(a)da\right). \quad (5.22)$$

It follows that:

$$O_\xi(f_2, f'_2) =$$
$$|\xi|^{s+1} \cdot \mathcal{G}\left(c \mapsto \int_{A\backslash G}\int_{F^\times} f_1\left(\begin{pmatrix}1 & ca\\ & 1\end{pmatrix} g\right) f'_1\left(\begin{pmatrix}1 & a^{-1}\\ & 1\end{pmatrix} wg\right)\eta(a)dadg\right) =$$
$$|\xi|^{s+1}\mathcal{G}\left(c \mapsto \int_{A\backslash G}\int_{F^\times} f_1\left(\begin{pmatrix}1 & c\\ & 1\end{pmatrix}\begin{pmatrix}a^{-1} & \\ & 1\end{pmatrix} g\right)\cdot\right.$$
$$\left. f'_1\left(\begin{pmatrix}1 & 1\\ & 1\end{pmatrix}\begin{pmatrix}a & \\ & 1\end{pmatrix} wg\right)dadg\right) =$$
$$|\xi|^{s+1}\mathcal{G}\left(c \mapsto \int_G f_1\left(\begin{pmatrix}cc1 & c\\ & 1\end{pmatrix} g\right) f'_1\left(\begin{pmatrix}1 & 1\\ & 1\end{pmatrix} wg\right)dg\right) =$$
$$|\xi|^{s+1}\mathcal{G}\left(c \mapsto O_c(f_1 \otimes f'_1)\right).$$

This proves the theorem.

□

6 Variation With a Parameter and Explicit Calculations

For global applications we will not be able to use the space of nonstandard sections for the Kuznetsov quotient directly. The reason is that, spectrally, they correspond to values of L-functions on the critical line, where global Euler products are non-convergent. We therefore need to introduce variations of this space, corresponding to the parameter s in:

$$L(\pi, \frac{1}{2}+s)L(\pi\otimes\eta, \frac{1}{2}+s).$$

We conclude with this, and some explicit calculations.

6.1 Nonstandard Whittaker space depending on s.

We generalize the definitions of §4.5 to an arbitrary parameter $s \in \mathbb{C}$ (the previous case corresponding to $s = 0$), borrowing freely notation from there.

We let $\mathcal{M}^s(\bar{X}\times X, \mathcal{L}_\psi \boxtimes \mathcal{L}_\psi^{-1})$ (resp. $\mathcal{S}^s(\bar{X}\times X, \mathcal{L}_\psi \boxtimes \mathcal{L}_\psi^{-1})$) denote the Schwartz cosheaf over $\bar{X}\times X$ consisting of smooth measures (resp. functions) on $X\times X$, valued in $\mathcal{L}_\psi \boxtimes \mathcal{L}_\psi^{-1}$, with the following properties:

- the restriction of the cosheaf to $X\times X$ coincides with the standard cosheaf of Schwartz measures (resp. functions) valued in $\mathcal{L}_\psi \boxtimes \mathcal{L}_\psi^{-1}$;
- in a neighborhood of $\mathbb{P}^1\times X$ they are finite sums of the form $\sum_i f_i F_i$, where:

  1. the f_i's are $\mathcal{L}_\psi \boxtimes \mathcal{L}_\psi^{-1}$-valued Schwartz functions on $\bar{X}\times X$;
  2. the F_i are scalar-valued measures (resp. functions) on $X\times X$ which are G-invariant in the second coordinate, and in the first coordinate are annihilated asymptotically by the operator:

$$\left(1-\delta^{-\frac{1}{2}-s}(a)\mathscr{L}_a\right)\cdot\left(1-\eta_{E/F}\delta^{-\frac{1}{2}-s}(a)\mathscr{L}_a\right). \qquad (6.1)$$

We let $\mathcal{M}(\mathcal{W}^s)$ denote the G-coinvariants of $\mathcal{M}^s(\bar{X}\times X, \mathcal{L}_\psi \boxtimes \mathcal{L}_\psi^{-1})$. Again, using the trivializations of Lemma 4.2, we have a map from $\mathcal{M}(\mathcal{W}^s)$ to measures on $\mathcal{B}^\times$. Finally, we identify those with functions on $\mathcal{B}^\times$, by dividing them by $|\xi|^{-2}d\xi$ (see the discussion following Lemma 4.3), and get the local Schwartz space $\mathcal{S}(\mathcal{W}^s)$ of the Kuznetsov trace formula with parameter s, consisting of functions on $\mathcal{B}^\times$. This is the space of orbital integrals of elements of $\mathcal{S}^s(\bar{X}\times X, \mathcal{L}_\psi \boxtimes \mathcal{L}_\psi^{-1})$.

6.2 Basic vector

We now come to the setting of §5.3, adopting (until the end of the section) all the conventions and notation from there. In particular, F is nonarchimedean and we have good integral models, measures, and isomorphisms for everything. We only denote here by X what was denoted there by X_2; namely, the space $N\backslash\mathrm{PGL}_2$. We defined in 5.3 certain sections Φ_s^0 of $\mathcal{L}_\psi \boxtimes \mathcal{L}_\psi^{-1}$ over $X \times X$. In analogy with Lemma 5.3 we have:

Lemma 6.1 *The section Φ_s^0 belongs to $\mathcal{S}^s(\bar{X} \times X, \mathcal{L}_\psi \boxtimes \mathcal{L}_\psi^{-1})$.*

The proof is identical to that of Lemma 5.3. We define f_s^0 to be the image of Φ_s^0 in $\mathcal{S}(\mathcal{W}^s)$. (In comparison to §5.3, we omit the index $_\mathcal{W}$ since here we only work on the Kuznetsov space, and introduce the index $_s$ so that the previous $f_\mathcal{W}^0$ is now f_0^0.)

6.3 Orbital integrals for the characteristic sections

Recall that $1_{x_m K}$ denotes a certain compactly supported section of $\mathcal{L}_\psi$ defined in §5.3, and $1_{y_m K}^-$ a compactly supported section of $\mathcal{L}_\psi^{-1}$. Now we compute the orbital integral:

$$O_\xi(1_{x_m K} \otimes 1_{y_0 K}^-)$$

for $\xi \in F^\times$. We also identify ξ with the representative $\begin{pmatrix} \xi & \\ & 1 \end{pmatrix}$ of $N\backslash G/N^-$, according to §4.3.

We have:

$$O_\xi(1_{x_m K} \otimes 1_{y_0 K}) = \int_{N^-\backslash G}\int_{N^-} 1_{x_m K}(\xi n g)\psi^{-1}(n)dn \cdot 1_{y_0 K}^-(g)dg =$$

$$\mathrm{Vol}(X(\mathfrak{o}))\int_{N^-} 1_{x_m K}(\xi n)\psi^{-1}(n)dn.$$

Let $n = \begin{pmatrix} 1 & \\ x & 1 \end{pmatrix} \in N$, then ξn admits the following Iwasawa decomposition ($G = NAK$):

- if $|x| \leq 1$: then $\xi \in A$, $n \in K$;
- if $|x| > 1$: then $\xi n = \begin{pmatrix} 1 & \xi x^{-1} \\ & 1 \end{pmatrix}\begin{pmatrix} -\xi x^{-1} & \\ & x \end{pmatrix}\begin{pmatrix} & 1 \\ 1 & x^{-1} \end{pmatrix}$.

Therefore,

$$\int_{N^-} 1_{x_mK}(\xi n)\psi^{-1}(n)dn =$$

$$1_{x_mK}\begin{pmatrix}\xi & \\ & 1\end{pmatrix} + \sum_{i=1}^{\infty} 1_{x_mK}\begin{pmatrix}\xi\varpi^i & \\ & \varpi^{-i}\end{pmatrix}\int_{\mathfrak{p}^{-i}\smallsetminus\mathfrak{p}^{-i+1}} \psi(\xi x^{-1} - x)dx.$$

Thus we get:

- if $|\xi| = q^{-m}$: $O_\xi(1_{x_mK}\otimes 1^-_{y_0K}) = \mathrm{Vol}X(\mathfrak{o})$;
- if $|\xi| = q^{2i-m}$ for some $i > 0$:

$$O_\xi(1_{x_mK}\otimes 1^-_{y_0K}) = \mathrm{Vol}X(\mathfrak{o})\int_{\mathfrak{p}^{-i}\smallsetminus\mathfrak{p}^{-i+1}} \psi(\xi x^{-1} - x)dx;$$

- zero otherwise.

For the integral in the second case, we have $|\xi x^{-1}| = |x|q^{-m} \le |x|$. Hence:

- If $m \ge 1$ and $i > 1$ then as x varies in a ball of radius q, ξx^{-1} varies in a ball of radius ≤ 1, therefore the integral is zero.
- If $m \ge 1$ and $i = 1$, i.e. $|\xi| = q^{2-m}$ then as x varies in $\mathfrak{p}^{-1}\smallsetminus\mathfrak{o}$, $\psi(\xi x^{-1}) = 1$ and we get: $O_\xi 1_{x_mK} = -\mathrm{Vol}X(\mathfrak{o})$.
- Finally, if $m = 0$ and $|\xi| > 1$ then we get (with a change of variables $x \mapsto -x$): $O_\xi 1_{x_mK} = \mathrm{Vol}X(\mathfrak{o})\int_{|x|^2=|\xi|}\psi(x - \xi x^{-1})dx$.

To summarize:

$$O_\xi(1_{x_mK}\otimes 1^-_{y_0K}) =$$

$$\mathrm{Vol}X(\mathfrak{o})\cdot\begin{cases}1 & \text{if } |\xi| = q^{-m};\\ -1 & \text{if } |\xi| = q^{2-m}, m\ge 1;\\ \int_{|x|^2=|\xi|}\psi(x-\xi x^{-1})dx & \text{if } |\xi| > 1, m = 1.\end{cases} \quad (6.2)$$

Remark 6.2 We notice that for any ξ and sufficiently large m we have $O_\xi(1_{x_mK}\otimes 1_{y_0K}) = 0$. Thus, for an element $\Phi \in \mathcal{S}^s(\bar{X}\times X, \mathcal{L}_\psi \boxtimes \mathcal{L}_\psi^{-1})^{K\times K}$ which can be written as a series:

$$\Phi = \sum_{m\ge 0} c(m)(1_{x_mK}\otimes 1^-_{y_0K}),$$

a regular orbital integral $O_\xi(\Phi)$ can be written as an eventually stabilizing series (compare with Lemma 4.11):

$$O_\xi(\Phi) = \sum_{m\ge 0} c(m)O_\xi(1_{x_mK}\otimes 1^-_{y_0K}).$$

6.4 Orbital integrals of the basic function

Recall that the basic vector $f_s^0 \in \mathcal{S}(\mathcal{W}^s)$ is obtained by the orbital integrals of $\Phi_s^0 = (H_s \star 1_{x_0K}) \otimes 1^-_{y_0K}$, where H_s is the formal series in the Hecke algebra corresponding to the unramified L-factor $L(\pi, \frac{1}{2}+s)L(\pi\otimes\eta, \frac{1}{2}+s)$. We compute its regular orbital integrals, according to the previous remark.

Lemma 6.3 *We have:*

$$f_s^0(\xi) = O_\xi(H_s \star 1_{x_0K} \otimes 1_{y_0K}) =$$
$$\mathrm{Vol}(X(\mathfrak{o}))L(\eta, 2s+1)\left(|\xi|^{s+1}\cdot(I - q^{-2s-1}\varpi^2\cdot)f(\xi) + 1_{|\xi|=q^2} + \mathcal{K}(\xi)\right), \tag{6.3}$$

where:

- $\mathcal{K}(\xi)$ *denotes the function which is supported on* $|\xi| > 1$ *and equal to:* $\int_{|x|^2=|\xi|} \psi(x - \xi x^{-1})dx$ *there. ($\mathcal{K}$ stands for "Kloosterman".)*
- f *is the function supported on* $|\xi| \le 1$ *and equal, there, to:*

$$\begin{cases} 1 - \log_q|\xi| & \text{in the split case,} \\ \frac{1+\eta(\xi)}{2} & \text{in the non-split case;} \end{cases}$$

- *the action of* ϖ^2 *is normalized as in (2.14).*

Proof Indeed, for given ξ with $|\xi| = q^{-n}$, the first term expresses the contributions of 1_{x_nK} and $1_{x_{n+2}K}$ whenever those are nonzero, according to the first two cases of (5.16). However, there is no contribution from 1_{x_0K} when $|\xi| = q^2$, and this is what the second term is correcting. The third term expresses the contribution of 1_{x_0K} when $|\xi| > 1$.

□

A Almost Smooth Functions, Schwartz and Tempered Functions

A.1 Almost smooth functions

The space of smooth functions on a real manifold has the structure of a Fréchet space. We would like, for the purpose of uniformity, to define a similar Fréchet space of functions for a p-adic manifold X, i.e. for a topological space equipped with an atlas of "p-adic analytic functions", which is locally isomorphic to the ring of p-adic analytic functions on $\mathfrak{o}^n$ (where, as usual $\mathfrak{o}$ denotes the ring of integers of a local nonarchimedean field F). The usual notions of "locally constant" and "uniformly locally constant" (when there is some uniform structure) functions do not lead to Fréchet spaces. We are going to define a new class of functions, which in this appendix will be called "almost smooth" and in the rest of the paper, for simplicity, just "smooth". Also, in this appendix we will be denoting the space of these functions by $C^{\infty+}$, but in the rest of the paper just by C^∞. Finally, for any statement about "almost smooth" functions in this appendix, when applied to real manifolds, the word "almost" should be disregarded; and moreover, complex manifolds and varieties will be considered as real manifolds/varieties.

Almost smooth functions will form a sheaf for the usual Hausdoff topology on X, and therefore it is enough to describe them locally around each point $x \in X$.

We choose an analytic chart for a neighborhood U of x, so that it becomes isomorphic to $\mathfrak{o}^n$ with its ring of analytic functions. Then we identify $C^{\infty+}(U)$ with $C^{\infty+}(\mathfrak{o}^n)$, the space of *almost smooth* functions on $\mathfrak{o}^n$, defined as those complex-valued functions of the form:

$$f = \sum_{i\geq 0} f_i \tag{A.1}$$

on $\mathfrak{o}^n$, where f_i is invariant under $\mathfrak{p}^i \times \cdots \times \mathfrak{p}^i$ and for every $N > 0$ there is a scalar C such that $\|f_i\|_\infty < Cq^{-iN}$ for all i. In other words, the supremum norms vanish faster than any power of the level. It is a Fréchet space under any of the following equivalent systems of seminorms:

Lemma A.1.1 *On the space of continuous functions on $\mathfrak{o}^n$, the following seminorms define* tamely equivalent[8] *Fréchet spaces:*

1. $\|f\|_N := \sup_{i\geq -1, x\in\mathfrak{o}^n} q^{iN}|f(x) - K_i \star f(x)|$;
2. $\|f\|_N := \sup_{i\geq -1, x\in\mathfrak{o}^n} q^{iN}|K_{i+1} \star f(x) - K_i \star f(x)|$;
3. $\|f\|_\infty$, *and* $\|f\|_N := \sup_{x\in\mathfrak{o}^n} \sup_{y\in\mathfrak{o}^n\smallsetminus\{0\}} |y|^{-N}|f(x) - f(x+y)|$ *(where* $|(y_1, \ldots, y_n)| = \sup_i |y_i|$*).*

Here K_i is the characteristic measure of $\mathfrak{p}^i \times \cdots \times \mathfrak{p}^i$, convolution is in the additive group $\mathfrak{o}^n$, and by convention $K_{-1} \star f = 0$.

[8] Recall that a tame Fréchet space is a Fréchet space with a presentation as a countable inverse limit of Banach spaces B_n, and a map $T : \lim_{\leftarrow} B_n \to \lim_{\leftarrow} B'_n$ is tame if there are integers b and r so that for all $n \geq b$ the map T is continuous from B_{n+r} to B'_n.

It is clear that the Fréchet structure is preserved under analytic automorphisms, thus the notion of an almost smooth function on a p-adic manifold is well-defined. Notice that, like smooth functions, these "almost smooth" functions have vanishing derivatives, for any reasonable notion of "derivative", for instance for any $Z \in \mathfrak{o}^n$ we have:

$$\lim_{t \to 0} \frac{f(tZ) - f(0)}{|t|} = 0.$$

Therefore, *any statements about derivatives in the nonarchimedean case, throughout the paper, should be taken to concern only the zeroth derivative.* However, we will encode the issue of how fast a function varies in what we will call "pseudo-derivatives", a notion that is related to the seminorms defined above.

A.2 Semialgebraic sets and charts

We recall that a semialgebraic set on a real algebraic variety X is obtained by a boolean combination (i.e. by taking unions and complements a finite number of times) of subsets of $X(\mathbb{R})$ given by an inequality of the form $f \geq 0$, where f is a regular function. For a smooth algebraic variety X over a nonarchimedean field F, on the other hand, semialgebraic sets are defined as boolean combinations of sets of the form:

$$\{x \in X(F) | f(x) \in P_k\}, \text{ where } P_k = \{y^k | y \in F\},$$

f is a regular function, and $k \in \mathbb{N}_{\geq 2}$, cf. [Den86]. By definition, a map: $X \to Y$ between semialgebraic sets is called semialgebraic if its graph is semialgebraic.

The above sets are the basic closed sets for the *restricted topology* of semi-algebraic sets (restricted means: only finite unions of open sets are required to be open), and this is the topology we will be using when talking about "open" and "closed" sets and neighborhoods, unless otherwise specified.

Notice that, in general, the notion of closure is not well-behaved for restricted topologies. However, for semialgebraic sets the following is true: The closure of a semialgebraic set in the usual (Hausdoff) topology is closed semialgebraic; hence, the notion of closure is well behaved, and closure in semialgebraic topology coincides with closure in the Hausdorff topology.

By a smooth semialgebraic set we will mean an open semialgebraic subset of the points of a smooth variety. (One can more generally define "semialgebraic manifolds", but we will not need this.) For the description of tempered functions, we will need to introduce a notion of "semialgebraic chart" for smooth semialgebraic set U in the nonarchimedean case. By a *semialgebraic chart* of U we mean a finite partition into open-closed subsets: $U = \bigsqcup_j U_j$ and, for every j, a semi-algebraic isomorphism $\alpha_j : V_j \xrightarrow{\sim} U_j$ with an open semialgebraic subset V_j of F^n *which is* $\mathfrak{o}^n$*-stable* (under addition in F^n).

Lemma A.2.1 *Semialgebraic charts exist.*

Proof It is easy to see that any smooth semialgebraic set in the nonarchimedean case is isomorphic to a finite disjoint union of open semialgebraic subsets of F^n, so there remains to consider the case that $U \subset F^n$, in order to show that one can find an $\mathfrak{o}^n$-invariant chart.

For simplicity, we only show that this is the case for the basic open set $\{x | f(x) \notin P_k\}$ where P_k is the set of k-th powers of elements of F as above and f is a polynomial; the general case is only notationally more complicated. We may even restrict to the intersection of this set with $\mathfrak{o}^n$, by partitioning the set and inverting coordinates as appropriate. Away from any neighborhood of the zero set of f (in $\mathfrak{o}^n$) the condition: $f(x) \notin P_k$ is locally constant in x, hence uniformly locally constant, hence multiplying the coordinates by a suitable scalar will give the required chart. We are left with finding an $\mathfrak{o}^n$-invariant chart for a neighborhood of the zero set $Z \subset \mathfrak{o}^n$ of f. By a resolution of singularities (which will be recalled in the next appendix, Theorem B.1.2), we can replace a neighborhood of Z by a compact semialgebraic set V of the same dimension, so that the pullback of f is, in semialgebraically-local coordinates $(y_1, \ldots, y_n)$, of the form: $c \cdot y_1^{i_1} \cdots y_n^{i_n}$. Then the claim is easy to show.

□

A.3 Schwartz functions

If U is an open semialgebraic subset of (the points of) a real or p-adic variety, we will define the space $\mathcal{S}(U)$ of *Schwartz functions* on U as a space of smooth (in the archimedean case), resp. almost smooth functions (in the nonarchimedean). The definition in the archimedean case is well-known, but to construct its analog for the nonarchimedean case we need to take into account not only the growth of f, but also the growth of the summands f_i of an expression as in (A.1). For this, we will introduce the following analog of differential operators:

Definition A.3.1 Let U be an open subset of the points of a smooth p-adic variety and $\mathcal{C} := (U_i, V_i, \alpha_i)_i$ a semialgebraic chart of U. For each almost smooth function f on U and each $N \geq 0$ we define the *N-th pseudoderivative of f with respect to $\mathcal{C}$* to be equal to f if $N = 0$, and otherwise:

$$f^{\mathcal{C},(N)}(x) = \sup_{y \in \mathfrak{o}^n \smallsetminus \{0\}} |y|^{-N} |f(x) - f(x+y)|, \tag{A.2}$$

where the "sum" $x + y$ should be interpreted in terms of the chart $\mathcal{C}$ — i.e. it really means: $\alpha_i(\alpha_i^{-1} x + y)$ for $x \in U_i$.

It is easy to prove:

Lemma A.3.2 *If $\mathcal{C}$, $\mathcal{C}'$ denote two different charts, for every N there is a semialgebraic function T such that:*

$$|f^{\mathcal{C},(N)}(x)| \leq |T(x)| \cdot |f^{\mathcal{C}',(N)}(x)|.$$

As a corollary, the notions of Schwartz and tempered functions that we are about to define do not depend on the choice of chart; we will omit the chart from the notation for pseudoderivatives from now on.

Now we define Schwartz functions on U. We recall that a "Nash differential operator" is a "smooth semialgebraic" differential operator, cf. [AG08]. In particular, the growth of these operators is bounded, locally for the semialgebraic topology, by regular functions.

Definition A.3.3 The space $\mathcal{S}(U)$ of Schwartz functions on U consists of those smooth functions on U, in the archimedean case, resp. almost smooth in the nonarchimedean case, with the property:

- for every Nash differential operator D on U, in the archimedean case, and for every (equivalently: some) chart $\mathcal{C}$, every $N \geq 0$ and semialgebraic function T, in the nonarchimedean case, the function Df, resp. $Tf^{(N)}$, is bounded.

The space of Schwartz functions on U is naturally a *nuclear Fréchet algebra*; its topology is generated by the seminorms:

$$\sup_{x \in U} |Df(x)|,$$

in the archimedean case, where D varies over all Nash differential operators (evidently, a countable number of them suffices), and:

$$\sup_{x \in U} |T(x) f^{(N)}(x)|$$

in the nonarchimedean, where $N \in \mathbb{N}$ and T varies over all semialgebraic functions on U (again, a countable number suffices).

We will discuss in appendix B cosheaf-theoretic properties of Schwartz functions. The following will be a consequence of B.1.1:

Proposition A.3.4 *Schwartz functions on U are precisely those functions which for one, equivalently any, smooth compactification $\bar{U}$ of U extend to smooth functions (in the archimedean case) resp. almost smooth functions (in the non-archimedean case) all of whose derivatives vanish on $\bar{U} \smallsetminus U$.*

We remind that any statement about derivatives should be understood to apply only to the zeroth derivative in the nonarchimedean case.

A.4 Tempered functions

Definition A.4.1 If U is an open semialgebraic subset of (the points of) a smooth real or p-adic variety, we define the space $\mathcal{O}(U)$ of *tempered functions* on U as those smooth (in the archimedean case), resp. almost smooth (in the nonarchimedean) functions f on U with the property:

- In the archimedean case, for every Nash differential operator D on U there is a semialgebraic function T on U with $|Df| \le |T|$; in the nonarchimedean, for every (equivalently: one) chart $\mathcal{C}$ and any $N \in \mathbb{N}$ there is a semialgebraic function T on U with $|f^{(N)}| \le |T|$.

The space $\mathcal{O}(U)$ of tempered functions on U is an algebra which acts on the space of Schwartz functions:

$$\mathcal{O}(U) \otimes \mathcal{S}(U) \to \mathcal{S}(U).$$

Moreover, each $f \in \mathcal{O}(U)$ is a bounded operator on $\mathcal{S}(U)$. We endow $\mathcal{O}(U)$ with the *strong operator topology* on the Fréchet space $\mathcal{S}(U)$; this way it becomes a locally convex topological algebra. By definition, convergence to zero of a net $(f_\alpha)_\alpha \subset \mathcal{O}(U)$ in the strong topology means that $f_\alpha \phi \to 0$ for every $\phi \in \mathcal{S}(U)$. Since $\mathcal{S}(U)$ is a nuclear, and hence Montel Fréchet space (i.e. bounded sets are precompact), it is known that this topology coincides with the operator topology of uniform convergence on bounded/compact sets [Köt79, p. 139]. It is easy to describe sequential convergence in this topology:

Lemma A.4.2 *For a sequence $f_n \in \mathcal{O}(U)$ we have $f_n \to 0$ iff:*

1. *$f_n \to 0$ in $C^\infty(U)$ (with the usual Fréchet topology of locally uniform convergence of all derivatives), resp. in $C^{\infty+}(U)$, and*
2. *for each Nash differential operator D, resp. for each chart $\mathcal{C}$ and integer N, there is a semialgebraic function T such that:*

$$|Df_n| \le |T|, \quad \text{resp. } |f_n^{(N)}| \le |T| \text{ for all } n.$$

Proof It is clear that such a sequence is a null sequence. Vice versa, it is clear that a null sequence should converge to zero in $C^\infty(U)$, resp. $C^{\infty+}(U)$. The proof of the second condition is reduced to tempered functions on F by the resolution of singularities that will be recalled in the next appendix (Theorem B.1.2). We prove that all f_n should be bounded by some $|x|^N$, for some N, in a neighborhood of ∞ (the proof for derivatives/pseudoderivatives is similar): if not, there is a Schwartz function ϕ on F with $\sup_x |f_n(x)\phi(x)|$ bounded below. Thus, f_n cannot be a null sequence.

□

In particular, $\mathcal{O}(U)$ is sequentially complete. In fact, it can be shown that it is a complete, nuclear topological vector space, but we will not use this.

B Schwartz Cosheaves

In this appendix we formalize certain properties of Schwartz functions. These properties are obvious for the Schwartz functions themselves, but not totally obvious for their coinvariants, hence the language that we are introducing is helpful in analyzing orbital integrals.

From now on, as in the main body of the text, **"smooth" function means "almost smooth" at nonarchimedean places**. Throughout this section, X denotes the F-points of a smooth algebraic variety over a local field F, "closed" and "open" refer to the restricted topology of semialgebraic sets.

Finally, to avoid repeating the same dichotomy again and again, any mention of **"a Nash differential operator D" should be understood, in the nonarchimedean case, as** the data consisting of:

1. a semialgebraic chart $\mathcal{C}$;
2. an integer $N \geq 0$;
3. a semialgebraic function T.

Then, a statement about the function Df should be replaced by the analogous statement about the function $|T|f^{(N)}$, as in the definition of Schwartz functions in Appendix A. On the other hand, we keep our convention that **any statement about derivatives should be understood to apply only to the zeroth derivative in the nonarchimedean case**. By consistently using the phrases "Nash differential operator" and "derivative", this should cause no confusion.

B.1 The sheaf of tempered functions

The association $U \to \mathcal{O}(U)$, where $\mathcal{O}(U)$ denotes the space of tempered functions on U (§A.4), is a sheaf of topological algebras on X. We will consider it as the "structure sheaf", in the sense that all other sheaves will be modules for it.

Except for the general sheaf properties, what is interesting for us now is the following relation between topology and algebra structure: Any closed $S \subset X$ gives rise to the sheaf of ideals $\mathcal{J}_S \subset \mathcal{O}$ of functions vanishing on S. We denote by $\mathcal{J}_S^n(U)$ the *closed* ideal of $\mathcal{O}(U)$ generated by n-fold products of elements in $\mathcal{J}_S(U)$; we write $\mathcal{K}_S(U) = \cap_n \mathcal{J}_S^n(U)$. Both $\mathcal{J}_S^n$ and $\mathcal{K}_S$ are sheaves on X.

Lemma B.1.1 *The sheaf $\mathcal{K}_S$ is the sheaf of tempered functions which vanish, together with all their derivatives, on S; in particular, in the nonarchimedean case $\mathcal{K}_S = \mathcal{J}_S$. Equivalently, for each open $U \subset X$ the space $\mathcal{K}_S(U)$ consists of those tempered functions f on U with the property that for any Nash differential operator D on $U \smallsetminus S$ the function Df is bounded in a neighborhood of $S \cap U$.*

Before we prove the lemma, we mention a basic tool for our proofs, namely the embedded resolutions of singularities, in the following sense:

Theorem B.1.2 *For every smooth semialgebraic set X and a closed semialgebraic subset $S \subset X$, there is a smooth semialgebraic set $\tilde{X}$ and a proper morphism $p : \tilde{X} \to X$, such that:*

1. *p is an isomorphism away from S;*
2. *there is a finite open cover $\tilde{X} = \bigcup U_i$ and, on each U_i, semialgebraic coordinates $(y_1, \dots, y_n)$ such that $p^{-1}(S) \cap U_i$ is given by finite intersections and unions of sets of the form:*

$$\{x | f(x) \geq 0\},$$

in the (real) archimedean case, and:

$$\{x | f(x) \in P_k\},\ P_k = \{y^k | y \in F\},$$

in the nonarchimedean case, where $f = cy_1^{i_1} \cdots y_n^{i_n}$, $i_j \in \mathbb{N}$.

This follows from Hironaka's embedded resolution of singularities [Hir64, Corollary 3, p. 146].

Along with the previous lemma, we will also prove the following, which will be useful elsewhere:

Lemma B.1.3 *Consider a resolution $\tilde{X} \to X$ as in Theorem B.1.2. Then for every open $U \subset X$ with preimage $\tilde{U} \subset \tilde{X}$, the pullback gives rise to an equality:*

$$\mathcal{K}_S(U) \simeq \mathcal{K}_{\tilde{S}}(\tilde{U}).$$

Proof (of Lemmas B.1.1 and B.1.3) We first prove that an element $f \in \mathcal{O}(U)$ belongs to all $\mathcal{J}_S^n(U)$ if and only if it vanishes on $S \cap U$ together with all its derivatives. One direction is easy: it is clear that any element of $\mathcal{J}_S^n(U)$ has vanishing i-th derivatives, for $i \leq n$, on all points of S.

Vice versa, consider a resolution $\tilde{X} \to X$ as in Theorem B.1.2. If $\tilde{U}$ is the preimage of U, it is easy to see (by reduction to $\tilde{U} = F^n$ with $\tilde{S} =$ a "standard" semialgebraic set defined by conditions on the coordinate functions) that if a smooth function vanishes with all its derivatives on $\tilde{S}$, it coincides locally around each point of $\tilde{S}$ with the restriction of a Schwartz function on $\tilde{U} \smallsetminus \tilde{S}$. (We remind again that in the nonarchimedean case there are no higher derivatives, so the statement is about almost smooth functions vanishing on $\tilde{S}$.) Let V denote the (closed) subspace of $\mathcal{O}(U)$ consisting of such functions; we will eventually prove that it coincides with $\mathcal{K}_{\tilde{S}}(\tilde{U})$.

To do so, we use two well-known results in the archimedean case, which can be similarly be proven for the nonarchimedean case, for sets as in Theorem B.1.2:

1. For every Schwartz function ϕ on $\tilde{U}$ there is a real-valued, positive Schwartz function ψ on the same space such that $\frac{\phi}{\psi}$ is a Schwartz function.
2. Every Schwartz function ϕ on $\tilde{U} \smallsetminus \tilde{S}$ is a product of two Schwartz functions on the same space.

From this it can immediately be deduced that any $f \in V$ can be written as a limit of $f\psi_\alpha$, where ψ_α runs over a suitable system of positive Schwartz functions on $\tilde{U}$, directed by majorization; and that every $f\psi_\alpha$ is the product of two Schwartz functions on $\tilde{U} \smallsetminus \tilde{S}$. But Schwartz functions on $\tilde{U} \smallsetminus \tilde{S}$ belong to V, hence the multiplication map: $V \otimes V \to V$ has dense image. Since $V \subset \mathcal{J}_{\tilde{S}}(\tilde{U})$, it follows that $V \subset \mathcal{J}_{\tilde{S}}^2(\tilde{U})$; repeating this argument, $V \subset \mathcal{J}_{\tilde{S}}^n(\tilde{U})$ for all n. Hence $V \subset \mathcal{K}_{\tilde{S}}(\tilde{U})$, therefore these spaces are equal.

Now, there is a sequence of natural numbers $k_n \to +\infty$ such that a smooth function on $\tilde{U}$ which vanishes with all its first n derivatives on $\tilde{S} \cap \tilde{U}$ descends to a function on U which vanishes with all its first k_n derivatives on $S \cap U$ (vice versa, if the first n derivatives of a function on U vanish on S then they also vanish for the pullback to $\tilde{U}$). This implies that elements of $\mathcal{K}_S(\tilde{U})$ descend to smooth functions on U with vanishing derivatives on S. Notice that the pullback of $\mathcal{O}(U)$ is closed in $\mathcal{O}(\tilde{U})$, and the topologies coincide (indeed, since the map $\tilde{X} \to X$ is proper, it suffices for defining the topology on $\mathcal{O}(\tilde{U})$ to consider only those Schwartz functions on $\tilde{U}$ which are pullbacks of Schwartz functions on U). Again by the fact that $\mathcal{K}_{\tilde{S}}(\tilde{U})^2$ is dense in $\mathcal{K}_{\tilde{S}}(\tilde{U})$ we deduce that $\mathcal{K}_{\tilde{S}}(\tilde{U}) \subset \mathcal{J}_S^n(U)$ for all n, hence $\mathcal{K}_{\tilde{S}}(\tilde{U}) = \mathcal{K}_S(U)$.

□

Now notice that we have an injective map: $\mathcal{K}_S(U) \to \mathcal{O}(U \smallsetminus S)$.

Lemma B.1.4 *There is a sequence* $(u_n)_n \subset \mathcal{K}_S(U)$ *with:*

$$u_n \to 1 \text{ in } \mathcal{O}(U \smallsetminus S).$$

We will call such a sequence an *approximate identity*, despite the fact that it is only bounded in the weaker topology of $\mathcal{O}(U \smallsetminus S)$, because it satisfies:

$$u_n f \to f \text{ for all } f \in \mathcal{K}_S(U).$$

Proof We may choose a countable, increasing filtration of $U \smallsetminus S$ by open sets U_n with the property that the complement of U_n contains a neighborhood of $U \cap S$. Then we can find a sequence of tempered functions $u_n \in \mathcal{K}_S(U)$, with $u_n|_{U_n} \equiv 1$ and the property that for every Nash differential operator D on $U \smallsetminus S$ there is a semialgebraic function T on $U \smallsetminus S$ such that $|Du_n| \le |T|$ for all n. (Again, this is easier to see with a blowup as in Theorem B.1.2.) By Lemma A.4.2, the limit of this sequence in $\mathcal{O}(U \smallsetminus S)$ is the constant function 1.

□

We denote the quotient sheaf $\mathcal{O}_S := \mathcal{O}/\mathcal{K}_S$. It is supported on S. Notice that for every quotient of $\mathcal{O}$ by an ideal subsheaf $\mathcal{I}$, the map: $\mathcal{O}(U) \to (\mathcal{O}/\mathcal{I})(U)$ is surjective for every U; thus, no sheafification is needed. Indeed, using a partition of unity as in [AG08, Theorem 5.2.1], we can patch functions f_i on a finite cover $U = \bigcup_i U_i$, which agree on intersections modulo $\mathcal{I}$, to a function $f \in \mathcal{O}(U)$ whose restriction to U_i is $\equiv f_i$ mod $\mathcal{I}(U_i)$.

Lemma B.1.5 *The sheaf* $\mathcal{O}_S$ *can also be described as the completion of* $\mathcal{O}$ *over the closed subset* S*:*

$$\mathcal{O}_S = \lim_{\substack{\leftarrow \\ n}} \mathcal{O}/\mathcal{J}_S^n. \tag{B.1}$$

Proof In the nonarchimedean case we have seen that $\mathcal{K}_S = \mathcal{J}_S$, so the statement is trivial. We discuss the archimedean case.

First of all, we use a resolution $\tilde{X} \to X$ as in Theorem B.1.2. Then we have, for every U, a commutative diagram of injective maps:

$$\begin{array}{ccc} \mathcal{O}/\mathcal{K}_S(U) & \hookrightarrow & \lim_{\overleftarrow{n}} \mathcal{O}/\mathcal{J}_S^n(U) \\ \downarrow & & \downarrow \\ \mathcal{O}/\mathcal{K}_{\tilde{S}}(\tilde{U}) & \hookrightarrow & \lim_{\overleftarrow{n}} \mathcal{O}/\mathcal{J}_{\tilde{S}}^n(\tilde{U}) \end{array}$$

Let us fix a sequence D_i of Nash differential operators which generate all Nash differential operators over the ring of Nash (i.e. smooth semialgebraic) functions, for the set $\tilde{U}$. An element of $\lim_{\overleftarrow{n}} \mathcal{O}/\mathcal{J}_{\tilde{S}}^n(\tilde{U})$ can be represented (non-uniquely) by a sequence $f_n \in \mathcal{O}(\tilde{U})$ with the property: for all $i \leq n$ we have $D_i f_n|_{\tilde{S}} = D_i f_i|_{\tilde{S}}$. The topology on $\lim_{\overleftarrow{n}} \mathcal{O}/\mathcal{J}_{\tilde{S}}^n(\tilde{U})$ is given by seminorms:

$$\sup_{x \in \tilde{S}} |T(x) D_i(\phi f_n)(x)|,$$

where T varies over semialgebraic functions on $\tilde{S}$ and ϕ varies over elements of $\mathcal{S}(\tilde{U})$.

It is then elementary to construct (by appropriate Taylor series) a smooth, tempered function f on $\tilde{U}$ with $D_i f|_{\tilde{S}} = D_i f_i|_{\tilde{S}}$ for all i. Moreover, for every $\phi \in \mathcal{S}(\tilde{U})$ the construction of f can be made such that $\|D_i(f\phi)\|_{L^\infty(U)}$ is bounded in terms of a finite number of seminorms of the form:

$$\sup_{x \in \tilde{S}} |T(x) D_j(\phi_j f_n)(x)|,$$

where T is semialgebraic and $\phi_j \in \mathcal{S}(\tilde{U})$. Thus, the bottom horizontal arrow of the above diagram is an isomorphism.

But if the given element of $\mathcal{O}/\mathcal{J}_{\tilde{S}}^n(\tilde{U})$ comes from $\mathcal{O}/\mathcal{J}_S^n(U)$, all the derivatives of the constructed function f on $\tilde{S}$ descend to S; therefore, the function descends to an element of $\mathcal{O}(U)$. In other words, the vertical arrows are closed embeddings into the same subspace, and this proves the claim.

□

B.2 Schwartz cosheaves

By a *Schwartz cosheaf* on X we will mean a cosheaf $\mathcal{F}$ of nuclear Fréchet spaces on X, satisfying certain axioms. The extension maps will be denoted by e_V^U (where $V \subset U$ are open subsets), or simply by e when the source and target are clear. We will sometimes call the extension maps "extension by zero", to emphasize their geometric meaning. The axioms are expressed in terms of an arbitrary open subset U (as I do not see a way to make them "sheaf-theoretic" combining the presence of a sheaf and a cosheaf), and are the following:

1. The extension maps $e_V^U : \mathcal{F}(V) \to \mathcal{F}(U)$ are closed.
2. $\mathcal{F}(U)$ is a continuous $\mathcal{O}(U)$-module, i.e. there is a continuous bilinear map:

$$\mathcal{O}(U) \times \mathcal{F}(U) \to \mathcal{F}(U)$$

   compatible with multiplication on $\mathcal{O}(U)$.

3. If $S \subset X$ is closed, sections of $\mathcal{F}$ vanishing to arbitrary degree on S are extensions by zero of sections on the complement of S, that is: if $\mathcal{J}_S$ denotes the sheaf of ideals of tempered functions vanishing on S as before then:

$$\bigcap_n \overline{\mathcal{J}_S^n(U)\mathcal{F}(U)} \subset e\left(\mathcal{F}(U \smallsetminus S)\right). \tag{B.2}$$

   (We will prove equality in the next lemma.)

4. Obvious compatibility conditions: If $V \subset U$ are open then the diagram commutes:

$$\begin{array}{ccccc}
\mathcal{O}(U) & \otimes & \mathcal{F}(U) & \longrightarrow & \mathcal{F}(U) \\
\big\downarrow r_V^U & & \big\uparrow e_V^U & & \big\uparrow e_V^U \\
\mathcal{O}(V) & \otimes & \mathcal{F}(V) & \longrightarrow & \mathcal{F}(V).
\end{array}$$

Lemma B.2.1 *Let $(u_n)_n$ be a weak approximate identity in $\mathcal{K}_S(U)$ as in Lemma B.1.4, then $u_n f \to f$ for every $f \in e_V^U\left(\mathcal{F}(U \smallsetminus S)\right)$. Consequently:*

$$\bigcap_n \overline{\mathcal{J}_S^n \mathcal{F}(U)} = e\left(\mathcal{F}(U \smallsetminus S)\right) = \overline{\mathcal{K}_S \mathcal{F}(U)}.$$

Proof Let $V = U \smallsetminus S$. Since the map $\mathcal{O}(V) \times \mathcal{F}(V) \to \mathcal{F}(V)$ is continuous, and $u_n \to 1$ in $\mathcal{O}(V)$, we have $u_n f \to f$ for every $f \in \mathcal{F}(V)$. By the compatibility of restrictions and corestrictions (fourth axiom), $u_n e_V^U(f)$ is the same as $e_V^U(u_n f)$, and since e_V^U is continuous, this tends to $e_V^U(f)$ for all $f \in \mathcal{F}(V)$.

But $u_n e_V^U(f) \in \mathcal{K}_S(e_V^U \mathcal{F}(V)) \subset \mathcal{J}_S^n \mathcal{F}(U)$ for all n, hence the claim. □

B.3 Functoriality

In what follows, we consider only morphisms between smooth semialgebraic sets. For a morphism $\pi : X \to Y$, and a cosheaf $\mathcal{F}$ on X, the push-forward $\pi_*\mathcal{F}$ is simply the cosheaf: $V \mapsto \mathcal{F}(\pi^{-1}V)$.

Lemma B.3.1 *The push-forward of a Schwartz cosheaf is a Schwartz cosheaf.*

Proof The only nontrivial verification is that of the third axiom. Let $\pi : X \to Y$ be a morphism and denote for clarity by $\mathcal{O}_X$ and $\mathcal{O}_Y$ the corresponding sheaves of tempered functions. Let $S \subset Y$ be a closed subset, and $\mathcal{K}_S, \mathcal{K}_{\pi^{-1}S}$ the corresponding sheaves on Y and X, respectively.

The third axiom for $\mathcal{F}$ implies that for every open $V \subset Y$ we have an equality:

$$\bigcap_n \overline{\mathcal{J}^n_{\pi^{-1}S}(\pi^{-1}V) \cdot \pi_*\mathcal{F}(V)} = e\left(\pi_*\mathcal{F}(V \smallsetminus S)\right).$$

In particular, since $\mathcal{J}_S(V) \subset \mathcal{J}_{\pi^{-1}S}(\pi^{-1}V)$, we get:

$$\bigcap_n \overline{\mathcal{J}^n_S(V) \cdot \pi_*\mathcal{F}(V)} \subset e\left(\pi_*\mathcal{F}(V \smallsetminus S)\right).$$

But by the proof of Lemma B.2.1, we have:

$$e\left(\pi_*\mathcal{F}(V \smallsetminus S)\right) \subset \overline{\mathcal{K}_S(V) \cdot \pi_*\mathcal{F}(V)},$$

and of course $\overline{\mathcal{K}_S(V) \cdot \pi_*\mathcal{F}(V)} \subset \bigcap_n \overline{\mathcal{J}^n_S(V) \cdot \pi_*\mathcal{F}(V)}$. Hence, all these three spaces coincide.

□

B.4 Stalks, fibers and a Nakayama-type lemma

Given a closed $S \subset X$ we define the *stalk* of a Schwartz cosheaf over S to be the cosheaf:

$$\mathcal{F}_S = \mathcal{F}/e\left(\mathcal{F}_{X \smallsetminus S}\right),$$

where $\mathcal{F}_{X \smallsetminus S}$ is the cosheaf: $\mathcal{F}_{X \smallsetminus S}(U) = \mathcal{F}(U \smallsetminus S)$. Clearly, $\mathcal{F}_S$ is zero on $X \smallsetminus S$; in that sense, it is supported on S.

Let now $\bar{\mathcal{O}}_S$ be the quotient $\mathcal{O}/\mathcal{J}_S$ — it is the sheaf of restrictions to S of (smooth) tempered functions on $\mathcal{O}$, and it satisfies $\bar{\mathcal{O}}_S(U) = \mathcal{O}(U)/\mathcal{J}_S(U)$. The *fiber* of $\mathcal{F}$ will be the cosheaf: $\bar{\mathcal{F}}_S(U) = \mathcal{F}(U)/\overline{\mathcal{J}_S(U)\mathcal{F}(U)}$. Of course, in the nonarchimedean case the fiber and the stalk coincide. In the archimedean case, we will use the following version of Nakayama's lemma:

Proposition B.4.1 *Let S =a point (hence $\bar{\mathcal{O}}_S(U) = \mathbb{C}$ for every $U \supset S$), and assume that a* finite-dimensional *subspace $N \subset \mathcal{F}(U)$ spans the fiber $\bar{\mathcal{F}}_S(U)$. Then the same subspace algebraically (i.e. without taking closures) generates $\mathcal{F}_S(U)$ as an $\mathcal{O}_S(U)$-module.*

Proof For simplicity of notation, $M := \mathcal{F}_S(U)$, $R = \mathcal{O}_S(U)$, $J = \mathcal{J}_S(U)$, $M_n = \overline{J^n M}$. Notice that by the third axiom for Schwartz cosheaves, M is separated with respect to the J-adic topology, i.e. we have an embedding:

$$M \hookrightarrow \lim_{\substack{\leftarrow \\ n}} M/M_n.$$

We claim that for every n the map $J^n \otimes N \to M_n/M_{n+1}$ is surjective. Indeed, we have a natural map with dense image:

$$J^n/J^{n+1} \otimes_{R/J} N \to M_n/M_{n+1},$$

but since the space on the left is finite dimensional, the map is surjective.

This implies that the map $(R/J^n) \otimes N \to M/M_n$ is surjective, and hence so is the map:

$$\lim_{\substack{\leftarrow \\ n}} ((R/J^n) \otimes N) \to \lim_{\substack{\leftarrow \\ n}} M/M_n.$$

Since N is finite-dimensional, the left hand side is equal to $(\lim_{\substack{\leftarrow \\ n}} R/J^n) \otimes N$, which is equal to $R \otimes N$ by Lemma B.1.5. Therefore N generates $\lim_{\substack{\leftarrow \\ n}} M/M_n$ algebraically over R, and in particular $M = \lim_{\substack{\leftarrow \\ n}} M/M_n$.

□

B.5 Group actions

Let $\mathcal{F}$ be a Schwartz cosheaf on a smooth F-variety X, and assume that it carries an action of a group G. (The group G is assumed to act trivially on X; for example, if Y is an affine G-variety, with G reductive, $X = Y /\!/ G$ and $\mathcal{G}$ is a cosheaf on Y with a compatible G-action, then $\mathcal{F}$ could be the push-forward of $\mathcal{G}$.)

We let $\mathcal{F}_G$ denote the *cosheaf of G-coinvariants*, that is:

$$\mathcal{F}_G(U) = \mathcal{F}(U)_G,$$

where $\mathcal{F}(U)_G$ denotes the quotient of $\mathcal{F}(U)$ by the closed subspace generated by vectors of the form $v - gv$, $v \in \mathcal{F}(U)$.

Proposition B.5.1 *The cosheaf $\mathcal{F}_G$ is a Schwartz cosheaf. For every two open sets $V \subset U$, the map: $\left(e_V^U \mathcal{F}(V)\right)_G \to \mathcal{F}_G(U)$ is a closed embedding.*

Proof The only nontrivial axiom to check is the closedness of extension maps, therefore it suffices to prove the second statement. Let $M_1 = e_V^U \mathcal{F}(V)$, $M_2 = \mathcal{F}(U)$, and let $N_i, i = 1, 2$, be the subspace of M_i algebraically spanned by elements of the form $v - g \cdot v$, $v \in M_i$. We need to prove that for a sequence $f_i \to f$, where $f_i \in N_2$ and $f \in M_1$, we have $f \in \overline{N_1}$.

Choose an approximate identity $u_n \in \mathcal{K}_{U \smallsetminus V}(U)$ (Lemma B.1.4). Then we have: $\lim_n u_n f = f$ (Lemma B.2.1). We also have $\lim_i u_n f_i = u_n f$ for every i; thus, $f \in \overline{(u_n f_i)_{n,i}} \subset \overline{N_1}$.

□

In applications to the present paper, all Schwartz cosheaves that we will encounter are flabby, i.e. the extension maps are monomorphisms (hence closed embeddings, by the first axiom). The last proposition implies:

Corollary B.5.2 *In the above setting, the cosheaf of G-coinvariants of a flabby Schwartz cosheaf is flabby.*

References

[AG08] A. Aizenbud and D. Gourevitch, *Schwartz functions on Nash manifolds*, Int. Math. Res. Not. IMRN **2008** (5):Art. ID rnm 155, 37, 2008. doi:10.1093/imrn/rnm155.

[CT] W. Casselman and Y. Tian, *Orbital integrals on the real hyperbolic plane*, Preprint, available at: http:/www.math.ubc.ca/~cass/research/pdf/Hyperbolic.pdf.

[Den86] J. Denef, *p-adic semi-algebraic sets and cell decomposition*, J. Reine Angew. Math. **369** (1986), 154–166. doi:10.1515/crll.1986.369.154.

[Hir64] H. Hironaka, *Resolution of singularities of an algebraic variety over a field of characteristic zero. I*, Ann. of Math. (2) **79** (1964), 109–203.

[Igu78] J. Igusa, *Forms of higher degree*, Lectures on Mathematics and Physics, **59**, Tata Institute of Fundamental Research, Bombay, 1978.

[II10] A. Ichino and T. Ikeda, *On the periods of automorphic forms on special orthogonal groups and the Gross-Prasad conjecture*, Geom. Funct. Anal. **19** (5), (2010), 1378–1425. doi:10.1007/s00039-009-0040-4.

[Jac86] H. Jacquet, *Sur un résultat de Waldspurger*, Ann. Sci. École Norm. Sup. (4) **19** (2) (1986), 185–229. URL: http://www.numdam.org/item?id=ASENS_1986_4_19_2_185_0.

[Jac05] H. Jacquet, *Kloosterman integrals for* GL$(2, \mathbb{R})$, Pure Appl. Math. Q. **1** (2, part 1) (2005), 257–289.

[Köt79] G. Köthe, *Topological vector spaces. II*, Grundlehren der Mathematischen Wissenschaften, **237**, Springer-Verlag, New York, 1979.

[Lan04] R. P. Langlands, *Beyond endoscopy*, In Contributions to automorphic forms, geometry, and number theory, pages 611–697. Johns Hopkins Univ. Press, Baltimore, MD, 2004.

[Saka] Y. Sakellaridis, *Beyond endoscopy for the relative trace formula II: global theory*, In preparation.

[Sakb] Y. Sakellaridis, *Spherical functions on spherical varieties*, Submitted for publication. arXiv:0905.4244.

[SV] Y. Sakellaridis and A. Venkatesh, *Periods and harmonic analysis on spherical varieties*, Preprint. arXiv:1203.0039.

[Tat67] J. T. Tate, *Fourier analysis in number fields, and Hecke's zeta-functions*, Algebraic Number Theory (Proc. Instructional Conf., Brighton, 1965), pages 305–347. Thompson, Washington, D.C., 1967.

[Ven04] A. Venkatesh, *"Beyond endoscopy" and special forms on GL(2)*, J. Reine Angew. Math. **577** (2004), 23–80. doi:10.1515/crll.2004.2004.577.23.

[Wal85] J.-L. Waldspurger, *Sur les valeurs de certaines fonctions L automorphes en leur centre de symétrie*, Compositio Math. **54** (2) (1985), 173–242. URL: http://www.numdam.org/item?id=CM_1985__54_2_173_0.

Department of Mathematics & Computer Science Rutgers, The State University of New Jersey, Smith Hall, Room 216, 101 Warren Street, Newark, NJ 07102, USA.

E-mail: sakellar@rutgers.edu

Automorphic Representations and L-Functions
Editors: D. Prasad, C.S. Rajan, A. Sankaranarayanan, J. Sengupta

Publisher: Hindustan Book Agency, New Delhi, India

On Interactions Between Harmonic Analysis and the Theory of Automorphic Forms

Marko Tadić[1]

Abstract

In this paper we review some connections between harmonic analysis and the modern theory of automorphic forms. We indicate in some examples how the study of problems of harmonic analysis brings us to the important objects of the theory of automorphic forms, and conversely. We consider classical groups and their unitary, tempered, automorphic and unramified duals. The most important representations in our paper are the isolated points in these duals.

1 Introduction

We start by recalling the very well known concept, due to Gelfand, of harmonic analysis on a locally compact group. The main problem of harmonic analysis on a group G is to understand some important unitary representations of G, such as $L^2(G)$. One approach to this problem is to break it into two parts:

(G1) Describe conveniently (possibly fully classify) the unitary dual $\hat{G}$ of G, i.e., the set of all equivalence classes of irreducible unitary representations of G.

(G2) Decompose important unitary representations of G in terms of $\hat{G}$ (as direct integrals, for example).

Observe that the main importance of (G1) comes from (G2).

Some very important problems of the modern theory of automorphic forms are typical problems of non-commutative harmonic analysis (in a broad sense), and progress in harmonic analysis has consequences in the theory of automorphic forms.

[1]The author was partly supported by Croatian Ministry of Science, Education and Sports grant #037-0372794-2804.

From the other side, in building harmonic analysis on reductive groups, automorphic forms are very useful. They are a very rich source of relevant ideas and concepts.

We review some of these connections in this paper, giving the picture from our point of view, which is closely related to harmonic analysis. We review only connections that have appeared primarily related to our work on problems of harmonic analysis. Therefore, a number of other very interesting connections are omitted (among others, the work of F. Shahidi contains very nice interactions). In this paper, we concentrate on relatively simple, but still interesting cases. Since both fields are very technical and often have very complicated notation, we try to keep the technical part as simple as possible and often give suggestive examples rather than the full results. We often simplify the situation as much as possible (trying not to oversimplify).

We deal only with classical groups in this paper. These groups have been generators of progress at some crucial stages in the development of harmonic analysis as well as the modern theory of automorphic forms. We now very briefly review the topics that we cover in the paper (one can find more details in the paper).

The unitary dual $\hat{G}$ carries a natural topology (defined in terms of approximation of matrix coefficients on compact subsets; see section 2.). For the commutative groups, the Pontryagin dual is formulated in this topology. In the non-commutative case, $\hat{G}$ does not need to be topologically homogenous. Of particular interest are the most singular representations of $\hat{G}$, the isolated points of this space, i.e., the isolated representations. In the case of unitary duals of reductive groups over local fields which are classified, usually isolated representations (of the group and its Levi subgroups) generate the whole unitary dual using some very standard constructions.

The part of the unitary dual of G which takes part in the decomposition of a unitary representation Π is called the support of Π, and it is denoted by

$$\text{supp}(\Pi).$$

Depending on the Π which one considers, $\text{supp}(\Pi)$ can be a relatively small part of $\hat{G}$. Therefore, we may be able to avoid general problem (G1), at least for dealing with such unitary representations (and focus our attention only on the classification of representations in $\text{supp}(\Pi)$). This was the case in Harish-Chandra's fundamental work on the explicit decomposition of the regular representation of semi-simple real group G on $L^2(G)$ (at the time when Harish-Chandra decomposed those representations for general semi-simple groups, unitary duals of simple groups were classified only for very low real ranks). There are also examples of unitary representations of different type. We shall later consider the unitary representations of general linear groups which have the most delicate parts of the unitary duals in their supports, all the isolated representations.

The support of $L^2(G)$, which was classified by Harish-Chandra, is called the tempered dual of G, and it is denoted by

$$\hat{G}_{\text{temp}}.$$

The support of a unitary representation Π can have its own isolated representations (which do not need to be isolated in $\hat{G}$), and the fact that they are isolated here may indicate their relevance for the unitary representation that one studies. The following fact is a nice example of this. For $\pi \in \hat{G}_{\text{temp}}$, we have

$$\pi \text{ is isolated in } \hat{G}_{\text{temp}} \iff \pi \text{ is square integrable.}$$

Let G be a reductive group defined over a number field k. Besides the regular representation (and the isolated points in its support), we also consider the representation of a local factor $G(k_v)$ of the adelic group $G(\mathbb{A}_k)$ on the space of square integrable automorphic forms $L^2(G(k)\backslash G(\mathbb{A}_k))$. The support of this representation is called the automorphic dual of G at v (see [Cl07] or the fourth section for a precise definition). Automorphic duals are very important, since they are related to a number of very hard questions in number theory (starting with Selberg $\frac{1}{4}$-conjecture). We do not know them even in the simplest cases, like $SL(2)$ or $GL(2)$. Nevertheless, we know pretty well what we can expect, at least in some cases. Further, we can prove a number of not-trivial facts about them. Representations in the automorphic dual which are unramified[2] with respect to a fixed maximal compact subgroup of $G(k_v)$ (i.e., containing a non-trivial vector fixed by the maximal compact subgroup) may be of particular interest. They are called Ramanujan duals[3] (following [BuLiSn92]). We also denote k_v by F. In what follows, we take F to be non-archimedean[4] (although a number of facts that we discuss hold or are expected to hold also in the archimedean case).

Now we move to the case of general linear groups, which are better understood then the other reductive groups. Take an irreducible square integrable representation σ of $GL(n, F)$. Equivalently, we might say that we took an isolated representation[5] from the tempered dual of $GL(n, F)$.

[2]The term spherical is also often used instead of unramified.

[3]The unramified classes in the unitary dual will be called the unramified unitary dual.

[4]In this case, we consider maximal compact subgroups $G(\mathcal{O}_F)$, where $\mathcal{O}_F$ is the maximal compact subring of F.

[5]In what follows, by isolated representation we shall always mean isolated modulo center (see the third section for the definition).

Fix a positive integer m and consider the representation

$$\mathrm{Ind}_P^{GL(mn,F)}(|\det|_F^{(m-1)/2}\sigma \otimes |\det|_F^{(m-1)/2-1}\sigma \otimes \cdots \otimes |\det|_F^{-(m-1)/2}\sigma), \tag{1.1}$$

parabolically induced from the appropriate parabolic subgroup which is standard with respect to the minimal parabolic subgroup consisting of upper triangular matrices. The above representation has a unique irreducible quotient which is denoted by

$$u(\sigma, m),$$

and called a Speh representation. Then each isolated representation in the unitary dual of a general linear group is a Speh representation, and Speh representations are almost always isolated ($u(\sigma, m)$ is isolated if and only if $m \neq 2$ and if σ is isolated[6] in $\widehat{GL(n,F)}$). It is an important fact that Speh representations are in the automorphic dual ([Jc84]). A further very important fact is that they are always isolated in the automorphic dual ([MüSp04] and [LoRuSn99]). Moreover, each isolated representation in the automorphic dual is expected to be some Speh representation (this would hold if we assume the generalized Ramanujan conjecture).

We have seen that the condition of being isolated in the tempered dual has a precise (and important) representation theoretic meaning. The precise (arithmetic) meaning of the condition of being isolated in the automorphic dual is less clear. Let us recall a very important and elegant paper [Ka67] of D. Kazhdan, where he proves that the trivial representation is isolated in the unitary dual of a simple algebraic group of rank $\neq 1$ over local field, and from this he derives some important arithmetic consequences.

Related to the result of D. Kazhdan, it is interesting to note that for $SL(n, F)$, $n \neq 2$, the trivial representation is also the only isolated representation in the unramified unitary dual for these groups. Clearly, the trivial representation is automorphic and isolated in the Ramanujan dual (also for $n = 2$; this follows from a general fact proved in [Cl03]). Further, it is also expected to be the only isolated representation there. Therefore, the set of isolated representations in the unramified unitary dual and the isolated representations in the Ramanujan dual are expected to coincide for $SL(n)$, with the exception $n = 2$. We have a surprisingly different situation for other classical groups, as we see next. We consider the example of $Sp(340, F)$.

The first surprise is that the number of isolated representations in the unramified unitary dual of $Sp(340, F)$ is

$$11\ 322\ 187\ 942$$

([MuTd11] and [Td10]). Further, G. Muić has proved an important fact

[6]Equivalently, σ does not correspond to a segment of cuspidal representations of length two in the Bernstein-Zelevinsky theory.

that these representations are all automorphic ([Mu07]; this is also how he proved their unitarity). In this way we get a huge number of isolated representations in the Ramanujan dual (isolated points there consist of at least 11 322 187 942 above representations). The following surprise is that [Mu07], [MuTd11] and Conjecture "Arthur + ϵ" of L. Clozel from [Cl07], would imply that the number of isolated representations in the Ramanujan dual is

$$568\ 385\ 730\ 874,$$

which is substantially bigger number than the number of isolated representations in the unramified unitary dual[7].

All this shows that compared to the $SL(n)$-case, new phenomena happen here (or are expected to happen). A new fact is that we have a huge number of isolated representations in the unramified unitary dual, and also in the Ramanujan dual. A new phenomenon here is that the isolated unramified representations in the unitary dual are expected to form a very small portion of isolated representations in the Ramanujan dual.

The above example also raises some questions. The first one is of an arithmetic nature. Having in mind D. Kazhdan's paper [Ka67], one may ask if the above stunning difference regarding the number of isolated points for $SL(n)$ and $Sp(2n)$ groups has some arithmetic explanation or consequence?

The second question is related to harmonic analysis: why does such a small portion of representations which are expected to be isolated in the Ramanujan dual remain isolated in the whole unramified dual (this was not the case for $SL(n)$)? The reason is that for each of 568 385 730 874 above (strongly negative) unramified representations, excluding 11 322 187 942 of them (i.e., the isolated ones in the unramified dual), there is a complementary series at whose end this representation lies (and complementary series are not expected to be in the Ramanujan dual). Complementary series representations start with an irreducible representation parabolically induced from a unitary one. Therefore, in the example of $Sp(340, F)$, for the 557 063 542 932 parabolically induced representations which are involved (where corresponding complementary series start), we need to know their irreducibility.

The above short discussion indicates that very often we have complementary series which end with representations that are expected to be isolated in the Ramanujan dual. But we have still a huge number of isolated representations in the unramified dual. Therefore, the following question arises: why are there still 11 322 187 942 isolated representations in the

[7]We have an intrinsic characterisation of the tempered dual and its isolated points among all the irreducible unitary representations (thanks to Harish-Chandra and W. Casselman). A similar situation might be the case regarding the Ramanujan dual and its isolated points (see the fifth section).

unramified dual? The answer is roughly: no complementary series ends with them (which is related to the question of reducibility of parabolically induced representations).

The above discussion suggests that if we want to know explicit answers regarding harmonic analysis or automorphic forms, we need also to have very explicit knowledge of complementary series (not just on some algorithmic level). This implies that we need to have a very explicit understanding of the question of irreducibility/reducibility of parabolically induced representations by unitary ones (unramified ones in this case). Such an understanding of the required irreducibility/reducibility is obtained by G. Muić in [Mu06]. We are not going to explain it here, but rather we go to a different (and dual) setting, where such an understanding is also crucial, and explain how one can deal with the question of irreducibility/reducibility there. We shall see how these basic questions of harmonic analysis lead to some deep problems in the number theoretic setting. The majority of this paper is devoted to this case. Below, we sketch only very basic idea.

Let us return to the fundamental problem (G1). A standard strategy for (G1) is to classify the non-unitary dual of G (formed of equivalence classes of irreducible representations of G), and then classify the unitarizable classes in it (i.e., $\hat{G}$). The Langlands classification of the non-unitary dual reduces this problem to the problem of tempered duals of its Levi subgroups. A very significant step in classifying the tempered dual is classifying the irreducible square integrable representations.

We now restrict to the case of classical groups (symplectic, orthogonal or unitary). For simplicity, in this discussion we consider only the case of symplectic groups (in the paper we also consider the case of groups $SO(2n+1,F)$). Here the structure of Levi subgroups (which are direct products of general linear groups and a symplectic group), and the existing classification of tempered duals of general linear groups, reduce the problem of the non-unitary dual to the problem of classifying the tempered duals of symplectic groups. To get irreducible tempered representations from the square integrable ones, one needs to classify all irreducible subrepresentations of the representations parabolically induced from the irreducible square integrable ones. We get the simplest example of such induced representations if we take irreducible square integrable representations δ and π of a general linear group and a symplectic group, and consider the representation

$$\mathrm{Ind}(\delta \otimes \pi) \tag{1.2}$$

of a symplectic group, parabolically induced from a maximal parabolic subgroup. Actually, the theory of R-groups reduces the general case to the question of whether the representations (1.2) reduce (see [Go94]). The representation (1.2) can be reducible only if δ is self dual (i.e., equivalent to its own contragredient). Therefore, we assume this in what follows.

One way to try to understand the reducibility of (1.2) is the following. Suppose σ in the induced representation (1.1) is an irreducible cuspidal representation of a general linear group with unitary central character (cuspidal representations are characterized by the property that their matrix coefficients are compactly supported modulo center - they can be characterized as isolated representations of the non-unitary dual, which also carries a natural topology). Then the representation (1.1) contains a unique irreducible subrepresentation, which we denote by

$$\delta(\sigma, m).$$

This representation is square integrable, and one gets all square integrable representations in this way ([Ze80]). Now we can slightly reinterpret the question of reducibility of (1.2). It is equivalent to the question of reducibility of

$$\mathrm{Ind}(\delta(\sigma, m) \otimes \pi), \tag{1.3}$$

when σ is a self dual irreducible cuspidal representation of a general linear group. If we fix σ and π as above, then there is one parity of positive integers (even or odd), such that for representations $\delta(\sigma, m)$, with m from that parity, the representation (1.3) is always irreducible (this parity depends only on σ). For the representations $\delta(\sigma, m)$ with m from the other parity, the representation (1.3) is always reducible, with finitely many exceptions. Denote by

$$\mathrm{Jord}(\pi)$$

the set of all such exceptions $\delta(\sigma, m)$ (for fixed π; we let σ run over all equivalence classes of self dual irreducible cuspidal representations of general linear groups)[8]. In other words, roughly speaking $\mathrm{Jord}(\pi)$ takes care of all the singularities of the parabolic induction of (1.3). Therefore, it is a crucial object for understanding the tempered representations which can be obtained from π. We illustrate the importance of $\mathrm{Jord}(\pi)$ with the following

Example 1.1 In this example, we consider odd orthogonal groups. A direct consequence of [Sh92] and [Td98] is that

$$\mathrm{Jord}(1_{SO(1,F)}) = \emptyset.$$

[8] C. Mœglin has defined Jordan blocks (slightly differently; one can find her original definition in [Mœ02]). Here we use a different notation than in [Mœ02] and the other papers, where elements of $\mathrm{Jord}(\pi)$ are pairs (σ, m) instead of square integrable representations $\delta(\sigma, m)$ (recall that $(\sigma, m) \leftrightarrow \delta(\sigma, m)$ is a bijection by [Ze80]).

Thus, for m from one parity (depending on σ), the representation

$$\mathrm{Ind}(\delta(\sigma, m) \otimes 1_{SO(1,F)})$$

is always reducible, while for m from the other parity it is always irreducible (σ is an irreducible self dual cuspidal representation of a general linear group). Now, for a self dual square integrable representation[9] δ we have

$$\mathrm{Ind}(\delta \otimes \pi) \text{ is reducible} \iff \mathrm{Ind}(\delta \otimes 1_{SO(1,F)}) \text{ is reducible and } \delta \notin \mathrm{Jord}(\pi).$$

In other words, roughly $\mathrm{Jord}(\pi)$ measures the difference between tempered induction of π and the trivial representation $1_{SO(1,F)}$[10]. This difference is not very big since $\mathrm{Jord}(\pi)$ must be finite. For the Steinberg representation, we have

$$\mathrm{Jord}(\mathrm{St}_{SO(2n+1,F)}) = \{\delta(1_{F^\times}, 2n)\} = \{\mathrm{St}_{GL(2n,F)}\}.$$

It is interesting and very important that $\mathrm{Jord}(\pi)$ has arithmetic meaning. We describe this briefly. The Langlands program predicts a natural parameterization of an irreducible representation τ of a split reductive groups G over F by a homomorphism

$$\Phi_G(\tau)$$

of $W_F \times SL(2, \mathbb{C})$ into the complex dual Langlands group, satisfying certain requirements (such homomorphisms are called admissible). The parameterization $\tau \mapsto \Phi_G(\tau)$ is called the local Langlands correspondence for G (these correspondences are expected to be instances of a more general phenomenon, called functoriality). Such correspondences can be viewed as generalizations of the local Artin reciprocity law from class field theory. Representations with the same parameter $\Phi_G(\tau)$ are called L-packets (these sets are expected to be finite). The representations inside L-packets are expected to be parameterized by equivalence classes of irreducible representations of the component group of $\Phi_G(\tau)$ (see the sixth section for more details). More then a decade ago, the existence of such a correspondence was established for general linear groups in full generality ([LmRpSu93], [HaTy01] and [He00]). We denote them by Φ_{GL} (here L-packets are singletons).

One of the very big breakthroughs in the theory of automorphic forms, obtained by J. Arthur in his recent book [Ar], is a classification of irreducible

[9]Recall that $\mathrm{Ind}(\delta \otimes \pi)$ is irreducible if δ is not self dual.

[10]For symplectic groups, the above discussion also holds if we exclude the trivial representation of $GL(1, F)$. This difference is caused by the fact that $\mathrm{Jord}(1_{SO(1,F)}) = \{1_{GL(1,F)}\}$, which again directly follows from [Sh92] and [Td98].

tempered representations of classical p-adic groups[11]. This classification can be viewed as an instance of a local Langlands correspondences. In the case of unitary groups, such a classification was obtained earlier by C. Mœglin in [Mœ07]. A fundamental result, which tells us that crucial objects of harmonic analysis are directly related to the fundamental objects of the number theory, is the following theorem of C. Mœglin:

Theorem 1.2 *The admissible homomorphism that J. Arthur has attached to a square integrable representation* π *is*

$$\bigoplus_{\sigma \in Jord(\pi)} \Phi_{GL}(\sigma). \tag{1.4}$$

Therefore, Arthur's classification singles out crucial information for the tempered induction related to π.

Besides this information, there are a number of other questions important for harmonic analysis which still remain to be answered. One of them is how irreducible square integrable representations are built from cuspidal ones. In the case of general linear groups, this question is answered by the Bernstein-Zelevinsky theory. For other classical groups, this question is directly related to understanding the internal structure of packets[12].

C. Mœglin has characterized the parameters corresponding to the cuspidal representations in the Arthur classification. They have a very simple description (see the seventh section). Now we can use the classification of irreducible square integrable representations of classical p-adic groups modulo cuspidal data (obtained in [Mœ02] and [MœTd02]), to obtain a different description of irreducible square integrable representations of classical p-adic groups. This also gives representation theoretic information on how packets are build.

There are a number of questions that one can further consider regarding the structure of the packets. We give one example. A new feature showing up in the Arthur classification (which was not present for groups like $GL(n)$ or $SL(n)$), is the existence of square integrable packets containing both cuspidal and non-cuspidal representations at the same time. The extreme instance of this phenomenon is (square integrable) packets containing at the same time a representation supported on the minimal parabolic subgroup and a cuspidal representation. Such packets are called packets with antipodes. A simple application of the representation theoretic description

[11] Arthur classification is still conditional, but this is expected to be removed soon (when the facts on which [Ar] relies become available). Therefore, we shall use [Ar] in what follows without mentioning that there is still a piece to be completed.

[12] They are not called L-packets, since [Ar] does not address the question of L-functions.

of packets is the fact that $SO(2n+1,F)$ has a packet of this type if and only if n is even.

We are very thankful to C. Mœglin for discussions and for providing us with some references in the last sections of the paper. G. Savin has read the first version of this paper, and gave us a number of useful suggestions. I. Matić also gave suggestions to the following version of the paper. C. Jantzen's numerous suggestions helped us a lot to improve the style of the paper. I. Badulescu gave us a number of useful mathematical remarks to improve the paper. Discussions with M. Hanzer, A. Moy and G. Muić were helpful during the preparation of this paper. The referee's numerous corrections and general remarks helped a lot to improve the exposition of the paper. We are very thankful to all of them.

We now briefly discuss the contents of the paper section by section. The second section reviews very basic notions related to the natural topology on representations. In the third section, we consider the example of $GL(n)$ and the isolated representations in this case. The fourth section introduces the automorphic duals, and follows the case of $GL(n)$. The fifth section studies the unramified duals of classical groups and automorphicity in this setting. In the sixth section, we follow how questions of harmonic analysis bring us to the square integrable packets of classical groups, and the recent work of J. Arthur and C. Mœglin. The seventh section recalls Mœglin's description of cuspidal representations in the Arthur classification, which is followed up in the eighth section by a description of the internal structure of packets.

2 Topology, Isolated Representations, Support

The unitary dual $\hat{G}$ is a topological space in a natural way: π is in the closure of $X \subseteq \hat{G}$ if and only if diagonal matrix coefficients of π on compact subsets can be approximated by finite sums of diagonal matrix coefficients of representations from X. One can find more details in [Di62] and [Fe62], or [Mi73] (the non-archimedean case is in [Td88]).

If G is commutative then the Pontryagin duality is formulated in this topology. In this case $\hat{G}$ is a group and hence topologically homogenous. In the non-commutative case $\hat{G}$ does not need to be homogenous. Of particular importance are isolated representations (or isolated modulo center[13]). They are pretty mysterious, but very important objects. Their unitarity

[13] If G has non-compact center, the role of isolated representations is played by isolated representations modulo center. Let ω_π be the central character of $\pi \in \hat{G}$ and denote $\hat{G}_{\omega_\pi} = \{\tau \in \hat{G}; \omega_\tau = \omega_\pi\}$. Then we say that π is isolated modulo center if π is an isolated point of the topological space $\hat{G}_{\omega_\pi}$. These representations we often simply call isolated representations.

is usually a very non-trivial fact. Local components of square integrable automorphic forms are a big source of isolated (or isolated modulo center) representations.

For performing step (G2) of the Gelfand concept on a fixed unitary representation Π of G, we usually do not need the whole $\hat{G}$, but only the representations which are in the support of the measure on $\hat{G}$ which decomposes Π into a direct integral of elements of $\hat{G}$ (we shall not go into detail here regarding direct integrals). This support of the measure is denoted by

$$\mathrm{supp}(\Pi).$$

One can describe the support of Π without finding the measure, but using only the topology. For $\pi \in \hat{G}$, we have: $\pi \in \mathrm{supp}(\Pi)$ if and only if it is weakly contained in Π, i.e., if diagonal matrix coefficients of π on compact subsets can be approximated by finite sums of diagonal matrix coefficients of Π (see [Di62] or [Fe62] for more details).

Clearly, $\mathrm{supp}(\Pi)$ inherits the topology from $\hat{G}$. Even for $\mathrm{supp}(\Pi)$, isolated representations can be again very distinguished. Let us illustrate this with the following:

Example 2.1 Let G be a semi-simple algebraic group over a local field F. Then the support of the regular representation of G on $L^2(G)$ by right translations is

$$\mathrm{supp}(L^2(G)) = \hat{G}_{\mathrm{temp}},$$

where $\hat{G}_{\mathrm{temp}}$ denotes the set of all tempered representations in $\hat{G}$, i.e., those whose matrix coefficients are in $L^{2+\epsilon}(G)$ for each $\epsilon > 0$. A very important class of tempered representations (for harmonic analysis, as well as for the theory of automorphic forms) are square integrable representations (i.e., those ones whose matrix coefficients are in $L^2(G)$). Now for $\pi \in \hat{G}_{\mathrm{temp}}$, we have

$$\pi \text{ is square integrable} \iff \pi \text{ is isolated in } \hat{G}_{\mathrm{temp}}.$$

There is a very effective criterion of Harish-Chandra and of W. Casselman, for checking square integrability of an irreducible representation.

Let us recall that the essential part of the monumental work of Harish-Chandra [HC83] was related to the regular representation of a semi-simple group G. In the real case, Harish-Chandra has constructed all the isolated points of $\hat{G}_{\mathrm{temp}}$'s. This was enough for him to explicitly decompose $L^2(G)$. The tempered dual $\hat{G}_{\mathrm{temp}}$ was classified by A. Knapp and G. Zuckermann later in [KnZu82], based on the work of Harish-Chandra. Here isolated representations in the tempered duals were crucial for the construction of the whole tempered dual (we shall see similar examples later).

In the book [GfNa50] of I.M. Gelfand and M.A. Naimark on harmonic analysis on complex classical groups, the authors wrote the lists of irreducible unitary representations of those groups. They expected that the representations from the lists form unitary duals of complex classical groups (their lists were very simple). In the case of symplectic and orthogonal groups, the incompleteness of the lists was clear pretty soon. For the special linear groups, E.M. Stein constructed in [St67], in a relatively simple way, representations (complementary series) which were not in the lists of Gelfand and Naimark (for $SL(2n,\mathbb{C})$, $n \geq 2$).

The above simple construction of E.M. Stein, the lack of significant progress in giving an explicit classification of the whole unitary dual for a long time (even for the groups like $SL(n,\mathbb{C})$) and very complicated approaches to the problem, have sometimes resulted in doubts that the unitary dual is the right object for harmonic analysis, that it may be too big and complicated, consisting perhaps mainly of non-relevant representations for harmonic analysis. That very successful strategy of Harish-Chandra might be the right way for the general approach: to go directly to (G2) for a specific important unitary representation (bypassing the general problem (G1)), and concentrate only on the part of $\hat{G}$ which is relevant for the unitary representation that we consider.

Considering some important groups, we shall see below why this strategy is not likely to be very successful for the general case. Already some important unitary representations that show up in the theory of automorphic forms indicate this. For example, we shall see for some groups that the most delicate part of the unitary duals, the isolated representations in $\hat{G}$, all show up. These representations are very distinguished representations, and important for number of other problems.

In the unitary duals appear very big and very complicated families of non-isolated representations (complementary series), which are not expected to show up in the automorphic setting, at least not in the split case. Therefore, from the point of view of automorphic forms, unitary duals may look too big, with significant parts which do not seem relevant. We shall see that even this part of the unitary duals can be interesting for the theory of automorphic forms.

To get an idea of isolated representations, their relation to automorphic forms, and the role of complementary series, we go to the relatively well understood case of $GL(n)$:

3 The Example of $GL(n)$

All parabolic subgroups of the groups that we consider in this paper, are assumed to be standard with respect to the minimal parabolic subgroup consisting of upper triangular matrices in the group.

Let F be a local field (or the ring of adeles of a global field). Let σ be an irreducible square integrable representation[14] of $GL(n, F)$ (in the adelic case we take an irreducible cuspidal representation of adelic $GL(n)$). Fix a positive integer m. Let P be the parabolic subgroup of $GL(nm, F)$ whose Levi subgroup is in a natural way isomorphic to

$$GL(n, F) \times \cdots \times GL(n, F).$$

Consider the parabolically induced representation

$$\mathrm{Ind}_P^{GL(mn,F)}(|\det|_F^{(m-1)/2}\sigma \otimes |\det|_F^{(m-1)/2-1}\sigma \otimes \cdots \otimes |\det|_F^{-(m-1)/2}\sigma) \tag{3.1}$$

($|\ |_F$ denotes the normalized absolute value on F). The above representation has a unique irreducible quotient, which is denoted by

$$u(\sigma, m),$$

and called a Speh representation. Observe that if σ is a unitary character of $F^\times$, then $u(\sigma, m) = \sigma \circ \det$ is also a character of $GL(m, F)$. This is the reason that for archimedean F, one gets Speh representations which are not characters only if $F = \mathbb{R}$ and σ is a square integrable representation of $GL(2, \mathbb{R})$ (this is where the name comes from; see [Sp83]).

Speh representations are very important in the theory of automorphic forms. It is interesting that we first came (in [Td86]) to the Speh representations and their unitarity without knowing their role in the automorphic forms, studying complementary series (which are not expected to show up in the setting of automorphic forms in this case; we comment on this later). We briefly sketch below how we came to the Speh representations.

Obviously $u(\sigma, 1) = \sigma$ is unitary (square integrable representations are unitary). Suppose that $u(\sigma, m)$ is unitary. Consider the family (complementary series)

$$\mathrm{Ind}_{P'}^{GL(2mn,F)}(|\det|_F^{\alpha}u(\sigma, m) \otimes |\det|_F^{-\alpha}u(\sigma, m)), \quad 0 \le \alpha < 1/2, \tag{3.2}$$

where P' is the appropriate parabolic subgroup. This is irreducible for $\alpha = 0$ by [Be84] (and also it is unitary). For other α's as above, it is

[14]These representations are also called square integrable representations modulo center, since the requirement is that the absolute value of their matrix coefficients be square integrable functions modulo center.

also irreducible, which is easy to see. Further, these representations are Hermitian. From this it easily follows that all the representations in (3.2) are unitary. If we put in (3.2) $\alpha = 1/2$ (the induced representation is then no more irreducible), we get for a subquotient the representation

$$\mathrm{Ind}_{P''}^{GL(2mn,F)}(u(\sigma, m+1) \otimes u(\sigma, m-1)), \tag{3.3}$$

if we can prove that (3.3) is irreducible. Then the representation (3.3) is unitary , since it is at the end of the complementary series (this follows from [Mi73]). Now a simple construction of unitary representations, which we call unitary parabolic reduction (since it is opposite to the unitary parabolic induction), implies that $u(\sigma, m+1) \otimes u(\sigma, m-1)$ is unitary, and thus also $u(\sigma, m+1)$ is unitary.

Now we shall see that Speh representations are distinguished from the point of view of harmonic analysis. We assume in the rest of the paper that F is a local non-archimedean field (although some facts also hold in the archimedean case). Now, let σ be a unitary irreducible cuspidal representation[15] of $GL(n, F)$. Then the representation (3.1) has a unique irreducible subrepresentation, which will be denoted by

$$\delta(\sigma, m).$$

This representation is square integrable, and J. Bernstein has shown that one gets all such representations in this way. An old result from [Td87] gives the following characterization of isolated points in the unitary dual of general linear groups:

Theorem 3.1 *Let $\pi \in \widehat{GL(k, F)}$. Then π is isolated (modulo center) if and only $\pi \cong u(\delta(\rho, l), m)$ for some irreducible unitary cuspidal representation ρ of $GL(k/(lm), F)$ and some positive integers $l \neq 2$ and $m \neq 2$ (clearly then lm divides k).*

In other words, if we have an isolated representation, then it is always a Speh representation. In the converse direction, a Speh representation is almost always isolated. The condition $l \neq 2$ and $m \neq 2$ spoils the picture given by the theorem a little bit[16] (the corresponding representations are subquotients of ends of complementary series, which easily implies that they can not be isolated). We shall soon see that we get a completely regular

[15] These representations are very distinguished square integrable representations. Their matrix coefficients are compactly supported modulo center. Irreducible cuspidal representations can be characterized as isolated points in the non-unitary dual (see [Td88]).

[16] We have similar "irregularity" in [Ka67] (rank must be $\neq 1$). This "irregularity" was removed in the automorphic dual by L. Clozel (see [Cl03]).

picture in the automorphic dual, which will be discussed in the following section.

Classification:

1. It is very easy to state the classification of $\widehat{GL(n,F)}$ modulo square integrable representations: each representation parabolically induced by a tensor product of Speh representations and complementary series (3.2) with $\alpha > 0$, is in $\widehat{GL(n,F)}$, and each representation π in $\widehat{GL(n,F)}$ is obtained in this way. Further, π determines the Speh representations and complementary series inducing to π, up to a permutation. The classification holds also in the archimedean case in exactly the same form and there irreducible square integrable representations are very simple (see [Td86] and [Td] or [Td09] for both cases)[17].

2. The above result (covering the non-archimedean as well as archimedean case, with the proofs along the same strategy) is just "the tip of the iceberg". It is only a special case of a much more general result for any local (finite dimensional) central division algebra[18] A over F. For an irreducible square integrable representation σ of $GL(n,A)$, let $s(\sigma)$ be the minimal positive exponent such that

$$\mathrm{Ind}_{P'''}^{GL(2n,A)}(|\det|_F^{s(\sigma)/2}\sigma \otimes |\det|_F^{-s(\sigma)/2}\sigma)$$

   reduces ($s(\sigma)$ is an integer dividing the rank of A). Then, thanks mainly to the recent work of I. Badulescu, D. Renard and V. Sécherre, if we define now representations $u(\sigma, m)$ and complementary series for general linear groups over A putting $|\det|_F^{s(\sigma)}$ instead of $|\det|_F$ in (3.1) and (3.2), the above classification holds in the same form[19] for all groups $GL(n,A)$.

[17]The complex case is particularly simple: each irreducible unitary representation is parabolically induced by tensor products of characters and complementary series starting with characters (constructed by E.M. Stein). This list differs from the list of I.M. Gelfand and M.A. Naimark in [GfNa50] only in that they have omitted complementary series of $GL(2n,\mathbb{C})$ for $n \geq 2$ (i.e., Steins's complementary series). The book of Gelfand and Naimark was much ahead of its time when it was published (1950), in particular regarding the intuition. It had strong influence on a number of further developments in harmonic analysis, as well as out of it.

[18]Recall that in the non-archimedean case the Brauer group is $\mathbb{Q}/\mathbb{Z}$ (the field case corresponds to the neutral element).

[19]Here representations $u(\sigma, m)$ are usually not in the automorphic dual (see [Ba08] and [BaRn10]), but their unitarity easily follows using closely related automorphic representations (see [BaRn04])

The above classification of $\widehat{GL(n,A)}$ is obtained along the same strategy in the non-archimedean and archimedean case (the outline of that strategy is in [Td85]). This strategy reduces the classification to proving the unitarity of Speh representations, and the irreducibility of unitary parabolic induction. We have described one possibility to prove the unitarity of Speh representations using complementary series (this strategy was used in [Td86], [BaHeLeSé10] and [Ba]). In the following section we comment on the possibility of proving unitarity of Speh representations using automorphic forms (used first in [Sp83], and then in [Td86] and [BaRn04]; this approach does not distinguish between the archimedean and non-archimedean cases). In the field case, the idea outlined in [Ki62] was a basis of proofs of irreducibility of unitary parabolic induction in [Be84] and [Bu03] (see also [AiGu09], [SuZh12] and [AiGuRlSf10]).

Remark 3.2 1. In the (non-commutative) division algebra case, D. Vogan proved in [Vo86] the irreducibility of unitary parabolic induction for Hamiltonian quaternions (see [BaRn10] for explanation how to get this irreducibility from [Vo86]) and V. Sécherre proved this irreducibility in [Sé09] for the case of non-archimedean division algebras (here we have plenty of division algebras). The two above proofs for division algebras are completely different. A uniform proof for both cases would be very desirable (the idea of Kirillov from [Ki62] does not work here, neither can the approach of [AiGu09], [SuZh12] and [AiGuRlSf10] be used here). A uniform proof in the division algebra case would also shed a new light on the field case. We expect such proof to be of a functional analytic nature.

2. To complete this discussion, let us mention that there is the classification of D. Vogan in the archimedean case. His classification is given by Theorem 6.18 of [Vo86], and is completely different from the classification that we have presented above. Stating his classification here would require a number of technical and combinatorial notions (and results), starting with several kinds of K-types. This is the reason that we do not present his classification here. His classification is equivalent to the specialization to the archimedean case of the classification that we presented above[20]. This equivalence is a non-trivial fact, and it is proved in [BaRn10].

[20]Vogan's classification gives the complete answer in the archimedean case, while the classification that we presented above (holding in both cases) only reduces unitary duals to square integrable representations. Despite this, the above classification in the archimedean case directly yields the complete classification, since the required square integrable representations were known already in 1950's (and even earlier). In the non-archimedean case, this was a very hard problem (see [Ze80], [BsKu93], [LmRpSu93], [HaTy01] and [He00]).

4 The Automorphic Dual

Definition 4.1 Let G be a reductive group defined over a number field k, v a place of k, k_v the completion of k at v (which we often denote by F), and $\mathbb{A}_k$ the ring of adeles of k. Then the automorphic dual

$$\hat{G}_{v,\mathrm{aut}}$$

is defined to be the support of the representation of k_v-rational points $G(k_v)$ of G on the space of square integrable automorphic forms

$$L^2(G(k)\backslash G(\mathbb{A}_k)).$$

If an appropriate maximal compact subgroup K of $G(k_v)$ is fixed, we denote by $\hat{G}^{\mathbf{1}}_{v,\mathrm{aut}}$ the representations in $\hat{G}_{v,\mathrm{aut}}$ which possess a non-trivial vector invariant for K (i.e., the unramified part[21]). This part will be called the Ramanujan dual (as in [BuLiSn92]).

One can find the original (more general) definition, and much more detail in [Cl07] (see also [BuLiSn92]).

The automorphic duals are very hard. They are related to such hard problems as Selberg's $\frac{1}{4}$-conjecture, and more generally Ramanujan's conjecture for Maass forms, generalized Ramanujan conjecture, etc. We do not know the classification of automorphic duals even in the simplest cases. Despite this, we know rather interesting facts for classical groups, as we shall see, and have more precise expectations than in the case of unitary duals. This may be good for both the duals.

H. Jacquet has proved in [Jc84] that Speh representations are automorphic (being local factors of global Speh representations, for which he proved that they are in the residual spectrum of the representation on the space of square integrable forms). B. Speh has proved this earlier for $u(\sigma, m)$'s when σ is a square integrable representation of $GL(2,\mathbb{R})$. Jacquet's proof clearly implies the unitarity of (local) Speh representations. The division algebra case requires an additional step (using unitary parabolic reduction, see [BaRn04]).

In this way, thanks to Theorem 3.1 and H. Jacquet's result that we mentioned above, we get plenty of isolated (modulo center) representations in the automorphic dual (although this term was not yet formally defined when the both results were available).

Further, the estimate in [MüSp04] of B. Speh and W. Müller of local components of irreducible representations in the residual spectrum (which

[21] The superscript $\mathbf{1}$ on a set of representations will always denote the unramified representations in that set.

is based on earlier estimates of local unramified components of irreducible representations in the cuspidal spectrum in [LoRuSn99]) and Langlands' description of automorphic spectra ([Ar79]) imply that the Speh representations excluded by Theorem 3.1 are also isolated in the automorphic dual.

Actually, the generalized Ramanujan conjecture (stating that each local component of an irreducible representation in the cuspidal spectrum of $GL(n)$ should be tempered), would imply that the Speh representations are the only isolated representations in the automorphic dual.

Moreover, the generalized Ramanujan conjecture also tells us what the automorphic dual should be. Let us briefly comment on this. In J. Bernstein's work on unitarity [Be84], a notion of rigid representations of $GL(n, F)$ naturally arose. We can define these representations as those for which the essentially tempered representation $|\det|_F^{\alpha_1}\tau_1 \otimes \cdots \otimes |\det|_F^{\alpha_k}\tau_k$ (of a Levi subgroup) corresponding to it by the Langlands classification of the non-unitary duals (τ_i are tempered and $\alpha_i \in \mathbb{R}$) have all the exponents α_i in $(1/2)\mathbb{Z}$. Denote by $\widehat{GL(n, F)}_{\mathrm{rig}}$ the subset of rigid representations in $\widehat{GL(n, F)}$. Then the classification theorem ([Td86], [Td09], or see our previous brief description of the classification) implies that $\widehat{GL(n, F)}_{\mathrm{rig}}$ consists of representations parabolically induced by tensor products of Speh representations (no complementary series). A. Venkatesh denotes $\widehat{GL(n, F)}_{\mathrm{rig}}$ by $\widehat{GL(n, F)}_{\mathrm{Ar}}$, since these representations come from the work of J. Arthur ([Ar89]).

A. Venkatesh observed in [Ve05] that the generalized Ramanujan conjecture would imply equality of the automorphic and the rigid duals, and conversely. Namely, Langlands' description of automorphic spectra implies that $\widehat{GL(n, F)}_{\mathrm{rig}}$ is contained in the automorphic dual, and further, the generalized Ramanujan conjecture would imply equality. In the other direction, knowledge of equality at all places would imply (using [Sa74]) the generalized Ramanujan conjecture.

One can find plenty of interesting facts, estimates and conjectures in [Cl07] and [Sn05] (and papers cited there).

5 Unramified Duals and Ramanujan Duals of Classical Groups

In this section, we consider unramified irreducible unitary representations (F is a non-archemedean field). For this section, it is convenient to introduce notion of negative and strongly negative representations. These terms were introduced by G. Muić in [Mu06]. We have already mentioned the Casselman square integrability criterion ([Ca]). His criterion for square integrability (resp., temperedness) modulo center, is given by some inequalities $<$ (resp., $\leq$). Reversing these inequalities, one gets the definition of strongly negative (resp., negative) representations. We do not go into further detail here (see [Mu06] or Definition 1.1 in [Td10]). Irreducible negative (resp., strongly negative) representations are dual to the tempered (resp., square integrable) irreducible representations by duality of A.-M. Aubert ([Au95]) and P. Schneider and U. Stuhler ([SdSl97]). Unramified irreducible negative representations are always unitary by [Mu07] (we comment on this later). The simplest example of a strongly negative representation of G is the trivial representation 1_G.

We first consider the simple example of $SL(n)$:

Example 5.1 Below, k is an algebraic number field, v a place of k and $F = k_v$, the completion of k at v. We have

1. Isolated representations in $\widehat{SL(n,F)}^{\mathbf{1}} = \begin{cases} \{1_{SL(n,F)}\} & n \neq 2, \\ \emptyset & n = 2. \end{cases}$
2. Strongly negative representations in $\widehat{SL(n,F)}^{\mathbf{1}} = \{1_{SL(n,F)}\}$.
3. Known (to us) isolated representations in $\widehat{SL(n)}^{\mathbf{1}}_{v,\mathrm{aut}} = \{1_{SL(n,F)}\}$.

Therefore, the above three sets coincide, with one exception (the case $n = 2$ in (1)).

Now we discuss the case of other classical groups. To simplify exposition, we concentrate on symplectic groups. Here we use a classification of unramified unitary duals obtained in [MuTd11]. By [MuTd11], an unramified irreducible unitary representation is either negative, or a complementary series starting with a negative irreducible representation. Clearly, complementary series can not be isolated. Therefore, we are left with irreducible negative representations. Further by [Mu06], each negative representation is a subrepresentation of a representation parabolically induced by a strongly negative irreducible one. From this it follows easily that an isolated unramified representation must be strongly negative (recall that for $SL(n)$, $n \neq 2$, the converse also holds).

Now we describe the parameters of unramified irreducible strongly negative representations of $Sp(2n, F)$. They are pairs of partitions (p_1, p_2), where each p_i is a partition of k_i into different odd positive integers such that

- $k_1 + k_2 = 2n + 1$;

- p_2 has even number of terms (i.e., p_1 has odd number of terms)[22].

In [Td10] it is described how one constructs in a simple way representations attached to these parameters.

Further, by [MuTd11] isolated representation in $\widehat{Sp(2n,F)}^{\mathbf{1}}$ are parameterized by pairs of partitions (p_1, p_2), which satisfy the above two conditions, and also the following two:

- neither p_1 nor p_2 contains consecutive odd numbers;
- 3 is neither in p_1 nor in p_2[23].

In [Td10], we have obtained combinatorial formulas for the number of above two classes of representations (i.e., irreducible unramified strongly negative and isolated ones). Instead of writing these formulas, we write here the numbers that we get for $Sp(340)$:

Example 5.2 We have

1. Number of isolated representations in $\widehat{Sp(340,F)}^{\mathbf{1}} = 11\,322\,187\,942$.
2. Number of strongly negative representations in
$$\widehat{Sp(340,F)}^{\mathbf{1}} = 568\ 385\ 730\ 874.$$
3. G. Muić has proved in [Mu07] that each unramified irreducible strongly negative representation is automorphic (in this way he also proved their unitarity). Therefore, 1. provides us with a huge number of isolated representations in the Ramanujan dual. From this it follows that in the Ramanujan dual $\widehat{Sp(340)}^{\mathbf{1}}_{v,\mathrm{aut}}$ we have at least 11 322 187 942 isolated representations (and this is the number of isolated representations in $\widehat{Sp(340)}^{\mathbf{1}}_{v,\mathrm{aut}}$ known to us).

This example shows several new phenomena when compared to the $SL(n)$-case, and raises some questions. A new fact is that we have a huge number of isolated representations in the Ramanujan dual, and also in the unramified unitary dual; also we have a huge number of unramified irreducible strongly negative representations. Further new fact is that the number of unramified irreducible strongly negative representation is much bigger then the number of isolated unramified representations (and also then the number of known isolated representations in the Ramanujan dual).

[22]These parameters are in bijection with discrete unramified admissible homomorphism of the Weil-Deligne group - see [Mu07].

[23]We have similar description for the special odd-orthogonal groups, where we deal with partitions into different even positive integers.

It is interesting that Conjecture "Arthur + ϵ" of L. Clozel from [Cl07] would imply that the set of isolated representations in $\widehat{Sp(340)}^{\mathbf{1}}_{v,\mathrm{aut}}$ is equal to the set of strongly negative representations in $\widehat{Sp(340,F)}^{\mathbf{1}}$ (see [Td10]).

The first question which arises related to the above example is of an arithmetic nature. D. Kazhdan's proof in [Ka67] that the trivial representation is isolated in the unitary dual of a simple group of rank different from 1 had important arithmetic consequences. Does the above stunning difference regarding the number of isolated points for $SL(n)$ and $Sp(2n)$ groups have some arithmetic explanation, or consequence?

The second question which arises is related to harmonic analysis: why is such a small portion of strongly negative representations isolated (this was not the case for $SL(n)$)? The reason is that for each of 568 385 730 874 strongly negative unramified representations, excluding 11 322 187 942 of them (i.e., the isolated ones in the unramified dual), there is a complementary series at whose end this representation lies. As it is well known, complementary series representations start with an irreducible representation parabolically induced from a unitary one (and which satisfies additional symmetry conditions with respect to the Weyl group). Therefore, in the example of $Sp(340, F)$, for the 557 063 542 932 parabolically induced representations which are involved (where corresponding complementary series start), we need to know their irreducibility.

The following question is related: why are there still 11 322 187 942 isolated representations? The answer is roughly: no complementary series ends with them.

All this tells us that we need to have very explicit knowledge of complementary series (not only on some algorithmic level). This means that we need also to have a very explicit understanding of the question of irreducibility/reducibility of parabolically induced representations from unramified unitary ones. Such an understanding is obtained by G. Muić in [Mu06]. Instead of explaining it here, we shall go to a different (and dual) setting in the following section, where we explain how one can get a similar understanding.

Note that for general linear groups we have a perfect understanding of the question of irreducibility/reducibility of parabolically induced representations from unitary ones. It is given by J. Bernstein in [Be84]. The answer is very simple: we always have irreducibility.

For other classical groups, the above answer is far from being true. Roughly, in the cases that we consider and when reducibility can happen, we have reducibility in about "half" the cases. It can be different from "half" and the portion for which the reducibility is different from "half" will be crucial information. We try to explain this in the following section.

6 Controlling Tempered Reducibility ⇝ Packets of Square Integrable Representations

We shall start this section with some very basic questions of pure harmonic analysis, and see how they bring us to some deep questions of the theory of automorphic forms. We deal with classical groups like $Sp(2n, F)$ or (split) $SO(2n+1, F)$ (F is a non-archimedean field). The main object of this section will be square integrable representations. Recall that they can be characterized as isolated representations in the tempered dual.

Suppose that we are interested in the unitary duals of these groups. Standard strategy to classify them is to use the non-unitary duals. The non-unitary dual of a reductive group G over F is the set of all equivalence classes of irreducible smooth representations of G[24]. The second part of this strategy is to identify in the non-unitary dual unitarizable classes (i.e. irreducible smooth representations which admit G-invariant inner products).

Langlands' classification of the non-unitary duals reduces the non-unitary duals to the tempered duals of Levi subgroups. Since a Levi subgroup in a classical group is a direct product of general linear groups and of a classical group, and since tempered duals of general linear groups are classified (we have commented this earlier in this paper), the non-unitary duals are in this way reduced to the tempered duals of classical groups.

Further, the theory of R-groups ([Go94]) reduces the problem of tempered duals of classical groups to the question of reducibility of the representations

$$\mathrm{Ind}_P^G(\delta \otimes \pi), \tag{6.1}$$

where δ and π are irreducible square integrable representations of a general linear group and a classical group respectively, and to the problem of classification of irreducible square integrable representations of classical groups (in (6.1), G is an appropriate classical group, and P is an appropriate parabolic subgroup of G)[25]. We concentrate now on the first question, the question of reducibility of (6.1). We can have reducibility only if δ is selfdual, i.e., if it is equivalent to its own contragredient. Therefore, we always assume that δ is selfdual in what follows.

---

[24] Some authors use the term smooth dual or admissible dual for the non-unitary dual. We prefer (more traditional) term non-unitary dual. This term has been used more often earlier (among others, by J. M. G. Fell in [Fe65]). Here the term "non-unitary" is used to stress that unitarity of representations is not required in the definition of this dual (the non-unitary dual contains the unitary dual).

[25] Actually, for a more explicit understanding of tempered duals we would need a little bit more than what the R-groups give (see [Jn], [Td+] where this is obtained). We do not go into detail here.

Each selfdual irreducible square integrable representation of a general linear group is of the form $\delta(\rho, m)$ for some cuspidal selfdual representation ρ and some positive integer m (see section 3 for notation). In what follows, we assume that ρ is selfdual. Therefore, we need to understand reducibility of

$$\mathrm{Ind}_P^G(\delta(\rho, m) \otimes \pi). \tag{6.2}$$

To get an idea of how the reducibility of (6.2) behaves, we shall look at some simple examples. For this, one of our old results from [Td98] will be useful. To state this result, we need to introduce some notation.

Let ρ be irreducible cuspidal representation of $GL(n, F)$ and m a non-negative integer. Set

$$[\rho, |\det|_F^m \rho] = \{\rho, |\det|_F \rho, \dots, |\det|_F^m \rho\}.$$

Then $\Delta = [\rho, |\det|_F^m \rho]$ is called a segment in cuspidal representations of general linear groups. The representation

$$\mathrm{Ind}_P^{GL((m+1)n,F)}(|\det|_F^m \rho \otimes |\det|_F^{m-1} \rho \otimes \cdots \otimes \rho) \tag{6.3}$$

contains a unique irreducible subrepresentation, which is denoted by

$$\delta(\Delta)$$

(P is the appropriate standard parabolic subgroup). This is an essentially square integrable representation[26], and J. Bernstein has shown that one gets all such representations in this way. Note that our former notation (for ρ unitary) becomes

$$\delta(\rho, m) = \delta([|\det|_F^{-(m-1)/2} \rho, |\det|_F^{(m-1)/2} \rho]). \tag{6.4}$$

To simplify discussion, we assume char$(F) = 0$ below. We now recall Theorem 13.2 from [Td98]:

Theorem 6.1 *Let $\delta(\Delta)$ be an essentially square integrable representation of a general linear group and σ an irreducible cuspidal representation of a classical group. Then*

$$Ind_P^G(\delta(\Delta) \otimes \sigma) \tag{6.5}$$

reduces if and only if

$$Ind_{P'}^{G'}(\rho \otimes \sigma) \tag{6.6}$$

reduces for some $\rho \in \Delta$ (G and G' are appropriate classical groups, and P and P' are appropriate parabolic subgroups in G and G', respectively).

[26] I.e., it becomes square integrable after twisting by an appropriate character.

The above theorem was proved under a condition in [Td98] (which was fulfilled if σ is generic thanks to the fundamental results of [Sh90]; see also [Sh92]). This condition now follows from [Ar]. Note that this book is still conditional, but this is expected to be removed very soon (when the facts on which [Ar] relies become available). Therefore, we shall use [Ar] in what follows without mentioning that there is still a piece to be completed.

In what follows, ρ will always be an irreducible selfdual cuspidal representation of a general linear group and π will be always an irreducible square integrable representation of a classical group (a series of classical groups will be fixed). Below we shall fix ρ and π, and consider the reducibility of (6.2) depending on m. We first look at some very simple examples.

The first example gives a very regular picture. We shall consider (for a moment) special odd-orthogonal groups. The simplest setting is if we take for the square integrable representation π the trivial representation $1_{SO(1,F)}$ of the trivial group, and for ρ the trivial representation $1_{F^\times}$ of $GL(1,F)$.

Example 6.2 Recall that $\mathrm{Ind}_{P_\emptyset}^{SO(3)}(|\ |_F^\alpha) = \mathrm{Ind}_{P_\emptyset}^{SO(3)}(|\ |_F^\alpha \otimes 1_{SO(1,F)})$, $\alpha \in \mathbb{R}$, reduces $\iff \alpha = \pm 1/2$. Now Theorem 6.1 implies

$$\mathrm{Ind}_P^{SO(2m+1,F)}(\delta(1_{F^\times}, m) \otimes 1_{SO(1,F)}) \text{ is } \begin{cases} \text{reducible for all even } m, \\ \text{irreducible for all odd } m, \end{cases}$$

since $|\ |_F^{\pm 1/2} \in [\,|\ |_F^{-(m-1)/2}, |\ |_F^{(m-1)/2}] \iff m$ is even (P is the Siegel parabolic subgroup above).

We now go to the symplectic counterpart, where we get a slightly less regular picture.

Example 6.3 Theorem 6.1 gives

$$\mathrm{Ind}_P^{Sp(2m,F)}(\delta(1_{F^\times}, m) \otimes 1_{Sp(0,F)}) \text{ is}$$
$$\begin{cases} \text{reducible for all odd } m \textbf{ except } m = 1, \\ \text{irreducible for all even } m, \end{cases}$$

since $|\ |_F^{\pm 1} \in [\,|\ |_F^{-(m-1)/2}, |\ |_F^{(m-1)/2}] \iff m$ is odd and $m \neq 1$ (reducibility of principal series of $SL(2,F)$ that we consider is at ± 1).

Therefore, $\boxed{1_{F^\times} = \delta(1_{F^\times}, 1)}$ is an exception among the representations $\delta(1_{F^\times}, m)$.

Remark 6.4 F. Shahidi had proved in [Sh92] that if $\rho \not\cong 1_{F^\times}$, and if $\mathrm{Ind}_P^G(|\det|_F^\alpha \rho \otimes 1)$ reduces, then either $a = 0$ or $\alpha = \pm 1/2$. Now Theorem 6.1 implies

$$\mathrm{Ind}_P^G(\delta(\rho, m) \otimes 1) \quad \text{is} \quad \begin{cases} \text{reducible for all } m \text{ from one parity,} \\ \text{irreducible for all } m \text{ from the other parity.} \end{cases}$$

The parity of reducibility is even (resp., odd) if Shahidi's reducibility is 1/2 (resp., 0). We do not further discuss here parity of reducibility/irreducibility (see [Sh92]), but it is related to the local Langlands correspondence for general linear groups (or analytic properties of corresponding L-functions). Clearly, the reducibility/irreducibility depends on the series of classical groups that we consider (F. Shahidi has obtained a duality among these reducibilities for the series of groups that we consider; see [Sh92]).

Observe that the picture that we have gotten from two last examples is pretty nice. We have only one exception which does not fit the general even - odd pattern. Actually, this exception is the first of a family of examples, when one takes for π the Steinberg representation $St_{Sp(2k,F)}$ of $Sp(2k, F)$. For $\pi = St_{Sp(2k,F)}$, we can not get the reducibility of (6.2) from Theorem 6.1, but one can get it for example, using some simple principles from [Td98] based on Jacquet modules (which were used in the proof of Theorem 6.1)[27]. We get

Example 6.5

$$\mathrm{Ind}_P^{Sp(2(m+k),F)}(\delta(1_{F^\times}, m) \otimes St_{Sp(2k,F)}) \text{ is}$$
$$\begin{cases} \text{reducible for all odd } m \textbf{ except } m = 2k+1, \\ \text{irreducible for all even } m. \end{cases}$$

So, among representations $\delta(1_{F^\times}, m)$, $\boxed{\delta(1_{F^\times}, 2k+1)}$ is an exception. For $\rho \not\cong 1_{F^\times}$ we have the same situation as in Remark 6.4, i.e., there are no exceptions.

In general, for general square integrable π, there can be more than one exception. For example, let ψ be a character of order two of $F^\times$. Then

$$\mathrm{Ind}_{P_\emptyset}^{Sp(4,F)}(|\ |_F \psi \otimes \psi) \tag{6.7}$$

[27]For dealing with Jacquet modules, the structure obtained in [Td95+] simplifies considerations.

contains precisely two irreducible square integrable subquotients. For each of them, the set of exceptional representations (as above) is

$$\{\delta(1_{F^\times},1),\delta(\psi,1),\delta(\psi,3)\} \tag{6.8}$$

(so they show up for $\rho = 1_{F^\times}$ and for $\rho = \psi$).

For general square integrable π, the set of such exceptional representations in the above sense (with respect to π) is denoted by[28]

$$\mathrm{Jord}(\pi)^{29}.$$

In other words:

Definition 6.6 For an irreducible square integrable representation π of a classical group, $\mathrm{Jord}(\pi)$ is the set of all selfdual irreducible square integrable representations $\delta(\rho,m)$ of general linear groups such that

$$\mathrm{Ind}_{P'}^{G'}(\delta(\rho,m)\otimes\pi)$$

is irreducible, and

$$\mathrm{Ind}_{P''}^{G''}(\delta(\rho,m+2k)\otimes\pi)$$

reduces for some positive integer k.

Above, G' and G'' are appropriate classical groups, while P' and P'' are appropriate parabolic subgroups of G' and G'' respectively.

After the above definition of $\mathrm{Jord}(\pi)$ we can describe the reducibility of (6.2) in the following way:

1. If $\delta(\rho,m')\in\mathrm{Jord}(\pi)$ for some ρ and m', then we have reducibility of (6.2) for m's of that parity, excluding $\delta(\rho,m)$'s from $\mathrm{Jord}(\pi)$, and we have irreducibility in the other parity.
2. For the remaining ρ's (not showing up in $\mathrm{Jord}(\pi)$), we have reducibility in one parity and irreducibility in the other parity. The parity does not depend on π. The reducibility parity is even if and only if $\mathrm{Ind}_P^G(|\det|_F^{1/2}\rho\otimes 1)$ reduces (P is the Siegel parabolic subgroup).

Therefore, it is crucial to know $\mathrm{Jord}(\pi)$.

[28] Here we use a slightly different notation than in [MœTd02] and the other papers, where elements of $\mathrm{Jord}(\pi)$ were pairs (ρ,m) instead of square integrable representations $\delta(\rho,m)$ (recall that $(\rho,m)\leftrightarrow\delta(\rho,m)$ is a bijection by [Ze80]). One can find the original definition of C. Mœglin in [Mœ02].

[29] Roughly, the set of "singularities" which happen in tempered induction related to π is just $\mathrm{Jord}(\pi)$.

Remark 6.7 Theorem 6.1 directly implies the following three consequences for an irreducible **cuspidal** representation π:

1. $$\delta(\rho, m) \in \mathrm{Jord}(\pi), m \geq 3 \implies \delta(\rho, m-2) \in \mathrm{Jord}(\pi). \tag{6.9}$$

2. We can read from $\mathrm{Jord}(\pi)$ the cuspidal reducibility points which are different from $0, \pm 1/2$. Namely if $\{m; \delta(\rho, m) \in \mathrm{Jord}(\pi)\} \neq \emptyset$, let
$$a_{\rho,\pi} = \max\{m; \delta(\rho, m) \in \mathrm{Jord}(\pi)\}.$$
Then
$$\mathrm{Ind}_P^G(|\det|_F^{(a_{\rho,\pi}+1)/2}\rho \otimes \pi) \text{ reduces.} \tag{6.10}$$

3. We can read information the other way, i.e., if $\mathrm{Ind}_P^G(|\det|_F^x\rho \otimes \pi)$ reduces for some $x \in (1/2)\mathbb{Z}$, $x \geq 1$, then
$$\delta(\rho, 2x-1), \delta(\rho, 2x-3), \ldots, \delta(\rho, \epsilon) \in \mathrm{Jord}(\pi), \tag{6.11}$$
where $\epsilon = 1$(resp., 2) if x is an integer (resp., not an integer). Further, these are the only members of the form $\delta(\rho, m)$ in $\mathrm{Jord}(\pi)$.

Therefore, for cuspidal π, knowing the cuspidal reducibilities ≥ 1 (which are in $(1/2)\mathbb{Z}$)[30] is equivalent to knowing the Jordan blocks of π. Other cuspidal reducibilities (i.e., those at 0 and 1/2) do not depend on π, but only on the series of groups (and clearly on ρ).

In the drawing below, x is the cuspidal reducibility exponent, as in (3) above (in our example below, x is integral), bold segments represent $\delta(\rho, m)$'s in the Jordan blocks (since for them $[|\det|_F^{-(m-1)/2}\rho, |\det|_F^{(m-1)/2}\rho]$ does not contain $|\det|_F^x\rho$; see (6.4)), and dashed segments represent $\delta(\rho, m)$'s which give reducibility (since they contain $|\det|_F^x\rho$), and therefore are not in the Jordan blocks of π:

[30] This is always the case by the recent results of J. Arthur, C. Mœglin and J.-L. Waldspurger.

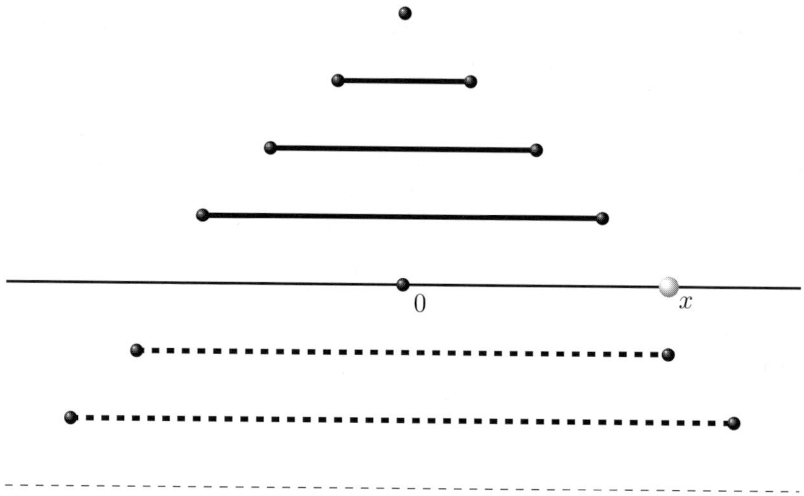

The above drawing is a graphical interpretation of Remark 6.7. The cuspidal reducibility point (exponent) x determines which segments are "bold" (i.e., belong to Jord(π)), and also conversely: if "bold" segments are given (i.e., Jord(π) is given), then we can read from this where the cuspidal reducibility point x is placed.

Definition 6.8 Jordan blocks (of not necessarily cuspidal representations) which satisfy (6.9) will be called Jordan blocks without gaps.

Remark 6.9 1. Observe that (1) of Remark 6.7 above tells us that Jordan blocks of cuspidal representations do not have gaps.

2. Theorem 6.1 and [Sh90] imply that Jordan blocks of cuspidal generic representations consist only of cuspidal representations (since the cuspidal reducibilities in this case can be only in $\{0, \pm 1/2, \pm 1\}$ by [Sh90]; see the above drawing).

Up to now in this section, our point of view was completely that of harmonic analysis. From the other side, square integrable representations are important for a number of problems in automorphic forms. One of the problems is to determine the local Langlands correspondence for the irreducible square integrable representations that we considered.

Briefly, the local Langlands correspondence (for the cases that interest us) attaches to an irreducible square integrable representation π of a split connected semi-simple group G over F, a conjugacy class of continuous homomorphisms

$$\varphi : W_F \times SL(2, \mathbb{C}) \to {}^L G^0,$$

such that φ maps elements of the first factor to semi simple elements, it is algebraic on the second factor, and its image is not contained in any proper Levi subgroup of ${}^LG^0$ (such homomorphisms are called discrete admissible homomorphisms). Above, W_F denotes the Weil group of F and ${}^LG^0$ the complex dual Langlands L-group. The mapping which sends π to φ will be denoted by Φ_G, or simply by Φ. Dual Langlands L-groups that interest us are

$$\begin{aligned} {}^LSp(2n,F)^0 &= SO(2n+1,\mathbb{C}), \\ {}^LSO(2n+1,F)^0 &= Sp(2n,\mathbb{C}). \end{aligned}$$

Further, elements with the same admissible homomorphism φ should be parameterized by the equivalence classes of irreducible representations of the component group

$$\mathrm{Cent}_{{}^LG^0}(\mathrm{Im}(\varphi))/\mathrm{Cent}_{{}^LG^0}(\mathrm{Im}(\varphi))^0\ Z({}^LG^0)$$

($\mathrm{Cent}_{{}^LG^0}X$ denotes the centralizer of $X \subseteq {}^LG^0$, $(\mathrm{Cent}_{{}^LG^0}X)^0$ the connected component of the identity and $Z({}^LG^0)$ the center of ${}^LG^0$). For classical groups, component groups are commutative. Therefore, their irreducible representations are characters. We shall also use local Langlands correspondences for general linear groups, where

$${}^LGL(n,F)^0 = GL(n,\mathbb{C})$$

and where component groups are trivial.

Recall that to square integrable representation π of a classical group, we have explained how to attach the set $\mathrm{Jord}(\pi)$ of square integrable representations of general linear groups. For general linear groups, the local Langlands correspondences are known for a while (they were established by G. Laumon, M. Rapoport and U. Stuhler in the positive characteristic, and by M. Harris and R. Taylor and by G. Henniart in the characteristic 0). Therefore, we can try to apply it to $\mathrm{Jord}(\pi)$, and see what we get.

Let us go to some examples. Consider $\pi = St_{Sp(2k,F)}$, for which we have observed that $\mathrm{Jord}(\pi) = \{\delta(1_{F^\times}, 2k+1)\}$. Applying the local Langlands correspondence Φ for general linear groups to the only term of $\mathrm{Jord}(\pi)$, we get an admissible homomorphism

$$1_{W_F} \otimes E_{2k+1} : W_F \times SL(2,\mathbb{C}) \to GL(2k+1,\mathbb{C}),$$

where E_{2k+1} denotes the irreducible $2k+1$-dimensional algebraic representation of $SL(2,\mathbb{C})$. Since irreducible algebraic odd-dimensional representations of $SL(2,\mathbb{C})$ are orthogonal, the above admissible homomorphism actually goes into

$$1_{W_F} \otimes E_{2k+1} : W_F \times SL(2,\mathbb{C}) \to SO(2k+1,\mathbb{C}).$$

Note that this is exactly where the admissible homomorphism given by the local Langlands correspondence for symplectic groups should go.

Consider now irreducible square integrable subquotients of (6.7). The admissible homomorphism corresponding to them should go to $SO(5,\mathbb{C})$. The Jordan blocks here are given by (6.8). Observe that neither of the homomorphisms $\Phi(\delta(1_{F^\times},1)), \Phi(\delta(\psi,1)), \Phi(\delta(\psi,3))$ goes into $SO(5,\mathbb{C})$. On the other side, their direct sum

$$\Phi(\delta(1_{F^\times},1)) \oplus \Phi(\delta(\psi,1)) \oplus \Phi(\delta(\psi,3))$$

goes into $SO(5,\mathbb{C})$. Actually, in the third section of [Mu98], G. Muić has conjectured that the admissible homomorphism corresponding to an irreducible square integrable generic representation π should be the direct sum $\oplus\ \Phi(\sigma)$, where the sum runs over $\sigma \in \mathrm{Jord}(\pi)$.

In Theorem 1.5.1 of [Ar], J. Arthur has obtained a classification of irreducible square integrable representations of classical groups (actually, he has obtained classification of tempered representations, but it is easy to single out the square integrable ones). He has attached to an irreducible square integrable representation π of a classical group a pair of an admissible homomorphism and a character of the component group of that homomorphism.

A fundamental result of C. Mœglin (Theorem 1.3.1 of [Mœ11]) is the following:

Theorem 6.10 *The admissible homomorphism that J. Arthur has attached to square integrable representation π is*

$$\bigoplus_{\sigma\in Jord(\pi)} \Phi(\sigma). \tag{6.12}$$

The above homomorphism is discrete, i.e., its image is not contained in any proper Levi subgroup of the (complex) Langlands dual group. Using the above formula, it is equivalent to work with discrete admissible homomorphisms and Jordan blocks. In what follows, we work with Jordan blocks.

In other words, one of the parameters by which J. Arthur classifies square integrable representations gives crucial information for tempered induction in a simple way. Roughly, we can define a packet as the representations which have the same tempered reducibility properties. Despite the same tempered reducibility, the representations can be pretty different (see the examples in section 8).

We end this section (in which we were considering square integrable representations, which can be characterized as isolated points in the tempered duals) with a short note about the Speh representations (which are isolated representations in the automorphic duals). Note that we have not mentioned Speh representations in this section up to now. Nevertheless, they play a significant role in Arthur's book [Ar], which is crucial to us in this and the following sections. The formula expressing Speh representation in terms of standard modules plays important role there (see page 427 of [Ar])[31]. The formula is surprisingly simple, and we briefly present it here.

Remark 6.11 The Speh representation $u(\delta, m)$ is the unique quotient of (3.1). Write each tensor factor of the inducing representation in (3.1) as

$$|\det|_F^{(m-1)/2+k-1}\sigma = \delta\big([|\det|_F^{b_k}\rho, |\det|_F^{e_k}\rho]\big), \quad k = 1, \dots, m,$$

where ρ is an irreducible unitary cuspidal representation. Then in the Grothendieck group of the category of smooth representations we have

$$u(\sigma, m) = \det\left(\left[\delta([|\det|_F^{b_i}\rho, |\det|_F^{e_j}\rho])\right]_{1\leq i,j\leq m}\right), \tag{6.13}$$

with additional convention that if $b_i = e_j + 1$ (resp., $b_i > e_j + 1$), we drop the corresponding term $\delta([|\det|_F^{b_i}\rho, |\det|_F^{e_j}\rho])$ (resp., we take it to be 0). The multiplication showing up in the determinant is given by parabolic induction (see [Td95], [ChRn08], [Ba] or [LpMi] for more details)[32].

A formula equivalent to the above one was obtained in [Td95]. The above simple interpretation is observed in [ChRn08], which opened way for substantial simplifications of the proof (in [ChRn08], and in particular in [Ba]), and a substantial generalization in [LpMi] to the completely non-unitary setting (to the ladder representations; see [LpMi] for the definition). It is interesting to note that this very recent non-unitary generalization has shown already to be very useful even in the unitary setting. Namely, it gives explicit formulas for the derivatives and Jacquet modules of irreducible unitary representations in a simple way .

7 Cuspidals

We have briefly discussed Arthur's classification (from [Ar]) of irreducible square integrable representations of classical groups in the last section. For a number of questions, it is important to understand what exactly happens in the packets and how the tempered, square integrable and cuspidal

[31]This formula is also important for the global (and the local) Jacquet-Langlands correspondences (which are instances of functoriality; see [Ba08], [BaRn10] and [Td06])

[32]The above formula directly gives an expression for each irreducible unitary representation of a general linear group in terms of standard modules

representations are related. In the case of general linear groups, all this is solved by the Bernstein-Zelevinsky theory (see [BeZe77] and [Ze80]; some of the main results are mentioned in the previous part of this paper). For other classical groups we have very briefly discussed the relation between tempered and square integrable representations in the last section.

We shall concentrate here on the relation between irreducible cuspidal and square integrable representations. A classification of irreducible square integrable representations modulo cuspidal data is obtained in [Mœ02] and [MœTd02], i.e., modulo cuspidal representations and cuspidal reducibilities (which is related to the knowledge of Jordan blocks by (2) and (3) of Remark 6.7). We do not go into details of this classification here (which can be found in [Mœ02] and [MœTd02]). Let us say only that C. Mœglin has attached to an irreducible square integrable representation π a triple

$$(\mathrm{Jord}(\pi), \epsilon_\pi, \pi_{cusp}). \tag{7.1}$$

We have defined $\mathrm{Jord}(\pi)$ in the previous section. The partial cuspidal support π_{cusp} is defined as an (equivalence class of) irreducible cuspidal representation(s) of a classical group such that there exists a representation θ of a general linear group so that $\pi \hookrightarrow \mathrm{Ind}_P^G(\theta \otimes \pi_{cusp})$. We do not go here into the definition of the third invariant, a partially defined function ϵ_π (see [Mœ02] or [MœTd02] or the introduction of [Td11]).

These triples satisfy certain conditions (short overview can be found in the introduction of [Td11]), and triples satisfying these conditions are called admissible triples. Now admissible triples classify irreducible square integrable representations of the series of classical groups that we consider. We do not go here into the definition of admissible triples. Let us only note that admissible triples are purely combinatorial objects modulo cuspidal data. Therefore, if in Arthur's classification we can single out cuspidal representations, this will not only imply the classification of cuspidal representations, but also give cuspidal reducibilities and further imply an understanding of square integrable representations in term of cuspidal representations (giving in this way an understanding of the internal structure of packets). We now explain how C. Mœglin has described the cuspidal representations in Arthur's classification.

Consider an irreducible square integrable representation π of a classical group. Then the admissible homomorphism corresponding to π is given by (6.12). For $\sigma \in \mathrm{Jord}(\pi)$, denote by z_σ the linear mapping on the space of (6.12), acting on the space of σ as the scalar -1, and as the identity on the spaces of σ' for all other $\sigma' \in \mathrm{Jord}(\pi)$. We very often use the identification

$$\sigma \leftrightarrow z_\sigma. \tag{7.2}$$

Now, we relate the component group and its characters to Jord(π). If we consider $SO(2m+1,F)$ (resp., $Sp(2m,F)$), then the centralizer in $Sp(2m,\mathbb{C})$ (resp., in $O(2m+1,\mathbb{C})$) of the image of (6.12) is a multiplicative group consisting of elements

$$\prod_{\sigma\in\mathrm{Jord}(\pi)} z_\sigma^{a_\sigma}, \tag{7.3}$$

where $a_\sigma \in \{0,1\}$. Further, such a_σ's are uniquely determined by (7.3). Therefore, we identify (7.3) with the formal (commutative) product

$$\prod_{\sigma\in\mathrm{Jord}(\pi),\, a_\sigma=1} \sigma. \tag{7.4}$$

Observe that (7.4) determines a subset of Jord(π) in an obvious way. In this way, the above centralizer is in a natural bijection with the set $2^{\mathrm{Jord}(\pi)}$ of all subsets of Jord(π) (it is an isomorphism when we consider the operation of symmetric difference on $2^{\mathrm{Jord}(\pi)}$). This is the reason that we shall denote the centralizer by

$$2^{\mathrm{Jord}(\pi)}.$$

Observe that the characters of $2^{\mathrm{Jord}(\pi)}$ are in a natural bijection with all the functions $\mathrm{Jord}(\pi)\to\{\pm1\}$.

If we consider the special odd-orthogonal group $SO(2m+1,F)$, then $\Phi(\sigma)$'s in (6.12) are symplectic. Further, the component group is the quotient of $2^{\mathrm{Jord}(\pi)}$ by the subgroup $\{1,\prod_{\sigma\in\mathrm{Jord}(\pi)}\sigma\}$. Therefore, the characters of the component group can be identified with the characters of $2^{\mathrm{Jord}(\pi)}$ which are trivial on

$$\prod_{\sigma\in\mathrm{Jord}(\pi)} \sigma. \tag{7.5}$$

Now consider $Sp(2m,F)$. Then $\Phi(\sigma)$'s in (6.12) are orthogonal. The component group is the subgroup of $2^{\mathrm{Jord}(\pi)}$ of all elements of determinant one, i.e., it consists of all $Y\subseteq \mathrm{Jord}(\pi)$ satisfying

$$\prod_{\sigma\in Y}\det(z_\sigma)=1. \tag{7.6}$$

This implies that if $\delta(\rho,2)\in\mathrm{Jord}(\pi)$ (resp., $\delta(\rho,a),\delta(\rho,b)\in\mathrm{Jord}(\pi)$), then $\delta(\rho,2)$ (resp., $\delta(\rho,a)\delta(\rho,b)$) is in the component group (we deal here with symplectic groups).

Therefore for both series of groups, after the above identification of characters of the component group, if $\delta(\rho,2)\in\mathrm{Jord}(\pi)$ (resp., $\delta(\rho,a),\delta(\rho,b)\in\mathrm{Jord}(\pi)$), then we can evaluate the characters on elements

$$\delta(\rho,2)\quad(\text{resp., }\delta(\rho,a)\delta(\rho,b)). \tag{7.7}$$

Observe that for both series of groups, the component group (which we write multiplicatively) is isomorphic in a natural way to a vector space over $\mathbb{Z}/2\mathbb{Z}$. Therefore, we can talk about a basis of the component group (or $2^{\mathrm{Jord}(\pi)}$), having in mind this vector space structure.

Let $\delta(\rho, a) \in \mathrm{Jord}(\pi)$. We define

$$a_- = \max\{b; \delta(\rho, b) \in \mathrm{Jord}(\pi), b < a\}$$

if $\{b; \delta(\rho, b) \in \mathrm{Jord}(\pi), b < a\} \neq \emptyset$ (otherwise, a_- is not defined).

Definition 7.1 A character φ of the component group corresponding to π will be called cuspidal[33], if it holds

1. $\mathrm{Jord}(\pi)$ is without gaps;
2. $\varphi(\delta(\rho, 2)) = -1$ whenever $\delta(\rho, 2) \in \mathrm{Jord}(\pi)$;
3. $\varphi(\delta(\rho, a)\delta(\rho, a_-)) = -1$ whenever $\delta(\rho, a) \in \mathrm{Jord}(\pi)$ and a_- is defined.

Now, Theorem 1.5.1 of [Mœ11] tells

Theorem 7.2 (C. Mœglin) *An irreducible square integrable representation π is cuspidal if and only if Jord(π) is without gaps, and if the character of the component group corresponding to π is cuspidal.*

Recall that the above result is again about isolated representations, since (as we have already mentioned) irreducible cuspidal representations can be characterized as the ones that are isolated in the non-unitary dual (see [Td88]).

8 Inside Packets

If $\sigma_1, \ldots, \sigma_k$ are nonequivalent irreducible square integrable representations of general linear groups such that all $\Phi(\sigma_1), \ldots, \Phi(\sigma_k)$ are symplectic, then $\Phi(\sigma_1) \oplus \cdots \oplus \Phi(\sigma_k)$ is also symplectic.

Let $\sigma_1, \ldots, \sigma_k$ be nonequivalent irreducible square integrable representations of general linear groups. Suppose that all $\Phi(\sigma_1), \ldots, \Phi(\sigma_k)$ are orthogonal. Then $\Phi(\sigma_1) \oplus \cdots \oplus \Phi(\sigma_k)$ is orthogonal. We want to know when the image goes into the special orthogonal group. Write $\sigma_i = \delta(\rho_i, n_i)$. Then $\Phi(\sigma_i) = \Phi(\rho_i) \otimes E_{n_i}$ (recall that by E_{n_i} we denote the irreducible n_i-dimensional algebraic representation of $SL(2, \mathbb{C})$). Further, $\Phi(\rho_i)$ is orthogonal if and only if n_i is odd (and $\Phi(\rho_i)$ is symplectic if and only if n_i is

[33] C. Mœglin uses term alternate. Since the same term is used in [Mœ02] and [MœTd02] (in a slightly different setting), we have rather chosen the term cuspidal (used by G. Lusztig).

even). If $\Phi(\rho_i)$ is symplectic, then $\Phi(\rho_i) \otimes E_{n_i}$ goes into the special orthogonal group. For orthogonal $\Phi(\rho_i)$, one has $\det(\Phi(\rho_i) \otimes E_{n_i}) = \det(\Phi(\rho_i))$ (by this, we simply mean that the determinant is trivial on E_{n_i}). By the properties of Φ, we know $\det(\Phi(\rho_i)) = \Phi(\omega_{\rho_1})$. Denote by

$$X$$

the set of all i such that $\Phi(\rho_i)$ is orthogonal. Therefore, $\Phi(\sigma_1) \oplus \cdots \oplus \Phi(\sigma_k)$ goes into the special orthogonal group if and only if $\prod_{i\in X} \Phi(\omega_{\rho_i}) \equiv 1$, which is equivalent to $\Phi(\prod_{i\in X} \omega_{\rho_i}) \equiv 1$, which is further equivalent to

$$\prod_{i\in X} \omega_{\rho_i} \equiv 1. \tag{8.1}$$

The next question that we try to explain is how to get the structure of a general packet. A packet is determined by its Jordan blocks (recall (6.12)). Therefore, in the symplectic (resp., orthogonal) case, we choose a finite set of non-equivalent orthogonal (resp., symplectic) irreducible square integrable representations of general linear groups, which we denote by

$$\mathrm{Jord}.$$

In the symplectic case, we also require that condition (8.1) be satisfied. It is easy to write all the characters of the component group. We fix one such character (see the previous section), and denote it by

$$\varphi.$$

Now to get elements of the packet, it is enough to know how to attach the corresponding square integrable representation to a character. We do this following [MœTd02] [34]. The representation that will be attached recursively in this way to the pair (Jord, φ) below, will be denoted by

$$\lambda_{\mathrm{Jord},\varphi}.$$

Recursive construction:

(1) If the character φ is cuspidal (see Definition 7.1), then we take $\lambda_{\mathrm{Jord},\varphi}$ to be the cuspidal representation attached by Arthur to (Jord, φ) (see Theorem 6.10).

In what follows, we assume that φ is not cuspidal.

[34] We do not know if at the moment it is proved in general that the representation that we attach is the same as the one that Arthur attaches. C. Mœglin has a proof in [Mœ07] for the unitary groups.

(2) Suppose that there exists some $\delta(\rho, a) \in \mathrm{Jord}$ for which a_- is defined, and also that

$$\varphi(\delta(\rho, a)\delta(\rho, a_-)) = 1. \tag{8.2}$$

Set $\mathrm{Jord}' = \mathrm{Jord}\backslash\{\delta(\rho, a), \delta(\rho, a_-)\}$. The character φ defines a character of Jord' in a natural way (by restriction), which we denote by φ'. Denote by π' the square integrable representation attached recursively to $(\mathrm{Jord}', \varphi')$, i.e., $\pi' = \lambda_{\mathrm{Jord}', \varphi'}$. Now the representation

$$\mathrm{Ind}_P^G(\delta([|\det|_F^{-(a_- -1)/2}\rho, |\det|_F^{(a-1)/2}\rho]) \otimes \pi')$$

has precisely 2 irreducible subrepresentations. Denote them by π_1 and π_2. They are not equivalent, and one of them corresponds to φ. We need to specify which one.

(a) Suppose that there exists $\delta(\rho, b) \in \mathrm{Jord}'$ such that $b_- = a$ (b_- is considered with respect to Jord). We attach to (Jord, φ) the representation π_i which embeds[35] into a representation of the form

$$\mathrm{Ind}_P^G(\delta([|\det|_F^{(a-1)/2+1}\rho, |\det|_F^{(b-1)/2}\rho]) \otimes \tau)$$

if and only if $\varphi(\delta(\rho, b)\delta(\rho, a)) = 1$.

(b) Suppose that there exists $\delta(\rho, b) \in \mathrm{Jord}'$ such that $b = (a_-)_-$ ($(a_-)_-$ is considered with respect to Jord). We attach to (Jord, φ) the representation π_i which embeds into a representation of the form

$$\mathrm{Ind}_P^G(\delta([|\det|_F^{(b-1)/2+1}\rho, |\det|_F^{(a_- -1)/2}\rho]) \otimes \tau)$$

if and only if $\varphi(\delta(\rho, b)\delta(\rho, a_-)) = 1$.

It remains to consider the case when we have no $\delta(\rho, b)$ in Jord'.

(c) If a is even, then we attach to (Jord, φ) the representation π_i which embeds into a representation of the form

$$\mathrm{Ind}_P^G(\delta([|\det|_F^{1/2}\rho, |\det|_F^{(a_- -1)/2}\rho]) \otimes \tau)$$

if and only if $\varphi(\delta(\rho, a)) = 1$.

(d) Suppose that a is odd. Then $\mathrm{Ind}_{P'}^{G'}(\rho \otimes \pi_{cusp})$ reduces. In [Mœ02] and [MœTd02] an indexing of the irreducible subrepresentations is fixed: $\mathrm{Ind}_{P'}^{G'}(\rho \otimes \pi_{cusp}) = \tau_1 \oplus \tau_{-1}$ (when π_{cusp} is generic, we

[35] Instead of the embedding requirement here and below, thanks to [Jn] or [Td+] we can use a Jacquet module requirement (in this case, the Jacquet module requirement is that $\delta([|\det|_F^{(a-1)/2+1}\rho, |\det|_F^{(b-1)/2}\rho]) \otimes \tau$ is a subquotient of the appropriate Jacquet module of π_i).

shall always take τ_1 to be generic). We attach to (Jord, φ) the representation π_i determined by the fact that it embeds into

$$\mathrm{Ind}_P^G(\theta \otimes \delta([|\det|_F\rho, |\det|_F^{(a-1)/2}\rho]) \otimes \tau_1)$$

for some irreducible representation θ of a general linear group, if and only if $\varphi(\delta(\rho, a)) = 1$.

In what follows, we can assume that (8.2) *does not occur for* φ.

(3) Suppose that Jord has gaps. Then there exists $\delta(\rho, a) \in \mathrm{Jord}$, $a \geq 3$ and $k \in \mathbb{Z}_{>0}$ such that $\delta(\rho, b) \notin \mathrm{Jord}$ for any $b \in [a-2k, a-2]$. Set $\mathrm{Jord}' = \mathrm{Jord}\backslash\{\delta(\rho, a)\} \cup \{\delta(\rho, a-2k)\}$. Define the character φ' of the component group of Jord' in a natural way from φ (putting $\delta(\rho, a-2k)$ instead of $\delta(\rho, a)$). Let π' be the representation recursively attached to $(\mathrm{Jord}', \varphi')$. Then the representation

$$\mathrm{Ind}_P^G(\delta([|\det|_F^{(a-2k+1)/2}\rho, |\det|_F^{(a-1)/2}\rho]) \otimes \pi')$$

contains a unique irreducible subrepresentation. This subrepresentation is square integrable, and we attach this subrepresentation to (Jord, φ) (this is compatible[36] with the construction in [MœTd02] by Theorem 8.2 of [Td+], or by Corollary 2.1.3 of [Jn]).

In what follows, we can assume that Jord *has no gaps.*

(4) The only possibility which remains is that we have some $\delta(\rho, 2) \in \mathrm{Jord}$ such that $\varphi(\delta(\rho, 2)) = 1$. Set $\mathrm{Jord}' = \mathrm{Jord}\backslash\{\delta(\rho, 2)\}$. Define the character φ' of the component group of Jord' in a natural way from φ (restricting). Let π' be the representation recursively attached to $(\mathrm{Jord}', \varphi')$. Then the representation

$$\mathrm{Ind}_P^G(|\det|_F^{1/2}\rho \otimes \pi')$$

has a unique irreducible subrepresentation. This subrepresentation is square integrable, and we attach this subrepresentation to (Jord, φ) (this is compatible with the construction in [MœTd02] by Lemma 9.1 in the appendix of this paper, or by Lemma 3.2.1 of [Jn]).

To illustrate what packets look like, we consider some very simple examples, first in the case of special odd-orthogonal groups, and later in the

[36]This means that restricting φ, we get the partially defined function attached to the representation in [Mœ02]

case of symplectic groups. We pay special attention to the packets simultaneously containing both cuspidal representations as well as representations supported by the minimal parabolic subgroup[37]. Such packets will be called packets with antipodes (cuspidal representations in such packets have simple parameters, and simple cuspidal reducibilities ≥ 1).

Below, we call packets containing an Iwahori-spherical representation simply Iwahori packets (clearly, in such a packet Iwahori-spherical representations are precisely the ones supported by the minimal parabolic subgroup).

The simple examples below are presented to give a flavor of what can happen in the packets: we can have representations supported on a number of different parabolic subgroups (often including a relatively small number of cuspidal representations when Jordan blocks are without gaps), and we can have also packets containing only cuspidal representations.

Example 8.1 In this example we consider packets for special odd-orthogonal groups.

1. Let ψ_1, ψ_2 be different characters of $F^\times$ satisfying $\psi_i^2 \equiv 1$ for $i = 1, 2$. Consider the packet determined by Jordan blocks

$$\text{Jord} = \{\delta(\psi_1, 2),\ \delta(\psi_2, 2)\}.$$

Obviously, the Jordan blocks are without gaps. We shall write a character of the centralizer 2^{Jord} as

$$\varphi_{(\epsilon_1, \epsilon_2)},$$

which means that this character sends the first element of the basis to ϵ_1 and the second one to ϵ_2 (we shall use this notation for characters in what follows, whenever a basis is fixed). If a character $\varphi_{(\epsilon_1,\epsilon_2)}$ of 2^{Jord} is a character of the component group, it must be trivial on $\delta(\psi_1, 2)\delta(\psi_2, 2)$. Therefore, we have two characters of the component group. One of them, $\varphi_{(-1,-1)}$, is a cuspidal character. The corresponding cuspidal representation we have denoted by

$$\lambda_{\{\delta(\psi_1,2),\delta(\psi_2,2)\},\varphi_{(-1,-1)}}$$

(here we apply step (1) of the recursive construction).

[37]i.e., which are subquotients of representations parabolically induced from the minimal parabolic subgroups.

It remains to consider the trivial character $\varphi_{(1,1)}$. To construct the corresponding representation, we apply step (4) of the recursive construction two times. In this way, we get that the corresponding representation is the unique irreducible subrepresentation of

$$\mathrm{Ind}_{P_\emptyset}^{SO(5,F)}(|\ |_F^{1/2}\psi_1 \otimes |\ |_F^{1/2}\psi_2),$$

which is also the unique irreducible square integrable subquotient of the above representation.

2. Let ψ_1, ψ_2, ψ_3 be different characters of $F^\times$ satisfying $\psi_i^2 \equiv 1$ for $i = 1, 2, 3$. Consider the packet determined by Jordan blocks

$$\delta(\psi_1, 2),\ \delta(\psi_2, 2),\ \delta(\psi_3, 2).$$

Evidently, Jordan blocks are without gaps but we do not have cuspidal characters, since the character $\varphi_{(-1,-1,-1)}$ is not trivial on $\delta(\psi_1, 2)\delta(\psi_2, 2)\delta(\psi_3, 2)$.

One gets three elements of the packet corresponding to the non-trivial characters in the following way (applying step (4) of Recursive construction once for each of these representations). Choose $i \in \{1, 2, 3\}$, and denote remaining two indices by $j_{i,1}$ and $j_{i,2}$. Now

$$\mathrm{Ind}_P^{SO(7,F)}(|\ |_F^{1/2}\psi_i \otimes \lambda_{\{\delta(\psi_{j_{i,1}},2),\delta(\psi_{j_{i,2}},2)\},\varphi_{(-1,-1)}}) \tag{8.3}$$

contains a unique irreducible square integrable subquotient. One gets the character corresponding to (8.3) from $\varphi_{(-1,-1,-1)}$ by putting 1 instead of -1 at i-th place.

Applying step (4) of the recursive construction, and the previous example, we get that an irreducible square integrable subquotient of

$$\mathrm{Ind}_{P_\emptyset}^{SO(7,F)}(|\ |_F^{1/2}\psi_1 \otimes |\ |_F^{1/2}\psi_2 \otimes |\ |_F^{1/2}\psi_3) \tag{8.4}$$

corresponds to the trivial character $\varphi_{(1,1,1)}$. Further, (8.4) contains a unique irreducible square integrable subquotient (it is the unique irreducible subrepresentation).

3. Take irreducible cuspidal representations $\rho_1, \dots, \rho_k$ of general linear groups. Suppose that all $\Phi(\rho_1), \dots, \Phi(\rho_k)$ are symplectic and non-equivalent. Then $\Phi(\rho_1) \oplus \cdots \oplus \Phi(\rho_k)$ is a discrete admissible homomorphism. The corresponding set of Jordan blocks is $\{\rho_1, \dots, \rho_k\}$. Observe that these Jordan blocks do not have gaps, and each character of the component group is cuspidal. Therefore, here one gets a packet consisting of 2^{k-1} cuspidal representations (here we apply only step (1) of the recursive construction).

4. If a packet (of $SO(2n+1,F)$) contains a representation supported by the minimal parabolic subgroup, then each element in the Jordan blocks of such packet has the form $\delta(\psi,2k)$ for some character ψ of $F^\times$ satisfying $\psi^2\equiv 1$, and some $k\in\mathbb{Z}_{\geq 1}$. The converse also holds (consider the trivial character of the component group).
5. We now consider a more general packet then the one in (1). Let ψ_1,ψ_2 be different characters of $F^\times$ satisfying $\psi_i^2\equiv 1$, $i=1,2$, and k a positive integer. Consider the packet of $SO(2k(k+1)+1,F)$ determined by the Jordan blocks
$$\{\delta(\psi_1,2i),\delta(\psi_2,2i); i=1,2,\ldots,k\}. \tag{8.5}$$
This packet obviously has one cuspidal representation. This cuspidal representation is not generic, which follows from [Sh92] since the above representation has two cuspidal reducibilities >1 (they correspond to ψ_1 and ψ_2; both reducibilities are at $k+1/2$).
6. Suppose that $SO(2n+1,F)$ has an Iwahori packet with antipodes. Then the fact that Jordan blocks of packets containing cuspidal representations have no gaps, implies that $2n=k_1(k_1+1)+k_2(k_2+1)$ for some $k_1,k_2\in\mathbb{Z}_{\geq 0}$ (i.e., n is a sum of two triangular numbers). Such a packet without gaps contains a cuspidal representation if and only if $\lfloor (k_1+1)/2\rfloor+\lfloor (k_2+1)/2\rfloor\in 2\mathbb{Z}$, where here $\lfloor x\rfloor$ denotes the largest integer not exceeding x.

Very often we shall consider a slightly different problem then the above one: for a given irreducible square integrable representation π, describe the remaining representations of the packet to which π belongs. To solve this, one possibility is to find the Jordan blocks of π, and apply the recursive construction. Proposition 2.1 of [MœTd02] is useful for this (see also Proposition 3.1 of [Td+]). We do not go into the details here.

Example 8.2 In this example we consider packets for symplectic groups.

1. We consider the packet given by Jordan blocks
$$\mathrm{Jord}=\{\delta(1_{F^\times},1),\delta(1_{F^\times},3),\delta(1_{F^\times},5)\}.$$
For a basis B of the component group we can take
$$\delta(1_{F^\times},1)\delta(1_{F^\times},3),\ \delta(1_{F^\times},3)\delta(1_{F^\times},5).$$
As before, we write characters of the component group as $\varphi_{(\epsilon_1,\epsilon_2)}$, which means that this character sends the first element of the basis to ϵ_1 and the second one to ϵ_2.

To attach representations to the characters $\varphi_{(\pm 1,1)}$, we apply step (2) of the recursive construction: the representation

$$\mathrm{Ind}_{P_\emptyset}^{Sp(8,F)}(\delta([|\ |_F^{-1}, |\ |_F^{2}]) \otimes 1_{Sp(0,F)})$$

has two irreducible subrepresentations. They are square integrable. Only one of them has a subquotient of the form $|\ |_F \otimes *$ in its Jacquet module. The character corresponding to this one is $\varphi_{(1,1)}$, while $\varphi_{(-1,1)}$ corresponds to the other subrepresentation (here, to attach characters to representations, we have applied (b) of step (2) in the recursive construction).

To attach representations to the characters $\varphi_{(1,\pm 1)}$, we consider the representation $\mathrm{Ind}_P^{Sp(8,F)}(\delta([|\ |_F^{0}, |\ |_F]) \otimes St_{Sp(4,F)})$. This representation has two irreducible subrepresentations (here we have applied step (2) and then step (3) of the recursive construction). They are square integrable. Only one of them does not have subquotient of the form $|\ |_F^2 \otimes *$ in its Jacquet module. The character corresponding to this one is $\varphi_{(1,-1)}$ (here, to attach character to the representation, we have applied (a) of step (2) in the recursive construction).

The fourth representation in the packet corresponds to the character $\varphi_{(-1,-1)}$. This character is cuspidal, so the corresponding representation is cuspidal (here we apply step (1) of the recursive construction). We comment on this representation later.

Observe that the principal series

$$\mathrm{Ind}_{P_\emptyset}^{Sp(8,F)}(|\ |_F^{-1} \otimes |\ |_F^{0} \otimes |\ |_F^{1} \otimes |\ |_F^{2})$$

has exactly 3 irreducible non-equivalent square integrable subquotients.

2. The following example is related to (6.8). We shall consider the packet determined by Jordan blocks given by $\delta(1_{F^\times}, 1), \delta(\psi, 1), \delta(\psi, 3)$ (i.e., by (6.8)). For the basis B of the component group, we can take

$$\delta(1_{F^\times}, 1)\delta(\psi, 1),\ \ \delta(\psi, 1)\delta(\psi, 3).$$

Now the square integrable subquotients of (6.7) correspond to the characters $\varphi_{(\pm 1,1)}$ by step (2) of the recursive construction ($\varphi_{(1,1)}$ corresponds to the generic one by (d) of step (2) of the recursive construction). The cuspidal characters are $\varphi_{(\pm 1,-1)}$. Thus, here we have two cuspidal representations in the packet (we apply step (1) of the recursive construction).

3. Let ρ be an irreducible selfdual cuspidal representation of $GL(2,F)$ with trivial central character[38]. We now consider the packet given by Jordan blocks

$$\delta(1_{F^\times},1),\ \delta(\rho,2).$$

For the basis B of the component group, we can take

$$\delta(\rho,2).$$

Then $\mathrm{Ind}_P^{Sp(4,F)}(|\det|_F^{1/2}\rho)$ contains precisely one irreducible square integrable subquotient. It is the unique irreducible subrepresentation. The above square integrable representation corresponds to the trivial character (by step (3) of the recursive construction).

There is one cuspidal character, denoted by $\varphi_{(-1)}$ (it sends $\delta(\rho,2)$ to -1). The corresponding cuspidal representation is denoted by $\lambda_{\{\delta(\rho,2)\},\varphi_{(-1)}}$ (this is step (1) of the recursive construction). Then

$$\mathrm{Ind}^{Sp(8,F)}(|\det|_F^{3/2}\rho\otimes\lambda_{\delta(\rho,2),\varphi_{(-1)}}) \tag{8.6}$$

reduces.

4. Let ρ be an irreducible selfdual cuspidal representation of $GL(2,F)$ with trivial central character (as in (3)). Consider Jordan blocks

$$\delta(1_{F^\times},1),\ \delta(\rho,2),\ \delta(\rho,4).$$

For the basis B of the component group, we can take

$$\delta(\rho,2),\ \delta(\rho,4).$$

Then $\mathrm{Ind}^{Sp(12,F)}(\delta([|\det|_F^{-1/2}\rho,|\det|_F^{3/2}\rho]))$ contains precisely two irreducible square integrable subquotients. Moreover, they are the unique irreducible subrepresentations. They correspond to the characters $\varphi_{(1,1)}$ and $\varphi_{(-1,-1)}$. Only one of these representations has a subquotient of the form $|\det|_F^{1/2}\rho\otimes *$ in its Jacquet module, and this one corresponds to $\varphi_{(1,1)}$ (here we have applied step (2) of the recursive construction, and then (c) in that step).

Further $\varphi_{(-1,1)}$ is the only cuspidal character here. It corresponds to a cuspidal representation (here we apply step (1) of the recursive construction).

[38]For this and the following example, we could take ρ to be any irreducible cuspidal representation of $GL(k,F)$ such that $\Phi(\rho)$ is symplectic.

The fourth element of the packet, corresponding to the character $\varphi_{(1,-1)}$, is the unique irreducible subrepresentation of

$$\mathrm{Ind}_P^{Sp(12,F)}(|\det|_F^{1/2}\rho \otimes |\det|_F^{3/2}\rho \otimes \lambda_{\{\delta(\rho,2)\},\varphi_{(-1)}}). \qquad (8.7)$$

This representation is not only square integrable, but moreover strongly positive (in the terminology of [MœTd02]). Here we have first applied step (4) of the recursive construction, then step (3) which brought us to the cuspidal character, where we apply step (1).

5. Let $\rho_1, \ldots, \rho_l$ be non-equivalent irreducible cuspidal representations of general linear groups, such that all $\Phi(\rho_1), \ldots, \Phi(\rho_l)$ are orthogonal and $\prod_{i=1}^{l} \omega_{\rho_i} \equiv 1$. Then $\Phi(\rho_1) \oplus \cdots \oplus \Phi(\rho_l)$ is an admissible homomorphism. The corresponding set of Jordan blocks is $\{\rho_1, \ldots, \rho_l\}$. These Jordan blocks do not have gaps, and each character of the component group is cuspidal. Therefore, here one gets a packet consisting of 2^{l-1} cuspidal representations.
6. Denote by ψ_{un} the unramified character of $F^\times$ of order 2. Suppose that $2n+1$ is a sum of two squares. Then one of them must be even and the other odd. Denote them by n_e and n_o, respectively. Then

$$\{\delta(1_{F^\times}, 2i-1); i = 1, 2, \ldots, n_o\} \cup \{\delta(\psi_{un}, 2i-1); i = 1, 2, \ldots, n_e\}$$

are parameters of a packet of $Sp(2n, F)$. This packet contains two cuspidal representations. Further, the representation corresponding to the trivial character of the component group is supported on the minimal parabolic subgroup, and it is Iwahori-spherical. Thus, this is an Iwahori packet with antipodes.

From this, it follows that $Sp(2n, F)$ has an Iwahori packet with antipodes if and only if $2n+1$ is a sum of two squares.
7. If a packet (of $Sp(2n, F)$) contains a representation supported on the minimal parabolic subgroup, then all the Jordan blocks of such a packet have the form $\delta(\psi, 2k-1)$ for some characters ψ of $F^\times$ satisfying $\psi^2 \equiv 1$, and some $k \in \mathbb{Z}_{\geq 1}$.

The converse does not hold. For example, take any three characters χ_1, χ_2, χ_3 of $F^\times$ or order two such that $\chi_1\chi_2\chi_3 = 1_{F^\times}$, and take any three odd positive integers k_1, k_2, k_3 satisfying $k_1 + k_2 + k_3 = 2n+1$. Then the packet of $Sp(2n, F)$ determined by Jordan blocks $\delta(\chi_1, k_1), \delta(\chi_2, k_2), \delta(\chi_3, k_3)$ consists of 4 representations. All these representations are supported on parabolic subgroups whose Levi factors are isomorphic to $GL(1, F)^{n-1} \times Sp(2, F)$.

Related to the above discussion, consider an irreducible cuspidal representation ρ of $GSp(2, F) = GL(2, F)$ which splits into 4 pieces after restriction to $Sp(2,F) = SL(2,F)$. Then a packet of $Sp(2,F)$ consists of all irreducible pieces of the restriction $\rho|_{Sp(2,F)}$. The Jordan blocks are three non-trivial characters ψ_1, ψ_2, ψ_3, characterized by the condition $(\psi_i \circ \det)\rho \cong \rho$ (all this follows directly from [Wa87] and Theorem 6.1). One can now consider the packet of $Sp(2n,F)$ determined by Jordan blocks $\delta(\psi_1,k_1), \delta(\psi_2,k_2), \delta(\psi_3,k_3)$ and easily describe its elements in terms of the elements of the packet of $Sp(2,F)$ considered above.

Some questions related to the packets arise naturally. One question is to determine, for some irreducible cuspidal representations obtained or constructed by other methods, the packets to which they belong.

From the other side, we have seen in the above examples that starting from very simple Jordan blocks, we can have cuspidal representations in the packet. In general, such cuspidal representations will be degenerate. Also, they will be rare in the packet. The question is, can one describe (at least some of) those representations in a different way? Clearly, these two questions are related. The Howe correspondences are a great source of representations for the second question. Related to this, we give some simple examples.

Let us first go to the packet in (1) of Example 8.2 (which has very simple Jordan blocks). Using the Howe correspondence for $Sp(8,F)$ and $O(Y)$, where Y is a totally anisotropic orthogonal space of dimension 4, C. Mœglin has obtained from the signum character of $O(Y)$ an irreducible cuspidal representation σ of $Sp(8,F)$. She has also gotten that $\mathrm{Ind}_P^{Sp(10,F)}(|\ |_F^3 \otimes \sigma)$ reduces (considering the Howe correspondence for $Sp(10,F)$ and $O(Y)$). Now (3) of Remark 6.7 and Theorem 6.10 imply directly that σ is the cuspidal representation which is in the packet in (1) of Example 8.2.

Consider for a moment the packet of $SO(2k(k+1)+1,F)$ in Example 8.1, (5), when one takes ψ_1 and ψ_2 unramified (then one of them is $1_{F^\times}$ and the other one is the unramified signum character; this is an Iwahori packet). Clearly, it is natural to expect that the cuspidal unipotent $SO(2k(k+1)+1,F)$-representation of G. Lusztig belongs to this packet.

Now consider the packet in Example 8.2, (6), when one takes $2n+1$ to be the sum of two consecutive squares k^2 and $(k+1)^2$ (then n is twice a triangular number). We get an Iwahori packet of $Sp(2k(k+1),F)$. Again, it is natural to expect that the cuspidal unipotent $Sp(2k(k+1),F)$-representation of G. Lusztig belongs to this packet.

Recall that the existence of Iwahori packets with antipodes for special odd-orthogonal (resp., symplectic) groups is related to the sums of triangular numbers (resp., sums of squares of integers). We have discussed above only the case of two equal triangular numbers (resp., sum of two consecutive squares). It is interesting to find other descriptions of the cuspidal representations in the remaining Iwahori packets. C. Mœglin's construction, which we have discussed above, gives a description of such a representation of $Sp(8, F)$ (where 9 is the sum of 3^2 and 0^2).

These are only some simple questions which arise related to the packets that we have considered.

Remark 8.3 Irreducible representations of compact Lie groups are classified by highest weights ([We68]). Other information about these representations may be obtained from the corresponding highest weight (dimension, character, representation itself, etc.). Similarly, for the irreducible square integrable representations of classical p-adic groups, using their parameters discussed at the beginning of this section, it would be interesting to get other relevant information about representations, in particular, for the representations with simple parameters. For this, other descriptions of the representations might be useful.

We end this section with a simple result:

Proposition 8.4 *The group $SO(2n+1, F)$ has a packet with antipodes if and only if n is even.*

Suppose that the residual characteristic of F is odd. Then $Sp(2n, F)$ has a packet with antipodes if and only if n is even[39].

Proof First we consider special odd orthogonal groups.

Let X be a packet with antipodes of $SO(2n+1, F)$. Since the packet contains a representation supported on the minimal parabolic subgroup, all elements of the Jordan blocks must be of the form $\delta(\psi, k)$, where ψ is a character of $F^\times$ satisfying $\psi^2 \equiv 1$ and k is even. Let ψ_i, $i = 1, \dots, m$, be all such different characters that show up in Jordan blocks, and let $k_i = \max\{l; \delta(\psi_i, 2l) \in X\}$. Since we have a cuspidal representation in the packet, there are no gaps. Therefore

$$X = \cup_{i=1}^{m} \{\delta(\psi_i, 2i); i = 1, \dots, k_i\} \tag{8.8}$$

and thus

$$\sum_{i=1}^{m} k_i(k_i + 1) = 2n. \tag{8.9}$$

Observe that there can be only one cuspidal character, and it must be trivial on (7.5). This implies

$$\sum_{i=1}^{m} \lfloor (k_i + 1)/2 \rfloor \in 2\mathbb{Z}, \tag{8.10}$$

[39] Odd residual characteristic is used only for the implication $\Longrightarrow$.

where, as before, $\lfloor x \rfloor$ denotes the largest integer not exceeding x. Observe that

$$\lfloor (k_i+1)/2 \rfloor \in 1+2\mathbb{Z} \iff k_i \equiv 1,2 \pmod 4 \iff k_i(k_i+1) \in 2+4\mathbb{Z}. \tag{8.11}$$

From the other side, if we have (different) ψ_i satisfying $\psi_i^2 \equiv 1$, and $k_i \in \mathbb{Z}_{\geq 1}$ satisfies (8.9) and (8.10), then if we define X by (8.8), we get a packet with antipodes.

From the above considerations, we see that in the case of packet with antipodes, (8.10), (8.11) and (8.9) imply that n must be even.

Suppose now that n is even. Then $n-1$ is odd. Recall that we have at least 4 different characters ψ satisfying $\psi^2 \equiv 1$. Denote them by $\psi_i, i = 1,\dots,4$. A classical result of Gauss says that $n-1$ is a sum of three triangular numbers, i.e., $2(n-1) = \sum_{i=1}^3 k_i(k_i+1)$. Now since we have proved that $SO(2(n-1)+1,F)$ does not have packets with antipodes, (8.10) cannot hold. Thus $\sum_{i=1}^3 \lfloor (k_i+1)/2 \rfloor \in 1+2\mathbb{Z}$. Now taking $k_4 = 1$, we get

$$\sum_{i=1}^4 k_i(k_i+1) = 2n \ \text{ and } \ \sum_{i=1}^4 \lfloor (k_i+1)/2 \rfloor \in 2\mathbb{Z}.$$

The above discussion now implies that $SO(2n,F)$ has a packet with antipodes.

It remains to consider the symplectic groups. Observe that we have now 4 characters ψ satisfying $\psi^2 \equiv 1$ (because the residual characteristic is odd).

Let X be a packet with antipodes of $Sp(2n,F)$. Since the packet contains a representation supported on the minimal parabolic subgroup, all elements of the Jordan blocks are of the form $\delta(\psi,k)$, where ψ is a character of $F^\times$ satisfying $\psi^2 \equiv 1$ and k is odd. Let ψ_i, $i = 1,\dots,m$, be all such characters that show up in the Jordan blocks, and let $k_i = \max\{l; \delta(\psi_i, 2l-1) \in X\}$. Since we have a cuspidal representation in the packet, there are no gaps. Therefore

$$X = \cup_{i=1}^m \{\delta(\psi_i, 2l-1); l = 1,\dots,k_i\}, \tag{8.12}$$

which implies

$$\sum_{i=1}^m k_i^2 = 2n+1. \tag{8.13}$$

Now, condition (7.6) tells us that

$$\prod_{i=1}^m \psi_i^{k_i} \equiv 1. \tag{8.14}$$

This can happen exactly in two ways:

1. if $\psi_i \neq 1$, then k_i is even;
2. if $\psi_i \neq 1$, then k_i is odd, and all the three non-trivial characters show up as ψ_i's.

We now show that (2) cannot happen. Suppose that 2. holds. Let φ be a character of the component group which corresponds to a representation supported on the minimal parabolic subgroup. Now perform the step (2) of the recursive construction as long as possible. We come to a packet of a (possibly smaller) group, where all three non-trivial ψ still show up. Now we perform step (3) as long as possible. We shall come to a packet of a (possibly smaller) group, where still all the three non-trivial ψ will show up (which implies that it is a packet of some $Sp(2\ell, F)$ with $\ell \geq 1$), but without gaps. The character that we get in this way must be cuspidal. The representation corresponding to this character is cuspidal. This implies that the representation corresponding to the initial character cannot be supported on the minimal parabolic subgroup. This finishes the proof of the claim.

Therefore, (1) holds for the packets with antipodes. Because of this and (8.13), $1_{F^\times}$ must always show up in X. We denote $1_{F^\times}$ by ψ_1. Then k_1 must be odd.

Now (8.13) implies that n must be even.

It remains to show that for each even n, we can find a packet with antipodes. The above discussion implies that for this, it is enough to show that for each $l \in \mathbb{Z}_{\geq 0}$ we can find $k_1 \in 1 + 2\mathbb{Z}_{\geq 0}$ and $k_2, k_3, k_4 \in 2\mathbb{Z}_{\geq 0}$ such that $4l + 1 = \sum_{i=1}^{4} k_i^2$. To prove this, it is enough to show that for each $l \in \mathbb{Z}_{\geq 0}$ we can find $m_1, \ldots, m_4 \in \mathbb{Z}_{\geq 0}$ such that

$$l = m_1(m_1 + 1) + \sum_{i=2}^{4} m_i^2. \tag{8.15}$$

If l is not of the form $4^a(8b + 7)$, then a classical result of Gauss tells us that we can do this with $m_1 = 0$. If l is of the form $4^a(8b + 7)$, then we take $m_1 = 1$ and apply the above classical result to $l - 2$, which is now not of the form $4^a(8b + 7)$. This again gives the representation (8.15). This completes the proof of the proposition.

□

9 Appendix

A lemma that we prove in this appendix is essentially Lemma 3.2.1 of [Jn]. It covers a case not covered by Lemma 7.1 of [Td+]. The proof that we include here uses methods (and notation) of the proof of Lemma 7.1 of [Td+]. We need this simple result to know that the last step of the construction of elements in packets described in the previous section is compatible with [MœTd02]. The claim is about partially defined functions. We do not recall their definition here (one can find it in [MœTd02]).

Lemma 9.1 *Let π be an irreducible square integrable representation of a classical group. Suppose $\delta(\rho,2) \in Jord(\pi)$ and $\epsilon_\pi(\delta(\rho,2)) = 1$. Then there exists an irreducible representation π' of a classical group of the same series, such that*

$$\pi \hookrightarrow Ind(|det|_F^{1/2}\rho \otimes \pi'). \tag{9.1}$$

Further, any such π' is square integrable and $Jord(\pi') = Jord(\pi)\backslash\{\delta(\rho,2)\}$. The representation $Ind(|det|_F^{1/2}\rho\otimes\pi')$ has a unique irreducible subrepresentation and one gets $\epsilon_{\pi'}$ from ϵ_π by restriction.

Proof The definition of $\epsilon_\pi(\delta(\rho,2)) = 1$ implies that there exists an embedding of type (9.1). Further, π' is square integrable by Remark 3.2 of [Mœ02]. Now $\mathrm{Jord}(\pi') = \mathrm{Jord}(\pi)\backslash\{\delta(\rho,2)\}$ by (i) in Proposition 2.1 of [MœTd02].

The fact that the representation $\mathrm{Ind}(|\mathrm{det}|_F^{1/2}\rho \otimes \pi')$ has a unique irreducible subrepresentation follows directly applying the structure formula of [Td95+] (Theorems 5.4 and 6.4 there), using $\delta(\rho,2) \notin \mathrm{Jord}(\pi')$ and Lemma 3.6 of [MœTd02] (one shows that $|\mathrm{det}|_F^{1/2}\rho \otimes \pi'$ has multiplicity one in the Jacquet module of $\mathrm{Ind}(|\mathrm{det}|_F^{1/2}\rho\otimes\pi')$, and then apply Frobenius reciprocity).

The proof that one gets $\epsilon_{\pi'}$ by restricting ϵ_π can be more or less extracted from the proof of Lemma 8.1 of [Td+]. We explain this below.

Let $\delta(\rho',c) \in \mathrm{Jord}(\pi')$ and suppose that c_- is defined. Then (C) of the proof of Lemma 8.1 of [Td+] proves also that $\epsilon_\pi(\delta(\rho',c)\delta(\rho',c_-)) = \epsilon_{\pi'}(\delta(\rho',c)\delta(\rho',c_-))$ (one needs to take $a=2$ there).

Let $\delta(\rho',c) \in \mathrm{Jord}(\pi')$. Suppose that c is odd and that $\epsilon_{\pi'}(\delta(\rho',c))$ is defined (then $\rho \not\cong \rho'$). Let b be the maximal element among such c's for this ρ'. Now (F) of the proof of Lemma 8.1 of [Td+] also proves that $\epsilon_\pi(\delta(\rho',b)) = \epsilon_{\pi'}(\delta(\rho',b))$ (again, one takes $a=2$ there).

Let $\delta(\rho',c) \in \mathrm{Jord}(\pi')$ and suppose that c is even. Let b be the minimal element among such c's (for this ρ').

First, consider the case $\rho \not\cong \rho'$. Then (A) of the proof of Lemma 8.1 of [Td+] also proves that $\epsilon_\pi(\delta(\rho',b)) = \epsilon_{\pi'}(\delta(\rho',b))$ (one takes $a=2$ there).

It remains to consider the case $\rho \cong \rho'$. Suppose $\epsilon_{\pi'}(\delta(\rho,b)) = 1$. Then

$$\pi' \hookrightarrow \mathrm{Ind}(\delta([|\mathrm{det}|_F^{1/2}\rho, |\mathrm{det}|_F^{(b-1)/2}\rho]) \otimes \tau)$$

for some τ, which implies

$$\pi \hookrightarrow \mathrm{Ind}(\delta([|\det|_F^{1/2}\rho, |\det|_F^{(b-1)/2}\rho]) \otimes |\det|_F^{1/2}\rho \otimes \tau).$$

This implies $\epsilon_\pi(\delta(\rho, b)) = 1$.

Suppose now $\epsilon_\pi(\delta(\rho, b)) = 1$. Then $\epsilon_\pi(\delta(\rho, b)\delta((\rho, 2)) = 1$. This implies that

$$\pi \hookrightarrow \mathrm{Ind}(\delta([|\det|_F^{-1/2}\rho, |\det|_F^{(b-1)/2}\rho]) \otimes \sigma)$$

for some irreducible square integrable representation σ. This and (9.1) imply that

$$\delta([|\det|_F^{-1/2}\rho, |\det|_F^{(b-1)/2}\rho]) \otimes \sigma$$

is in the Jacquet module of $\mathrm{Ind}(|\det|_F^{1/2}\rho \otimes \pi')$. Now, a simple analysis based on the structure obtained in [Td95+] implies that a subquotient of the form $\delta([|\det|_F^{-1/2}\rho, |\det|_F^{(b-1)/2}\rho]) \otimes *$ is in the Jacquet module of π'. Using this, section 7 of [Td+] implies $\epsilon_{\pi'}(\delta(\rho, b)) = 1$. This completes the proof.

□

References

[AiGu09] A. Aizenbud and D. Gourevitch, *Multiplicity one theorem for $(GL(n+1,\mathbb{R}), GL(n,\mathbb{R}))$*, Selecta Mathematica **15** no. 2 (2009), 271–294.

[AiGuRlSf10] A. Aizenbud, D. Gourevitch, S. Rallis and G. Schiffmann, *Multiplicity One Theorems*, Ann. of Math. **172** no. 2 (2010), 1407–1434.

[Ar79] J. Arthur, *Eisenstein series and the trace formula*, in "Automorphic forms, representations and L-functions", Proc. Sympos. Pure Math. **33**, Part 1, Amer. Math. Soc., Providence, Rhode Island, 1979, 253–274.

[Ar89] J. Arthur, *Unipotent automorphic representations: conjectures*, Astérisque **171-172** (1989), 13–71.

[Ar] J. Arthur, *The Endoscopic Classification of Representations: Orthogonal and Symplectic Groups*, preprint (`http://www.claymath.org/cw/arthur/pdf/Book.pdf`).

[Ar+] J. Arthur, *The endoscopic classification of representations*, this proceedings.

[Au95] A.M. Aubert, *Dualité dans le groupe de Grothendieck de la catégorie des représentations lisses de longueur finie d'un groupe réductif p- adique*, Trans. Amer. Math. Soc. **347** (1995), 2179–2189; *Erratum*, Trans. Amer. Math. Soc **348** (1996), 4687–4690.

[Ba08] A.I. Badulescu, *Global Jacquet-Langlands correspondence, multiplicity one and classification of automorphic representations (with an appendix by Neven Grbac)*, Invent. Math. **172** no. 2 (2008), 383–438.

[Ba] A.I. Badulescu, *On p-adic Speh representations*, Bulletin de la SMF, to appear.

[BaHeLeSé10] A.I. Badulescu, G. Henniart, B. Lemaire and V. Sécherre, *Sur le dual unitaire de $GL_r(D)$*, Amer. J. Math. **132** no. 5 (2010), 1365–1396.

[BaRn04] A.I. Badulescu and D.A. Renard, *Sur une conjecture de Tadić*, Glasnik Mat. **39** no. 1 (2004), 49–54.

[BaRn10] A.I. Badulescu and D.A. Renard, *Unitary dual of GL_n at archimedean places and global Jacquet-Langlands correspondence*, Compositio Math. **146** no. 5 (2010), 1115–1164.

[BbMo96] D. Barbasch and A. Moy, *Unitary spherical spectrum for p-adic classical groups*, Representations of Lie groups, Lie algebras and their quantum analogues, Acta Appl. Math. **44** no. 1-2 (1996), 3–37.

[Bu03] D. Baruch, *A proof of Kirillov's conjecture*, Ann. of Math. (2) **158** no. 1 (2003), 207–252.

[Be74] J. Bernstein, *All reductive p-adic groups are tame*, Functional Anal. Appl. **8** (1974), 3–6.

[Be84] J. Bernstein, *P-invariant distributions on $GL(N)$ and the classification of unitary representations of $GL(N)$ (non-archimedean case)*, Lie Group Representations II, Lecture Notes in Math. **1041**, Springer-Verlag, Berlin, 1984, 50–102.

[BeZe77] J. Bernstein and A.V. Zelevinsky, *Induced representations of reductive p-adic groups I*, Ann. Sci. École Norm Sup. **10** (1977), 441–472.

[BuLiSn92] M. Burger, J.-S. Li and P. Sarnak, *Ramanujan duals and automorphic spectrum*, Bull. Amer. Math. Soc. **26** no. 2 (1992), 253–257.

[BsKu93] C.J. Bushnell and P.C. Kutzko, *The admissible dual of $GL(N)$ via compact open subgroups.* Annals of Math. Studies **129**, Princeton University Press, Princeton, NJ, 1993.

[Ca] W. Casselman, *Introduction to the theory of admissible representations of p-adic reductive groups*, preprint (`http://www.math.ubc.ca/~cass/research/pdf/p-adic-book.pdf`).

[ChRn08] G. Chenevier and D. Renard, *Characters of Speh representations and Lewis Caroll identity*, Represent. Theory **12** (2008), 447–452.

[Cl03] L. Clozel, *Démonstration de la Conjecture τ*, Inventiones Math. **151** (2003), 297–328.

[Cl04] L. Clozel, *Combinatorial consequences of Arthurs conjectures and the Burger-Sarnak method*, Int. Math. Res. Not. **11** (2004), 511–523.

[Cl07] L. Clozel, *Spectral Theory of Automorphic forms*, in "Automorphic forms and applications", IAS/Park City Math. Ser. **12** (2007), 41–93.

[Cl11] L. Clozel, *The ABS principle : consequences for $L^2(G/H)$*, in "On certain L-functions", Clay Math. Proc. **13** (2011), 99–115.

[ClUl04] L. Clozel and E. Ullmo, *Equidistribution des points de Hecke*, in "Contributions to Automorphic Forms, Geometry and Arithmetic", Johns Hopkins University Press, 2004, 193–254.

[CoKiPSSh04] J. Cogdell, H. Kim, I. Piatetski-Shapiro and F. Shahidi, *Functoriality for the classical groups*, Publ. Math. Inst. Hautes Études Sci. **99** (2004), 163–233.

[De74] P. Deligne, *La conjecture de Weil I*, Publ. Math. Inst. Hautes Études Sci. **43** (1974), 273–307.

[DeKaVi84] P. Deligne, D. Kazhdan and M.-F. Vignéras, *Représentations des algèbres centrales simples p-adiques*, in book "Représentations des Groupes Réductifs sur un Corps Local" by J.-N. Bernstein, P. Deligne, D. Kazhdan, and M.-F. Vignéras, Hermann, Paris, 1984.

[Di62] J. Dixmier, *Les C^*-algebras et leurs Représentations,* Gauthiers-Villars, Paris, 1969.

[Fe62] J. M. G. Fell, *Weak containment and induced representations of groups*, Canad. J. Math. **14**, (1962), 237–268.

[Fe65] J.M.G. Fell, *Non-unitary dual space of groups,* Acta Math. **114** (1965), 267–310.

[GaGrPr] W.T. Gan, B.H. Gross and D. Prasad, *Symplectic local root numbers, central critical L-values, and restriction problems in the representation theory of classical groups*, Asterisque **346** (2012), 1–109.

[GaTk10] W.T. Gan and S. Takeda, *The local Langlands conjecture for $Sp(4)$*, Int. Math. Res. Not. **2010** no. 15 (2010), 2987–3038.

[Gb75] S.S. Gelbart, *Automorphic forms on adele groups*, Annals of Math. Studies **83**, Princeton University Press, Princeton, 1975.

[Gb84] S.S. Gelbart, *An elementary introduction to the Langlands program*, Bulletin Amer. Math. Soc. **10** (1984), 177–219.

[GfGrPS69] I.M. Gelfand, M. Graev and I. Piatetski-Shapiro, *Representation theory and automorphic functions*, Saunders, Philadelphia, 1969.

[GfNa50] I.M. Gelfand and M.A. Naimark, *Unitäre Darstellungen der Klassischen Gruppen* (German translation of Russian publication from 1950), Akademie Verlag, Berlin, 1957.

[Go94] D. Goldberg, *Reducibility of induced representations for $Sp(2n)$ and $SO(n)$*, Amer. J. Math. **116** (1994), 1101–1151.

[HC83] Harish-Chandra, *Collected papers*, Springer-Verlag, Berlin, 1983.

[HaTy01] M. Harris and R. Taylor, *On the geometry and cohomology of some simple Shimura varieties*, Annals of Math. Studies **151**, Princeton University Press, 2001.

[He00] G. Henniart, *Une preuve simple des conjectures de Langlands pour $GL(n)$ sur un corps p-adique*, Invent. Math. **139** (2000), 439–455.

[HiIcIk08] H. Hiraga, A. Ichino and T. Ikeda, *Formal degrees and adjoint γ-factors*, J. Amer. Math. Soc. **21** (2008), 283–304; *Erratum*, J. Amer. Math. Soc. **21** (2008), 1211–1213.

[Ho79] R. Howe, *θ-series and invariant theory*, Symp. Pure Math. **33**, part 1, Amer. Math. Soc., Providence, R.I., 1979, pp. 275–286.

[HoMo85] R. Howe, with collaboration of A. Moy, *Harish Chandra Homomorphisms for p-adic groups*, CBMS Regional Conference Series **59**, Amer. Math. Soc., Providence, R.I., 1985.

[Jc71] H. Jacquet, *Représentations des groupes linéaires $\mathfrak{p}$-adiques*, in "Theory of group representations and Fourier analysis", C.I.M.E. pp. 119–220, Edizioni Cremonese, Rome, 1971.

[Jc84] H. Jacquet, *On the residual spectrum of $GL(n)$*, Lie Group Representations II, Lecture Notes in Math. **1041**, Springer-Verlag, Berlin, 1984, 185–208.

[JcLn70] H. Jacquet and R.P. Langlands, *Automorphic Forms on $GL(2)$*, Lecture Notes in Math. **114**, Springer-Verlag, Berlin, 1970.

[Jn97] C. Jantzen, *On supports of induced representations for symplectic and odd-orthogonal groups*, Amer. J. Math. **119** (1997), 1213–1262.

[Jn] C. Jantzen, *Tempered representations for classical p-adic groups*, preprint.

[JiSo04] D. Jiang and D. Soudry, *Generic representations and local Langlands reciprocity law for p-adic $SO(2n+1)$*, in "Contributions to automorphic forms, geometry, and number theory", Johns Hopkins Univ. Press, Baltimore, MD, 2004, 457–519.

[Ka67] D. Kazhdan, *Connection of the dual space of a group with the structure of its closed subgroups*, Functional Anal. Appl. **1** (1967), 63–65.

[KaLu87] D. Kazhdan and G. Lusztig, *Proof of the Deligne–Langlands conjecture for Hecke algebras,* Invent. Math. **87** (1987), 153–215.

[KaSv90] D. Kazhdan and G. Savin, *The smallest representation of simply-laced groups*, in Israel Math. Conference Proceedings, Piatetski-Shapiro Festschrift, **2** (1990), 209–233.

[Ki62] A.A. Kirillov, *Infinite dimensional representations of the general linear group*, Dokl. Akad. Nauk SSSR **114** (1962), 37–39; Soviet Math. Dokl. **3** (1962), 652–655.

[Ki76] A.A. Kirillov, *Elements of the Theory of Representations*, Springer-Verlag, New York, 1976.

[KnZu82] A.W. Knapp and G.J. Zuckerman, *Classification of irreducible tempered representations of semisimple groups*, Ann. of Math. **116** (1982), 389–455.

[Ln70] R.P. Langlands, *Problems in the theory of automorphic forms*, Lecture Notes in Math. **170**, Springer-Verlag, Berlin, 1970, 18–86.

[Ln89] R.P. Langlands, *On the classification of irreducible representations of real algebraic groups*, in "Representation Theory and Harmonic Analysis on Semisimple Lie Groups", AMS Mathematical Surveys and Monographs, **31**, 1989, 101–170.

[LpMi] E. Lapid and A. Minguez, *On a determinantal formula of Tadić, Amer. J. Math.*, to appear.

[LpRo] E. Lapid and J. Rogawski, *On a result of Venkatesh on Clozel's conjecture*, in "Automorphic forms and L-functions II. Local aspects", Contemp. Math. **489** (2009), 173–178.

[LpMuTd04] E. Lapid, G. Muić and M. Tadić, *On the generic unitary dual of quasisplit classical groups*, Int. Math. Res. Not. **26** (2004), 1335–1354.

[LmRpSu93] G. Laumon, M. Rapoport and U. Stuhler, *P-elliptic sheaves and the Langlands correspondence*, Invent. Math. **113** (1993), 217–338.

[LoRuSn99] W. Luo, Z. Rudnick and P. Sarnak, *On the generalized Ramanujan conjecture for $GL(n)$*, in "Automorphic forms, automorphic representations, and arithmetic (Fort Worth,

TX, 1996)", Symp. Pure Math. **66**, part 2, Amer. Math. Soc., Providence, R.I., 1999, 301–310.

[Ls95] G. Lusztig, *Classification of unipotent representations of simple p-adic groups*, Int. Math. Res. Not. (1995), 517–589.

[LbZu] A. Lubotzky and A. Zuk, *On property* (τ), preprint.

[Ms69] H. Maass, *Nichtanalytishe Automorphe Funktionen*, Math. Ann. **121** (1949), 141–183.

[Mc52] G.W. Mackey, *Induced representations of locally compact groups I*, Ann. of Math. **55** (1952), 101–139.

[Mi73] D. Miličić, *On C^*-algebras with bounded trace*, Glasnik Mat. Ser. III **8(28)** (1973), 7–21.

[Mœ00] C. Mœglin, *Normalisation des opérateurs d'entrelacement et réductibilité des induites de cuspidales; le cas des groupes classiques p-adiques*, Annals of Math. **151** no. 2 (2000), 817–847.

[Mœ02] C. Mœglin, *Sur la classification des séries discrètes des groupes classiques p-adiques: paramètres de Langlands et exhaustivité*, J. Eur. Math. Soc. **4** (2002), 143–200.

[Mœ03] C. Mœglin, *Points de réductibilité pour les induites de cuspidales*, J. Algebra **268** (2003), 81–117.

[Mœ07] C. Mœglin, *Classification et Changement de base pour les séries discrètes des groupes unitaires p-adiques*, Pacific J. Math. **233** (2007), 159–204.

[Mœ07+] C. Mœglin, *Classification des séries discrètes pour certains groupes classiques p-adiques*, in "Harmonic analysis, group representations, automorphic forms and invariant theory", 209–245, Lect. Notes Ser. Inst. Math. Sci. Natl. Univ. Singap. 12, World Sci. Publ., Hackensack, NJ, 2007.

[Mœ11] C. Mœglin, *Multiplicité 1 dans les paquets d'Arthur aux places p-adiques*, in "On certain L-functions", Clay Math. Proc. **13** (2011), 333–374.

[MœTd02] C. Mœglin and M. Tadić, *Construction of discrete series for classical p-adic groups*, J. Amer. Math. Soc. **15** (2002), 715–786.

[MœViWa87] C. Mœglin, M.-F. Vignéras and J.-L. Waldspurger, *Correspondances de Howe sur un corps p-adique*, Lecture Notes in Math. **1291**, Springer-Verlag, Berlin, 1987.

[MœWa89] C. Mœglin and J.-L. Waldspurger, *Le spectre residuel de* $GL(n)$, Ann. Sci. École Norm. Sup **22** (1989), 605–674.

[MœWa06] C. Mœglin and J.-L. Waldspurger, *Sur le transfert des traces tordues d'un group linéaire à un groupe classique p-adique*, Selecta Mathematica **12** (2006), 433–516.

[Mu97] G. Muić, *The unitary dual of p-adic* G_2, Duke Math. J. **90** (1997), 465–493.

[Mu98] G. Muić, *Some results on square integrable representations; Irreducibility of standard representations*, Int. Math. Res. Not. **14** (1998), 705–726.

[Mu98+] G. Muić, *On generic irreducible representation for* $Sp(n, F)$ *and* $SO(2n+1, F)$, Glasnik Mat. Ser III **33(53)** (1998), 19–31.

[Mu06] G. Muić, *On the Non-Unitary Unramified Dual for Classical p-adic Groups*, Trans. Amer. Math. Soc. **358** (2006), 4653–4687.

[Mu07] G. Muić, *On Certain Classes of Unitary Representations for Split Classical Groups*, Canadian J. Math. **59** (2007), 148–185.

[MuTd11] G. Muić and M. Tadić, *Unramified unitary duals for split classical p–adic groups; the topology and isolated representations*, in "On Certain L-functions", Clay Math. Proc. **13**, 2011, 375–438.

[MüSp04] W. Müller and B. Speh, *Absolute convergence of the spectral side of the Arthur trace formula for* GL_n. *With an appendix by E. M. Lapid*, Geom. Funct. Anal. **14** no. 1 (2004), 58–93.

[Ng] B.C. Ngô, *Le lemme fondamental pour les algèbres de Lie*, Publ. Math. Inst. Hautes Études Sci **111** (2010), 1–169.

[Rm16] S. Ramanujan, *On certain arithmetical functions*, Transactions of the Cambridge Philosophical Society XXII (9) (1916), 159–184.

[Rd97] M. Reeder, *Hecke algebras and harmonic analysis on p-adic groups*, Amer. J. Math. **119** no.1 (1997), 225–249.

[Sn05] P. Sarnak, *The generalized Ramanujan Conjectures*, in "Harmonic analysis, the trace formula, Shimura varieties", Clay Math Proc. **4** (2005), 659–685.

[Sk66] I. Satake, *Spherical functions and Ramanujan's conjecture*, Algebraic Groups and Discontinuous Subgroups , Proc. Symp. Pure Math. **9**, Amer. Math. Soc. (1966), 258–264.

[SdSl97] P. Schneider and U. Stuhler, *Representation theory and sheaves on the Bruhat-Tits building*, Publ. Math. Inst. Hautes Études Sci. **85** (1997), 97–191.

[Sé09] V. Sécherre, *Proof of the Tadić conjecture (U0) on the unitary dual of* $GL_m(D)$, J. Reine Angew. Math. **626** (2009), 187–203.

[Sb65] A. Selberg, *On the estimation of Fourier coefficients of modular forms*, Proc. Symp. Pure Math. **8** Amer. Math. Soc. (1965), 1–15.

[Sr95] J.-P. Serre, *Représentations linéaires et espaces homogènes kählériens des groupes de Lie compacts* (d'après Armand Borel et André Weil), Séminaire Bourbaki (Paris: Soc. Math. France) 2 (100), 1995, 447–454.

[Sh90] F. Shahidi, *A proof of Langlands conjecture on Plancherel measures; complementary series for p-adic groups*, Ann. of Math. **132** (1990), 273–330.

[Sh92] F. Shahidi, *Twisted endoscopy and reducibility of induced representations for p-adic groups*, Duke Math. J. **66** (1992), 1–41.

[Sa74] J. Shalika, *The multiplicity one theorem for* $GL(n)$, Ann. of Math. **100** (1974), 171–193.

[Si80] A. Silberger, *Special representations of reductive p-adic groups are not integrable*, Ann. of Math. **111** (1980), 571–587.

[Sp83] B. Speh, *Unitary representations of* $GL(n,\mathbb{R})$ *with non-trivial* (g, K)*- cohomology*, Invent. Math. **71** (1983), 443–465.

[St67] E.M. Stein, *Analysis in matrix spaces and some new representations of* $SL(N, \mathbb{C})$, Ann. of Math. **86** (1967), 461–490.

[SuZh12] B. Sun and C.B. Zhu, *Multiplicity one theorems: the Archimdean case*, Ann. of Math. **175** (2012), 23–44.

[Td83] M. Tadić, *The topology of the dual space of a reductive group over a local field*, Glas. Mat. Ser. III **18(38)** (1983), 259–279.

[Td] M. Tadić, *Unitary representations of general linear group over real and complex field,* preprint MPI/SFB 85-22 Bonn (1985).

[Td85] M. Tadić, *Unitary dual of p-adic* $GL(n)$*, Proof of Bernstein Conjectures*, Bulletin Amer. Math. Soc. **13** (1985), 39–42.

[Td86] M. Tadić, *Classification of unitary representations in irreducible representations of general linear group (non-archimedean case)*, Ann. Sci. École Norm. Sup. (4) **19** (1986), 335–382.

[Td87] M. Tadić, *Topology of unitary dual of non-Archimedean* $GL(n)$*,* Duke Math. J. **55** no. 2 (1987), 385–422.

[Td88] M. Tadić, *Geometry of dual spaces of reductive groups (non-archimedean case)*, J. Analyse Math. **51** (1988), 139–181.

[Td90] M. Tadić, *Induced representations of* $GL(n, A)$ *for p-adic division algebras* A, J. Reine Angew. Math. **405** (1990), 48–77.

[Td93] M. Tadić, *An external approach to unitary representations,* Bull. Amer. Math. Soc. (N.S.) **28** no. 2 (1993), 215–252.

[Td95] M. Tadić, *On characters of irreducible unitary representations of general linear groups*, Abh. Math. Sem. Univ. Hamburg **65** (1995), 341–363.

[Td95+] M. Tadić, *Structure arising from induction and Jacquet modules of representations of classical p-adic groups*, J. of Algebra **177** (1995), 1–33.

[Td98] M. Tadić, *On reducibility of parabolic induction,* Israel J. Math. **107** (1998), 29–91.

[Td99] M. Tadić, *Square integrable representations of classical p-adic groups corresponding to segments*, Represent. Theory **3** (1999), 58–89.

[Td02] M. Tadić, *A family of square integrable representations of classical p-adic groups in the case of general half-integral reducibilities*, Glas. Mat. Ser. III **37(57)** no. 1 (2002), 21–57.

[Td04] M. Tadić, *On classification of some classes of irreducible representations of classical groups*, in book "Representations of real and p-adic groups", Singapore University Press and World Scientific, Singapore, 2004, 95–162.

[Td06] M. Tadić, *On the representation theory of $GL(n)$ over a p-adic division algebra and unitarity in the Jacquet-Langlands correspondences*, Pacific J. Math. **223** (2006), 167–200.

[Td07] M. Tadić, *An exercise on unitary representations in the case of complex classical groups*, Functional analysis IX (Dubrovnik, 2005), 91–102., Various Publ. Ser. 48, Univ. Aarhus, Aarhus, 2007.

[Td09] M. Tadić, $GL(n,\mathbb{C})\hat{}$ *and* $GL(n,\mathbb{R})\hat{}$, in "Automorphic Forms and L-functions II, Local Aspects", Contemp. Math. **489** (2009), 285–313.

[Td09+] M. Tadić, *On reducibility and unitarizability for classical p-adic groups, some general results*, Canad. J. Math. **61** (2009), 427–450.

[Td10] M. Tadić, *On automorphic duals and isolated representations; new phenomena*, J. Ramanujan Math. Soc. **25** no. 3 (2010), 295–328.

[Td11] M. Tadić, *On invariants of discrete series representations of classical p-adic groups*, Manuscripta Math. **136** (2011), 417–435.

[Td12] M. Tadić, *Reducibility and discrete series in the case of classical p-adic groups; an approach based on examples*, in "Geometry and Analysis of Automorphic Forms of Several Variables", World Scientific, Singapore, 2012, 254–333.

[Td+] M. Tadić, *On tempered and square integrable representations of classical p-adic groups*, Sci. China Math., to appear.

[Ve05] A. Venkatesh, *The Burger-Sarnak method and operations on the unitary dual of $GL(n)$*, Represent. Theory **9** (2005), 268–286.

[Vo86] D. A. Vogan, *The unitary dual of $GL(n)$ over an archimedean field*, Invent. Math. **82** (1986), 449–505.

[Wa87] J.-L. Waldspurger, *Un exercice sur $GSp(4, F)$ et les représentations de Weil*, Bull. Soc. Math. France **115** (1987), 35–69.

[Wa90] J.-L. Waldspurger, *Démonstration d'une conjecture de dualité de Howe dans le cas p-adique, $p \neq 2$*, in "Festschrift in honor of I. I. Piatetski-Shapiro on the occasion of his sixtieth birthday", Part I, 267–324, Israel Math. Conf. Proc., 2, Weizmann, Jerusalem, 1990.

[Wa97] J.-L. Waldspurger, *Le lemme fondamental implique le transfert*, Compos. Math. **105** (1997), 153–236.

[Wa03] J.-L. Waldspurger, *La formule de Plancherel pour les groupes p-adiques*, J. Inst. Math. Jussieu **2** (2003), 235–333.

[We68] H. Weyl, *Theorie der Darstellung kontinuierlicher halbeinfacher Gruppen durch lineare Transformationen*, Gesammelte Abhandlungen, Bd. II, **68**, 543–647. Springer-Verlag, Berlin, 1968.

[Ze80] A.V. Zelevinsky, *Induced representations of reductive p-adic groups II. On irreducible representations of $GL(n)$*, Ann. Sci. École Norm. Sup. **13** (1980), 165–210.

Department of Mathematics, University of Zagreb, Bijenička 30, 10000 Zagreb, Croatia.

E-mail: tadic@math.hr

Automorphic Representations and L-Functions
Editors: D. Prasad, C.S. Rajan, A. Sankaranarayanan, J. Sengupta

Publisher: Hindustan Book Agency, New Delhi, India

On the Rank of Selmer Groups for Elliptic Curves Over $\mathbb{Q}$

Eric Urban

Abstract

The following are extended notes of a lecture given by the author at the international colloquium on L-functions and Automorphic Representation held at TIFR in january 2012. This lecture reported on some joint work of Chris Skinner and the author on the link between central L-values and Selmer groups of elliptic curves. The detailed proofs of our results will appear in [SU13]. The author presents here the main lines of the arguments.

1 Introduction

Let E be an elliptic curve over the rational. By the works initiated by Wiles, it is known that E is modular and therefore that its L-function $L(E,s)$ is entire on the whole complex plane. The Birch and Swinnerton-Dyer conjecture predicts that

$$\mathrm{ord}\, L(E,s)|_{s=1} = \mathrm{rank}_{\mathbb{Z}}\, E(\mathbb{Q}).$$

This conjecture is proved thanks to the works of Kolyvagin and Gross-Zagier when the order of vanishing is 0 or 1. When the order of vanishing is higher, very little is known in general for the Mordell-Weil rank. However studying the co-rank of the Selmer group of E seems more accessible. Let p be a rational prime and let $\mathrm{Sel}_p(\mathbb{Q},E)$ be the p-Selmer group of E over $\mathbb{Q}$. Recall that it is a subgroup of $H^1(\mathbb{Q},E[p^\infty])$ fitting in the Kümmer exact sequence:

$$0 \to E(\mathbb{Q}) \otimes \mathbb{Q}_p/\mathbb{Z}_p \to \mathrm{Sel}_p(\mathbb{Q},E) \to \text{Ш}_p(\mathbb{Q},E) \to 0$$

where $\text{Ш}_p(\mathbb{Q},E) \subset H^1(\mathbb{Q},E)$ stands for the p-part of the Tate-Shafarevitch group of E over $\mathbb{Q}$. Birch and Swinnerton-Dyer conjecture also that this later is finite. The corank of the Selmer group should therefore be equal to the Mordel-Weil rank of E. A special case of our result is the following:

Theorem 1.1 ([SU13]) *Let E be a semi-stable elliptic curve over $\mathbb{Q}$ having good reduction at p. If $L(E,1)=0$ then $\mathrm{Sel}_p(\mathbb{Q},E)$ is infinite. Furthermore if $L(E,s)$ vanishes at $s=1$ with a positive even order then the corank of $\mathrm{Sel}_p(\mathbb{Q},E)$ is at least 2.*

This result is valid for a larger class of elliptic curves. More generally, we prove a similar result for the Bloch-Kato Selmer group attached to an elliptic cuspidal eigenform of trivial nebentypus. In this note, we explain the main steps of the strategy to prove such a result. In Section 2, one recalls the definition of the Bloch-Kato Selmer groups and states the main result. The basic strategy is to construct and use a certain deformation of a reducible Galois representation like in our previous work [SU06a, SU06b] but in a different way than in [SU10] where we proved the Iwasawa conjecture for p-ordinary elliptic curves. In Section 3, we explain how the existence of such a deformation leads to the construction of a non trivial extension in the Bloch-Kato Selmer group. There isn't much novelty in that part of the argument, except that the Hodge-Tate weights of the reducible Galois representation that is deformed have multiplicities and a slightly different argument is necessary to prove the first part of Theorem 1.1. From this, the even order situation can be resolved like in [SU06b]. In Section 4, we explain how to construct a p-adic deformation of a certain *critical* p-adic Eisenstein series whose Galois representation is isomorphic to the Galois representation we want to deform. This will give rise to the deformation of the Galois representation studied in Section 3. In [SU06b], we treated the case of the Bloch-Kato Selmer group for a cuspidal eigenform f of weight $k \geq 4$, because this condition on the weight is necessary to construct the appropriate holomorphic Eisenstein series. The case of weight 2 we treat here[1] is obtained by constructing a p-adic Eisenstein series that we can think of as an overconvergent automorphic form with a non arithmetic p-adic weight. By overconvergent here, we mean a p-adic modular form which defines a point of the Eigenvariety for the quasi-split unitary group $U(2,2)$. This can be achieved under the condition that $L(f,1)=0$. To construct this p-adic Eisenstein series, we put the modular eigenform f into a Coleman family. For each member of this family of weight ≥ 4, we can construct an Eisenstein series with the arithmetic appropriate weight but that is holomorphic only if the central L-value of this member vanishes. However, this Eisenstein series is always nearly holomorphic in a sense defined originally by Shimura in the seventies for elliptic modular forms and generalized later for symplectic and unitary groups (see [Sh04] for an unified treatment of his theory). We are therefore led to study the arithmetic of

[1]We actually treat the general case $k \geq 2$.

nearly holomorphic forms and give an algebraic definition[2] of those. This naturally leads us to define the notion of nearly overconvergent forms which can be seen naturally as special p-adic modular forms. We have given a taste of this notion here although it is not really necessary for our goal but it gives a good idea of the kind of objects we are dealing with in this work. This notion of nearly overconvergence have other applications in particular for the construction of p-adic L-function. We hope to come back to this in the future.

The author would like to thank Michael Harris, Haruzo Hida, Benoit Stroh, Vincent Pilloni and Jacques Tilouine for their interest in this work. The author wishes also to thank the organizers of the International Colloquium on Automorphic Representations and L-functions that was held at the TIFR for their kind invitation to give a lecture and for their hospitality. The author have been founded by the CNRS and the NSF during periods when this research was undertaken and is grateful to these institutions for their financial support.

Notations and conventions

Throughout this paper p is a fixed prime. We denote by $\mathbb{Z}$ and $\mathbb{Z}_p$ the rings of integers and p-adic integers with respective field of fractions $\mathbb{Q}$ and $\mathbb{Q}_p$. We denote respectively by $\overline{\mathbb{Q}}$ and $\overline{\mathbb{Q}}_p$ the algebraic closures of $\mathbb{Q}$ and $\mathbb{Q}_p$ and by $\mathbb{C}$ the field of complex numbers. We fix embeddings $\iota_\infty : \overline{\mathbb{Q}} \hookrightarrow \mathbb{C}$ and $\iota_p : \overline{\mathbb{Q}} \hookrightarrow \overline{\mathbb{Q}}_p$ and we fix an identification $\overline{\mathbb{Q}_p} \cong \mathbb{C}$ compatible with these embeddings. Throughout we implicitly view $\overline{\mathbb{Q}}$ as a subfield of $\mathbb{C}$ and $\overline{\mathbb{Q}}_p$ via the embeddings ι_∞ and ι_p. All number fields will be considered as subfield of $\overline{\mathbb{Q}}$ and of $\overline{\mathbb{Q}}_p$ or $\mathbb{C}$ via the above embeddings. We denote respectively by $\mathbb{A}$ and $\mathbb{A}_f$ the rings of adeles and finite adeles of $\mathbb{Q}$. For each place v of a number field K, we denote K_v the completion of K with respect to the norm $|\cdot|_v$ associated to v. If $\mathfrak{X}$ is a rigid analytic variety over an extension of $\mathbb{Q}_p$, we denote by $A(\mathfrak{X})$ the ring of analytic functions on $\mathfrak{X}$. If H is a reductive group over $\mathbb{Z}$, an algebraic representation of H is seen a functorial pair (ρ, V) in the sense that for any ring R, we have a group homomorphism $\rho : \; H(R) \rightarrow \mathrm{GL}_R(V_R)$ where $V_R = V_{\mathbb{Z}} \otimes R$ is free over R satisfying the obvious base change property for any ring homomorphism $R \rightarrow S$.

[2]We learned after this was achieved that Michael Harris had given an equivalent definition in [Ha85, Ha86].

2 Bloch-Kato Selmer Groups

2.1 Some definitions

We recall the definition of the Bloch-Kato Selmer group attached to a Galois representation and precise our conventions for L-functions. Let K be a number field and $G_K = \mathrm{Gal}(\overline{\mathbb{Q}}/K)$ be its absolute Galois group. Let L be a finite extension of $\mathbb{Q}_p$ and let V be a finite dimensional L-vector space equipped with a continuous linear action of G_K. We assume that this Galois representation is geometric in the sense of Fontaine. For such a representation, we denote by $V(n)$ the n-th Tate twist of V.

We denote by $H^1(K, V)$ the continuous cohomology of G_K with coefficients in V. This space parametrizes the isomorphisms classes $[E]$ of extensions E of the form

$$0 \to V \to E \to L \to 0.$$

where we understand L as the one dimensional L-vector space with trivial G_K-action. Then $H^1_g(K, V)$ is the subset of $H^1(K, V)$ classifying classes of extensions $[E]$ such that E is also geometric. We now assume that the action of the decomposition subgroups of G_K at the places above p are crystalline. Then, $H^1_f(K, V)$ denotes the subspace of $H^1(K, V)$ of extensions classes $[E]$ such that E is crystalline at all places above p and such that for all places v not divising p, we have

$$0 \to V^{I_v} \to E^{I_v} \to L \to 0,$$

where I_v stands for an inertia subgroup at v. In general, we have

$$H^1_f(K, V) \subset H^1_g(K, V). \tag{2.1}$$

We now recall the definition of the L-function attached to a Galois representation. For any finite place v not dividing p, we put

$$P_v(V, X) := \det(1 - X\mathrm{Frob}_v; V^{I_v})$$

where Frob_v stands for a geometric Frobenius at v and

$$P_v(V, s) := \det(1 - X\varphi_v; D_{\mathrm{crys},v}(V))$$

if v divides p and where φ_v stands for the geometric crystalline Frobenius induced on

$$D_{\mathrm{crys},v}(V) := (B_{\mathrm{crys}} \otimes V)^{D_v}.$$

For each finite place v, let q_v be the cardinal of the residue field of K at v. Then we put $L_v(V,s) := P_v(V, q_v^{-s})^{-1}$ and the L-function of V is defined as

$$L(V,s) = \prod_{v<\infty} L_v(V,s).$$

It is conjectured that $L(V,s)$ has a meromorphic continuation to the complex plane. This fact is of course established when we know that V is attached to an automorphic representation.

Remark 2.1 The inclusion (2.1) is an equality if $P_v(V, q_v) \neq 0$ for all finite place v.

2.2 Selmer groups for modular forms

Let f be a new cuspidal elliptic eigenform of even weight $k = 2m$ with trivial nebentypus and conductor N. Assume f is normalized and let us write its Fourier expansion

$$f(q) = \sum_{n=1}^{\infty} a(n,f)q^n.$$

Let L be a finite extension of $\mathbb{Q}_p$ containing the Hecke eigenvalues of f. By Eichler-Shimura and Deligne, there exists a two dimensional L-vector space V_f with a continuous $G_{\mathbb{Q}}$-linear action such that

$$L(V_f,s) = L(f,s) := \sum_{n=1}^{\infty} a(n,f)n^{-s}.$$

Recall that $L(f,s)$ satisfies a functional equation of the form:

$$L(f,s) = \varepsilon(f,s)L(f,2m-s).$$

Our main result is the following theorem.

Theorem 2.2 ([SU13]) *Assume N is prime to p. If $k=2$ and $\varepsilon(f,1)=1$, we further assume that $a_\ell \neq 0$ for some prime $\ell | N$. Then,*

a) *if $L(f,m) = 0$, we have $H^1_f(\mathbb{Q}, V_f(m)) \neq 0$,*

b) *if $L(f,m) = 0$ and $\varepsilon(f,m) = 1$, then $\mathrm{rank}_L\, H^1_f(\mathbb{Q}, V_f(m)) \geq 2$.*

Remark 2.3 Theorem 1.1 follows from Theorem 2.2 because for f_E the weight 2 square free level cusp form associated to E, the rank of $H^1_f(\mathbb{Q}, V_{f_E}(1))$ is equal to the co-rank of $\mathrm{Sel}_p(\mathbb{Q}, E)$. Notice that we have $V_{f_E}(1) \cong V_p(E) = T_p(E) \otimes \mathbb{Q}_p$ where $T_p(E)$ stands for the Tate module of E.

Remark 2.4 If f is ordinary at p, the part a) of this Theorem follows from [SU10]. It follows also from [SU02] if the order of vanishing is odd. The method used in [SU02] works *mutatis mutandis* if p is a supersingular prime for E. The construction of the deformation of the corresponding Saito-Kurokawa lift follows from Example 5.5.3 of [Ur11].

Remark 2.5 If $k > 2$, this Theorem is proved using the strategy outlined in [SU06b]. It is based on the construction of an holomorphic Eisenstein series on the quasi-split unitary group $U(2,2)$ attached to an imaginary quadratic field $\mathcal{K}$. To include the case $k = 2$, we use an arithmetic theory of nearly holomorphic forms to construct a specific p-adic (overconvergent) Eisenstein series (having a non-arithmetic weight).

Remark 2.6 The condition on the conductor N when $k = 2$ is a simplified assumption making sure f is in the image of the Jacquet-Langlands correspondence for a definite quaternion algebra. This is necessary to show that the p-adic (overconvergent) Eisenstein series that will construct as a p-adic limit is non-trivial. See (iv) in Theorem 4.8.

2.3 The basic strategy

In this note, we explain the main steps of the basic strategy for proving the part a) of this theorem. The main idea is an extension and generalization of a method introduced and developed by C. Skinner and the author in a series of papers[3] [SU02, SU06a, SU06b]. We introduce an imaginary quadratic field $\mathcal{K}$ and construct a generically irreducible deformation of the $G_{\mathcal{K}}$-representation

$$W_f := L \oplus V_f(m) \oplus L(1).$$

From this deformation, we are able to construct a non-split extension of the following form by using a version of Ribet's lemma

$$0 \to L(1) \to E_f \to V_f(m) \to 0.$$

Since $V_f^\vee \cong V_f(k-1)$, we get a non trivial class $[E_f] \in H^1(\mathcal{K}, V_f(m))$. We further show that it is contained in $H^1_f(\mathcal{K}, V_f(m))$. Let denote by $\chi_{\mathcal{K}}$ the quadratic character attached to the extension $\mathcal{K}/\mathbb{Q}$. If $\mathcal{K}$ is chosen so that

$$L(f, \chi_{\mathcal{K}}, m) \neq 0. \tag{2.2}$$

We know by results of Kato or Kolyvagin that $H^1_f(\mathbb{Q}, V_f(m) \otimes \chi_{\mathcal{K}}) = 0$. We deduce that $[E_f] \in H^1_f(\mathbb{Q}, V_f(m))$ and the part a) of Theorem 1.1 follows. For the part b), the strategy is already explained in [SU06b].

[3] In [SU02, SU06a], one does not need to introduce an imaginary quadratic field.

3 An Analytic Family of Trianguline Galois Representations

Using the theory of p-adic families of automorphic forms of finite slopes, we will construct a certain family of trianguline Galois representations. We describe the family in Theorem 3.1 and show how we deduce the part a) of Theorem 1.1

3.1 Polarized Galois representations

We fix $\mathcal{K} \subset \overline{\mathbb{Q}}$ an imaginary quadratic field. We denote by c the complex conjugation of $\mathbb{C}$ (and hence of $\mathcal{K}$ induced by the embedding ι_∞). Sometimes we write $\bar{a}$ instead of a^c for $a \in \mathbb{C}$. Let $\mathcal{O}_\mathcal{K}$ the ring of integers of $\mathcal{K}$. We assume p splits in $\mathcal{K}$, i. e. $p.O_\mathcal{K} = \wp.\wp^c$ where $\wp$ stands for the prime ideal of $\mathcal{O}_\mathcal{K}$ induced by ι_p. We denote by $\mathcal{O}_{(\wp)}$ the localization of $\mathcal{O}_\mathcal{K}$ at $\wp$ and by $\mathcal{O}_\wp$ its completion.

We will consider Galois representations (ρ, W) of $G_\mathcal{K}$. For such a representation, we denote by (ρ^c, W^c) the representation on the space W with the conjugate action by c (i.e. $\rho^c(g) = \rho(cgc)$, $\forall g \in G_\mathcal{K}$) and by $(\rho^\vee, W^\vee)$ the contragredient representation. We will say W is polarized if it satisfies:

$$W^c \cong W^\vee(1).$$

Notice that the representation W_f that we have defined in the previous section satisfies this condition. We will consider families of such representations of dimension 4. For this, it is convenient to define the parametrizing space of Hodge-Tate-Sen weights. Let $\mathfrak{W}$ be the rigid analytic space over $\mathbb{Q}_p$ such that $\mathfrak{W}(L) = \mathrm{Hom}_{\mathrm{cont}}((\mathbb{Z}_p^\times)^4, L^\times)$. An element $\underline{\kappa} = (\kappa_1, \kappa_2, \kappa_3, \kappa_4) \in \mathfrak{W}(\overline{\mathbb{Q}}_p)$ is called a weight. If $k_1, \ldots, k_4$ are integers, we write (k_1, k_2, k_3, k_4) for the weight defined by

$$(t_1, t_2, t_3, t_4) \mapsto t_1^{k_1} t_2^{k_2} t_3^{k_3} t_4^{k_4}.$$

Such a weight is called arithmetic regular if $k_1 < k_2 < k_3 < k_4$. Using the theory of p-adic families of automorphic forms, one shows the following theorem.

Theorem 3.1 ([SU13]) *Assume that $L(f, m) = 0$ and that the conditions of Theorem 2.2 are satisfied. Let α be an eigenvalue of $X^2 - a(p, f)X + p^{k-1}$. Then there exist:*

(i) *$\mathfrak{V}$ an irreducible finite cover of a one-dimensional affinoid subdomain $\mathfrak{U}$ of $\mathfrak{W}$ with structural map $\underline{\kappa}$,*

(ii) *a point* $x_0 \in \mathfrak{V}(L)$ *such that* $\underline{\kappa}(x_0) = \underline{\kappa}_0 = (-m, -1, 0, m-1) \in \mathfrak{U}(L)$,

(iii) *A pseudo-character*[4] $T: G_{\mathcal{K}} \to A(\mathfrak{V})$ *of dimension* 4,

(iv) *An infinite set* $\Sigma \subset \mathfrak{V}(\overline{\mathbb{Q}}_p)$ *sitting over arithmetic regular weights,*

(v) *Analytic functions* $\underline{\varphi} = (\varphi_1, \varphi_2, \varphi_3, \varphi_4) \in A(\mathfrak{V})^4$

For $x \in \mathfrak{V}(\overline{\mathbb{Q}}_p)$, *let us write* T_x *for the evaluation of* T *at* x. *The data* T, Σ, $\underline{\varphi}$ *and* x_0 *satisfy the following properties:*

(a) T_{x_0} *is the character associated to* W_f *and* $\underline{\varphi}(x_0) = (\alpha p^{-m}, 1, p^{-1}, \alpha^{-1}p^{m-1})$,

(b) *For all* $x \in \Sigma$, T_x *is the trace of a semi-simple polarized representation* ρ_x *such that*

$$\dim(\rho_x)^{I_v} = \dim V_f^{I_v} + 2$$

for all finite places v *not dividing* p.

(c) *For all* $x \in \Sigma$, *the restriction*[5] *of* ρ_x *at* $G_\wp$ *is crystalline with Hodge-Tate weight* $\underline{\kappa}(x)$ *and Frobenius eigenvalues* $(\varphi_1(x)p^{\kappa_1(x)}, \varphi_2(x)p^{\kappa_2(x)}, \varphi_3(x)p^{\kappa_3(x)}, \varphi_4(x)p^{\kappa_4(x)})$.

(d) *For all integers* N, *the subset* $\Sigma_N \subset \Sigma$ *of points* $x \in \Sigma$ *such that* $\kappa_{i+1}(x) - \kappa_i(x) > N$ *for* $i = 1, 2, 3$ *is infinite.*

This theorem is proved by constructing a p-adic family of cuspidal representations for $U(2,2)$ that specializes to an Eisenstein representation at the point x_0 and by looking at corresponding family of Galois representations. The strategy to construct this cuspidal family is explained in Section 4.

Lemma 3.2 *The pseudo representation* T *is (generically) irreducible.*

The proof of this generical irreducibility property goes along the same lines as the one of Theorems 3.3.12 and 4.2.7 of [SU06a]. It is done by contradiction. In loc. cit. one uses the fact that there are finitely many units in $\mathbb{Q}$, here the same fact for imaginary quadratic fields is crucial.

3.2 Construction of the desired extension

We keep the hypothesis and notations of Theorem 3.1. After replacing $\mathfrak{V}$ by a finite cover, we may assume that the representation attached to the pseudo-character T is defined over the fraction field of $A(\mathfrak{V})$. We may also assume that $A(\mathfrak{V})$ is a Dedekind domain. Then, we consider a lattice $\mathcal{L}$

[4] See the papers [Ro96] and [Ta91] for the definitions and properties of pseudo-characters.

[5] A similar property holds for the restriction of ρ_x to $G_{\wp^c}$ thanks to the polarization property satisfied by ρ_x.

of the representation of dimension 4 with trace given by T such that its localization $\mathcal{L}_{(x_0)}$ at the maximal ideal corresponding to x_0 has a unique irreducible quotient, this quotient being isomorphic to $V_f(m)$. To see how to construct such a lattice see [SU06b]. Let $\mathcal{L}_{x_0}$ the reduction of $\mathcal{L}$ modulo the maximal ideal corresponding to x_0. Notice that by condition (a) the semi-simplification of $\mathcal{L}_{x_0}$ is isomorphic to W_f. By construction, $V_f(m)$ is the unique irreducible quotient of $\mathcal{L}_{x_0}$. An important fact is the following.

Lemma 3.3 *$\mathcal{L}_{x_0}$ contains the trivial representation as a subrepresentation. The quotient E_f of $\mathcal{L}_{x_0}$ by this trivial subrepresentation is a nontrivial extension of the form*

$$0 \to L(1) \to E_f \to V_f(m) \to 0.$$

Proof Again the argument to prove this fact is already present in [SU06a] and [SU06b] at least when $k > 2$. If the first assertion were not true, then the representation $\mathcal{L}_{x_0}$ would contain a non trivial extension:

$$0 \to L(1) \to E' \to L \to 0.$$

By the condition (b) of the theorem, this extension would be unramified away from p. It remains to prove that this representation is crystalline at $\wp$ and $\wp^c$ which would give the contradiction we are seeking since $H^1_f(\mathcal{K}, L(1)) = 0$. By a result of B. Perrin-Riou, this extension is semi-stable at $\wp$ because it is ordinary. Let us call N' the monodromy operator on[6] $D_{st,\wp}(E')$. Let us consider the exterior square $\wedge^2\mathcal{L}_{x_0}$. It contains the representation $E' \otimes V_f(m)$ as a subquotient. This latter representation is also semi-stable since $V_f(m)$ is crystalline[7] and

$$D_{st,\wp}(E' \otimes V_f(m)) = D_{st,\wp}(E') \otimes D_{\mathrm{crys},\wp}(V_f(m))$$

Moreover, its monodromy operator is given by $N = N' \otimes Id_{D_{\mathrm{crys},\wp}(V_f(m))}$. On the other hand using a result of Kisin, we know that

$$D_{\mathrm{crys},\wp}(\wedge^2\mathcal{L}_{x_0})^{\Phi=\alpha p^{-m}} \neq 0.$$

This implies clearly that

$$\mathrm{rank}_L\, D_{\mathrm{crys},\wp}(E' \otimes V_f(m))^{\Phi=\alpha p^{-m}} = 1. \tag{3.1}$$

[6]For a representation W of $G_{\mathcal{K}}$, we put $D_{st,\wp}(W) := (B_{st} \otimes W)^{\mathrm{Gal}(\overline{\mathbb{Q}}_p/\mathcal{K}_\wp)}$ where B_{st} stands for the ring of semi-stable periods of Fontaine. It is equipped with a filtration and an action of Frobenius ϕ.

[7]This is because the conductor N of f is prime to p.

From the relation, $N\Phi = p\Phi N$ in $D_{st,\wp}(E')$, it is easy to see that if E' is not crystalline there exists $v' \in D_{st,\wp}(E')^{\Phi=1}$ such that $N'v' \neq 0$. If $v_\alpha \in D_{\mathrm{crys},\wp}(V_f(m))$ is an eigenvector for the eigenvalue αp^{-m}, therefore $v = v' \otimes v_\alpha$ is an eigenvector for the eigenvalue αp^{-m} and $N.v = Nv' \otimes v_\alpha \neq 0$ which contradicts (3.1) since $D_{\mathrm{crys},\wp}(E \otimes V_f(m))$ is the kernel of N. Therefore E' is crystallline at $\wp$. A similar argument applies for $\wp^c$ using the fact that the polarization property implies that the conditions satisfied for the restriction to the decomposition subgroup $D_\wp$ at $\wp$ are also satisfied for the restriction to the decomposition subgroup $D_{\wp^c}$ at $\wp^c$. The second assertion of the lemma is clear from the construction.

□

Assuming the hypothesis and the conclusions of Theorem 3.1, we are now in position to finish the proof of the part a) of Theorem 2.2.

Lemma 3.4 *If $v_p(\alpha) < k - 1$, then $[E_f] \in H^1_f(\mathcal{K}, V_f(m))$; in particular $H^1_f(\mathcal{K}, V_f(m)) \neq 0$.*

Proof By Remark 2.1, in order to prove that the extension class $[E_f]$ belongs to $H^1_f(\mathcal{K}_\wp, V_f(m))$, we only need to show that the restriction to $D_\wp$ of E_f is de Rham. Indeed, since $P_\wp(V_f(m), X) = (1-\alpha p^{-m}X)(1-\alpha^{-1}p^{m-1}X)$ and α is a Weil number of weight $2m-1$, we see that $P_\wp(V_f(m), p) \neq 0$ and $H^1_f(\mathcal{K}_\wp, V_f(m)) = H^1_g(\mathcal{K}_\wp, V_f(m))$. In order to prove that E_f is de Rham at $\wp$, we use Lemma 4.2.3 of [SU06b]. Let g be the projection map from E_f onto $V_f(m)$. We need to show there exists $D' \subset D_{dR,\wp}(E_f)$ such that

$$g \otimes id_{B_{dR}}(D') \oplus Fil^0 D_{dR,\wp}(V_f(m)) = D_{dR,\wp}(V_f(m)). \qquad (3.2)$$

Let D' be the image in $D_{dR,\wp}(E_f)$ of $D_{\mathrm{crys},\wp}(E_f)^{\Phi=\alpha p^{-m}}$. We know that $D' \neq 0$ by the Corollary 5.3 of [Ki03] (see also Proposition 4.2.2 of [SU06b]). Moreover its image by $g \otimes id_{B_{dR}}$ is the image D of $D_{\mathrm{crys},\wp}(V_f(m))^{\Phi=\alpha p^{-m}}$ in $D_{dR,\wp}(V_f(m))$. Since $v_p(\alpha.p^{-m}) < m-1$ and the Hodge-Tate numbers of $V_f(m)$ are $-m$ and $m-1$, we deduce that $D \cap Fil^0 D^{dR,\wp}(V_f(m)) = \{0\}$ by weak admissibility of $D_{\mathrm{crys},\wp}(V_f(m))$. This finishes the proof of (3.2) and that $[E_f] \in H^1_f(\mathcal{K}_\wp, V_f(m))$. Similarly we have $[E_f] \in H^1_f(\mathcal{K}_{\wp^c}, V_f(m))$.

One need to prove a similar fact at places w not dividing p. More precisely, one has to show that the following sequence is right exact:

$$0 \to L(1)^{I_w} \to E_f^{I_w} \to V_f(m)^{I_w} \to 0$$

but this follows easily by the property (b) of Theorem 3.1. This finishes the proof of our lemma.

□

4 Nearly Overconvergent Eisenstein Series on $U(2,2)$

In this section, we explain how the vanishing of the L-value $L(f,m)$ implies the existence of a certain overconvergent automorphic forms for the quasi-spplit unitary group $U(2,2)$. We will start by recalling standard facts on unitary automorphic forms mainly due to Shimura and then give their algebraic interpretations which enables us to have an arithmetic theory for nearly holomorphic forms. This will be the cornerstone of the construction of p-adic families of nearly holomorphic forms and the notion of nearly overconvergent forms. Strictly speaking, the theory of nearly overconvergent forms could be avoided as it will appear in the last subsection of these notes. However, we have introduced them because when $k=2$, the Eisenstein series which is deformed is obtained as a p-adic limit of nearly holomorphic Eisenstein series and cannot be seen as a nearly holomorphic form since its (p-adic) weight $(1,2,-2,1)$ is not dominant. So it is a purely p-adic object. We have not defined it as a section of some p-adic sheaf but this could be done easily by using the techniques of Andreatta-Iovita-Pilloni [AIP].

4.1 Nearly holomorphic unitary automorphic forms

4.1.1 Unitary automorphic forms

We consider the skew-hemitian form on $\mathcal{K}^4$ given by the matrix

$$J = \begin{pmatrix} 0_2 & 1_2 \\ 1_2 & 0_2 \end{pmatrix}.$$

Let $\mathbb{G} = GU(2,2) \subset \mathrm{GL}_{4/\mathcal{O}_\mathcal{K}}$ be the group scheme of unitary similitudes preserving the skew-hermitian form on $\mathcal{K}^4$ given by J . That is for any ring R, we put

$$\mathbb{G}(R) := \{g \in \mathrm{GL}_d(\mathcal{O}_\mathcal{K} \otimes_\mathbb{Z} R)\ : gJ^t\bar{g} = \nu(g)J\}.$$

with $\nu(g) \in \mathbb{G}_m(R) = R^\times$. We denote by $G = U(2,2)$ the unitary group defined as the kernel of $\nu : \mathbb{G} \to \mathbb{G}_m$. We similarly define $GU(1,1)$ and $U(1,1)$. A matrix $\gamma \in G(R)$ will be written by blocs of size 2×2 in the following way:

$$\gamma = \begin{pmatrix} a_\gamma & b_\gamma \\ c_\gamma & d_\gamma \end{pmatrix}.$$

The hermitian tube domain associated to G is the four dimensional complex analytic manifold $\mathcal{D}$ defined by

$$\mathcal{D} := \{z \in M_{2\times 2}(\mathbb{C})\ : i.(z^* - z) > 0\}$$

where we write $z^* = {}^t\bar{z}$. The identity component $G^+(\mathbb{R})$ of $G(\mathbb{R})$ acts transitively on $\mathcal{D}$ by the usual Möbius transformation

$$\gamma.z = (a_\gamma z + b_\gamma) \cdot (c_\gamma z + d_\gamma)^{-1} \text{ for } \gamma \in G^+(\mathbb{R}).$$

We consider the automorphic factor

$$j(\gamma, z) := (c_\gamma z + d_\gamma, (\bar{c}_\gamma {}^t z + \bar{d}_\gamma))$$

taking values in $H(\mathbb{C})$ with $H := \mathrm{GL}_2 \times \mathrm{GL}_2$. We also define

$$\Xi(z) := (i(\bar{z} - {}^t z), i(z^* - z)) \text{ and } r(z) := i(\bar{z} - {}^t z)^{-1}.$$

Let (ρ, V) be an algebraic representation of H and let $K \subset G(\mathbb{A}_f)$ be an open compact subgroup. We denote by $\mathcal{A}_\rho(K, \mathbb{C})$ the space of $V_{\mathbb{C}}$-valued real analytic functions on $G(\mathbb{A}_f) \times \mathcal{D}$ such that

$$f(\gamma.g_f.k, \gamma.z) = \rho(j(\gamma, z)).f(g_f, z)$$

for all $\gamma \in G(\mathbb{Q})$ and $k \in K$. Following Shimura [Sh04], we now define some differential operators on the space of automorphic forms. We first introduce a few more notations. Let $(S, M_2(-))$ be the representation of $H = \mathrm{GL}_2 \times \mathrm{GL}_2$ on the space of 2×2 matrices $M_2(-)$ given $S(g_1, g_2).M = g_1 M \, {}^t g_2$ for $M \in M_2(R)$ and $(g_1, g_2) \in H(R)$. Let St^+ (resp. St^-) be the standard representation of the first (resp. second) copy of GL_2 in H then we have $S \cong St^+ \otimes St^-$.

Let $(\frac{\partial}{\partial r_{ij}})_{i,j}$ be the differential operators on the differentiable functions on $\mathcal{D}$ defined by the relation

$$\frac{\partial}{\partial \bar{z}_{kl}} = \sum_{i,j} \frac{\partial r_{ij}}{\partial \bar{z}_{kl}} \frac{\partial}{\partial r_{ij}}$$

where for $1 \leq i, j \leq 2$, the function $r_{ij}(z)$ stands for the (i, j) entry of the function $r(z)$. We consider the differential operator ϵ_ρ from $\mathcal{A}_\rho(K, \mathbb{C})$ taking value in the space of real analytic $\mathrm{Hom}_{\mathbb{C}}(M_2(\mathbb{C}), V_{\mathbb{C}})$-valued functions on $G(\mathbb{A}_f) \times \mathcal{D}$ defined by:

$$(\epsilon_\rho f)(g_f, z)(u_{ij}) := \sum_{i,j} u_{ij} \cdot \frac{\partial}{\partial \bar{r}_{ij}} f(g_f, z)$$

for $(u_{ij}) \in M_2(\mathbb{C})$. If there is no possible confusion, we sometimes just write ϵ. The image of ϵ_ρ is contained in $\mathcal{A}_{S^\vee \otimes \rho}(K, \mathbb{C})$. The space of nearly holomorphic forms $\mathcal{N}_\rho^r(K, \mathbb{C})$ of order $\leq r$ is by definition the kernel of the differential operator of degree $r + 1$ defined by

$$\epsilon_\rho^{r+1} = \epsilon_{(S^\vee)^{\otimes r} \otimes \rho} \circ \cdots \circ \epsilon_\rho.$$

For $r = 0$, one obtains the usual space of holomorphic unitary automorphic forms of weight ρ.

We now recall the generalized Maass-Shimura differential operators. For $f \in \mathcal{N}_\rho^r(K, \mathbb{C})$, one defines $\delta_\rho.f \in \mathcal{N}_{\rho\otimes S}^{r+1}(K, \mathbb{C})$ as the function taking values in $V_\mathbb{C} \otimes M_2(\mathbb{C})$ defined by the formula

$$(\delta_\rho.f)(g_f, z) := \sum_{ij} \left(\rho(\Xi(z))^{-1} \frac{\partial}{\partial z_{ij}} [\rho(\Xi(z)).f(g_f, z)] \right) \otimes E_{ij}$$

where E_{ij} stands for the elementary matrix of $M_2(\mathbb{C})$ having the (i, j)-entry equal to 1 and zero elsewhere. If $\rho = \det^k \otimes 1$, we denote by δ_k the Maass-Shimura operator. We record the following lemma which follows from a simple direct computation.

Lemma 4.1 *Let f be an holomorphic form of scalar weight k. Then we have:*

$$\delta_k(f) = \sum_{i,j} \frac{\partial f}{\partial z_{ij}} E_{ij} + kf(z)\mathrm{tr}({}^t r(z) E_{**})$$

where E_{**} stands for the formal matrix with entries E_{ij}. Moreover

$$\epsilon_{\det^k \otimes S}(\delta_k f)(z) = kf(z) \sum_{ij} E_{ij}^* \otimes E_{ij}$$

where E_{ij}^* stands for the dual basis of E_{ij}. In particular $\epsilon_{\det^k \otimes S}(\delta_k f)$ takes values in the canonical invariant line of $S^\vee \otimes S$.

It is possible to define algebraic and arithmetic versions of these spaces and differential operators. We will do that in the next sections, we first need to introduce the unitary Shimura variety attached to $\mathbb{G}$.

4.1.2 Unitary Shimura variety

We fix a neat open compact subgroup $K^p \subset \mathbb{G}(\mathbb{A}_f^p)$ and we denote by $X = X_K$ the Shimura variety of level $K = K^p.\mathbb{G}(\mathbb{Z}_p)$ given by the usual Shimura data so that its complex points are given by

$$X_K(\mathbb{C}) = \mathbb{G}(\mathbb{Q}) \backslash (\mathcal{D} \times \mathbb{G}(\mathbb{A}_f)/K).$$

The Shimura variety X_K is a smooth quasi-projective scheme defined over $\mathcal{K}$ which is not geometrically connected in general. By the work of Kottwitz, we know it has a canonical model over $O_{(\wp)}$ that we denote by $\mathcal{X}_K$. The scheme $\mathcal{X}_K$ represents the functor which sends a $\mathcal{O}_{(\wp)}$-scheme S to the set of the equivalent classes of certain quadruplets $(f : A \to S, \iota, \lambda, \alpha)$ where

- $f : A \to S$ is an abelian scheme over S of relative dimension 4,
- ι is a a ring homomorphism $\iota : \mathcal{O}_{\mathcal{K}} \to \mathrm{End}_S(A)$,
- $\lambda : A \to A^t$ is a polarization of degree prime to p,
- α is a K^p-level structure, that is to say an isomorphism modulo K^p :

$$\alpha : H_1(A/S, \mathbb{A}_f^p) \cong (\mathbb{A}_f \otimes K)^4_{/S}.$$

Let $\omega_{A/S} = f_*\Omega_{A/S}$. It is a locally free sheaf over S. Let $\omega^+_{A/S}$ (resp. $\omega^-_{A/S}$) the sub-sheaf of sections on which the complex multiplication by $O_{(\wp)}$ coincides with the one (resp. with the conjugate of the one) induced by the $O_{(\wp)}$-scheme structure of S. We denote by α^t the K-level structure of A^t induced by λ and α. These quadruplets are required to satisfy the following conditions.

- $\omega^+_{A/S}$ and $\omega^-_{A/S}$ are locally free of rank 2 over S;
- The pairing on $(\mathbb{A}_f^p \otimes \mathcal{K})^4$ induced by α, α^t and the Weil pairing has matrix J.

We also consider the relative de Rham cohomology $\mathcal{H}^1_{dR}(A/S) := R^1 f_*\Omega^\bullet_{A/S}$. It is a locally free sheaf of rank 8 and it fits in the canonical exact sequence

$$0 \to \omega_{A/S} \to \mathcal{H}^1_{dR}(A/S) \to \omega^\vee_{A/S} \to 0$$

after we have identified $R^1 f_*\mathcal{O}_A$ with $\omega^\vee_{A/S}$ using Poincaré duality and the polarization λ. One defines its $+$ part $\mathcal{H}^1_{dR}(A/S)^+$ as we did for $\omega_{A/S}$ and for which we have

$$0 \to \omega^+_{A/s} \to \mathcal{H}^1_{dR}(A/S)^+ \to (\omega^-_{A/S})^\vee \to 0.$$

We define $\mathcal{J}(A/S)$ the sheaf obtained by making the following diagramm commutative and the bottom short sequence exact:

$$\begin{array}{ccccccccc}
0 \longrightarrow & \omega^+_{A/S} \otimes \omega^-_{A/S} & \longrightarrow & \mathcal{H}^1_{dR}(A/S)^+ \otimes \omega^-_{A/S} & \longrightarrow & (\omega^-_{A/S})^\vee \otimes \omega^-_{A/S} & \longrightarrow 0 \\
& \| & & \uparrow & & \uparrow & \\
0 \longrightarrow & \omega^+_{A/S} \otimes \omega^-_{A/S} & \longrightarrow & \mathcal{J}(A/S) & \longrightarrow & \mathcal{O}_S & \longrightarrow 0
\end{array}$$

We denote by $\pi : \mathcal{A} \to X$ the universal abelian scheme over $X = X_K$. We consider the sheaves $\omega^\pm := \omega^\pm_{\mathcal{A}/X}$ and $\mathcal{H}^\pm := \mathcal{H}^1_{dR}(\mathcal{A}/X)^\pm$. Let $\bar{X}$ be a smooth toroidal compactification of X over $\mathcal{O}_{(\wp)}$ constructed by K.W. Lan in his thesis [La08]. The boundary $\partial\bar{X} = \bar{X}\backslash X$ is a normal crossing divisor of $\bar{X}$. We denote by $\Omega_{X/O_{(\wp)}}(\log \partial\bar{X})$ the sheaf of Khäler differential having logarithmic poles along $\partial\bar{X}$. Recall we have the Gauss-Manin connexion

$$\nabla : \mathcal{H}^1_{dR}(\mathcal{A}/X) \to \mathcal{H}^1_{dR}(\mathcal{A}/X) \otimes \Omega_{X/O_{(\wp)}}(\log \partial\bar{X}).$$

It induces on the + part a map

$$\omega^+ \hookrightarrow \mathcal{H}^1_{dR}(\mathcal{A}/X)^+ \to \mathcal{H}^1_{dR}(\mathcal{A}/X)^+ \otimes \Omega_{X/O_{(\wp)}}(\log \partial\bar{X}) \\ \to (\omega^-)^\vee \otimes \Omega_{X/O_{(\wp)}}(\log \partial\bar{X})$$

which yields the Kodaira-Spencer isomorphism ([La08]):

$$\omega^+ \otimes \omega^- \xrightarrow{\sim} \Omega_{X/O_{(\wp)}}(\log \partial\bar{X}).$$

We will identify these sheaves. Notice that we therefore have

$$0 \longrightarrow \Omega_{X/O_{(\wp)}}(\log \partial\bar{X}) \longrightarrow \mathcal{J} \longrightarrow \mathcal{O}_S \longrightarrow 0$$

with $\mathcal{J} = \mathcal{J}(\mathcal{A}/X)$ and it could be seen that $\mathcal{J}^\vee$ is isomorphic to the sheaf of 1-jets on X.

4.1.3 Automorphic sheaves

We define now locally free coherent sheaves on our unitary Shimura varieties in order to get rational and integral structures on the spaces of automorphic forms we have introduced above. Since ω^+ and ω^- are locally free of rank 2, we may consider the following $H = \mathrm{GL}_2 \times \mathrm{GL}_2$-torsor over X:

$$\mathcal{T} := \mathrm{Isom}(\omega^+ \oplus \omega^-, (O_X)^2 \oplus (O_X)^2).$$

For any algebraic representation (ρ, V) of $H = \mathrm{GL}_2 \times \mathrm{GL}_2$, we denote by ω_ρ the coherent sheaf on X defined as the contracted product

$$\omega_\rho := V \times^H \mathcal{T}.$$

In particular we have $\omega_{St^\pm} = \omega^\pm$ and

$$\omega_S = \omega^+ \otimes \omega^- \cong \Omega_{X/\mathcal{O}_{(\wp)}}(\log \partial\bar{X}).$$

Let $\omega^1_\rho := \omega_\rho \otimes \mathcal{J}^\vee$. By the isomorphism above we have the short exact sequence of locally free sheaves:

$$0 \to \omega_\rho \to \omega^1_\rho \to \omega_{\rho\otimes S^\vee} \to 0.$$

The Hodge decomposition provides the isomorphisms:

$$H^0(X_K, \omega_{\rho/\mathbb{C}}) \cong \mathcal{N}_\rho^0(K, \mathbb{C}) \text{ and } H^0(X_K, \omega^1_{\rho/\mathbb{C}}) \cong \mathcal{N}_\rho^1(K, \mathbb{C}).$$

Moreover the map $\omega_\rho^1 \to \omega_{\rho\otimes S^\vee}$ induces the differential operator ϵ_ρ we have defined in Section 4.1.1. Using Leibnitz rule, there is a canonical way to define a connexion on $\mathcal{H}^1_{dR}(\mathcal{A}/X)^{\otimes^r}$, from which we deduce a connexion[8] :

$$\nabla_\rho : \omega_\rho \to \omega^1_{\rho\otimes S}$$

in the sense that it satisties the relation[9] :

$$\nabla_\rho(f.\omega) = df \otimes \omega + f\nabla\omega$$

for $f \in \mathcal{O}_X(U)$ and $\omega \in \omega_\rho(U)$ for any Zariski open set $U \subset X$

Remark 4.2 We can more generally define $\omega_\rho^r := \mathrm{Hom}_{\mathcal{O}_X}(\mathrm{Sym}^r(\mathcal{J}), \omega_\rho)$ and verify that

$$H^0(X_K, \omega^r_{\rho/\mathbb{C}}) = \mathcal{N}_\rho^r(K, \mathbb{C}).$$

But we will not use this generalization in this note.

We now give a more concrete definition of the sections of ω^1_ρ à la Katz. Let Q be the standard parabolic of GL_4 stabilizing a plane. Then we identify H with the Levi subgroup of Q by the map $(g_1, g_2) \mapsto \mathrm{diag}(g_1, {}^tg_2^{-1})$. Let X_{ij} be formal variables with $i, j \in \{1, 2\}$ and let $R[X_{ij}]_r$ be the polynomials in X_{ij} of total degree at most r with coefficients in a ring R. We consider the representation ρ^r_V of Q on $V_R[X_{ij}]_r := V_R \otimes R[X_{ij}]_r$ given by

$$(\rho_V^r(g).P)(\underline{X}) := \rho_V(\mathrm{diag}(a_g, {}^td_g^{-1}).P((a_g^{-1}.\underline{X}d_g - a_g^{-1}b_g)$$

where

$$\underline{X} = \begin{pmatrix} X_{11} & X_{12} \\ X_{21} & X_{22} \end{pmatrix} \text{ and } g = \begin{pmatrix} a_g & b_g \\ 0 & d_g \end{pmatrix} \in Q(R)$$

A global section φ of ω^r_ρ can be seen as a functorial rule defined as follows. We consider the quintuplet $(f : A \to \mathrm{Spec}(R), \iota, \lambda, \alpha, \psi)$ where $(f : A \to \mathrm{Spec}(R), \iota, \lambda, \alpha)$ is an in Section 4.1.2 and ψ is an isomorphism $\mathcal{H}^1_{dR}(A/\mathrm{Spec}(R)^+ \cong R^4$ inducing $\omega^+_{A/\mathrm{Spec}(R)} \cong R^2 \oplus \{0\} \subset R^4$. Then φ can be seen as a functor on such quintuplets taking values in $V_R[X_{ij}]_r = V \otimes R[X_{ij}]_r$ and such that

$$\varphi(f : A \to \mathrm{Spec}(R), \iota, \lambda, \alpha, g \circ \psi) = \rho_V^r(g)\varphi(f : A \to \mathrm{Spec}(R), \iota, \lambda, \alpha, \psi)$$

for $g \in Q(R)$.

---

[8]The fact that the image of ω_ρ is contained in $\omega^1_{\rho\otimes S}$ follows from Griffith transversality.

[9]Notice that $df \otimes \omega \in \Omega_{X/\mathcal{O}_{(\wp)}}(\log \partial\bar{X}) \otimes \omega_\rho = \omega_{\rho\otimes S} \subset \omega^1_{\rho\otimes S}$.

4.1.4 Polynomial q-expansions

We define the polynomial q-expansion of a nearly holomorphic form by evaluating it on a Mumford-Tate object. The general theory of the q-expansion of holomorphic forms is written in great details by K.W. Lan in [La08, La12]. We extend here the definition in the case of nearly holomorphic forms. For simplicity, we now assume that $K = K_f \subset \mathbb{G}(\hat{\mathbb{Z}})$. We consider the lattice $\mathcal{H} = \mathcal{H}_{K_f}$ of hermitian matrices inside $\mathcal{H}er_2(\mathcal{K}) \subset M_2(\mathcal{K})$ such that

$$\mathcal{H} := \left\{ h \in M_2(\mathcal{K}) | \begin{pmatrix} 1_2 & h \\ 0_2 & 1_2 \end{pmatrix} \in K_f \right\}.$$

We denote by $\mathcal{H}^\vee$ the dual lattice for the pairing on $\mathcal{H}er_2(\mathcal{K})$ defined by $(h, h') = tr(hh') \in \mathbb{Q}$. We will denote $\mathcal{H}^\vee_{\geq 0}$ the submonoid of $\mathcal{H}^\vee$ of positive hermitian matrices. For any ring A and a monoid M with a neutral element 0, we denote by $A[[q^M]]$ the formal power series ring with coefficient in A over the monoid M. We denote element q^h of H inside $A[[q^M]]$ multiplicatively as q and q^0 is just denoted 1. We will consider the case A is a $\mathcal{O}_\mathcal{K}$-algebra and $M = \mathcal{H}^\vee_{\geq 0}$.

We consider the decomposition of $\mathcal{K}^4 = W \oplus W'$ where W (respectively W') is the standard totally isotropic subspace of vectors whose last (respectively first) two coordinate entries with respect to the standard basis of $\mathcal{K}^4$ are zero. Let L and L' be respectively the free $\mathcal{O}_\mathcal{K}$-lattices of W and W' such that $L \oplus L' = \mathcal{O}_\mathcal{K}^4$. For any $h \in \mathcal{H}$, h induces a canonical $\mathcal{O}_\mathcal{K}$-linear map $h : L \to L'$. Let $\underline{q}$ the map

$$\underline{q} : L \to L' \otimes \mathbb{G}_{m/\mathcal{O}_{(\wp)}[[q^{\mathcal{H}_{\geq 0}}]]}$$

defined by the composition

$$L \to \mathrm{Hom}(H, L') = H^\vee \otimes L' \to L' \otimes \mathbb{G}_{m_{/\mathcal{O}_{(\wp)}[[q^{\mathcal{H}^\vee_{\geq 0}}]]}}$$

where the first map is the obvious map and the last one is defined by $h \otimes l' \mapsto l' \otimes q^h$. By the work of Mumford, there exists a abelian variety $\mathrm{Mum}(q)$ over $\mathcal{O}_{(\wp)}((q^\mathcal{H}))$ endowed with a canonical complex multiplication noted ι_{can} by $\mathcal{O}_\mathcal{K}$ which can be described as the quotient $L' \otimes \mathbb{G}_m / \underline{q}(L)$. In particular, the formal completion along the origin gives a canonical isomorphism:

$$\widehat{\mathrm{Mum}(q)} \cong L' \otimes \widehat{\mathbb{G}}_{m/\mathcal{O}_{(\wp)}((q^\mathcal{H}))}$$

which induces a canonical isomorphism

$$\omega_{\mathrm{Mum}(q)/\mathcal{O}_{(\wp)}((q^\mathcal{H}))} \cong L \otimes_\mathbb{Z} \mathcal{O}_{(\wp)}((q^\mathcal{H})).$$

From the decomposition $\mathcal{O}_{(\wp)}\otimes\mathcal{O}_{(\wp)} \cong \mathcal{O}_{(\wp)}\oplus\mathcal{O}_{(\wp)}$ given by $z\otimes a \mapsto (za, \bar{z}a)$, we deduce an isomorphism

$$\omega^+_{\mathrm{Mum}(q)/\mathcal{O}_{(\wp)}((q^{\mathcal{H}}))} \cong L \otimes_{\mathcal{O}_{(\wp)}} \mathcal{O}_{(\wp)}((q^{\mathcal{H}})).$$

We define $\omega^+_{\mathrm{can}} = (\omega_{1,\mathrm{can}}, \omega_{2,\mathrm{can}})$ the basis of the left hand side of this isomorphism induced by the canonical basis of L. We complete it into a basis of $\mathcal{H}^1_{dR}(\mathrm{Mum}(q)/\mathcal{O}_{(\wp)}((q^{\mathcal{H}})))^+ \cong L \otimes_{\mathbb{Z}_p} \mathcal{O}_{(\wp)}((q^{\mathcal{H}}))$ using the Gauss-Manin connection. Let D_{ij} be the derivation of $\mathcal{O}_{(\wp)}((q^{\mathcal{H}}))$ such that $D_{ij}(q^h) = h_{ij}q^h$. Then we define[10]

$$\delta_{i,\mathrm{can}} = \nabla(D_{ii})(\omega_{i,\mathrm{can}}).$$

Then $(\omega_{1,\mathrm{can}}, \omega_{2,\mathrm{can}}, \delta_{1,\mathrm{can}}, \delta_{2,\mathrm{can}})$ is a basis of $\mathcal{H}^1_{dR}(\mathrm{Mum}(q)/\mathcal{O}_{(\wp)}((q^{\mathcal{H}}))^+$ which defines an isomorphim

$$\psi_{\mathrm{can}} : \mathcal{H}^1_{dR}(\mathrm{Mum}(q)/\mathcal{O}_{(\wp)}((q^{\mathcal{H}}))^+ \cong \mathcal{O}_{(\wp)}((q^{\mathcal{H}}))^4$$

as in the end of the previous section. Finally, it is not difficult to define a canonical polarization λ_{can} and a canonical level structure α_{can}. We are now ready to define the polynomial q-expansion[11] of a global section of ω^1_ρ. For any $\mathcal{O}_{(\wp)}$-algebra R, we consider the map

$$H^0(X_K, \omega^1_{\rho/R}) \to R((q^{\mathcal{H}})) \otimes_R V_R[X_{ij}]_1$$

defined by

$$f \mapsto f(q, X_{ij}) := f(\mathrm{Mum}(q)/\mathcal{O}_{(\wp)}((q^{\mathcal{H}})), \lambda_{\mathrm{can}}, \iota_{\mathrm{can}}, \alpha_{\mathrm{can}}, \psi_{\mathrm{can}}).$$

Moreover, it can be shown as usual that

$$f(q, X_{ij}) \in V_R[[q^{\mathcal{H}^\vee_{\geq 0}}]][X_{ij}]_1.$$

One can see the action of ϵ_ρ on the polynomial q-expansion is given by the following formula:

$$(\epsilon_\rho f)(q, X_{ij}) = \sum_{ij} E^*_{ij} \otimes \frac{\partial}{\partial X_{ij}} f(q, X_{ij}).$$

[10]It can be checked easily using the complex uniformization that it defines an horizontal section.

[11]In fact, we can (and need to) define a polynomial expansion for each connected component of X_K to be able to formulate a polynomial q-espansion principle.

4.2 p-adic unitary automorphic forms

4.2.1 p-adic unitary automorphic forms

Let X_{rig} be the rigid space obtained as the generic fiber of the formal scheme obtained by taking the formal completion of $\mathcal{X}_K$ along its special fiber at p. We consider $X_{\text{ord}} \subset X_{\text{rig}}$ the ordinary locus (i.e. the open rigid analytic subvariety of points $(f : A \to \mathrm{Spec}\overline{\mathbb{Q}}_p, \iota, \lambda, \alpha)$ for abelian varieties A having good ordinary reduction). The space of p-adic forms of weight ρ is defined as

$$M_\rho^{p-adic}(K, \mathbb{Q}_p) := H^0(X_{\text{ord}}, \omega_{\rho/\mathbb{Q}_p}).$$

We defined the spaces of overconvergent and nearly overconvergent forms of degree at most 1 by:

$$\mathcal{N}_\rho^{r,\dagger}(K, \mathbb{Q}_p) := \lim_{V \supset X_{\text{ord}}} H^0(V, \omega^r_{\rho/\mathbb{Q}_p}).$$

Here the V's in the injective limit run in the set of strict neighborhood of X_{ord} inside X_{rig}. Recall that Dwork defines a canonical splitting $\mathcal{H}_{dR}(\mathcal{A}/X_{\text{ord}}) \cong \omega \oplus \mathcal{U}$ where $\mathcal{U}$ is the unit root crystal of $\mathcal{H}_{dR}(\mathcal{A}/X_{\text{ord}})$. For any representation ρ of H, it induces a splitting

$$\omega^r_{\rho/X_{\text{ord}}} \overset{split_p}{\longrightarrow} \omega_{\rho/X_{\text{ord}}} \to 0.$$

The following proposition follows from the fact that the canonical Dwork splitting does not extend to any strict neighborhood of X_{ord}. We will write a proof in a subsequent paper (See [Ur12] in the case GL(2)). It shows nearly overconvergent forms or just nearly holomorphic forms can be seen as special p-adic forms. It will not be used in this paper.

Proposition 4.3 *For any strict neighborhood V of X_{ord}, the compositum of the two following maps*

$$H^0(V, \omega^r_{\rho/\mathbb{Q}_p}) \to H^0(X_{\text{ord}}, \omega^r_{\rho/\mathbb{Q}_p}) \overset{split_p}{\longrightarrow} H^0(X_{\text{ord}}, \omega_{\rho/\mathbb{Q}_p})$$

induces a canonical injection

$$\mathcal{N}_\rho^{r,\dagger}(K, \mathbb{Q}_p) \hookrightarrow M_\rho^{p-adic}(K, \mathbb{Q}_p).$$

Let $I \subset \mathbb{G}(\mathbb{Z}_p)$ be the Iwahori subgroup associated to the standard Borel subgroup of GL_4. We can define the unitary Shimura variety of Iwahori level above X we denote by X^I_{rig} the corresponding rigid analytic space and X^I_{ord} its ordinary part. We can define similarly the space of p-adic , overconvergent and nearly overconvergent forms of Iwahori level in a similar way as in [PS11]. We denote them respectively $\mathcal{M}_\rho^{p-adic}(K^pI, \mathbb{Q}_p)$, $\mathcal{M}_\rho^\dagger(K^pI, \mathbb{Q}_p)$ and $\mathcal{N}_\rho^{r,\dagger}(K^pI, \mathbb{Q}_p)$. The proposition above extends to these spaces without difficulty.

4.2.2 Families of finite slope nearly overconvergent forms

Weights. For any decreasing quadruplet of integers $\underline{k} := (k_1, k_2; k_3, k_4)$, we write $\omega_{\underline{k}}$ and $\omega^1_{\underline{k}}$ for the sheaves attached to the representation $\rho_{\underline{k}}$ of H given by $V^+_{k_1,k_2} \otimes V^-_{-k_4,-k_3}$ where we denote by $V^+_{a,b}$ (resp. $V^-_{a,b}$) the representation of the first (resp. second) copy of GL_2 in H of highest weight (a, b) for any pair of integers (a, b) with $a \geq b$. A unitary automorphic form of one of the types we have defined before will be said of weight $\underline{k}$ if the corresponding representation of H is $\rho_{\underline{k}}$. It is well-known that an holomorphic eigenform of weight $\underline{k}$ is attached to an automorphic representation which archimedean component is a discrete series when $k_2 - 2 \geq k_3 + 2$. It will be convenient to consider the coordinate function functor on $V_{\underline{k}/\mathbb{Z}}$ along the highest weight vector. We denote this function by

$$\pi_{\underline{k}} : V_{\underline{k}}(R) \to R.$$

It satisfies

$$\pi_{\underline{k}}(tn.v) = t_1^{k_1} t_2^{k_2} t_3^{-k_4} t_4^{-k_3} \pi_{\underline{k}}(v)$$

for any diagonal $t = \mathrm{diag}(t_1, t_2, t_3, t_4) \in H$ and n in the standard unipotent subgroup of H. It is defined up to sign.

Weight space. Let $\mathfrak{X}$ be the rigid analytic space such that

$$\mathfrak{X}(L) := \mathrm{Hom}_{\mathrm{cont}}((\mathbb{Z}_p^\times)^4, L^\times).$$

The points of $\mathfrak{X}(\overline{\mathbb{Q}}_p)$ are called p-adic weights. If $\underline{k} = (k_1, k_2, k_3, k_4) \in \mathbb{Z}^4$, we write $[\underline{k}]$ the point of $\mathfrak{X}(\mathbb{Q}_p)$ corresponding to the continuous character $(x_1, x_2, x_3, x_4) \mapsto \prod_{i=1}^4 x_i^{k_i}$. Those points are called algebraic weights if $k_1 \geq k_2 \geq k_3 \geq k_4$. We denote by $\mathfrak{X}(\mathbb{Q}_p)^{\mathrm{alg}}$ the subset of algebraic weights of $\mathfrak{X}$.

Slopes. For each $t = \mathrm{diag}(t_1, t_2, t_3, t_4) \in T(\mathbb{Q}_p)$ such that

$$v_p(t_1) \leq v_p(t_2) \leq v_p(t_3) \leq v_p(t_4), \tag{4.1}$$

we consider the Hecke operators u_t attached to the double class ItI and acting on the various spaces automorphic forms of level $K^p.I$ we have defined. It is important to remember that there are two way to normalize the action of these operators on the spaces $\mathcal{N}^r_{\underline{k}}$. We call them the algebraic and p-adic normalizations. The difference between the two normalization is given by the following formula:

$$(u_t)^{p-adic} = |\lambda_{\underline{k}}(t)|_p (u_t)^{\mathrm{alg}} \tag{4.2}$$

where $\lambda_{\underline{k}}$ is the algebraic weight[12] $(k_1 - 2, k_2 - 2, k_3 + 2, k_4 + 2)$. Here the algebraic normalization is defined as usual by

$$(f|u_t)(g_f, z) := \sum_i f(g_f \xi_i^{-1}, z)$$

where $I\xi_i$ are the left coset representatives such that $ItI = \sqcup_i I\xi_i$. The p-adic normalization extends to an action on the space of overconvergent and p-adic forms and this action preserves the integrality of those. We refer to [Hi04] for this fact and to [PS11] for the definition of the action on the overconvergent forms using the theory of the canonical subgroup. This definition extends easily to nearly overconvergent forms. Again we don't really need this fact here but it is an important feature of the general theory which is good to keep in mind.

Let $\mathcal{U}_p$ the Hecke algebra generated by the u_t's for t satisfying (4.1). It is isomorphic to a polynomial algebra in 4 variables. More precisely if θ is a character of $\mathcal{U}_p$, one can find a quadruple $(\alpha_1, \alpha_2, \alpha_3, \alpha_4) \in \overline{\mathbb{Q}}_p^4$ such that

$$\theta(t) = \prod_{i=1}^{4} \alpha_{5-i}^{v_p(t_i)}.$$

We say that θ is of finite slope if the α_i's are all non zero and we define the slope of θ as the quadruplet $\underline{s} = (s_1, s_2, s_3, s_4)$ with $s_i = v_p(\alpha_i)$ for $i = 1, 2, 3, 4$. A p-adic form is said of finite slope if it belongs to the sum of generalized $\mathcal{U}_p$-eigenspaces of finite slope characters.

Remark 4.4 If θ is a character of $\mathcal{U}_p$ acting on the I-invariants of an unramified principal series, then the corresponding Hecke polynomial is

$$\prod_{ii=1}^{4} (X - p^{i-1}\alpha_i).$$

Definition 4.5 Let $\mathfrak{V}$ be an affinoid over a finite extension of $\mathbb{Q}_p$ and $w : \mathfrak{V} \to \mathfrak{X}$ be a finite morphism. We assume that the subset $\Sigma_{\mathfrak{V}} = \mathfrak{V}(\overline{\mathbb{Q}}_p) \cap w^{-1}(\mathfrak{X}^{\mathrm{alg}}(\mathbb{Q}_p))$ is Zariski dense. A $\mathfrak{V}$-family[13] of nearly overconvergent forms F of degree at most r is a polynomial q-expansion[14] of degree at

[12] This character is dominant exactly when the weight correspond to holomorphic discrete series and is the cohomological weight of this holomorphic discrete series.

[13] It is possible to give a definition using a theory similar to [AIP] for unitary groups. Instead we use a shortcut here which is sufficient for our application.

[14] In fact, the correct definition is to take a formal q-expansion for each connected component of X_K. We will not do it for the sake of the notations.

most r with coefficient in $A(\mathfrak{V})$ (i.e. $F \in A(\mathfrak{V})[X_{ij}]_r[[q^{\mathcal{H}^\vee_{\geq 0}}]]$) such that there exists a Zariski dense set $\Sigma_F \subset \Sigma_{\mathfrak{V}}$ satisfying:

$\forall x \in \Sigma_F$, there exist a *finite slope* nearly holomorphic form F_x of weight $\underline{k}_x$ with $w(x) = [\underline{k}_x]$ and level $K^p.I$ such that $\iota_p(\pi_{\underline{k_x}}(F_x(q, X_{ij}))) \in \overline{\mathbb{Q}}_p[X_{ij}][[q^{\mathcal{H}^\vee_{\geq 0}}]]$ is equal to the evaluation of the formal polynomial q-expansion F at the point x.

4.3 Eisenstein series

4.3.1 Klingen-type Eisenstein series

We recall some results of [SU06b] on certain Eisenstein series for G. Let P be the stabilizer in G of the line $\{(0,*,0,0) \in \mathcal{K}^4 \;: * \in \mathcal{K}\}$. Then P is a standard, maximal $\mathbb{Q}$-parabolic subgroup of G with standard Levi subgroup L isomorphic to $U(1,1) \times \mathrm{Res}_{\mathcal{K}/\mathbb{Q}}\mathbb{G}_m$. A pair $(g,t) \in U(1,1) \times \mathrm{Res}_{\mathcal{K}/\mathbb{Q}}\mathbb{G}_m$ is identified with

$$m(g,t) = \begin{pmatrix} a & 0 & b & 0 \\ 0 & \bar{t}^{-1} & 0 & 0 \\ c & 0 & d & 0 \\ 0 & 0 & 0 & t \end{pmatrix} \in G.$$

with $g = \left(\begin{smallmatrix} a & b \\ c & d \end{smallmatrix}\right) \in U(1,1)$ and $t \in \mathbb{G}_{m/\mathcal{K}}$. We write N for the unipotent radical of P. The modulus function giving the determinant of the action of L on the Lie$N$ is given by

$$\delta(m(g,t)) = |t|^{-3}_{\mathbb{A}_\mathcal{K}}.$$

Let f be an elliptic cusp form of weight $k = 2m > 2$ for $\Gamma_0(N)$. We denote by ϕ_f be the automorphic form on $U(1,1)$ such that:

$$\varphi_f(\gamma g_\infty k z) = (c_\infty i + d_\infty)^{-m}(\bar{c}_\infty i + \bar{d}_\infty)^{-m} f(g_\infty.i)$$

for $\gamma \in U(1,1)(\mathbb{Q})$, $g_\infty \in U(1,1)(\mathbb{R})$, z in the center of $U(1,1)(\mathbb{A})$ and $k \in U(1,1)(\hat{\mathbb{Z}})$ such that c_k is divisible by N. We denote by (π, V_f) the irreducible cuspidal representation of $U(1,1)$ generated by φ_f. For any $s \in \mathbb{C}$, we then consider the induced representation $I(s)$ of smooth functions $\phi_s : G(\mathbb{A}) \to V_f$ satisfying

$$\phi_s(h.m(g,t).n) = \delta(m(g,t))^{1/2}|t|_\mathbb{A}^{-s}\pi(g).\phi_s(h)$$

for all $h \in G(\mathbb{A}), g \in U(1,1)(\mathbb{A})$ and $t \in \mathbb{A}_\mathcal{K}^\times$. Let us decompose π into the restricted tensor product of its local component $\pi_f = \bigotimes'_v \pi_v$. For each place v, we consider the local induction $I_v(s) = \mathrm{Ind}_{P(\mathbb{Q}_c)}^{G(\mathbb{Q}_v)} \pi_v \otimes \delta^{s/3}$. Then we have

$$I(s) = \bigotimes_v{}' I_v(s).$$

For each finite place v, we denote $L(\pi_v)$ the Langlands quotient of $I_v(\pi_v) := I_v(s_0)$ for $s_0 = 1/2$.

Let w be a finite place of $\mathcal{K}$ above v. Let us write W_{K_w} for the Weil group of $\mathcal{K}_w$ and denote rec_w the isomorphism of the local class field theory $W_{\mathcal{K}_w}^{ab} \xrightarrow{\sim} \mathcal{K}_w^\times$ sending arithmetic Frobenius onto uniformizor and by $N_w = |\cdot|_w \circ \mathrm{rec}_w$. By the local Langlands correspondence, the parameter of the base change Π_w of $L(\pi_v)$ to $\mathrm{GL}_4(\mathcal{K}_w)$ with w a finite place of $\mathcal{K}$ above v is given by:

$$\mathrm{rec}(\Pi_w)(x) = \begin{pmatrix} N_w(x)^{1/2} & & \\ & \mathrm{rec}(\pi_w)(x) & \\ & & N_w(x)^{-1/2} \end{pmatrix} \quad \forall x \in W_{\mathcal{K}_w} \tag{4.3}$$

where π_w stands for the base change of π_v to $\mathrm{GL}_2(K_w)$ and $\mathrm{rec}(\pi_w)$ is the image of π_w by the local Langlands correspondence for $\mathrm{GL}_2(\mathcal{K}_w)$. In particular, the local L-function of Π_w is given by:

$$L(\Pi_w, s) = (1 - q_w^{s+1/2})^{-1} L(\pi_w, s)(1 - q_w^{s-1/2})^{-1}.$$

Let $\Phi \in I(\pi)$ such that $\Phi(1) \in \Pi \otimes \mathbb{C}(m, -m)]^{U(1)\times U(1)}$. Here we denote $\mathbb{C}(a, b)$ is the one dimensional representaion of $U(1) \times U(1)$ defined by the character $(u_1, u_2) \mapsto u_1^a u_2^{-b}$. Then, we consider the form f_Φ on $\mathfrak{H}$ defined by:

$$f_\Phi(z) = (ci + d)^m (\bar{c}i + \bar{d})^m \Phi(1)(\left(\begin{smallmatrix} a & b \\ c & d \end{smallmatrix}\right))$$

with $z = \frac{ai+b}{ci+d}$ belongs to the Poincaré upper half plane $\mathfrak{H}$. It is an holomorphic form of weight k. Let N an integer divisible by the conductor of π and $\phi \in \prod_{v|N} I(\pi(f)_v)$, we consider

$$\Phi = \phi \otimes (\bigotimes_{\substack{v \text{ such that} \\ v(N)=0}} \phi_v^0) \otimes \phi_\infty$$

where ϕ_v^0 is a canonical spherical section of $I(\pi_v)$ for $v \not| N$ and $\phi_\infty(1)$ is a basis of $[\pi_\infty \otimes \mathbb{C}(m, -m)]^{U(1)\times U(1)}$. We now define the Eisenstein series attached to ϕ. For $z \in \mathcal{D}$, $g_f \in G(\mathbb{A}_f)$ and $s \in \mathbb{C}$ such that $\mathrm{Re}(s)$ is sufficiently large, we put

$$E(\phi, s)(g_f, z) := \sum_{\gamma \in P(\mathbb{Q})\backslash G(\mathbb{Q})} \rho_{\mathcal{K}_m}(j(\gamma, z)) |\det(c_\gamma z + d_\gamma)|^{-s} . f_{\gamma_f g_f . \Phi}([\gamma . z]) v_{\mathcal{K}_m}$$

where the map $[\cdot] : \mathcal{D} \to \mathfrak{H}$ is defined by $[z] = z_{11}$ for $z = \left(\begin{smallmatrix} z_{11} & z_{12} \\ z_{21} & z_{22} \end{smallmatrix}\right) \in \mathcal{D}$, the weight κ_m is defined by

$$\kappa_m := (m, 2; -2, -m),$$

v_{κ_m} is an highest weight vector of ρ_{κ_m}, γ_f is the image of γ in $G(\mathbb{A}_f)$ and $\gamma_f g_f.\Phi$ is defined by the action of $G(\mathbb{A}_f)$ on $I(\pi)$.

Proposition 4.6 *Let f be an eigenform of weight $k = 2m > 2$ for $\Gamma_0(N)$ and $\phi \in \prod_{v|N} I(\pi(f)_v)$, then $E_{\kappa_m}(\phi, s)$ has no pole at $s = 0$ and its evaluation at $s = 0$ is a nearly holomorphic form $E_{\kappa_m}(f, \phi)$ of weight κ_m of order at most 1. It is an holomorphic form if $L(f, m) = 0$. This latter condition is necessary if ϕ projects non trivially in the Langlands quotient $\otimes_{v|N} L(\pi_v)$. Moreover, after an adequate normalization, $E_{\kappa_m}(f, \phi)$ is defined over $\overline{\mathbb{Q}}$.*

Proof This was proved in [SU06b] except the fact that it is nearly holomorphic in general. However this point follows easily from the computation of the constant term realized in loc. cit. This will be explained in greater details in [SU13]. The algebraicity follows from a general result due to M. Harris [Ha81]. His argument is written only for holomorphic Eisenstein series but it extends easily to nearly holomorphic ones.

□

Remark 4.7 Let $E_{\kappa_m}(f)$ be the image of the $G(\mathbb{A}_f)$-representation generated by the $E_{\kappa_m}(f, \phi)$'s in the product of the local Langlands quotients at the finite places. It is an irreducible representation and we have

$$L(BC(E_{\kappa_m}(f)), s) = L(f, s + m - 1/2)\zeta_{\mathcal{K}}(s + 1/2)\zeta_{\mathcal{K}}(s - 1/2)$$

with $\zeta_{\mathcal{K}}$ the Dedekind zeta function of $\mathcal{K}$. According to the conventions of [SU06b, §4], its associated Galois representation is

$$\rho_{E_{\kappa_m}(f)} = L(-1) \oplus L(-2) \oplus V_f(m - 2).$$

Theorem 3.1 establishes the existence of a deformation of $\rho_{E_{\kappa_m}(f)}(2)$ when $L(f, m) = 0$. It will follow from the existence of a p-adic family of cusp forms degenerating into a p-adic version of $E_{\kappa_m}(f)$. The construction is outlined in the next paragraphs. Notice that when $k = 2$, the previous proposition does not hold and we will need to replace $E_{\kappa_m}(f)$ by a finite slope p-adic automorphic representation having the Galois representation $W_f(-2)$.

4.3.2 Families of Eisenstein series

Let $\mathfrak{X}_1$ be the rigid rigid variety over $\mathbb{Q}_p$ such that

$$\mathfrak{X}_1(\overline{\mathbb{Q}}_p) = \mathrm{Hom}_{\mathrm{cont}}(\mathbb{Z}_p^{\times}, \overline{\mathbb{Q}}_p^{\times}).$$

Let $\mathfrak{U}$ be an affinoid, $k : \mathfrak{U} \to \mathfrak{X}_1$ be a finite morphism and

$$F = \sum_{n=1}^{\infty} a(n, F)q^n \in A(\mathfrak{U})[[q]]$$

be a Coleman family[15] of normalized new cuspidal eigenforms for $\Gamma_0(N)$ of slope $s_0 \in \mathbb{Q}_{\geq 0}$. This means that for each $x \in \mathfrak{V}(\mathbb{Q}_p)$ such that $k(x) = [k_x]$ with $k_x \in \mathbb{Z}_{\geq 2}$ and $k_x > s_0 + 1$, $F_x = \sum_{n=1}^{\infty} a(n, F)(x)q^n = \iota_p(f_x(q))$ where $f_x(q)$ is the q-expansion of a normalized N-new eigenform of even weight $k_x = 2m_x$ and level $\Gamma_0(Np)$.

Theorem 4.8 ([SU13]) *We keep the hypothesis and notations as above. Let* $\kappa = i \circ k$ *with* i *the closed immersion of* $\mathfrak{X}_1$ *into* $\mathfrak{X}$ *given by* $i(\xi) = (\xi^{1/2}, [2], [-2], \xi^{-1/2})$*. Then there exists a* $\mathfrak{U}$*-family of nearly holomorphic automorphic forms* $E(F)$ *such that for all* $x \in \mathfrak{V}(\mathbb{Q}_p)$ *such that* $k(x) = [k_x]$ *and* f_x *is of trivial nebentypus and weight* $k_x = 2m_x \in \mathbb{Z}$*, we have:*

(i) *If* $k_x \in \mathbb{Z}_{\geq 4}$ *and* $k_x > s_0+1$*, then* $E(F)_x$ *is the polynomial* q*-expansion of a nearly holomorphic Klingen-Eisenstein series* $E_{\kappa_{m_x}}(f_x, \phi_x)$ *of weight* $\kappa_x = \kappa_{m_x}$ *for some section* $\phi_x \in \otimes_{v|Np} I_v(\pi(f_x)_v)$ *projecting non trivially on the Langlands quotient* $\otimes_{v|Np} L(\pi(f_x)_v)$.

(ii) $E(F)_x$ *is an eigenform of finite slope for a character of* $\mathcal{U}_p$ *with normalized eigenvalues given by the quadruplet*

$$(a(p, F)(x), p, p^{-1}, a(p, F)(x)^{-1}.$$

In particular, its slope is $\underline{s}_0 = (s_0, 1, -1, -s_0)$.

(iii) *If* $k_x \geq 2$, $\epsilon(E(F)_x) = 0$ *when* $L(f_x, k_x/2) = 0$.

(iv) *If* $k_x \geq 2$*, then* $E(F)_x$ *is non trivial.*

The general construction of this family is done using the doubling method and the use of differential operators on the Siegel Eisenstein series on $U(3, 3)$. It uses an explicit description of certain harmonic polynomials and the effect of the Maass-Shimura differential operators on the polynomial q-expansion. The fact that we can make an analytic interpolation is easy

[15] This corresponds to an affinoid of the Eigencurve of Coleman-Mazur on which the slope is constant.

from these facts by using the standard technique of the p-adic Petersson inner product due to Hida.

The point (ii) is done by making a good choice of families of sections at p, that is, Iwahori invariant sections which are proper for the Hecke operators u_t with eigenvalues given by the same quadruplet after renormalization. Notice that the slope of the Eisenstein series $E(F)_x$ is critical with respect to the weight $\lambda_{\kappa_x} = (m_x - 2, 0, 0, 2 - m_x)$ in the sense of [Ur11]. Therefore any larger deformation of this family will no longer be Eisenstein. The point (iii) follows from Proposition 4.6 when $k_x > 2$. For the case $k_x = 2$, it follows by showing that $\epsilon(E(F)_x)$ is divisible by the p-adic L-function interpolating the central values $L(f_x, m_x)$. This fact is another consequence of the doubling method. The point (iv) is easy when $k_x > 2$, it follows from the computation of the Fourier coefficients of the Eisenstein series. When $k_x = 2$, one needs[16] to show that some p-sdic limit is non trivial and we use crucially that f is in the image of the Jacquet-Langlands correspondence for a definite quaternion algebra.

When $k_x = 2$, $E(F)_x$ is the q-expansion of a p-adic form as it is obtained as a limit of p-adic (since nearly holomorphic) forms. Its weight is $(1, 2; -2, -1)$, so it is not arithmetic since not dominant. Therefore $E(F)_x$ is even not nearly holomorphic in general. However it can be proved it is nearly overconvergent for a suitable p-adic sheaf as those constructed in [AIP]. In the next section, we show that when $L(f, 1) = 0$, it defines a point of the Eigenvariety so it can be considered as overconvergent in that sense. To do this we will show that it is a p-adic limit of holomorphic forms of arbitrary regular weight and fixed slope $\underline{s}_0$.

Remark 4.9 The corresponding family of Galois representations $\rho_{E_{\kappa_{m_x}}(F_x)}$ is a finite slope deformation in the sense of the Section 3. In that case the function $\underline{\varphi}$ is given by

$$\underline{\varphi}(x) = (a(p, F)(x), p, p^{-1}, a(p, F)(x)^{-1}).$$

However, it does not satisfy the property (d) of Theorem 3.1 and this is why this family is reducible.

Remark 4.10 In fact there are two ways to do the construction of this family of nearly holomorphic Eisenstein series. The first one is by making the construction directly from the pull-back to obtain a nearly holomorphic form of weight κ_m. The second one is by constructing a family of Eisenstein series of weights $(a, b; -b, -a)$ with slope $(s_0, 0, 0, -s_0)$ and apply one time

[16]This fact is more delicate but much less difficult than proving that the ordinary Eisenstein series appearing in our previous work [SU10] is non zero modulo p.

a differential operator $\delta^*_{(a,b;-b,-a)}$ to obtain a family of slope $(s_0, 1, -1, -s_0)$ and weights $(a, b+1; -b-1, -a)$ and then evaluate at $a = m$ and $b = 1$. Here $\delta^*_{\underline{k}}$ is the differential operator for $\underline{k} = (k_1, k_2, k_3, k_4)$ obtained as the composition of the generalized Maass-Shimura operator $\delta_{V_{\underline{k}}}$ and the projection $H^0(X_K, \omega^1_{V_{\underline{k}}\otimes S}) \to H^0(X_K, \omega^1_{k_1,k_2+1;k_3-1,k_4})$ coming from the decomposition when $\underline{k}$ is regular:

$$V_{\underline{k}} \otimes S \cong$$
$$V_{k_1+1,k_2;k_3-1,k_4} \oplus V_{k_1,k_2+1;k_3-1,k_4} \oplus V_{k_1+1,k_2;k_3,k_4-1} \oplus V_{k_1,k_2+1;k_3,k_4-1}.$$

For GL_2, the similar construction gives the critical Eisenstein series E_2^{crit} from the ordinary family of Eisenstein series E^{ord}_{k-2} by applying one time the Maass-Shimura operator δ_{k-2} and then evaluate the result at $k = 2$ (see [Ur12]).

4.3.3 Cuspidal deformation of critical Eisenstein series

We sketch the construction of a generically cuspidal deformation of the p-adic Eisenstein series[17] $E_{\kappa_m}(\phi, f)$. More precisely, we show that the Eisenstein $E_{\kappa_m}(\phi, f)$ when f is as in Theorem 3.1 can be seen as a p-adic limit of finite slope holomorphic forms of very regular weights. From the theory of the Eigenvariety as developed in [Ur11], this implies our Theorem 3.1 and therefore would conclude the first part of the main theorem stated in these notes. The details of this argument will appear in [SU13].

A common feature to construct p-adic families of modular forms is to use high p-powers of a lifting of the Hasse invariant. This is insufficient to construct families of very regular weight in general but combining this with our previous family of Eisenstein series it will be sufficient for our goal. Before sketching our argument, we therefore start by explaining how this technic can be extended in the context of nearly holomorphic forms. Let A be a lifting of some power of the Hasse invariant. It is an holomorphic form of weight $k_0 \in (p-1)\mathbb{Z}_{>0}$. For any integer $s > 0$, we write

$$B_s := \frac{1}{4k_0 s}\delta_{sk_0}(A^s).$$

By Lemma 4.1, B_s is a global section of $\det(\omega^+)^{\otimes^{k_0 s}} \otimes \mathcal{J} \subset \omega^1_{\rho\otimes S}$ for $\rho = \det(St_+)^{\otimes^{k_0 s}}$, because its image $\epsilon(B_s)$ by ϵ in $H^0(X_K, \det(\omega^+)^{\otimes^{k_0 s}} \otimes \omega_{S\otimes S^\vee})$ is A^s. In other words, we have:

$$\epsilon(B_s) = A^s \in H^0(X_K, \det(\omega^+)^{\otimes^{k_0 s}}) \subset H^0(X_K, \det(\omega^+)^{\otimes^{k_0 s}} \otimes \omega_{S\otimes S^\vee}).$$

[17] This Eisenstein series is purely p-adic only when $m = 1$.

By computing the polynomial q-expansion of B_s, it is also very easy to verify that B_s is a p-adic analytic family in the variable s.

We now return to our goal. We treat the general case $k \geq 2$ but the main reason we have to work with nearly holomorphic forms is to treat the special case $k = 2$. Let f be as in Theorem 3.1 and let us choose a p-stabilization f_α such that its U_p-eigenvalue α satisfies $s_0 := v_p(\alpha) < k-1$. Let F be a Coleman family as in Theorem 4.8 passing through f_α at a point $x_0 \in \mathfrak{U}(\mathbb{Q}_p)$. For a point x in $\mathfrak{U}$ such that $k_x \in \mathbb{Z}_{>2}$ with $[k_x] = k(x)$, we consider

$$G'_{x,s} := A^s E(F)_x - B_s \epsilon(E(F)_x).$$

Since $E(F)_x \in H^0(X_K, \omega_{\kappa_{m_x}} \otimes \mathcal{J}^\vee)$, this is a well defined section of $\omega_{\kappa_{k_x}} \otimes \omega_{S^\vee} \otimes \mathcal{J} \otimes \det(\omega^+)^{\otimes^{k_0 s}}$ because $\mathcal{J}^\vee \subset \omega_{S^\vee} \otimes \mathcal{J}$. Moreover since $\epsilon(E(F)_x)$ is an holomorphic[18] form, we have

$$\epsilon(G'_{s,x}) = A^s \epsilon(E(F)_x) - \epsilon(B_s)\epsilon(E(F)_x) = 0.$$

Therefore $G'_{x,s}$ is a global section of $\omega_{S^\vee} \otimes \omega_S \otimes \det(\omega^+)^{\otimes^{k_0 s}} \otimes \omega_{\kappa_{m_x}}$. We define $G_{x,s}$ as the projection of $G'_{x,s}$ onto $H^0(X, \det(\omega^+)^{\otimes^{k_0 s}} \otimes \omega_{\kappa_{m_x}})$. Therefore $G_{s,x}$ is an holomorphc form of weight $\kappa_{x,s} = (sk_0 + m_x, sk_0 + 2; -2, -m_x)$. When $s \gg 0$ and x varies, we can make the weight of $G_{s,x}$ arbitrary regular. Since $L(f,m) = 0$, by the point (iii) of Theorem 4.8, we have

$$G'_{x_0,0} = E(F)_{x_0}. \tag{4.4}$$

This implies that $G_{x_0,0} = E(F)_{x_0}$ and in particular that $G_{x,s}$ is non-trivial thanks to the point (iv) of Theorem 4.8. We now make use of the construction of the Eigenvariety in [Ur11] to prove the existence of a deformation that will lead us to the proof of Theorem 3.1. Let $\kappa_0 = (m, 2; -2, -m) \in \mathfrak{X}(\overline{\mathbb{Q}}_p)$ and $\underline{s} = (s_0, 1, -1, -s_0) \in \mathbb{Q}^4$. For each x as in Theorem 4.8, $E(F)_x$ is of slope $\underline{s}$. From the results in [Ur11], one can deduce there exists a neighborhood $\mathfrak{V}$ of κ_0 and a polynomial $Q(\kappa, X) \in A(\mathfrak{V})[X]$ such that $Q(\kappa, u_0)$ projects the space of nearly holomorphic forms of weight κ, order ≤ 1 and level $K^p.I$ onto its subspace of slope $\underline{s}_0$ for any sufficiently regular weight inside $\mathfrak{V}$. Here u_0 is the Hecke operator $u_{\mathrm{diag}(1,p,p^2,p^3)}$. In particular, that implies that

$$Q(\kappa_{m_x}, u_0).E(F)_x = E(F)_x \tag{4.5}$$

when $k_x = 2m_x > 2$ and therefore for all $x \in \mathfrak{U}(\overline{\mathbb{Q}}_p) \cap i^{-1}(\mathfrak{V}(\overline{\mathbb{Q}}_p))$ by analytic continuation and in particular for $x = x_0$. Now, we consider the family of forms given by:

$$K_{x,s} := Q(\kappa_{x,s}, u_0).G_{x,s}$$

[18] This is because $E(F)_x$ is nearly holomorphic of degree ≤ 1.

for (x, s) such that $\kappa_{x,s} \in \mathfrak{V}$. Then $K_{x,s}$ is a well defined holomorphic form for all but finitely many pairs (x, s) for which $\kappa_{x,s}$ is algebraic dominant. Moreover by (4.4) and (4.5), we have

$$\lim_{(x,s)\to(x_0,0)} K_{x,s} = E(F)_{x_0}. \tag{4.6}$$

From this, it is easy to see that $E(F)_{x_0}$ gives a point of the cuspidal Eigenvariety for $U(2,2)$ with slope $\underline{s}_0$. Then, by using the theory developed in [Ur11], one shows that there is a 4-dimensional p-adic family of automorphic representations of finite slope specializing to the representation $E_{\kappa_0}(f)$ in the sense of [SU06b, §2, Thm 2.3.2] and [Ur11, Thm 5.4.4.].

From this, we construct the corresponding families of Galois representations using the theory of pseudo-characters and the existence of Galois representations for cuspidal representations of $U(2,2)$ and their properties[19] with respect to the local-global compatibility with the Langlands correspondence. We can deduce easily Theorem 3.1 by taking a suitable 1-dimensional subfamily of it. The fact that the sections used for constructing the Eisenstein series project subjectively on the Langlands quotient implies the crucial local property (b) in Theorem 3.1 by (4.3) and local-global compatibility of the Galois representation with the Langlands correspondence.

References

[AIP] F. Andreatta, A. Iovita and V. Pilloni, *Families of Siegel modular forms*, preprint 2011.

[Ha81] M. Harris, *Eisenstein series on Shimura varieties*, Ann. of Math. **119** (1984), 59–94.

[Ha85] M. Harris, *Arithmetic vector bundles and automorphic forms on Shimura varieties. I*, Invent. Math. **82** (1985), 151–189.

[Ha86] M. Harris, *Arithmetic vector bundles and automorphic forms on Shimura varieties. II*, Compositio Math. **60** no 3 (1986), 323–378.

[Hi04] H. Hida, *p-adic Automorphic Forms on Shimura Varieties*, Springer Monographs in Mathematics, Springer-Verlag, 2004.

[Ki03] M. Kisin, *Overconvergent modular forms and the Fontaine-Mazur conjecture*. Invent. Math. **153** (2003), no. 2, 373–454.

[19] This uses essentially the works of Sophie Morel, Chris Skinner and Shrenik Shah.

[La08] K.W. Lan, *Arithmetic compactifications of PEL-type Shimura varieties*, Ph.D. thesis, Harvard University, May 20, 2008.

[La12] K.W. Lan, *Comparison between analytic and algebraic constructions of toroidal compactifications of PEL-type Shimura varieties*, J. Reine Angew. Math. **664** (2012), 163–228.

[PS11] B. Stroh and V. Pilloni, *Surconvergence et classicité: le cas déployé.*, preprint 2011.

[Sh04] G. Shimura, *Arithmeticity in the Theory of Automorphic Forms*, American Mathematical Society, Mathematical Surveys and Monographs, Vol. 82, (2004).

[SU02] C. M. Skinner and E. Urban, *Sur les déformations p-adiques des formes de Saito-Kurokawa*, C. R. Math. Acad. Sci. Paris **335** (2002), no. 7, 581–586.

[SU06a] C. M. Skinner and E. Urban, *Sur les déformations p-adiques de certaines représentations automorphes*, J. Inst. Math. Jussieu **5** no. 4 (2006), 629–698.

[SU06b] C. M. Skinner and E. Urban, *Vanishing of L-functions and ranks of Selmer groups*, 28 pages, Proc. International Congress of Mathematicians, Vol. II, 473–500, Eur. Math. Soc., Zurich, 2006.

[SU10] C. M. Skinner and E. Urban, *The Main conjecture for* GL_2, to appear in Invent. Math..

[SU13] C. M. Skinner and E. Urban, *Nearly overconvergent automorphic forms and applications*, in preparation.

[Ro96] R. Rouquier, *Caractérisation des caractères et Pseudo-caractéres*, J. Algebra **180** (1996), 571–586.

[Ta91] R. Taylor, *Galois representations associated to Siegel modular forms of low weight*, Duke Math. J. **63** (1991), 281–332.

[Ur11] E. Urban, *Eigenvarieties for reductive groups*, Ann. of Math. **174** (2011), 1685–1784.

[Ur12] E. Urban, *Nearly overconvergent modular forms*, preprint 2012.

Current Address: Eric Urban, Department of Mathematics, Columbia University, 2990 Broadway, New York, NY 10027, USA

E-mail: urban@math.columbia.edu, urban@math.jussieu.fr

Automorphic Representations and L-Functions
Editors: D. Prasad, C.S. Rajan, A. Sankaranarayanan, J. Sengupta

Publisher: Hindustan Book Agency, New Delhi, India

Harmonic Analysis For Relative Trace Formula

Wei Zhang[1]

This is an expository article on some local harmonic analysis related to relative trace formula.

1 An Overview of the Relative Trace Formula

Let G be a reductive group and H a subgroup both defined over a number field F. Let $\mathbb{A}$ denote the ring of adeles of F. Let π be a cuspidal automorphic representation of $\mathrm{G}(\mathbb{A})$. Then the automorphic period integral is a linear functional on π:

$$\mathscr{P}_{\mathrm{H}}(\phi) :\coloneqq \int_{\mathrm{H}(F)\backslash \mathrm{H}(\mathbb{A})} \phi(h)dh, \quad \phi \in \pi.$$

Many questions on special values of L-functions are tied to the study of period integrals of automorphic forms. A notable example is the formula of Waldspurger ([22]) that relates the toric period on GL_2 or its inner form to some central critical L-value. The conjecture of Gan, Gross and Prasad ([4]) started with the Gross-Prasad conjecture ([5]), as well as the refinement of Ichino and Ikeda ([8]) of the Gross-Prasad conjecture, vastly generalizes the Waldspurger formula to higher rank groups. We briefly state their conjecture. Let E be either F or a quadratic extension of F. If $E = F$ (resp., if E is a quadratic extension of F), let W_{n+1} be a quadratic space over $E = F$ (resp., a Hermitian space associated to E/F) of E-dimension $n + 1$. Let $W_n \subset W_{n+1}$ be a non-degenerate subspace of codimension one. Let G_i be $\mathrm{SO}(W_i)$ or $\mathrm{U}(W_i)$ for $i = n, n+1$. The Gan-Gross-Prasad period is attached to the pair (H, G) where $\mathrm{G} = \mathrm{G}_n \times \mathrm{G}_{n+1}$ and $\mathrm{H} \subset \mathrm{G}$ is the diagonal embedding of G_n. Assume further that π is tempered. The conjecture of Gan-Gross-Prasad ([4, Conjecture 24.1]) asserts that the following two statements are equivalent:

[1]The work is supported by NSF # 1204365.

(i) For some automorphic representation in the Vogan L-packet (cf. [4]) of π, the linear functional $\mathscr{P}_{\mathrm{H}}$ does not vanish.

(ii) The central value $L(1/2, \pi, \mathrm{R})$ does not vanish, where R is a certain representation of the L-group ${}^L G$.

In the Hermitian case, the author in [29] proves the conjecture for π satisfying a certain local condition. The tool is the relative trace formula first introduced by Jacquet to study period integrals. We give an outline of this strategy (cf. the survey articles [11], [14], [15]). We start with a triple $(\mathrm{G}, \mathrm{H}_1, \mathrm{H}_2)$ consisting of a reductive group G and two suitable subgroups $\mathrm{H}_1, \mathrm{H}_2$. Here the two subgroups H_1 and H_2 are possibly the same. We associate to a test function $f \in \mathscr{C}_c^\infty(\mathrm{G}(\mathbb{A}))$ a kernel function:

$$K_f(x, y) := \sum_{\gamma \in \mathrm{G}(F)} f(x^{-1}\gamma y), \quad x, y \in \mathrm{G}(\mathbb{A}).$$

Then we consider the linear functional on $\mathscr{C}_c^\infty(\mathrm{G}(\mathbb{A}))$ defined by the double integral:

$$I(f) = \int_{\mathrm{H}_1(F)\backslash \mathrm{H}_1(\mathbb{A})} \int_{\mathrm{H}_2(F)\backslash H_2(\mathbb{A})} K_f(h_1, h_2) dh_1 dh_2. \tag{1.1}$$

We may also insert some "small" representation as a weight factor: for example, a character of $\mathrm{H}_i(F)\backslash \mathrm{H}_i(\mathbb{A})$, or the Weil representation. The relative trace formula attached to the triple $(\mathrm{G}, \mathrm{H}_1, \mathrm{H}_2)$ is an identity between two different expansions of $I(f)$, known as the "*spectral expansion*" and the "*geometric expansion*" of $I(f)$, are equal. A cuspidal automorphic representation π contributes a term to the spectral expansion, which we will call a (global) *"spherical character"*, or *"relative character"*. This term is a distribution I_π on $\mathrm{G}(\mathbb{A})$ defined by

$$I_\pi(f) := \sum_{\phi \in \mathscr{B}(\pi)} \mathscr{P}_{\mathrm{H}_1}(\pi(f)\phi) \overline{\mathscr{P}_{\mathrm{H}_2}(\phi)},$$

where $\mathscr{B}(\pi)$ denotes an orthonormal basis of π. The terms in the geometric expansion are parameterized by double cosets $\gamma \in \mathrm{H}_1(F)\backslash \mathrm{G}(F)/\mathrm{H}_2(F)$. Given such a double coset γ, its contribution is the distribution $\mathrm{Orb}(\gamma, \cdot)$ defined by the orbital integral:

$$\mathrm{Orb}(\gamma, f) := \tau((\mathrm{H}_1 \times \mathrm{H}_2)_\gamma) \int_{(\mathrm{H}_1 \times \mathrm{H}_2)_\gamma(\mathbb{A})\backslash (\mathrm{H}_1 \times \mathrm{H}_2)(\mathbb{A})} f(h_1^{-1}\gamma h_2) dh_1 dh_2,$$

where $(\mathrm{H}_1 \times \mathrm{H}_2)_\gamma$ denotes the stabilizer of γ, and $\tau((\mathrm{H}_1 \times \mathrm{H}_2)_\gamma)$ is the volume of $(\mathrm{H}_1 \times \mathrm{H}_2)_\gamma(F)\backslash(\mathrm{H}_1 \times \mathrm{H}_2)_\gamma(\mathbb{A})$. Note that typically there is problem with convergence; in this expository article we will ignore such convergence issues. When we take the triple

$$(\mathrm{H} \times \mathrm{H}, \Delta_{\mathrm{H}}, \Delta_{\mathrm{H}}),$$

where $\Delta_{\mathrm{H}} \subset \mathrm{H} \times \mathrm{H}$ is the diagonal embedding of H, the associated relative trace formula is equivalent to the Arthur-Selberg trace formula associated to H. In general, Sakellaridis and Venkatesh ([19]) have initiated a conjectural framework which includes the case where each of the homogeneous spaces G/H_1 and G/H_2 is a *spherical variety* under G.

We will use RTF to stand for "relative trace formula". In application we usually need to compare two RTFs that are close to each other. In general the problem of identifying 'comparable' RTFs is a subtle one. To attack the Gan-Gross-Prasad conjecture in the Hermitian case, Jacquet and Rallis ([12]) constructed a pair of RTFs. The first RTF deals with the period integral and is associated to the triple $(\mathrm{G}, \mathrm{H}, \mathrm{H})$ where

$$\mathrm{H} = \Delta_{\mathrm{U}(W_n)}, \quad \mathrm{G} = \mathrm{U}(W_n) \times \mathrm{U}(W_{n+1}).$$

The second one deals with the L-values and is associated to the triple $(\mathrm{G}', \mathrm{H}_1, \mathrm{H}_2)$ where

$$\mathrm{G}' = \mathrm{Res}_{E/F}(\mathrm{GL}_n \times \mathrm{GL}_{n+1}),$$

and

$$\mathrm{H}_1 = \mathrm{Res}_{E/F}\mathrm{GL}_n, \quad \mathrm{H}_2 = \mathrm{GL}_n \times \mathrm{GL}_{n+1}.$$

Moreover it is necessary to insert a quadratic character $\eta_{n,n+1}$ of $\mathrm{H}_2(\mathbb{A})$:

$$\eta_{n,n+1} : \mathrm{H}_2(\mathbb{A}) \ni (h_n, h_{n+1}) \mapsto \eta^{n-1}(\det(h_n))\eta^n(\det(h_{n+1})),$$

where η is the quadratic character associated to E/F by class field theory.

We should take this discussion as an opportunity to mention a parallel version of the Jacquet–Rallis construction. As one way to generalize the Gross-Zagier formula ([6], [25]), there is also an *arithmetic* [2] version of the Gan-Gross-Prasad conjecture and the Ichino–Ikeda refinement ([27]). Inspired by the Jacquet-Rallis construction above, the author was led to a comparison of two '*arithmetic*' RTFs ([28]) to attack the problem. Now we resume with the same notation as in the Jacquet-Rallis construction above. For simplicity, we assume that $F = \mathbb{Q}$ and E is an imaginary quadratic field. We further assume that the Hermitian spaces W_n, W_{n+1} are of signatures $(n-1, 1)$ and $(n, 1)$ respectively. From the embedding

[2]Though it may be misleading, we will use the word "arithmetic" to indicate the case involving algebraic cycles.

$\mathrm{H} \subset \mathrm{G}$ one can construct an embedding of the associated Shimura varieties $\mathrm{Sh}_{\mathrm{H}} \subset \mathrm{Sh}_{\mathrm{G}}$. Then to a test function f in a *suitable* subspace of $\mathscr{C}_c^\infty(\mathrm{G}(\mathbb{A}))$, we may associate a Hecke correspondence $\mathrm{R}(f)$ on Sh_{G}. We then define an analogue of the distribution $I(f)$ in (1.1):

$$J(f) = \langle \mathrm{Sh}_{\mathrm{H}}, \mathrm{R}(f)\mathrm{Sh}_{\mathrm{H}} \rangle,$$

where $\langle \cdot, \cdot \rangle$ is the Beilinson-Bloch height pairing between two algebraic cycles. When Sh_{G} is a curve and Sh_{H} is a divisor, the paring is the Neron-Tate height pairing. This serves as the first arithmetic RTF that deals with the height of algebraic cycles. For the second one, we modify the RTF associated to the triple $(\mathrm{G}', \mathrm{H}_1, \mathrm{H}_2)$ in the Jacquet–Rallis construction as follows. For $f' \in \mathscr{C}_c^\infty(\mathrm{G}'(\mathbb{A}))$, we may define a family of distributions parameterized by $s \in \mathbb{C}$:

$$I(f', s) = \int_{\mathrm{H}_1(F)\backslash \mathrm{H}_1(\mathbb{A})} \int_{\mathrm{H}_2(F)\backslash H_2(\mathbb{A})} K_{f'}(h_1, h_2) |\det(h_1)|^s \eta_{n,n+1}(h_2) dh_1 dh_2.$$

Its first derivative at $s = 0$ is intimately tied to the central derivative of the L-function $L(s, \pi, \mathrm{R})$. Then we expect that, for *suitable* 'matching' test functions f and f', we have an equality

$$\frac{d}{ds} I(f', s)|_{s=0} = J(f).$$

When $\dim \mathrm{Sh}_{\mathrm{G}} > 1$, substantial work needs to be done before we can prove this identity in any non-trivial case. Some partial progress is presented in [28].

Back to general RTFs, as in the case of Arthur-Selberg trace formula, the comparison of two RTFs leads us to at least two local questions: the *fundamental lemma* and the *existence of smooth transfer*. Due to the author's limited knowledge, we will not discuss the first one. Instead, we would like to focus on the second one and present some questions in related topics of harmonic analysis. Examples are provided by Waldspurger's work in the endoscopic case ([24]), Jacquet's work on the Jacquet-Ye conjecture for quadratic base change ([9], [10]), and the author's work on the Jacquet-Rallis RTFs ([29]) which leads to a proof of some cases of the Gan-Gross-Prasad conjecture for Hermitian spaces. The emphasis on the Jacquet–Rallis construction is due simply to the author's ignorance of the other cases.

2 A Little Bit of Invariant Theory

We will restrict ourselves to the case where $\mathrm{H}_1, \mathrm{H}_2$ and G are all reductive.[3] We first introduce some notation in a more abstract setting. Let F be a field of characteristic zero. Let H be a reductive group acting on an algebraic variety X over F. The *categorical quotient* of X by H (cf. [1], [16], [17]) consists of a pair (Y, π) where Y is an algebraic variety with the trivial action by H and $\pi : X \to Y$ is an H-morphism such that for any pair (Y', π') with $\pi' : X \to Y'$ an H-morphism, there exists a unique morphism $\phi : Y \to Y'$ such that $\pi' = \phi \circ \pi$. If such a pair exists, then it is unique up to a unique isomorphism. When X is affine, which we assume from now on, the categorical quotient always exists. Indeed we may construct this categorical quotient as follows. Consider the affine variety

$$X_{/\mathrm{H}} := \operatorname{Spec} \mathcal{O}(X)^{\mathrm{H}}$$

together with the obvious morphism

$$\pi = \pi_{X,\mathrm{H}} : X \to X_{/\mathrm{H}}.$$

Then $(X_{/\mathrm{H}}, \pi)$ defines a categorical quotient of X by H. By abuse of notation, we will also let π denote the induced map $X(F) \to X_{/\mathrm{H}}(F)$. We say that a point $x \in X(F)$ is

- H-semisimple if Hx is Zariski closed in X (when F is a local field of characteristic zero, this is equivalent to requiring that $\mathrm{H}(F)x$ be closed in $X(F)$ for the analytic topology, cf. [18, p.109]).
- H-regular if the stabilizer H_x of x achieves the minimal dimension $\dim \mathrm{H}_y$ among all $y \in X$.

If no confusion can arise, we will simply use the words "semisimple" and "regular".We say that x is *regular semisimple* if it is regular and semisimple. We are interested in the following two cases

- For a triple $(\mathrm{G}, \mathrm{H}_1, \mathrm{H}_2)$ above, we consider $X = \mathrm{G}$, and the product $\mathrm{H} = \mathrm{H}_1 \times \mathrm{H}_2$ where H_1 (H_2, resp.) acts by left (right, resp.) multiplication.
- An algebraic representation: $X = \mathrm{V}$ is a vector space (considered as an affine variety) with an action by a reductive group H. We will use (H, V) to denote the representation to emphasize the dependence on the group and the vector space on which the group acts. The action is then self-evident in our examples below.

[3] But there are indeed many interesting cases which involve non-reductive subgroups, for example, the Jacquet-Ye construction ([9], [10]).

Given an action of a reductive group H on an affine variety X, there is a procedure to obtain an algebraic representation from each semisimple point $x \in X(F)$. Indeed if $x \in X(F)$ is H-semisimple, the stabilizer H_x is a reductive subgroup of H defined over F. It preserves the normal space $\mathrm{N}^X_{\mathrm{H}x,x}$ [4] at x of the orbit $\mathrm{H}x$ in the ambient space X. The induced action of H_x on the vector space $\mathrm{N}^X_{\mathrm{H}x,x}$ is called the *sliced representation* at x. The sliced representation is a generalization of the Harish-Chandra's 'semisimple descent' ([13, §16]) in which case $X = \mathfrak{h}$ is the Lie algebra of a reductive group H endowed with the adjoint action of H, and the H_x-representation $\mathrm{N}^X_{\mathrm{H}x,x}$ is equivalent to the Lie algebra centralizer $\mathfrak{h}_x$ of $x \in \mathfrak{h}$ on which the stabilizer H_x operates by the adjoint action.

Roughly speaking, the sliced representation $(\mathrm{H}_x, \mathrm{N}^X_{\mathrm{H}x,x})$ models the action of H on X around the semisimple point x, at least when X is smooth. More precisely, if X is a smooth affine variety and $x \in X(F)$ is a semisimple point, there exists an *étale Luna slice* (cf. [1, A.2], [17, §6]), namely there exists a locally closed smooth H_x-invariant subvariety Z of X together with a strongly étale [5] H_x-morphism $Z \to \mathrm{N}^X_{\mathrm{H}x,x}$ such that the induced H-morphism $\mathrm{H} \times_{\mathrm{H}_x} Z \to X$ [6] by $(h, z) \mapsto hz$ is strongly étale. Because of this, we usually try to transfer questions on X to questions on the normal space $\mathrm{N}^X_{\mathrm{H}x,x}$. It is therefore important to study the harmonic analysis attached to various algebraic representations (H, V).

We list some examples of spliced representations. In the case of the RTF associated to a triple $(\mathrm{G}, \mathrm{H}_1, \mathrm{H}_2)$, we will be interested in the sliced representations at $1 \in \mathrm{G}$ for the action of $\mathrm{H}_1 \times \mathrm{H}_2$ on G. We will call this sliced representation the '*Lie algebra*' of the RTF associated to the triple $(\mathrm{G}, \mathrm{H}_1, \mathrm{H}_2)$, or simply the '*Lie algebra*' of the triple $(\mathrm{G}, \mathrm{H}_1, \mathrm{H}_2)$.

1. Let $\mathfrak{h}$ be the Lie algebra of H. Then the adjoint representation $(\mathrm{H}, \mathfrak{h})$ is the sliced representation at $1 \in \mathrm{H}$ for the adjoint action of H on itself. We may also obtain some representations of similar flavor if we

[4] The vector space $\mathrm{N}^X_{\mathrm{H}x,x}$ is the quotient of the tangent space T^X_x of X at x by the tangent space $\mathrm{T}^{\mathrm{H}x}_x$ of the orbit $\mathrm{H}x$ at x.

[5] Let X and Y be two affine varieties with H-action. We say an H-morphism $\phi : X \to Y$ strongly étale if the induced morphism $\phi_{/\mathrm{H}} : X_{/\mathrm{H}} \to Y_{/\mathrm{H}}$ is étale and the following induced diagram is Cartesian:

$$\begin{array}{ccc} X & \xrightarrow{\phi} & Y \\ \downarrow & & \downarrow \\ X_{/\mathrm{H}} & \xrightarrow{\phi_{/\mathrm{H}}} & Y_{/\mathrm{H}}. \end{array}$$

[6] The notation $\mathrm{H} \times_{\mathrm{H}_x} Z$ means the quotient of $\mathrm{H} \times Z$ by H_x via the free action $h \cdot (g, z) = (gh^{-1}, hz)$.

consider symmetric spaces and Vinberg's θ-group ([18, §6]).

2. The Gan–Gross–Prasad case: for the triple $(\mathrm{G}, \mathrm{H}, \mathrm{H})$, we obtain an $\mathrm{H} \times \mathrm{H}$-action on $\mathrm{G} = \mathrm{G}_n \times \mathrm{G}_{n+1}$ where $\mathrm{H} \subset \mathrm{G}$ is the diagonal embedding of G_n. The sliced representation at $1 \in \mathrm{G}$ can be identified with the pair $(\mathrm{H}, \mathfrak{g}_{n+1})$ where $\mathfrak{g}_{n+1}$ is the Lie algebra of G_{n+1} and the action of $\mathrm{H} \simeq \mathrm{G}_n$ is via the restriction of the adjoint action of G_{n+1} on $\mathfrak{g}_{n+1}$ to G_n.
3. The Jacquet–Rallis case: for the triple $(\mathrm{G}', \mathrm{H}_1, \mathrm{H}_2)$, the sliced representation at $1 \in \mathrm{G}'$ is given by the pair $(\mathrm{GL}_n, \mathfrak{s}_{n+1})$ where $\mathfrak{s}_{n+1}$ is the F-vector space consisting of $X \in M_{n+1}(E)$ ($(n+1) \times (n+1)$-matrices with coefficients in E) satisfying

$$X + \overline{X} = 0,$$

where $\overline{X}$ is the Galois conjugate of X (entry-wise). The action of GL_n on $\mathfrak{s}_{n+1}$ is by conjugation. If we write the quadratic extension E as $F[\sqrt{\delta}]$ with $\delta \in F$, we have a non-canonical isomorphism $\mathfrak{s}_{n+1} \simeq \mathfrak{gl}_{n+1,F}$ (the Lie algebra of $\mathrm{GL}_{n+1,F}$) by $X \mapsto X/\sqrt{\delta}$.

In the case of an algebraic representation (H, V), there is a distinguished point in V as well as in $\mathrm{V}_{/\mathrm{H}}$, namely the zero vector in V and its image in $\mathrm{V}_{/\mathrm{H}}$ denoted still by 0. We define the H-nilpotent cone to be the fiber $\pi^{-1}(0)$, and denote it by $\mathcal{N}$. Elements in $\mathcal{N}$ are called H-nilpotent. An element is H-nilpotent if and only if the Zariski closure of the orbit $\mathrm{H}x$ contains $0 \in \mathrm{V}$. The notions of H-semisimplicity, regularity and nilpotency coincide with the usual ones when $(\mathrm{H}, \mathrm{V}) = (\mathrm{H}, \mathfrak{h})$ is the adjoint representation of H on its Lie algebra $\mathfrak{h}$. The geometry of the H-nilpotent cone in some sense dictates many aspects of the quotient morphism π. For example, the number of H-orbits in each fiber is bounded above by the number of H-orbits in the nilpotent cone.

It is an interesting question how to extend the above discussion to the case where H_1 and/or H_2 is non-reductive.

3 Local Relative Trace Formula for (H, V)

Let (H, V) be an algebraic representation. Imitating the convention of Harish-Chandra in [7] for the adjoint action, we will write the action in an exponential way: $h \cdot X = X^h$ for $h \in \mathrm{H}, X \in \mathrm{V}$. From now on we assume that F is a non-archimedean local field of characteristic zero (i.e., finite extension of $\mathbb{Q}_p$). We abuse notation and use V, H to denote the F-points of V, H. Let $\mathscr{C}_c^\infty(\mathrm{V})$ be the space of locally constant and compactly supported functions on V. For $h \in \mathrm{H}$, we denote by ${}^h f$ the function ${}^h f(X) := f(X^h)$. Let $\mathscr{D}(\mathrm{V})$ be the space of distributions on V, i.e.:

$$\mathscr{D}(\mathrm{V}) = \mathrm{Hom}(\mathscr{C}_c^\infty(\mathrm{V}), \mathbb{C}).$$

By an H-*polarization*, we mean a non-degenerate symmetric F-bilinear pairing

$$\langle \cdot, \cdot \rangle : \mathrm{V} \times \mathrm{V} \to F,$$

that is H-invariant:

$$\langle X^h, Y^h \rangle = \langle X, Y \rangle, \quad X, Y \in \mathrm{V}, h \in \mathrm{H}.$$

Then for a fixed nontrivial character $\psi : F \to \mathbb{C}^\times$ we may define an automorphism (of order 4) of $\mathscr{C}_c^\infty(\mathrm{V})$, i.e., the Fourier transform:

$$\widehat{f}(X) = \int_{\mathrm{V}} f(Y)\psi(\langle Y, X \rangle) dX.$$

We use the self-dual measure on V. The H-invariance of the pairing implies that the Fourier transform commutes with the H-action on $\mathscr{C}_c^\infty(\mathrm{V})$:

$$\widehat{{}^h f} = {}^h \widehat{f}, \quad h \in \mathrm{H}.$$

By duality each h also defines an automorphism of $\mathscr{D}(\mathrm{V})$.

To illustrate the idea of the local trace formula, we first assume that $\mathrm{H}(F)$ is compact (cf.[2] for the case of the usual local trace formula on groups). In particular, we may simply define the orbital integral as

$$\mathrm{Orb}(X, f) := \int_{\mathrm{H}} f(X^h) dh, \quad X \in \mathrm{V}.$$

The Parseval-Plancherel theorem asserts that the Fourier transform preserves the L^2-norm:

$$\int_{\mathrm{V}} f_1(X)\widehat{f_2}(X) dX = \int_{\mathrm{V}} \widehat{f_1}(X) f_2(X) dX.$$

We may replace f_1 by ${}^h f_1$:

$$\int_{\mathrm{V}} f_1(X^h)\widehat{f_2}(X) dX = \int_{\mathrm{V}} \widehat{f_1}(X^h) f_2(X) dX.$$

Since H is compact we may integrate the above identity over H and interchange the order of integration to obtain

$$\int_{\mathrm{V}} \mathrm{Orb}(X, f_1)\widehat{f_2}(X) dX = \int_{\mathrm{V}} \mathrm{Orb}(X, \widehat{f_1}) f_2(X) dX. \tag{3.1}$$

We will call this *the local relative trace formula*, or simply *the local trace formula*, associated to (H, V). Equivalently, under a suitable choice of a measure on the quotient $V_{/H}$ [7]:

$$\int_{V_{/H}} \mathrm{Orb}(X, f_1)\mathrm{Orb}(X, \widehat{f_2})dX = \int_{V_{/H}} \mathrm{Orb}(X, \widehat{f_1})\mathrm{Orb}(X, f_2)dX.$$

Clearly if $H = \{1\}$, the formula reduces to the Parseval-Plancherel theorem on Fourier transform. In general, one may view the local trace formula as an equivariant version of the Parseval-Plancherel theorem.

When the group H is non-compact, the formula needs to be regularized and some weighted orbital integrals may appear. It is much more difficult to establish such a formula. For example, in the case of the adjoint action $(\mathrm{H}, \mathfrak{h})$, Waldspurger ([23]) proves a local trace formula for Lie algebras which is closely related to Arthur's local trace formula for groups ([3]). In the case of symmetric spaces, Sparling has made some progress towards deriving a local trace formula for Lie algebras ([20], [21]).

It is plausible that the local trace formula for (H, V) may retain the simple form as in the compact case above when the stabilizer of every regular semisimple element of V is compact. Then we may expect the situation to be similar to the *elliptic part* of the local trace formula of Waldspurger for Lie algebras. Two examples where this happens are the local trace formulae associated to the representations in the second (the Gan-Gross-Prasad case) and the third example (the Jacquet-Rallis case) in §2. In these cases, the stabilizer of a regular semisimple element is always trivial. The local trace formula turns out to have the same shape as (3.1) above and is proved in [29].

There are some examples of sliced representations in the Vinberg theory of θ-group where the stabilizer of every regular semisimple element is always finite (hence compact). It may be interesting to investigate the local trace formula for these sliced representations.

In the context of the RTF for a triple $(\mathrm{G}, \mathrm{H}_1, \mathrm{H}_2)$, we will consider the sliced representation at the identity. In general it may be also necessary to consider the sliced representations at *all* semisimple elements. But in our examples above the sliced representations at other semisimple elements of G are the same as some sliced ones at the identity of some other RTFs. Therefore it suffices to study the local trace formula for sliced representations at the identity element for many RTFs.

One may also wonder about the Arthur's local trace formula for groups in the context of RTFs. The problem has not receive much attention and

[7]Strictly speaking, there may be an issue of "stability": there may be regular semisimple elements that are not $\mathrm{H}(F)$-equivalent but have the same image under $\mathrm{V} \to \mathrm{V}_{/\mathrm{H}}$.

the author does not know any such example. However, for the study of existence of smooth transfer, it suffices to use the local trace formula for sliced representations.

4 Fourier Transform of Orbital Integrals

Representability question. We retain the notations from the previous section. We say that an H-invariant function κ on V is *nice* if it is locally constant on the regular semisimple locus V_{rs} and locally integrable on V. It is important to consider the Fourier transform of orbital integrals for regular semisimple $X \in \mathrm{V}$:

$$\widehat{\mathrm{Orb}}(X, f) := \mathrm{Orb}(X, \widehat{f}), \quad f \in \mathscr{C}_c^\infty(\mathrm{V}).$$

We denote this distribution also by $\widehat{\mathrm{Orb}}_X$. When $\mathrm{V} = \mathfrak{h}$ is the Lie algebra of H, Harish-Chandra discovered that the Fourier transform of an orbital integral on Lie algebra $\mathfrak{h}$ plays the role of the character of an irreducible representation on the group H.

Question 4.1 Let $X \in \mathrm{V}$ be H-regular semisimple. Is the distribution $\widehat{\mathrm{Orb}}_X$ representable by a nice function?

In other words, the questions asks whether there is a nice function $\kappa(X, \cdot)$ such that for all $f \in \mathscr{C}_c^\infty(\mathrm{V})$:

$$\widehat{\mathrm{Orb}}(X, f) = \int_{\mathrm{V}} f(Y)\kappa(X, Y)dY.$$

The two-variable function κ is then a sort of kernel function which contains a great deal of the harmonic analysis of (H, V). In the case of the adjoint representation $(\mathrm{H}, \mathfrak{h})$, a theorem of Harish-Chandra ([7, Theorem 1.1]) answers the question affirmatively. The key ingredients are the local trace formula for $(\mathrm{H}, \mathfrak{h})$ [8] and *Howe's finiteness hypothesis* ([13, §26]). We recall the statement of Howe's finiteness hypothesis. Let ω be a compact subset of V and let ω^{H} be the set of elements $X^h, X \in \omega, h \in \mathrm{H}$. We denote by $J(\omega)$ the set of H-invariant distributions with support in ω^{H}. Let Λ be a lattice in V and let $\mathscr{C}_c^\infty(\mathrm{V}/\Lambda)$ be the subspace of $\mathscr{C}_c^\infty(\mathrm{V})$ consisting of Λ-translation invariant functions. We denote by $j_\Lambda^* J(\omega)$ the image of $J(\omega)$ under the homomorphism $j_\Lambda^* : \mathrm{Hom}(\mathscr{C}_c^\infty(\mathrm{V}), \mathbb{C}) \to \mathrm{Hom}(\mathscr{C}_c^\infty(\mathrm{V}/\Lambda), \mathbb{C})$. Then Howe's

[8] Harish-Chandra's original proof did not use the full strength of the local trace formula. He essentially used the elliptic part of the local trace formula, cf. [7] and [13, §26].

finiteness hypothesis for (H, V) states that

For any compact subset ω of V and any lattice Λ of V, the space $j_\Lambda^* J(\omega)$ is finite dimensional.

This was shown to be true for the adjoint representation $(\mathrm{H}, \mathfrak{h})$ by Howe in the case where $\mathrm{H} = \mathrm{GL}_n$ and in general by Harish-Chandra ([7]). Rader and Rallis ([18]) also proved the finiteness in the case of (H, V) arising from Vinberg's θ-group (hence, including the case of symmetric spaces).

However, when Howe's finiteness hypothesis fails, it seems to be a challenging question how to prove (or disprove) the representability of $\widehat{\mathrm{Orb}}_X$. Note that a counterexample to Howe's finiteness hypothesis already shows up in the Jacquet–Rallis case for $n = 2$: $\mathrm{H} = \mathrm{GL}_2$, $\mathrm{V} = \mathfrak{gl}_3$. Consider the lattice $\Lambda = \mathrm{V}(\mathcal{O}_F)$ and the characteristic functions

$$f_t = 1_{t^{-1}\Lambda}, \quad t \in F^\times, |t| \leq 1.$$

Then $f_t \in \mathscr{C}_c^\infty(\mathrm{V}/\Lambda)$. Let ω be any compact open neighborhood of 0 in V. Then ω^{H} contains the H-nilpotent cone $\mathcal{N}$. [9] In $\mathcal{N}$, there are infinitely many nilpotent orbits. A continuous family can be given as

$$n(u) := \begin{bmatrix} 0 & u & 1 \\ 0 & 0 & 0 \\ 0 & 1 & 0 \end{bmatrix}, \qquad u \in F.$$

The stabilizer of $n(u)$ is the nilpotent group N consisting of matrices of the form $\begin{pmatrix} 1 & * \\ 0 & 1 \end{pmatrix}$. The naive definition of orbital integral turns out to be absolutely convergent and defines an H-invariant distribution on V:

$$\mathrm{Orb}(n(u), f) = \int_{N\backslash \mathrm{H}} f(n(u)^h) dh.$$

Indeed, by the Iwaswa decomposition $\mathrm{H} = NAK$ (A being the diagonal subgroup, $K = \mathrm{GL}_2(\mathcal{O}_F)$), we have (for a suitable choice of measures)

$$\mathrm{Orb}(n(u), f) = \int_{F^2} f_K \begin{bmatrix} 0 & uxy & x \\ 0 & 0 & 0 \\ 0 & y & 0 \end{bmatrix} dxdy, \quad f_K(X) := \int_K f(X^k) dk.$$

This is clearly absolutely convergent. We see immediately

$$\mathrm{Orb}(n(u), f_t) = |t|^{-2} \mathrm{Orb}(n(t^{-1}u), f_1),$$

[9] To see this, note that if $X \in \mathcal{N}$, the closure of $\{X\}^{\mathrm{H}}$ contains zero. Hence $\{X\}^{\mathrm{H}} \cap \omega$ is non-empty.

and, if we normalize the measures such that $\mathrm{vol}(\mathcal{O}_F) = \mathrm{vol}(\mathrm{GL}_2(\mathcal{O}_F)) = 1$:

$$\mathrm{Orb}(n(u), f_1) = \begin{cases} 1, & |u| \leq 1; \\ (-v_F(u)\zeta_F(1)^{-1} + 1)|u|^{-1}, & |u| > 1. \end{cases}$$

Here v_F is the valuation on F and $\zeta_F(s)$ is the local zeta function of F. From these facts we may deduce that the subspace of $j_\Lambda^* J(\omega)$ generated by the image of the various $\mathrm{Orb}_{n(u)}$ is not finite dimensional.

Compatibility between smooth transfer and Fourier transform. In the comparison of two RTFs, we may first consider an infinitesimal version, namely the comparison between sliced representations at semisimple points, especially at 1. This simplifies the question and usually one can deduce the original questions from the infinitesimal comparison when we vary x in all semisimple points.

We again consider the Jacquet-Rallis construction for the Gan-Gross-Prasad conjecture in the Hermitian case. There is a canonical isomorphism between the categorical quotients

$$\mathfrak{u}(W_{n+1})_{/\mathrm{U}(W_n)} \simeq \mathfrak{s}_{n+1/\mathrm{GL}_n}. \tag{4.1}$$

For each F-point in the quotient, the fiber (possibly empty) under the map

$$\pi : \mathfrak{u}(W_{n+1})(F) \to \mathfrak{u}(W_{n+1})_{/\mathrm{U}(W_n)}(F)$$

contains at most *one* regular semisimple $\mathrm{U}(W_n)(F)$-orbit (note that there is no 'stability' issue here). The same holds for $\mathfrak{s}_{n+1/\mathrm{GL}_n}$. The isomorphism allows us to match orbits: given a regular semisimple $\mathrm{U}(W_n)(F)$-orbit in $\mathfrak{u}(W_{n+1})(F)$, say of $X \in \mathfrak{u}(W_{n+1})(F)$, it turns out that there exists a unique regular semisimple $\mathrm{GL}_n(F)$-orbit of Y in $\mathfrak{s}_{n+1}(F)$ whose image in the quotient is the same as that of X, under the isomorphism (4.1). Conversely, given a regular semisimple $\mathrm{GL}_n(F)$-orbit of Y in $\mathfrak{s}_{n+1}(F)$, there exists a pair of Hermitian spaces W_n, W_{n+1} and a unique $\mathrm{U}(W_n)(F)$-orbit in $\mathfrak{u}(W_{n+1})(F)$ with the same image in the quotient. In each of the two cases above, the stabilizer of a regular semisimple element is trivial. We thus simply take the orbital integrals as

$$\mathrm{Orb}(X, f) = \int_{\mathrm{U}(W_n)(F)} f(X^h)dh, \quad f \in \mathscr{C}_c^\infty(\mathfrak{u}(W_{n+1}(F)), X \in \mathfrak{u}(W_{n+1})(F),$$

and

$$\mathrm{Orb}(Y, f') = \int_{\mathrm{GL}_n(F)} f'(Y^h)\eta(h)dh, \quad f' \in \mathscr{C}_c^\infty(\mathfrak{s}_{n+1}(F)), Y \in \mathfrak{s}_{n+1}(F),$$

where $\eta(h)$ stands for $\eta(\det(h))$ and η is the quadratic character of $F^\times$ attached to the quadratic extension E/F. We say that f' and f are *smooth transfers* of each other if for all matching regular semisimple X and Y, we have

$$\mathrm{Orb}(X, f) = \eta'(Y)\mathrm{Orb}(Y, f'),$$

where $\eta'(Y) \in \{\pm 1\}$ is a certain transfer factor satisfying $\eta'(Y^h) = \eta(h)\eta'(Y)$ for all regular semisimple Y and $h \in \mathrm{GL}_n(F)$ ([29]). We may define Fourier transforms that are compatible in a suitable sense. The following compatibility result is proved in [29], basing partially on the local trace formula.

Theorem 4.2 *If f, f' are smooth transfers of each other, so are $\widehat{f}$ and $\lambda \cdot \widehat{f'}$, where λ is an explicit constant independent of f, f'.*

This is a relative variant of Waldspurger's result on the compatibility between the endoscopic transfer and Fourier transform ([24]). But in our case the proof is purely local and does not require the fundamental lemma, which has been proved by Yun ([26]). Together with other ingredients, this compatibility result implies the existence of smooth transfer for all f and f':

Theorem 4.3 *Given $f \in \mathscr{C}_c^\infty(\mathfrak{u}(W_{n+1}(F))$, there exists $f' \in \mathscr{C}_c^\infty(\mathfrak{s}_{n+1}(F))$ that is a smooth transfer of f. Conversely, given $f' \in \mathscr{C}_c^\infty(\mathfrak{s}_{n+1}(F))$, there exists $f \in \mathscr{C}_c^\infty(\mathfrak{u}(W_{n+1}(F))$ that is a smooth transfer of f'.*

Roughly speaking, Fourier transforms provide a way to generate more pairs of functions that are smooth transfers of each other, starting from certain basic pairs (for example, pairs of (f, f') with each of f, f' supported in the regular semisimple locus).

We would like to give a heuristic as to why we should expect the compatibility result of Theorem 4.2 to hold. Let F be a number field. For (H, V) being one of the two representations above, we may consider a θ-function attached to $f \in \mathscr{C}_c^\infty(\mathrm{V}(\mathbb{A}))$:

$$K_f(h) = \sum_{X \in \mathrm{V}(F)} f(X^h), \quad h \in \mathrm{H}(\mathbb{A}).$$

It is clearly absolutely convergent and $\mathrm{H}(F)$-invariant on the left. We consider the integral

$$I(f) := \int_{\mathrm{H}(F)\backslash \mathrm{H}(\mathbb{A})} K_f(h)dh.$$

We proceed formally to ignore the convergence problem. We may expand it as

$$I(f) = \sum_X \mathrm{Orb}(X, f) + \cdots.$$

where the sum is over a set of representatives for the regular semisimple orbits X, and where the non-regular-semisimple contribution needs to be regularized. On the other hand, by the Poisson summation formula we have

$$K_f(h) = K_{\widehat{f}}(h).$$

Therefore we have an alternative expansion of $I(f) = I(\widehat{f})$ and we conclude

$$\sum_X \mathrm{Orb}(X, f) + \cdots = \sum_X \mathrm{Orb}(X, \widehat{f}) + \cdots . \tag{4.2}$$

This can be viewed as an infinitesimal version, or a 'Lie algebraic version', of the global RTF. In the case of an adjoint representation $(\mathrm{H}, \mathfrak{h})$, if we compare the Lie algebraic version with the Arthur-Selberg trace formula for the group H, we may view the Fourier transform of an orbital integral on the Lie algebra $\mathfrak{h}$ as an analogue of an irreducible character of the group H. Now if we compare the equality (4.2) for two pairs of (H, V), we may naturally expect that the Fourier transform should respect the smooth transfer (possibly up to some local constant whose product over all places is equal to one).

In the endoscopic case too, such a compatibility result holds and was reduced to the fundamental lemma for Lie algebras by Waldspurger ([24]). In this case, the kernel function κ representing the Fourier transform of orbital integrals is known to exist but hard to evaluate. The fundamental lemma is used to apply a global argument to deduce an identity between the kernel functions. There is another example due to Jacquet ([9]), where the kernel function κ is computed explicitly and the fundamental lemma can then be deduced from the kernel function. In our case, it is not known whether the kernel function κ exists but a certain special feature of $(\mathrm{GL}_n, \mathfrak{s}_{n+1})$ allows us to circumvent the difficulty and prove the compatibility result of Theorem 4.2.

Acknowledgement. The author thanks the anonymous referee for many helpful suggestions, and Rahul Krishna for some useful comments.

References

[1] A. Aizenbud, D. Gourevitch, *Generalized Harish-Chandra descent, Gelfand pairs, and an Archimedean analog of Jacquet-Rallis's theorem.* With an appendix by the authors and Eitan Sayag. Duke Math. J. **149** (2009), no. 3, 509–567.

[2] J. Arthur, *Towards a local trace formula*, Algebraic analysis, geometry, and number theory (Baltimore, MD, 1988), 1–23, Johns Hopkins Univ., Baltimore, MD, 1989.

[3] J. Arthur, *A local trace formula*, Inst. Hautes Études Sci. Publ. Math. **73** (1991), 5–96.

[4] W. Gan, B. Gross and D. Prasad, *Symplectic local root numbers, central critical L-values, and restriction problems in the representation theory of classical groups*, To appear in Asterisque.

[5] B. Gross and D. Prasad, *On irreducible representations of $SO(2n+1) \times SO(2m)$*. Canad. J. Math. **46** (1994), no. 5, 930–950.

[6] B. Gross and D. Zagier, *Heegner points and derivatives of L-series*, Invent. Math. **84** (1986), no. 2, 225–320.

[7] Harish-Chandra, *Admissible invariant distributions on reductive p-adic groups*, Preface and notes by Stephen DeBacker and Paul J. Sally, Jr. University Lecture Series, 16. American Mathematical Society, Providence, RI, 1999. xiv+97 pp. ISBN: 0-8218-2025-7.

[8] A. Ichino and T. Ikeda, *On the periods of automorphic forms on special orthogonal groups and the Gross-Prasad conjecture*, Geom. Funct. Anal. **19** (2010), no. 5, 1378–1425.

[9] H. Jacquet, *Smooth transfer of Kloosterman integrals*, Duke Math. J. **120** (2003), no. 1, 121–152.

[10] H. Jacquet, *Kloosterman identities over a quadratic extension*. Ann. of Math. (2) **160** (2004), no. 2, 755–779.

[11] H. Jacquet, *A guide to the relative trace formula*, in Automorphic representations, L-functions and applications: progress and prospects, 257–272, Ohio State Univ. Math. Res. Inst. Publ., 11, de Gruyter, Berlin, 2005.

[12] H. Jacquet and S. Rallis, *On the Gross-Prasad conjecture for unitary groups* in *On certain L-functions*, 205–264, Clay Math. Proc., 13, Amer. Math. Soc., Providence, RI, 2011.

[13] R. Kottwitz, *Harmonic analysis on reductive p-adic groups and Lie algebras*, Harmonic analysis, the trace formula, and Shimura varieties, 393–522, Clay Math. Proc., **4**, Amer. Math. Soc., Providence, RI, 2005.

[14] E. Lapid, *The relative trace formula and its applications*, Automorphic Forms and Automorphic L-Functions (Kyoto, 2005), Surikaisekikenkyusho Kokyuroku **1468** (2006), 76–87.

[15] E. Lapid, *Some applications of the trace formula and the relative trace formula*, Proceedings of the International Congress of Mathematicians, Volume III, 1262–1280, Hindustan Book Agency, New Delhi, 2010.

[16] D. Mumford, F. Fogarty and F. Kirwan, *Geometric invariant theory*, Third edition, Ergebnisse der Mathematik und ihrer Grenzgebiete (2), **34**, Springer-Verlag, Berlin, 1994. xiv+292 pp. ISBN: 3-540-56963-4 .

[17] V.L. Popov and B. Vinberg, *Invariant theory*, in Algebraic geometry, 4, 137–314, 315 Itogi Nauki i Tekhniki, Akad. Nauk SSSR, Vsesoyuz. Inst. Nauchn. i Tekhn. Inform., Moscow, 1989.

[18] C. Rader and S. Rallis, *Spherical characters on p-adic symmetric spaces.* Amer. J. Math. **118** (1996), no. 1, 91–178.

[19] Y. Sakellaridis and A. Venkatesh, *Periods and harmonic analysis on spherical varieties*, arXiv:1203.0039.

[20] J. Sparling, A Local Relative Trace Formula for $F^*\backslash \mathrm{SL}(2,F)$, arXiv:0811.4462

[21] J. Sparling, On the θ-split side of the local relative trace formula, arXiv:0905.0034

[22] J.-L. Waldspurger, *Sur les valeurs de certaines fonctions L automorphes en leur centre de symétrie.* Compositio Math. **54** (1985), no. 2, 173–242.

[23] J.-L. Waldspurger, *Une formule des traces locale pour les algèbres de Lie p-adiques*, J. Reine Angew. Math. **465** (1995), 41–99.

[24] J.-L. Waldspurger, *Le lemme fondamental implique le transfert*, Compositio Math. **105** (1997), no. 2, 153–236.

[25] X. Yuan, S. Zhang and W. Zhang, *The Gross–Zagier formula on Shimura curves*, Ann. of Math. Studies #184, ISBN: 9780691155920.

[26] Z. Yun, *The fundamental lemma of Jacquet and Rallis*, With an appendix by Julia Gordon. Duke Math. J. **156** (2011), no. 2, 167–227.

[27] S. Zhang, *Linear forms, algebraic cycles, and derivatives of L-series*, preprint, 2010.

[28] W. Zhang, *On arithmetic fundamental lemmas*, Invent. Math. **188** Number 1 (2012), 197–252.

[29] W. Zhang, *Fourier transform and the global Gan–Gross–Prasad conjecture for unitary groups*, submitted, 2011.

Wei Zhang, Department of Mathematics, Columbia University, MC 4423, 2990 Broadway, New York, NY 10027, USA.
E-mail: wzhang@math.columbia.edu.